Nanotechnology Applications in Environmental Engineering

Rabia Nazir
Pakistan Council of Scientific and Industrial Research Laboratories Complex, Pakistan

A volume in the Advances in Environmental Engineering and Green Technologies (AEEGT) Book Series

IGI Global
DISSEMINATOR OF KNOWLEDGE

Published in the United States of America by
IGI Global
Engineering Science Reference (an imprint of IGI Global)
701 E. Chocolate Avenue
Hershey PA, USA 17033
Tel: 717-533-8845
Fax: 717-533-8661
E-mail: cust@igi-global.com
Web site: http://www.igi-global.com

Library of Congress Cataloging-in-Publication Data

Names: Nazir, Rabia, 1977- editor.
Title: Nanotechnology applications in environmental engineering / Rabia Nazir, editor.
Description: Hershey, PA : Engineering Science Reference (an imprint of IGI Global), [2019] | Includes bibliographical references and index.
Identifiers: LCCN 2017055226| ISBN 9781522557456 (hardcover) | ISBN 9781522557463 (ebook)
Subjects: LCSH: Nanostructured materials--Environmental aspects. | Environmental engineering--Materials.
Classification: LCC TD196.N36 N368 2019 | DDC 628.028/4--dc23 LC record available at https://lccn.loc.gov/2017055226

This book is published in the IGI Global book series Advances in Environmental Engineering and Green Technologies (AEEGT) (ISSN: 2326-9162; eISSN: 2326-9170)

British Cataloguing in Publication Data
A Cataloguing in Publication record for this book is available from the British Library.

For electronic access to this publication, please contact: eresources@igi-global.com.

Advances in Environmental Engineering and Green Technologies (AEEGT) Book Series

Sang-Bing Tsai
University of Electronic Science and Technology of China Zhongshan Institute, China
Ming-Lang Tseng
Lunghwa University of Science and Technology, Taiwan
Yuchi Wang
University of Electronic Science and Technology of China Zhongshan Institute, China

ISSN:2326-9162
EISSN:2326-9170

Mission

Growing awareness and an increased focus on environmental issues such as climate change, energy use, and loss of non-renewable resources have brought about a greater need for research that provides potential solutions to these problems. Research in environmental science and engineering continues to play a vital role in uncovering new opportunities for a "green" future.

The **Advances in Environmental Engineering and Green Technologies (AEEGT)** book series is a mouthpiece for research in all aspects of environmental science, earth science, and green initiatives. This series supports the ongoing research in this field through publishing books that discuss topics within environmental engineering or that deal with the interdisciplinary field of green technologies.

Coverage

- Electric Vehicles
- Water Supply and Treatment
- Policies Involving Green Technologies and Environmental Engineering
- Pollution Management
- Renewable energy
- Alternative Power Sources
- Industrial Waste Management and Minimization
- Biofilters and Biofiltration
- Waste Management
- Radioactive Waste Treatment

IGI Global is currently accepting manuscripts for publication within this series. To submit a proposal for a volume in this series, please contact our Acquisition Editors at Acquisitions@igi-global.com or visit: http://www.igi-global.com/publish/.

The Advances in Environmental Engineering and Green Technologies (AEEGT) Book Series (ISSN 2326-9162) is published by IGI Global, 701 E. Chocolate Avenue, Hershey, PA 17033-1240, USA, www.igi-global.com. This series is composed of titles available for purchase individually; each title is edited to be contextually exclusive from any other title within the series. For pricing and ordering information please visit http://www.igi-global.com/book-series/advances-environmental-engineering-green-technologies/73679. Postmaster: Send all address changes to above address.

Titles in this Series

For a list of additional titles in this series, please visit: www.igi-global.com/book-series

Food Systems Sustainability and Environmental Policies in Modern Economies
Abiodun Elijah Obayelu (Federal University of Agriculture – Abeokuta (FUNAAB), Nigeria)
Engineering Science Reference • copyright 2018 • 371pp • H/C (ISBN: 9781522536314) • US $195.00 (our price)

Microbial Biotechnology in Environmental Monitoring and Cleanup
Pankaj (G. B. Pant University of Agriculture and Technology, India) and Anita Sharma (G. B. Pant University of Agriculture and Technology, India)
Engineering Science Reference • copyright 2018 • 427pp • H/C (ISBN: 9781522531265) • US $235.00 (our price)

Handbook of Research on Renewable Energy and Electric Resources for Sustainable Rural Development
Valeriy Kharchenko (Federal Scientific Agroengineering Center VIM, Russia) and Pandian Vasant (Universiti Teknologi PETRONAS, Malaysia)
Engineering Science Reference • copyright 2018 • 672pp • H/C (ISBN: 9781522538677) • US $345.00 (our price)

Green Production Strategies for Sustainability
Sang-Binge Tsai (University of Electronic Science and Technology of China (Zhongshan Institute), China & Civil Aviation University of China, China) Bin Liu (Shanghai Maritime University, China) and Yongian Li (Nankai University, China)
Engineering Science Reference • copyright 2018 • 325pp • H/C (ISBN: 9781522535379) • US $235.00 (our price)

Innovative Strategies and Frameworks in Climate Change Adaptation Emerging Research and Opportunities
Alexander G. Flor (University of the Philippines Open University, Philippines) and Benjamina Gonzalez Flor (University of the Philippines, Philippines)
Engineering Science Reference • copyright 2018 • 165pp • H/C (ISBN: 9781522527671) • US $165.00 (our price)

Utilizing Innovative Technologies to Address the Public Health Impact of Climate Change Emerging Research and Opportunities
Debra Weiss-Randall (Florida Atlantic University, USA)
Engineering Science Reference • copyright 2018 • 295pp • H/C (ISBN: 9781522534143) • US $185.00 (our price)

Economical and Technical Considerations for Solar Tracking Methodologies and Opportunities for Energy Management
S. Soulayman (Higher Institute for Applied Sciences and Technology, Syria)
Engineering Science Reference • copyright 2018 • 647pp • H/C (ISBN: 9781522529507) • US $245.00 (our price)

IGI Global
DISSEMINATOR OF KNOWLEDGE

701 East Chocolate Avenue, Hershey, PA 17033, USA
Tel: 717-533-8845 x100 • Fax: 717-533-8661
E-Mail: cust@igi-global.com • www.igi-global.com

Table of Contents

Foreword xiv

Preface xvi

Acknowledgment xxi

Section 1

Chapter 1
Introduction to Environmental Nanotechnology: E-Nano 1
Rabia Nazir, Pakistan Council of Scientific and Industrial Research, Pakistan

Chapter 2
Properties of Nanomaterials and Environment 28
Phool Shahzadi, Pakistan Council of Scientific and Industrial Research (PCSIR) Labs, Pakistan

Chapter 3
Characterization of Nanomaterials 44
Mintu Maan Dutta, Gauhati University, India

Chapter 4
Green Approaches to Environmental Sustainability 81
Lubna Tahir, Pakistan Council of Scientific and Industrial Research, Pakistan

Section 2

Chapter 5
Nanomembrane Applications in Environmental Engineering 103
Sumaira Nosheen, Pakistan Council of Scientific and Industrial Research Lahore, Pakistan

Chapter 6
Nanotechnology for Air Remediation 121
Shafaq Mubarak Mubarak, Pakistan Council of Scientific and Industrial Research, LLC, Pakistan

Chapter 7
Nanomaterial Surface Modifications for Enhancement of the Pollutant Adsorption From Wastewater: Adsorption of Nanomaterials 143
Hamidreza Sadegh, West Pomeranian University of Technology Szczecin, Poland
Gomaa A. M. Ali, Al-Azhar University, Egypt
Hamid Jafari Nia, Islamic Azad University, Khorramabad Branch, Iran
Zahra Mahmoodi, Islamic Azad University, Science and Research Branch, Iran

Chapter 8
Nanotechnology for Water Environmental Application: Functionalized Silica Hybrids as Nano-Sorbents 171
Saima Nasreen, The Women University Multan, Pakistan
Uzaira Rafique, Fatima Jinnah Women University, Pakistan

Chapter 9
Application of Magnetic Nanomaterials for Water Treatment 201
Ambreen Lateef, University of the Punjab, Pakistan
Rabia Nazir, Pakistan Council of Scientific and Industrial Research, Pakistan

Chapter 10
Wastewater Treatment in Removal of Heavy Metals: Nanotechnology Applications in Environmental Engineering 220
Tehseen Yaseen, Pakistan Council of Scientific and Industrial Research, Pakistan
Anum Yaseen, University of Punjab, Pakistan

Chapter 11
Nanoparticles for Degradation of Organic Pollutants 241
Sana Aslam, Government College Women University Faisalabad, Pakistan
Matloob Ahmad, Government College University Faisalabad, Pakistan

Chapter 12
Wastewater Treatment and Role of Green Synthesized Metal Oxide Nanocomposites 268
Durr-e-Shahwar Noman, University of South Australia, Australia
Gary Owens, University of South Australia, Australia

Chapter 13
Recent Progresses in Membranes for Proton Exchange Membrane Fuel Cell (PEMFC) for Clean and Environmentally Friendly Applications 308
Jay Pandey, Indian Institute of Technology Delhi, India

Chapter 14
Toxicocinetic and Mechanisms of Action of Nanoparticles 344
Abderrezak Khelfi, National Center of Toxicology, Algeria

Compilation of References 369

About the Contributors 464

Index 467

Detailed Table of Contents

Foreword xiv

Preface xvi

Acknowledgment xxi

Section 1

Chapter 1
Introduction to Environmental Nanotechnology: E-Nano 1
Rabia Nazir, Pakistan Council of Scientific and Industrial Research, Pakistan

With the advent of industrialization, there are new technological developments going on which have triggered the new researches and fields. Nanotechnology is one such field that is marked as twenty-first century industrial revolution and led to the development of various products that have impacted life in almost every field. Among these fields, nanotechnology has contributed significantly to the environment protection and its remediation by addressing issues that have been threatening the human for long. In this regard, several nanomaterials have been synthesized and many more are still in pipeline. No doubt the field has made great contributions and has bestowed human with various amenities by enhancing atom efficiency, use of fewer chemicals, less energy and other resource, reduction in waste, providing better materials, and technology for environmental applications promising environmental sustainability, however there are certain negative impacts that still need to be explored.

Chapter 2
Properties of Nanomaterials and Environment 28
Phool Shahzadi, Pakistan Council of Scientific and Industrial Research (PCSIR) Labs, Pakistan

The chapter provides a timely review of the various properties of nonmaterial and their applications into environmental compartments. An extensive variety of poisonous chemicals is discharged into the environment because of globalization and industrialization. The dimensional, compositional, geometric, and structural properties are fundamental to convey usefulness of the nanomaterials. The controlled sizes and shapes of nanoparticles are anticipated to yield unique catalytic, electrochemical, and photochemical properties. The electrochemical properties of monolayer-functional metal nanoparticles are expected to be controlled by the particle sizes. Metal nanomaterials have interesting optical properties due to strong surface plasmon absorption and field enhancement effects; metal oxides lack visible absorption due to very large bandgap. Nanocomposites have complex optical properties. Nanomaterials present gigantic advantages on diverse applications, catalysis, imaging, biotechnological, and sensor applications due to their improved properties.

Chapter 3
Characterization of Nanomaterials 44
Mintu Maan Dutta, Gauhati University, India

This chapter provides a general introduction to the new learners in the field of the synthesis and characterization of nanoparticles. The people who are familiar with this field but are unable to interpret the data will learn how to interpret from various data available. Keeping in the mind the usefulness of numerical data, this chapter includes a number of tables and graphs for the help and easy understanding of the students.

Chapter 4
Green Approaches to Environmental Sustainability... 81
Lubna Tahir, Pakistan Council of Scientific and Industrial Research, Pakistan

The development of different eco-friendly technologies in material synthesis is playing its role to expand their biological applications. Nanoparticles are synthesized chemically, but because of their hazardous effects, the researchers have now turned to biological systems for inspiration. There is a growing concern about the possible results of effects of chemical synthesis of nanoparticles. Therefore, the aim of this chapter is to provide an overview of how nanomaterials can be synthesized using green route to play its role in environmental sustainability as green synthesis processes are considered to be a safe alternative to routine processes due to their cost effectiveness, easy to handle, and environmentally friendly nature. In fact, it opens a new scope for phyto-chemists by discouraging the use of toxic chemicals.

Section 2

Chapter 5
Nanomembrane Applications in Environmental Engineering 103
Sumaira Nosheen, Pakistan Council of Scientific and Industrial Research Lahore, Pakistan

This chapter delivers an outline of the strategies and techniques that are used and developed for the fabrication of membrane techniques/methods for application in environmental engineering. Human activities are the cause of increased hazardous gases in atmosphere mainly in soil and water. Nanotechnology deals the ability to control matter at the nano-scale level. Materials prepared are in nano level and thus they will possess some special properties to deal with specific functions. One of the major aspects of nanotechnology is nanomembrane fabrication which is mainly employed for water purification plants. The chapter is specifically offers a full understanding of the technologies and laws used to synthesize membranes. The chapter also provides an introduction to techniques to characterize nanomembranes.

Chapter 6
Nanotechnology for Air Remediation 121
Shafaq Mubarak Mubarak, Pakistan Council of Scientific and Industrial Research, LLC, Pakistan

The world around us is a gift of God. A thrust to know and reconnoiter the environment around us is innate. Man has drastically explored and utilized the resources hidden in nature, but unfortunately in this sprint of development, the natural environment is severely affected. It is the need of the hour to focus on methodologies for environmental remediation. Many technologies have been developed to reduce the pollution causing factors. Use of nanotechnology for the sake of saving environment is an emerging field. Nano-technology is based on nano-sized (smaller than 1 micron) materials. Nanosize particles have initiated the advancement in new and low cost techniques for environmental pollution control including air pollution.

Chapter 7
Nanomaterial Surface Modifications for Enhancement of the Pollutant Adsorption From Wastewater: Adsorption of Nanomaterials 143
Hamidreza Sadegh, West Pomeranian University of Technology Szczecin, Poland
Gomaa A. M. Ali, Al-Azhar University, Egypt
Hamid Jafari Nia, Islamic Azad University, Khorramabad Branch, Iran
Zahra Mahmoodi, Islamic Azad University, Science and Research Branch, Iran

With the development of dyeing, textile, leather, paper, and other chemical industries, an increasing amount of dye wastewater containing refractory organic dyes is discharged. Undoubtedly, much high content dye wastewater will lead to serious environmental issues such as color pollution, light penetration interference, and virulence to aquatic organisms, even endanger human health. Therefore, it is an imminent problem and has become a global concern to degrade dye wastewater efficiently. So far, many techniques have been used to degrade dyeing wastewater, such as chemical degradation, biological degradation, photochemical degradation, coagulation, membrane filtration, and combined methods. These methods have certain impacts on the degradation of dye wastewater, but usually with slow degradation rate, complex and high operation costs, as well as easily causing secondary pollution. The adsorption process is a simple, effective, and low-cost way to remove dyes.

Chapter 8
Nanotechnology for Water Environmental Application: Functionalized Silica Hybrids as Nano-Sorbents 171
Saima Nasreen, The Women University Multan, Pakistan
Uzaira Rafique, Fatima Jinnah Women University, Pakistan

In developing countries like Pakistan, industries do not allocate budget for control and treatment of pollution due to limited resources. Most of the pollutants emitted from these industries are usually reusable but their capture from waste is highly uneconomical. This goal can be attained by synthesizing the advanced materials such as silica hybrids, the class of silica hybrids consisting of inorganic silica linked with organic ligands to form network structure. Their high porosity, surface area, and crystallinity provide a way for removing large amount of pollutants with small quantity of silica hybrids. These applications include the use of powder hybrids as adsorbent for removal of different pollutants such as heavy metals, persistent organic pollutants, and toxic organic chemicals. These functionalized hybrids can be further fabricated into thin films. This new form will be further used for catalytic degradation of various pollutants. Besides all these applications, these hybrids will have the potential to be used as sensors for detecting various pollutants in industries.

Chapter 9
Application of Magnetic Nanomaterials for Water Treatment 201
Ambreen Lateef, University of the Punjab, Pakistan
Rabia Nazir, Pakistan Council of Scientific and Industrial Research, Pakistan

The availability of clean drinking water becomes a critical issue for all the people of the world due to a rapid increase in population and industrialization. The water bodies get contaminated due to the discharge of wastewater, that will not only disturb the aquatic life but also badly affect human health. Therefore, different methods are adopted to treat the contaminated water to make it clean and safe for people. In last few years, the nanomaterials have gained much attention for water treatment because of their unique properties. Among all nanomaterials, magnetic nanomaterials are considered more efficient and attractive because of their easy separation and reusable property. In this chapter, a brief review related to synthesis and characterization of MNM was studied along with their application in removal of dyes, heavy metals, and microbes from wastewater through simple adsorption processes.

Chapter 10
Wastewater Treatment in Removal of Heavy Metals: Nanotechnology Applications in Environmental Engineering 220
Tehseen Yaseen, Pakistan Council of Scientific and Industrial Research, Pakistan
Anum Yaseen, University of Punjab, Pakistan

Nanotechnology is the area of nano science that shows great potential to establish a new process for wastewater treatment. It has been applied on a nanometer scale level. Currently, limited water resources and real treatment of wastewater is a chief requirement for the growing economy. It is in great demand to introduce the progressive wastewater treatment technologies. Therefore, the modern innovative processes in nanomaterial sciences have been appealing the target of scientists. The chapter addresses the developments in nanotechnology with respect to wastewater treatment, especially the removal of heavy metals and to the environmental applications. It will discuss the application of different classes of nanomaterials for wastewater treatment in removal of heavy metals and its possible effects to the environment. Therefore, the scope is to offer an overview of how nanomaterials are causing concerns related to heavy metal removal for water and in the surrounding environment.

Chapter 11
Nanoparticles for Degradation of Organic Pollutants 241
Sana Aslam, Government College Women University Faisalabad, Pakistan
Matloob Ahmad, Government College University Faisalabad, Pakistan

Water pollution is becoming a serious threat to the environment. The access to safe drinking water, especially for the poor, is reaching an alarmingly low level, and the number of people suffering from kidney, stomach, and liver diseases, etc. is increasing every day. Among various pollutants, organic pollutants have major contribution in the contamination of water. These pollutants include toxic dyes, pharmaceutical compounds, industrial chemicals, and organic solvents, etc. Various routes are being employed for the removal/degradation of these pollutants. The use of nanoparticles (i.e., metal nanoparticles, nanocomposites, carbon nanotubes, and graphene oxide-based nanocomposite materials) possesses significant position as photocatalysts. This chapter has a focus on the environment remediation using these nanomaterials and contains the recent work in this field.

Chapter 12
Wastewater Treatment and Role of Green Synthesized Metal Oxide Nanocomposites 268
Durr-e-Shahwar Noman, University of South Australia, Australia
Gary Owens, University of South Australia, Australia

Water pollution by metalloids is a global environmental concern. Owing to their propensity for bioaccumulation, water solubility, and interaction with environment, they are threatening both human and ecosystem health. Inherent limitations like low efficiency, sensitive operating conditions, and high capital and operating costs are associated with conventional removal methods which restricts adoption of these technologies on large scale. While adsorption is commonly recognized as both an effective and affordable remediation technology, many common adsorbents often have inherited limitations including non-renewability and high operating costs. Thus, limitations in conventional remediation technologies have headed to the rapid progression of new avenues for advanced treatment technologies for metalloid pollutant removal such as green nanotechnology. In contrast to many of the currently available adsorbents, nanoparticles often have unique properties such as tiny size, more active sites and big surface area, easy separation, and high reactivity that enhance removal efficiencies.

Chapter 13
Recent Progresses in Membranes for Proton Exchange Membrane Fuel Cell (PEMFC) for Clean and Environmentally Friendly Applications 308
Jay Pandey, Indian Institute of Technology Delhi, India

Fuel cell has become an emerging renewable energy device with potential to meet energy demand by portable and transport applications with zero-emission, easy operation, and compact design. The chapter provides an insight into design and development of membranes for PEMFCs and recent progresses made in membranes so far. Although majority of research has focused on fluorinated and non-fluorinated membranes, these polymeric membranes have showed deteriorated properties at elevated temperature (>80oC) and lower relative humidity (30%). Considering the major issues with polymeric membranes, the authors have reviewed inorganic-organic nanocomposite membranes showing improved physical and electrochemical properties at elevated temperature and lower relative humidity. Recently, metal-organic framework (MOF), a novel and unique material, has attracted tremendous attention due to their enhanced proton conductivity, easy functionality, and stability. MOFs have also exhibited excellent compatibility with different polymeric materials that are also discussed in this chapter.

Chapter 14
Toxicocinetic and Mechanisms of Action of Nanoparticles 344
Abderrezak Khelfi, National Center of Toxicology, Algeria

Human exposure to nanoparticles has been dramatically increased in the past 25 years as a result of the rapidly developing field of nanotechnology. Many have recognized the importance of identifying potential effects on human health associated with the manufacture and use of these important technology.

Many questions remain unanswered regarding the short- and long-term effect, systemic toxicity, and carcinogenicity. Engineered nanoparticles can be taken up by the human body via inhalation, ingestion, dermal uptake, and injection. They can reach the bloodstream and ultimately affect multiple body organs such as liver and spleen or even transcend the blood-brain barrier. Because of the huge diversity of materials used and the wide range in size of nanoparticles, these effects will vary a lot. Local and systemic adverse effects consist of primarily inflammatory reactions. Other observed effects include generation of reactive oxygen species and subsequent oxidative stress, disruption of proteins, DNA, mitochondria and membrane structures, as well as changes in cell signaling pathways.

Compilation of References 369

About the Contributors 464

Index 467

Foreword

Environmental Nanotechnology or E-Nano is sprouted as a new branch of Nanotechnology that addresses the multifarious applications nanotechnology play in environmental engineering. This field has a strong potential to address the current issues that are being faced by human like reducing the already entered pollutants in the system and also avoiding the more pollutants to enter into the system. The E-nano is a combination of technologies and processes that are used to put a step forward towards sustainable development and hence consequently to sustainable environment. These technologies and processes can be summarized into three main heads that can become and portray the true essence of environmental nanotechnology: 1) environmental monitoring, 2) environmental remediation, and 3) pollution prevention.

Environmental monitoring deals with the detection of contaminants at low detection limits so that the proper measures can be taken to remove these pollutants before they can enter in environment in larger amounts. In this regard nano-sensors have played a major role that enable low level detections and with high sensitivity that was earlier not possible with the use of traditional sensors.

Environmental remediation includes both the treatment of effluents so that the influx of pollutants into the environment can be controlled. In addition to that it also includes the technologies and materials used to remove the pollutants already present in water, air and soil so as to diminish the negative impacts caused by their occurrence. This includes the use of nanoparticles, nanocomposites, nanomembanes and nanofilters along with associated technologies that use these materials to purify the environment.

Pollution prevention deals with the modification of processes and technologies that ensure use of green approaches to avoid the generation of contaminants in first place so that all the further steps might not be required that resulted in the cost saving, resource conservation and better efficacy of the system.

This book provides a valuable information and overview about these three aspects of the environmental nanotechnology covering all the necessary details required to cover this broader aspect as aforementioned. The challenges faced by today's researchers and engineers in developing new materials and designing new processes have made their endeavor much more complicated. The book in a short way caters the entire topic that can help in getting know how of the multifaceted aspects of nanotechnology that can help in resolving these issues and challenges. In this new age of industrial revolution marked by nanotechnology, it is necessary to provide simple and basic information on different aspects of environmental nanotechnology that can benefit students, professionals and researchers. This book is a good step in that direction.

Shahzad Alam
Pakistan Council of Scientific and Industrial Research, Pakistan

Foreword

Shahzad Alam *has been working as Chairman PCSIR since September 2015 and prior to that he had been performing duties as Director General, PCSIR Lahore. He had paved the way for establishment of material science and nanotechnology innovation culture in the country by keeping focus on development of advanced engineering products with industry focused application. He received his Ph.D. degree from University of Leeds, UK in 1991 and did three Post doctorates from Mechanical Engineering Laboratory Tsukuba, Japan. His current field of specialization is nanomaterials, laser and plasma hybrid sprayed functional coatings and biomaterials. Dr. Shahzad Alam has more than 100 research publications published in journals, conferences/ symposiums local and abroad and 22 technical reports. He is the co-author of three proceedings of international conferences and member of prestigious national and international societies and associations engaged for the promotion of materials science. Bearing in mind the global advancements in Science and Technology, Dr. Alam envisaged the idea of creating an infrastructure for Materials Research in the country and ultimately succeeded in establishing Pakistan Institute of Technology for Minerals and Advanced Engineering Materials (PITMAEM) in 2002, under a PSDP project of the Ministry of Science & Technology (MoST). The establishment of such a facility was indispensable for the country to keep abreast with the technological revolutions worldwide. The institute under his supervision is contributing a lot for industrial capacity building, export enhancement, import substitution and human resource development in the various fields of materials science.*

Preface

INTRODUCTION

Nanotechnology is the twenty-first century revolution that has impacted each and every aspect of life by its small size. The industry by manipulating the small size and its associated properties has resulted in many diverse amounts of applications ranging from simple house hold things to complex engineering products. The nanotechnology utilization is the field of environment only has taken many directions and is helping in different directions like monitoring, treatment, remediation and prevention. Hence, to focus on this upcoming role of nanotechnology and to give an overview of the areas of environment where nanotechnology has played critical role, the current book is launched.

The book is themed at highlighting the role of nanotechnology in environment. The main objective of the book is to provide a thorough and comprehensive look at the nanomaterials and their impact on environment. The recent advances in the field of Nanotechnology have shown that the nano-adsorbents, nano-membranes, nano-catalysts and nano-sensors etc. can effectively deal with the challenges of today's environment. The book will also provide the fundamental basis needed for understanding of nanomaterials synthesis and their unique properties and characterization.

The book targets the audience composed of professionals and researchers working in the field of information and knowledge in field of nanotechnology and its applications in environmental engineering. Moreover, the book is a good source to provide insights and support Environmental scientists, Nanotechnologists Chemists and Engineers concerned and working in field of environmental nanotechnology.

ORGANIZATION OF THE BOOK

The book is divided into two main sections consisting of total 14 chapters, written by the authors who are working in their field of expertise since long time and have sufficient know-how about the topics they cater. The first section presents the underlying science behind nanotechnology, the tools used to create and characterize nanostructures and the various unique properties of nanomaterials. The second section discusses the current and potential applications of nanomaterials in environmental science and engineering; showcasing how nanomaterials can be tailored to address some of the environmental remediation and sensing/detection problems faced today. Areas covered include air and water remediation. The book will also highlight few aspects of toxicology related impacts to the environment and human.

The first chapter deals with general introduction to the field of environmental nanotechnology famously known as E-nano. Three main sub fields are engrossed in E-nano i.e. Environmental monitoring,

Environmental remediation and Pollution prevention. Nano-sensors play important role in environmental monitoring owning to their high sensitivity and low detection limits. By use of these nano-sensors it becomes much easier to detect the contaminants level at much lower concentrations and in different types of setups. Different types of nanomaterials are used to synthesize these nanosensors and are utilized based on their sensing properties towards different pollutants. Second important side of E-nano is environmental remediation that basically focuses on the clean-up technologies that helps in removal of pollutants from the water, air and soil. In this area the role of nano-adsorbents and nano-catalysts is highlighted in the chapter with focus on three types of pollutants i.e. organic, inorganic and microbes. The third aspect of E-nano is pollution prevention that is basically a way to support green technologies by use of which the sources of pollution are mitigated. This is made possible by having innovations in products and processes as well as enhancing energy and resource efficacy. This was made possible by modifying these aspects in the industrial processes and especially the ones that earlier on lead to generation of huge waste. The chapter also focuses on the balance between the benefits and risks this field caters.

The second chapter focuses on the properties of the nanomaterials. Nanomaterials having small size ranging from 1-100nm and high surface area give them properties that are different from their bulk counterparts. These materials can show different physical and chemical properties that can be manipulated by incorporating other materials or by modifying their size and shape. The chapter briefly describes the electrical, magnetic, catalytic, mechanical and biological properties of the nanomaterials. The chapter also discusses the impacts of these properties on environment by altering the ways these nanomaterials interacts with the pollutants. Further these properties can play significant role in deciding the faith and behavior of nanomaterials in the environment.

The nanomaterials are characterized by using different techniques that are overviewed in third chapter. Owing to their small size, these materials require special instruments for their characterization. The chapter details about the spectroscopy, X-ray diffraction, microscopy, thermal analysis, Vibrating sample magnetometry etc. Spectroscopy is widely used to determine the geometry, aggregation and surface composition of the nanomaterials while X-ray diffraction is helpful in analyzing crystallite size, phase and material identification. The technique of microscopy includes scanning electron microscopy (SEM), transmission electron microscopy (TEM) and atomic force microscopy (AFM). All these microscopic techniques use different underlying working principles to determine the nanoparticle shape, size, particle size distribution, composition, crystallographic information, thickness, phases etc. in the samples. Thermal analysis techniques are helpful in determining the thermal stability of the sample along with the phase transformations that can occur as a result of heating. It also gives information about the decomposition, dehydration, adsorption, desorption behavior of the nanomaterials. Such techniques are specifically helpful in getting multiple information in case of polymeric nanocomposites. Magnetic properties of the nanomaterials are judged with help of Vibrating sample magnetometer giving information about type of magnetism, magnetization, remanence, coercivity, hysteresis etc. The adsorption behavior of the nanomaterials is studied by applying various kinetic models that are also detailed in the chapter.

These nanomaterials are fabricated or synthesized using various approaches like sol-gel, chemical vapour deposition, ball milling, hydrothermal, micro-emulsion, sonication and many more. One context of the environmental aspect of nanotechnology is to focus on the designing of novel and new synthetic approaches that help in pollution prevention by using principals of green chemistry. Fourth chapter of this book focus light on the green synthetic approaches of nanomaterials and the need to promote such approaches. These green methods are usually based on utilization of plant extracts of different origin prepared in various kinds of solvents including the aqueous ones. The chapter details the preparation of some silver nanoparticles that is the most studied nanoparticle by using this green approach. The chapter also focuses on the various techniques peculiarly required for the characterization of silver nanoparticles prepared by green route followed by their detailed application overview in medicine/ drug delivery, water treatment, agriculture and food. The chapter also briefly points out the environmental aspects of the nanoparticles.

The membrane technology is one of the promising technologies of the nano field that have played a major role in environmental pollution control, remediation and treatment. The fifth chapter deals with nano-membranes detailing their introduction, types, synthetic approaches, characterization and applications. The different kinds of membranes are isotropic, anisotropic, micro-porous, nonporous-dense, electrically charged, ceramic, metal and liquid. The synthetic techniques used for their preparation are mainly focused on keeping the barrier perm selective which means that it restricts the movement of molecules in it in a very specific manner. These membranes are characterized using two main approaches i.e. static and dynamic techniques which can further be sub-divided into different categories. The characteristic tests for these membranes are mostly similar to the ones required by simple nanomaterials but in addition to that specific characterization techniques are also required to get insight into the properties and behavior of these membranes. These membranes have found ample applications in the environmental engineering field and are frequently used in industrial setups to treat the effluents.

The sixth chapter deals with the E-nano implication in the air remediation. The chapter smoothly runs around the criteria for air pollutants and the ways that are used to mitigate these pollutants from the environment. In this aspect the chapter discussed the role of nanomaterials as nano-sensors, nano-adsorbents, nano-filters, nano-catalysts, antibacterial agents and dendimers. The role of nano-sensors like solid-state and gas used to monitor the pollutants in air with high precision is detailed with respect to their efficacy and types of materials used for detecting. Another role of nanomaterials in air remediation is by use of nano-adsorbents. The role of different nanomaterials like carbon nanotubes both single and multi-walled, Fullerenes, Graphene as nano-adsorptive materials has been discussed for absorbing pollutants like metal oxide, volatile organic compounds and microbes. Special attention is also given to highly carcinogen pollutants removal that includes mercury and formaldehyde using different kinds of nanomaterials. Another aspect of utilization of nanotechnology in air pollution control is through use of nano-filters that have been effectively designed for the filtration of particulate matter, industrial emissions and vehicle exhausts. The chapter also highlights the use of nano-catalysts in convertor reactions of automobile vehicles and industrial exhaust which convert the toxic gases to harmless gases. The role of various catalysts like titanium and carbon based nano-catalyst have been emphasized with details about their mode of working and practical applications. The role of these nanomaterials as bacterial decontaminant is also being addressed in this chapter.

The water and waste treatment was also tried by using surface modifications of different nanomaterials. The seventh chapter of the book deals with different ways that surface modification can be made effective and its impact on enhancement of adsorption capacity of these modified nanomaterials. The various types of nano-adsorbents are discussed that is metal oxide based nanomaterials, carbon based nanomaterials (including carbon nanotubes, activated carbon) and polymer based nanomaterials. There are several factors that affect the adsorption phenomena are also briefed in this chapter.

Water treatment applications of nanomaterials are widely studied by using simple single system based small sized particles as well as complex ones like functionalized silica hybrids that have been discussed in the eighth chapter of this book. The chapter briefly describes the complexity and present status of water pollution and prevailing contaminants and the advantages of using nanomaterials for water purification. Functionalized silica hybrids are chosen as potential nano-adsorbents based on the ample advantages offered by silica and especially the ones that are mesoporous. Surface area, particle size and surface morphology are actually basic features of nano-sorbents that make them unique for specific sorbates and hence owing to their high adsorption capacity and high catalytic activity as well they can enjoy high efficacy by use of their multi-action approach towards removal of various kinds of contaminants. There are different approaches to synthesize these materials which are briefly discussed in the chapter. The contaminants that have been removed by these functionalized silica hybrids are metal ions, phenols and polycyclic Aromatic Hydrocarbons. Sorption of gas molecule like nitrogen have also been achieved with the help of these advanced engineered nanomaterials that is comprehensively been addressed in the chapter.

In addition to functionalized nanomaterials, magnetic nanoparticles and their nano-composites have also been widely used to treat water and remove contaminants by providing additional advantage of easy separation of these magnetic nanomaterials by simple application of applied magnetic field. The role of these magnetic nanomaterials in water and wastewater treatment is comprehensively been addressed in ninth chapter of this book. These magnetic nanomaterials are characterized by having magnetic properties owing to presence of iron and other magnetically active metals making these materials preferred over the non-magnetic ones as the later required additional separation technologies leading to extra cost and time. The chapter also highlights different methodologies for the preparation of these magnetic materials which are sol-gel, microemulsion, co-precipitation, chemical polymerization and many others. These magnetic nanoparticles are effectively used for treatment of dyes, heavy metals and even microbes. The types of magnetic materials used for each of the case and the resulting removal capacity is briefly detailed in the chapter giving readers a bird-eye view about the mechanism and advantages of these magnetic systems.

Removal of heavy metals from waste and waste water using nanomaterials is discussed in tenth chapter of the book. Heavy metals are considered to be one of the most toxic contaminants that can even in very small amount portray serious health impacts to the aquatic flora and fauna as well as human. Hence there effective removal from the water bodies is highly desirable and also in very high amounts and less time. Nanotechnology had played immense role in this regard and has always remained as a topic of interest and attention to their high removal rates. The chapter briefs about the synthetic approaches used to prepare the nanomaterials that have been used as heavy metal adsorbents followed by their characterization techniques. The heavy metal removal was discussed in respect with nano-adsorbents, nano-catalysts and nano-membranes. The merger of biological-nanotechnology processes for better and enhanced removal of heavy metals was also elaborated in the chapter.

The organic pollutants are other hazardous substances which have adverse effects on human health like dysfunction of liver, adverse effects on brain, reproductive system, kidney and central nervous system. Nanotechnology is considered as one of the advanced technologies that can help in the removal of these complex organic molecules. The eleventh chapter focuses on use of various nanomaterials like carbon nanotubes, graphene oxide, metal organic frame works, and metal based. The organic contaminates removed by using these nanomaterials are pharmaceutical products, dyes, phenols etc. The detailed mechanism of photocatalytic degradation of these organic pollutants in presence of photocatalysts was also discussed in the chapter.

The twelfth chapter of the book deals with specific applications of nanomaterials prepared by using green approach as already catered in chapter fourth of the book. The chapter goes smoothly covering various aspects of these green synthesized materials like different synthetic approaches, characterization, surface chemistry, plant extracts used, mechanism of synthesis and factors effecting the growth of these nanoparticles. The main focus of the chapter was kept on magnesium and zinc oxide nanoparticle and their use in treatment technologies.

Addressing their power demand through new and renewable sources can reduce demand on the large stationary power generation systems and help in environment sustainability through non-renewable resources conservation. Chapter 13 of the book is based on the this aspect of energy consumption through use of fuel cell, which utilizes large range of hydrocarbon fuels, has emerged as a potential candidate to fulfill the energy demand of portable applications largely. Performance of fuel cells depends on the judicious design and selection of effective and robust materials like solid proton exchange membranes (PEMs), electro-catalysts and bipolar plates. The chapter focuses on polymer electrolyte membrane fuel cell and its comparison with earlier developed and frequently used Nafion membrane and other commercially available membranes. The alternative membranes that were prepared in order to get better properties were also discussed. The membranes made up of inorganic-organic nanocomposite that comprise of stable polymeric support doped with submicron to nano-sized inorganic ion-exchangers are discussed in comprehensive way. The impact of different kinds of ion-exchangers and polymeric support is also discussed along with their role in defining the properties of these membranes. The applications of different cells like polymer electrolyte membrane fuel cell, vanadium redox flow battery, microbial fuel cell etc. were also discussed along with a brief note on challenges for the next generation fuel cell membranes.

The nanotechnology has no doubt bestowed human with multifarious facilities and have given a new revolution to the industry but this technology like all other technologies has certain negative aspects associated with it. The last chapter i.e. fourteenth chapter of the book deals with the toxicological impact of nanomaterials when they become part of the environment either through food chain or direct contact. With nanomaterials becoming an integral part of the economy and a relevant component of consumer products the identification of risk these materials can pose hence must be identified and studied. The chapter discusses the various routes to exposure of these nanoparticles through absorption and distribution followed by their fate in metabolism and excretion. Mechanisms of action of nanoparticles is also discussed that can play a pivotal role in deciding the toxicological impact and potential of these small particles.

CONCLUSION

Nanotechnology has played a pivotal role in refining the industry and amenities of life by giving new definition to the materials, their designing, processing, fabrication and even to the characterization techniques. The vast domain of nanotechnology cannot be catered in this small book of 14 chapters but the editor and authors tried to present an overview of this large horizon. The potential this field of nanotechnology has still not being fully explored and there is a lot at the bottom that still can been discovered. The simple synthetic approaches to the complex novel ones have been used to prepare and fabricate nanomaterials. These approaches can use green ways to prevent pollution and the generation of waste. The nanotechnology field has also aided in treatment of water, air and soil resources. These materials enable easy removal of organic, inorganic and biological contaminant. The pros and cons of every technique are in the hands of human and so is the case with this nanotechnology. Being small size facilities these particles to show diverse properties that are different from their bulk counterparts but on other hand these properties can result in negative impacts to the environment as well as human.

Acknowledgment

The editors would like to acknowledge the help of all the people involved in this project and, more specifically, to the authors and reviewers that took part in the review process. Without their support, this book would not have become a reality.

First, the editors would like to thank each one of the authors for their contributions. Our sincere gratitude goes to the chapter's authors who contributed their time and expertise to this book.

Second, the editors wish to acknowledge the valuable contributions of the reviewers regarding the improvement of quality, coherence, and content presentation of chapters. Most of the authors also served as referees; we highly appreciate their double task.

Section 1

Chapter 1
Introduction to Environmental Nanotechnology: E-Nano

Rabia Nazir
Pakistan Council of Scientific and Industrial Research, Pakistan

ABSTRACT

With the advent of industrialization, there are new technological developments going on which have triggered the new researches and fields. Nanotechnology is one such field that is marked as twenty-first century industrial revolution and led to the development of various products that have impacted life in almost every field. Among these fields, nanotechnology has contributed significantly to the environment protection and its remediation by addressing issues that have been threatening the human for long. In this regard, several nanomaterials have been synthesized and many more are still in pipeline. No doubt the field has made great contributions and has bestowed human with various amenities by enhancing atom efficiency, use of fewer chemicals, less energy and other resource, reduction in waste, providing better materials, and technology for environmental applications promising environmental sustainability, however there are certain negative impacts that still need to be explored.

INTRODUCTION

Nanotechnology according to American Society for Testing and Materials (ASTM) International is defined as "a term referring to a wide range of technologies that measure, manipulate, or incorporate materials and/or features with at least one dimension between approximately 1 and 100 nanometers (nm); such applications exploit the properties, distinct from bulk/macroscopic systems, of nanoscale components" (Sellers et al., 2008). Owing to the manipulation of the properties and materials the field offers, Nanotechnology is considered 21st century industrial revolution. The term was initiated with an idea of Richard Feynman and is understood as a "technology at nanoscale" (Ramsden, 2011). With its multidisciplinary nature the field encompasses production and development of materials at atomic and

DOI: 10.4018/978-1-5225-5745-6.ch001

molecular scales and their integration into large nanostructures. The resulting nanomaterials show different properties from their bulk counterpart which can be tailored as well due to their small size and high surface to volume ratio (Figure 1) (Karkare, 2008; Ratner et al., 2003).

The power of this small technology can be envisioned by the fact that it not only offers improved products but has also resulted in better manufacturing and technological processes in almost every field of life starting from simple house hold products to complex engineered materials (Karkare, 2008). In the environmental arena nanotechnology has contributed in multiplex ways and has significant impact on its sustainability. The chapter presents a brief overview of the role of nanotechnology in environment field.

E-NANO

According to the U.S. Environmental Protection Agency defines nanotechnology as "research and technology development at the atomic, molecular, or macromolecular levels using a length scale of approximately one to one hundred nanometers in any dimension; the creation and use of structures, devices and systems that have novel properties and functions because of their small size; and the ability to control or manipulate matter on an atomic scale" (Sellers et al., 2008; Tratnyek et al., 2006). The three major areas of environment in which nanotechnology plays its role are:

1. Environmental monitoring
2. Environmental remediation
3. Pollution prevention

These three traits of E-Nano along with the famous kinds of nanomaterials that have really bloomed these aspects are briefly overviewed in Figure 2.

Figure 1.

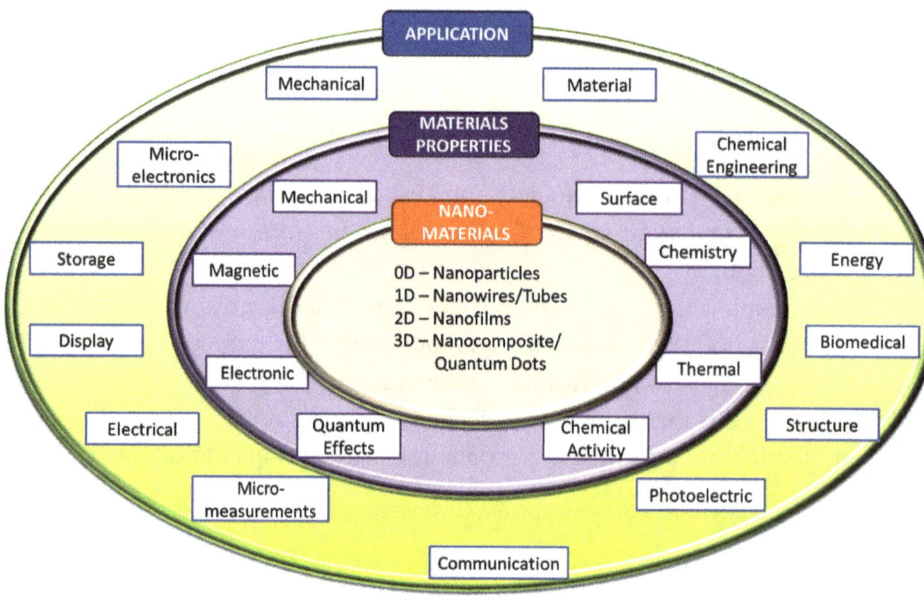

Figure 2.

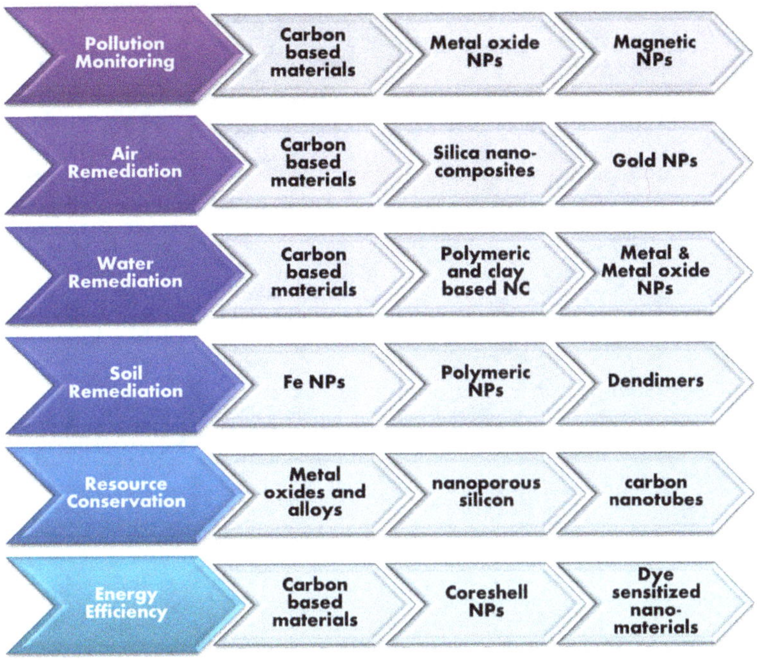

Environmental Monitoring

Environmental monitoring has become much more complex in terms of quality and quantity of pollutants that are generated with hike in world population and rapid urbanization. The generation of toxic wastes is at high levels and its presence at minute levels has raised the question on current available technologies and have surged the scientists to look for more viable alternatives that can detect these pollutants at very small levels with high sensitivity and selectivity and in shortest possible time (Rogers, 2006). Nanotechnology has helped in this regard by development of several nanomaterials and hence nanosensors which have the desired properties.

Nanosensors are defined as "*devices having size, sensitivity and spatial distance between sensor and object in nanoscale*". The general attributes which help in distinguishing the nanosensors are depicted in Figure 3a (Khanna, 2016; Wang et al., 2015) along with the nanomaterials used to synthesize these sensors (Figure 3b). These nanomaterials can be in form of nano-fibers, nano-tubes, nano-belts, nano-probes, nano-wires, nano-grains, quantum dots and thin films (Khanna, 2016). Nanosensors usually make use of any specific property to work on and based on these they are classified into different types which are given in Table 1 and the working principle of these is given in Figure 4 (Akyildiz et al., 2010; Khanna, 2016). Additional classes have also been described by other researchers as optical, electromagnetic, mechanical, vibrational, magnetic, thermal and bio nanosensors (Khanna, 2016; Lim et al., 2006).

Nanosensors enjoy the special advantage in environmental monitoring as they facilitate high sensitivity, lower detection limits and real-time analysis without need to undergo extensive, tedious and lengthy sample preparation procedures (Long et al., 2013).

Several materials are used to prepare and fabricate nanosensors among them the few important ones are discussed.

Figure 3.

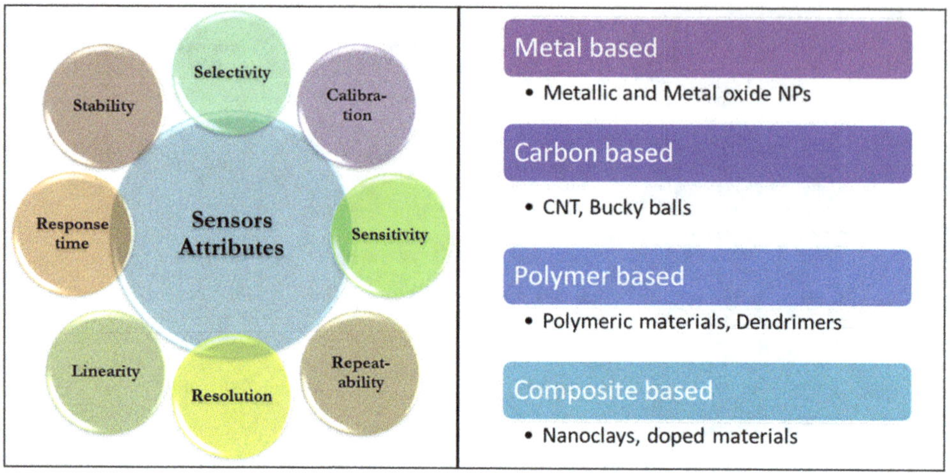

(a) (b)

Table 1. Classification of sensors

Class	Description	Function	Measurand
Physical Sensor	Sensors used to measure physical properties like temperature, force, stress, flow, position, pressure etc.	The working principle is based on the fact that the electronic properties of both nanomaterials change when these are bent or deformed.	Mechanical, acoustic, thermal, radiation, magnetic, electric charge
Chemical Sensor	Sensors used to determine the nature and quantity of the chemical substance like gases, pesticides etc.	The functioning is based on the fact that the electronic properties of nanomaterials change when different types of molecules are adsorbed on top of them.	Identity, concentration, states of elements or compounds
Biological Sensor	Sensors used to monitor bio-molecular processes like antibody -antigen interaction, DNA interaction, protein etc.	The functioning is either based on the fact that the electronic properties of nanomaterials change when different types of biological molecules interact with them (electrochemical biological sensors) or based on the use of noble metal nanoparticles and the excitation using optical waves, the resonant frequency of the surface plasmons resulting from light irradiation changes when different materials are adsorbed on and in between the particles (photometric biological nanosensors).	Identity, concentration, states of bio molecules

Metal Based Nanomaterials

These include metal and metal oxide based materials that have been employed in nanosensors preparation and have offered advantages of being low cost, better life expectancy, robust, lightweight, high sensitivity and swift response time with high accuracy for detection of toxic gases i.e. CO_x, NO_x and NH_3 (Fine et al., 2010). Gold NPs functionalized with fluorescent dye molecules have showed sensitivity for the

Figure 4.

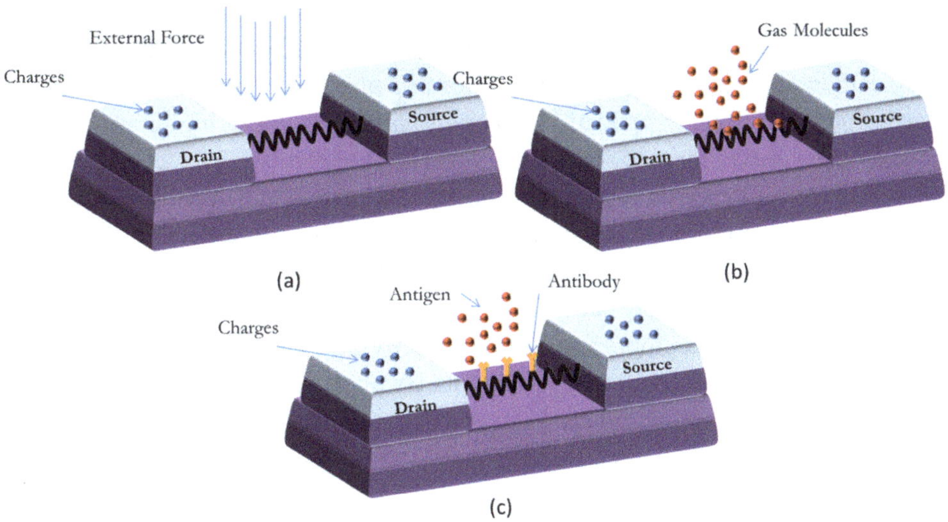

mercury ions in aqueous system. The mechanism of these is based on generation of fluorescent signals based on the released dye molecules from the surface of Au as they interact with the Hg at a distance of 10nm. Improved selectivity was achieved when the dye molecules surface was modified using thiol ligands. Even better approach and sensitivity for mercury in water was attained by use of un-modified gold nanowires (Selid et al., 2009). Colorimetric sensors of Ag or Au NPs were developed for the detection of heavy metals, radioactive metals and isomeric aromatic compounds (Su et al., 2012). The Au NP - enzyme complex changed color from blue to red on exposure to Pb (II) which is based on aggregation of Au NPs that is inhibited by presence of Pb (II) and consequently complex formation with enzyme (Riu et al., 2006). The Ur (VI) radionuclide had been detected in ppb levels by superparamagnetic core (Fe_3O_4) –shell (Au) magnetic NPs functionalized with organophosphorous ligand by exploiting the uranium natural affinity for the phosphate molecules. (Banerjee et al., 2010).

Metal oxides with their semiconducting nature have depicted variations in conductance response owing to their stoichiometry, defects and interacting target species. n-type materials (e.g. ZnO, SnO_2) exhibit decrease in conductance on interaction with oxidizing gases (e.g. NO_2) while the p-type materials such as CuO demonstrate increase in conductance. Exposure to reducing gases (i.e. H_2S, CO, H_2, H_2O vapors) impart opposite behavior in n- and p-type materials. Surface morphology has also played its role in defining the sensitivity of the nanosensors e.g. ZnO brushes depicted detection of ethanol upto 5 ppm level while the ZnO flowers can go as low as 1 ppm. Similarly, better sensing performance was shown by thin SnO_2 nanorods in contrast to thick ones (Huang et al., 2009). SnO_2 nano-ribbons, synthesized by thermal deposition method, have shown considerable change in electrical conductance with adsorption of NO_2. The sensor can be cleaned up by simple application of UV light unlike SnO_2 thin films that underwent desorption at high temperatures with risk of explosion. The UV light triggered the conductance to increase due to photocurrent generation and number of free carriers leading to photo-desorption (Law et al., 2002).

Carbon Based Nanomaterials

Carbon-based nanomaterials usually designated as CBNs include carbon nanotubes (both single and double walled), carbon nanodots, graphene, fullerenes and nano-diamonds have been in spotlight due to their properties and have found applications in environmental monitoring in areas of heavy metals detection, gas molecules, pesticides etc.(Hu et al., 2017).

Among these, carbon nanotubes have gained ample attention due to their high sensitivity and ability to detect very minute concentration of gas molecules at room temperature (Hefzollah Mohammadian et al., 2013) with added advantage of low-cost, power saving, large life span and regeneration (McNicholas et al., 2011; Mubeen et al., 2009). The changes in transport and electronic properties is recorded when these CNT interact with gas molecules either directly through physio- or chemi-sorption or by indirect interaction through CNT bonded contaminants By altering these parameters the specificity of the Single-walled CNT (SWNT) can be altered and employed for the monitoring of toxic gases (Goldoni et al., 2003). The mode of synthesis of CNT had considerable impact on the electrical response generated on exposure to contaminant. The CNT mats fabricated by plasma enhanced CVD and the SWCNT synthesized by drop casting resulted in detection limits of 10 ppb NO_2 at 165°C and 44 ppb NO_2 at room temperature respectively (Zhang et al., 2008).

The electrical properties of CNTs showed enhanced sensitivity to chemical doping and transfer of charge owing to doping or presence of various molecules (Rogers, 2006). The decoration of SWNT with gold NPs have also enabled the detection of H_2S (Limit of detection (LOD) 3 ppb) by triggering the conduction mechanism working across the nanotubes and electron exchange occurring between the defect sites of gold and SWNT (Mubeen et al., 2009). These Au loaded CNT are able to monitor the levels of both oxidizing and reducing gases at much lower level and better sensitivity as compared to simple CNTs by undergoing decrease and increase in electrical resistance, respectively upon exposure. Better mean sensitivity was observed in case of NO_2, NH_3 and H_2S at 200°C with 5% Au loading (Penza et al., 2009). Similar Au-SWNT have enabled detection of Hg vapors in gas phase at even 2ppb level by utilization of Surface Enhanced Raman Scattering phenomena that operates through transfer of Au NPs electron density to SWNT on Hg exposure resulting in resistance changes and hence measurements (McNicholas et al., 2011). The resistance response in both the cases is found to be directly proportional to the size and density of Au NPs while desorption phenomena is simply affected by lowering the temperature resulting in weakening the bond affinity between the contaminant and Au NPs (McNicholas et al., 2011; Mubeen et al., 2009). Direct label-free analyte detection was made possible by employing thin films of biological receptors (aptamers, antibodies and binding proteins) modified SWNT enabled the detection of biological contaminants (*E. coli*, hepatitis C virus, *Aureococcus anophagefferens* (algae), *Bacillus anthracis*) as well as Trinitrotoluene and organophosphate nerve agents (Sarkar et al., 2013).

Modified or unmodified SWCNT have been extensively used for the detection of volatile organic carbons (VOC) like nitrotoluene, toluene, ethyl acetate, ethanol, acetone, methanol, methyl ethyl ketone, dichloromethane, water vapor, propanol, hexane, trichloroethylene, chloroform, methane etc. (Li et al., 2003; Lu et al., 2004; Penza, Michele et al., 2006; Shirsat et al., 2012; Tasaltin et al., 2014). Moreover, Multi-walled CNT (MWNTs) based sensor exhibited early response with appreciable sensitivity towards the detection of pesticides (0.04 ppb Dichlorvos and 0.1 ppb carbofuran) in aqueous media (Su et al., 2012).

Environmental Remediation

Environmental remediation deals with the clean-up strategy that enables removal or degradation of pollution from the water, air and soil.

Approaches to Environmental Remediation

There are two different aspects that are dealt in this area:

- **Ex-situ vs. In-situ approach**
- **Absorptive vs. Reactive approach**

A general comparison of both approaches under one roof is presented in Table 2. The basic essence in both the terminologies is cost-effectiveness that poses a major challenge to the current treatment methods available. In this context nanotechnology offers improved and energy efficient processes for environmental remediation. In this regard extensive work is done to reach out nanomaterials that have increasing ability to degrade or remove the pollutants from environment.

1. Ex-Situ and In-Situ Environmental Remediation

The former technique entails the pollutant removal from site followed by its treatment while the later involves treatment of pollutants on site directly (Lofrano et al., 2017; Tratnyek et al., 2006).

Ex-situ treatment can be explained in a way that it involves the pumping of the contaminated water from the place of its occurrence to another place like surface and then treating it at the treatment plant (Mulvaney, 2011). The famous example of such technique is the photo-oxidation of organic contaminants using semiconducting nanomaterials as this involves the illumination which is conducted in specially designed reactor (Tratnyek et al., 2006).

In-situ technologies incorporate one of the three ways that are illustrated in detail in Figure 5.

- The first *in-situ* approach is based on the construction of barrier made up of nanomaterials incorporated in that. This barrier is usually named as "Permeable reactive membrane" (Figure 5a).
- The second one is named as "Reactive treatment zone" which involves formation of a reactive region made by injecting the low mobility nanomaterials when flowing water comes in contact with this zone the clean-up phenomena takes place (Figure 5b)

Table 2. Environmental remediation approaches

	In-situ	Ex-situ
Adsorptive	Sequestering of pollutants in-situ by addition of nano-adsorbents	Nano-filtration which involves removal of contaminants by filtration on filtration plant
Reactive	Reactive degradation of organic contaminants on site	Photo-oxidation of organic contaminants in especially designed UV reactors

Figure 5a.

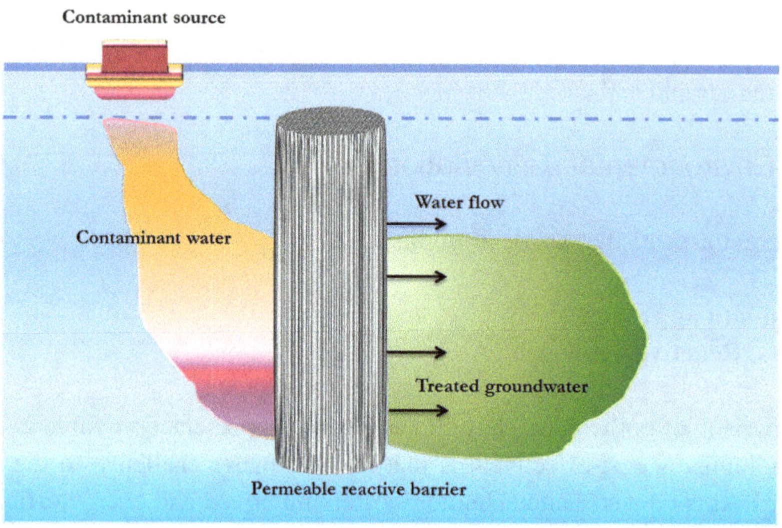

Figure 5b.

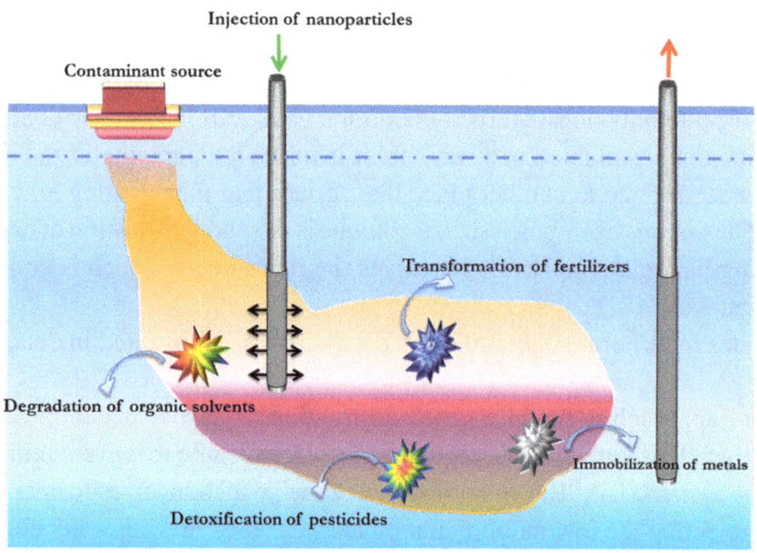

- Third way is to develop "reactive nanoparticles plumes" with aid of mobile nanoparticles that migrate to contaminated areas and come in contact with the non-aqueous phase liquid (DNAPL) to ensure their clean-up as shown in Figure 5c.

The zero-valent iron are usually considered the successful candidates for in-situ treatment (Tratnyek et al., 2006).

Both techniques have their own pros and cons; ex-situ techniques inspite of being costly offer fast and timely solution to the pollution problem, on other hand in-situ approach are quite effective for long term

Figure 5c.

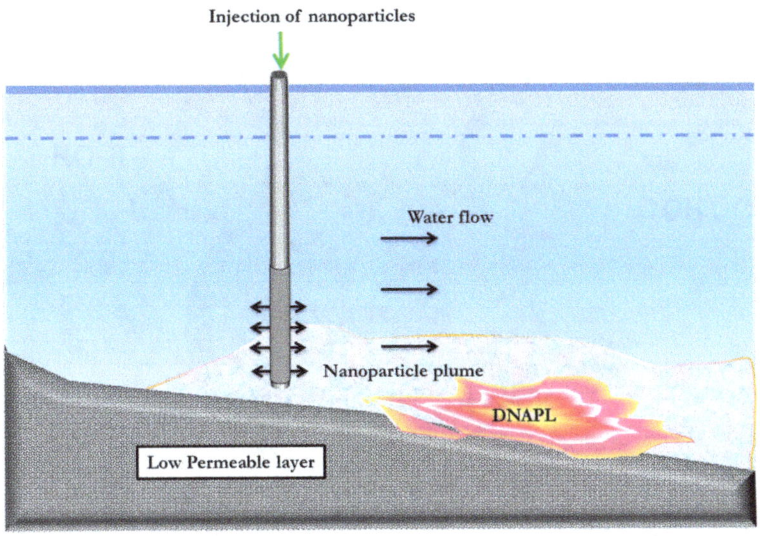

treatment but they usually face threats of effectiveness risk and residual liability in addition to delivery difficulties (Mulvaney, 2011). Another short-coming of the ex-situ methods is the generation of hazardous waste concentrates that need further treatment and disposal protocols to be followed (Grieger et al., 2015).

2. Absorptive vs. Reactive Approach

Absorptive remediation approach is the removal of contaminants by sequestration using different nanomaterials. On other hand reactive remediation technologies undergo degradation of contaminants to less benign and sometimes to completely harmless products i.e. carbon dioxide and water. The earlier method is usually targeted to the heavy metals removal while the later refers to organic pollutants (Tratnyek et al., 2006) by involvement of the different processes as depicted in Figure 6. Hence, according to this approach, nanomaterials can be broadly categorized as nano-adsorbent and nano-catalysts. In this context many nanomaterials have been prepared and checked for their efficacy for the treatment of pollutants in water, soil and air. These nanomaterials with large surface to volume ratio show high reactivity that can be utilized well in adsorption and catalyzation (Tratnyek et al., 2006). This part of chapter will present an overview of the developments that have been taken place over the years in the field of adsorption and catalysis.

Nano-Adsorbents

The simple definition of nano-adsorbents is "*Nanoadsorbents are nanoscale particles from organic or inorganic materials that have a high affinity to adsorb substances*" (Bhatnagar et al., 2014). These materials owing to their small size, high surface reactivity, porosity, high binding capacities, and regeneration ability are able to adsorb variety of contaminants with different physical and chemical attributes. A simple layout of the potential these materials cater is presented in Table 3 which presents just few of the contaminants that can be removed by using nano-adsorbents. Various types of nanomaterials that can

Figure 6.

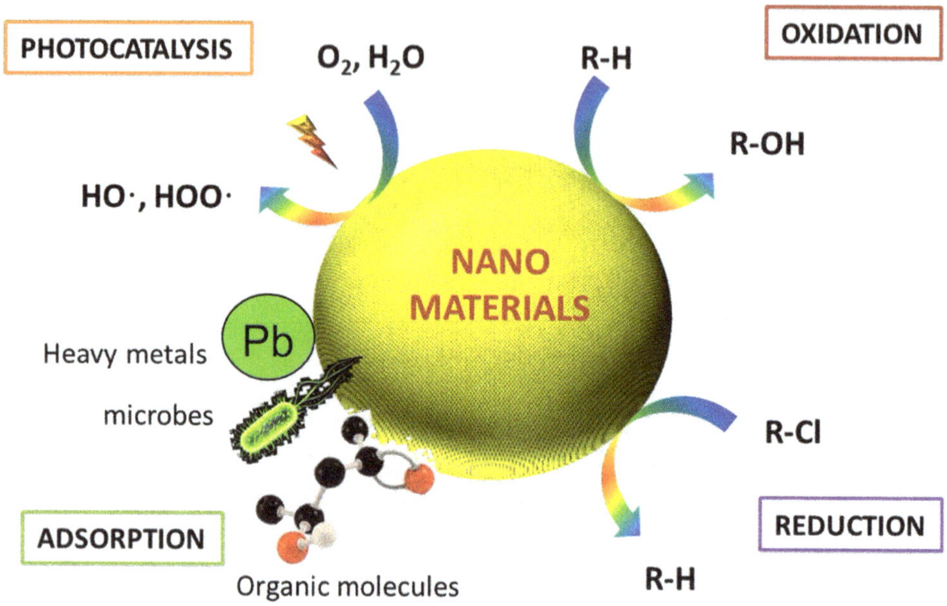

Table 3. Different types of contaminants treated by using nano-adsorbents

Chlorinated Methanes	Chlorinated benzenes	Trihalomethanes	
Carbon tetrachloride Chloroform Dichloromethane Chloromethane	Hexachlorobenzene Pentachlorobenzene Tetrachlorobenzene Trichlorobenzene Dichlorobenzene Chlorobenzene	Bromoform Dibromochloromethane Dichlorobromomethane	
Chlorinated Ethenes	**Polychlorinated Hydrocarbons**	**Organic Contaminants**	
Tertachloroethene Trichloroethene *cis*-Dichloroethene *trans*-Dichloroethene 1,1- Dichloroethene Vsinyl chloride	PCBs Pentachlorophenol Dioxins	N-nitrosodimethylamine TNT	
Pesticides	**Dyes**	**Inorganic and Heavy Metal Ions**	
Lindane DDT	Orange II Chrysoidine Tropaeolino Acid Orange Acid Red	Dichromate Arsenic Perchlorate Nitrate	Mercury Silver Cadmium Arsenic

be used in this regard are given in Figure 7. The materials can be subdivided into many other types like metallic NPs, metallic nanoalloys, metallic oxide NPs, mixed metal oxide NPs, single, double and multi walled CNT, carbon NPs, carbon nandots, carbon nanosheets, metal-clay nancomposites, polymer-clay nanocomopsite, polymer-metal nanocomposite etc. (Khajeh et al., 2013)

Figure 7.

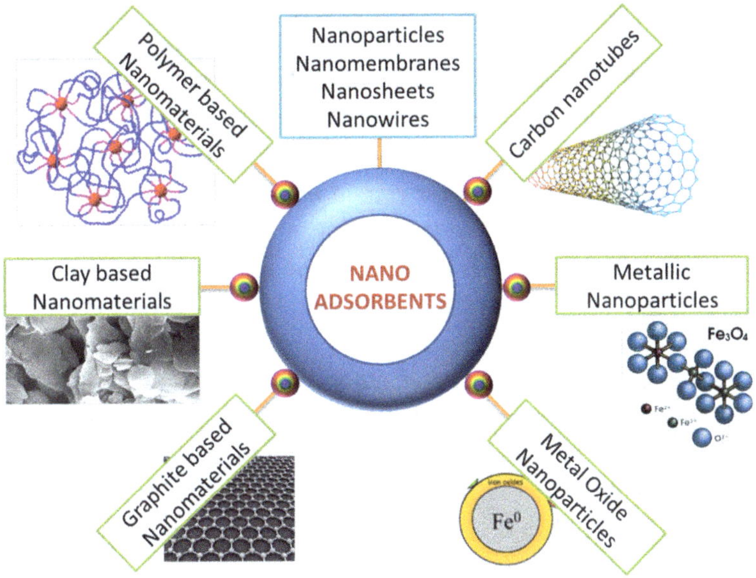

All these materials have been utilized in water, soil and air remediation by the exploitation of two main properties' i.e.

1. Innate Surface Properties
2. External Functionalization

- **Innate Surface Properties:** These properties are basically determined by the nanomaterials' composition, size and surface structure. The nature of active sites and their distribution together with the location of surface atoms, surface binding energy and surface chemical activity overall affect the adsorption potential of the nano-adsorbent.
- **External Functionalization***:* These are the changes that are induced into the material owing to the functionalization induced into the material. Various functional groups are added on the surface as a result of that they impart several impacts to the adsorbent properties. Few of these can be listed in terms of enhanced adsorption capacity, changes in hydrophilicity, increased selectivity, resistance to agglomeration and oxidation prevention (Khajeh et al., 2013).

Nano-Catalyst

Nano-catalysts, the catalysts on nanoscale, have advantages over the traditional catalyst because of three attributes which are:

- **High Activity:** Large surface area that helps in providing more catalytic sites for enhancing reaction rates.
- **High Selectivity:** Special tailored properties that enhance selectivity and reduction of byproducts.

- **Long Life Time:** Lesser usage of catalysts lead to longer life time reducing the cost of catalysts and generation of waste.

Owing to last two traits, the nano-catalysts are famously known as environment catalyst. Further, these catalysts exhibit the properties that are in between homogeneous (*catalysts existing in same phase as reactants; easily accessible; high surface area*) and heterogeneous catalysts (*catalysts existing in insoluble and exist as solid particles; stable; easy handling and isolation).* They are divided into four classes i.e. Nanoparticulate, Nanoporous, Nanocrystalline and Supramolecular in broad sense (Anastas et al., 2012; Lofrano et al., 2017; Polshettiwar et al., 2013; Sanghi et al., 2012). Several researches based on applications of these nanocatalysts towards environmental remediation were done over the years and have shown the broad spectrum aspects of these nano-catalysts, some of these are outlined in Table 4.

Photocatalyst has received an exceptional place among the catalysts that undergo degradation of organic pollutants by getting activated in presence light. These catalysts are usually solid semiconducting materials having large and gap and are readily available owing to their low cost. The materials that have gained attention in this regard are TiO_2 (Yu et al., 2007), ZnO (Hariharan, 2006), clay based nanocomposites (Chong et al., 2009), ZnS (Chen et al., 2010) and Graphite nanocomposites (Bai et al., 2012).

Types of Environmental Remediation

Considering the three dimensions of the environment, environmental remediation is categorized broadly into water, soil and air remediation (as already defined in Table 4) which will be discussed briefly in this chapter:

Water Remediation

Water remediation including drinking water, ground water and wastewater is one of the most researched areas of environmental remediation. The "pump and treat" i.e. *ex-situ* treatment methods have gained much attention in this regard and there are a number of research publications available that have reviewed the role of different types of nanoparticles and materials in this field. In case of on-site (*in-situ*) approaches zero-valent iron nanoparticles (nZVI) have enjoyed the attention of majority of the environmental scientists (Grieger et al., 2015; Mueller et al., 2012). The plenty of contaminants have been removed from the drinking and wastewater by adopting absorptive and reactive approaches including heavy metals, organic contaminants, anionic inorganic pollutants and microbiological impurities using nanoparticles, nanomembanes etc. Nano-filtration has recently gained much attention owing to its effectiveness in removing hardness, natural organic material, pesticides, VOCs, dyes, metals, inorganic anions and more over the viruses and bacteria (Van der Bruggen et al., 2003).

Soil Remediation

Soil remediation approaches are also categorized into *ex-situ* and *in-situ* and can have very strong but indirect influence on the groundwater quality and vice versa. Owing to this reason the soil and water remediation are usually dealt simultaneously (Mueller et al., 2010; Mulvaney, 2011). Most promising nano approach towards remediating soil is the use of zero-valent iron nanoparticles (nZVI). Several studies are done regarding the pesticides decontamination of soil using nZVI utilizing in-situ techniques mostly. Malathion degradation was found to be inversely depended on the particle size and NP dose

Table 4. Application of nanocatalysts towards air, water and soil remediation

Nanocatalyst	Target Contaminants	Reference
For Air		
Metal (Cu, Co, Fe, Ni) NPs supported on Activated carbon	Catalytic incineration of volatile organic compounds (VOC) i.e. Xylene, toluene, benzene	(Lu et al., 2007)
Carbon based nanocatalysts	Reduction in CO_2 emissions	(Khavarian et al., 2013)
TiO2	Formaldehyde photocatalytic degradation indoor environment	(Yu et al., 2007)
$Ti_{1-x}Zr_xO_2$ Solid Solutions	Photocatalytic Degradation of Acetone in Air	(Yu et al., 1998)
TiO2	degradation of NOx in air	(Cappelletti et al., 2008)
For water		
TiO_2-V and Ce	3,4-Dichloroaniline degradation	(Padmini et al., 2013)
Mn/Iranian hematite	Oxidation of SO_2 pollutant in aqueous environment	(Shokri et al., 2017)
nano-gold composite	4-chlorophenol degradation in water	(Narayanan et al., 2011)
nZVI	Ground water remediation	(Lofrano et al., 2017; Tratnyek et al., 2006)
carboxymethyl cellulose stabilized iron NP	perchloroethylene (PCE) and trichloroethylene (TCE) and polychlorinated biphenyls (PCBs) degradation in groundwater	(He et al., 2010)
nZVI/palygorskite composite materials	Bisphenol A degradation by advanced oxidation	(Xi et al., 2014)
nZVI/Cu bimetallic nanoparticles with activated carbon support	γ-HCH degradation in water system	(Chang et al., 2011)
TiO_2	Dyes degradation in water	(Aarthi et al., 2007; Nagaveni et al., 2004a, 2004b; Sivalingam et al., 2003)
nZVI	Dyes degradation in water	(Hoag et al., 2009)
crosslinked chitosan/nano-CdS composite	Dye degradation in water	(Zhu et al., 2009)
TiO_2	Nitrobenzene degradation in water	(Yang et al., 2007)
Magnesium and barium doped TiO_2	4-chlorophenol degradation in water	(Venkatachalam et al., 2007)
Sn (IV)/TiO_2/AC	Dye degradation in water	(Sun et al., 2006)
ZnO	Dye degradation in water	(Jang et al., 2006; Kumar et al., 2015)
Grafted TiO2	Polystyrene degradation in water	(Zan et al., 2006)
nano-TiO2 coated films	Organochlorine pesticides degradation in water	(Yu et al., 2007)
In Soil		
nZVI	Pesticides degradation in soil	(El-Temsah et al., 2013; Reddy et al., 2012; Satapanajaru et al., 2008; Singhal et al., 2012)
nZVI	Poly aromatic Hydrocarbon degradation in soil	(Chang et al., 2007)
nZVI	Degradation of soil-sorbed trichloroethylene	(Zhang et al., 2011)
CMC-Pd/nFe 0	Pesticides in soil	(Singh et al., 2013)
nZVI	PCB degradation in soil	(Varanasi et al., 2007)

(Singhal et al., 2012) whereas DDT degradation rate was also impacted by mode of nZVI synthesis (El-Temsah et al., 2013; El-Temsah et al., 2016). Atrazine, another pesticide, was also degraded by two way step i.e. reductive dechlorination followed by acid hydrolysis in presence nZVI. The degradation was facilitated by lowering the soil pH and use of additives like sulfates of iron (II and III) and Al (III) and alloying with palladium (Satapanajaru et al., 2008). Similar studies have shown the potential of nZVI towards destruction of polychlorinated biphenyl present in soil which has shown dependence on mixing temperature, mixing time, NP dose, type of mixing and presence of air while no effect was imparted by pH, particle size and surfactant addition (Varanasi et al., 2007). Pyrene, a polycyclic aromatic hydrocarbon, has been degraded under ambient conditions with 62% efficacy within an hour by using optimized dose of 0.15 g g^{-1} soil (Chang et al., 2007). Moreover, the drug (ibuprofen) contaminated soil was successfully affected by the green synthesized nZVI. The studies also highlighted the role of soil composition and pH in determining the dominant mechanism of degradation i.e. aggregation, deposition and particle-soil interactions. Humic acid levels can change NPs' physiochemcial properties by promoting agglomeration, reducing surface reactivity by humic acid accumulation, readsorption inhibition and enhancing NP dissolution. The article also points out that the degradation starts at the supercritical level of soil and afterwards as the reaction progresses nZVI moves downwards and get oxidized by soil oxygen or dissolved by soil moisture hence losing their reactivity with passage of time (Machado et al., 2013).

In addition to organic decontamination of soil, nZVI have played an effective role in immobilization of heavy metals from soil. This in-situ immobilization of heavy metals like Pb and Zn has the advantages of overcoming the transport issues of NP. Treatment of Cr and As through same approach helped in reducing the leaching of these metals from soil by utilization of 2% nZVI which is enhanced many times by synergy action of using bio compost with nZVI (Galdames et al., 2017). The importance and effectiveness of this technique can be well understood by considering the fact that some commercial products have been launched as well e.g. NanoFer 5S (i.e. nZVI), Fe-oxide (nano goethite). The former showed 60% decrease in Zn, Pb and Co levels in the soil leachlets and the level was also maintained to a constant value even after several weeks indicating reductive action of nZVI (Gonçalves, 2016).

In addition to nZVI, many other nanomaterial have also been researched for their role in soil remediation. Few of these are amphiphilic polyurethane NP for treatment of polynuclear aromatic hydrocarbon (phenanthrene) (Tungittiplakorn et al., 2004), alumina with 10% ferrihydate and nano carbon black for Cu decontamination (McBride et al., 2000; Wang et al., 2009), nano-hydroxyapatite for Cu and Pb (Zhang et al., 2010), magnetite NPs for phosphate immobilization and calcium peroxide NPs for absorptive removal of organic pollutants (gasoline, methyl tertiary butyl ether, ethylene glycol etc.) (Mueller et al., 2010).

Air Remediation

Air remediation techniques are usually defined in context with point source and non-point source. The former is based on the removal of targeted contaminants coming from either the cars or industrial stacks into harmless gases (by use of reactive approach) or restricting their release into atmosphere (by absorptive approach).

The exhaust nano-catalyst (named as convertor catalyst) have played a major role in catalytic abatement of toxic automotive exhaust which due to fuels' combustion releases harmful gases into the atmosphere which include NO_x, hydrocarbons (HC), CO, CO_2 and SO_X in addition to particulate matters (Kašpar et al., 2003). Among these NO_x, nitric oxide (NO) which can be converted successfully to N_2 by use of gold based nanocatalyst with the added advantage of enhanced selectivity towards Nitrogen formation, resistant to changes affected by moisture and utilization of other unburnt gases (like CO, HC) in catalytic conversion reaction (Cortie et al., 2002). These gold based nanocatalysts are also helpful in complete combustion of methane by oxidative destruction (Choudhary et al., 2008). Inspite of these advantages these catalysts face the issues with thermal instability at high temperatures. The use of ceria-zirconium nanocatalyst can not only resolve the thermal stability issues in addition to offering simultaneous reduction of CO, NOx and HC (Masui et al., 2000). Industrial stacks polluted stream has been effectively decontaminated from VOCs by use of MnO_2 nanofibers (Das et al., 2015). Further developments in the field of air nano-remediation led to development of nanostructured membranes, made up of CNT, capable of separating toxic gases like methane or CO_2 from the industrial smoke stack (Kowalczyk et al., 2008). Metallic organic frameworks, marks another class of nonporous materials that have been able to efficiently sorb CO_2 and desorb by application of vacuum (Ondreiovic et al., 1995).

In-door or non-point source pollution has been tackled by the introduction of photocatalytic reactors, the active catalyst being the nanomaterials. The polluted air passes through these reactors which work in continuous but turbulent flow mode enabling maximum contact of the polluted air with the pores of membrane type filter resulting in transfer of organic pollutants to the surface of nanomaterial followed by their catalytic degradation (Luque et al., 2012). TiO_2 in addition to CNT have been used to mitigate air pollutants including phenolic compounds, airborne microbes and odorous chemicals (Khin et al., 2012). A novel development in this aspect is the self-cleaning buildings materials coated with TiO2 nanoparticle having photocatalytic properties. These particles adsorb harmful VOC from the air and in presence of sunlight covert them into other less toxic degradable products. NP incorporated in paints and cement have shown to clean the indoor environment from various pollutants (Rickerby, 2007).

Pollution Prevention

The other terminology, pollution prevention, portrays the pathways by which pollution is not created in any of these three major areas: 1) Products (less toxic, less polluting), 2) processes (more efficient, less waste generation) and 3) energy and resource efficient (Products or processes that use less energy and fewer raw materials) (Sellers et al., 2008). The pollution prevention marks the "green nano vision" approach that will help in revolutionizing the manufacturing approaches by making them more energy efficient, less hazardous, use of fewer chemicals, less byproducts generation and use of less toxic chemicals.

In industrial setup the nanotechnology has paved the way towards revolutionizing in environmental friendly way (Elnashaie et al., 2015), few of the developments are discussed below in context to major fields:

1. **Chemical Manufacturing:** The use of nanocatalysts in chemical reactions that enable the faster reactions with better product yield due their high mobilization within the reactor leading to increase in contact between the reactants. Further high stability of these nanocatalysts prevents frequent replacing of the catalysts and consequently frequent shut down of the machines that led to energy loses as well as time restrains (Fihri et al., 2011; Tsang et al., 2004). Magnetically active nanocatalysts facilitates easy separation of catalyst used in various organic reactions saving extra and tedious process requirements that helps in saving energy (Polshettiwar et al., 2011)
2. **Petroleum Industry:** The in-situ preparation of nanocatalysts helped in enhanced recovery of bitumen by converting it into lighter products that can be easily extracted and pumped from the surface. This modification in the steam-assisted gravity drainage process has not only helped in

cost reduction but also tackling the environmental concerns (Almao, 2012). Moreover, the majority of processes in petroleum industry like separation of olefin/paraffin and light solvents, solvent dewaxing, recover of phenol and aromatic compounds, dehydrogenation, oxidative coupling of methane and steam reforming of methane have all been facilitated by use of nanomembranes that have advantages of high selectivity and permeability over the traditional membranes making all these processes cost-effective in terms of energy and material consumption (Ravanchi et al., 2009).

3. **Material Industry:** Nanocomposites with improved mechanical properties including stability, fire resistance, scratch resistance makes them a viable alternate to conventional nanofillers as they can be used in small amounts to impart same effect and hence cut the cost and waste generation (Elnashaie et al., 2015). Self-cleaning glass made up of TiO_2 NPs deposited over soda-lime glass substrates can degrade organic contaminants from in-door and out-door with no acid residues (Cedillo-González et al., 2014).
4. **Transport Industry:** Use of nanosilica and CNT in tires have considerably deceased the rolling resistance and durability which helped in overcoming fuel losses, waste generation and raw materials excessive use (De et al., 2013; Sui et al., 2008). The nanocomposite have also resulted in vehicle weight reduction that reduce fuel consumption and energy saving (Watarai, 2006). More advancement in the field of transportation are in area of automotive catalytic converters, by the use of these vehicle exhaust emissions can be treated at point-of-source helping in automotive CO oxidation (Seyfi et al., 2009).
5. **Textile and Paint Industries:** Textiles with nanocoatings that are dirt resistant or easy-to-clean facilitating cleaning procedures and consumption of water, chemicals and energy. Paint industry also boosted with nanochemistry by use of nanomaterials and nanoemulsions that lead to reduction in use of organic content and hence effective in saving the petroleum resources. Also use of nanoparticle can prevent the walls from microbial growth hence restricting the use of pesticides etc. (Li et al., 2003; Mizutani et al., 2006).
6. **Construction Industry:** Construction material prepared by using nanotechnology have better barrier properties hence improving the buildings' energy requirements needed for heating and cooling these structures (Elnashaie et al., 2015). In addition to that the buildings constructed from photocatalytic nanomaterials can help in environmental pollution remediation, self-cleaning and self-disinfecting by taking advantage of sunlight and rain water for getting activated (Chen et al., 2009).
7. **Energy Storage:** Lithium-ion batteries (LIB) consisting of CNT showed improved life span, enhancement in storage capacity, size reduction, less weight, reduction in material consumption, frequent recycling and charging and less energy losses. In case of Fuel cells (considered as alternate to fossil fuels and reducing dependence on petroleum based products), super capacitors (store electrical energy) and solar cells better performance in terms of efficacy have been achieved by use nanomaterials (Cui et al., 2009; Moniz, 2010; Serrano et al., 2009; Zäch et al., 2006).
8. **Agriculture and Food Industry:** The benefits of nanotechnology can be best seen in case of pest management that facilities better absorption capacity of pesticides by reducing their size (El-Bendary et al., 2013) or in total eradication of pesticide use (Rai et al., 2012) as that is a direct threat to environment (Miller et al., 2008). The controlled release of nutrients facilitated with nanofertilizers have helped in reducing the excessive use of fertilizers and hence leaching of these into the envi-

ronment (Lateef et al., 2016). Moreover, the development of bio-degradable packaging materials have assisted easy degradation of plastics by micro-organisms and enzymes hence avoiding their piling up (Bordes et al., 2009).

E-NANO: A BALANCE BETWEEN BENEFITS AND RISKS

E-Nano – a technology that has played a pivotal role in reshaping the environmental technologies and concept. From the old concept where the focus was just on designing new treatment ways, E-Nano has stimulated the older technologies with innovative approaches by designing ways to remediate environment by mitigating the potential causes of pollution thereby preventing its occurrence. As every technology has potential benefits and risks associated with it, hence currently there is controversial aspect linked to the E-Nano as well in terms of NPs safety and their influence on environment itself and consequently on flora and fauna. This debate between positive and negative impacts is presented in Figure 8 (Bernd, 2010; Pavličević et al., 2016).

Nanotechnology have helped in sustaining and refining the environment by designing novel and improved ways to enhance the quality of three elements of life i.e. water, soil and air through either product development or process designing. The three aspects of environment sustainability i.e. monitoring, remediation and pollution prevention have been well catered by nanotechnology in terms of designing excellent sensors, adsorbents, catalysts and energy and atom efficient products and processes. Nanosensors have reformed the older sensors by enhancing the sensitivity, use of less material and enabling early detection of contaminants. On other hand the environmental cleanup was made possible by development of efficient nano-catalysts and nano-adsorbents that have resulted in not only enhanced removal of the pollutants from the environment but have resulted in conversion of these pollutants to non-toxic products. Further these catalysts and adsorbents are energy-efficient as they require less stringent conditions to give better performance than their bulk counterparts (Philippot et al., 2012). The execution and employment of green principles can be seen in context of nanotechnology either in case of nanoparticles synthesis or facilitation in processes or product formation by use of nanomaterials, have given this technique an

Figure 8.

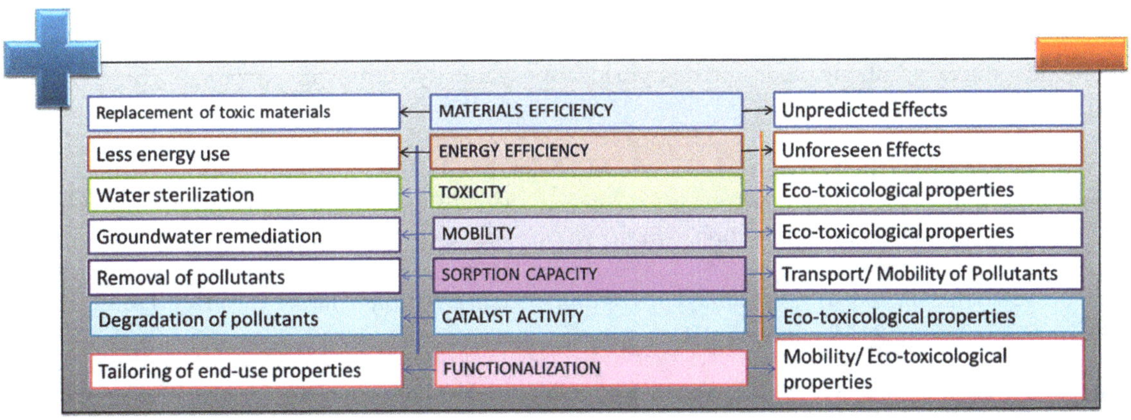

edge over the existing technologies (Sanghi et al., 2012). Nanotechnology has also reformed the energy production, energy storage and energy saving by providing alternatives to current practices (Zang, 2011).

In-spite of all the benefits, the nanotechnology has always been under debate and facing the controversies regarding its eco-toxicological impacts. The behavior of nanomaterials in terms of their bioavailability, toxicity, mobility and stability has raised questions on the safe use of these materials. The inhalation of nanoparticles, absorption from the skin and oral intake are the areas of concern. Some studies have shown some potential impacts on cellular functions, DNA functioning etc. but equal number of studies have negated the probability of the release of nanoparticles into the environment (Bernd, 2010). The careful review of the NT in context with its overall impacts can be well assessed while carrying out full life assessment studies and careful use of the products and processes developed and used under this umbrella of technology.

CONCLUSION

Despite of the negative impacts associated with the nanotechnology, the field caters great potential to benefit the lives of humans by overflow of numerous commercial and industrial applications in almost every field of life; the most promising and impactful being observed in field of environments.

REFERENCES

Aarthi, T., & Madras, G. (2007). Photocatalytic degradation of rhodamine dyes with nano-TiO2. *Industrial & Engineering Chemistry Research*, *46*(1), 7–14. doi:10.1021/ie060948n

Akyildiz, I. F., & Jornet, J. M. (2010). Electromagnetic wireless nanosensor networks. *Nano Communication Networks*, *1*(1), 3–19. doi:10.1016/j.nancom.2010.04.001

Almao, P. P. (2012). In situ upgrading of bitumen and heavy oils via nanocatalysis. *Canadian Journal of Chemical Engineering*, *90*(2), 320–329. doi:10.1002/cjce.21646

Anastas, P. T., & Zimmerman, J. B. (2012). Innovations in Green Chemistry and Green Engineering: Selected Entries from the Encyclopedia of Sustainability Science and Technology. Springer.

Bai, S., Shen, X., Zhong, X., Liu, Y., Zhu, G., Xu, X., & Chen, K. (2012). One-pot solvothermal preparation of magnetic reduced graphene oxide-ferrite hybrids for organic dye removal. *Carbon*, *50*(6), 2337–2346. doi:10.1016/j.carbon.2012.01.057

Banerjee, R., Katsenovich, Y., Lagos, L., Senn, M., Naja, M., Balsamo, V., ... Li, C. (2010). Functional magnetic nanoshells integrated nanosensor for trace analysis of environmental uranium contamination. *Electrochimica Acta*, *55*(27), 7897–7902. doi:10.1016/j.electacta.2010.05.060

Bernd, N. (2010). *Pollution prevention and treatment using nanotechnology*. Wiley Online Library. doi:10.1002/9783527628155.nanotech010

Bhatnagar, A., & Sillanpää, M. (2014). Application of nanoadsorbents in water treatment. *Nanomaterials for Environmental Protection*, 237-247.

Bordes, P., Pollet, E., & Avérous, L. (2009). Nano-biocomposites: Biodegradable polyester/nanoclay systems. *Progress in Polymer Science*, *34*(2), 125–155. doi:10.1016/j.progpolymsci.2008.10.002

Cappelletti, G., Ardizzone, S., Bianchi, C. L., Gialanella, S., Naldoni, A., Pirola, C., & Ragaini, V. (2008). Photodegradation of Pollutants in Air: Enhanced Properties of Nano-TiO2Prepared by Ultrasound. *Nanoscale Research Letters*, *4*(2), 97–105. doi:10.100711671-008-9208-3 PMID:20596293

Cedillo-González, E. I., Riccò, R., Montorsi, M., Montorsi, M., Falcaro, P., & Siligardi, C. (2014). Self-cleaning glass prepared from a commercial TiO 2 nano-dispersion and its photocatalytic performance under common anthropogenic and atmospheric factors. *Building and Environment*, *71*, 7–14. doi:10.1016/j.buildenv.2013.09.007

Chang, C., Lian, F., & Zhu, L. (2011). Simultaneous adsorption and degradation of γ-HCH by nZVI/Cu bimetallic nanoparticles with activated carbon support. *Environmental Pollution*, *159*(10), 2507–2514. doi:10.1016/j.envpol.2011.06.021 PMID:21763039

Chang, M.-C., Shu, H.-Y., Hsieh, W.-P., & Wang, M.-C. (2007). Remediation of soil contaminated with pyrene using ground nanoscale zero-valent iron. *Journal of the Air & Waste Management Association*, *57*(2), 221–227. doi:10.1080/10473289.2007.10465312 PMID:17355083

Chen, D., Huang, F., Ren, G., Li, D., Zheng, M., Wang, Y., & Lin, Z. (2010). ZnS nano-architectures: Photocatalysis, deactivation and regeneration. *Nanoscale*, *2*(10), 2062–2064. doi:10.1039/c0nr00171f PMID:20689875

Chen, J., & Poon, C. (2009). Photocatalytic construction and building materials: From fundamentals to applications. *Building and Environment*, *44*(9), 1899–1906. doi:10.1016/j.buildenv.2009.01.002

Chong, M. N., Vimonses, V., Lei, S., Jin, B., Chow, C., & Saint, C. (2009). Synthesis and characterisation of novel titania impregnated kaolinite nano-photocatalyst. *Microporous and Mesoporous Materials*, *117*(1), 233–242. doi:10.1016/j.micromeso.2008.06.039

Choudhary, V. R., Patil, V. P., Jana, P., & Uphade, B. S. (2008). Nano-gold supported on Fe2O3: A highly active catalyst for low temperature oxidative destruction of methane green house gas from exhaust/waste gases. *Applied Catalysis A, General*, *350*(2), 186–190. doi:10.1016/j.apcata.2008.08.008

Cortie, M., & Van der Lingen, E. (2002). *Catalytic gold nano-particles.* Paper presented at the Materials forum.

Cui, L.-F., Yang, Y., Hsu, C.-M., & Cui, Y. (2009). Carbon– silicon core– shell nanowires as high capacity electrode for lithium ion batteries. *Nano Letters*, *9*(9), 3370–3374. doi:10.1021/nl901670t PMID:19655765

Das, S., Sen, B., & Debnath, N. (2015). Recent trends in nanomaterials applications in environmental monitoring and remediation. *Environmental Science and Pollution Research International*, *22*(23), 18333–18344. doi:10.100711356-015-5491-6 PMID:26490920

De, D., Panda, P. K., Roy, M., Bhunia, S., & Jaman, A. I. (2013). Reinforcing effect of nanosilica on the properties of natural rubber/reclaimed ground rubber tire vulcanizates. *Polymer Engineering and Science*, *53*(2), 227–237. doi:10.1002/pen.23255

El-Bendary, H., & El-Helaly, A. (2013). First record nanotechnology in agricultural: Silica nano-particles a potential new insecticide for pest control. *App. Sci. Report*, *4*(3), 241–246.

El-Temsah, Y. S., & Joner, E. J. (2013). Effects of nano-sized zero-valent iron (nZVI) on DDT degradation in soil and its toxicity to collembola and ostracods. *Chemosphere*, *92*(1), 131–137. doi:10.1016/j.chemosphere.2013.02.039 PMID:23522781

El-Temsah, Y. S., Sevcu, A., Bobcikova, K., Cernik, M., & Joner, E. J. (2016). DDT degradation efficiency and ecotoxicological effects of two types of nano-sized zero-valent iron (nZVI) in water and soil. *Chemosphere*, *144*, 2221–2228. doi:10.1016/j.chemosphere.2015.10.122 PMID:26598990

Elnashaie, S. S., Danafar, F., & Rafsanjani, H. H. (2015). *From Nanotechnology to Nanoengineering Nanotechnology for Chemical Engineers*. Springer. doi:10.1007/978-981-287-496-2_2

Fihri, A., Bouhrara, M., Nekoueishahraki, B., Basset, J.-M., & Polshettiwar, V. (2011). Nanocatalysts for Suzuki cross-coupling reactions. *Chemical Society Reviews*, *40*(10), 5181–5203. doi:10.1039/c1cs15079k PMID:21804997

Fine, G. F., Cavanagh, L. M., Afonja, A., & Binions, R. (2010). Metal oxide semi-conductor gas sensors in environmental monitoring. *Sensors (Basel)*, *10*(6), 5469–5502. doi:10.3390100605469 PMID:22219672

Galdames, A., Mendoza, A., Orueta, M., de Soto García, I., Sánchez, M., Virto, I., & Vilas, J. (2017). *Development of new remediation technologies for contaminated soils based on the application of zero-valent iron nanoparticles and bioremediation with compost*. Resource-Efficient Technologies.

Goldoni, A., Larciprete, R., Petaccia, L., & Lizzit, S. (2003). Single-wall carbon nanotube interaction with gases: Sample contaminants and environmental monitoring. *Journal of the American Chemical Society*, *125*(37), 11329–11333. doi:10.1021/ja034898e PMID:16220955

Gonçalves, J. R. (2016). The Soil and Groundwater Remediation with Zero Valent Iron Nanoparticles. *Procedia Engineering*, *143*, 1268–1275. doi:10.1016/j.proeng.2016.06.122

Grieger, K. D., Hjorth, R., Rice, J., Kumar, N., & Bang, J. (2015). *Nano-remediation: tiny particles cleaning up big environmental problems.* Blog entry for IUCN.

Hariharan, C. (2006). Photocatalytic degradation of organic contaminants in water by ZnO nanoparticles: Revisited. *Applied Catalysis A, General*, *304*, 55–61. doi:10.1016/j.apcata.2006.02.020

He, F., Zhao, D., & Paul, C. (2010). Field assessment of carboxymethyl cellulose stabilized iron nanoparticles for in situ destruction of chlorinated solvents in source zones. *Water Research*, *44*(7), 2360–2370. doi:10.1016/j.watres.2009.12.041 PMID:20106501

Hoag, G. E., Collins, J. B., Holcomb, J. L., Hoag, J. R., Nadagouda, M. N., & Varma, R. S. (2009). Degradation of bromothymol blue by 'greener'nano-scale zero-valent iron synthesized using tea polyphenols. *Journal of Materials Chemistry*, *19*(45), 8671–8677. doi:10.1039/b909148c

Hu, Q., Wujcik, E. K., Kelarakis, A., Cyriac, J., & Gong, X. (2017). Carbon-Based Nanomaterials as Novel Nanosensors. *Journal of Nanomaterials*.

Huang, J., & Wan, Q. (2009). Gas sensors based on semiconducting metal oxide one-dimensional nanostructures. *Sensors (Basel)*, *9*(12), 9903–9924. doi:10.339091209903 PMID:22303154

Jang, Y. J., Simer, C., & Ohm, T. (2006). Comparison of zinc oxide nanoparticles and its nano-crystalline particles on the photocatalytic degradation of methylene blue. *Materials Research Bulletin*, *41*(1), 67–77. doi:10.1016/j.materresbull.2005.07.038

Karkare, M. (2008). *Nanotechnology: Fundamentals and Applications: I.K.* International Publishing House Pvt. Limited.

Kašpar, J., Fornasiero, P., & Hickey, N. (2003). Automotive catalytic converters: Current status and some perspectives. *Catalysis Today*, *77*(4), 419–449. doi:10.1016/S0920-5861(02)00384-X

Khajeh, M., Laurent, S., & Dastafkan, K. (2013). Nanoadsorbents: Classification, preparation, and applications (with emphasis on aqueous media). *Chemical Reviews*, *113*(10), 7728–7768. doi:10.1021/cr400086v PMID:23869773

Khanna, V. K. (2016). *Nanosensors: Physical, Chemical, and Biological*. CRC Press.

Khavarian, M., Chai, S.-P., & Mohamed, A. R. (2013). Carbon dioxide conversion over carbon-based nanocatalysts. *Journal of Nanoscience and Nanotechnology*, *13*(7), 4825–4837. doi:10.1166/jnn.2013.7569 PMID:23901504

Khin, M. M., Nair, A. S., Babu, V. J., Murugan, R., & Ramakrishna, S. (2012). A review on nanomaterials for environmental remediation. *Energy & Environmental Science*, *5*(8), 8075–8109. doi:10.1039/c2ee21818f

Kowalczyk, P., & Holyst, R. (2008). Efficient adsorption of super greenhouse gas (tetrafluoromethane) in carbon nanotubes. *Environmental Science & Technology*, *42*(8), 2931–2936. doi:10.1021/es071306+ PMID:18497146

Kumar, R., Kumar, G., Akhtar, M., & Umar, A. (2015). Sonophotocatalytic degradation of methyl orange using ZnO nano-aggregates. *Journal of Alloys and Compounds*, *629*, 167–172. doi:10.1016/j.jallcom.2014.12.232

Lateef, A., Nazir, R., Jamil, N., Alam, S., Shah, R., Khan, M. N., & Saleem, M. (2016). Synthesis and characterization of zeolite based nano–composite: An environment friendly slow release fertilizer. *Microporous and Mesoporous Materials*, *232*, 174–183. doi:10.1016/j.micromeso.2016.06.020

Law, M., Kind, H., Messer, B., Kim, F., & Yang, P. (2002). Photochemical sensing of NO2 with SnO2 nanoribbon nanosensors at room temperature. *Angewandte Chemie*, *114*(13), 2511–2514. doi:10.1002/1521-3757(20020703)114:13<2511::AID-ANGE2511>3.0.CO;2-N PMID:12203608

Li, J., Lu, Y., Ye, Q., Cinke, M., Han, J., & Meyyappan, M. (2003). Carbon nanotube sensors for gas and organic vapor detection. *Nano Letters*, *3*(7), 929–933. doi:10.1021/nl034220x

Li, Y.-f., Wang, B.-h., & Huang, W.-x. (2003). Study on bactericidal performance of nano-scale ZnO modified interior wall paint. *Paint and Coatings Industry-China*, *33*(8), 3–5.

Lim, T.-C., & Ramakrishna, S. (2006). A conceptual review of nanosensors. *Zeitschrift fur Naturforschung. Section A. Physical Sciences*, *61*(7-8), 402–412. doi:10.1515/zna-2006-7-815

Lofrano, G., Libralato, G., & Brown, J. (2017). *Nanotechnologies for Environmental Remediation: Applications and Implications*. Springer International Publishing. doi:10.1007/978-3-319-53162-5

Long, F., Zhu, A., & Shi, H. (2013). Recent advances in optical biosensors for environmental monitoring and early warning. *Sensors (Basel)*, *13*(10), 13928–13948. doi:10.3390131013928 PMID:24132229

Lu, C.-Y., Wey, M.-Y., & Chen, L.-I. (2007). Application of polyol process to prepare AC-supported nanocatalyst for VOC oxidation. *Applied Catalysis A, General*, *325*(1), 163–174. doi:10.1016/j.apcata.2007.03.030

Lu, Y., Li, J., Han, J., Ng, H.-T., Binder, C., Partridge, C., & Meyyappan, M. (2004). Room temperature methane detection using palladium loaded single-walled carbon nanotube sensors. *Chemical Physics Letters*, *391*(4), 344–348. doi:10.1016/j.cplett.2004.05.029

Luque, R., Varma, R. S., & Clark, J. H. (2012). *Sustainable Preparation of Metal Nanoparticles: Methods and Applications*. Royal Society of Chemistry. doi:10.1039/9781849735469

Machado, S., Stawiński, W., Slonina, P., Pinto, A., Grosso, J., Nouws, H., ... Delerue-Matos, C. (2013). Application of green zero-valent iron nanoparticles to the remediation of soils contaminated with ibuprofen. *The Science of the Total Environment*, *461*, 323–329. doi:10.1016/j.scitotenv.2013.05.016 PMID:23738986

Masui, T., Ozaki, T., Machida, K.-i., & Adachi, G.-y. (2000). Preparation of ceria–zirconia sub-catalysts for automotive exhaust cleaning. *Journal of Alloys and Compounds*, *303-304*(Supplement C), 49–55. doi:10.1016/S0925-8388(00)00603-4

McBride, M. B., & MartÍnez, C. E. (2000). Copper Phytotoxicity in a Contaminated Soil: Remediation Tests with Adsorptive Materials. *Environmental Science & Technology*, *34*(20), 4386–4391. doi:10.1021/es0009931

McNicholas, T. P., Zhao, K., Yang, C., Hernandez, S. C., Mulchandani, A., Myung, N. V., & Deshusses, M. A. (2011). Sensitive detection of elemental mercury vapor by gold-nanoparticle-decorated carbon nanotube sensors. *The Journal of Physical Chemistry C*, *115*(28), 13927–13931. doi:10.1021/jp203662w PMID:21922036

Miller, G., van Schaik, D., & Adelaide, R. (2008). *Nanotechnology in food and agriculture*. Radio Adelaide.

Mizutani, T., Arai, K., Miyamoto, M., & Kimura, Y. (2006). Application of silica-containing nanocomposite emulsion to wall paint: A new environmentally safe paint of high performance. *Progress in Organic Coatings*, *55*(3), 276–283. doi:10.1016/j.porgcoat.2005.12.001

Mohammadian, H., Heidari, M. B., & Esfandiarijahromi, E. (2013). *Analysis the Different Types of Nano Sensors on Based of Structure and it's Applications on Nano Electronics*. Paper presented at the 2nd International Conference on Computer and Communication Technologies (ICCCT'2013), Pattaya, Thailand.

Moniz, E. J. (2010). *Nanotechnology for the energy challenge*. John Wiley & Sons.

Mubeen, S., Zhang, T., Chartuprayoon, N., Rheem, Y., Mulchandani, A., Myung, N. V., & Deshusses, M. A. (2009). Sensitive detection of H2S using gold nanoparticle decorated single-walled carbon nanotubes. *Analytical Chemistry, 82*(1), 250–257. doi:10.1021/ac901871d PMID:19968250

Mueller, N. C., Braun, J., Bruns, J., Černík, M., Rissing, P., Rickerby, D., & Nowack, B. (2012). Application of nanoscale zero valent iron (NZVI) for groundwater remediation in Europe. *Environmental Science and Pollution Research International, 19*(2), 550–558. doi:10.100711356-011-0576-3 PMID:21850484

Mueller, N. C., & Nowack, B. (2010). Nanoparticles for remediation: Solving big problems with little particles. *Elements, 6*(6), 395–400. doi:10.2113/gselements.6.6.395

Mulvaney, D. (2011). *Green Technology: An A-to-Z Guide*. SAGE Publications. doi:10.4135/9781412973793.n84

Nagaveni, K., Sivalingam, G., Hegde, M., & Madras, G. (2004a). Photocatalytic degradation of organic compounds over combustion-synthesized nano-TiO2. *Environmental Science & Technology, 38*(5), 1600–1604. doi:10.1021/es034696i PMID:15046366

Nagaveni, K., Sivalingam, G., Hegde, M., & Madras, G. (2004b). Solar photocatalytic degradation of dyes: High activity of combustion synthesized nano TiO 2. *Applied Catalysis B: Environmental, 48*(2), 83–93. doi:10.1016/j.apcatb.2003.09.013

Narayanan, K. B., & Sakthivel, N. (2011). Synthesis and characterization of nano-gold composite using Cylindrocladium floridanum and its heterogeneous catalysis in the degradation of 4-nitrophenol. *Journal of Hazardous Materials, 189*(1), 519–525. doi:10.1016/j.jhazmat.2011.02.069 PMID:21420237

Ondreiovic, G., & Sirota, A. (1995). Selective binding and removal of guests in a microporous metal-organic framework. *Nature, 378*, 14.

Padmini, E., & Miranda, L. R. (2013). Nanocatalyst from sol–sol doping of TiO2 with Vanadium and Cerium and its application for 3,4 Dichloroaniline degradation using visible light. *Chemical Engineering Journal, 232*(Supplement C), 249–258. doi:10.1016/j.cej.2013.07.083

Pavličević, J., Ikonić, B., Bera, O., Vukić, N., & Budinski-Simendić, J. (2016). Environmental nanotechnology: benefits and risk assessment. *Conference: 10. International Conference Risk and safety engineering, 3*, 129-134.

Penza, M., Cassano, G., Aversa, P., Cusano, A., Consales, M., Giordano, M., & Nicolais, L. (2006). Acoustic and optical VOCs sensors incorporating carbon nanotubes. *IEEE Sensors Journal, 6*(4), 867–875. doi:10.1109/JSEN.2006.877974

Penza, M., Rossi, R., Alvisi, M., Cassano, G., & Serra, E. (2009). Functional characterization of carbon nanotube networked films functionalized with tuned loading of Au nanoclusters for gas sensing applications. *Sensors and Actuators. B, Chemical, 140*(1), 176–184. doi:10.1016/j.snb.2009.04.008

Philippot, K., & Serp, P. (2012). *Concepts in nanocatalysis*. Nanomaterials in Catalysis.

Polshettiwar, V., Asefa, T., & Hutchings, G. (2013). *Nanocatalysis: Synthesis and Applications*. Wiley. doi:10.1002/9781118609811

Polshettiwar, V., Luque, R., Fihri, A., Zhu, H., Bouhrara, M., & Basset, J.-M. (2011). Magnetically Recoverable Nanocatalysts. *Chemical Reviews*, *111*(5), 3036–3075. doi:10.1021/cr100230z PMID:21401074

Rai, M., & Ingle, A. (2012). Role of nanotechnology in agriculture with special reference to management of insect pests. *Applied Microbiology and Biotechnology*, *94*(2), 287–293. doi:10.100700253-012-3969-4 PMID:22388570

Ramsden, J. (2011). *Nanotechnology: An Introduction*. William Andrew.

Ratner, M. A., & Ratner, D. (2003). *Nanotechnology: A Gentle Introduction to the Next Big Idea*. Prentice Hall.

Ravanchi, M. T., Kaghazchi, T., & Kargari, A. (2009). Application of membrane separation processes in petrochemical industry: A review. *Desalination*, *235*(1), 199–244. doi:10.1016/j.desal.2007.10.042

Reddy, A., Madhavi, V., Reddy, K. G., & Madhavi, G. (2012). Remediation of chlorpyrifos-contaminated soils by laboratory-synthesized zero-valent nano iron particles: Effect of pH and aluminium salts. *Journal of Chemistry*.

Rickerby, D. G. (2007). Nanotechnology Nanotechnologies and nanoparticles in remediation and pollution prevention. *Chemistry Today, 25*(6), 24.

Riu, J., Maroto, A., & Rius, F. X. (2006). Nanosensors in environmental analysis. *Talanta*, *69*(2), 288–301. doi:10.1016/j.talanta.2005.09.045 PMID:18970568

Rogers, K. (2006). Recent advances in biosensor techniques for environmental monitoring. *Analytica Chimica Acta*, *568*(1), 222–231. doi:10.1016/j.aca.2005.12.067 PMID:17761264

Sanghi, R., & Singh, V. (2012). *Green Chemistry for Environmental Remediation*. Wiley.

Sarkar, T., Gao, Y., & Mulchandani, A. (2013). Carbon nanotubes-based label-free affinity sensors for environmental monitoring. *Applied Biochemistry and Biotechnology*, *170*(5), 1011–1025. doi:10.100712010-013-0233-z PMID:23653139

Satapanajaru, T., Anurakpongsatorn, P., Pengthamkeerati, P., & Boparai, H. (2008). Remediation of atrazine-contaminated soil and water by nano zerovalent iron. *Water, Air, and Soil Pollution*, *192*(1-4), 349–359. doi:10.100711270-008-9661-8

Selid, P. D., Xu, H., Collins, E. M., Striped Face-Collins, M., & Zhao, J. X. (2009). Sensing mercury for biomedical and environmental monitoring. *Sensors (Basel)*, *9*(7), 5446–5459. doi:10.339090705446 PMID:22346707

Sellers, K., Mackay, C., Bergeson, L. L., Clough, S. R., Hoyt, M., Chen, J., & Hamblen, J. (2008). *Nanotechnology and the Environment*. CRC Press.

Serrano, E., Rus, G., & Garcia-Martinez, J. (2009). Nanotechnology for sustainable energy. *Renewable & Sustainable Energy Reviews*, *13*(9), 2373–2384. doi:10.1016/j.rser.2009.06.003

Seyfi, B., Baghalha, M., & Kazemian, H. (2009). Modified LaCoO 3 nano-perovskite catalysts for the environmental application of automotive CO oxidation. *Chemical Engineering Journal*, *148*(2), 306–311. doi:10.1016/j.cej.2008.08.041

Shirsat, M. D., Sarkar, T., Kakoullis, J. Jr, Myung, N. V., Konnanath, B., Spanias, A., & Mulchandani, A. (2012). Porphyrin-functionalized single-walled carbon nanotube chemiresistive sensor arrays for VOCs. *The Journal of Physical Chemistry C, 116*(5), 3845–3850. doi:10.1021/jp210582t PMID:22393460

Shokri, A., Rabiee, F., & Mahanpoor, K. (2017). Employing a novel nanocatalyst (Mn/Iranian hematite) for oxidation of SO2 pollutant in aqueous environment. *International Journal of Environmental Science and Technology, 14*(11), 2485–2494. doi:10.100713762-017-1346-7

Singh, R., Manickam, N., Mudiam, M. K. R., Murthy, R. C., & Misra, V. (2013). An integrated (nano-bio) technisque for degradation of γ-HCH contaminated soil. *Journal of Hazardous Materials, 258*, 35–41. doi:10.1016/j.jhazmat.2013.04.016 PMID:23692681

Singhal, R., Gangadhar, B., Basu, H., Manisha, V., Naidu, G., & Reddy, A. (2012). Remediation of malathion contaminated soil using zero valent iron nano-particles. *American Journal of Analytical Chemistry, 3*(01), 76–82. doi:10.4236/ajac.2012.31011

Sivalingam, G., Nagaveni, K., Hegde, M., & Madras, G. (2003). Photocatalytic degradation of various dyes by combustion synthesized nano anatase TiO 2. *Applied Catalysis B: Environmental, 45*(1), 23–38. doi:10.1016/S0926-3373(03)00124-3

Su, S., Wu, W., Gao, J., Lu, J., & Fan, C. (2012). Nanomaterials-based sensors for applications in environmental monitoring. *Journal of Materials Chemistry, 22*(35), 18101–18110. doi:10.1039/c2jm33284a

Sui, G., Zhong, W., Yang, X., Yu, Y., & Zhao, S. (2008). Preparation and properties of natural rubber composites reinforced with pretreated carbon nanotubes. *Polymers for Advanced Technologies, 19*(11), 1543–1549.

Sun, J., Wang, X., Sun, J., Sun, R., Sun, S., & Qiao, L. (2006). Photocatalytic degradation and kinetics of Orange G using nano-sized Sn (IV)/TiO 2/AC photocatalyst. *Journal of Molecular Catalysis A Chemical, 260*(1), 241–246. doi:10.1016/j.molcata.2006.07.033

Tasaltin, C., & Basarir, F. (2014). Preparation of flexible VOC sensor based on carbon nanotubes and gold nanoparticles. *Sensors and Actuators. B, Chemical, 194*, 173–179. doi:10.1016/j.snb.2013.12.063

Tratnyek, P. G., & Johnson, R. L. (2006). Nanotechnologies for environmental cleanup. *Nano Today, 1*(2), 44–48. doi:10.1016/S1748-0132(06)70048-2

Tsang, S. C., Caps, V., Paraskevas, I., Chadwick, D., & Thompsett, D. (2004). Magnetically Separable, Carbon-Supported Nanocatalysts for the Manufacture of Fine Chemicals. *Angewandte Chemie, 116*(42), 5763–5767. doi:10.1002/ange.200460552 PMID:15495206

Tungittiplakorn, W., Lion, L. W., Cohen, C., & Kim, J.-Y. (2004). Engineered polymeric nanoparticles for soil remediation. *Environmental Science & Technology, 38*(5), 1605–1610. doi:10.1021/es0348997 PMID:15046367

Van der Bruggen, B., & Vandecasteele, C. (2003). Removal of pollutants from surface water and groundwater by nanofiltration: Overview of possible applications in the drinking water industry. *Environmental Pollution, 122*(3), 435–445. doi:10.1016/S0269-7491(02)00308-1 PMID:12547533

Varanasi, P., Fullana, A., & Sidhu, S. (2007). Remediation of PCB contaminated soils using iron nanoparticles. *Chemosphere*, *66*(6), 1031–1038. doi:10.1016/j.chemosphere.2006.07.036 PMID:16962632

Venkatachalam, N., Palanichamy, M., & Murugesan, V. (2007). Sol–gel preparation and characterization of alkaline earth metal doped nano TiO 2: Efficient photocatalytic degradation of 4-chlorophenol. *Journal of Molecular Catalysis A Chemical*, *273*(1), 177–185. doi:10.1016/j.molcata.2007.03.077

Wang, H., Wang, Y., Chen, J., Wang, S., Cheng, J., & Zhou, D. (2009). Application of modified nanoparticle black carbon for the remediation of soil heavy metal pollution. *Zhongguo Huanjing Kexue*, *29*(4), 431–436.

Wang, Q., Xiong, Y., & Lou, L. (2015). Biosensor for detection of pesticide residue. *American Journal of Nano Research and Application*, *3*(1), 18-22.

Watarai, H. (2006). *Trend of Research and Development for Magnesium Alloys-Reducing the Weight of Structural Materialsin Motor Vehicles*. NISTEP Science & Technology Foresight Center.

Xi, Y., Sun, Z., Hreid, T., Ayoko, G. A., & Frost, R. L. (2014). Bisphenol A degradation enhanced by air bubbles via advanced oxidation using in situ generated ferrous ions from nano zero-valent iron/palygorskite composite materials. *Chemical Engineering Journal*, *247*, 66–74. doi:10.1016/j.cej.2014.02.077

Yang, Y., Ma, J., Qin, Q., & Zhai, X. (2007). Degradation of nitrobenzene by nano-TiO 2 catalyzed ozonation. *Journal of Molecular Catalysis A Chemical*, *267*(1), 41–48. doi:10.1016/j.molcata.2006.09.010

Yu, B., Zeng, J., Gong, L., Zhang, M., Zhang, L., & Chen, X. (2007). Investigation of the photocatalytic degradation of organochlorine pesticides on a nano-TiO 2 coated film. *Talanta*, *72*(5), 1667–1674. doi:10.1016/j.talanta.2007.03.013 PMID:19071814

Yu, H., Zhang, K., & Rossi, C. (2007). Experimental study of the photocatalytic degradation of formaldehyde in indoor air using a nano-particulate titanium dioxide photocatalyst. *Indoor and Built Environment*, *16*(6), 529–537. doi:10.1177/1420326X07083513

Yu, J. C., Lin, J., & Kwok, R. W. (1998). Ti1-x Zr x O2 Solid Solutions for the Photocatalytic Degradation of Acetone in Air. *The Journal of Physical Chemistry B*, *102*(26), 5094–5098. doi:10.1021/jp980332e

Zäch, M., Hägglund, C., Chakarov, D., & Kasemo, B. (2006). Nanoscience and nanotechnology for advanced energy systems. *Current Opinion in Solid State and Materials Science*, *10*(3), 132–143. doi:10.1016/j.cossms.2007.04.004

Zan, L., Wang, S., Fa, W., Hu, Y., Tian, L., & Deng, K. (2006). Solid-phase photocatalytic degradation of polystyrene with modified nano-TiO 2 catalyst. *Polymer*, *47*(24), 8155–8162. doi:10.1016/j.polymer.2006.09.023

Zang, L. (2011). *Energy efficiency and renewable energy through nanotechnology*. Springer.

Zhang, M., He, F., Zhao, D., & Hao, X. (2011). Degradation of soil-sorbed trichloroethylene by stabilized zero valent iron nanoparticles: Effects of sorption, surfactants, and natural organic matter. *Water Research*, *45*(7), 2401–2414. doi:10.1016/j.watres.2011.01.028 PubMed

Zhang, T., Mubeen, S., Myung, N. V., & Deshusses, M. A. (2008). Recent progress in carbon nanotube-based gas sensors. *Nanotechnology*, *19*(33), 332001. doi:10.1088/0957-4484/19/33/332001 PMID:21730614

Zhang, Z., Li, M., Chen, W., Zhu, S., Liu, N., & Zhu, L. (2010). Immobilization of lead and cadmium from aqueous solution and contaminated sediment using nano-hydroxyapatite. *Environmental Pollution*, *158*(2), 514–519. doi:10.1016/j.envpol.2009.08.024 PMID:19783084

Zhu, H., Jiang, R., Xiao, L., Chang, Y., Guan, Y., Li, X., & Zeng, G. (2009). Photocatalytic decolorization and degradation of Congo Red on innovative crosslinked chitosan/nano-CdS composite catalyst under visible light irradiation. *Journal of Hazardous Materials*, *169*(1), 933–940. doi:10.1016/j.jhazmat.2009.04.037 PMID:19477069

Chapter 2
Properties of Nanomaterials and Environment

Phool Shahzadi
Pakistan Council of Scientific and Industrial Research (PCSIR) Labs, Pakistan

ABSTRACT

The chapter provides a timely review of the various properties of nonmaterial and their applications into environmental compartments. An extensive variety of poisonous chemicals is discharged into the environment because of globalization and industrialization. The dimensional, compositional, geometric, and structural properties are fundamental to convey usefulness of the nanomaterials. The controlled sizes and shapes of nanoparticles are anticipated to yield unique catalytic, electrochemical, and photochemical properties. The electrochemical properties of monolayer-functional metal nanoparticles are expected to be controlled by the particle sizes. Metal nanomaterials have interesting optical properties due to strong surface plasmon absorption and field enhancement effects; metal oxides lack visible absorption due to very large bandgap. Nanocomposites have complex optical properties. Nanomaterials present gigantic advantages on diverse applications, catalysis, imaging, biotechnological, and sensor applications due to their improved properties.

INTRODUCTION

Nanotechnology is based on materials that fall in size range of 1-100nm.This exceptionally small size makes their surface to volume ratio very high which in turn is responsible for nanomaterial properties that are different from their bulk counterparts. By manipulating their size and shape these properties can be tailored and hence can be utilized for various commercial and domestic applications in different areas like; medical, energy resource research, agricultural and food, electronics, environmental applications etc. At this nano-scale the bonding is different from that of the bulk materials and so is the atom that is available on the surface to that of the one available inside the bulk. Further the differential behavior is shown by atoms that is present on a smooth surface to the ones available on rough surface; the ones present on cluster to the ones occurring on support and the atoms present in doped and pure form (Roduner, 2006).

DOI: 10.4018/978-1-5225-5745-6.ch002

NANOMATERIALS' SIZE MATTERS

Nanomaterials have large surface to volume ratio and that ratio changes dramatically as the size reduces. For example, when the size of iron cube reduces from $1cm^3$ to $10\ nm^3$ the 10% increase in surface atoms is noticed; with further decrease in size to $1nm^3$ all the atoms present will be surface atoms only. Similarly, the total surface energy hypes from 7.2×10^{-5} J/g to 560 J/g with decrease in particle side from 0.77cm (total surface area 3.6 cm^2) to 1nm (total surface area $2.8 \times 10^7\ cm^2$).This kind of pronounced impact of surface to volume ratio is held responsible for changes in physical and chemical properties of the nanomaterials (Cao, 2004).

These size based effects can be divided into two types:

1. Surface effects
2. Quantum effects

In the first case atoms at the surface have less neighboring atoms and hence less coordination number and consequently the bond energy per atom (cohesive energy). This is the reason of high stability of atoms preset in the bulk which increases with increase in coordination number and cohesive energy. The order of atom' stability for a cube follows the order: corner> edge > in-plane surface > interior atoms. Thus, owing to highest instability of the corner atom and maximum stability of the interior atoms reverse order of is followed for affinity for bond formation and absorption of atoms. Same factor goes for less well defined phase transitions in nanomaterials having lower number of atoms in a cluster. This surface driven effect is responsible for variations in material's properties like melting point, catalytic reactivity, adsorption capacity etc. (Roduner, 2006).

The second type of effect, quantum confinement, is based on delocalization of electron states. The density of states (DOS) refers to the band formed by contribution of a number of atoms and width of DOS is directly proportional to the number of atoms. This number amounts to the Avogadro's number in case of bulk and very low in case of nanomaterials thereby reducing the DOS. All this contributes to modification in distance between the highest and lowest occupied levels in case of nanomaterials resulting in size-induced metal-insulator transitions leading to conversion of insulator to semiconductor to metal. This phenomenon is also responsible for change in magnetic and optical properties in case of nano-size materials (Roduner, 2006).

The DOS unequivocally impacts the electronic and optical properties of semiconductors and metals. The adjustments in DOS, alongside changes in electronic and vibrational levels, modify the properties of the nanomaterials. This in turn brings changes in both their dynamic and harmony properties; for example changes in the DOS of the electrons and phonons influence the electron–phonon coupling and along these lines the electronic unwinding due to electron–phonon communication.

The density of state of a diminished measurement framework additionally changes significantly with diminishing size. For instance, for a three-dimensional mass material, the DOS is relative to the square foundation of vitality E. For a system with control in one measurement (a two-dimensional material or quantum well), the DOS is a stage work. For frameworks with imprisonment in two measurements (a one-dimensional material or quantum wire), the DOS has an idiosyncrasy. For frameworks with repression in three measurements (zero-dimensional material or supposed quantum dabs, QDs), the DOS has the state of a δ-work (Yang et al., 2009). The adjustment in DOS in decreased measurement frameworks is delineated in Figure 1.

Figure 1. Illustration of change in DOS as a function of physical dimension of the system, from 3D (bulk) to 2D, 1D, and 0D

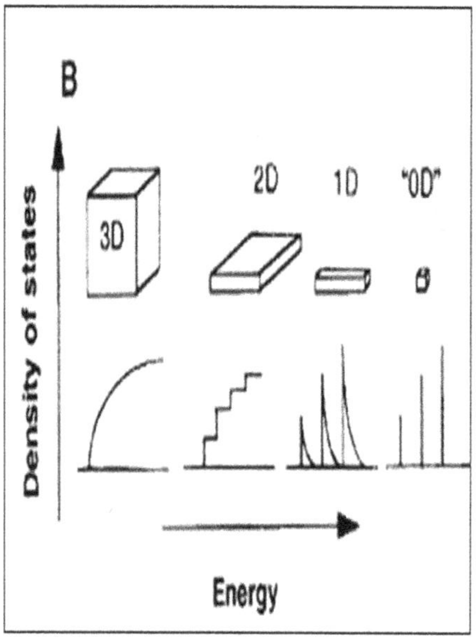

PROPERTIES OF NANOMATERIALS

The materials at nanoscale behave differently when compared to their larger counterparts. They show unique physical, chemical, electronic and optical properties owing to the modification in particle size that ultimately impacts their behavior and interaction with the adjacent atoms. The group of nanoparticles can have altogether different properties as from the individual ones as a result of assembling of these nanoparticles in different shapes and structures. This behavior of changing and altering in properties hence can be exploited in multiple applications (Altavilla et al., 2016). Properties of nanomaterials are influenced by their chemical composition, size, geometry, crystal structure and surface properties like charge, organic and inorganic coating.

The excellent optical, magnetic, electrical and catalytic properties signify the profits over the macro-scale materials, revealing the notable specific properties (Weiner et al., 2015). The physical and catalytic properties of the nanomaterials can be easily tuned or altered by the reduction of the spatial dimension or the confinement of the structures in a precise crystallographic direction. In particular, the properties of the nanomaterials are mainly associated to different origins via large fraction of surface atoms, large surface energy, spatial confinement, and the reduced imperfections. These can be tuned simply by adjusting the size, shape or extent of agglomeration (Gilroy et al., 2016). Due to the enormous fraction of surface atoms in the total amount of atoms, nanomaterials possess lower melting point and reduced lattice constants. The reduction of the materials dimension has obvious effects on the optical properties and the size and structure dependence can be categorized into surface plasmon resonance (SPR) and the quantum size effects (Plant, S. R., et al. 2014). The electrical conductivity decreases by increased surface scattering. The mechanism of electrical conductivity on the thin film can be extensively explained by a simple tunneling between localized insulating states, which imply a high resistivity at low temperatures (Kramer, I. J., et al. 2013). For the catalytic property of the nanomaterials, geometry, composition, oxidation state, and chemical/physical environment can play a vital role in determining catalytic activity and reactivity of the nanomaterials, while particle size and shape is an important consideration. Thus, the relationship between these parameters and the catalytic performance of the nanomaterials may be system dependent. Moreover, a systematic understanding of the factor that control catalyst reactivity and selectivity is a vital (Howes et al., 2014; Zhou et al., 2015).

The size based effects, as discussed earlier in the chapter, contribute to the different properties like physical, chemical, magnetic, electronic, optical, mechanical, thermal, biological etc. as outlined in Table 1. Few of these are detailed as under:

Magnetism in Small Particles

Magnetic nanoparticles are of great curiosity for investigators from an eclectic range of disciplines, which include heterogeneous and homogenous catalysis, biomedicine, magnetic fluids, data storage magnetic resonance imaging (MRI), and environmental remediation such as water decontamination. Magnetic nanoparticles demonstrate an assortment of irregular magnetic behavior when contrasted with the bulk materials, mostly due to surface or interface effects, including symmetry breaking, electronic environment or charge transfer and magnetic interactions. The literature revealed that nanoparticles perform best when the size is <critical value i.e. 10–20 nm, at such low scale the magnetic properties of nanoparticles conquered effectively, which make these particles priceless and can be used in different applications. The uneven electronic distribution in nanoparticles leads to magnetic property. These properties are also dependent on the synthetic protocol and various synthetic methods such as co-precipitation, micro-emulsion, thermal decomposition, flame spray synthesis and solvothermal Magnetostriction, the procedure in which attractive material distorted because of essence of attractive field, can help in development of various properties in materials like nano-scale films making them useful in different applications which are impossible when utilizing existing bulk materials. For instance these constitute driving components of small scale robots, pumps, engines and so forth. These can likewise be utilized for attractive control of flexible properties or reliance of stress or strain on attractive porousness to create

Table 1. Properties of nanoparticles

Types	Properties
Electrical	Higher electrical conductivity in ceramics and higher electric resistance in metals.
Magnetic	Increased magnetic, coercivity, super paramagnetic behavior
Catalytic	Higher surface to volume ratio improves catalytic efficiency
Mechanical	Improved hardness of metals and alloys, super-plasticity and ductility of ceramics.
Optical	Spectral shift of optical and fluorescence properties and increased quantum efficiency of semiconductors.
Sterical	Selectivity transportation and controlled release of drugs.
Biological	Biocompatibility and permeability increased through biological barriers.
Sensors	Pollutants sensors that able to detect lower limits with low cost
Carbon capture	Photocatalyst consisting of silica Nanosprings coated with a combination of titanium dioxide
Energy	Heat distribution e.g. ceramic-like materials that provide sufficient reliability and durability of the entire structure

different electronic gadgets like a resonator with attractively (Qi et al., 2016). Two main features impact the magnetic properties of nanoparticles: a) Finite-size effects and b) surface effects while coercivity is determined by spin rotation and shape anisotropy (Issa et al., 2013).

Decreasing particle size also affects magnetic behavior of the individual nanoparticles which is different in core and surface. Core usually exhibits magnetic behavior similar to that of bulk while owing to higher magnetic disorder in the surface, different types of magnetic behaviors are observed, and it's the competition between surface and core that determines the ground state of particles (Blanco et al., 2008). Interaction and size effects in magnetic nanoparticles and enhanced magnetism are observed in nanosized particles of famous ferromagnetic materials i.e. Fe and Co (Brechignac et al., 2008; Roduner, 2006) Amazingly the materials those are non-magnetic in bulk state have shown tremendous magnetism when their particle size is reduced. The Rhodium cluster consisting of 32 atoms only showed high magnetic moment equivalent of $1\mu_B$/atom while that of Pd (13 atoms) exhibited magnetic moment of 0.65 μ_B/atom; similar results are observed for Pt and Au clusters. This magnetic polarization was attributed to presence of unpaired electrons on the cluster, increase in Fermi level owing to reduction in interatomic distances and higher orbital overlap. A new concept of super paramagnetism is also peculiar to nanoparticles that are characterized by blocking temperature that is dependent on particle size e.g. the blocking temperature of Co having size of 6nm is around room temperature which reduces 30K when particle size reaches 3nm. In case of thin films, high coercivity is shown as compared to the bulk materials e.g. in $Fe_x(SiO_2)_{1-x}$ based thin films with x = 0.5 depicted coercivity value of 2500 Oe at 2K which magnitude is two order high as compared to bulk Fe (Edelstein et al., 1998).

Generally, the metallic nanoparticles have large magnetization as compared to their oxides but they can easily lose their magnetization due to easy oxidation of these nanoparticles (Gubin, 2009).

Another phenomenon that is peculiar of the nanomaterials is superparamagnetism – it happens that below a critical size the nanoparticles behave as single domain instead of the typical multi-domain. At this size the magnetic properties are drastically changed (Koksharov, 2009). For iron the critical size is 20 nm (Gubin, 2009).

Optical and Electronic Properties of Small Particles

Optical properties of nanoparticles depend strongly on the size and shape e.g. bulk Au looks yellowish in reflected light but its thin film looks blue which changes to different tones of purple and red as particle size decreases. This effect is attributed to surface-plasmon resonance ("the frequency at which conduction electrons oscillates in response to the alternating electric field of incident electromagnetic radiation") (Liz-Marzan, 2004, 26-31). For the size affects the color of CdSe-CdS core shell nanoparticles with size varying from 1.7 – 6nm the cooler changes from blue to red owing to the decrease in band gap (distance between conduction and valance band). The gold nanoparticles having size in range of 10 nm exhibit absorbance of green light and hence appearing red (Roduner, 2006).The shape effect is also prominent on absorption as well as emission properties of the material. The extreme case is nanowires (NW) that can be considered as a limiting case of nanoparticle in the general sense. Due to the extended dimension in one direction for NW or nanorod (NR) that is comparable or even longer than the wavelength of light, scattering becomes more significant than for nanoparticles with all three dimensions much smaller than the wavelength of light (Alén et al., 2001). In addition to size and shape, nature of surrounding media also affects the optical properties of nanoparticles as when particles are sufficiently close, interactions between neighboring particles come into action.

There is a strong link between optical and electronic properties of the nanoparticles (Khan et al., 2017). It is observed that with change in structural geometry of individual particles and aggregation the optical responses of metal-dielectric nano-composites change to considerably. For example the network of Au nanoparticles showed varied dielectric function and the electrical conductivities which strongly depend on of the particle size, volume fraction, and interaction among the particles (Su et al., 2009). In case of Ti_2O_3 nanoparticles the colour shift from dark brown to bluish brown was accompanied by the 70K reduction in temperature of insulator-metal transition which was attributed to the reduction in band gap (Tsujimoto et al., 2015). The size related metal-to-insulator transitions also occur in nanomaterials that is related to the HOMO-LUMO energy difference (Bréchignac et al., 2008) as depicted in Figure 2. (Roduner, 2006).

Mechanical Properties of Small Particles

For novel applications in many important fields such as tribology, surface engineering, nanofabrication and nano-manufacturing the distinct mechanical properties of nanoparticles play very important role. Different mechanical parameters such as elastic modulus, hardness, stress and strain, adhesion and friction can be surveyed to know the exact mechanical nature of nanoparticles. Beside these parameters surface coating, coagulation, and lubrication also aid to mechanical properties of nanoparticles. Nanoparticles show different mechanical properties as compared to microparticles and their bulk materials. Moreover, in a lubricated or greased contact, the contrast in the stiffness between nanoparticles and the contact-

Figure 2. The size related metal-to-insulator transitions occurring in nanomaterials

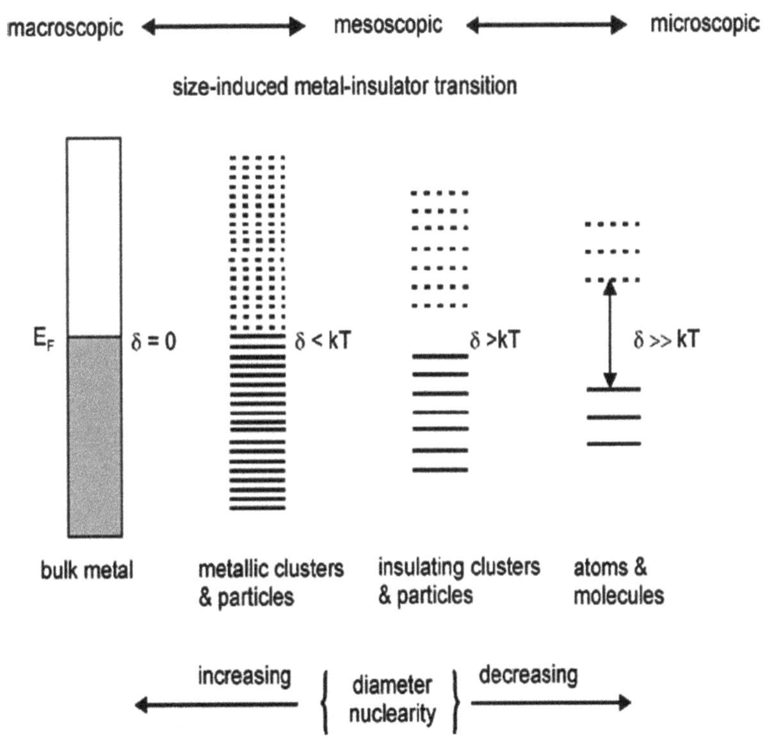

ing external surface controls whether the NPs are indented into the plan surface or deformed when the pressure at contact is significantly large. The main applications of nanoparticles that result from their special mechanical properties; these applications include a lubricant with nanoparticles as additives, nanoparticles in nano-manufacturing and nanoparticle reinforced composite coating.

This important information could divulge how nanoparticles perform in the contact situation. Decent controls over mechanical features of nanoparticles and their interactions with any kind of surface are vital for enlightening the surface quality and elevating material removal. Fruitful outcomes in these fields generally need a deep insight into the basics of the mechanical properties such as elastic modulus and hardness, movement law, friction and interfacial adhesion and their size dependent characteristics (Guo et al., 2013).

Thermal Properties of Small Particles

Nanomaterials having fewer numbers of atoms don't show well pronounced phase transitions and hence are no longer sharp. It is verifiable truth that metals nanoparticles have warm conductivities higher than those of liquids in strong shape. For instance, the warm conductivity of copper at room temperature is around 700 times more prominent than that of water and around 3000 times more noteworthy than that of motor oil. Indeed, even oxides, for example, alumina (Al_2O_3) have warm conductivity higher than that of water. In this way, the liquids containing suspended strong particles are required to show altogether upgraded warm conductivities in respect to those of traditional warmth exchange liquids. Nanofluids are created by scattering the nanometric scales strong particles into fluid, for example, water, ethylene glycol or oils. Nanofluids are required to display better properties relative than those of regular warmth exchange liquids and liquids containing tiny estimated particles. Since the warmth exchange happens at the surface of the particles, it is alluring to utilize the particles with vast aggregate surface zone. The expansive aggregate surface territory likewise expands the soundness suspension. As of late it has been shown that the nanofluids comprising of CuO or Al_2O_3 NPs in water or ethylene display propel warm conductivity (Cao, 2002).

Nano-solids' melting temperature is determined by size-dependent cohesive energy. The melting point os nanomaterials decrease with reduction in particle size. This cohesive energy is also the basics used to govern nanomaterials' melting enthalpy, melting entropy and specific heat. The studies on melting entropy of nanoparticle showed to have high impact with small change of nanoparticles' size. Similar increase was observed in case of melting enthalpy of Ag, Cu and In spherical nanoparticles as their size increases. For Al and Se NPs the melting entropy as well as enthalpy varies with shape following the order Nanospheres < Nanowires < Nanofilms (Singh et al., 2017).

Catalytic Properties of Small Particles

As for other properties, in case of catalytic properties as well, the enhanced catalytic activities owing to high reactivity and better selectivity. The main contributor in this regard is considered to be size but other factors such as geometry, composition, oxidation state, and chemical or physical environment also play significant role. The famous example in this case is of gold which in bulk form is inert material but its nanoparticles are highly reactive. In addition to the earlier stated factors that contribute to high catalytic activity surface roughness of the nanoparticles and transformation between different energetically favorable isomers of the clusters also contribute (Cuenya, 2010). The catalytic activity of nanoparticle is six times that of conventional catalysts. The iron nanoparticles showed higher selectivity towards methane synthesis because of their finite size (Huber, 2005).

Generally, smaller sized nanomaterials show high catalytic activity. In case of nanoclusters of gold, the electro-oxidation potential depends strongly on the size of cluster (Cuenya et al., 2003). The similar impact of size was observed for degradation of phenol and COD reduction by use of Co_3O_4, the order followed is 70 > 19 > 3.5nm (Yuming et al., 2007).

The studies on 2-propanol oxidation via Pt supported Al_2O_3 nanoparticles showed the control of catalytic properties via shape change (Mostafa et al., 2010). Another studies related to shape effect on catalytic activity of nanomaterials showed that rate of CO oxidation when compared to nanoparticles is about three times higher for nanobelts and six times higher for nanoplatelets (Zhou et al., 2006). In case of CeO_2 different chemical reactivity is observed for nanowires, nanorods and nanoparticles; highest being offered by nanowires as they offer much more exposed surface area (Zhang et al., 2009). Alkyne coupling reactions catalyzed by copper nanoparticles and nanorods showed dependence on shape of catalyst as well (Wei et al., 2010).

The nanomaterials prepared with support materials have shown support dependent catalytic activity (Prieto et al., 2013). In such case two effects dominate the catalytic potential of these materials: 1) transfer of electron to or from support and 2) an epitaxial stress that can alter the particle structural attributes. The nano-gold (3nm) deposited on three different supports powders (TiO_2, ZrO_2 and Al_2O_3) depicted different catalytic reactivities towards CO oxidation owing to the electronic effects induced by interaction between gold and support particles. Similarly the metals alloyed together impart electronic effects in different way that contribute differently than the support type materials (Bréchignac et al., 2008). Conducting substrate (ITO-coated glass) and semiconducting (TiO_2) substrates perform better in respect to catalytic reactivity as compared to insulating ones (SiO_2) (Cuenya, et al., 2003). Another studies carried out on MnO_x nanoparticles as potential catalyst for CO oxidation demonstrated to depend on the precursor used for the synthesis. The lower catalytic reactivity was recorded by nanoparticles prepared by using manganese nitrate as compared to the ones prepared by manganese acetyl acetonate (Han et al., 2006). The synthetic approach also influence the catalytic power of the nanomaterials; the catalytic reforming of methane using Nickel ferrite nanocatalyst prepared by sol-gel and hydrothermal approach was greater than the one prepared by co-precipitation (Benrabaa et al., 2013).

Antimicrobial Properties of Small Particles

The antimicrobial activity of nanoparticles is also considered as a function of particle size. Other factors that aid this activity are high surface area, crystal morphologies and reactive sites. The small size has larger number of particles present on the surface that helps in gaining the maximum interaction of these atoms with the microorganism. This plus size makes these nanomaterials potent antimicrobial agents against many microbial strains like bacteria and fungi. The metals studied in this regard are Ag, Cu, Au, Al, Ti, Fe and Zn (Santos et al., 2013). The species sensitivity to the nanoparticles is sometimes traced to the cell wall characteristics of the microbial strains but certain other factors might also contribute e.g. growth rate and biofilm formation. The contributing factors in case of nano are their composition, surface modifications (Hajipour et al., 2012), shape (Pal et al., 2007) and size (Azam, Ahmed, Oves, Khan, & Memic, 2012). Silver nanoparticles having truncated shape depicted better antimicrobial activity against *E. coli* as compared to the ones having rod and spherical shape. In case of CuO nanoparticles the particles with smaller size (4.6 nm) exhibited zone of inhibition twice as that of larger size particles (4.7nm) against three bacterial strains (Azam, Ahmed, Oves, Khan, & Memic, 2012). Similar size dependent activity was observed in presence of visible light against broad spectrum of microorganisms for ZnO nanoparticles (Azam, Ahmed, Oves, Khan, Habib, & Memic, 2012).

In addition to above mentioned factors, the surfactants and polymeric materials added to enhance stabilization of silver nanoparticles have shown strong correlation with the antibacterial activity (Kvitek et al., 2008). Addition of support materials or doping with nanoparticles also enhances the antimicrobial activities. In case of chitosan embedded with copper nanoparticles the presence of chitosan has positive impact by impeding the bacterial growth rate (Usman et al., 2013). Further increase in activity of nanocomposite was achieved by increase in amount of nanoparticles doped. The reduction in MIC value from 1.9 to 3.9 μg/ml to 62.5 μg/ml was noticed with increase in concentration of silver in ssilver-zeolite nanocomposite from 1.9% to 38% Ag (Egger et al., 2009).

The mechanism of action of these nanoparticles is still not clear, but it is proposed that the nanoparticles enter the bacterial cell by rupturing the cell wall which later on impact the respiratory chain and cell division and hence leading to cell death. (Li et al., 2010). In case of fungal activity the nanoparticles bind to the fungal membrane leading to formation of transmembrane pores followed by leakage of cell constituents that eventually result in cell death (Kim et al., 2009).

Properties of Small Particles and Environment

The properties of nanomaterials no doubt have played important role in delineating their use and applicability for different systems. These properties like size, shape, surface area, surface morphology, surface charge, chemical composition etc. can play significant role in deciding their faith and behavior in environment.

Interactions of Nanomaterials With Pollutants

It has been proposed that nanomaterials may collaborate with contaminants which may bring about harmful impacts on biota demonstrated the capability of nanomaterials to upgrade the lethal impacts of natural contaminants. Researchers reported that AgNPs produce size dependent mortality, whereas, interestingly but not surprisingly, the behavior of Au NPs was independent of size (Chen et al., 2006).

Furthermore, as the attributes of nanoparticles such as surface roughness, hydrophobicity, and charge of nanoparticles influence the phenomena of cellular uptake of nanoparticles indeed influence the toxicity associated with nanoparticles.

Surface coarseness dictates the strength of nanoparticle-cell interactions and promotes cell adhesion. Pore structure is critical in cell-nanoparticle interactions. It has been demonstrated that size dependent hemolysis effect of mesoporous silica nanoparticles is only observed when the nanoparticles have long range ordered porous structure (De Angelis et al., 2010) showed that nanoporous silicon NPs with a pore size of about 2 nm do not have any toxicity in mouse-models with no histological evidence of tissue pathology (Park et al., 2009) observed that luminescent porous silicon nanoparticles did not show any toxicity in animal models.

Studies showed that instillation of Ir192-particles of 80 nm resulted in accumulation in the rat liver with an extent of 0.1% of total amount, while particles of 15 nm size displayed increased accumulation to an extent of 0.3–0.5%. Moreover, it has been observed that when smaller particles are retained in the respiratory tract for longer duration it leads to increased translocation to the pulmonary interstitium with impairment of alveolar macrophages function. Redistribution of NPs from their site of deposition or deposition into renal tissues and escape from normal phagocytic defences may also lead to toxicity (Kreyling et al., 2002; Seaton et al., 2005). The respiratory system represents a unique target for the potential toxicity of nanoparticles due to the fact that in addition to being the portal of entry for inhaled particles, it also receives the entire cardiac output (Ferreira et al., 2013).

Due to high surface to mass ratio natural NPs play an important role in the solid/water partitioning of contaminants can be absorbed to the surface of NPs, co-precipitated during the formation of natural NPs or trapped by aggregation of NPs which had contaminants adsorbed to their surface. The interaction of contaminants with NPs is dependent on characteristics, such as size, composition, morphology, porosity, aggregation/disaggregation and aggregate structure. The luminophores are not safe in the environment and are protected from the environmental oxygen when they are doped inside the silica network (Santra et al., 2001).

As nanoparticles are directly related to their release in the environment. The effects of NPs have been described in a wide variety of organisms, such as micro-organisms protozoa, invertebrates and vertebrates. However, interactions of NPs with plants and other organisms that share similarities with plant cells, such as algae, have been poorly studied, remaining unclear the general consequences of NPs exposure for plant cells (Qian et al., 2013).

Vapor Pressure and Solubility

Vapor pressure and water dissolvability of traditional chemicals are utilized to anticipate air/water parceling. These properties may not be exceptionally valuable in foreseeing the degree to which nanoparticles may segment from water into air. Be that as it may, dissolvability of nanomaterials in water, or rather the rate of disintegration of nanoparticles in water is essential for an altogether unique reason. Lethal impacts of the nearness of nanomaterials may well outcome, at any rate to some degree, from the nearness of broke up species that start from disintegration of the nanomaterials. To date, it is misty to what degree the impacts watched can be credited to the broken down shape or to the nanoformulation the impact being a mix of part and size have proposed that phytoxicity saw on exposures to ZnO nanoparticles may not be ascribed exclusively to broke down zinc (Navarro et al., 2008).

Shapes of Nanoparticles

Nanomaterials are of different shapes including planes, tubes, fibers and rings. Toxicity also depends upon shape of nanoparticles for example carbon nanotubes, silica, allotropies, nickel, gold, and titanium nanomaterials. Major advancement in the understanding of interplay between particle size and shape for development of more efficacious nanomaterial based targeted delivery system; nevertheless, this also reinforced that their untoward effects should also be examined.

Surface Charge

The high surface area to volume ratio of the nanomaterials allows enhanced catalytic and sensing response by the rapid movement of analytes through nanomaterials based electrodes or sensors. The size and surface area are important factors in determining toxicity of nanoparticles other factors such as chemical nature of the constituents may also contribute to the intrinsic toxicity of the nanoparticles. As size decreases surface area increase relative to volume. Nanoparticle's size also shows pharmacological behaviors. It has been observed that nanoparticles smaller than 50 nm slanting quickly to all tissues and pass on potentially toxic effects in various tissues. Nanoparticles greater than 50 nm are readily taken up by reticuloendothelial system which refrain their path to other tissues. Gold nanoparticles behavior is independent of size while silver nanoparticles produce size dependent mortality in the size range of 3, 10, 50, and 100 nm.

Effect of Surface Coating and Surface Roughness

The surface properties of particles have significant role on toxicity of nanoparticles as they play a critical role in determining the outcome of their interaction with the cells and other biological entities. Surface coating can affect the cytotoxic properties of nanoparticles by changing their physicochemical properties such as magnetic, electric, and optical properties and chemical reactivity and can alter the pharmaco kinetics, distribution, accumulation, and toxicity of nanoparticles. In general, surface coating can mitigate or eliminate the adverse effects of nanoparticles. In particular, proper surface coating can lead to stabilization of nanoparticles as well as elude release of toxic ions from nanomaterials. Moreover, it has also been demonstrated that spherical gold nanoparticles with various surface coatings have been found to be nontoxic to human cells.

Effect of Solvents

Medium/solvent conditions have been known to affect particle dispersion and agglomeration state of nanoparticles, which in turn have effect on their particle size, thereby influencing the toxicity associated with nanoparticles. It has been observed that particles of TiO_2, ZnO, or carbon black have significantly greater size in PBS than in water; moreover, it is also in general consensus that nanoparticles display different diameters in biological environment. Accordingly, the toxic effects of nanoparticles show variation depending upon the medium composition in which the nanoparticles are suspended. Although, the dispersing agent may improve the physicochemical and solution properties of nanomaterials formulations, they may also adversely affect the toxicity of nanomaterials.

CONCLUSION

Nanotechnology is arising field with multiple potentialities and applications. Nanotechnology is being envisaged as burgeoning field with many potential human health benefits and with rapid upsurge in the field, it becomes increasingly imperative to evaluate the toxicities issues associated with these nanomaterial based products. The increasing area of engineered nanoparticles in industrial and household applications leads to the release of such materials into the environment. Assessing the risk of these nanoparticles in the environment requires on understanding of their mobility, reactivity, Eco toxicity and persistency.

Physicochemical properties such as size, surface area, surface chemistry, surface roughness, dispersion medium, and ability to agglomerate play vital role in determining their toxicity. With newer nanomaterials based products being introduced in the market on daily bases, there is urgent need to reduce the knowledge gap between the physicochemical properties and their influence on the manifestation of toxicities issues. This will certainly pave ways towards maneuvering these physicochemical properties for their safer implementation in diverse fields. In this review, we presented a review about NPs properties. It was revealed that NPs have size ranges from a few nanometers to 500 nm. Due to their tiny size, NPs have large surface area, which make them suitable candidate for various applications. Beside this, the optical properties are also dominant at that size, which further increase the importance of these materials in photocatalytic applications. Synthetic techniques can be useful to control the specific morphology, size and magnetic properties of NPs. Though NPs are useful for many applications, but still there are some health hazard concerns due to their uncontrollable use and discharge to natural environment, which should be consider for make the use of NPs more convenient and environmental friendly. Nanoparticles also have different toxicity profiles. Nanoparticles have always been present in the environment, with the increase of NPs industry and multiple applications more information is needed regarding the potential impacts that NPs release may have on environmental and animals' health. The uptake, bioaccumulation, biotransformation, and risks for food crops are still not well understood. The possible biomagnification in the food chain is also unknown.

SOLUTIONS AND RECOMMENDATIONS

More importantly environmental issued should be taken into account before using these materials for any applications, especially in case of heavy metals, which are prone to environmental hazards and can also affect the livings as well. For future work it is suggested that different reaction parameters such as temperature, pressure, time, and pH. play important role in controlling the shape and morphology of the nanoparticles, so that should be optimize for achieving specific characteristic product. Beside this for good implications and properties study specific characterization techniques should be used.

REFERENCES

Alén, B., Martınez-Pastor, J., Garcıa-Cristobal, A., González, L., & Garcıa, J. M. (2001). Optical transitions and excitonic recombination in InAs/InP self-assembled quantum wires. *Applied Physics Letters, 78*(25), 4025–4027. doi:10.1063/1.1379991

Altavilla, C., & Ciliberto, E. (2016). *Inorganic nanoparticles: synthesis, applications, and perspectives.* CRC Press.

Azam, A., Ahmed, A. S., Oves, M., Khan, M., & Memic, A. (2012). Size-dependent antimicrobial properties of CuO nanoparticles against Gram-positive and-negative bacterial strains. *International Journal of Nanomedicine, 7*, 3527. doi:10.2147/IJN.S29020 PMID:22848176

Azam, A., Ahmed, A. S., Oves, M., Khan, M. S., Habib, S. S., & Memic, A. (2012). Antimicrobial activity of metal oxide nanoparticles against Gram-positive and Gram-negative bacteria: A comparative study. *International Journal of Nanomedicine, 7*, 6003. doi:10.2147/IJN.S35347 PMID:23233805

Benrabaa, R., Löfberg, A., Rubbens, A., Bordes-Richard, E., Vannier, R. N., & Barama, A. (2013). Structure, reactivity and catalytic properties of nanoparticles of nickel ferrite in the dry reforming of methane. *Catalysis Today, 203*, 188–195. doi:10.1016/j.cattod.2012.06.002

Bréchignac, C., Houdy, P., & Lahmani, M. (2008). *Nanomaterials and Nanochemistry*. Springer Berlin Heidelberg.

Cao, G. (2004). *Nanostructures & Nanomaterials: Synthesis, Properties & Applications*. Imperial College Press. doi:10.1142/p305

Chen, Z., Meng, H., Xing, G., Chen, C., Zhao, Y., Jia, G., ... Zhao, F. (2006). Acute toxicological effects of copper nanoparticles in vivo. *Toxicology Letters*, *163*(2), 109–120. doi:10.1016/j.toxlet.2005.10.003 PMID:16289865

Cuenya, B. R. (2010). Synthesis and catalytic properties of metal nanoparticles: Size, shape, support, composition, and oxidation state effects. *Thin Solid Films*, *518*(12), 3127–3150. doi:10.1016/j.tsf.2010.01.018

Cuenya, B. R., Baeck, S.-H., Jaramillo, T. F., & McFarland, E. W. (2003). Size-and support-dependent electronic and catalytic properties of Au0/Au3+ nanoparticles synthesized from block copolymer micelles. *Journal of the American Chemical Society*, *125*(42), 12928–12934. doi:10.1021/ja036468u PMID:14558841

De Angelis, F., Pujia, A., Falcone, C., Iaccino, E., Palmieri, C., Liberale, C., ... De Laurentiis, A. (2010). Water soluble nanoporous nanoparticle for in vivo targeted drug delivery and controlled release in B cells tumor context. *Nanoscale*, *2*(10), 2230–2236. doi:10.1039/c0nr00161a PMID:20835434

Egger, S., Lehmann, R. P., Height, M. J., Loessner, M. J., & Schuppler, M. (2009). Antimicrobial properties of a novel silver-silica nanocomposite material. *Applied and Environmental Microbiology*, *75*(9), 2973–2976. doi:10.1128/AEM.01658-08 PMID:19270121

Ferreira, A., Cemlyn-Jones, J., & Cordeiro, C. R. (2013). Nanoparticles, nanotechnology and pulmonary nanotoxicology. [English Edition]. *Revista Portuguesa de Pneumologia*, *19*(1), 28–37. doi:10.1016/j.rppneu.2012.09.003 PMID:23265236

Gilroy, K. D., Ruditskiy, A., Peng, H.-C., Qin, D., & Xia, Y. (2016). Bimetallic nanocrystals: Syntheses, properties, and applications. *Chemical Reviews*, *116*(18), 10414–10472. doi:10.1021/acs.chemrev.6b00211 PMID:27367000

Gubin, S. P. (2009). *Magnetic Nanoparticles*. Wiley. doi:10.1002/9783527627561

Guo, D., Xie, G., & Luo, J. (2013). Mechanical properties of nanoparticles: Basics and applications. *Journal of Physics. D, Applied Physics*, *47*(1), 013001. doi:10.1088/0022-3727/47/1/013001

Hajipour, M. J., Fromm, K. M., Ashkarran, A. A., de Aberasturi, D. J., de Larramendi, I. R., Rojo, T., ... Mahmoudi, M. (2012). Antibacterial properties of nanoparticles. *Trends in Biotechnology*, *30*(10), 499–511. doi:10.1016/j.tibtech.2012.06.004 PMID:22884769

Han, Y.-F., Chen, F., Zhong, Z., Ramesh, K., Chen, L., & Widjaja, E. (2006). Controlled Synthesis, Characterization, and Catalytic Properties of Mn_2O_3 and Mn_3O_4 Nanoparticles Supported on Mesoporous Silica SBA-15. *The Journal of Physical Chemistry B*, *110*(48), 24450–24456. doi:10.1021/jp064941v PMID:17134200

Howes, P. D., Chandrawati, R., & Stevens, M. M. (2014). Colloidal nanoparticles as advanced biological sensors. *Science*, *346*(6205), 1247390. doi:10.1126cience.1247390 PMID:25278614

Huber, D. L. (2005). Synthesis, properties, and applications of iron nanoparticles. *Small*, *1*(5), 482–501. doi:10.1002mll.200500006 PMID:17193474

Issa, B., Obaidat, I. M., Albiss, B. A., & Haik, Y. (2013). Magnetic nanoparticles: Surface effects and properties related to biomedicine applications. *International Journal of Molecular Sciences*, *14*(11), 21266–21305. doi:10.3390/ijms141121266 PMID:24232575

Khan, I., Saeed, K., & Khan, I. (2017). Nanoparticles: Properties, applications and toxicities. *Arabian Journal of Chemistry*. doi:10.1016/j.arabjc.2017.05.011

Kim, K.-J., Sung, W. S., Suh, B. K., Moon, S.-K., Choi, J.-S., Kim, J. G., & Lee, D. G. (2009). Antifungal activity and mode of action of silver nano-particles on Candida albicans. *Biometals*, *22*(2), 235–242. doi:10.100710534-008-9159-2 PMID:18769871

Koksharov, Y. (2009). *Magnetism of Nanoparticles: Effects of Size, Shape, and Interactions*. Wiley-VCH Verlag GmbH & Co. KGaA

Kreyling, W., Semmler, M., Erbe, F., Mayer, P., Takenaka, S., Schulz, H., ... Ziesenis, A. (2002). Translocation of ultrafine insoluble iridium particles from lung epithelium to extrapulmonary organs is size dependent but very low. *Journal of Toxicology and Environmental Health. Part A.*, *65*(20), 1513–1530. doi:10.1080/00984100290071649 PMID:12396866

Kvitek, L., Panáček, A., Soukupova, J., Kolář, M., Večeřová, R., Prucek, R., ... Zbořil, R. (2008). Effect of surfactants and polymers on stability and antibacterial activity of silver nanoparticles (NPs). *The Journal of Physical Chemistry C*, *112*(15), 5825–5834. doi:10.1021/jp711616v

Li, W.-R., Xie, X.-B., Shi, Q.-S., Zeng, H.-Y., You-Sheng, O.-Y., & Chen, Y.-B. (2010). Antibacterial activity and mechanism of silver nanoparticles on Escherichia coli. *Applied Microbiology and Biotechnology*, *85*(4), 1115–1122. doi:10.100700253-009-2159-5 PMID:19669753

Mostafa, S., Behafarid, F., Croy, J. R., Ono, L. K., Li, L., Yang, J. C., ... Cuenya, B. R. (2010). Shape-Dependent Catalytic Properties of Pt Nanoparticles. *Journal of the American Chemical Society*, *132*(44), 15714–15719. doi:10.1021/ja106679z PMID:20949968

Navarro, E., Piccapietra, F., Wagner, B., Marconi, F., Kaegi, R., Odzak, N., ... Behra, R. (2008). Toxicity of silver nanoparticles to Chlamydomonas reinhardtii. *Environmental Science & Technology*, *42*(23), 8959–8964. doi:10.1021/es801785m PMID:19192825

Pal, S., Tak, Y. K., & Song, J. M. (2007). Does the antibacterial activity of silver nanoparticles depend on the shape of the nanoparticle? A study of the gram-negative bacterium Escherichia coli. *Applied and Environmental Microbiology*, *73*(6), 1712–1720. doi:10.1128/AEM.02218-06 PMID:17261510

Park, J.-H., Gu, L., Von Maltzahn, G., Ruoslahti, E., Bhatia, S. N., & Sailor, M. J. (2009). Biodegradable luminescent porous silicon nanoparticles for in vivo applications. *Nature Materials*, *8*(4), 331–336. doi:10.1038/nmat2398 PMID:19234444

Prieto, G., Zečević, J., Friedrich, H., De Jong, K. P., & De Jongh, P. E. (2013). Towards stable catalysts by controlling collective properties of supported metal nanoparticles. *Nature Materials*, *12*(1), 34–39. doi:10.1038/nmat3471 PMID:23142841

Qi, M., Zhang, K., Li, S., Wu, J., Pham-Huy, C., Diao, X., ... He, H. (2016). Superparamagnetic Fe 3 O 4 nanoparticles: Synthesis by a solvothermal process and functionalization for a magnetic targeted curcumin delivery system. *New Journal of Chemistry*, *40*(5), 4480–4491. doi:10.1039/C5NJ02441B

Qian, H., Peng, X., Han, X., Ren, J., Sun, L., & Fu, Z. (2013). Comparison of the toxicity of silver nanoparticles and silver ions on the growth of terrestrial plant model Arabidopsis thaliana. *Journal of Environmental Sciences (China)*, *25*(9), 1947–1956. doi:10.1016/S1001-0742(12)60301-5 PMID:24520739

Roduner, E. (2006). Size matters: Why nanomaterials are different. *Chemical Society Reviews*, *35*(7), 583–592. doi:10.1039/b502142c PMID:16791330

Santos, C., Albuquerque, A., Sampaio, F., & Keyson, D. (2013). Nanomaterials with antimicrobial properties: applications in health sciences. *Microbial Pathogens and Strategies for Combating Them: Science, Technology and Education, 4*(2).

Santra, S., Zhang, P., Wang, K., Tapec, R., & Tan, W. (2001). Conjugation of biomolecules with luminophore-doped silica nanoparticles for photostable biomarkers. *Analytical Chemistry*, *73*(20), 4988–4993. doi:10.1021/ac010406+ PMID:11681477

Seaton, A., & Donaldson, K. (2005). Nanoscience, nanotoxicology, and the need to think small. *Lancet*, *365*(9463), 923–924. doi:10.1016/S0140-6736(05)71061-8 PMID:15766981

Singh, M., & Tlali, S. (2017). Effects of size and shape on the specific heat, melting entropy and enthalpy of nanomaterials. *Journal of Taibah University for Science*, *11*(6), 922–929. doi:10.1016/j.jtusci.2016.09.011

Su, H., Li, Y., Li, X.-Y., & Wong, K. S. (2009). Optical and electrical properties of Au nanoparticles in two-dimensional networks: An effective cluster model. *Optics Express*, *17*(24), 22223–22234. doi:10.1364/OE.17.022223 PMID:19997469

Tsujimoto, Y., Matsushita, Y., Yu, S., Yamaura, K., & Uchikoshi, T. (2015). Size dependence of structural, magnetic, and electrical properties in corundum-type Ti2O3 nanoparticles showing insulator–metal transition. *Journal of Asian Ceramic Societies*, *3*(3), 325–333. doi:10.1016/j.jascer.2015.06.007

Usman, M. S., El Zowalaty, M. E., Shameli, K., Zainuddin, N., Salama, M., & Ibrahim, N. A. (2013). Synthesis, characterization, and antimicrobial properties of copper nanoparticles. *International Journal of Nanomedicine*, *8*, 4467. PMID:24293998

Wei, Y., Chen, S., Kowalczyk, B., Huda, S., Gray, T. P., & Grzybowski, B. A. (2010). Synthesis of Stable, Low-Dispersity Copper Nanoparticles and Nanorods and Their Antifungal and Catalytic Properties. *The Journal of Physical Chemistry C*, *114*(37), 15612–15616. doi:10.1021/jp1055683

Weiner, R. G., & Skrabalak, S. E. (2015). Metal Dendrimers: Synthesis of Hierarchically Stellated Nanocrystals by Sequential Seed-Directed Overgrowth. *Angewandte Chemie International Edition*, *54*(4), 1181–1184. doi:10.1002/anie.201409966 PMID:25424302

Yang, Z., Ye, Z., Xu, Z., & zhao, B. (2009). Effect of the morphology on the optical properties of ZnO nanostructures. *Physica E, Low-Dimensional Systems and Nanostructures*, *42*(2), 116–119. doi:10.1016/j.physe.2009.09.010

Yuming, D., Kun, H., Lin, Y., & Aimin, Z. (2007). A facile route to controlled synthesis of Co_3O_4 nanoparticles and their environmental catalytic properties. *Nanotechnology*, *18*(43), 435602. doi:10.1088/0957-4484/18/43/435602

Zhang, M., Li, J., Li, H., Li, Y., & Shen, W. (2009). Morphology-dependent redox and catalytic properties of CeO2 nanostructures: Nanowires, nanorods and nanoparticles. *Catalysis Today*, *148*(1-2), 179–183. doi:10.1016/j.cattod.2009.02.016

Zhou, K., Wang, R., Xu, B., & Li, Y. (2006). Synthesis, characterization and catalytic properties of CuO nanocrystals with various shapes. *Nanotechnology*, *17*(15), 3939–3943. doi:10.1088/0957-4484/17/15/055

Zhou, X., Lee, S., Xu, Z., & Yoon, J. (2015). Recent progress on the development of chemosensors for gases. *Chemical Reviews*, *115*(15), 7944–8000. doi:10.1021/cr500567r PMID:25651137

Chapter 3
Characterization of Nanomaterials

Mintu Maan Dutta
Gauhati University, India

ABSTRACT

This chapter provides a general introduction to the new learners in the field of the synthesis and characterization of nanoparticles. The people who are familiar with this field but are unable to interpret the data will learn how to interpret from various data available. Keeping in the mind the usefulness of numerical data, this chapter includes a number of tables and graphs for the help and easy understanding of the students.

CHARACTERIZATION OF NANOMATERIALS

Nanotechnology is an emerging interdisciplinary area with wide application in all fields of science and technology. The basic of nanotechnology lies with the fact that with decrease in the particle size to nanometer range the properties of the materials changes noticeably (Joshi et al., 2008). But measuring the particle size is not a very easy task and they are posing challenges to the scientists working in this field. Thus, the discovery of various sophisticated nano characterization techniques has helped in better control of the size, morphology of the materials in nano range.

There are various characterization techniques; some of them are surface area analyzer, characterization by spectroscopy, characterization by X-Ray, characterization by microscopy, thermal analyzer and magnetic property analyzer.

Nanomaterial Characterization by Spectroscopy

Use of spectroscopic methods for the characterization began in late 1950s. These methods can provide information about the chemical nature of the surfaces as well as help in determining their concentration. Sometimes it is seen that the chemical nature of the surface of a solid is different from the interior of the solid. So, one shouldn't focus completely on the interior bulk composition, because the composition of the surface layer may be more important.

DOI: 10.4018/978-1-5225-5745-6.ch003

Ultraviolet and Visible Radiation (UV-Vis)

Principle

Ultraviolet-Visible spectroscopy works with the principle of Beer Lambert's law.

"When a beam of monochromatic light is passed through a solution of an absorbing substance, the rate of decrease of intensity of radiation with thickness of the absorbing solution is proportional to the incident radiation as well as the concentration of the solution."

The expression of Beer-Lambert law is- $A = \log (I_0/I) = \varepsilon cl$

Where, A = absorbance, I_0 = intensity of light incident upon sample cell, I = intensity of light leaving sample cell, C = molar concentration of solute, L = length of sample cell (cm.), ε= molar absorptivity

Ultraviolet light: wavelengths in the range of 190 and 400 nm, Visible light: wavelengths in the range of 400 and 800 nm

- **Light Source**

Tungsten filament lamps: Rich in red radiations, emit the radiations of 375 nm.

Hydrogen-Deuterium: Emits the radiations below 375 nm

Ultraviolet-Visible (UV-Vis) spectroscopy is a characterization technique that is used to quantify the amount of light that is absorbed and scattered by a sample. Absorption is associated with the **excitation of electrons** from lower to higher energy levels. The Double beam UV-Vis spectrophotometer is shown in figure 1.

Working

The cell holding the sample is transparent, and is usually made up of quartz or fused silica. The thickness of cell is usually 1 cm. Cell containing the sample is placed between a light source and a photo detector (figure 1), and the intensity of a beam of light is measured before and after passing through the sample based on Beer- Lamberts law.

Figure 1.

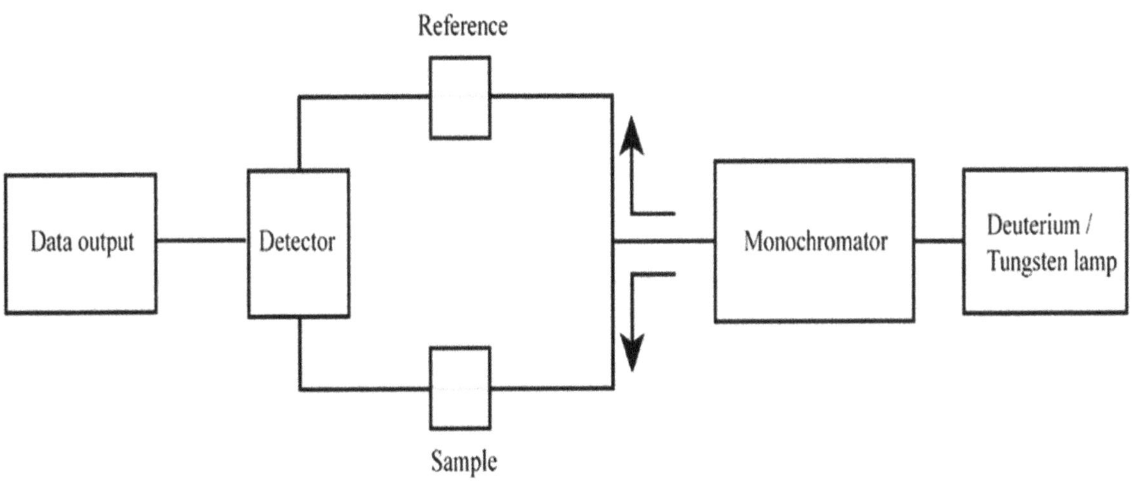

1. UV-Vis Spectroscopy is Used to Predict Nanoparticles Geometry

Figure 2 below shows that the UV-Vis absorbance spectrum is highly dependent nanoparticles geometry (Moddy, 2015). Two spectra differ in their shapes despite of the two types of nanoparticles being composed of the same material and having similar dimensions.

2. UV-Vis Spectroscopy is Used to Determine Nanoparticles Aggregation States

When nanoparticles are close to each other, their Plasmon's couple, which affects their LSPR (localized surface Plasmon resonance) and thus absorption of light takes place. Dimerization of nanospheres causes a "red shift," i.e. a shift towards longer wavelengths. Unlike dimerization, aggregation of nanoparticles causes a decrease in the intensity of the peak absorbance without shifting the wavelength at which the peak occurs (λmax). Figure 3 (A) illustrates the UV-Vis. absorbance spectrum of 50 nm gold nanosphere dimers with a reference spectrum of single gold nanospheres (Moddy, 2015) and (B) illustrates the increase in nanoparticles aggregation with increased salt concentrations based on the decreased absorbance peak intensity.

3. UV-Vis Spectroscopy is Used to Determine Nanoparticle Surface Composition

The UV-Vis absorbance spectrum of noble metal nanoparticles is highly influenced by the environment surrounding the nanoparticles. Because of this, shifts in λ_{max} are observed and can be used to detect changes in the surface composition of the nanoparticles. Figure 4 shows the red shift in the λ_{max} of the UV-Visible absorbance spectrum. It is due to the addition of human serum albumin protein. It indicates that the protein is binding to the surface of the nanoparticles.

Figure 2.

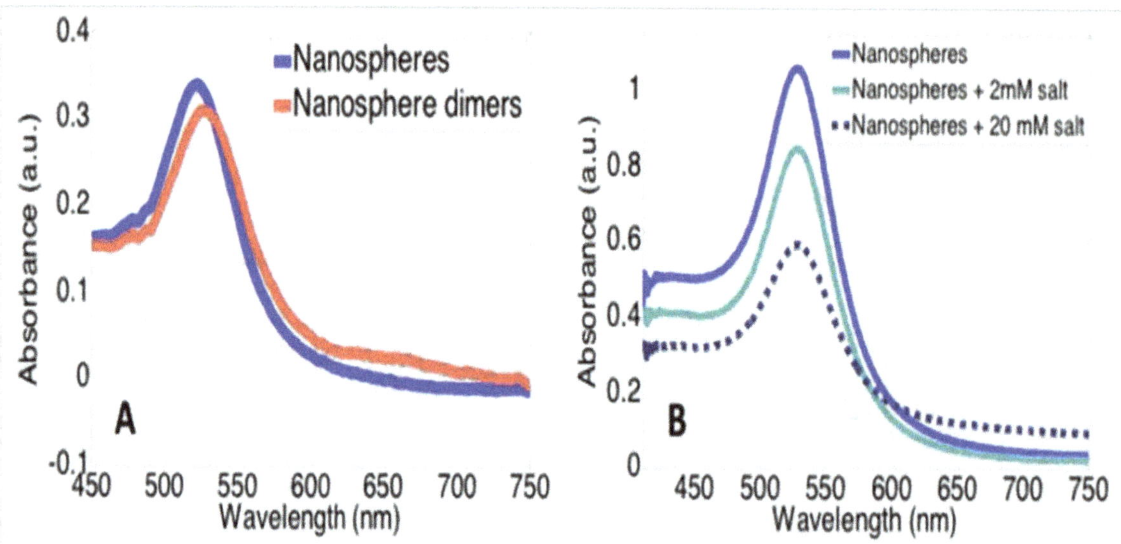

Figure 3.

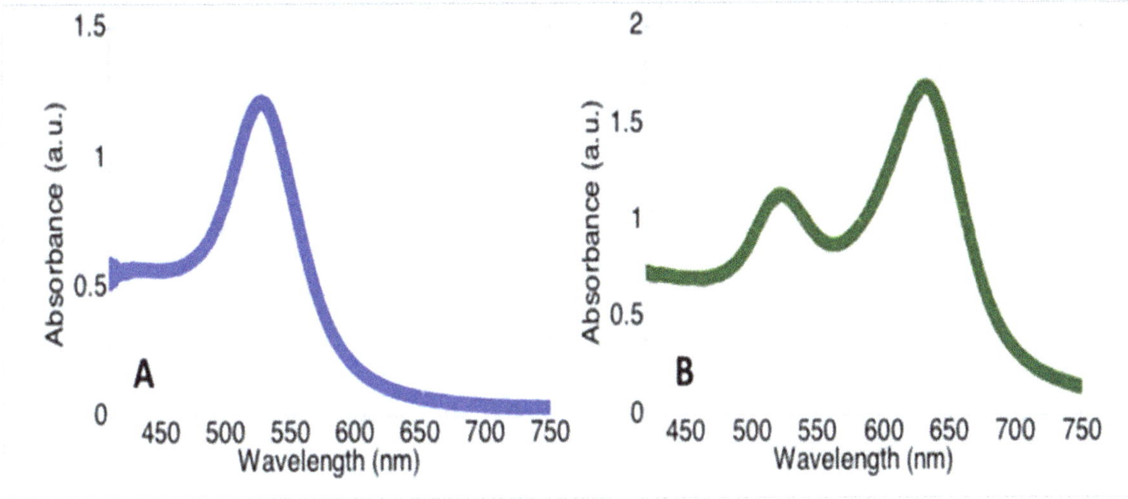

Figure 4.

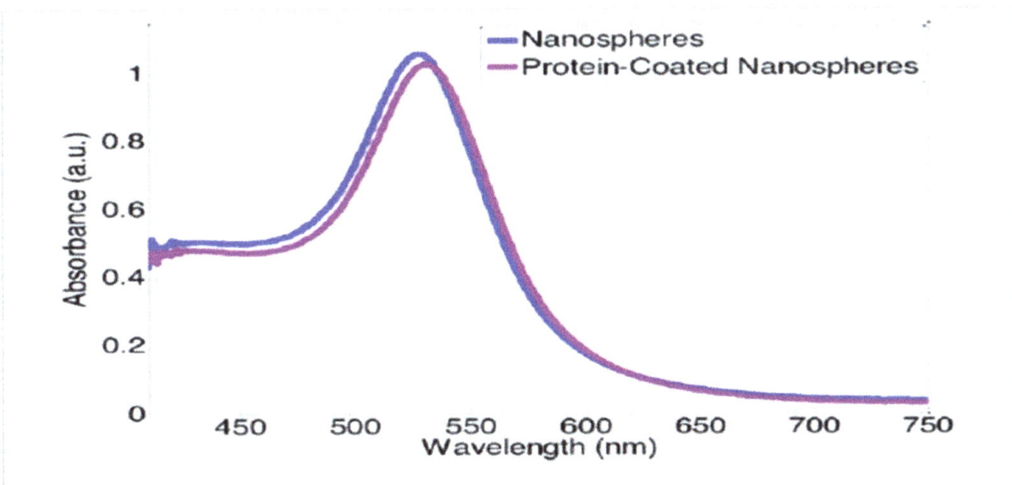

X-Ray Diffraction (XRD)

Introduction

Powder XRD is an analytical technique that uses finely ground, homogenized material to explain the crystalline nature of the materials. It can also provide information regarding unit cell dimensions of the materials.

Principle

In 1912, Max Von Laue discovered that crystals could be used as diffraction gratings for X-rays. Thus the diffraction pattern has been used for the study of crystal structure. Crystals are formed as a result of regular arrangement of atoms and X-rays are considered as electromagnetic radiation. Diffraction occurs when X-rays are scattered by the atoms with long range order, thus X-Ray striking an atom's electron produces secondary wave from the electron. This phenomenon is referred to as elastic scattering. This produces constructive interferences at some specified angles. This is determined by Bragg's Law:

$$n\lambda = 2d\sin\theta$$

Where d= spacing between two diffracting planes, θ= incident angle, n= any integer, λ= wavelength of the incident X-ray

The atoms in crystals diffract X-ray only when the incident wavelength overcomes the distance between atoms. These diffracted X-rays are detected and proper study of the diffraction pattern gives all the information regarding the atomic arrangement of the powdered material. Diffraction pattern can be converted to d-spacing via. mathematical calculation and it allows identifying the mineral by comparing the d-spacing with the standard pattern as available in XRD data base (JCPDS). Significant peak in the diffraction pattern is not observed in case of amorphous materials like glass because the periodic arrangement of the atoms is of short-range order.

Information in a Diffraction Pattern

- **Peak Position:** It describes about the translational symmetry i.e., most precisely the shape and size of the unit cell.
- **Peak Intensities:** Gives information about the position of atoms.
- **Peak Shape & Width:** Gives information about the particle size using Sherrer formula (FWHM). It also describes the deviation from the perfect crystal.

Figure 5. A PXRD (Bruker D8 Advance) showing the goniometer and its drives inside the cabinet (Chauhan & Chauhan, 2014)

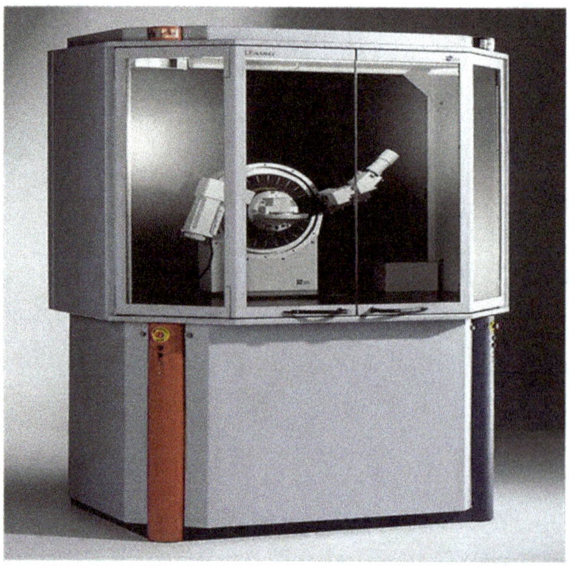

1. Steps for Mathematical Calculation

1. Identify the peak (**2θ**) values from the XRD data.
2. Obtain the θ values from 2θ, determine the value for **$Sin^2θ$.**
3. Calculate the ratio of $Sin^2θ / Sin^2θ_{min}$ and multiply with appropriate integers.
4. Round off the ratio to the nearest integer that yields the value for **$h^2+k^2+l^2$.**
5. Identify the Bravais lattice. Eg: **For BCC**: 2,4,6,8……, **For Primitive**: 1,2,3,4,5,6,8,9,10,11,12…….., **For FCC**: 3, 4,8, 11, 12, 16, 19, 20…….
6. Determine the lattice parameter **(hkl)**.

2. Qualitative Analysis

Diffraction pattern of known samples are maintained by database known as ICDD (International Centre for Diffraction Data), this are easily available online www.icdd.com. The diffraction data obtained for the sample can be matched with the diffraction pattern of the known samples. Presently there are 115,000 pattern of which 95, 000 are Inorganic compounds and 20,000 are organic compounds.

a. XRD Crystallite Size Calculator

Using Scherer formula XRD data can be interpreted and one can calculate the crystallite size:

Table 1. Allowed list of $h^2+k^2+l^2$ for cubic crystals

Forbidden Numbers	Primitive, P	Face Centered, F	Body Centered, B	Corresponding hkl
	1			100
	2		2	110
	3	3		111
	4	4	4	200
	5			210
	6		6	211
7				
	8	8	8	220
	9			221,300
	10		10	310
	11	11		311
	12	12	12	222
	13			320
	14		14	321
15				
	16	16	16	400

$Dp = K\lambda/\beta_{1/2}\text{Cos}\theta$

Where Dp = Average crystallite size, β = Line broadening in radians (FWHM- Full Width Half Maximum)

K =Shape factor

– 0.94 (for FWHM of spherical crystallites with cubic symmetry)
– 0.89 (for integral breadth of spherical crystallites with cubic symmetry)

λ = X-ray wavelength. (For Cu anode λ = 1.54056 Å)

For the above XRD data, if we calculate the Average crystallite size, it would be as follows:

For peak at 2θ= 29.567 °, $\beta_{1/2}$ = 0.06 °, Cos 29.567 ° = 0.8697
Thus, Dp = 1.44812/0.052182 = 27.75 Å = **2.75 nm**
For peak at 2θ= 42.2309 °, $\beta_{1/2}$ = 0.05 °, Cos 42.2309 ° = 0.7404
Thus Dp = 1.44812/0.03702 = 39.117 Å = **3.91 nm**
For peak at 2θ= 49.96454 °, $\beta_{1/2}$ = 0.05 °, Cos 49.96454 ° = 0.64326
Thus Dp = 1.44812/0.03216 = 45.024 Å = **4.50 nm**

Mean of the three crystallite gives the Avaerage crystallite size = (2.75 + 3.91 + 4.50)/3 = 3.72 nm
The crystallite size may in size in few Å over the range of 25 to 90°.

b. Strengths of XRD

a. Sample analysis time is < 20 minutes.
b. Data interpretation is quite simple.
c. Sample preparation time is minimum.
d. Non-destructive sample.

c. Limitations of XRD

1. This techniques best suits for homogeneous and single phase materials.
2. The data must be matched with the standard references to draw an inference.
3. For non-isometric crystal system indexing (i.e. unit cell determination) is complicated.
4. In case of high angle reflection, peak may overlap.

d. Application of PXRD in Science

Most important application includes the identification of unknown crystalline materials. It has wide application in several branches of science such as geology, material science, engineering and biology.

1. PXRD is the basic tool for differentiating the crystalline compounds (metal salt sample) from semi-crystalline (polymer E.g.: cotton) and amorphous compounds (phenol formaldehyde complex resin).

Figure 6 (a), (b) Contains three XRD plots: First shows a narrow, sharp, and significant peaks that are obtained for crystalline compounds, whereas amorphous compounds and semi-crystalline compounds does not give significant peaks rather they look like noise signals and shows broad peaks respectively (Chauhan & Chauhan, 2014).

Figure 6.

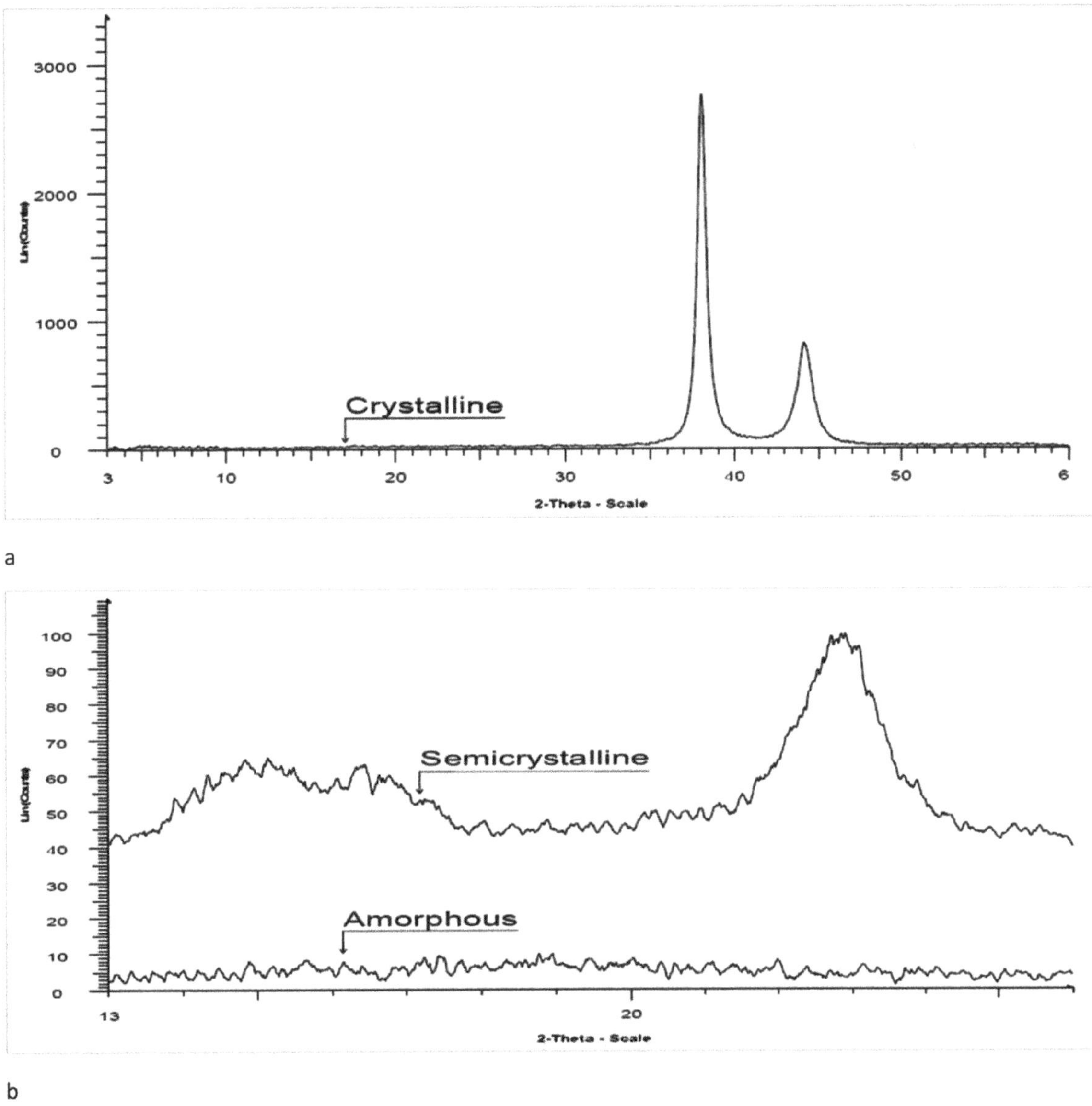

2. PXRD in Nano and Material science

PXRD makes an important distinction between particle and crystal size, particles are made of a large number of small crystallites. Thus PXRD can measure millions of crystals that are distributed uniformly in the nanomaterial and accurately determine the size. Determination of the particle size is done using Scherer equation.

3. Crystal structure determination using Lattice parameters:

XRD diffraction pattern gives information about lattice parameters. Position of the diffraction peaks helps in determining the shape and size of the unit cell, peaks also refers to the lattice plane and they are characterized by Miller indices. These are done via. Mathematical calculation that are explained previously.

4. In Pharmaceutical industry

PXRD plays a major role in drug development, testing and production.

5. In Glass industry

Diffraction pattern of glass does not contain any significant peaks as it is amorphous in nature. It helps in the identification of crystalline particles that create defects in glass.

6. In Mineral exploration:

Geological science used PXRD for mineral exploration. Every mineral explored are identified by its unique diffraction pattern.

7. Variable temperature and Relative humidity study:

A compound analyzed under various temperature helps to study the Crystallinity as a function of temperature. Several times polymorph interconversion is observed.

It is seen that most of the inorganic metal salts contain water of hydration in non-stoicheometric ratio within the crystal lattice. Thus PXRD helps to study the amount of water present as a function of relative humidity in the environment of sample.

Nanomaterial Characterization by Microscopy

In many fields of science and technology such as chemistry, material science, geology, biology etc. deals with the detailed knowledge of the physical nature of the solid. Thus the use of microscopic methods for the characterization came into existence. Classical methods made use of the optical microscope for obtaining the information but they are limited to the micron level materials with reasonable resolutions. Hence, some new imaging techniques such as scanning electron microscope (SEM), transmission electron microscope (TEM), atomic force microscopy (AFM) has been developed. These methods can observe the materials up to sub micron size. Thus nanomaterials can be observed through this imaging techniques which was not possible with the human eye and optical microscope.

Transmission Electron Microscope (TEM)

Introduction

The microscopic technique uses a beam of electron at 100-400kv accelerating voltage that is transmitted through an ultra- thin specimen, thereby interacting with the sample which is immersed in the objective lens of microscope. An image is formed as a result of interaction of transmitted electron and diffracted electron that are recombined by the objective lens to form the image (diffraction pattern). The image is magnified and focused onto a fluorescent screen for observation.

Principle

Electron possess a wave like character. Electrons emitted into vaccum from a heated filament with increased accelerating potential will have small wavelength. Such higher energy electron can penetrate distances of several microns into a solid. If these transmitted electrons could be focused-images with better resolution may be obtained. Focusing is dependent on the fact that electron behave as negatively charged particles and are therefore deflected by electric or magnetic field.

Properties of Electron

Based on de-Broglie's idea we can relate the particle momentum p, with its wavelength λ, and it given by the equation (Shmeliov): $\boldsymbol{\lambda = h/p}$

Electrons are accelerated by potential E, thus kinetic energy $\mathbf{eE = 1/2m_0v^2}$ with momentum $\mathbf{p = m_0v = \sqrt{(2m_0eE)}}$. Substituting the value of $\lambda = h/\sqrt{(2m_0eE)}$.

Thus including relavistics effects, the above equation can be written as: $\boldsymbol{\lambda = h/\sqrt{(2m_0eE)(1+Ee/2m_0C^2)}}$

Thus by increasing the acceleration voltage, the wavelength of the electron decreases, thereby increasing the resolution.

From Table 2 it is clear that when the electrons are at a accelerating potential of 300 keV have a wavelength of ~2pm and have a diffraction limited resolution of ~1pm.

- **Source Information:** There are two types of emission sources (Langfeldt):
  - **Thermionic Source:** The therminonic emission sources most commonly a tungsten filament (or) a lanthanum hexaboride (LaB_6).

Table 2.

E(keV)	λ Relativistic (pm)
50	5.36
80	4.18
100	3.70
200	2.51
300	1.97

        - **Tungsten Filament:** It looks like either a hairpin- style filament (or) a small spike-shape filament.
        - **Lab$_6$ Source:** It utilizes small single crystals.
    - **Field Emission Source:** The field emission sources most commonly a tungsten filament and a ZnO/W. The gun begins to emit electron by thermionic (or) field electron emission into the vaccum once the gun are connected to a high voltage source (typically 100-400kv)
- **Scattering Cross Section From the Specimen**

Total scattering is given by number of scattering taking place per unit distance after the electron travels through the specimen.

$$\sigma_{total} = N\sigma_{atom} = N_0\sigma_{atom}\varrho/A$$

N= atoms/ unit volume, $\mathbf{N_0}$=Avogadros number, ϱ= density of specimen, A= atomic weight of scattering atom, A specimen with thickness ‘t’, the above equation is modified as:

$$t\sigma_{total} = N\sigma_{atom} = N_0\sigma_{atom}\varrho t/A$$

Diffraction

Figure 7. Shows the TEM image for polycrystalline thin films/ amorphous solid material which produces an image with a series of rings, whereas in case of single crystal thin films, image consists of a pattern of dots that is shown in Figure 7 (a).

The images obtained provide the user with information about the space group symmetries in the crystal. This is done by utilizing information of the observed image symmetry at which the diffraction spots appear.

Figure 7.

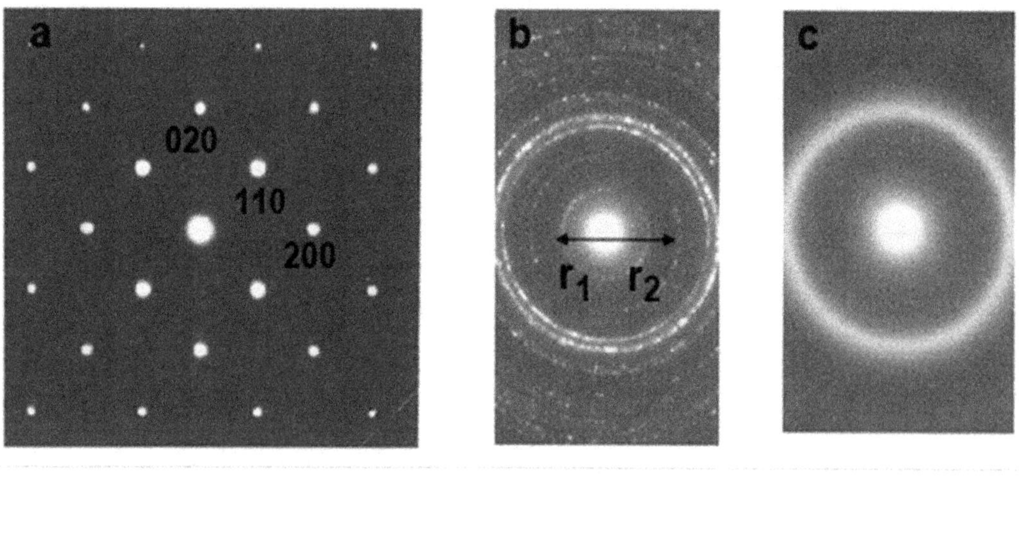

1. Indexing Diffraction Pattern

Diffraction data from TEM analysis are determined using near plane-sections through diffraction space. The angle formed between the transmitted and diffracted beams gives the magnitude of the diffraction vector (Δk). Two degree of freedom were previously required for obtaining diffraction pattern that were obtained with a double-slit specimen holder, where the two tilt axes lie perpendicular to each other. Modern Transmission electron microscope uses SAED and CBED method for obtaining the diffraction pattern. Single crystal diffraction spot obtained can be highly symmetrical when they are oriented along zone axis.

2. SAED (Selected Area Electron Diffraction)

To obtain SAED pattern in a two-lens TEM. The intermediate lens is focused to the back focal plane of the objective lens. Transmitted beam and the diffracted beams are viewedon screen. Selected area aperture can be placed in the image plane of the objective lens and the diffraction pattern of the selected area of the specimen can be obtained. SAED is generally performed on region as small as micrometer in diameter.

3. CBED (Convergent Beam Electron Diffraction)

To obtain selected area diffraction pattern for smaller region, as small as 1nm probe technique i.e. CBED is used.

Indexing Rules

1. For indexing we have to first identify the transmitted beam i.e. (000) diffraction spot. Diffraction spot is the brightest spot at the center of the diffraction pattern.
2. Thereafter we have to index two non collinear diffraction spot near the brightest spot.
3. After we get two vector, we can make linear combination of it to determine the position and miller indices of all the other neighboring spots. To complete the indexing, we have to specify the zone axis. Most commonly the zone axis points towards the electron gun.
4. 4. Indexing involves measuring the distance between diffraction spots and angle between two spots and comparing it with geometrical calculation of the distances and angle. From camera equation $rd = \lambda L$. It is used to measure the distance 'r' of diffraction spot from transmitted beam (bright spot)

$$R = \lambda L/a \sqrt{(h^2+k^2+l^2)}$$

5. Initially we measure the vertical and horizontal distances between spots near to the diffraction spot, as reference distance. (say 0.65 and 1.10nm). We then find the ratio of $\sqrt{(h^2+k^2+l^2)}$ from the allowed (hkl) values of the FCC crystal structure. Eg: 0.65/1.10= 0.591 By trial and error method we can find the value that matches with the above calculated ratio.

Transmission electron microscopy (TEM) is the first method used to determine the size and particle size distribution of nanoparticles samples. Once a representative group of TEM images is obtained, the next important task is to count particles as many as possible, ideally a few thousand, so that good statistics

Table 3. FCC lattice plane and their interplanar spacings

hkl	$\sqrt{(h^2+k^2+l^2)}$	Relative Spacing
(111)	$\sqrt{3}$	1.732
(200)	$\sqrt{4}$	2.000
(220)	$\sqrt{8}$	2.828
(311)	$\sqrt{11}$	3.316
(222)	$\sqrt{12}$	3.464
(400)	$\sqrt{16}$	4.000
(331)	$\sqrt{19}$	4.358
(420)	$\sqrt{20}$	4.472
(422)	$\sqrt{24}$	4.898

For (hkl)= (111)= $\sqrt{3}$; (hkl)= (220)= $\sqrt{8}$, thus $\sqrt{3}/\sqrt{8}$=0.61
(hkl)=(200)= $\sqrt{4}$; (hkl)= (311)= $\sqrt{11}$, thus $\sqrt{4}/\sqrt{11}$=0.60
(hkl) = (200)= $\sqrt{4}$; (hkl)= (222)= $\sqrt{12}$, thus $\sqrt{4}/\sqrt{12}$=0.58

on the size and size distribution present can be obtained. This counting step can be either by eye and by hand" method, those are common in the literature. It has several problems, as such these methods are tedious and often result in a few hundred or fewer particles being counted, and hence yield poor statistics.

To overcome those drawbacks, computer image processing methods are of great interest for counting a relatively large number of particles, with high accuracy as possible (Woehrle, 2006). This image processing need is recognized in a report on commercial software for nanoparticles counting. Some commercial software packages that have been used for this purpose, such as SPIP (Image Metrology ApS), LUCIA (Laboratory Imaging) and KONTRON KS 400 (Zeiss-Kontron). Amongst other readily available, free, public-domain program for particle counting includes NIH-*Image* (and 2 others derived from it, Scion Image and Image J).

Limitations

1. This microscope is very large and expensive.
2. Samples are limited to those that are electron transparent, able to tolerate the vaccum chamber and small enough to fit in the chamber.
3. Sampling is done for a very small area of the specimen.
4. TEM images represent 2D projection of a 3D specimen.

Advantages

1. It offers the most powerful magnifications with value ranging over one million times or more.
2. It has wide applications in various scientific, educational and industrial fields.
3. Provides information about chemical composition, crystal structure (symmetry, unit cell parameters).
4. Atomically resolved imaging about the structure, surface, interfaces, defects.
5. Provides information about the structural order i.e. crystalline, semi-crystalline or amorphous.

Application of TEM

1. Conventional TEM

In the determination of microstructure, morphology, phase distribution and types of defects.

2. Analytical TEM

In the determination of chemical composition of the specimen – EDX (Energy Dispersive X-ray analysis), EELS (Electron Energy Loss spectrometry), ELNES, EXELFS.

3. High Resolution TEM

Lattice imaging, structural order of complex material (crystalline, polycrystalline or amorphous).

Thus, Transmission Electron Microscopy (TEM) is a very important characterization instrument for directly imaging nanomaterials to obtain quantitative measures of particle and/or grain size, size distribution, and morphology.

Scanning Electron Microscope (SEM)

In these days SEM has become one of the most widely used instruments in the field of material science and fabrication plants. SEM provides information related to morphology, compositional differences, topographical features, crystal structure and many more. A Scanning Electron Microscope is a type of electron microscope that produces image of a sample by scanning it with a focused high energy beam of electrons in a raster scan pattern. The electrons interact with atoms in the sample, thereby producing various signals that can be detected and that contain information about the sample's surface topography and composition. The most common SEM mode is detection of secondary electron emitted by atoms excited by the electron beam. The number of secondary electron (SE) depends on the angle at which beam meets the surface of the specimen. By scanning the sample and collecting the SE with a special detector, an image is obtained that displays the topography of the surface. SEM is able to determine the elemental composition of a species with the addition of X-ray (commonly known as Energy Dispersive Spectrometry).

The sole purpose of the lens in the SEM is to demagnify and focus the electron beam towards the sample surface, which give rises to two major benefits:

- Increase in the range of magnification
- Increase in depth of field in the image.

The SEM has around 300 times more depth of field than the light microscope. It is due to the use of a very narrow beam of electron and a large amount of sample can be focused at one time that gives a characteristic 3D-view of the sample.

Characteristics That Can Be Analyzed on SEM

1. **Topography:** Describes about the surface property of the sample i.e. how it looks, its texture.
2. **Morphology:** Describes the shape and size of the particles that the specimen/sample is made up of.
3. **Composition:** Able to determine the elements and compounds that are present in the sample.
4. **Crystallographic Information:** Describes the arrangement of atoms making up the object.

Principle and Capacities

- **Data Analysis and Interpretation**

For the complete interpretation of geometry of probe, specimen and detector are extremely important.

The signals produced by SEM includes the secondary electron (SE), back scattered electron (BSE), characteristic X-rays, cathode luminescence (CL), specimen current and transmitted electron. Thus SE provides information on surface morphology. When the probe is at different angles with respect to the sample surface different brightness levels are observed. Holes remain dark while the raised area appears brighter.

  - **Back Scattered Electron:** These are the beam of electron that is reflected from the sample by elastic scattering. Theses electrons can give information about the distribution of various elements in the sample. The area with higher average atomic number produces more BSE's. Thus they appear brighter. Samples that exhibit multiple phases may exhibit contrast because of their different crystal orientation or magnetic domain.

- **Sample Preparation Steps**
  - **Cleaning the Surface of Specimen:** Proper cleaning of the surface of sample so as to remove the unwanted deposits, dusts, silt or other contaminants depending upon the type of specimen.
  - **Stabilizing the Specimen:** Fixatives are used as stabilizers. Common fixatives includes aldehydes, osmium tetroxide, tannic acid etc and it is achieved via. Perfusion and microinjection, immersions or vapors.
  - **Rinsing the Specimen:** Rinsing is done, so as to remove the excess fixative.
  - **Dehydrating the Specimen:** Dehydration to be performed very careful (especially for biological sample) using acetone or ethanol.
  - **Drying the Specimen:** As the SEM operates with high vaccum, it needs the sample to be dried, otherwise the sample may be destroyed inside the electron microscope chamber.
  - **Coating the Specimen:** Coating is done using a conductive metal (usually gold, palladium or platinum) so that to increase the conductivity in SEM and to prevent high voltage charges on the sample by transferring the charge to ground. Coating is usually 20 to 30nm thick.
  - **Mounting the Specimen:** After following the above steps, the specimen must be mounted on a holder usually aluminum stubs using a double-sticky tape. Then it is mounted on a holder that is inserted into the electron microscope.

- **Charging and its Influence**

The incoming electron to the sample/specimen loses their energy and gets absorbed in the specimen. If the specimen is conductive the electron flow through the specimen stage takes place, however if the specimen is non-conductive, the electron stops in the specimen, and results in charging. Charging causes the image to distort i.e. once the charging occurs, the electron probe that scans over the specimen is deflected by a repulsive force from a charged potential, resulting in the shift in the position of the electron probe.

- **How to Prevent Charging**

Conductive coating over the sample surface helps to prevent charging. In this method a highly conductive thin material is coated over the non conductive sample. Then the sample is subjected to Ion sputtering and vaccum evaporation. Most commonly noble metal is used for coating, it is because they are highly stable and its secondary electron yield is considerably high.

- **Electron Guns**
    - **Thermionic Guns:** Thermo electrons are emitted from a filament (cathode) made up of thin tungsten wire (about 0.1mm) by heating the filament at high temperature (about 2800 K). These thermo electrons are gathered as an electron beam flowing into the metal plate (anode) by applying a positive voltage (1 to 30kv) to the anode.
    - **Field Emission Guns:** A strong electric field is created to pull the electron away from the atoms to which they were associated with. Field emission electron guns are also called as Schottky-emission electron gun.
- **Vaccum System**

The electron optical system and the specimen chamber operates at a high vaccum, thus a high vaccum of 10^{-3} to 10^{-4} Pa is essential. Thus, these components are evacuated generally by a diffusion pump.

Advantages of SEM

1. Most SEM samples require minimal preparation actions.
2. Gives detailed 3D and topographical imaging and the versatile information from different detectors.
3. Modern SEM allows for the generation of data in digital form.

Table 4. Advantages of SEM

Microscope	Magnification	Depth of Field	Resolutions
OM	4X -1400X	0.5mm	0.2mm
SEM	10X-500KX	30mm	1.5nm

Disadvantages of SEM

1. SEM is quite expensive and large.
2. It is limited to solid samples, inorganic samples and the samples that can handle moderate vacuum pressure.
3. In some cases electron are scattered from the beneath of the sample surface, this carry a small risk of radiation exposure.

Application of SEM

1. Identification of metals and materials.
2. Particle contamination identification and elimination.
3. Examination of surface morphology.
4. Biological application includes 3D tissue imaging.
    a. **Virology:** For the investigations of virus structures.
    b. **Cryoelectron microscopy:** Helps in the determination of the surface of the frozen materials.
5. Forensics.

Thus, SEM has proved as an important characterization technique for the nanoparticles which provides information related to topographical features, morphology, crystal structure and compositional differences.

Differences Between SEM and TEM

Shown in Table 5.

Energy Dispersive X-Ray Spectrometer (EDS)

It is an analytical technique that is used for the elemental analysis of a sample. It relies on the interaction of high energy beam of charged particle i.e. electron, proton, or a beam of X-ray to stimulate the emission of X-ray from the sample.

Table 5. Differences between SEM and TEM

TEM (Transmission Electron Microscope)	SEM (Scanning Electron Microscope)
Based on the principle of transmitted electron i.e. electron beam passes through the thin sample.	Based on the principle of scattered electron i.e. electron scans over the surface of the sample.
Specially prepared thin films on Copper grids	Sample of any thickness mounted on an aluminum stub.
Specimen stage is halfway down the column	Specimen stage is at the bottom of the column
Image is shown on a fluorescent screen	Image is shown on TV monitor
Image is a 2D projection of the sample	Image is a 3D view of the sample
Small amount of the sample can be analyzed at a time	Large amount of sample can be analyzed at a time
It is used to analyze the particle size.	It is used to determine the surface morphology of the sample

- *Detectors Used*
- **Crystal Si (Li):** Previously used detectors. These detectors require cooling using liquid nitrogen, generate and collect electron hole pairs.
- **Silicon Drift Detector:** It consists of high-resistivity silicon chips with a large area contact on entrance side. On the opposite side there is a small collecting anode, which is surrounded by concentric drift electrodes. This anode has very low capacitance. The electrons are drifted down the field gradient that are applied between drift rings and collected at anode.
    - **Advantages of SDD**
        - High processing and count rates.
        - At high count rates SDD has better resolution than Si (Li) detectors.
        - Faster analytical capabilities.
        - More precise X-ray maps
        - Can be operated at high temperatures, thereby eliminating the use of liquid nitrogen for cooling.

In some cases when the X-ray energies of interest exceeds ~30keV, commonly used silicon-based detectors suffers from poor quantum efficiency, which may be due to the reduction in detectors stop-

Figure 8. Schematic cross-section of Silicon Drift Detector

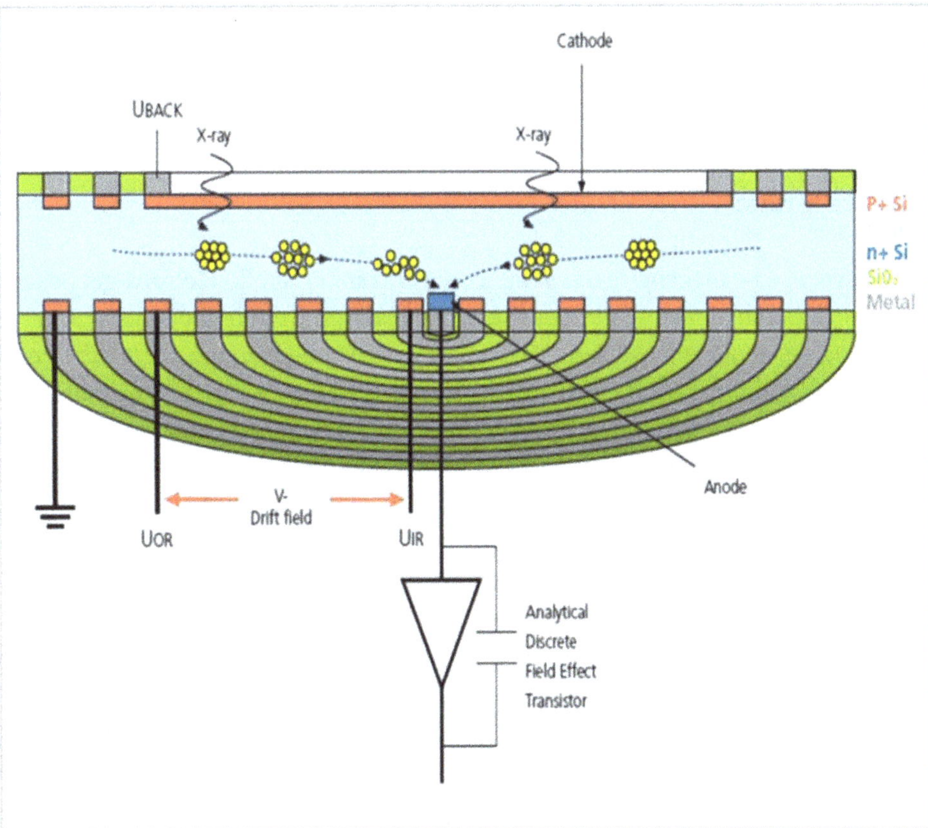

Figure 9.

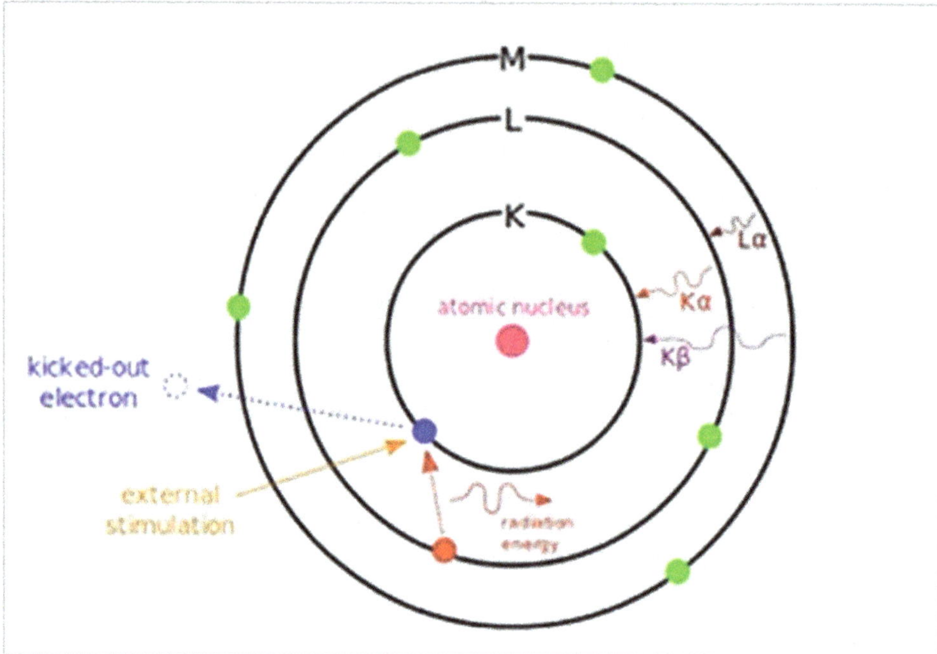

ping power. Thus high density semiconductors have been used for the production of alternate detectors (cadmium telluride-CdTe, cadmium zinc telluride (CdZnTe). These detectors are highly efficient for higher energy X-rays and are able to operate under room temperature.

- **Generation of X-rays:**

When the high energy beams enters the sample, various inner shell electrons gets excited, and are ejected from the shell thereby creating electron hole. An electron from higher energy (outer) shell then fills the hole, and the substance release X-ray. This X-ray corresponds to the energy difference between the outer shell electron and inner shell electron. The energy and number of X-ray emitted from a specimen can be determined by energy dispersive-spectrometer. These X-rays are called as 'Characteristic X-rays" because their energies corresponds to individual elements. Thus EDS allows measuring the elemental composition of the specimen. EDS can be used to detect all the elements from atomic number 4 (Be) to 92 (U).

Figure 9 shows the X-ray lines that are designated by capital Roman letter that indicates the shell with inner vacancy (K, L,M…), a Greek letter denotes the group to which the lines belong, they are arranged in order of their decreasing importance (α,β, etc.) and a number denotes the intensity of the line, they are also arranged in order of their decreasing order (1,2,etc.). Most intense line of K-series is Kα1. Similarly most intense line of L-series is Lα1. As L shell is further splitted into three sub shell, L spectrum is more complicated and contains as many as 12 lines, out of which most of them are weak.

X-ray emitted from the specimen enters the semi-conductor detector. The detector is cooled by liquid nitrogen, in order to reduce the electric noise. Thus amount of X-ray energy is plotted along X-axis and X-ray counts are plotted along Y-axis (Figure 10).

Advantages of EDS

1. Rapid identification of contaminant and source
2. Full control of environmental factors, emissions etc.

Application of EDS

1. To determine the composition of the coating
2. Rapid material alloy identification
3. Small component material analysis
4. Catalyst quality, poisoning and elemental distribution
5. Product imperfections and defect analysis

Atomic Force Microscopy (AFM)

AFM is a very high resolution (of the order of fraction of nanometer) scanning probe microscope (SPM), which has 1000 times better resolution than the optical diffraction limit.

Figure 10.

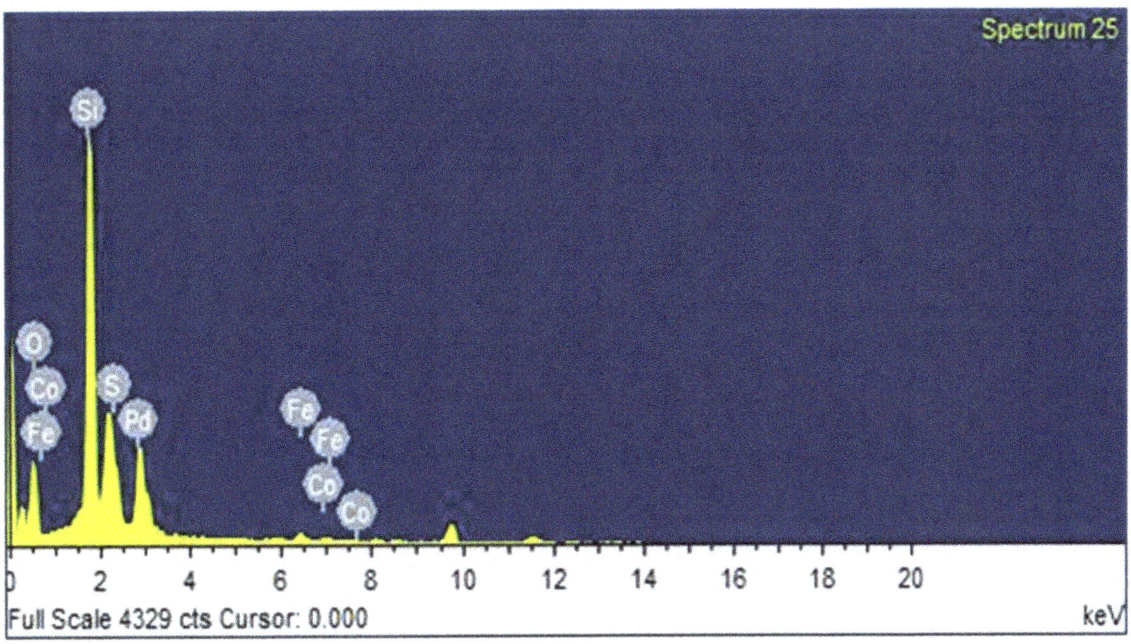

History

In 1986, Binning and Quate developed AFM based on the STM (Scanning Tunneling Microscope), which consists of a cantilever with a small probe tip (Binning, 1986). In 1987, Wickramsinghe et al. developed a modified version of AFM with a vibrating cantilever technology (Wickramsinghe, 1987). AFM gathers the information by "feeling" the surface with a small probe tip, piezoelectric elements used facilitates the accurate and precise movement and enable very precise scanning over the sample surface.

Principle

AFM comprises of a probe (sharp tip) that is attached at one of the end of the cantilever, and is used to scan the specimen surface. Cantilever is generally made up of Silicon (or) Silicon Nitride (Si_4N_4) with the tip radius of curvature of few nanometers. A variety of silicon and silicon nitride cantilevers are commercially available with

- Micron-scale dimensions,
- Spring constants ranging from 0.01 to 100N/m, and
- Resonant frequencies ranging from 5 kHz to over 300 kHz.

When the tip is brought close to the sample surface deflection in the cantilever is observed according to the Hooks law i.e. due to the force acting between the tip and the sample (Figure 11). This deflection is measured using a laser spot i.e. reflected from the top of the cantilever into an array of photodiodes. AFM is able to measure various types of forces depending upon the situation that includes Vander Waal's force, electrostatic forces, magnetic forces, capillary forces, chemical bonding etc.

Depending upon application, AFM can be operated in various modes. Imaging mode is divided into three types (Peltonen):

Figure 11.

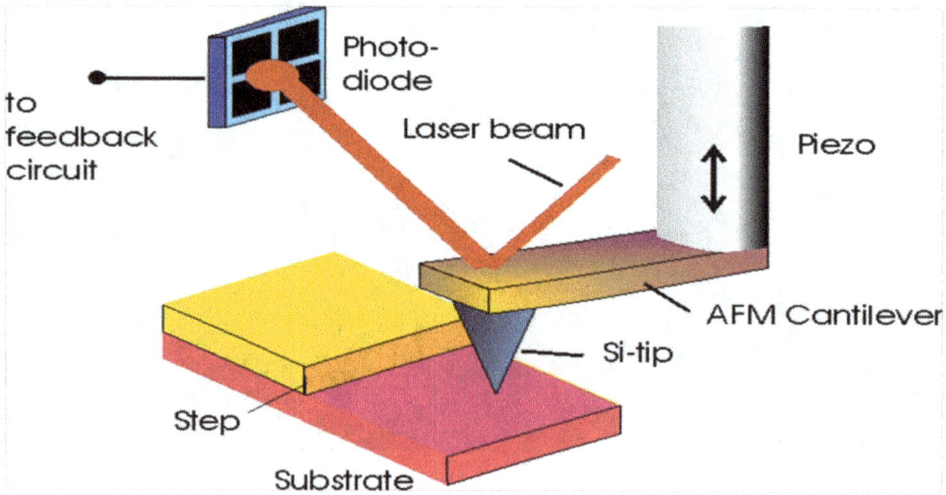

1. **Contact Mode (c-AFM):** In this mode, the probe tip is dragged across the sample surface, the contours of the surface is measured from the deflection of cantilever or from the feedback signal that is required to keep the position of cantilever constant. This mode is applicable for a solid surface i.e. below any adsorbed layer. In this mode the probe tip is very close to the sample surface, thus attractive forces are quite strong. This causes the tip to "snap-in" to the surface. Thus contact mode is always done at the depth.
2. **Intermittent Contact Mode (ic-AFM):** This mode is also called as tapping mode. Under normal condition it is seen that a liquid meniscus layer is developed over the sample. Due to this reason it becomes difficult to keep the probe tip very close to the sample for the short range forces to become detectable. To overcome this difficulty tapping mode was developed. In this mode, the cantilever with a small piezoelectric element probe tip oscillates up and town at near its oscillating frequency. The amplitude of oscillation is typically between 100-200 nm. Electrostatic forces, Vander Waal's forces, dipole-dipole interactions etc. are the different interaction forces that acts on the cantilever as the tip come close to the surface of the sample. This force decreases the amplitude of oscillation as the tip come closer to the sample. The height of the cantilever over the surface of the sample is controlled by an electronic servo that uses piezoelectronic actuator. This method reduces the damage that occurs to the sample and AFM tip.
    a. Advantages of Tapping Mode
        i. Helps in the visualization of supported lipid bilayers, adsorbed single polymer molecule.
        ii. It has true atomic resolution.
        iii. It can measure atomic forces (so called Atomic Force Microscopy)
        iv. It can measure mechanical response i.e. even elastic deformation.
3. **Non Contact Mode (nc-AFM):** In this mode, the tip of the cantilever is just oscillated at either its resonant frequency or just above it where the amplitude of oscillation is few nanometers i.e. < 10nm to few pm, such that the tip does not touch come in contact with the surface of the sample. Scanning software allows constructing the topographic image of the sample surface by measuring the tip to sample distance (x,y) from the data plot. This method reduces the damage that occurs to the surface of the sample and the probe tip. Thus this mode become useful for getting topographical information of soft samples, e.g. biological samples and thin organic films.
    a. Advantages of AFM
        i. Able to achieve a resolution of 10 pm with special tips (typical resolution ≈ 10 nm).
        ii. High resolution AFM is comparable with the resolution of TEM & STM.
        iii. Able to image samples in air and under liquids, thus biological macromolecules and living organisms can be studied.
        iv. Able to measure in 3D- surface profile (within limits)
        v. Able to measure non-conductive surface (unlike SEM or STM), i.e. sample do not require any special treatment (i.e. metal/ carbon coating)
        vi. Applicability of AFM can be extended by combining AFM with variety of optical microscopy techniques such as fluorescent microscopy.
    b. Disadvantages of AFM
        i. Incorrect tip choice, sample operating environment can lead to measurement artifacts or sample damage.
        ii. Depth of field limited by cantilever / z positioning PZ
        iii. Scan area limited by PZ scanners(i.e. 10-20μm height and 150x150μm in area)

iv. Slow scan rate compared to SEM. Slow rate may lead to thermal drift in the image.
v. AFM probes cannot measure steep walls or overhangs, thus especially made cantilevers are used that are expensive and they have low lateral resolution.

c. Application of AFM
   i. **Biochemistry Application:** Imaging cellular components, cells or tissues, topography of soft biological materials in their native environments, determining the mechanical properties of many tissues and cells.
   ii. **Chemistry, Material Science and Nanotechnology Application:** Imaging of polymer, nanostructure or other materials.
   iii. **Physics and Biophysics Application:** Measuring forces between AFM tip and the sample surface.

Nanomaterial Characterization by Thermal Analysis

Thermal analysis comprises of a group of techniques which measures the change in physical and chemical properties of the sample as a function of temperature. Solid samples are generally used for analysis. The samples are subjected to controlled heating and cooling program.

Thermo Gravimetric Analysis (TGA)

This analysis describes the changes in weight of the sample as a function of temperature under a controlled atmosphere (N_2, O_2, air or He).A derivative weight loss curve is obtained that is able to describe the point at which weight loss takes place.

The weight loss curve gives information about: The change in the sample composition, thermal stability of the sample, kinetic studies (Chatwal, 2002; Hatakeyama & Quinn, 1999; Coats & Redfren, 1963).

TGA can explain about the change in the physical and chemical properties of the sample.

Factors Affecting the TGA Curve

Various factors that affect the TGA curve can be broadly divided into two categories namely, instrumental related and sample related.

Types of TGA

The TGA instrument is initially loaded with sample and it is heated to the desired temperatures at a fixed heating rate under a controlled atmosphere. The heating rates are fixed between 0.5 and 50 K/min, often

Table 6. Properties of the sample

Physical Properties	Chemical Properties
Adsorption	Dehydration
Desorption	Decomposition
Phase transition • Sublimation • Vaporization	Solid-gas reaction • Oxidation • Reduction

Table 7. Factors affecting the TGA curve

Instrumental Related	Sample Related
Heating rate	Particle size of the sample
Furnace atmosphere and gas flow rate $CaCO_3 \rightarrow CaO + CO_2$ • In vacuum, sample decompose at ~500°C • In air, it decompose at ~700°C • In CO_2, ~900°C	Packing
Crucible shape	Thermal conductivity
Crucible material (Al, alumina, Pt, Cu, Au,etc.)	Heat of reaction

Figure 12.

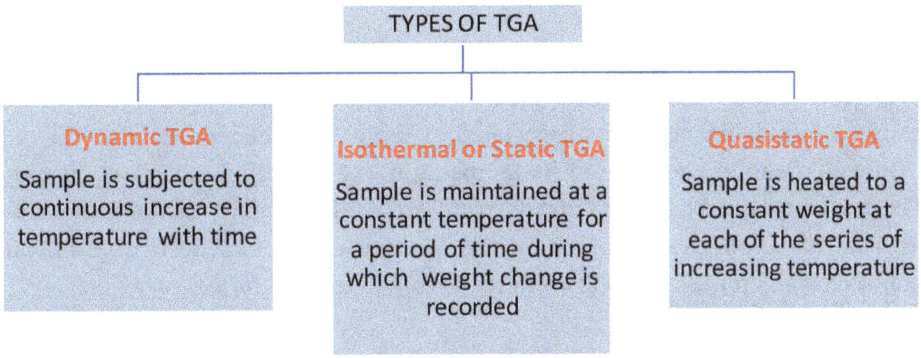

at 20K/min. starting temperature is usually 25 to 30 °C, in order to detect possible drying of the sample. The final temperature is relatively very high i.e., 600 °C for organic substances and > 1000 °C for inorganic compounds Various components of the sample starts decomposing with the rise in temperature, and the change in the mass of the sample can be measured. Results are plotted with temperature along X-axis and percentage weight loss along Y-axis. The data can be adjusted using curve smoothing and its first derivatives are often plotted to determine points of inflection (i.e., onset and offset points) for better interpretations.

Interpretation of TGA Curves

Several TGA curves are used in addition to the actual TGA curves are used for interpretation:

- First derivative curve.
- SDTA curve (it measures the exothermic and endothermic effects similar to the DSC curves).
- EGA (Evolved Gas Analysis).

TGA curves are plotted with the weight percentage along Y- axis and temperature (T) or time (t) along X-axis. There are two temperature in the curve, T_i (representing procedural decomposition temperature) and T_f representing the final temperature. T_i represents the temperature at which initial weight loss takes place and T_f represents the temperature at which the process has been completed.

Types of TGA curves

Based on their shapes TGA curves can be categorized into 7 types:

1. Curve A is shown by sample when there is no mass change over the entire range of temperature. Thus the decomposition temperature of the sample is greater than the instrument operating temperature.
2. Curve B represents the large mass loss, which occur due to the evaporation of volatile compounds.
3. Curve C represents single step decomposition of the sample.
4. Curve D represents multi-step decomposition reaction where the reaction is resolved.
5. Curve E represents multi-step decomposition reaction where the reaction is not resolved.
6. Curve F represents the increase in the of sample as a result of interacting atmosphere. E.g. Surface oxidation reaction.
7. Curve G represents the multiple reactions. E.g. Surface oxidation reaction followed by the decomposition of reaction product.

Derivative Thermogravimetry

TGA curve shows a sample's weight loss associated over a wide temperature range. TGA is unable to identify a unknown compound because of its temperature range. This problem is overcome by DTG (Derivative Thermo Gravimetry). In DTG first derivative of the TGA curve (dw/dt) is plotted with respect to temperature. Rate of weight change is along y-axis. The peak of the first derivative indicates the point of greatest rate of change on the weight loss curve. This is known as *Inflection Point.*

Applications of Thermo Gravimetric Analysis

1. Characterization of
   a. Thermal stability
   b. Purity of the sample
   c. Humidity content

Figure 13. Types of TGA curves

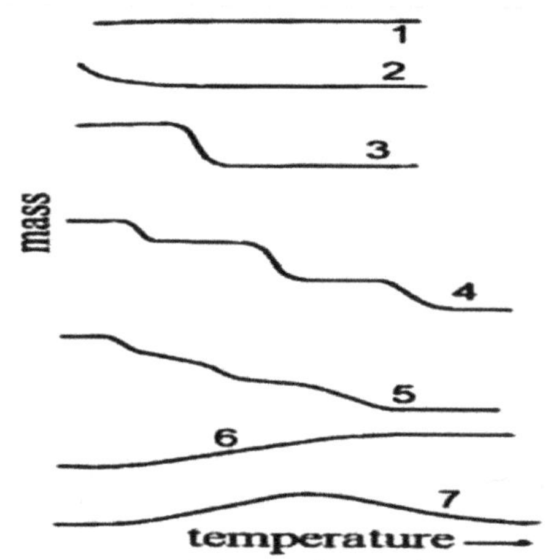

2. Examination of
    a. Corrosion studies (oxidation and reduction reactions)
    b. Gasification processes
    c. Kinetic processes

On decomposition of the inorganic materials following gaseous components are evolved: H_2O, CO, CO_2, SO_X, NO_X, Cl_2, F_2, CH_3OH etc. There are few solid phase chemical reactions where gaseous components are released:

$$Na_2SiO_3(s) + SiO_2(s) \rightarrow Na_2SiO_3(s) + CO_2(g)$$

$$CaC_2O_4H_2O \overset{-H_2O}{\rightarrow} CaC_2O_4 \overset{-CO}{\rightarrow} CaCO_3 \overset{-CO_2}{\rightarrow} CaO$$

Example:
Decomposition of calcium oxalate monohydrate (Figure 14)

- It is used as a standard material to demonstrate the working of TGA.
- It exhibits three weight losses at varying temperature under N_2 atmosphere.

$$CaC_2O_4H_2O \overset{-H_2O}{\rightarrow} CaC_2O_4 \overset{-CO}{\rightarrow} CaCO_3 \overset{-CO_2}{\rightarrow} CaO$$

Thermo gravimetric analysis measures the amount of weight change of material with increase in temperature as a function of time in an inert atmosphere or in vaccum. It can be used for both the amorphous and crystalline components. Thus, it has been widely used as an important characterizing technique for the nanoparticles.

Differential Scanning Calorimetry (DSC)

In the year 1962 E.S. Watson and M.L.O'Neill first developed this technique. DSC is a thermo analytical technique which measures the heat of the sample relative to the reference as a function of temperature. Throughout the experiment both the sample and reference are maintained at nearly same temperature.

Principle

The temperature programmed DSC analysis is designed in such a way that the sample holder temperature increases or decreases linearly as a function of time. During the process of heating two types of reaction namely endothermic and exothermic reaction can take place. If the sample undergoes a physical transformation such as phase transitions, and if some amount of heat need to flow into the sample so that to maintain both at the same temperature then the process is referred to as *endothermic reaction*. In this case large amount of energy is required so that zero temperature difference is maintained (e.g.: Melting, boiling, vaporization, sublimation).

Figure 14.

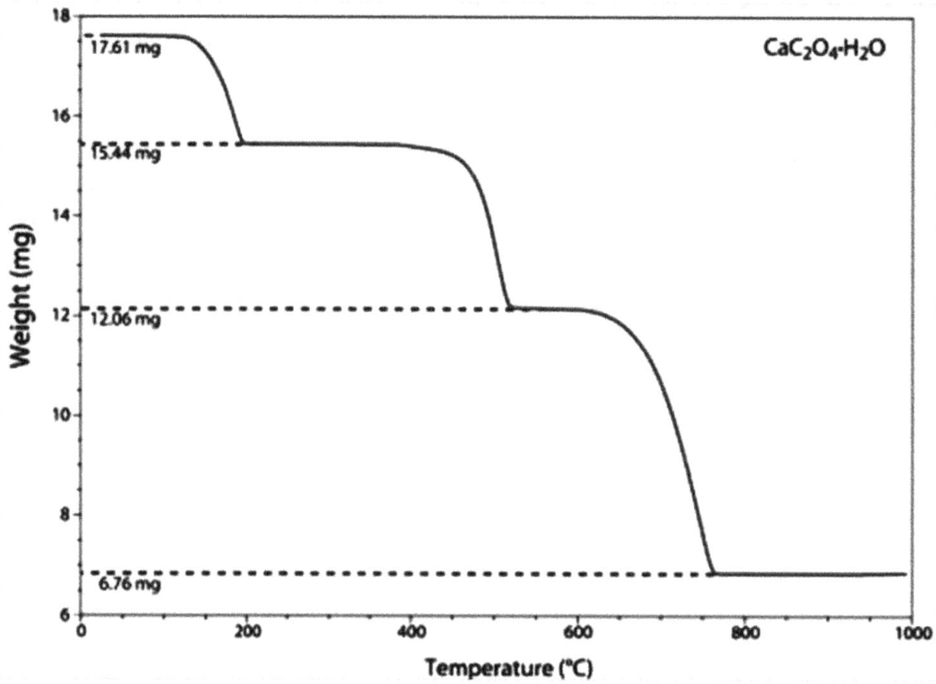

And if the sample undergoes a phase transitions, and if some amount of heat is released from the sample, so that to maintain zero temperature difference between the sample and the reference then the process is referred to as *exothermic reaction*. In this case less amount of energy is required so that zero temperature difference is maintained (e.g.: Crystallization, degradation, polymerization).

DSC is used to observe the subtle phase changes very easily, such as glass transitions. DSC has wide industrial application as a quality control instrument because of its applicability in evaluating sample purity. The result of a DSC experiment is a combination of heating and cooling curve.

DSC curves (figure 15 a,b) are plotted with the heat flux along Y- axis and temperature (T) or time (t) along X-axis. There are two conventions: a sample may show positive or negative peak, positive represents exothermic reaction while negative peak resembles endothermic reaction. By integrating the peak corresponding to a particular transition, enthalpies of transitions can be calculated. Enthalpy of transition can be expressed by using the following equation: $\mathbf{\Delta H = KA}$

Where, ΔH = enthalpy of transition, K = calorimetric constant (vary from instrument to instrument), and

A = area under the curve, A α heat absorbed or evolved by the reaction.

It is seen that that heaters attached to the two pans does not heat at the same rate. It is because one pan has sample in it and the other doesn't. The pan in which sample is present have extra materials in it, this means that pan takes more heat to raise the temperature of the sample pan at the same rate with the reference pan. So the heater under sample pan has to put more heat.

Figure 15.

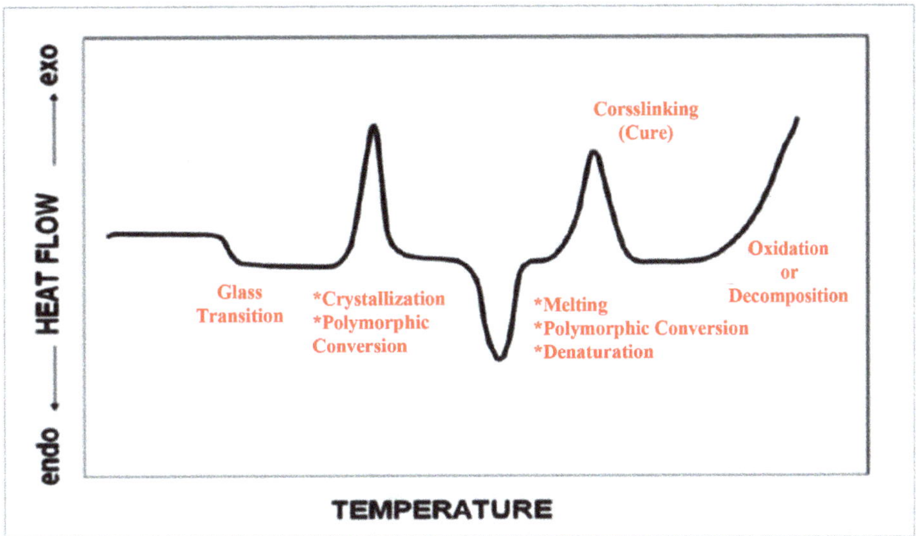

a

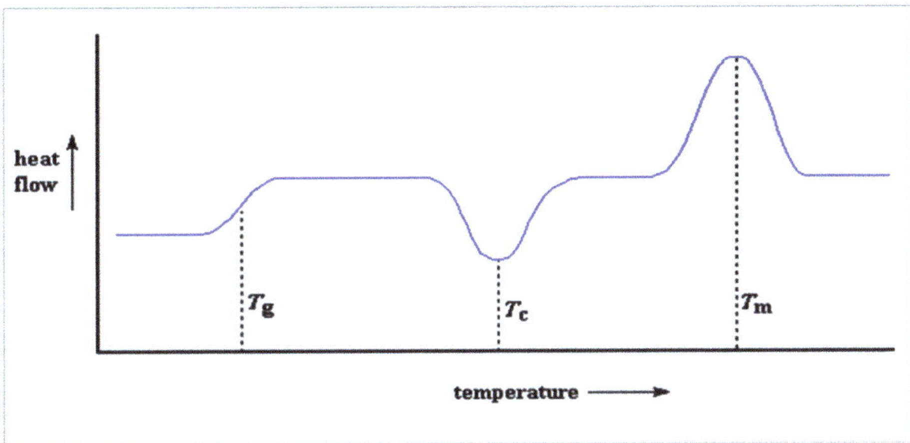

b

In DSC curve we get crystallization peaks and melting dip for samples that form crystals and melts, these peaks are not observed for completely amorphous samples.

Heat flow is given by the formula, heat flow= heat/time= q/t

Heating rate = increase in temperature/time= ΔT/t

Material's Heat capacity is given by: Cp= Heat flow/heating rate=q/ ΔT

A graph may be labeled with four critical points: the glass transition temperature (T_g), the crystallization temperature (T_c), the melting temperature (T_m) and the curing temperature.

1. **The Glass Transition Temperature (T_g):** It is an important characteristic of crystalline materials. In DSC curve, it is seen that continuous heating of the material shows a sudden downward shift, this permanent decrease in base line heat flow represents the Tg, and Tg is taken as inflection point in the curve. It is also an important property for amorphous and semi crystalline materials, below this temperature the materials tends to be hard and brittle, and above this temperature they become soft and ductile.
2. **The Crystallization Temperature (T_c):** After glass transition temperature, the materials attain a lot of mobility, when the crystallization temperature is obtained they will release energy to form crystals (a ordered arrangements), i.e. crystallization is an exothermic process. In a DSC curve onset point of crystallization curve represents the crystallization temperature. Amorphous materials do not undergo crystallization.
3. **The Melting Temperature (T_m):** If we further heat the material after attaining T_c, the material will reach another thermal transition, known as melting. At this temperature material changes its phase from solid to liquid. When the crystal melts, heat is absorbed i.e. its intermolecular bond begin to loosen and break. Once the material completes melting, heat flow attains the original baseline value. As this process involves absorption of heat, the process is endothermic. In a DSC curve onset point of melting curve represents the melting temperature.
4. **The Curing Temperature:** This curve is mainly observed in polymers, in which cross linking takes place. Like crystallization this is also a bond formation process, and the process is exothermic. In a DSC curve onset point of curing peak represents the curing temperature.

DSC data analysis and interpretation may be difficult for some samples. E.g. Some polymers may contain moisture, fillers, additives, plasticizers that behave differently than pure polymers.

Factors Affecting DSC Curves

DSC can explain about the following change in the physical and chemical properties of the sample.

1. Glass transitions
2. Melting and boiling points

Table 8. Factors affecting DSC curves

Instrumental Factors	Sample Characteristics
Furnace heating rate	Amount of sample
Recording speed	Nature of sample
Furnace atmosphere	Sample packing
Crucible shape	Particle size
Crucible composition	Heat of reaction
	Thermal conductivity

3. Heat of fusion
4. Percent Crystallinity
5. Specific heat capacity
6. Thermal stability
7. Kinetics study
8. Purity of sample

Advantages of DSC

1. Highly sensitive
2. Operates at very high temperature
3. Flexible in sample volume and form
4. Characteristic phase transition temperatures can be determined

Disadvantages of DSC

1. Not suitable for two phase mixtures
2. It is used for thermal screening of isolated intermediates and products.
3. Unable to detect gas generation

Application of DSC

1. Polymers:
    a. Composition of polymer can be determined.
    b. Glass transition temperature, crystallization temperature and melting point can be determined.
    c. Percentage Crystallinity.
    d. Thermal degradation of the sample and amount of impurity present in it.
2. Oxidative stability:
    a. Stability and optimum storage condition of the sample can be determined.
3. Drug analysis:
    a. Widely used in pharmaceutical industry and polymer industry (cross linking of samples via. curing process)
4. General chemical analysis:
    a. Purity analysis of the sample can be performed using freezing point depression as an important tool.
5. Functional application:
    a. Binding studies can be performed.

Differential Thermal Analysis (DTA)

It is a technique which continuously measures the temperature difference that is developed between the sample and thermally inert reference as a function of temperature or time.

In DTA both the samples and reference material (alumina) are subjected to controlled heating and cooling program. Zero temperature difference between the test sample and reference material indicates that there doesn't exists any physical or chemical change in the test sample. This differential temperature is then plotted against time, or against temperature. For the analysis of inorganic compounds in DTA, reference materials used are alumina (Al_2O_3), carborundum (SiC) or magnesium oxide (MgO) powder.

- *Physical changes* in the samples usually results in two different endothermic curves:
  - **Sharp Endothermic Curve:** Changes in Crystallinity (or) fusion.
  - **Broad Endothermic Curve:** Dehydration reaction.
- *Chemical reaction* usually results in exothermic curves.

Factors Affecting DTA

Shown in Table 9.

Advantages of DTA

1. Highly sensitive
2. Operates up to a very high temperature
3. Characteristic transitions can be determined accurately.
4. Flexibility in the use of crucible of any volume and form.

Application of DTA Curves

1. Qualitative and quantitative identification of minerals, it can be done by comparing the DTA curve of the test sample with the reference curve.
2. Purity assessment of sample: Impurities in the sample can be detected by depression of melting point.
3. Measurement of Crystallinity: Helps to determine mass fraction of crystalline material present in semi-crystalline materials.
4. Analysis of Biological materials: Date the bones and archaeological materials. Polymeric materials: Identification of thermo physical, thermo chemical, thermo elastic and thermo mechanical changes or transitions in materials.

Table 9. Factors affecting DTA

Factors	Effects
Heating Rate	Changes the position of peak and peak size
Furnace atmosphere	Change in the curve
Amount of sample	Changes the position of peak and peak size
Particle size	Irreproducible peaks
Packing density	Irreproducible peaks
Position of thermocouple	Irreproducible peaks

It is advantageous to use both DTA and TGA to draw a constructive conclusion, it is because the DTA events can then be classified into those which do or do not involve mass change. Example (Table 10).

Nanomaterial Characterization by Vibrating Sample Magnetometer (VSM)

VSM was invented in the year 1955 by Simon Foner at Lincoln Laboratory, MIT is a scientific instrument. It is a scientific instrument that measures the magnetic moment and hysteresis. VSM operates via. Faraday's law of Induction and lock in principle of measurement, which explains that change in magnetic field that produces electric field. By measuring the change in electric field we can get information about the magnetic field change. Thus VSM measures the magnetic behavior of magnetic materials.

Commercially available VSM magnetometers have sensitivity below $10^{-9}Am^{-2}$. VSM uses parallel configuration and transverse configuration of sample vibrations with respect to the external magnetic field. Figure 16 represents the basic setup of Vibrating Sample Magnetometer.

In VSM, first the sample is placed in a uniform external magnetic field. The magnetic domains or individual magnetic spins are aligned in the direction of magnetic field. This induces magnetization in the sample. A magnetic field around the sample is created due to the magnetic dipole moment; this field is referred to as *Magnetic Stray Field.* When the magnetized sample starts vibrating, magnetic stray field

Table 10.

Temperature	TGA	DTA
500-700 °C Change in mass	Dehydration of $Al_4(Si_4O_{10})(OH)_8$ (kaolin)	Dehydration - ENDOTHERM
950-980 °C No change in mass	No effect	Recrystallization- EXOTHERM

Figure 16.

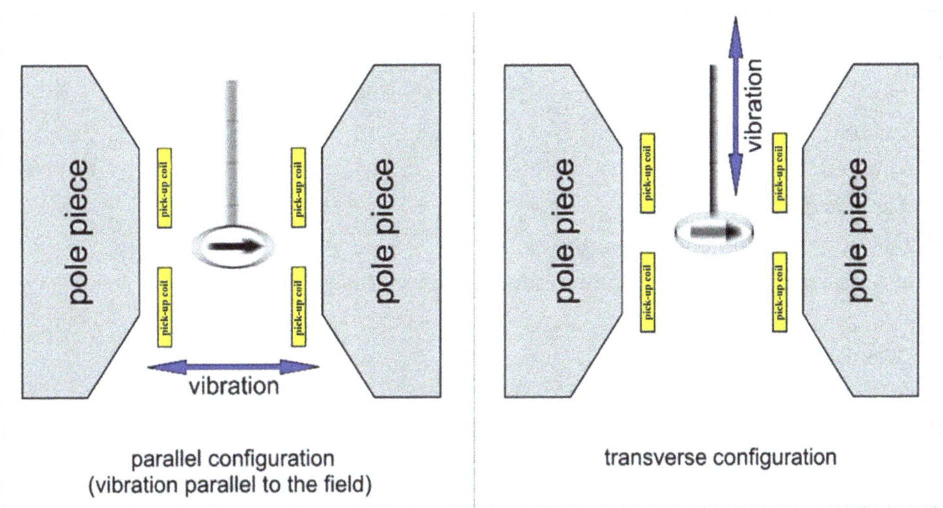

starts changing as a function of time, this change is measured by a set of pick up coils that are arranged around the sample. According to the Faraday's law of induction the alternating magnetic field will create an electric field in the coils.

The current generated in the coils depend upon:

- Amplitude and frequency of vibrations.
- External applied magnetic field.
- Magnetization of the sample.

Software attached to control and monitors the system, can predict the amount of sample magnetized and how the magnetization depends upon the strength of magnetic field. Thus a plot of magnetization (M) and magnetic field (H) is generated. This can be plotted using origin software to get a hysteresis loop.

The hysteresis loop can be explained using the following terms:

- **Coercivity (H_C):** It is the maximum magnetic field required to demagnetize a sample that has been fully magnetized. It measures the resistance of a ferromagnetic material to change its magnetization. It unit is Oersted or ampere/meter. Ferromagnetic materials with high coercivity are used for making permanent magnets. In hysteresis loop coercivity is represented as a horizontal intercept.
- **Retentivity or Remanence (B_C):** It is the capacity of a substance to retain magnetization after the removal of magnetizing force. Its unit is Tesla (T).
- **Magnetization (M_S):** It is the process of making a substance temporarily or permanently magnetic, as by insertion in a magnetic field. Its unit magnetic moment per unit volume (emu/g).

Magnetic domains walls are broad in case of magnetically soft materials, so their movement requires small field, whereas larger fields are required for magnetically hard materials (Askeland, 1994). Secondly, magnetization of sample increases with the increase in magnetization rotation within the domains. There are certain directions that are easier to magnetize, thus spontaneous magnetization occurs along that direction. If the applied field is not parallel to that direction, the domains rotate and aligned them along the directions of magnetic flux lines.

Advantages of VSM

1. Highly Sensitive up to 10^{-6} emu
2. Fully automated
3. Suitable for powders
4. Can operate under variable temperature
5. Sufficient high field possible- if semiconductor magnet is used

Disadvantages of VSM

1. Process is destructive
2. It undergoes self-demagnetization
3. Limited field dH/ dt.

Applications of VSM

1. Measurement of magnetic properties (Magnetization, Coercivity and Remanance) of the sample (liquids, powders, bulk)
2. Magnetic measurements under hydrostatic pressure.

Adsorption Process

Adsorption Isotherm Model

Adsorption is an extensively used technique for the decolorization of synthetic dyes from the industrial wastewater. It is a surface phenomenon and it occurs when the adsorbate comes into contact with a solid surface (adsorbent) which is porous in nature. The adsorption process is generally classified as physisorption or chemisorptions but in some cases it may be due to the electrostatic attraction. The nature of bonding and state of adsorbed molecule can be known by photoelectron spectroscopy. However Infrared and Raman spectra can be used to study the nature of adsorbed species.

During the adsorption process, equilibrium is established between the solution and adsorbent. Initially, aqueous solution of the dye was used for carrying out the adsorption experiments. The adsorption was carried out in batches in order to obtain rate and equilibrium data. Following equation can be used to determine the amount of adsorption of molecules (q_e) at the equilibrium time:

$$q_e = \frac{\left(C_0 - C_e\right)V}{W} \tag{1}$$

Where C_0 represents the initial concentration of the dye/metal solution and C_e represents the final concentration of the dye/metal solution. V represents the volume of the solution and W represents the mass of adsorbent.

Different adsorption isotherms were considered in order to determine the adsorption capacity and adsorption behavior of the adsorbent.

Langmuir adsorption isotherm model illustrates that the active sites are uniformly distributed over the adsorbent surface and adsorption occurs only on the active sites of the adsorbent surface (Langmuir, 1916). Once all the active sites are occupied by the adsorbate, further adsorption stops. The Langmuir adsorption equation is:

$$\frac{C_e}{q_e} = \frac{1}{Q_0 b} + \frac{C_e}{Q_0} \tag{2}$$

Where, q_e represents the amount of dye/metal adsorbed,

C_e represents the equilibrium solution concentration,

Q_0 and b are the Langmuir constants, and they represent the maximum adsorption capacity and energy of adsorption respectively.

The Freundlich isotherm model is based on the hypothesis that the adsorbate gets absorbed onto the surface of an adsorbent (Freundlich, 1906). Equation (3) expresses the model:

$$\log q_e = \log K_F + \frac{1}{n} \log C_e \tag{3}$$

Where q_e represents the amount adsorbed on adsorbent, C_e represents the equilibrium concentration of dye solution; adsorption capacity and adsorption intensity are represented by two Freundlich constants K_F and n respectively.

Various kinetic models such as the linearized form of pseudo first order, pseudo second order or intra particle diffusion model are used to investigate the adsorption process. Lagergren explains the pseudo first order kinetic model.

Temkin isotherm explains the relationship between the amount of dye adsorbed on adsorbent and the heat of adsorption. Chemical adsorption process, explains the endothermic or exothermic processes during the process of adsorption. Equation (4) represents the Temkin isotherm model:

$$\log q_e = B \ln A + B \ln C_e \tag{4}$$

Where A and B represents the equilibrium binding constant and the heat of adsorption.

The surface area of the nanocomposite can be was determined by nitrogen adsorption desorption method using BET *(Brunauer–Emmett Teller)* equation:

$$\frac{1}{V\left[\left(\mathrm{p0}-p\right)-1\right]} = \frac{c-1}{vmc}\left(\frac{p}{p0}\right) + \frac{1}{vmc}$$

Where p_0 = partial vapour pressure of adsorbate gas in equilibrium with the surface at 77.4 K (b.p. of liquid nitrogen), in pascals, p= saturated pressure of adsorbate gas, in pascals, V= volume of gas adsorbed at standard temperature and pressure (STP) [273.15 K and atmospheric pressure (1.013×10^5 Pa)], in millilitres, v_m= volume of gas adsorbed at STP to produce an apparent monolayer on the sample surface, in millilitres, c = dimensionless constant that is related to the enthalpy of adsorption of the adsorbate gas on the powder sample.

And the micro pore volume and external surface area (mesoporous surface area) was calculated using *t-plot method.* Before analysis, the samples were oven dried at 100 °C for 7 h and then degassed under vacuum at 120 °C for 2 h. At a relative pressure close to unity all the pores of the composite were filled with nitrogen under a specific pressure, thus pore volume was determined from the amount of nitrogen adsorbed.

CONCLUSION

The recent developments in the field of advanced materials, polymers etc shows a drift towards nano-materials. However the challenges and success of the researchers in this field depends on its availability, cost and ease of handling and the performance of the sophisticated instruments that are described throughout the chapter. Further, the extent of understanding of some particular instrumental techniques such as SEM, TEM, AFM, XRD etc. by the users, right sample preparation are important for deriving information from those techniques.

Thus, it can be summarized that nanotechnology research in various fields of Science and technology has a lot of potential in the future. But the developments basically depend on development of some newer, faster, simple and efficient characterization techniques for nanomaterials, nanocoatings and nanocomposite.

REFERENCES

Binning, G., & Quate, C. F. (1986). Article. *Physical Review Letters*, *56*, 930–933. doi:10.1103/PhysRevLett.56.930 PMID:10033323

Chatwal, G. R., & Anand, S. K. (2002). Instrumental method of Chemical Analysis: Thermal Methods (5th ed.). Himalaya Publishing House.

Chauhan, A., & Chauhan, P. (2014). Article. *Journal of Analytical & Bioanalytical Techniques*, *5*(5), 1–5.

Coats, A. W., & Redfren, J. P. (1963). Thermogravimetric Analysis. *RE:view*, *88*, 906–924.

Cullity, B. D., & Graham, C. D. (2009). *Introduction to Magnetic Materials*. John Wiley & Sons.

Freundlich, H. M. F. (1906). Über die adsorption in Lösungen. Z. *Journal of Physical Chemistry*, *57*, 385–470.

Hatakeyama, T., & Quinn, F. X. (1999). Thermal Analysis Fundamental and Applications to Polymer Science: Thermogravimetry (2nd ed.). John Wiley and Sons Publications.

Howe, J. M., Fultz, B., & Miao, S. (2012). Characterization of Materials. John Wiley & Sons, Inc.

Jiles, D. (1998). *Introduction to Magnetism and Magnetic Materials*. Chapman & Hall.

Langmuir, I. (1916). The constitution and fundamental properties of solids and liquids. *Journal of the American Chemical Society*, *38*(11), 2221–2295. doi:10.1021/ja02268a002

Moody, N., & Barron, A. R. (2015). *UV-Visible Spectroscopy of Noble Metal Nanoparticles*. OpenStax-CNX module: m55548.

Particle Sizing using image. (2013). Retrieved from http://mesa.ac.nz/2013/03/particle-sizing-using-imagej/

Pungor, E. (1995). A Practical Guide to Instrumental Analysis. Academic Press.

Reetz, M. T., Maase, M., Schilling, T., & Tesche, B. (2000). Computer Image Processing of Transmission Electron Micrograph Pictures as a Fast and Reliable Tool To Analyze the Size of Nanoparticles. *The Journal of Physical Chemistry B*, *104*(37), 8779–8781. doi:10.1021/jp000328e

Wickramasinghe, H. K., Martin, Y., & Williams, C. C. (1987). Article . *Journal of Applied Physics*, *61*, 4723. doi:10.1063/1.338807

Wikipedia. (2018a). *Differential Scanning Calorimetry*. Retrieved from https://en.wikipedia.org/wiki/Differential_scanning_calorimetry

Wikipedia. (2018b). *Vibrating Sample Magnetometer*. Retrieved from https://en.wikipedia.org/wiki/Vibrating_sample_magnetometer

Wikipedia. (2018c). *Ultraviolet Visible Spectroscopy*. Retrieved from https://en.wikipedia.org/wiki/Ultraviolet%E2%80%93visible_spectroscopy

Woehrle, G. H., Hutchison, J. E., Ozkar, S., & Finke, R. G. (2006). Article . *Turkish Journal of Chemistry*, *30*, 1–13.

Chapter 4
Green Approaches to Environmental Sustainability

Lubna Tahir
Pakistan Council of Scientific and Industrial Research, Pakistan

ABSTRACT

The development of different eco-friendly technologies in material synthesis is playing its role to expand their biological applications. Nanoparticles are synthesized chemically, but because of their hazardous effects, the researchers have now turned to biological systems for inspiration. There is a growing concern about the possible results of effects of chemical synthesis of nanoparticles. Therefore, the aim of this chapter is to provide an overview of how nanomaterials can be synthesized using green route to play its role in environmental sustainability as green synthesis processes are considered to be a safe alternative to routine processes due to their cost effectiveness, easy to handle, and environmentally friendly nature. In fact, it opens a new scope for phyto-chemists by discouraging the use of toxic chemicals.

INTRODUCTION

The word environment is usually associated with the impact of human on natural system and this then differentiates it from the commonly used word ecological, which is considered as a concept of interdependence of elements within the system. Basically the ecological definition of sustainability is a biological conservation. So, ecological sustainability is basically to meet the human needs without compromising the ecosystems health (Morelli, 2013). Recently, the importance of sustainable development and environmental issues has increased both in the developing and developed nations (Sharma & Gupta, 2015). The total material health of humanity was improved over the past century (Steffen et al., 2011), but in present century we are facing the scarcity of resources, erosion of our earth capability to absorb our wastes and degradation of ecosystem services and the issue of equity is still difficult to solve and the situation is a threat to flexibility of our earth.

The advance of eco-friendly technologies in material synthesis is significantly important in order to enhance their biological applications. Today, variety of inorganic nanoparticles have been synthesized with distinct chemical composition, morphology and size using different microorganisms along-with

DOI: 10.4018/978-1-5225-5745-6.ch004

their applications in various cutting edge technology have also been explored (Li, Xu, Chen, & Chen, 2011) . Although nanomaterials maybe synthesized using different chemical methods, but now it is possible to include the use of biological materials (Mohanpuria, Rana, & Yadav, 2008). To synthesize the nanoparticles using the green route really opens a new scope for the phyto-chemist and at the same rime discourages the use of different toxic chemicals (Kumar, Smita, Cumbal, & Debut, 2017). This green synthesis is regarded as safer alternative to usual chemical, microbial and physical methods due to easy to handle, environmental friendly and cost-effective nature (Saha, Begum, Mukherjee, & Kumar, 2017). In the past few years the experimental and theoretical research on biological nanomaterials has attained all the attention with emphasis on biophysics, biochemistry, electromagnetism and thermodynamics and biomechanics of protein and nucleic acid and its composites are playing the lead role in material and life sciences. (Qin, Liu, & Dong, 2017).

In the present chapter we will discuss the environmental sustainability, factors effecting the environment, comparison between the chemical and biosynthetic route of nanoparticle synthesis, synthesis of nanoparticles using green routes with brief light on different fields of applications.

BACKGROUND

Biosynthesis of nanoparticles by plant extracts is currently under exploitation. Nanoparticles are a class of materials with properties distinctively different from their bulk and molecular counterparts (Biswas & Wu, 2005). Plant extracts are cost effective and eco-friendly and they are the efficient alternative for the large scale production of nanoparticles (Dubey, Bhadauria, & Kushwah, 2009). With the development of different physical and chemical methods, the concern for environmental contaminations are heightened as the chemical procedures involved in the synthesis of nanomaterials generate a large amount of hazardous byproduct emphasizing a need for 'green chemistry' that includes a clean, nontoxic and environment-friendly method of nanoparticle synthesis (Mukherjee et al., 2001). Nanotechnology has become a popular technology in recent years and addresses nanoparticles that are atomic or molecular aggregates characterized by size of less than 100 nm (Fatimah, 2016). Due to the unique features and applications of the nanoparticles, they are very useful especially in the field of medical imaging, biotechnology, and catalysts (Bagherzade, Tavakoli, & Namaei, 2017).

In recent years different environmentally friendly methods have been employed in the synthesis of nanoparticles, and in this regard over the last few decades the use of plants has been growing increasingly in the world (Balayssac et al., 2009). The principles of green chemistry have been spreading since the mid-1990s, connected with advances in nanomaterial synthesis and lately these two groups have begun to significantly converge. Nanomaterial synthesis groups are developing greener, more sustainable production methods, while nanoparticle application groups are exploring sustainable energy sources and environmental remediation as end goals (Murphy, 2008).

There are different chemical and physical methods that have been exploited in the synthesis of several inorganic metal nanoparticles by using different methods like, ultraviolet irradiation, wet and dry approaches viz, lithography, aerosol technologies, laser ablation, ultrasonic fields, and photochemical reduction techniques. But these methodologies are still expensive and they involve the use of hazardous chemicals. Hence, there is a growing concern for the development of alternative environment friendly and sustainable methods. Increasing awareness towards green chemistry and biological processes has led to a necessity to develop simple, cost-effective and eco-friendly procedures (Narayanan & Sakthivel, 2011)

WHAT IS ENVIRONMENTAL SUSTAINABILITY AND HOW IT IS AFFECTED BY HUMAN ACTIVITIES

Basically the natural environment is a combination of different chemical, biological and physical components and different interactions among them. In this contest the term 'sustainability' infers the meeting of present demands nut at same time maintaining the environments ability to meet the future demand as well. The basic approach is to have a balance of competing demands through establishment of interconnections and pliability or some kind of link among different components (Lal, 2016). We need to pinpoint those factors, processes and different causes that are in actual affecting the sustainability of any environment under conditions that are site- specific and properly addressed by interrupting science into action using government policies. In this regard, the management and land use are two very important determinants of these factors, process and causes.

Morelli (2011) defined environmental sustainability as conservation of natural capital and also as a concept that is even though separate but still connected both with economic and social sustainability. The management of natural resources is basically a driver and an outcome of economic, ecological and social dynamics as are recognized by applied ecology and conservation biology. Protected areas actually offer a central approach to conservation of ecosystem but in actual they are also the socio-ecological systems whose sustainability and ecological management are affected by people (Cumming & Allen, 2017).

Globally, sustainability is progressively understood as a criterion to achieve the human development (GSP) at all levels, starting from local communities to different cities, to nations and then to world (Folke, Hahn, Olsson, & Norberg, 2005). It is because we entered into a new geological period, the Anthropocene, in which the global environmental risks are increasing because of human pressures and for the very first time it become the reason of the largest planetary changes (Steffen, Crutzen, & McNeill, 2007)

Various pollutants that are released in the environment because of human activities are briefly summarized in Table 1, which clearly indicates that the energy supply is very important both in the traditional and industrial, in the movement of toxic substances like cadmium, Sulphur oxides as well as the principle greenhouse gas the carbon dioxide. There is a picture of human disruption index as well for each substance, which is actually the ratio of amount release by human activities to the natural resources present. So, the energy systems along with human activities is seriously affecting the cycling of important chemicals in the ecosystem (Holdren et al., 2000). In fact the accidental and routine release of all the pollutants in the environment is the most important insult, and human activities are rated far beyond the natural flow of all the climatologically and biologically active elements in it. So, if these problems are taken under control, the world will move towards the sustainable energy future in ways of environment.

NANOTECHNOLOGY AND NANOMATERIALS

Nanotechnology, an emerging new field in research that deals with the synthesis of nanomaterials and nanoparticles and they find their applications in different field like catalysis, pharmaceutics, cosmetics, electrochemistry, food technology, biomedicine, space technology, textile industry, energy science, biomedicines etc (Abdelghany et al., 2017; Hussein, 2016). In fact nanotechnology has changed our expectations, vision, and abilities to control the material world. By definition the particles whose sizes are in the dimension of 1-100nm are known as nanoparticles and because of their size and morphology

Table 1. Environmental insults due to human activities by sector, mid-1990s

Insult	Natural Baseline Tones in Year	Human Disruption Index	Share of Human Disruption Caused by			
			Commercial Energy Supply	Traditional Energy Supply	Agriculture	Manufactures Others
Lead emission to atmosphere	12,000	18	41% (fossil fuel burning, including additives	Negligible	Negligible	59% (metal processing, manufacturing, refuse burning
Sulphur emission	31 million	2.7	85% (fossil fuel burning)	0.5% (traditional fuel burning)	1% (agricultural burning)	13% (smelting, refuse burning)
Nitrous oxide flow	33 million	0.5	12% (fossil fuel burning)	8% (traditional fuel burning)	80% (fertiliser, land clearing, aquifer disruption)	Negligible
Oil added to rivers	200,000	10	44% (petroleum harvesting, processing, and transport)	Negligible	Negligible	56% (disposal of oil wastes, including motor oil changes)
Carbon dioxide flow in air	150 billion	0.05	75% (fossil fuel burning)	3% (net deforestation for fuelwood)	15% (net deforestation for land clearing)	7% (net deforestation for lumber, cement manufacturing)
Mercury emission	2500	1.4	20% (fossil fuel burning)	1% (traditional fuel burning)	2% (agricultural burning)	77% (metals processing, manufacturing, refuse burning)
Methane flow	160 million	2.3	18% (fossil fuel harvesting and processing)	5% (traditional fuel burning)	65% (rice paddies, domestic animals, land clearing)	12% (landfills)
Nitrogen fixation	140 million (nitrogen)	1.5	30% (fossil fuel burning)	2% (traditional fuel burning)	67% (fertiliser, agricultural burning	1% (refuse burning)
Non-methane hydrocarbon emissions to atmosphere	1,000 million	0.12	35% (fossil fuel processing and burning)	5% (traditional fuel burning)	40% (agricultural burning)	20% (nonagricultural land clearing, refuse burning)
Cadmium emission	1400	5.4	13% (fossil fuel burning)	5% (traditional fuel burning)	12% (agricultural burning)	70% (metals processing, manufacturing, refuse burning)

(Holdren et al., 2000)

these particles show enhanced properties. Nanotechnology has been emerging strongly and plays a main role in daily human life because of the impact it is bringing due to its various applications (Saratale et al., 2017).

Nanoparticles can be synthesized by using different methodologies. Traditionally there are different methods like irradiation, aerosol technologies, ultrasound field's photochemical reduction, lithography for preparation of nanoparticles. Among them the chemical methods include, using a reducing agent, colloidal chemical method, sol-gel method and physical methods include sonication, thermal decomposition and sputtering (Ahmad, Senapati, Khan, Kumar, & Sastry, 2003; Sharma, Akhtar, Ameen, Srivastava, & Singh, 2015).The two very common approaches are the bottom up and top down. Following figure (Figure 1) can illustrate these two methods clearly

Nanocomponents are assimilated from building blocks of their respective precursors in a solid, liquid or gas phase by using either physical or chemical deposition processes in bottom up method (Wang & Xia, 2004), whereas in top-down method, the nanoscale structures are engraved onto a substrate with the help of electron beams and by applying proper etching and deposition procedure (Shamaila et al., 2016; Shankar, Jaiswal, Aparna, & Prasad, 2014).

The possible mechanism of synthesis of nanoparticle using bottom up approach is shown in Figure 2.

Nanotechnology, with great ability to improve water, air and soil quality in environment can help in the improvement of sensing and detection of different pollutants in the environment and help in developing different new technologies for their remediation. In fact understanding the process of nanoparticles synthesis will allow us to develop effective and efficient methods for reducing the formation of pollutants (Figure 3) in the very first place. (Masciangioli & Zhang, 2003)

Figure 1. Synthesis of nanoparticles using bottom up/top down approach

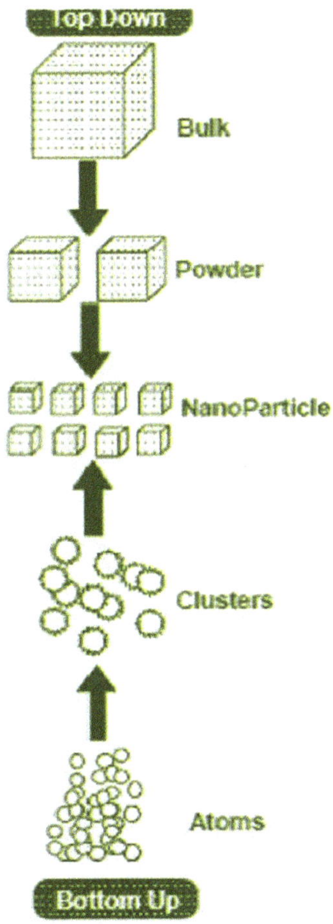

Figure 2. Mechanism of synthesis of nanoparticle (M+-metal ion)

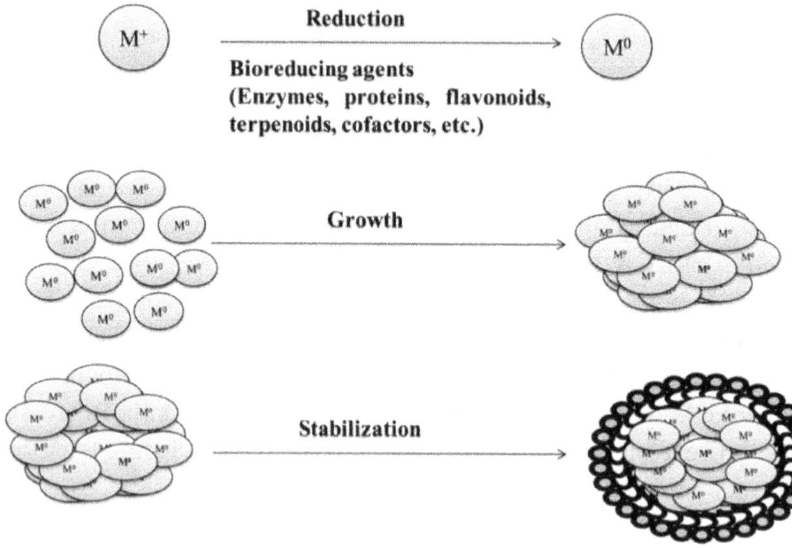

Figure 3. Nano particles in the environment (Biswas & Wu, 2005)

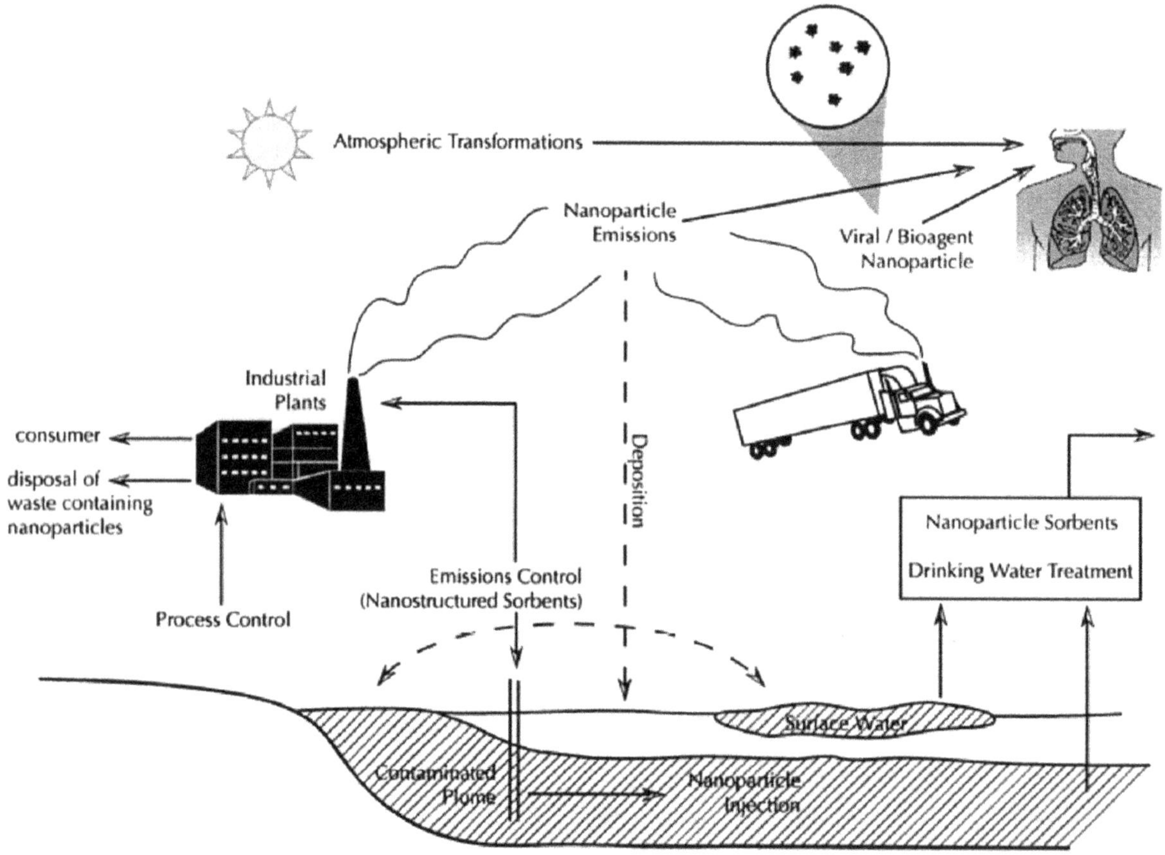

NANOTECHNOLOGY AND GREEN CHEMISTRY

As a result of worldwide problems that are linked with various environmental contaminations, the environmental friendly, greener processes in chemical technology and chemistry are getting popular as they are need of time. In the last two years many books on 'green chemistry' have been published which describe in general terms the green processes (Ahluwalia, 2006; Anastas, Boethling, & Voutchkova-Kostal, 2014; Patel, Patel, & Raval, 2012) and their specialized characteristics that includes microwaves, ultrasound, and different other methods involved in the green synthesis (D. Chen, Sharma, & Mudhoo, 2011). Different research work describes the green synthesis of α-Fe2O3, CuS, FeO, Pd, Ag, ZnO, Cu, graphene and Fe based nanoparticles which find their applications in parasitology, pharmacology and entomology.

Green nanotechnology, a multidisciplinary field, emerged as a rapidly developing area of research that serve as a way of developing procedures that are non-toxic, clean and environment friendly as compare to those of currently employed chemical and physical methods for synthesis of nanoparticles. These biogenic routes are called green routes because they do not involve any toxic chemicals or high energy inputs for their synthesis and can lead to production of nanoparticles with unique sizes, shape and different bioactivity. The beauty of the green fabricated nanoparticles have capacitated their possible applications in many sectors (Saratale et al., 2017).

Different aerosol technologies, laser ablation, UV irradiation, lithography and photochemical reduction techniques have been successfully employed to synthesize the nanoparticles but they involve the use of different harmful chemicals and they remain expensive. This, therefor demand and provoke a significant interest in developing some sustainable and environmental friendly method for their synthesis (Narayanan & Sakthivel, 2010). For this, closed reactors, low temperatures, non- toxic solvents (usually water) and different green techniques, without contacting the reaction media and air can be used. Green synthesis of nanomaterials using different plant materials is emerging as a field in nanoscience, emphasizing to avoid the use of various hazardous chemicals and help in supporting development of ecofriendly technique (Kumar et al., 2017). In developing less toxic chemical synthesis, the 12 principles of green chemistry are becoming a guideline for the chemical technologists worldwide (Anastas & Warner, 2000). Even though rapid development in nanotechnology is occurring but still the harmful effects of nanomaterials are somewhat unknown, and to synthesize such materials using some biocompatible and environmentally friendly reagents could in fact, lower the harmfulness of the resulting materials and their byproducts on the environment (Sanchez-Mendieta & Vilchis-Nestor, 2012; Varma, 2012). Therefore, there is a growing concern to synthesize the nanoparticles using some environmental friendly methods.

NANOPARTICLES: GREEN SYNTHESIS - WHY DO WE NEED IT?

In traditional eastern medicine, especially in India and China, plants have been used to cure different physiological disorders (Kuppusamy, Yusoff, Maniam, & Govindan, 2016).

Vitamins, plant extracts, microorganisms, sugar and biodegradable polymers are some of the naturally occurring reagents that are considered attractive as capping and reductant agents for obtaining nanoparticles and this had led to the assembly of limited number of different inorganic nanoparticles, mainly the metal nanoparticles even though different metals and salts nanoparticles are reported already (Siavash Iravani, 2011). Copper (Cu), nickel (Ni), gold (Au), platinum (Pt), selenium (Se), titanium (Ti), silver (Ag), and zinc nanoparticles (ZiNPs) are synthesized using plant resources (Kuppusamy et al., 2016).

Among these reagents the plant based materials are considered to be the best for nanoparticles synthesis especially for large-scale biosynthesis. Different biomolecules that are present in the plants can be used as reducing agents to reduce the metal ions to nanoparticles in a one- step green synthesis process. This biogenic reduction process is very rapid and can be conducted easily at room temperature and pressure and can be easily scaled up. This kind of plant base synthesis is environmentally benign. The reducing agents involved from plants include water soluble plants secondary metabolites (e.g. flavonoids, alkaloids, terpenoids, and phenolic compounds) and different co-enzymes. Gold (Au) and silver (Ag) nanoparticles are considered important for plant based syntheses. Extracts of different parts of different plants and even the live plants have been used in making nanoparticles (Mittal, Chisti, & Banerjee, 2013). Exploiting botanical extracts as chemical reagents in synthesis of nanoparticle sometimes complicates identification of reactive species, in situ surface-pacification usually occurs in order to give the nanoparticles that are biocompatible for use in biological, clinical and environment testing (Benelli & Lukehart, 2017).

The proteins and the nucleic acids are significant nanocomponents that perform different life functions and are considered to be the best natural biological nanomaterials. Their study not only involve the structure and function of genes and proteins but also the release of specific factors, their binding and recognition and release of biochemical and bio-electrochemical signals that are involved in the development of new technologies. Even though different issues still need to be evaluated but development of different new materials with both biological and nanotech qualities has really made a big progress (Qin et al., 2017).

Recent studies are mainly focused on greener methods using non-toxic aqueous medium for production of large amount of nanoparticles. Extracts of different parts of the plant like leaf, fruit, seed (Kumar, Pammi, Kollu, Satyanarayana, & Shameem, 2014), bark (Mehmood et al., 2014), peel (Kumar et al., 2015b), and root (Shameli et al., 2012) work very efficiently in the synthesis of green nanoparticles.

PREPARATION OF NANOPARTICLES USING PLANT EXTRACT

Many studies reported the synthesis of nanoparticles using physical and chemical approaches to produces different metal nanoparticles (Ahmad et al., 2003), but the synthesis of metal oxide and metal nanoparticles using plants and different phytochemicals present in plant are considered to be advantageous because of its simplicity, cost effectiveness, availability, using non-toxic chemicals, reproducibility, higher efficiency, environment friendly and production of stable materials as it does not require the strong maintenance of laboratory cultures used for synthesis of nanoparticles (Benelli & Lukehart, 2017; Saratale et al., 2017). Extracts from leaf, fruit (juices, pericarp), root and stems are used to synthesize the nanoparticles where they actually act as a reducing / stabilizing agents. In particular, the main role was observed by polyphenols from extracts used in nano-synthesis but other secondary metabolites like flavonoids, alkaloids, saponins, tannins. Carboxylic acids, ascorbic acid, amino acids were also found to be involved in the capping, reduction and stabilization of metal nanoparticles (Borase et al., 2014; S Iravani, Korbekandi, Mirmohammadi, & Zolfaghari, 2014; Prasad, Kambala, & Naidu, 2013) and because of these biologic constituents the synthesized nanoparticles showed significant antidiabetic, antioxidant radical scavenging and antimicrobial activities against different pathogens. Structures of some of the secondary metabolites found in extracts of different parts of plants used in nano-synthesis are shown in Figure 4.

Plants because of rich genetic variability have many interesting biomolecules in different forms like vitamin based intermediates, coenzymes and many others which then reduce the metal ions to nanopar-

Figure 4. Structure of compounds found in different parts of plant extracts

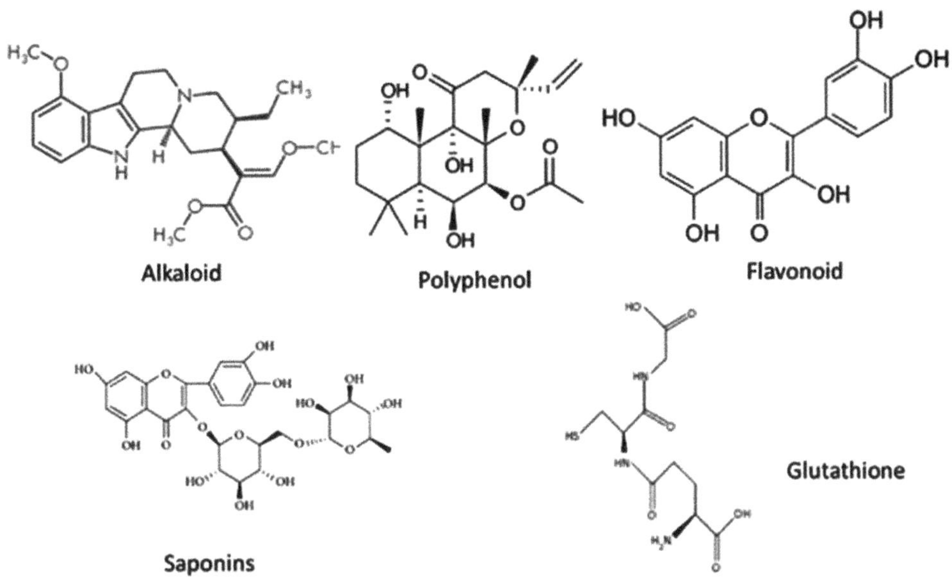

ticles in one step process. The advantage is that these methods can be conducted at room temperature and pressure without using any hard and fast technical requirements. Moreover these plant based approaches for nanoparticle synthesis are easy to scale up and are favored as they are environment friendly. The main driving force behind the attraction of synthesis of plant based nanoparticles is that this route of synthesis actually allows the products to be exploited for different applications especially those of nanomedicine based innovation. Another important feature of these nanoparticles that is playing its role in making them famous is that these materials are easy and inexpensive to be cultured as compare to those of microorganisms. The ease of procedural and result based advantages along with fairly quicker application administration which make these plants better and most favored destinations. There are studies that report the use of whole plant based material in the synthesis of nanoparticles (Mubarak Ali, Thajuddin, Jeganathan, & Gunasekaran, 2011; Park, Hong, Weyers, Kim, & Linhardt, 2011). Some of the plants that are commonly used in the synthesis of nanoparticles used for different purposes are like *Aloe vera, Ocimum tenuiflorum, Camellia sinensis, Diopyros kaki, Jatropha curcas L, Glycine max, Arachis hypogaea L, Solanum trilobatum, Pelargonium graveolens* etc. Figure 5 represent some of the plants used in nanoparticle synthesis.

The production of nanoparticles from plants involved some very common steps. The plant extract (organic solvent, aqueous) is mixed with a solution of desired metal salt at room temperature. The reaction completes within a few minutes and as a result of biochemical reduction the used metal is converted into zero-valent states from there mono or divalent oxidation states. This is the formation of nanoparticles which is also physically indicated by change of color that is observed in culture medium vessel. Similarly synthesis of silver, gold and many other metals are reported in the same way and these plant based reductions are so adaptable in nature that silver nanoparticles are produced from some of the most common plants like *Azadirachta indica* (neem) and *Ocimum tenuiflorum* (tulsi), that are common in almost every household. These bioreductions are brought about because of different compounds in plants extracts. The plant driven synthesis of nanoparticles is so selective and specific that it can either be in

Figure 5. Different plant species used to make metal nanoparticles

the form of aggregation mediated by self-assembly or stabilization of nanoparticle by their production in a control environment in which they can be without any alterations in their stabilities. This quality can be very useful when we are dealing with assembly of integrated nanostructures (Malik, Shankar, Malik, Sharma, & Mukherjee, 2014). As an example we will discuss in detail the green route synthesis of silver nanoparticle using plant extract.

Methodology

Preparation of Plant Extracts and Silver Nanoparticles

Collect any part of plant like roots, stem, leaves or whole plant. In order to remove any contamination or dust from the plant first wash the plant with tap water several times and then wash it carefully with de-ionized water. After washing weigh specific amount of Plant and cut them into small pieces and then soak them into water or solvent of your choice. Let the plant be soaked in the solvent for some time or place it on shaking bath at certain temperature to get maximum nutrients/ compounds in the solvent. After some time filtered the extract and can be stored in refrigerator at 4 °C for the further use. Then take some required amount of plant extract (5, 10, 15 mL) and add the prepared solution of silver nitrate (Ag NO3) drop by drop. Stirred the reaction mixture of the extract and the AgNO3 for some time depending upon the conditions (2-5min) and observed visually the color change of the reaction mixture from colorless to dark brown etc. Different conditions like reaction temperature, concentration and pH of the reaction mixture and the ratio of the reactants have to be considered to get the silver nanoparticles with tunable size and morphology(Tripathy, Raichur, Chandrasekaran, Prathna, & Mukherjee, 2010)

Characterization

The X-ray diffraction pattern will be recorded using an X-ray diffractometer (PANalytical-PW 340/60 Xpert PRO) and the optical absorption spectrum will be obtained using UV–vis–NIR spectrophotometer (Perkin Elmer Lambda 35 model). The morphology of the prepared samples was observed by using transmission electron microscope (TEM, Hitachi H-7100) and energy dispersive X-ray analysis (EDAX) was used to do elemental analyses. Fourier transform infrared (FTIR) spectrum was measured using Perkin-Elmer RX-I FTIR spectrophotometer(Gomathi, Rajkumar, Prakasam, & Ravichandran, 2017)

Antibacterial Activity of Ag NPs

The antimicrobial activities of the plant extract and silver nanoparticles was measured against any pathogenic microbes like *E.coli*, *P. aeruginosa*, *S. aureus*, *K. pneumoniae*, *M. tuberculosis* etc. using either the agar well method or the disc diffusion assay. In disc diffusion method, the discs were soaked in double distillation water, the plant extract and the silver nanoparticle solution separately. Different concentrations of silver nanoparticles were used in order to make sure the specification of antibacterial activity. These discs were air dried in sterile conditions before they are placed in different agar media that contain the microbial culture. After that the plates containing both media and cultures were then divided into four parts and already prepared discs were then placed on each plate. The water disc will be considered as negative control and as a positive control we can use any antibiotic for example chloramphenicol. The plates were then carefully incubated at 37ºC for 24 hours. After 24 hours the maximum zone of inhibition was observed and calculated for the analysis of activity against each pathogen. (Fatimah, 2016).

APPLICATIONS OF BIOSYNTHESIZED NANOPARTICLES

The unique and fascinating characteristics of different types of green synthesized nanoparticles have found numerous applications in different fields like biomedicines, food sciences, water treatment, agriculture and environmental engineering (Barabadi et al., 2017).we will discuss some of them in little detail below

Medicine/ Drug Delivery

Nanotechnology is playing its role in medicine and specifically in drug delivery and even in drug delivery the focus is mainly on cancer therapy. The ability of the nanoparticles to cross the blood brain barrier actually opens the new ways of drug delivery to brain and also to various cellular compartment and even nucleus (De Jong & Borm, 2008). Along with this the drug itself maybe formulated at a nanoscale and become its own carrier (Baran & Hasirci, 2002; Cascone, Lazzeri, Carmignani, & Zhu, 2002; Kipp, 2004). Recently green route in different substrates like plant extracts, enzymes etc is used to synthesize the zinc oxide nanoparticles (ZnO-NPs) to treat a range of different skin conditions and have anticancer properties.(Mirzaei & Darroudi, 2017). Different plant extracts are used to make an antiseptic bandages to treat the external wounds (Tahir, Nazir, & Khan, 2013). Similarly the antimicrobial efficacy of silver nanoparticles synthesized using leaf extract of *Corchorus capsularis* was effective against different resistant isolates from post-surgical wound infections (Kasithevar, Periakaruppan, Muthupandian, &

Mohan, 2017).The biologically synthesized metal nanoparticles have showed potent anticancer activity as well (Jacob, Finub, & Narayanan, 2012; Ramkumar et al., 2017; Saratale et al., 2017)

Water Treatment

New technologies for efficient disinfection and microbial control is needed to achieve the appropriate disinfection without any harmful byproducts and for this different natural and engineered nanomaterials have shown a strong antimicrobial activity through different mechanisms that destroy the cell components, affecting bacterial cell wall, and inhibition of enzymes and DNA synthesis (Li et al., 2008). The Fe3O4 nanocrystals and magnetic separations offer a promising method for arsenic removal (Mayo et al., 2007)

Removal of toxic materials from wastewater is necessary for health and environmental protection. A variety of efficient, low cost and eco-friendly nanomaterials using different plant extracts with different functionalities have been suggested for their applications in purification of ground water, industrial effluents, drinking and surface water (Dil, Ghaedi, & Asfaram, 2017; Savage & Diallo, 2005; Theron, Walker, & Cloete, 2008)

Agriculture/Food

Efforts for application of nanotechnology in agriculture began because of the realization that the traditional technologies would neither restore the damage to ecosystem due to use of fertilizers, irrigation and pesticides nor increase productivity (Mukhopadhyay, 2014). Nanoparticles synthesized using green routes are used to control the disease causing pathogens in plants by offering an easy, environmental friendly and cost effective approach. They can either be sprayed directly on plants or they can be developed as carriers for fertilizers, pesticides and herbicides (Alghuthaymi, Almoammar, Rai, Said-Galiev, & Abd-Elsalam, 2015; Bouwmeester et al., 2009). As the traditional strategies used for pest management are insufficient and affect the humans apart from affecting soil fertility. In this scenario the green nanotechnology is providing the efficient alternatives for the insect pest management without harming the environment (Rai & Ingle, 2012).

In food industry it includes the enhancement of shelf life, improved food storage, encapsulation and delivery of substances in targeted sites and brand protection etc. Similarly the nano barcodes are used for labelling and monitoring of different food products and nano supplements are also incorporated by different encapsulation techniques for the enhancement of nutrition (Chellaram et al., 2014). Figure 6 is reflecting the applications of nanotechnology in different sectors of food sciences and technology.

IMPLICATIONS

Sometimes nanoparticles are engineered using different physical and chemical methods which make them harmful (Gottschalk & Nowack, 2011) reported that even though the release of engineered nanoparticles (ENP) in the environment is much less than the natural nanoparticles that are released but still they pose great danger not to the environment but also to living organisms (Huang, Fan, Grusak, Sherrier, & Huang, 2014) because of the presences of harmful chemicals that actually are not present in the natural nanoparticles (NPs) (Handy, Owen, & Valsami-Jones, 2008) and presences of different organic pollut-

Figure 6. Applications of nanotechnology in food sciences ((Ravichandran, 2010)

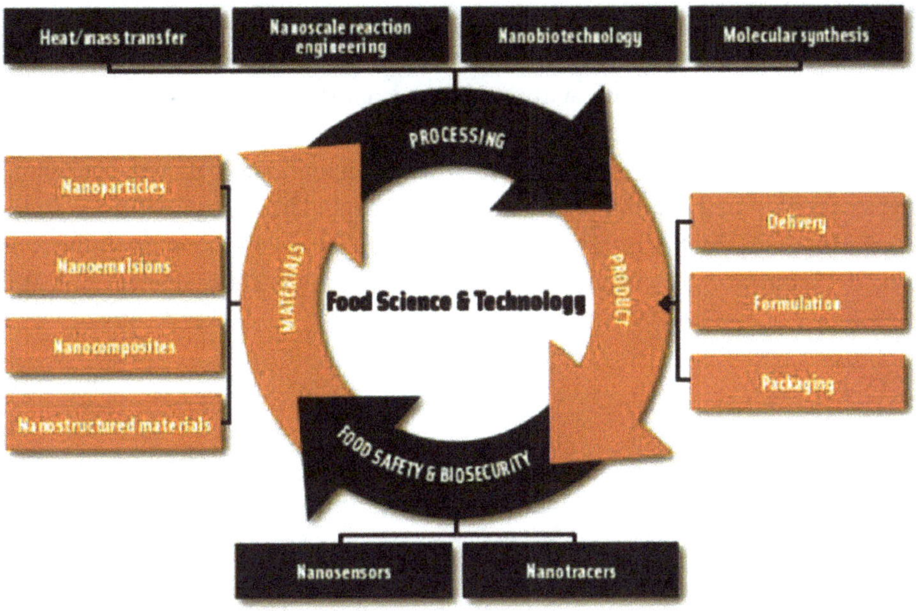

ants and toxic chemicals that are already present in the environment (Figure 7) make them more toxic as they(ENP) react with them (Brar, Verma, Tyagi, & Surampalli, 2010). Once in the environment these ENP, they also enter in plants using different routes like leaves, roots and stem (Cifuentes et al., 2010; Hong et al., 2014; Lin & Xing, 2007; X. Wang et al., 2012) and they can be transported inside plants through vascular pathways.

Figure 7. Release of ENPs from different sources into the environment

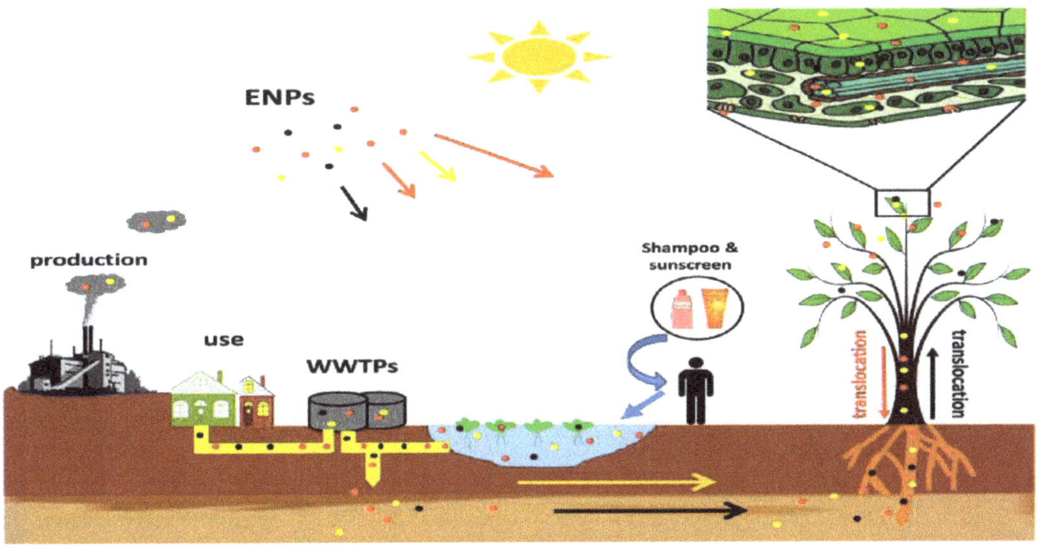

Although nanotechnology is very promising for different sectors but still scientist agree that the development in this field should be done in a very consistent manner and government should play a major role in this through existing regulations and the development authorities must have the new procedures to have the sustainability of the procedures (Bergeson, 2013). Researchers are studying the toxic effect for both the environment and organisms within that environment when these nanomaterials reach the nanobio interface. Focus is also on the release of ENP, how they accumulate and the types and number of ENP we are exposed to. Either the nanoparticles are used intentionally during agriculture process or their presences as a contaminate, important thing is to understand how they will affect the surroundings (Chen, Seiber, & Hotze, 2014). Currently there is low risk of exposure to nanoparticles (Batley, Kirby, & McLaughlin, 2012) but work is still needed to study the toxicity under real environmental conditions.

CONCLUSION

The use of different plant extracts to make metal nanoparticles is easily scaled up, cheap, and environmental friendly way. The plant based nanoparticles are of control size and morphology, but still more research is needed to understand and improve the applications of these nanoparticles. Further applications in the field of drug delivery and clinical diagnostics still need to be improved. The possible threats to environment and human health caused because of nanoparticles increases the concerns and therefore require the full understanding of the toxicity and mechanism of actions of these nanoparticles. In order to achieve the large scale nanoparticles production in a cost effective way research is needed to be done in the field of isolation and purification of plant metabolites in large scale production (Saratale et al., 2017) and lastly this is an important area of research and with constant efforts we can implement and utilize these technologies for scale up and commercial applications in different fields.

ACKNOWLEDGMENT

This research received no specific grant from any funding agency in the public, commercial or not for profit sector.

REFERENCES

Abdelghany, T. M., Al-Rajhi, A. M. H., Al Abboud, M. A., Alawlaqi, M. M., Ganash Magdah, A., Helmy, E. A. M., & Mabrouk, A. S. (2017). Recent Advances in Green Synthesis of Silver Nanoparticles and Their Applications: About Future Directions. A Review. *BioNanoScience*. doi:10.100712668-017-0413-3

Ahluwalia, V. (2006). *Green chemistry: Environmentally benign reaction*. Ane Books Pvt Ltd.

Ahmad, A., Senapati, S., Khan, M. I., Kumar, R., & Sastry, M. (2003). Extracellular biosynthesis of monodisperse gold nanoparticles by a novel extremophilic actinomycete, Thermomonospora sp. *Langmuir*, *19*(8), 3550–3553. doi:10.1021/la0267721

Alghuthaymi, M. A., Almoammar, H., Rai, M., Said-Galiev, E., & Abd-Elsalam, K. A. (2015). Myconanoparticles: Synthesis and their role in phytopathogens management. *Biotechnology, Biotechnological Equipment, 29*(2), 221–236. doi:10.1080/13102818.2015.1008194 PMID:26019636

Anastas, P. T., Boethling, R., & Voutchkova-Kostal, A. (2014). *Handbook of Green Chemistry, Green Processes, Designing Safer Chemicals* (Vol. 9). John Wiley & Sons.

Anastas, P. T., & Warner, J. C. (2000). *Green chemistry: theory and practice*. Oxford university press.

Bagherzade, G., Tavakoli, M. M., & Namaei, M. H. (2017). Green synthesis of silver nanoparticles using aqueous extract of saffron (Crocus sativus L.) wastages and its antibacterial activity against six bacteria. *Asian Pacific Journal of Tropical Biomedicine, 7*(3), 227–233. doi:10.1016/j.apjtb.2016.12.014

Balayssac, S., Trefi, S., Gilard, V., Malet-Martino, M., Martino, R., & Delsuc, M.-A. (2009). 2D and 3D DOSY 1 H NMR, a useful tool for analysis of complex mixtures: Application to herbal drugs or dietary supplements for erectile dysfunction. *Journal of Pharmaceutical and Biomedical Analysis, 50*(4), 602–612. doi:10.1016/j.jpba.2008.10.034 PMID:19108978

Barabadi, H., Honary, S., Mohammadi, M. A., Ahmadpour, E., Rahimi, M. T., Alizadeh, A., ... Saravanan, M. (2017). Green chemical synthesis of gold nanoparticles by using Penicillium aculeatum and their scolicidal activity against hydatid cyst protoscolices of Echinococcus granulosus. *Environmental Science and Pollution Research International, 24*(6), 5800–5810. doi:10.100711356-016-8291-8 PMID:28054267

Baran, E., & Hasirci, V. (2002). In vivo half life of nanoencapsulated L-asparaginase. *Journal of Materials Science. Materials in Medicine, 13*(12), 1113–1121. doi:10.1023/A:1021125617828 PMID:15348652

Batley, G. E., Kirby, J. K., & McLaughlin, M. J. (2012). Fate and risks of nanomaterials in aquatic and terrestrial environments. *Accounts of Chemical Research, 46*(3), 854–862. doi:10.1021/ar2003368 PMID:22759090

Benelli, G., & Lukehart, C. M. (2017). Special Issue: Applications of Green-Synthesized Nanoparticles in Pharmacology, Parasitology and Entomology. *Journal of Cluster Science, 28*(1), 1–2. doi:10.100710876-017-1165-5

Bergeson, L. L. (2013). *Sustainable nanomaterials: emerging governance systems*. ACS Publications.

Biswas, P., & Wu, C.-Y. (2005). Nanoparticles and the Environment. *Journal of the Air & Waste Management Association, 55*(6), 708–746. doi:10.1080/10473289.2005.10464656 PMID:16022411

Borase, H. P., Salunke, B. K., Salunkhe, R. B., Patil, C. D., Hallsworth, J. E., Kim, B. S., & Patil, S. V. (2014). Plant extract: A promising biomatrix for ecofriendly, controlled synthesis of silver nanoparticles. *Applied Biochemistry and Biotechnology, 173*(1), 1–29. doi:10.100712010-014-0831-4 PMID:24622849

Bouwmeester, H., Dekkers, S., Noordam, M. Y., Hagens, W. I., Bulder, A. S., De Heer, C., ... Sips, A. J. (2009). Review of health safety aspects of nanotechnologies in food production. *Regulatory Toxicology and Pharmacology, 53*(1), 52–62. doi:10.1016/j.yrtph.2008.10.008 PMID:19027049

Brar, S. K., Verma, M., Tyagi, R., & Surampalli, R. (2010). Engineered nanoparticles in wastewater and wastewater sludge–Evidence and impacts. *Waste Management (New York, N.Y.), 30*(3), 504–520. doi:10.1016/j.wasman.2009.10.012 PMID:19926463

Cascone, M. G., Lazzeri, L., Carmignani, C., & Zhu, Z. (2002). Gelatin nanoparticles produced by a simple W/O emulsion as delivery system for methotrexate. *Journal of Materials Science. Materials in Medicine*, *13*(5), 523–526. doi:10.1023/A:1014791327253 PMID:15348607

Chellaram, C., Murugaboopathi, G., John, A. A., Sivakumar, R., Ganesan, S., Krithika, S., & Priya, G. (2014). Significance of Nanotechnology in Food Industry. *APCBEE Procedia*, *8*(Supplement C), 109–113. doi:10.1016/j.apcbee.2014.03.010

Chen, D., Sharma, S. K., & Mudhoo, A. (2011). *Handbook on applications of ultrasound: sonochemistry for sustainability*. CRC Press.

Chen, H., Seiber, J. N., & Hotze, M. (2014). ACS Select on Nanotechnology in Food and Agriculture: A Perspective on Implications and Applications. *Journal of Agricultural and Food Chemistry*, *62*(6), 1209–1212. doi:10.1021/jf5002588 PMID:24479582

Cifuentes, Z., Custardoy, L., de la Fuente, J. M., Marquina, C., Ibarra, M. R., Rubiales, D., & Pérez-de-Luque, A. (2010). Absorption and translocation to the aerial part of magnetic carbon-coated nanoparticles through the root of different crop plants. *Journal of Nanobiotechnology*, *8*(1), 26. doi:10.1186/1477-3155-8-26 PMID:21059206

Cumming, G. S., & Allen, C. R. (2017). Protected areas as social-ecological systems: Perspectives from resilience and social-ecological systems theory. *Ecological Applications*, *27*(6), 1709–1717. doi:10.1002/eap.1584 PMID:28618079

De Jong, W. H., & Borm, P. J. A. (2008). Drug delivery and nanoparticles: Applications and hazards. *International Journal of Nanomedicine*, *3*(2), 133–149. doi:10.2147/IJN.S596 PMID:18686775

Dil, E. A., Ghaedi, M., & Asfaram, A. (2017). The performance of nanorods material as adsorbent for removal of azo dyes and heavy metal ions: Application of ultrasound wave, optimization and modeling. *Ultrasonics Sonochemistry*, *34*, 792–802. doi:10.1016/j.ultsonch.2016.07.015 PMID:27773307

Dubey, M., Bhadauria, S., & Kushwah, B. (2009). Green synthesis of nanosilver particles from extract of Eucalyptus hybrida (safeda) leaf. *Digest Journal of Nanomaterials and Biostructures*, *4*(3), 537–543.

Fatimah, I. (2016). Green synthesis of silver nanoparticles using extract of Parkia speciosa Hassk pods assisted by microwave irradiation. *Journal of Advanced Research*, *7*(6), 961–969. doi:10.1016/j.jare.2016.10.002 PMID:27857843

Folke, C., Hahn, T., Olsson, P., & Norberg, J. (2005). Adaptive governance of social-ecological systems. *Annual Review of Environment and Resources*, *30*(1), 441–473. doi:10.1146/annurev.energy.30.050504.144511

Gomathi, M., Rajkumar, P., Prakasam, A., & Ravichandran, K. (2017). *Green synthesis of silver nanoparticles using Datura stramonium leaf extract and assessment of their antibacterial activity*. Resource-Efficient Technologies.

Gottschalk, F., & Nowack, B. (2011). The release of engineered nanomaterials to the environment. *Journal of Environmental Monitoring*, *13*(5), 1145–1155. doi:10.1039/c0em00547a PMID:21387066

Handy, R. D., Owen, R., & Valsami-Jones, E. (2008). The ecotoxicology of nanoparticles and nanomaterials: Current status, knowledge gaps, challenges, and future needs. *Ecotoxicology (London, England)*, *17*(5), 315–325. doi:10.100710646-008-0206-0 PMID:18408994

Holdren, J. P., Smith, K. R., Kjellstrom, T., Streets, D., Wang, X., & Fischer, S. (2000). *Energy, the environment and health*. New York: United Nations Development Programme.

Hong, J., Peralta-Videa, J. R., Rico, C., Sahi, S., Viveros, M. N., Bartonjo, J., ... Gardea-Torresdey, J. L. (2014). Evidence of translocation and physiological impacts of foliar applied CeO2 nanoparticles on cucumber (Cucumis sativus) plants. *Environmental Science & Technology*, *48*(8), 4376–4385. doi:10.1021/es404931g PMID:24625209

Huang, Y. C., Fan, R., Grusak, M. A., Sherrier, J. D., & Huang, C. (2014). Effects of nano-ZnO on the agronomically relevant Rhizobium–legume symbiosis. *The Science of the Total Environment*, *497*, 78–90. doi:10.1016/j.scitotenv.2014.07.100 PMID:25124056

Hussein, A. K. (2016). Applications of nanotechnology to improve the performance of solar collectors–Recent advances and overview. *Renewable & Sustainable Energy Reviews*, *62*, 767–792. doi:10.1016/j.rser.2016.04.050

Iravani, S. (2011). Green synthesis of metal nanoparticles using plants. *Green Chemistry*, *13*(10), 2638–2650. doi:10.1039/c1gc15386b

Iravani, S., Korbekandi, H., Mirmohammadi, S., & Zolfaghari, B. (2014). Synthesis of silver nanoparticles: Chemical, physical and biological methods. *Research in Pharmaceutical Sciences*, *9*(6), 385. PMID:26339255

Jacob, S. J. P., Finub, J., & Narayanan, A. (2012). Synthesis of silver nanoparticles using Piper longum leaf extracts and its cytotoxic activity against Hep-2 cell line. *Colloids and Surfaces. B, Biointerfaces*, *91*, 212–214. doi:10.1016/j.colsurfb.2011.11.001 PMID:22119564

Kasithevar, M., Periakaruppan, P., Muthupandian, S., & Mohan, M. (2017). Antibacterial efficacy of silver nanoparticles against multi-drug resistant clinical isolates from post-surgical wound infections. *Microbial Pathogenesis*, *107*, 327–334. doi:10.1016/j.micpath.2017.04.013 PMID:28411059

Kipp, J. (2004). The role of solid nanoparticle technology in the parenteral delivery of poorly water-soluble drugs. *International Journal of Pharmaceutics*, *284*(1), 109–122. doi:10.1016/j.ijpharm.2004.07.019 PMID:15454302

Kumar, B., Smita, K., Cumbal, L., & Debut, A. (2017). Green synthesis of silver nanoparticles using Andean blackberry fruit extract. *Saudi Journal of Biological Sciences*, *24*(1), 45–50. doi:10.1016/j.sjbs.2015.09.006 PMID:28053570

Kumar, P. V., Pammi, S., Kollu, P., Satyanarayana, K., & Shameem, U. (2014). Green synthesis and characterization of silver nanoparticles using Boerhaavia diffusa plant extract and their anti bacterial activity. *Industrial Crops and Products*, *52*, 562–566. doi:10.1016/j.indcrop.2013.10.050

Kuppusamy, P., Yusoff, M. M., Maniam, G. P., & Govindan, N. (2016). Biosynthesis of metallic nanoparticles using plant derivatives and their new avenues in pharmacological applications–An updated report. *Saudi Pharmaceutical Journal, 24*(4), 473–484. doi:10.1016/j.jsps.2014.11.013 PMID:27330378

Lal, R. (2016). Environmental Sustainability. In R. Lal, D. Kraybill, D. O. Hansen, B. R. Singh, T. Mosogoya, & L. O. Eik (Eds.), *Climate Change and Multi-Dimensional Sustainability in African Agriculture: Climate Change and Sustainability in Agriculture* (pp. 3–11). Cham: Springer International Publishing. doi:10.1007/978-3-319-41238-2_1

Li, Q., Mahendra, S., Lyon, D. Y., Brunet, L., Liga, M. V., Li, D., & Alvarez, P. J. J. (2008). Antimicrobial nanomaterials for water disinfection and microbial control: Potential applications and implications. *Water Research, 42*(18), 4591–4602. doi:10.1016/j.watres.2008.08.015 PMID:18804836

Li, X., Xu, H., Chen, Z.-S., & Chen, G. (2011). Biosynthesis of Nanoparticles by Microorganisms and Their Applications. *Journal of Nanomaterials, 16*. doi:10.1155/2011/270974

Lin, D., & Xing, B. (2007). Phytotoxicity of nanoparticles: Inhibition of seed germination and root growth. *Environmental Pollution, 150*(2), 243–250. doi:10.1016/j.envpol.2007.01.016 PMID:17374428

Malik, P., Shankar, R., Malik, V., Sharma, N., & Mukherjee, T. K. (2014). Green Chemistry Based Benign Routes for Nanoparticle Synthesis. *Journal of Nanoparticles, 14*. doi:10.1155/2014/302429

Masciangioli, T., & Zhang, W.-X. (2003). *Peer reviewed: environmental technologies at the nanoscale*. ACS Publications.

Mayo, J. T., Yavuz, C., Yean, S., Cong, L., Shipley, H., Yu, W., ... Colvin, V. L. (2007). The effect of nanocrystalline magnetite size on arsenic removal. *Science and Technology of Advanced Materials, 8*(1-2), 71–75. doi:10.1016/j.stam.2006.10.005

Mirzaei, H., & Darroudi, M. (2017). Zinc oxide nanoparticles: Biological synthesis and biomedical applications. *Ceramics International, 43*(11, Part B), 907–914. doi:10.1016/j.ceramint.2016.10.051

Mittal, A. K., Chisti, Y., & Banerjee, U. C. (2013). Synthesis of metallic nanoparticles using plant extracts. *Biotechnology Advances, 31*(2), 346–356. doi:10.1016/j.biotechadv.2013.01.003 PMID:23318667

Mohanpuria, P., Rana, N. K., & Yadav, S. K. (2008). Biosynthesis of nanoparticles: Technological concepts and future applications. *Journal of Nanoparticle Research, 10*(3), 507–517. doi:10.100711051-007-9275-x

Morelli, J. (2011). Environmental Sustainability: A Definition for Environmental Professionals. *Journal of Environmental Sustainability, 1*(1), 9. doi: 10.14448/jes.01.0002

Morelli, J. (2013). Environmental sustainability: A definition for environmental professionals. *Journal of Environmental Sustainability, 1*(1), 2.

MubarakAli, D., Thajuddin, N., Jeganathan, K., & Gunasekaran, M. (2011). Plant extract mediated synthesis of silver and gold nanoparticles and its antibacterial activity against clinically isolated pathogens. *Colloids and Surfaces. B, Biointerfaces, 85*(2), 360–365. doi:10.1016/j.colsurfb.2011.03.009 PMID:21466948

Mukherjee, P., Ahmad, A., Mandal, D., Senapati, S., Sainkar, S. R., Khan, M. I., ... Kumar, R. (2001). Fungus-mediated synthesis of silver nanoparticles and their immobilization in the mycelial matrix: A novel biological approach to nanoparticle synthesis. *Nano Letters*, *1*(10), 515–519. doi:10.1021/nl0155274

Mukhopadhyay, S. S. (2014). Nanotechnology in agriculture: Prospects and constraints. *Nanotechnology, Science and Applications*, *7*, 63–71. doi:10.2147/NSA.S39409 PMID:25187699

Murphy, C. J. (2008). Sustainability as an emerging design criterion in nanoparticle synthesis and applications. *Journal of Materials Chemistry*, *18*(19), 2173–2176. doi:10.1039/b717456j

Narayanan, K. B., & Sakthivel, N. (2010). Biological synthesis of metal nanoparticles by microbes. *Advances in Colloid and Interface Science*, *156*(1), 1–13. doi:10.1016/j.cis.2010.02.001 PMID:20181326

Narayanan, K. B., & Sakthivel, N. (2011). Green synthesis of biogenic metal nanoparticles by terrestrial and aquatic phototrophic and heterotrophic eukaryotes and biocompatible agents. *Advances in Colloid and Interface Science*, *169*(2), 59–79. doi:10.1016/j.cis.2011.08.004 PMID:21981929

Park, Y., Hong, Y., Weyers, A., Kim, Y., & Linhardt, R. (2011). Polysaccharides and phytochemicals: A natural reservoir for the green synthesis of gold and silver nanoparticles. *IET Nanobiotechnology / IET*, *5*(3), 69–78. doi:10.1049/iet-nbt.2010.0033 PMID:21913788

Patel, J., Patel, O., & Raval, B. (2012). *Green Chemistry: New Avenues in Chemical Research: Focus in Healthcare*. LAP LAMBERT Academic Publishing.

Prasad, T. N., Kambala, V. S. R., & Naidu, R. (2013). Phyconanotechnology: Synthesis of silver nanoparticles using brown marine algae Cystophora moniliformis and their characterisation. *Journal of Applied Phycology*, *25*(1), 177–182. doi:10.100710811-012-9851-z

Qin, S., Liu, X., & Dong, L. (2017). *Advances in Application of Nanomaterials in Life Science*. Academic Press.

Rai, M., & Ingle, A. (2012). Role of nanotechnology in agriculture with special reference to management of insect pests. *Applied Microbiology and Biotechnology*, *94*(2), 287–293. doi:10.100700253-012-3969-4 PMID:22388570

Ramkumar, V. S., Pugazhendhi, A., Gopalakrishnan, K., Sivagurunathan, P., Saratale, G. D., Dung, T. N. B., & Kannapiran, E. (2017). Biofabrication and characterization of silver nanoparticles using aqueous extract of seaweed Enteromorpha compressa and its biomedical properties. *Biotechnology Reports (Amsterdam, Netherlands)*, *14*, 1–7. doi:10.1016/j.btre.2017.02.001 PMID:28459002

Ravichandran, R. (2010). Nanotechnology Applications in Food and Food Processing: Innovative Green Approaches, Opportunities and Uncertainties for Global Market. *International Journal of Green Nanotechnology: Physics and Chemistry*, *1*(2), 72–P96. doi:10.1080/19430871003684440

Saha, J., Begum, A., Mukherjee, A., & Kumar, S. (2017). A novel green synthesis of silver nanoparticles and their catalytic action in reduction of Methylene Blue dye. *Sustainable Environment Research*, *27*(5), 245–250. doi:10.1016/j.serj.2017.04.003

Sanchez-Mendieta, V., & Vilchis-Nestor, A. R. (2012). *Green synthesis of noble metal (Au, Ag, Pt) nanoparticles, assisted by plant-extracts. In Noble Metals*. InTech. doi:10.5772/34335

Saratale, R. G., Saratale, G. D., Shin, H. S., Jacob, J. M., Pugazhendhi, A., Bhaisare, M., & Kumar, G. (2017). New insights on the green synthesis of metallic nanoparticles using plant and waste biomaterials: Current knowledge, their agricultural and environmental applications. *Environmental Science and Pollution Research International*. doi:10.100711356-017-9912-6 PMID:28815433

Savage, N., & Diallo, M. S. (2005). Nanomaterials and water purification: Opportunities and challenges. *Journal of Nanoparticle Research*, *7*(4), 331–342. doi:10.100711051-005-7523-5

Shamaila, S., Sajjad, A. K. L., Farooqi, S. A., Jabeen, N., Majeed, S., & Farooq, I. (2016). Advancements in nanoparticle fabrication by hazard free eco-friendly green routes. *Applied Materials Today*, *5*, 150–199. doi:10.1016/j.apmt.2016.09.009

Shankar, S., Jaiswal, L., Aparna, R., & Prasad, R. (2014). Synthesis, characterization, in vitro biocompatibility, and antimicrobial activity of gold, silver and gold silver alloy nanoparticles prepared from Lansium domesticum fruit peel extract. *Materials Letters*, *137*, 75–78. doi:10.1016/j.matlet.2014.08.122

Sharma, J. K., Akhtar, M. S., Ameen, S., Srivastava, P., & Singh, G. (2015). Green synthesis of CuO nanoparticles with leaf extract of Calotropis gigantea and its dye-sensitized solar cells applications. *Journal of Alloys and Compounds*, *632*, 321–325. doi:10.1016/j.jallcom.2015.01.172

Sharma, R., & Gupta, N. (2015). Green HRM: An Innovative Approach to Environmental Sustainability. *Proceeding of the Twelfth AIMS International Conference on Management*.

Steffen, W., Crutzen, P. J., & McNeill, J. R. (2007). The Anthropocene: Are humans now overwhelming the great forces of nature. *Ambio*, *36*(8), 614–621. doi:10.1579/0044-7447(2007)36[614:TAAHNO]2.0.CO;2 PMID:18240674

Steffen, W., Persson, Å., Deutsch, L., Zalasiewicz, J., Williams, M., Richardson, K., ... Svedin, U. (2011). The Anthropocene: From Global Change to Planetary Stewardship. *Ambio*, *40*(7), 739–761. doi:10.100713280-011-0185-x PMID:22338713

Tahir, L., Nazir, R., & Khan, N. (2013). *Nanoherbal antiseptic medicine/bandage for effective and safe treatment of external wounds*. Paper presented at the Nanotechnology 2013: Bio Sensors, Instruments, Medical, Environment and Energy, Washington, DC.

Theron, J., Walker, J., & Cloete, T. (2008). Nanotechnology and water treatment: Applications and emerging opportunities. *Critical Reviews in Microbiology*, *34*(1), 43–69. doi:10.1080/10408410701710442 PMID:18259980

Tripathy, A., Raichur, A. M., Chandrasekaran, N., Prathna, T., & Mukherjee, A. (2010). Process variables in biomimetic synthesis of silver nanoparticles by aqueous extract of Azadirachta indica (Neem) leaves. *Journal of Nanoparticle Research*, *12*(1), 237–246. doi:10.100711051-009-9602-5

United Nations Secretary-General's High-Level Panel on Global Sustainability. (2012). Resilient People, Resilient Planet. A Future Worth Choosing. New York: UN GSP.

Varma, R. S. (2012). Greener approach to nanomaterials and their sustainable applications. *Current Opinion in Chemical Engineering*, *1*(2), 123–128. doi:10.1016/j.coche.2011.12.002

Wang, X., Han, H., Liu, X., Gu, X., Chen, K., & Lu, D. (2012). Multi-walled carbon nanotubes can enhance root elongation of wheat (Triticum aestivum) plants. *Journal of Nanoparticle Research*, *14*(6), 841. doi:10.100711051-012-0841-5

Wang, Y., & Xia, Y. (2004). Bottom-up and top-down approaches to the synthesis of monodispersed spherical colloids of low melting-point metals. *Nano Letters*, *4*(10), 2047–2050. doi:10.1021/nl048689j

Section 2

Chapter 5 Nanomembrane Applications in Environmental Engineering

Sumaira Nosheen
Pakistan Council of Scientific and Industrial Research Lahore, Pakistan

ABSTRACT

This chapter delivers an outline of the strategies and techniques that are used and developed for the fabrication of membrane techniques/methods for application in environmental engineering. Human activities are the cause of increased hazardous gases in atmosphere mainly in soil and water. Nanotechnology deals the ability to control matter at the nano-scale level. Materials prepared are in nano level and thus they will possess some special properties to deal with specific functions. One of the major aspects of nanotechnology is nanomembrane fabrication which is mainly employed for water purification plants. The chapter is specifically offers a full understanding of the technologies and laws used to synthesize membranes. The chapter also provides an introduction to techniques to characterize nanomembranes.

INTRODUCTION TO MEMBRANE TECHNOLOGY

In eighteenth century many scientists started studies on membrane science and technology and how they work. In start these membranes were only used in laboratories for experimental studies and they did not possess any commercial or daily life importance. For example, the behavior of solutions by using membranes was carried out by Traube and Pfeffer. Their experiments and work done was used by van't Hoff in 1887 which lead to limit law explaining behavior of ideal dilute solutions that contributed to the development of famous "Van't Hoff Equation". The work is further proceeded by other scientists who worked on different types of membranes including semipermeable membranes which resulted in development of another theory known as "Kinetic Theory of Gases" by Maxwell was introduced to the world. Early membrane investigators experimented with every type of diaphragm which was available to them including animal bladders. In 1907, Bechhold, a well know scientist, prepared nitrocellulose membranes having very small pore size to determine a bubble test (Bechhold, 1907). Thus, nowadays membranes importance have increased so much that many technologies are in use to synthesize nanomembranes. The major problems in membrane development are its pore size and its sustainability as

DOI: 10.4018/978-1-5225-5745-6.ch005

the real challenge is to allow specific materials to pass through the membrane and blocking the others depending upon the type of application (Baker, 2004) hence fabricating membranes targeted to specific application and vice versa. Thereby, for environmental issues, food and chemicals or wastewater treatment, the diverse types of membranes having different parameters will be employed. To permit separation process, specially designed synthetic membranes are widely used which separate particles on the basis of their molecular size and shape. Since early 1970s, membranes were not considered very important in separation industries but their importance rose suddenly in mid 1970s in that area resulting in dramatic increase in their successful use in industries. The area of synthetic membranes further flourished in middle of the twentieth century impacting positively both small and large scale industrial processes. Now an extensive variety of synthetic membranes is known to the world (Toyomoto et al., 1992) that include polymers, liquids or even inorganic materials. Among these the successful and commercially viable membranes are made up of polymeric materials mainly.

Loeb and Sourirajan are two well-known scientists who investigated the process of osmosis to prepare a synthetic membrane (Loeb & Sourirajan, 1962). Their experiments led to the successful completion of desalination plant which was later installed in some arid regions to purify sea water. Thus it is an influential discovery that transformed membrane separation from a laboratory to an industrial process and called as Loeb–Sourirajan process for making defect-free, high-flux, anisotropic reverse osmosis membranes. The characteristics of these membranes are that they are ultrathin, selective in nature and possess permeable microporous support for increased mechanical strength. The studies showed that membranes prepared by Loeb and coworker had increased mechanical strength and very good efficiency (10 times higher than that of any membrane then available).

Membranes can be classified based on their surface chemistry, bulk structure, and morphology and production method. Common types of membranes include porous membranes, asymmetric and dense membranes which are followed by separation industries (Freeman & Pinnau, 1999).

Types of Membranes

A membrane is nothing but a discrete, thin interface that moderates the permeation of chemical species in contact with it. This interface may be molecularly homogeneous (completely uniform in composition and structure) or it may be chemically or physically heterogeneous. For example, it may have holes and pores of restricted dimensions or it may have layering structures inside the membrane unit. Commonly it is called that membranes are just like filters; no doubt a normal filter meets the definition of a membrane, but in actual the filters are only those structures which act as separation unit. It means filters work on the rule that they will separate smaller and larger particle in the range of 1-10μm in a heterogeneous system. A large variety of membranes is available and Figure 1 schematically presents the principal types of membrane which are described briefly below.

Isotropic Membranes

Isotropic membranes are further categorized according to the specification of membranes. Some are discussed below:

Figure 1. General types of membranes (Baker, 2004)

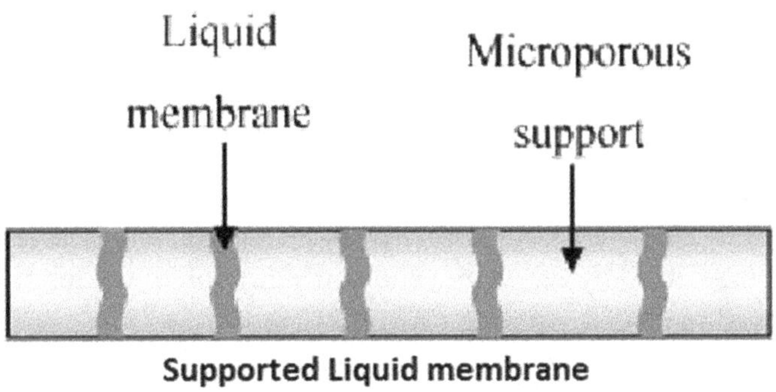

Microporous Membranes

Just like a conventional filter, microporous membrane works same as filters. The pores are distributed inside membrane in random manner having interconnections among them. The only difference in the pores of filter and membrane is that the membrane has extremely small pores in it the diameter of which lies in the range of 0.01 to 10 µm. This enables the retention of big particles that cannot pass through microporous membranes resulting in their rejection. On other hand the particles which are much smaller in size than the smallest pores of the specific membrane will pass through the membrane very easily. Thus, separation of solutes is ensured by defining the pore size in microporous membranes and is mainly a function of molecular size and pore size distribution. In general, only molecules that differ considerably in size can be separated effectively by microporous membrane. This can be observed in microfiltration and ultra-filtration processes (Baker, 2004).

Nonporous, Dense Membranes

When molecules travel from area of higher concentration towards area of lower concentration, the process is called as diffusion. Pressure and electro potential gradient are tools which help out in transport of particles across the membrane. Transport rate of particles across the membrane allows separation of various components which is determined by their diffusivity and solubility in the membrane materials. If the material has considerable similar size, then dense membranes having non porous nature can separate them. Most gas separation, pre-evaporation, and reverse osmosis membranes use dense membranes to perform the separation. The membrane structure is kept anisotropic in order to achieve as much flux as possible (Baker, 2004).

Electrically Charged Membranes

The membranes whose particles have charge on them (either positive or negative) are termed as electrically charged membranes. Their structure may be microporous or dense in nature.

Two types of membranes come under this category:

- Anion exchange membranes
- Cation exchange membranes

A positively charged ions membrane is called as an anion-exchange membrane because it binds anions in the surrounding fluid. In the same way the membrane having negative charge on its particles is called a cation-exchange membrane. Separation with charged membranes is achieved mainly by exclusion of ions of the same charge as the fixed ions of the membrane structure, and to a much lesser extent by the pore size. The separation is affected by the charge and concentration of the ions in solution. These types of membranes are employed in reverse osmosis (Wei, 2012),

Anisotropic Membranes

When efficiency of membrane is taken then the size and thickness of membrane is very important. Thin membranes are no doubt very difficult to synthesize but they have the highest efficiency in performing separation. In general, membrane thickness takes part in transport rate determination; thicker the membrane less will be the transport across membrane and vice versa. But thin membranes having good efficiency are not very strong mechanically therefore synthesis of these membranes is limited to minimum thickness of 20 μm and having no defect by following conventional fabrication techniques. Many experiments and studies have been conducted in last thirty years to attain desired properties in a membrane which is anisotropic in nature. These experiments led to discovery of more and more fabrication techniques in the field of membrane synthesis. Anisotropic membrane is actually very thin and supported by a thicker porous membrane parallel to it. The surface layer and its substructure may be formed in a single operation or separately. The benefits of the higher fluxes delivered by anisotropic membranes are so pronounced that almost all commercial processes adopt these membranes in their industrial applications.

Ceramic, Metal and Liquid Membranes

Depending upon the fabrication and synthesis process used, membranes are either polymeric in nature, or they may be organic/ inorganic in properties according to the method used for their preparation. But, in recent years, interest in membranes formed from less conventional materials has enlarged dramatically. Another type of membranes which are called ceramic membranes also exists. It is a special class of microporous membranes used in ultrafiltration and microfiltration applications for which solvent resistance and thermal stability is recommended.

Synthesis Parameters for Membranes

A membrane can be described as a semi permeable barrier between two phases which prevents near contact. This barrier must be perm selective which means that it restricts the movement of molecules in it in a very specific manner. This barrier can be solid, liquid or gas depending upon applications. Perm selectivity can be obtained by many mechanisms; some of which are given below:

1. Size exclusion process
2. Differences in diffusion coefficients (bulk as well as surface)
3. Electrical charge differences

4. Differences in solubility
5. Differences in adsorption and/or reactivity on inside surfaces

Pressure gradient is the tool to create flux of liquids or gases through the membrane. Sometimes the flux may be created by using an electric field gradient. Membranes can be used for many purposes like:

1. Separation of mixtures (liquids, gases or liquid-solid mixtures can be separated).
2. Handling of chemical reactions: shifting the equilibrium situation or management of the conversion or selectivity of catalytic reactions can be done.

The effectiveness of the membrane in any application depends on the detailed structure of the membrane system. Morphology is also very important in addition to the performance of the above mentioned physicochemical mechanisms in a membrane. These are critically resolute by the synthesis process and this is why details of the preparation procedures are so important in synthesis (Bhave, 2012).

Basic Principle of Membrane Synthesis

In this section a short introduction will be given on how the membranes are synthesized by sol-gel technique and how anodization is done on carbon membranes and glass membranes. Some new terminologies will be introduced here as well including symmetric and asymmetric membranes, in addition words like mesoporous and macroporous will be added. Symmetric membranes are those membranes which have homogeneous structure throughout the frame. Capillary glass membranes are example of symmetric membrane. On the other hand, asymmetric membranes have a gradual change in structure throughout the system. In majority of the cases they are composite membranes which have several layers with a gradual decrease in pore size to the feed side of the membrane. Ceramic alumina membranes which are prepared by using sol-gel technique are typical example of this type of membrane. Macropores is the term used when pore diameter is larger than 50 nm while Mesopores have a diameter between 2 and 50 nm and below a diameter of 2 nm the system is called microporous (Sing et al 1985).

Ceramic Asymmetric membrane

Ceramic asymmetric membrane synthesis demand very critical care for development of center layer and to develop support system of the membrane. Special technologies and processes are required for increasing the quality of intermediate layer. By far the most frequently used principle to meet these requirements is the formation of a layer consisting of a packing of well-ordered, uniform-sized particles. The size and shape of the particles determine the minimum obtainable mean size and pore size. To change porosity of membrane, heat treatment can be done. The main process for making ceramic membranes is to first prepare a dispersion of fine particles, also known as slip and then to deposit the particles which are present in the slip on a porous support by a slip-casting method. The capillary pressure drop created on letting the slip come into contact with the microporous support forces the dispersion medium of the slip to flow into the pores of the support. The slip particles are concentrated at the pore entrance to form a layer of particles or a gel layer (Bhave, 2012)

Polymeric Membranes

Polymeric membrane can also be synthesized if continuous efforts done. Many fabrication techniques are present which can be used to prepare membrane with desired parameter and applications. Some of the fabrication process includes the following:

- Template leaching
- Stretching of polymeric film
- Phase separation
- Interfacial polymerization

Template Leaching

In the leaching process, two different components are mixed together forming a heterogeneous structure. In the second step the undesired component is removed from the mixture by leaching process (Chau et al., 1989; Ichikawa et al.1987; Lopatin et al., 1989).

Stretching of Polymeric Films

It is very common technique to make semi permeable membranes in which the polymeric sheets are stretched in such a way that some pores are created in it (Baker 2004).

Figure 2. Simplest example of leaching process

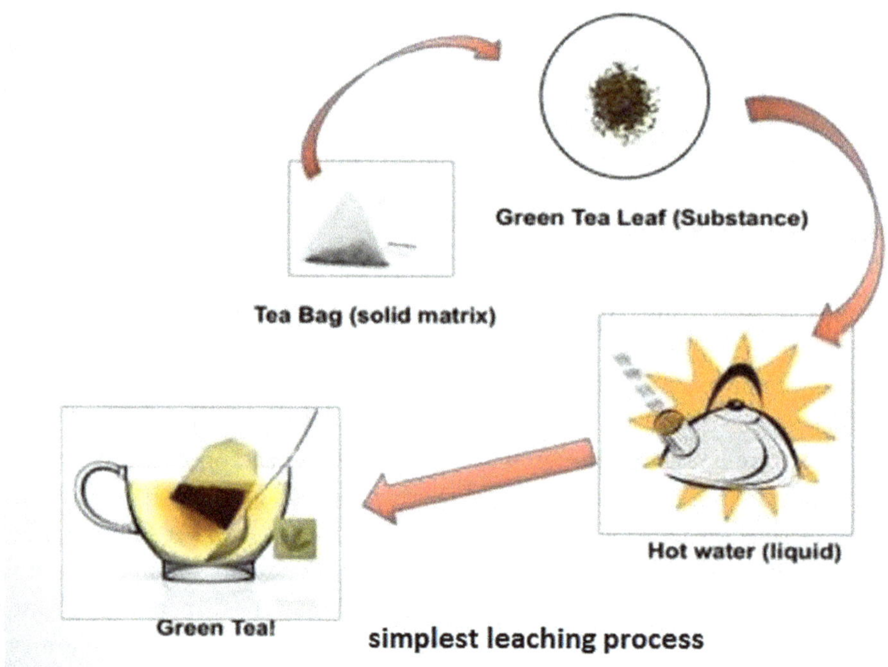

Figure 3. Figure showing SEM image of stretching of membrane

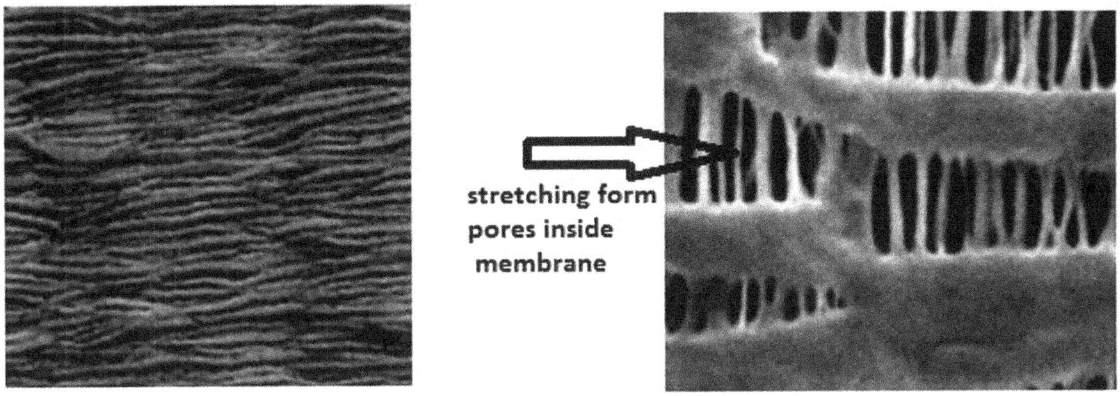

Phase Separation

By employing phase separation technique asymmetric membranes can be easily synthesized. In this process on phase is split up into two or more membranes and it is supported by thermodynamic reaction. The unwanted parts of the membrane can also be eluted out easily. During the phase separation, the prepared material may be heated up to vaporize the extra particles or may be solidified according to the requirements (Schnabel & Vaulont, 1978).

Phase separation technique is further categorized into six parts as follows:

- Vapor induced phase separation (VIPS)
- Liquid induced phase separation (LIPS)
- Thermally induced phase separation (TIPS)
- Colloidal induced phase separation (CIPS)
- Pressure induced phase separation (PIPS)
- Reaction induced phase separation (RIPS)

1. **Vapor Induced Phase Separation Technique**

The technique is based on vaporization principle in which the solvent is vaporized by heating or by inducing thermal reaction. The solvent used for vaporization may be organic in nature or aqueous. But aqueous solvents may remain inside the pores of membrane after vaporization and cause inefficiency of membrane (Baker, 2004)

2. **Liquid Induced Phase Separation Technique**

Wet casting is done by liquid induced separation. The synthesized membrane will have dense pores inside the structure. This technique can also be combined with vapor induced technique to enhance its effectiveness (Guillen, 2011).

3. **Thermally Induced Separation Technique**

Material is melted by using high temperature in a well-defined cast (Guillen, 2011).

Characterization of Membranes

Characterization techniques can be classified into two main processes:

1. Static technique
2. Dynamic technique

Both of these techniques are separated on the basis of membrane properties and characterization parameters. Static one deals with all synthesis process, structural features and changes related with their physical and chemical properties while the dynamic technique is specifically for explaining characterization parameters. Techniques may be destructive or non -destructive in nature. Except for bubble pressure all other reported techniques have not yet been standardized or harmonized (K.-L. Tung et al.). Table 1 shows the main characterization tests for membranes:

Bubble Pressure

This method was originated by Bechhold in 1908 (Tung, K. et al.) and it is based on the fact that the pressure (P) which is essential to blow a gas through a liquid-filled capillary is inversely proportional to the capillary radius (r). The pore size can be premeditated by following the Cantor equation (1):

Table 1. Characteristic tests of a synthesized membrane (Tung et al., 2014)

Methods	Characteristics	Topology
Bubble pressure	Maximum pore size	Non destructive
Gas and liquid displacement methods	Distribution of pore size	Non destructive
Mercury porosimetry	Distribution of pore size in membranes	Destructive
Atomic force microscopy (AFM)	Surface characterization	Non destructive
Scanning electron microscopy (SEM)	Surface characterization Pore size of membrane	Non destructive
Transmission electron microscopy (TEM)	Surface characterization Quantitative analysis of membrane	Destructive
Flux and retention measurements	Permeability and selectivity of membranes	Non destructive
Gas adsorption and desorption methods	Distribution of pore size in membrane	Non destructive
Permporometry	Distribution of pore size	Non destructive
Fourier transform infrared spectroscopy (FTIR)	Qualitative and quantitative analysis of functional group in membranes	Destructive
Contact angle measurements	Surface studies	Non destructive
Stress strain measurements	-	Destructive

$$P = \frac{2.v.\cos\theta}{r} \tag{1}$$

The bubble point test is a measure of the radius of the largest pore. According to the equation given above, gas will pass through it first. To calculate pore size in an air-water system where the surface tension (g) will be almost 72 mN/m following equation will be followed (2):

$$r(\mu m) = \frac{1.4}{P(bar)} \tag{2}$$

The main advantages of this test are:

- It is purely a non-destructive method,
- A very simple way with simple set up of apparatus,
- It is very beneficial for integrity test.

But the technique suffers from a drawback that it cannot determine the radius of pores. Moreover, the idea about how the pores are distributed inside a membrane system cannot be evaluated by following bubble test. An increased surface tension in the membrane will cause compression in the system as well. This increase surface tension is required for permeation of gas through pores. For example, the bubble point's method was used for calculation of pore size of membrane. These types of membrane are called as three hollow fiber polyphenylsulfone membranes (Cuperus & Smolders, 1991) The pore size range was from 53 to 237 nm and pressure calculated was between 2-9 bars respectively (Manesh et al., 2011)

Liquid-Liquid Displacement Porosimetry

The next technique was a little bit improved in that sense that it enables calculation of pore sizes as well. Same principles were followed like in air liquid displacement method or bubble point processes. In both the techniques inter-relation between pressure and pore sizes were taken (Ding & Bikson, 2010). These techniques are advantageous as very high potential is created and pore sizes of nano levels can be attained. In addition membrane efficiency is increased due to less surface tension. Because of favorable properties of this membrane, many trials were done and when polymeric membranes were forms, they gave outstanding results. Due to very small pore sizes, these membranes have wide range of applications as well (Calvo et al., 1995).

Some of the advantages of these membranes are given below:

- Membranes can be tested in their original state.
- It can give information very close to the normal operating conditions of the membrane
- There is advantage of this membrane that due to reduced pressure, there is no chance of membrane breaking
- It is very simple and easily handleable technique. Improved Due to process efficiency and timings, the technique is considered s simple and easily manageable

Liquid and Air Permeability

As concerned with properties of membranes, one of the important factors is permeability of membranes for specific liquids. It is important to develop a model or to follow an approach to estimate permeability values correctly. It should be taken in consideration that the model cannot differentiate between a system with few large pores or large number of small pores in a membrane. This method can be further upgraded by adding gas which will act as a permeating medium as a substitute of liquid. Thus it will be possible to differentiate between pore size distributions in a system as gas will be dependent on the overall pressure. These techniques have been used for pore size calculation in polymeric hollow fibers (Carretero et al., 2013), which give analogous trends in the observed mean pore size. The gas permeation has certain advantages considering the influence of the polymer concentration (Khayet & Matsuura, 2003). Consequently, gas permeation technique has showed to be very helpful in characterization of the thin layer in asymmetric membranes. The negative aspect of this technique is that it does not give any insight into the remaining membrane pore size structure (Feng et al., 2013).

Nitrogen Adsorption/Desorption (Pore Size Distribution)

The technique is based on adsorption and desorption isotherms of gas. The gas is usually nitrogen which is subjected to adsorption and cause capillary condensation in the pores. The nitrogen adsorption BET analysis is very beneficial to determine surface area and pore size distribution of ceramic membranes. It has found application in structures like glassy polymers (See-Toh et al., 2008), in characterization of the support membrane in case of thin film composite OSN membranes (Park et al., 2008), as well as in the characterization of micro- and meso-pore materials as composite amorphous membranes (Fukumoto et al., 2014) or semi-crystalline polymer membranes (Jimenez Solomon, M.F. et al.).

Permporometry (Pore Size Distribution)

Permporometry technique, first introduced by Eyraud and modified by Katz et al. (Tsuru et al., 2008; Tsuru et al., 2011; Mey-Maron & Katz, 1986) was based on the theory that the vapor pressure at the surface of a liquid depends on its curvature. The permeation of gases will be blocked due to condensation of membranes.

Electron Microscopy

Electron microscopy is a method used to obtain visual information of the membrane structure and porosity. The observation is obtained through magnification by scanning electron microscope (SEM) or a transmission electron microscope (TEM). But polymeric membranes are very sensitive in nature and high energy of electron beam can damage the membrane. For microscopic studies, sample preparation is relatively easy. In majority of the cases sample can be directly evaluated without any preparation. The sample for TEM analysis is prepared with high care and requires surface to be dry and thin enough to allow penetration of electrons easily. Due to this requirement, sample preparation for TEM requires extra care in case of membrane. SEM is used to characterize the predominant morphology of polymeric asymmetric and mixed matrix membranes also (Feng et al., 2013; Alexander et al., 2014; Valodez-Blanco & Livingston, 2009; Siddique et al., 2014; Bogdanowicz et al, 2013). These membranes are used in organic solvent nanofiltration and to calculate the thickness of the top- and the support layers. Their role in the control of the membrane mass-transport rates, as well as eventual changes in the membrane before and after nanofiltration is also studied (Freeman & Pinnau, 1999; Loeb & Sourirajan, 1962; Tylkowski et al., 2010;Tylkowski et al., 2011; Fan et al., 2008)

Stress - Strain Property

As membranes have critical surface properties, high surface tension weakens the membranes also. It is necessary to consider mechanical properties of membrane as well. Normally the materials are exposed to stress and the accompanying deformation throughout processing and further applications. The mechanical properties of the membranes characterized by stress-strain tests are very important to confirm the potential suitability of these membrane in applications including nanofiltration (Tylkowski et al., 2010;Tylkowski et al., 2011;in respect of mechanical strength of hollow fiber membranes (Tsuru et al., 2008; Tsuru et al., 2011) and in connection with the abrasive resistance of flat sheet membranes (Tylkowski et al., 2010;Tylkowski et al., 2011), mainly.

The stress-strain relationships can then be used to establish the compressive or tensile yielding strength, the modulus of elasticity and the ultimate strength.

Applications of Nansomembranes in the Environmental Engineering

In industrialized nations, the air is filled with numerous pollutants caused by human activity or industrial processes, such as carbon monoxide (CO), chlorofluorocarbons (CFC), heavy metals (arsenic, chromium, lead, cadmium, mercury, zinc), hydrocarbons, nitrogen oxides, organic chemicals (volatile organic compounds, known as VOCs, and dioxins), sulphur dioxide and particulates. Soil is contaminated by acid rains and increased nitrogen levels. The burning of oil, gas, coal, fuels are the main factors to increase hazardous gases in the atmosphere. Some amount of contaminants came from natural processes also including volcanic activity, paleontology process and soil decay etc. By using highly toxic pesticides and fertilizers, industrials effluents wastes are also cause of increased water pollution. All these contaminants are very hazardous if their level increases to a certain level. Their levels are usually measured in ppm or ppb levels. For example arsenic is a hazardous chemical. It's very small amount can cause serious affects. The toxic level for arsenic is 10 ppm in soil. Similarly the toxic level for mercury is 0.002 ppm in water. Contaminants may be present as a single form but in most cases they occur as mixtures. As a result, there is a need for technologies that should be proficient of monitoring, recognizing and, ideally, treating such small amounts of contaminants in air, water and soil. Nanomembranes are one of the state of the art field of science. Advanced materials including organic and inorganic polymers and ceramics are in practice to synthesize membranes. They have wide range of applications in resolving environmental issues. Applications of nanofiltration can be found, e.g. in drinking water treatment, in the food industry as well as in cleaning environment and natural resources as well. A great potential for nanotechnology is expected in the field of seawater desalination also (Martins et al., 2010).

Nanomembrane for Industrial Wastes Treatment

Microfiltration and ultrafiltration are archetypal pressure-driven membrane operations which along with reverse osmosis, to be ingrained industrial separation technologies. All these processes are based on the use of polymeric membranes and represent a valid approach to resolve separations problems involving particulate material and macromolecules. They are a valid alternative to competitive separation processes like distillation, extraction, fractionation, and adsorption, since the separation process is athermal and involves no phase change or chemical agents. Due to simplicity these techniques have additional advantages like low energy consumption, easy scale-up, low weight and space requirements, modularity and the possibility of carrying out the separation continuously (Bilstad, 1997).

The combination of bioreactors and UF membranes (MBRs) documents an inventive and effective cleaning process for both municipal and industrial wastewaters to be obtained. The bioreactor is an oxygenated feed tank in which the dissolved organic substances are disintegrated by activity of microorganisms and result in formation of nitrogen, water and carbon dioxide respectively. The biomass is separated from this effluent by the nanomembrane and finally recycled in the feed tank. By following this treatment high molecular weight substances are prohibited by the membrane and can be further decomposed. The biomass concentration in the system can reach values 10 times higher than those obtained in conventional plants (Zenon Environmental b.v., 1995).

The recovery of water, toxic and valuable compounds from water and wastewaters in the chemical industry, to elude pollution of the environment is another interesting field of applications (Reynolds, 1996).

The application of ED (electro dialysis) with MF(microfiltration), UF (ultrafiltration) and ion-exchange processes provides a good solution for recovering and recycling valuable compounds (nickel, cadmium, copper) from rinse solutions in the metal dispensation industry. An integrated membrane process based on the use of UF and NF has been proposed by Cassano *et al.* (2001) for the recovery of chromium from spent tanning liquors in the leather industry. Osmota GmbH (Germany) proposed a treatment scheme for the recovery and recycling of water and sulphuric acid from the rinse of a lead battery production line including MF, NF, ED and RO (Strathmann, 2004). Kim *et al.* (2006) reported a pilot-scale wastewater treatment and reuse system by integrating MF and ED. MF removes suspending solids and heavy metal ions (immobilized by separating materials) from sewage, while the following ED operation further purifies and desalinates the outlet. UF and RO can be also integrated to recover dyes from wastewaters produced by the textile industry (Short, 1993).

This interesting field offers the skill to control matter at the nanoscale level and to develop and fabricate specific materials with outstanding properties that can serve specific functions for remedy of environmental pollution. Nano-technology is a vast field of science that offers various trends to resolve in particular environmental issues where pollution often arises from the presence of a specific contaminant within a mixture of materials, in solid, liquid or gas form. This is the property of nano science that we deal with materials having very small size.

Assimilated processes which include water softening, MF, RO, UF, UV-sterilization and mixed bed ion-exchange have been anticipated for purifying well or surface waters (Strathmann *et al.*, 2006).

Due to extra ordinary dominant properties of nanomaterials, it allows the development of highly reduced, accurate and sensitive pollution-monitoring devices known as nano-sensors. Nanomaterials can also be engineered to actively interrelate with a pollutant and decompose it into less toxic species. Nanotechnology studies not only used to detect contaminated sites but also to treat or reduce them efficiently. This technology can be used to reduce the production of harmful wastes in manufacturing processes by reducing the amount of material used, and by engaging less toxic compounds.

1. Soil and groundwater contamination ascending from manufacturing processes are a matter of great complexity and concern. Many sites are affected including lakes and rivers in their neighborhood,

underground storage tank leakages, landfills and abandoned mines. Pollutants in these areas include heavy metals like mercury, lead and cadmium and organic compounds including benzene and chlorinated solvents. Nanotechnology can develop techniques that will allow for more specific tools to resolve that area. Many of the methods are used to remove toxic contaminants that involve laborious, time-consuming and expensive techniques. A pretreatment process and removal of the contaminated area is often required. This will eventually disturb the ecosystem also.

Nano science can also be used to develop different tools for remedy that are specific to a certain pollutant like metals or chemicals. Consequently increasing affinity and selectivity, as well as refining the sensitivity of the technique.

One more very important parameter is contamination of drinking water. Highly toxic elements like Mercury and arsenic are major health hazardous risk. Remediation methods that allow the fast, economic and effective treatment of water polluted with such contaminants is urgently needed. Nanotechnology can introduce new methods for the treatment and purification of water from pollutants, as well as new techniques for wastewater management and water desalination.

There is considerable reduction in formation of toxic components when nano powders are used. As a result of the increased reactivity and stability of the nanoparticles as compared to the granular iron powder, high efficiency is acquired. As a result of their stability, nano-iron particles may remain active in a location for more than six weeks before they become dispersed completely in the groundwater and become less concentrated than naturally occurring iron.

Semiconducting nanoparticles made of titania and zirconia is used in photo catalytic remediation. These materials produce an electron-hole pair when irradiated with a light having energy in the order of the material band gap because they are electrically conductive in nature. Both titania and zirconia are easily available and have been used for environmental issue for many years In recent studies, nano-sized titania and zirconia particles have been formed, as these have more active surface given the same volume of material. The aim of synthesizing these particles is to develop solar material which can be helpful for catalysis remediation system. In this system the nano particles will remove contaminants by the help of rays.

2. Nanotechnology could bring an improvement in two ways. Initially when noble metals like gold and platinum are chemisorbed to the TiO_2 and ZnO nanoparticles, the photo catalytic activity is enhanced. The reason is that the presence of the metal will help to keep the electrons and holes from recombining in the semiconductor and thereby it will enhance the efficiency of the photo catalysis.
3. Nanomaterials have also been found able to remove metal contaminants from air. For example, silica-titania nanocomposites are being investigated for use in the removal of elementary mercury from vapors such as those from combustion sources. In these nanocomposites, silica acts as a support material and titania transforms mercury to a less volatile form (mercury oxide). Remediation using dendrimers are highly branched polymers with controlled composition and nanoscale dimensions. Chelating agents in the form of dendrimers are also studied for the removal of metal contaminants. These can be designed so as to able to act as ‘cages’ and trap metal ions and zero-valent metals, making them soluble in appropriate media or able to bind to certain surfaces. The vision is to employ dendrimers as nanoscale chelating agents for polymers supported ultrafiltration systems.

Nanomembranes and Nanofilters

Nanotechnology can also be employed for the fabrication of nanofilters, nano-adsorbents and nanomembranes having required properties that can be used for decontaminating water and air. It is the facility to manipulate matter at a molecular level that makes nanotechnology so promising in this field, together with the small size and high surface-to-volume ratio of nanomaterials that are employed in the fabrication of these products. In principle, 'nano traps' designed for a certain contaminant can be produced, for example with a specific pore size and surface reactivity. An example is given by the work carried out at Rice's CBEN, where researchers are developing reactive iron oxide ceramic membranes (ferroxane membranes) that are capable of remediating organic waste in water. Filters and membranes can also be engineered to be 'active' in the sense of being able not only to trap a certain contaminant, but also to chemically react with it and convert it to a non-toxic product. For example, researchers at the University of Tennessee are investigating a new type of nanofibre for the removal of micro-organisms via filtration that can also kill them on contact. An interesting application of nanomembranes has been developed by researchers at the University of California Los Angeles (UCLA) in the form of a new reverse osmosis (RO) membrane for seawater desalination and wastewater remediation. The membrane is made of a uniquely cross-linked matrix of polymers and engineered nanoparticles designed to draw in water ions but repel contaminants. This is possible due to the nano-size of the holes forming the membrane, which are 'tunnels' accessible only to the water molecules. Another distinctive feature of this nanomembrane is its ability to repel organics and bacteria, as a result of the chemical composition of the nanoparticles embedded in the membrane. Compared with conventional RO membrane, these membranes are thus less prone to clogging, which increases the membrane lifetime with an obvious economic benefit.

Super Hydrophilic Filters

Access to safe drinking water to human life is becoming more and more critical day by day and in areas of the world clean water is not available to mankind. Nano filters are used to remove dangerous materials like arsenic and other heavy metals from water. The filter is inserted in a plastic bottle and enables contaminated water to be cleaned on-site in an easy way.

Pollution Prevention

Increased development in industries is also increasing different types of pollution day by day. Scientists are also working to resolve this issue side by side. Nanotechnology studies deal with many advanced approaches to reduce pollution. These are only a few of the many approaches that can be taken to reduce pollution of the environment. Nanotechnologies are already actively involved in this sector, either as a technology producing advanced materials that pollute less, or as a method to increase the efficiency of certain industrial processes.

CONCLUSION

Membrane processes may donate to explaining diverse problems of natural environment protection. They save raw materials and reduce energy consumption as well as allow understanding various recovery strategies which are linked with the subject matter. In addition, membranes can be easily combined in systems including economic hybrids that would permit free adjustment of product parameters and energy use. By their versatility and multiple functions, membrane methods can fulfill various aims and expectations of strategy of clean technologies. All membrane methods are very efficient, nevertheless high separation ability of each membrane process is limited to some specific range of separated particles. Reverse osmosis is capable to retain all dissolved species that includes monovalent ions, but it only applies high pressures which are not mandatory when thermally driven process like membrane distillation, which has similar separation capabilities is also in use (Zakrzewska-Trznadel, 2005).

REFERENCES

Alexander, J. V., Neely, J. W., & Grulke, E. A. (2014). Effect of chemical functionalization on the mechanical properties of polypropylene hollow fiber membranes. *Journal of Polymer Science. Part B, Polymer Physics*, *52*(20), 1366–1373. doi:10.1002/polb.23572

Baker, R. W. (2004). *Overview of membrane science and technology* (2nd ed.). Membrane Technology and Applications.

Bechhold, H. (1907). Kolloidstudien mit der Filtrationsmethode. *Zeitschrift für Physikalische Chemie*, *60*(1), 257–318. doi:10.1515/zpch-1907-0113

Bhave, R. (2012). *Inorganic Membranes Synthesis, Characteristics and Applications: Synthesis, characteristics, and applications*. Springer Science & Business Media.

Bilstad, T. (1997). Membrane operations. *Water Science and Technology*, *36*(2-3), 17–24.

Bogdanowicz, K. A., Tylkowski, B., & Giamberini, M. (2013). Preparation and characterization of light-sensitive microcapsules based on a liquid crystalline polyester. *Langmuir*, *29*(5), 1601–1608. doi:10.1021/la3038878 PMID:23245267

Calvo, J. I., Hernandez, A., Pradanos, P., Martınez, L., & Bowen, W. R. (1995). Pore size distributions in microporous membranes II. Bulk characterization of track-etched filters by air porometry and mercury porosimetry. *Journal of Colloid and Interface Science*, *176*(2), 467–478. doi:10.1006/jcis.1995.9944

Carretero, P., Molina, S., Lozano, A., de Abajo, J., Calvo, J. I., Prádanos, P., & Hernández, A. (2013). Liquid–liquid displacement porosimetry applied to several MF and UF membranes. *Desalination*, *327*, 14–23. doi:10.1016/j.desal.2013.08.001

Cassano, A., Donato, L., Conidi, C., & Drioli, E. (2008). Recovery of bioactive compounds in kiwifruit juice by ultrafiltration. *Innovative Food Science & Emerging Technologies*, *9*(4), 556–562. doi:10.1016/j.ifset.2008.03.004

Chau, C. C., & Im, J. H. (1989). *U.S. Patent No. 4,874,568*. Washington, DC: U.S. Patent and Trademark Office.

Cuperus, F. P., & Smolders, C. A. (1991). Characterization of UF membranes: Membrane characteristics and characterization techniques. *Advances in Colloid and Interface Science*, *34*, 135–173. doi:10.1016/0001-8686(91)80049-P

Darvish Manesh, S., Tasselli, F., Jansen, J. C., Tocci, E., Bazzarelli, F., Bernardo, P., & Van der Bruggen, B. (2011). Preparation of solvent stable polyphenyl sulfone hollow fiber nanofiltration membranes. *Journal of Membrane Science, 384*(1), 89-96.

Ding, Y., & Bikson, B. (2010). Preparation and characterization of semi-crystalline poly (ether ether ketone) hollow fiber membranes. *Journal of Membrane Science*, *357*(1), 192–198. doi:10.1016/j.memsci.2010.04.021

Fan, Q., Sirkar, K. K., & Michniak, B. (2008). Iontophoretic transdermal drug delivery system using a conducting polymeric membrane. *Journal of Membrane Science*, *321*(2), 240–249. doi:10.1016/j.memsci.2008.04.061

Feng, C. Y., Khulbe, K. C., Matsuura, T., & Ismail, A. F. (2013). Recent progresses in polymeric hollow fiber membrane preparation, characterization and applications. *Separation and Purification Technology*, *111*, 43–71. doi:10.1016/j.seppur.2013.03.017

Freeman, B. D., & Pinnau, I. (Eds.). (1999). *Polymer membranes for gas and vapor separation: chemistry and materials science*. American Chemical Society. doi:10.1021/bk-1999-0733

Fukumoto, T., Yoshioka, T., Nagasawa, H., Kanezashi, M., & Tsuru, T. (2014). Development and gas permeation properties of microporous amorphous TiO 2–ZrO 2–organic composite membranes using chelating ligands. *Journal of Membrane Science*, *461*, 96–105. doi:10.1016/j.memsci.2014.02.031

Guillen, G. R., Pan, Y., Li, M., & Hoek, E. M. (2011). Preparation and characterization of membranes formed by nonsolvent induced phase separation: A review. *Industrial & Engineering Chemistry Research*, *50*(7), 3798–3817. doi:10.1021/ie101928r

Ichikawa, T., Takahara, K., Shimoda, K., Seita, Y., & Emi, M. (1987). *U.S. Patent No. 4,708,800*. Washington, DC: U.S. Patent and Trademark Office.

Jimenez-Solomon, M. F., Gorgojo, P., Munoz-Ibanez, M., & Livingston, A. G. (2013). Beneath the surface: Influence of supports on thin film composite membranes by interfacial polymerization for organic solvent nanofiltration. *Journal of Membrane Science*, *448*, 102–113. doi:10.1016/j.memsci.2013.06.030

Katz, M. G., & Baruch, G. (1986). New insights into the structure of microporous membranes obtained using a new pore size evaluation method. *Desalination*, *58*(3), 199–211. doi:10.1016/0011-9164(86)87004-7

Khayet, M., & Matsuura, T. (2003). Determination of surface and bulk pore sizes of flat-sheet and hollow-fiber membranes by atomic force microscopy, gas permeation and solute transport methods. *Desalination*, *158*(1), 57–64. doi:10.1016/S0011-9164(03)00433-8

Lai, C. Y., Groth, A., Gray, S., & Duke, M. (2014). Preparation and characterization of poly (vinylidene fluoride)/nanoclay nanocomposite flat sheet membranes for abrasion resistance. *Water Research*, *57*, 56–66. doi:10.1016/j.watres.2014.03.005 PMID:24698723

Loeb, S., & Sourirajan, S. (1962). *Sea water demineralization by means of an osmotic membrane*. Academic Press.

Lopatin, G., Yen, L. Y., & Rogers, R. R. (1989). *U.S. Patent No. 4,874,567*. Washington, DC: U.S. Patent and Trademark Office.

Martins, Eggers, & Evertz. (2010). *Applications of nanomaterials in environmental engineering*. Environmental Research of the Federal Ministry of the Environment, Nature Conservation and Safety, UBA-FB001337.

Mey-Marom, A., & Katz, M. G. (1986). Measurement of active pore size distribution of microporous membranes-a new approach. *Journal of Membrane Science*, *27*(2), 119–130. doi:10.1016/S0376-7388(00)82049-9

Park, H. B., Jung, C. H., Lee, Y. M., Hill, A. J., Pas, S. J., Mudie, S. T., & Cookson, D. J. (2007). Polymers with cavities tuned for fast selective transport of small molecules and ions. *Science*, *318*(5848), 254–258. doi:10.1126cience.1146744 PMID:17932294

Reynolds, T. D. (1977). *Unit operations and processes in environmental engineering*. Brooks.

See-Toh, Y. H., Silva, M., & Livingston, A. (2008). Controlling molecular weight cut-off curves for highly solvent stable organic solvent nanofiltration (OSN) membranes. *Journal of Membrane Science*, *324*(1), 220–232. doi:10.1016/j.memsci.2008.07.023

Short, J. L. (1993). Membrane Industry News. Westford, MA: Academic Press.

Siddique, H., Rundquist, E., Bhole, Y., Peeva, L. G., & Livingston, A. G. (2014). Mixed matrix membranes for organic solvent nanofiltration. *Journal of Membrane Science*, *452*, 354–366. doi:10.1016/j.memsci.2013.10.012

Strathmann, H. (1985). *Production of microporous media by phase inversion processes*. Academic Press.

Szymanska, K., Zouboulis, A. I., & Zamboulis, D. (2014). Hybrid ozonation–microfiltration system for the treatment of surface water using ceramic membrane. *Journal of Membrane Science*, *468*, 163–171. doi:10.1016/j.memsci.2014.05.056

Toyomoto, K., Higuchi, A., Osada, Y., & Nakagawa, T. (1992). *Membrane Science and Technology*. Academic Press.

Tsibranska, I., Tylkowski, B., Kochanov, R., & Alipieva, K. (2011). Extraction of biologically active compounds from Sideritis ssp. L. *Food and Bioproducts Processing*, *89*(4), 273–280. doi:10.1016/j.fbp.2010.10.004

Tsibranska, I. H., & Tylkowski, B. (2013). Concentration of ethanolic extracts from Sideritis ssp. L. by nanofiltration: Comparison of dead-end and cross-flow modes. *Food and Bioproducts Processing*, *91*(2), 169–174. doi:10.1016/j.fbp.2012.09.004

Tsuru, T., Nakasuji, T., Oka, M., Kanezashi, M., & Yoshioka, T. (2011). Preparation of hydrophobic nanoporous methylated SiO 2 membranes and application to nanofiltration of hexane solutions. *Journal of Membrane Science*, *384*(1), 149–156. doi:10.1016/j.memsci.2011.09.018

Tsuru, T., Narita, M., Shinagawa, R., & Yoshioka, T. (2008). Nanoporous titania membranes for permeation and filtration of organic solutions. *Desalination*, *233*(1), 1–9. doi:10.1016/j.desal.2007.09.021

Tung, K. L., Chang, K. S., Wu, T. T., Lin, N. J., Lee, K. R., & Lai, J. Y. (2014). Recent advances in the characterization of membrane morphology. *Current Opinion in Chemical Engineering*, *4*, 121–127. doi:10.1016/j.coche.2014.03.002

Tylkowski, B., Carosio, F., Castañeda, J., Alongi, J., García-Valls, R., Malucelli, G., & Giamberini, M. (2013). Permeation behavior of polysulfone membranes modified by fully organic layer-by-layer assemblies. *Industrial & Engineering Chemistry Research*, *52*(46), 16406–16413. doi:10.1021/ie402942g

Tylkowski, B., Peris, S., Giamberini, M., Garcia-Valls, R., Reina, J. A., & Ronda, J. C. (2010). Light-induced switching of the wettability of novel asymmetrical poly (vinyl alcohol)-co-ethylene membranes blended with azobenzene polymers. *Langmuir*, *26*(18), 14821–14829. doi:10.1021/la101809r PMID:20799708

Tylkowski, B., Trusheva, B., Bankova, V., Giamberini, M., Peev, G., & Nikolova, A. (2010). Extraction of biologically active compounds from propolis and concentration of extract by nanofiltration. *Journal of Membrane Science*, *348*(1), 124–130. doi:10.1016/j.memsci.2009.10.049

Tylkowski, B., Tsibranska, I., Kochanov, R., Peev, G., & Giamberini, M. (2011). Concentration of biologically active compounds extracted from Sideritis ssp. L. by nanofiltration. *Food and Bioproducts Processing*, *89*(4), 307–314. doi:10.1016/j.fbp.2010.11.003

Valadez-Blanco, R., & Livingston, A. G. (2009). Solute molecular transport through polyimide asymmetric organic solvent nanofiltration (OSN) membranes and the effect of membrane-formation parameters on mass transfer. *Journal of Membrane Science*, *326*(2), 332–342. doi:10.1016/j.memsci.2008.10.016

Wei, Q., Yang, D., Fan, M., & Harris, H. G. (2013). Applications of nanomaterial-based membranes in pollution control. *Critical Reviews in Environmental Science and Technology*, *43*(22), 2389–2438. doi :10.1080/10643389.2012.672066

Zakrzewska-Trznadel, G. (2006). Membrane processes for environmental protection: Applications in nuclear technology. *Nukleonika*, *51*, 101–111.

Zenon Environmental, B. V. (1995). Ultrafiltration combined with a bioreactor for the treatment of tannery effluents. *World Leather*, 43-5.

Chapter 6
Nanotechnology for Air Remediation

Shafaq Mubarak Mubarak
Pakistan Council of Scientific and Industrial Research, LLC, Pakistan

ABSTRACT

The world around us is a gift of God. A thrust to know and reconnoiter the environment around us is innate. Man has drastically explored and utilized the resources hidden in nature, but unfortunately in this sprint of development, the natural environment is severely affected. It is the need of the hour to focus on methodologies for environmental remediation. Many technologies have been developed to reduce the pollution causing factors. Use of nanotechnology for the sake of saving environment is an emerging field. Nano-technology is based on nano-sized (smaller than 1 micron) materials. Nanosize particles have initiated the advancement in new and low cost techniques for environmental pollution control including air pollution.

INTRODUCTION

Latest trends in nanotechnology have been adopted for its use in environmental remediation. Nanomaterials are different from other macro and micro- molecules in their chemical and physical properties and are found to be more effective. Various nano-sized materials have been synthesized to reduce many upcoming health issues based on air pollution.

Air pollution comes from many different sources depending upon the mobility of the pollution emitting source. Static or stationary sources include power plants, factories, smelters and in-door sources such as dry cleaning operations and degreasing operations. Mobile sources which kept on enhancing toxic chemicals in air over a larger area include cars, buses, trucks, planes and trains. All these sources contribute in deteriorating the natural air quality. Pollutants from these sources range from green-house gases to volatile organic compounds and even solid particulate matter.

DOI: 10.4018/978-1-5225-5745-6.ch006

Attention to control the air quality in different regions has been made worldwide. Many countries have set limits in their regions for the emission of specific pollutants. Environmental protection agency (EPA), USA has categorized six major air pollutants as criteria air pollutants. This concept of "criteria air pollutants" is accepted worldwide.

A brief description of major air pollutants, their sources and health hazardous are discussed in Table 1.

Table 1. Criteria air pollutants

Pollutant	Description	Sources	Health hazard	Release	Effect on environment
Carbon Monoxide (CO)	CO is an odorless, colorless, and poisonous gas produced by the incomplete burning of fossil fuels (gasoline, oil, natural gas).	Cars, trucks, buses, small engines, and some industrial processes are major sources. Wood stoves, cigarette smoke, and forest fires are also sources of CO.	CO interferes with the blood's ability to carry oxygen, slowing reflexes and causing drowsiness. In high concentrations, CO can cause death. Headaches and stress on the heart can result from exposure to CO.	Direct	contribute to the formation of photochemical smog
Nitrogen Oxides (NOx)	Nitrogen and oxygen combine during combustion (burning) to form nitrogen oxides. Many nitrogen oxides are colorless and odorless gases.	NOx come from burning fuels in motor vehicles, power plants, industrial boilers and other industrial, commercial, and residential sources that burn fuels.	NOx can make the body vulnerable to respiratory infections, lung disease, and possibly cancer.	Direct	contribute to the formation of photochemical smog, formation of acid rain, visibility reduction, water quality deterioration,
Sulfur Dioxide (SO_2)	SO2 is a gas produced by chemical interactions between sulfur and oxygen.	SO2 comes largely from burning fossil fuels (gasoline, oil, natural gas). It is released from petroleum refineries, paper mills, chemical and coal burning power plants.	Eye irritant Damage to lungs	Direct	formation of acid rain, visibility reduction, plant damages
Particulate Matter (PM) also known as Particle Pollution	Particulate matter is a term used to describe very small solids. Smoke, ash, soot, dust, lead, and other particles from burning fuels are examples of some of the compounds that make up particulate matter.	Some particles are directly emitted from cars, trucks, buses, factories, construction sites, tilled fields, unpaved roads, and burning wood. Other particles are indirectly formed when gases from burning fuels react with sunlight and water vapor.	Particulate matter can reduce visibility and cause a variety of respiratory problems. Particulate matter has also been linked to cancer. It can also corrode metal; erode building and sculptures, and soil fabrics.	Direct and formed in the air	visibility impairment, impacts on trace gas cycles, cloud and fog formation, absorption and scattering radiation
Lead	Lead is a metal found naturally in the environment as well as in manufactured products. Small solid particles of lead can become suspended in the air. Lead can then be deposited on soil and in water.	The major source of lead is metal processing with the highest levels of lead generally found near land smelters. Other sources include waste incinerators, utilities, and leadacid battery manufacturers.	Exposure to lead can cause blood, organ and neurological damage in humans and animals. Lead can also slow down the growth rate in plants.	Direct	Contaminated crops and livestock Smog
Ozone (O3)	Ozone (O3) is a gas not usually emitted directly into the air. Ground level ozone is created by a chemical reaction between NOx and VOCs in the presence of heat and sunlight.	Motor vehicle exhaust, industrial emissions, gasoline vapors, and chemical solvents are some of the major sources of NOx and VOCs.	Ozone can irritate lung airways and cause wheezing and coughing. Repeated exposure can cause permanent lung damage. Ozone damages leaves of trees and other plants. It decreases the ability of plants to produce and store food, and reduces crop yield.	Formed in the air	plant and ecosystem damage, visible injury, decreased productivity, crop yield, indirect effect on global warming

(Source: EPA, USA)

Standards Values of Some Air Pollutants

USA has passed a Clean Air Act which states and limits the hazardous air pollutants present in environment. The Clean Air Act, helped in giving standard limits for exposure to the polluted air with respect to specific pollutant. The clean air act (amended in 1990) requires EPA to regulate and set National Ambient Air Quality Standards for harmful pollutants.

Two types of ambient air quality standards have been made for toxic pollutant exposure. Primary standards and secondary ambient air quality standards.

The standard which deals with the direct effect of the air pollutants to human health comes under the category of "primary ambient air quality standards". This standard also deal with setting limits for exposure of sensitive individual like children, elderly persons and children. Generalized effects of air quality like visibility, harmful effects on vegetation, buildings and animals are discussed under "Secondary standards for ambient air quality".

National Ambient Air Quality Standards (NAAQS) for criteria air pollutants has been set by EPA. These standards are reviewed periodically and may be revised when needed. The current standards are listed in Table 2.0.(Units of measure for the standards are parts per million (ppm) by volume, parts per billion (ppb) by volume, and micrograms per cubic meter of air (μg/m^3).

Table 2. NAAQS standards for air quality (Source: USA,EPA)

Pollutant		Primary/Secondary	Averaging Time	Level	Form
Carbon monoxide		Primary	8 hours	9ppm	Not to be exceeded once per year
			1 hour	35ppm	
Lead		Primary and Secondary	Rolling 3 months average	0.15ug/m3	Not to be exceeded
Nitrogen Dioxide		Primary	1 hour	100ppb	98th percentile of 1-hour daily maximum concentrations, averaged over 3 years
		Primary and Secondary	1 year	53ppb	Annual Mean
Ozone		Primary and Secondary	8 hours	0.070ppm	Annual Fourth-highest daily maximum 8-hour concentration, averaged over 3 years
Particle Pollution	PM 2.5	Primary	1 year	12.0ug/m3	Annual mean, averaged over 3 years
PM10		Secondary	1 year	15.0 ug/m3	Annual mean, averaged over 3 years
		Primary and Secondary	24 hours	35 ug/m3	98th percentile, averaged over 3 years
		Primary and Secondary	24 hours	150 ug/m3	Not to be exceeded more than once per year on average over 3 years
Sulphur Dioxide		Primary	1 hour	75ppb	99th Percentile of 1-hour daily maximum concentrations, averaged over 3 years
		Secondary	3 hour	0.5 ppm	Not to be exceeded more than once per year

* Standard levels are year specified as mention in NAAQS by USA-EPA

ROLE OF NANOTECHNOLOGY IN AIR POLLUTION CONTROL

Nanotechnology is an emerging field of science which has shown utilization in many scientific disciplines. Various compounds have been explored for their unique property at nano-scale level under the umbrella of nanotechnology. Maintenance and monitoring of air quality can be done by using numerous nano materials in different ways.

Nano bio-sensors, nano-adsorbents, nano-membranes/ thin films for air-filters, photo-catalysis, nano-dendrimers, and many others metal sensors can be used to regulated air quality.

Nano Solid State-Sensor

Sensors are used to monitor air quality by identifying specific compounds. Nano-particle based solid phase sensors can assess the pollutants in the air more precisely. Sensitivities of nano solid sensors are usually high for even small amount (ppm to ppb) of toxic pollutants. Detection studies of toxic particle in few environmental systems has been done via using nano-based solid sensor devices.(Zhou et al., 2015) .

Following ways could be used to utilize nanotechnology as a sensor technique for air pollution (Zaporotskova et al., 2016).

1. Coating of nano-particle on organic and biological ligands can be used to enhance specificity of the sensor.
2. The interaction of nano-material to the pollutant molecule can be promoted due to higher surface to volume ratio property of nano-particle.
3. By adsorption/coating on different metals the conduction and pollutant specificity of the nano-sensor increases.

Concept of designing multiplex sensor (to sense more than one pollutant) is gaining interest due to advances in nanotechnology resulted in cost reduction of analyzing and sensing for different air pollutants (Kim et al., 2016; Bhawana, et al.,2012).

Gas Nano-Sensors

In air various gaseous pollutants can be sensed using various gas sensors especially from industrial smoke or from vehicle exhaust. These are the coatings of nano-solid sensor materials mainly metal oxides (Zinc Oxide, Tungsten oxide, Tin oxide and titanium oxide) on solid surfaces with electrodes. These gas sensors are designed to operate in drastic thermal or chemical attack that's way find their utility at industrial chimneys and vehicle exhaust. Hence the solid state nano-gas sensor are getting importance at industrial scale and at large to determine air pollution due to specific pollutant over a large area (Pummakarnchanaa et al., 2005).

- *Mechanism of Gas Sensor*:

Gas sensor operates at the principle of change in conduction or resistance of the sensors fixed on certain places to monitor air pollution. The sensing material showed variation in voltage fluctuation of the electrodes by the attack of specific pollutant. Hence showing signals as compared to the background plane air.(Simon et al., 2001).Tungstate oxide based sensor is discussed here as it is the most suitable and economical nano-gas sensor.

Tungstate Oxide Based Nano-Gas Sensor

Studies have been carried out to synthesize and fabricate different kinds of nano metal oxide materials to sense the presence of noxious gases in air; among these TiO_2 has gained considerable attention. Micro-fabrication thin film technique can be use to coat tungsten oxide nanoparticle on solid surface. Tungsten oxide showed excellent sensing ability towards nitrogen dioxide presence in air as a part of thin film sensor (Shih-Han et al.,2003). The thin film was fabricated by crystalline nano-porous metal oxide from its chloride salt through sol-gel process. The synthesized nano-crystalline thin film calcinated on different temperature has been studied for its surface modification and sensitivity towards NO_2 gas in air. It has been observed that the nano-film calcinated at 550oC up to an hour, adhere properly to the sensor surface and showed higher sensitivity for NO_2 particles present in air . The sensor operating temperature was found to be best at 300oC. Further, experiments showed that the tungsten oxide nano-crystalline based sensor has quick response time i.e. three minutes to sense the NO_2 concentration for levels i.e-50 to 550ppb.

Surface modification of the developed tungsten oxide nano film on calcination at 550°C has been shown in figure 1. Sensitivity of the thin film depends upon the extent of fabrication of the tungsten oxide on the sensor solid surface. The structural modification played important role for the interaction of NO_2 . Proper calcination results in producing good thin film of metal oxide which depends upon grain size, surface geometry and heterogeneity of nano-particles. Hence, tungsten oxide calcinated at 550oC, depicts relatively higher sensitivity and response to NO_2 particles with an operating temperature of 300oC.

Figure 1. Surface morphologies of tungsten oxide thin films (on the highly polished alumina substrate) at calcination temperature 550oC
(Source: Wang,2013).

Nano Adsorbents

Another application of nanotechnology is the use of nanoparticles for the adsorption of pollutants from air. Many nano-adsorptive materials has been developed and categorized for detecting and absorbing pollutants like metal oxide, VOC (volatile organic compounds) and microbes. Some of the nano-adsorbents used to clean air has been discussed as follows.

Carbon Nano Tubes (CNT)

Studies have been conducted to explore carbon nano-tubes as an adsorbent for many metal ions. It has been noted that these CNT showed better adsorption than the use of simple activated carbon particles from different sources. Carbon nano tube possess graphitic surface of the carbon donated electrons. In a research study of nanomaterials, it has been found that electrons has been donated by CNT form chemical bond with organic functional groups like amide, hydroxyl etc.(Prachi et al., 2013; Pan & Xing, 2008). Carbon nano tubes are also use to control the emission of methane and carbon dioxide from exhausts of industrial chimneys and vehicles. It also helped to trap the green house gases from power generation and coal mines.(Chirag, 2015).

Contaminated air with radio nuclides and heavy metals can also be treated with nano-sized metal oxides adsorbent .these nano-adsorbents includes Titanium oxide, aluminium oxide, ferrous oxide etc. (Mayo et al.,2007).

- *Feature s of CNT:*

The properties which make CNT as a good nano-adsorptive material for the removal of many contaminants from air include their high affinity for the selective pollutant uptake. CNT possess various morphological features which enhance their capacity of pollutant uptake. These features include, the increased surface area of the nano adsorbent, enhanced pore volume and pore diameter . (Wang et al.,2013). These surface features increases the reactive sites which can bound with the specific pollutant particle present in air(Gupta & Saleh, 2013).

In addition to this, structural morphological changes the absorptive capacity can be increased by adding different functional groups to nano-adsorbents to increase active sites. Some nano-structures based on fullerene, Carbon nano tube, Graphene and graphite has shown in figure 2.

Carbon nanotubes exist in different forms structurally (Zhao & Stoddart, 2009) One is Singled walled CNT(SWCNT),, as result of rolling up of grapheme sheets in concentric manor. The other one is Multi-walled CNT(MWCNT), which is also made from layering up of graphen sheets (Figure 3).

Much work has been done for exploring Treatment techniques based on various carbon nano-adsorbents under different conditions . Studies of few researches have been summarized in table3.0. to monitor and eliminate the VOC from polluted air.

Removal of Specific Pollutants by Nano-Adsorbents

Various outdoor and indoor air pollutants from different industries and houses continuously emitted in the surrounding air. To decrease the level of these pollutants, nano-adsorptive materials are used. Mer-

Figure 2. Structural features of nano-adsorbents
(Source: Bergmann & Machado, 2015)

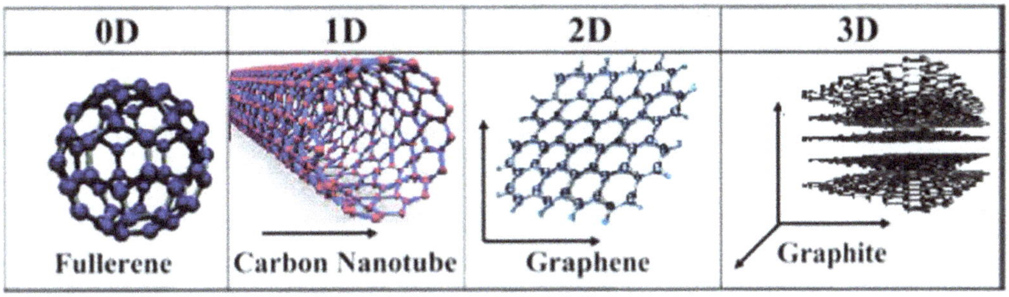

Figure 3. Structure representations of (a) MWCNT and (b) SWCNT
(Source: Zhao & Stoddart, 2009)

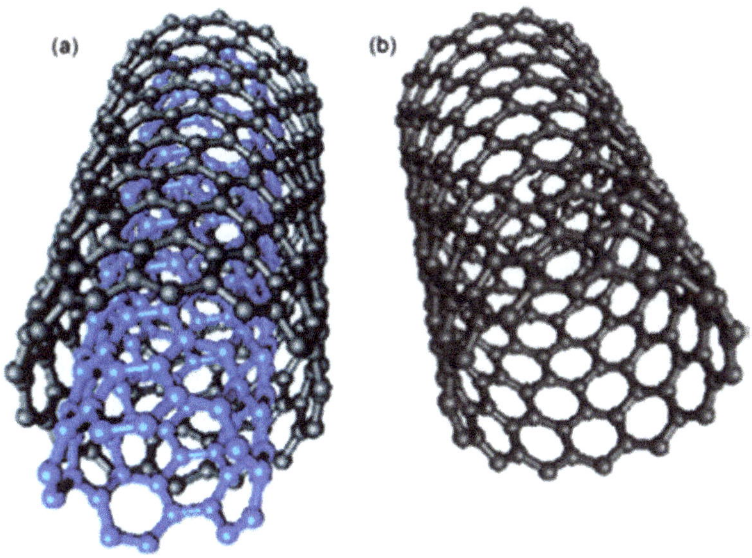

cury vapors, formaldehyde etc. are few of these air pollutants. Nano-adsorption studies on adsorption of mercury and formaldehyde has discussed here:

- *Mercury Adsorption*:

Flue gases from various industries cause the emission of mercury vapors in the form of mercuric chloride. It is need of the hour to bring the level of mercury in the air at optimum level to avoid any harmful effects. Nano-adsorbent material based on some chelating ligand for mercury has been developed (Malyuba et al.,2005) .Synthesizing an ionizing nano-layer surface with a stable chelating ligand is the basic concept of the research. Studies were carried out for the removal of mercuric chloride from gas phase. The nano-ligand surface was mobilized with metal (mercury) chloride, resulted in complexation of the metal with ligand. The nano-layer of the ligand has been developed on solid meso-porous silica.

Table 3. Nanomaterials for air remediation

Nano-Adsorptive Materials	Types of Nanoparticles	Target Pollutant Gases	Removal Mechanism	References
Carbon nanotubes (CNTs)	(SWNTs and MWNTs)	NO*x* (mixture of NO and NO2)	NO and O_2 pass through CNTs and NO is oxidized to NO_2 and then adsorbed on the surface of nitrate species.	(Long & Yang, 2001; Zhang et al., 2012)
	(CNTs-APTS), Modified CNTs using 3aminopropyltriethoxysilane (APTS).	CO_2	Surface of CNTs with abundant amine groups that provide numerous chemical sites for CO_2 adsorption which makes CNTs adsorb more CO_2 gases at low temperature range (20-100 °C).	(Su et al.,2009)
	SWNTs/NaClO	Isopropyl vapor	Physical adsorption by van derWaals forces and chemical adsorption onto adsorbent surface functional groups.	(Hsu & Lu,2007)
	CNTs deposited on quartz filters	VOCs	It carried out by π- π interactions.	(Amade et al., 2014)
	Si-doped and Boron-doped SWCNTs	CO and CH_3OH gases	Physisorption or chemisorption, the electronic properties of SWCNT improves significantly the gas adsorption.	(Azama et al., 2017)
Fullerene	fullerene B40	CO_2	high adsorption capacity for CO_2 by strong chemisorptions.	(Dong et al., 2015)
	fullerene-like boron nitride nanocage	N_2O	Adsorption and decomposition of N_2O.	(Esrafili, 2017)
Graphene	Graphene oxide (GO)/nanocomposits	CO_2, NH_3, SO_2, H_2S and N_2	Functional groups on GO are responsible for adsorption of gases, synergetic effect between the metal centers andGO surface promote the gases adsorption.	(Petit & Bandosz, 2009; Seredych & Bandosz, 2012; Wu et al., 2013)

Table source: (Elham, 2017)

This layer provides an excellent space for the adsorption of mercury chloride in vapor form . Optimization of adsorption parameters has done. Further, it has been observed from elemental analysis that mercury can be adsorbed upto 33mg/g of nano-adsorbent layer. Effect of temperature on adsorption revealed that it can be operated at lower temperatures with the highest possible temperature of 135°C. Minimum capacity for mercury vapors of the synthesized ligand found to be 12mg/g.

- *Formaldehyde Adsorption*:

In-door air quality can be enhanced by using various nano-adsorbents based on titanium and zinc oxide materials. Formaldehyde is one of the common indoor air pollutant,. In a detailed study for removing formaldehyde from air through nano-particle based adsorbent, various materials has been investigated (Chen et al.,2016). Solyo-thermal process to synthesized nano-adsorbent has been used. Reaction is mediated by diethylenetriamine (DETA) on one-pot synthesis to synthesized hierarchial titanate nanosphere(HTS). Surface area of the nano-material is enhanced which results in adsorption of many functional groups present on the VOC in air as pollutants.

Adsorptive studies of HTS for the removal of formaldehyde from in door air have shown 100% removal. This is based on the hierarchical structure of titnate adsorbent which increases the surface area of adsorption and also increases the pore size of adsorbent. This HTS has several amine binding groups for the efficient removal of formaldehyde. Use of DETA for reaction mediation played a vital role in inducing textural properties. This surface modification boosted the adsorption capacity of HTS for formaldehydes vapors from in-door air. Hence helped in controlling the air pollution.

Nano-Filters

To control air pollution another technique, based on using nano-materials, is the use of nano-air filters. These nano-filters based on the principle to clean the air from pollutant material via- adsorption or filtrating the pollutants from exhausts gases. This is done by using various nano-membranes in air filters and nanostructures to adhere the pollutants. Also, some nano-fiber coated air-filters has been installed for filtration of inlet gases in industrial plants (Muralikrishnan et al., 2014)

Filtration of Particulate Matter

Presence of small size particles in air categorized as particulate matter which resulted in polluting the air. (Zhang et al., 2013) . These particulate matter have various sizes . The size which could be harmful for human health is denoted as 2.5micron and 10micron. (Harrison, 2000). The pollution caused by PM2.5 and PM10 can cause health hazards related to breathing . These small particle adhere to the lung passage and can penetrate to the tissues(Betha et al., 2014; Wu et al.,2014; Brook et al.,2010; Anenberg, 2010; Timonen et al., 2006).

Many frame works have been introduce to capture the particulate matter from air. The nano-material can be used as filtration source in air filter or can adhere to the window panes to adhere particulate matter from in-door air resulted in adhesion . Figure 4 showed image of SEM of a PAN nano-fiber based air filter material capturing the particulate matter.

Another, approach to control particulate matter air pollution is to capture these via metal-organic frame work (MOF) technique. The fabricated nano-sized crystalline MOF possesses rich functionalities, increased porosity and good adherence for particulate matter (Zhang et al., 2016). Figure 5 showed SEM image of one of such MOF with and without adherence of particulate matter.

Membrane Filter for Industrial Emission and Vehicles

Use of nano membrane filters to remove various polluting gases from the air finds it's utility in controlling air quality. Incomplete catalytic conversion of fuel in vehicle resulted in emission of green house gases from the vehicle exhaust. The nano membrane material is fabricated on the lining of exhaust system. This is a very intricate process. Fabrication of the nano-membrane requires its selectivity with toxic gases. The choice of nano-meberane filter material should be based on considering the stringent operational temperature of vehicle exhaust and it should be heat resistance. (Sugg et al.,2012)

Following characteristics of the nano-membrane filters are important to control air pollution

Figure 4. a) SEM image showing the detailed morphologies of attached soft PM that formed a coating layer wrapping around the PAN nanofibre. Scale bar, 1 mm. (b) SEM image showing that the nanofibre junction has more PM aggregated to form bigger particles. Scale bar, 1 mm.
(Source: Liu et al., 2015)

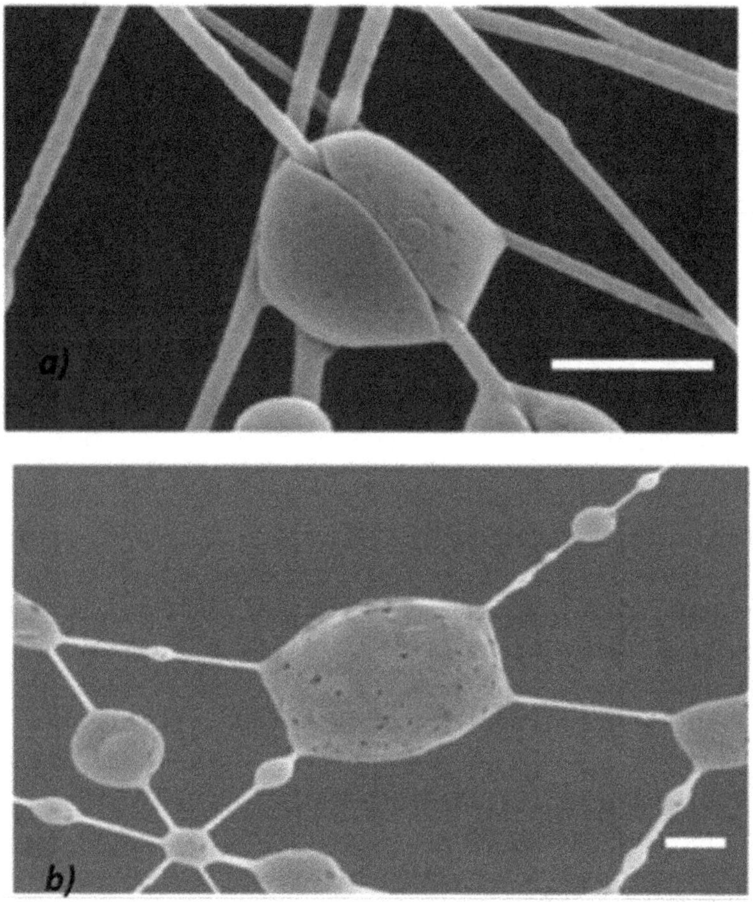

- *Pore Size*:

Since nano fabrication on nano-membrane surface involved the mechanism of adsorption of toxic materials from the polluted air, its pore size design is very crucial. Care must be taken in designing the nano-filter membrane according to the choice of the adsorption of pollutant. Some filters require small pore sized nano-membrane but some require large pore size to capture bigger size molecules. Surface pore size of nano-membrane can be controlled by using various surfactant at the time of synthesis. Similarly, symmetric and asymmetric pore size can be formed for various applications. Once synthesized, the membrane further be treated by a solution of surfactant or by physical means to set the size of it's pores. Use of acidic condition can resulted in the oxidation of the surface of nano-membrane resulting in a smaller pore size.(Muralikrishnan et al., 2014)

Figure 5. Photos and SEM images of the MO Filter before and after PM capture
(Source: Zhang et al., 2016)

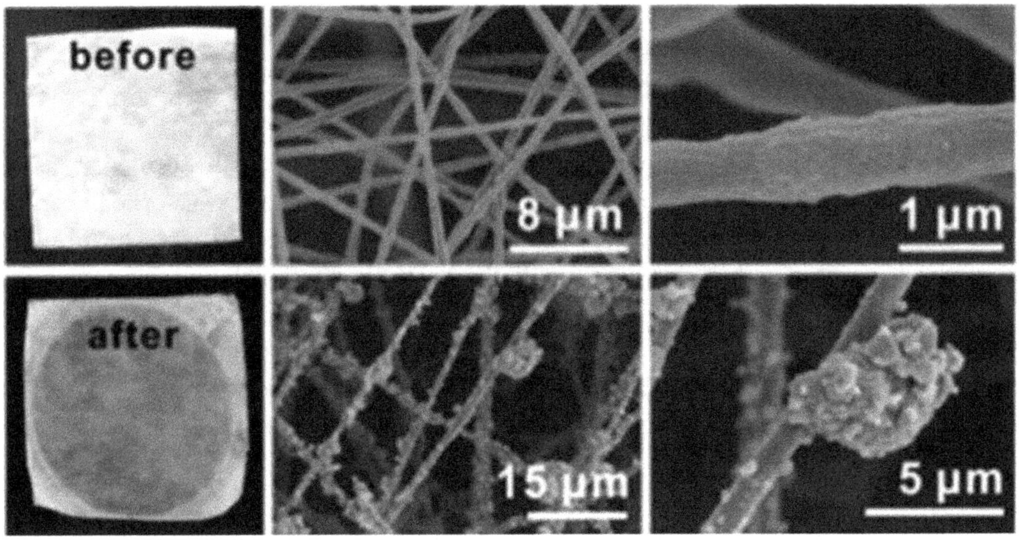

- *Suggested Nano-Absorptive Materials:*

Other important factor which can contribute for the characteristics of the nano-membrane is the type of absorptive material fabricated on the nano-memberane. Specificity of pollutant removal depends upon the nature of the nano-absorptive material. Concept of nano-sized molecular sieves to remove dust or particulate matter from air is gaining interest of many researchers. Zeolite based alumino-silicate molecular sieves has been developed and investigated for its efficacy for the adsorption of sand particles. Molecular sieves have capacity to absorb up to 24% of their mass. These sieve restrict the flow of bigger size molecule by absorbing directly on the membrane . Carbon dioxide gas can also be trapped from pollutes air by using molecular nano-sieves (Albo, 2010). Structural modification also have little effect on the adsorption as the empty spaces or pores hold up the smaller sized particles and acts as a trap for them. Further, these sieves worked on the principle of size only and does not restrict to the nature of the pollutant present in air. This property makes these nano-sized molecular sieves to be used in trapping various toxic compounds from air passed through the air filters.

Nano-Catalyst

Air pollution can not only be controlled by removing it from surrounding air, it can also be controlled by applying certain treatment techniques to reduce the emission of these toxic pollutants in air. The effectiveness of using nano-catalytic system for air remediation is acquiring attention of the scientists. Nano-catalyst help in fast and specific chemical transformation which resulted in the proper and fast conversion of chemical compounds to desired products with a lesser emission of fuel gases.

Also these nano-catalyst used in the convertor reactions of automobile vehicles and industrial exhaust which convert the toxic gases to harmless gases. One example is the use of manganese oxide based nano-fiber catalyst which help in reduction of VOC from smokestacks at industrial level.(Pandey et al.,2012). Use of different nano-catalyst in various reactions and air-purification techniques has been discussed here in detail.

Titanium and Carbon Based Nano-Catalyst

Conventional methods for purifying air uses simple activated carbon and oxide of titanium as an adsorbent for removing noxious gases. This technique is less specific and have low absorptive capacity. It is relatively expensive and does not work properly in ambient air, hence loses its popularity. In spite of this, use of nano-based activated carbon and titanium oxide showed exceptional air purification properties by photo-catalysis based purification principle.

A titanium based nanophotocatalyst system has been developed by compounding activated carbon nano particle on solid surface then the nano-sized titanium oxide is loaded on granules of carbon. This supported nano-catalyst based membrane filtration support is used in various air purification techniques (Gao, 2004).

Example: Use of Nano-Catalysts in Air Conditioning System

A practical example to understand the air purification application of nan-catalyst has been discussed in detail by taking into account the complete system of air-conditioning and purification of the air. This limits the use of a separated cleaner system for air cleaning. Moreover, the air supply of the conditioning system should itself be clean hence require air-purifier to clean the inlet air.

Generalized unit of air cleaning has been shown in figure 6. Consisting of five unit operations like filters, heat /wet system, nano-catalysts purifier and air machine (Hu et al., 2005).

Figure 6. Air-and-air conditioning system with purification section 1. First filter 2. Middle filter 3.Heat and wet handling facility 4.Active carbon and nano-titanium dioxide photocatalysis purification section 5.Air machine
Source (Hu et al., 2005)

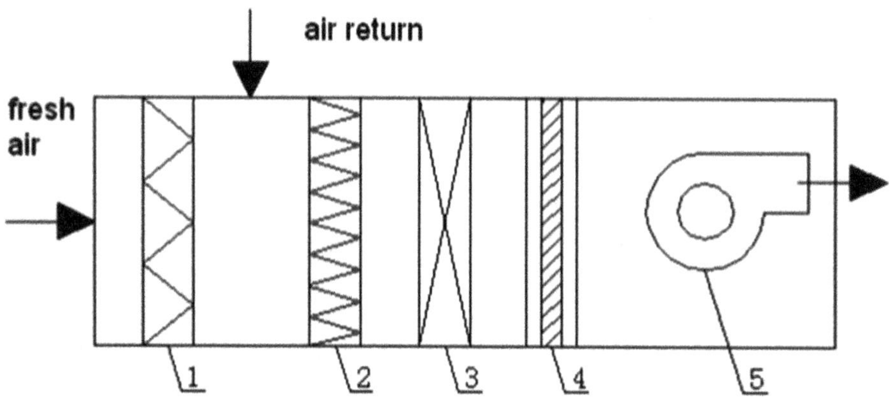

- *Mode of Working:*

The first filter receives air in natural form from surrounding environment. This filter clear the dust particles and granules present in air. This cleaner air circulates via air return valve to air filter and nano titanium oxide catalyst unit. Noxious gases get oxidized and adsorbed on the catalytic chamber. From here air goes again in circulation and passed through primary filter to remove larger sized particles. Then the flow of air continued to the heat and wet unit. After this process the air flow throw air blower outside the system in the room. Hence in this way nano-catalyst based air purification air conditioning system operates.

It is further concluded that this activated carbon –titanium oxide nano-photocatalyst technique is more effective than using single technique like for using adsorption of CO_2 and formaldehyde from contaminated air.

Common Nano-Photocatalyst

Various studies have been conducted to investigate the use of nano-catalyst to adsorbed gases from air. Titanium oxide film has used for degradation of ammonia and acetaldehyde in industrial flue gases (Sopyan et al., 2007). Surfaces are modified from titanium oxide film on solid surface with advance catalytic property. It is designed for self –sensing, quantifying and decontamination properties. Skubal et al. (2008) based on feedback monitoring system.

Certain toxic vapors in industries contaminates the air and resulted in its pollution .Exposure to these toxic VOC can cause skin disease and cancer.(Qi jin et al., 2008). Mixed metal oxides in the form of heterogeneous nano-catalyst have been explored for degradation of different pollutants from air. (Linda et al., 2006).

Oxidation through nano-catalyst of benzene and toluene present in industrial smoke has been studied on titanium and zinc oxide based heterogeneous nano-photoctalyst. In the study a glass surface is fixed with nano-catalyst. Removal efficiency for toluene and benzene has been extensively studied. Possible parameters to effect the removal of these VOC has investigated and concluded that the use of UV intensity enhance the degradation activity of these nano-catalyst. Maximum UV intensity on which experiments were conducted has found to be 10 w/m^2 and minimum removal efficiency is as low as UV intensity at 4 w/m^2 (Gholami et al., 2014). In another study only Zinc oxide nano-photocatalyst is employed with the exposure of UV radiation (Nassehinia et al.,2013).

Degradation of VOC(toluene and benzne) by nano-photocatlyst is increased by the exposure of UV radiation .Nano catalyst, titanium oxide has absorption on UV at $\leq$ 387 nm, which induces electrons from valence band to conduction band leaving behind Positive hole (McMurray et al., 2005) as shown in the following reaction mechanism.

$H_2O \rightarrow OH^- + H^+$

$H^+ + UV + OH^- \rightarrow OH^o$

$e- + O_2 \rightarrow O_2^o$

OH^o + Organic reactants + $O_2 \rightarrow$ Products (CO_2+ H_2O etc.)

Formation of hydroxyl ion causes the degradation of toluene and benzene. The more nano-catalyst surface exposed to UV light, more degradation off VOC occurred. Excellent results obtained by increased UV light intensity and time duration of exposure of these VOC in the reactor. TiO_2/ZnO based removal of VOC followed first order reaction kinetics. The catalyst showed more affinity to degrade toluene than benzene.

Nano-Photo Catalytic Removal of Dyes

In –door air pollutants not only consist of common VOC and exhaust gases but also some dye present in the air can effect the quality of air. Nano-photocatalyst found its application in sensitizing the dye particles in the air. Dye sensitized solar cells(DSSC) is a known technique to detect various natural and synthetic dyes. However, evaluation has been conducted to fabricate a glass surface with nano titanium oxide material to form a dye sensitized TiO_2(DST) device followed by calcination at elevated temperature (Lee et al., 2013).

Thorough studies to check the ability of the developed DST to sense natural and synthetic dyes has been done based on natural dyes (raspberry, blueberry, cranberry). Artificial dyes used were Ru complexes with N3 ($RuL_2(NCS)_2•2H_2O$), N719($C_{58}H_{86}O_8N_8S_2Ru$).

Removal efficiency of dye via catalytic degradation under different irradiation sources were conducted by using light bars with LEDi.e light emitting set up. Effect of different light wavelengths i.e., BLED (blueLED), RLED (red LED),WLED (white LED) and UVLED(ultraviolet LED).

Results showed 80% higher sensitization of DST to TiO_2 up to the level of 30times raspberry dye solution. Order of η for N_3 and N719 sensitization under different light bars, the sequence is as follows

UVLED>BLED>WLED>RLED

N719 removal through DST degradation was 248mg/g, this is 11 times increased than the use of simple TiO_2 catalyst. Same studies when done on UVLED, the removal was 720mg/g and its sensitization increased upto 95%. In addition to it, visible light always has effect on photo-catalytic reactions.

Photocatalytic Air Filter

Applying combination techniques to regulate air quality is an emerging filed. Combining PCO (photocatalystic oxidation) and ESF(electrostatic air filter pack) can reduce the air pollution efficiently. Both these techniques helped in maintaining the quality of air, indoor and outdoor.

ESF worked on the principle of electrical precipitation of air particles. This is based on a needle like electrode (corona wire), which charge the surrounding air particles by creating electromagnetic field. This ionized air particles get adhere to the collecting plates at air filters (Hinds 1998).Unfortunately, this ESF technique has limitation in removing VOC from polluted air, hence a technique to treat the coating of air filter based on Silica substrate introduced by coating photo catalytic oxidizing material, like titanium dioxide.(TiO_2). Additives, e.g PEG applied during fabrication of PCO (titanium dioxide) to solid glass fiber has been investigated and found better application for good adhesion of nano-catalyst to the substrate as compared to conventional surfactants (Uzunova-Bujnova et al., 2009; Yao & Yeung, 2011).

Use of combined ESF and PCO with structural modification has been investigated to remove VOCs., particularly xylene from polluted air. (Limmongkon et al., 2013). Different chemical and mechanical modification for air filter has been done and tested for the removal of xylene. Basic comparison was one between PCO coated glass filter surrounding with three layers of ESF air filter packs and other is PCO glass filler with four layers of ESF air filter pack. Further, these combinations were placed in air filter packs. In addition to it, PCO coating is done with varying titanium dioxide concentration, i.e., at 2% and 5% of TiO_2. Degradation studies on xylene showed that 5% TiO_2 coating with four layers of ESF have removal efficiency from 5-13% under irradiation.

Modification of the combination system with HEPA(high efficiency particulate air) glass filter mounted with 5% TiO_2 with four layer of ESF in similar studies showed very high degradation efficiency up to 41-49% for xylene removal.

PCO for Bacterial Decontamination

Few studies has been conducted on using different nano-techniques for purifying polluted air from bacterias/ microbes . Deactivation of *E.coli* in presence of visible light illumination of nitrogen doped Titanium dioxide by following sol-gel synthetic technique. This nanocatalyst based on PCO, resulted in decontamination of air from bacterial pollution. (Qi et al., 2007).

In another work degradation of two strains of bacteria, i.e. *bacillus subtilis* and *pseudomonas aeruginosa,* has been monitored by using nano structured nitrogen based TiO_2 films. The photocatalytic activity showed 75% and 95% removal of these strains from contaminated air respectively (Xu et al., 2006)

The limitation of operational efficiency of PCO technique under visible light only has been overcome by developing carbon and vanadium doped TiO_2 nano catalyst. Studies showed that coating with 2% carbon and vanadium has the capacity for bacterial removal irrespective of the irradiation process i.e., it can be operated both in night and day time (Yang et al., 2007)

Nano Dendrimers

Nano-technology not only deals with tiny particles but can found its application on larger structures as well in the form of nano-dendrimers. This complex structure is based on star-shaped molecules having nano-scale size with specific composition. These nano-dendrimers played a vital role in air remediation technologies (Vani Padmaja, 2016).

Basic structure of dendrimer consists of three components: interior dendritic structure with a central core and exterior surface mounted with various functional groups which need up in formation of supra nano-molecular structure (Figure 7).

Much work is done to explore the role of dendrimers in air remediation. Modification of dendrimers has been done via changing functional groups on dendrimers to chelating ligands for trapping various contaminants from polluted air. These structures can be used in ultra-filtration systems of air-purification to create a cage for capturing the pollutants from airVossmeyer et al., 2002; Yunus et al., 2012; Balogh et al., 2001)

Other researchers showed that addition of various functional groups to dendrimer globular molecules resulted in excellent adherence of small sized molecules. These additional functionalities at intra-

Figure 7. Dendrimer general structure
(Source: G. Vani Padmaja)

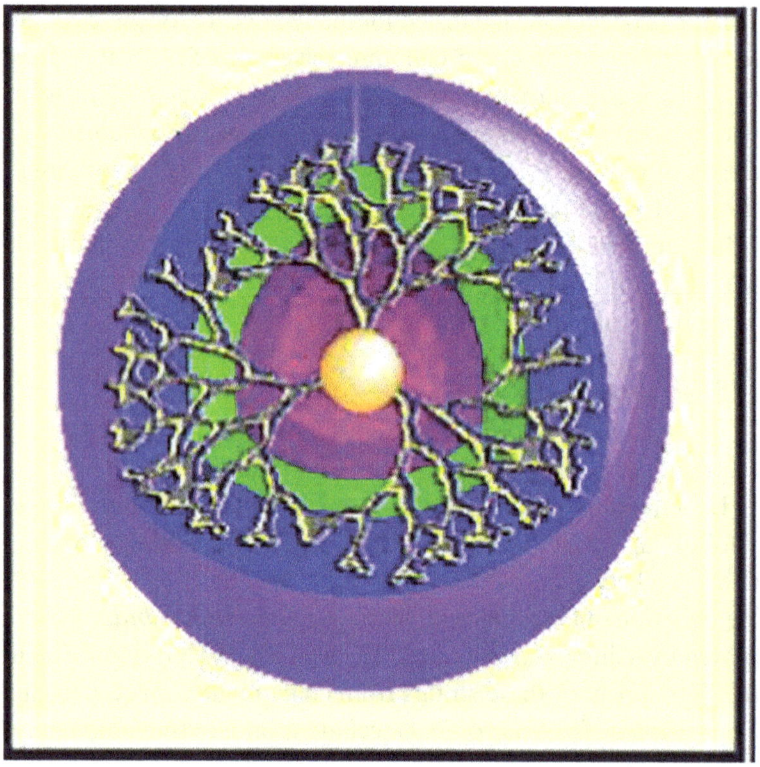

molecular level and at terminal level of dendrimers, make it a proper sized molecular structure to trap specific air pollutants (Balogh et al., 1999; Newkome et al., 1996; Bosman et al.,1999). Dendrimers are the highly sensitive and specific in their interaction with various gases and metals. Thin films in air filters can be mounted by these nano-dendrimers for high specificity and fast removal of contaminants from air (Schlupp et al., 2001).

For the removal of VOCs, from the contaminated air, dendrimer macro-structures has been doped with metal nano-particles. Research showed that by adding gold nano-parrticle to organic dendrimers i.e polyphenylene and propylene imine to form thin film resistors for sensitizing VOC., is achieved through layer by layer synthetic technique of thin film formation (Nadejda K., et al., 2002). Mechanism of trapping VOC involves induction of some electrical conductivity to the dendrimer core via metal-nano-particle which helped in selective adsorption of charged vapors of organic molecules like 1-propoanal and toluene on the dendrimer thin films. Selectivity is controlled by varying the metal type of nano-particle dopped on dendrimers.

CONCLUSION

Pollution prevention and control is one of the major targets of the countries worldwide. For this, application of nanotechnology in the field of air-remediation has been explored. Use of many nano-techniques for sensing and cleaning air from industrial smoke stacks and vehicle exhaust has been applied to industries. These techniques utilizes nano-materials like carbon nano tubes, nano-membranes etc. Use of nano-catalysis has also played important role in cleaning air from industrial smoke by speeding up the reaction involved for degradation of toxic pollutants. Moreover, certain surface modification of nano-adsorptive material has also been discussed for removal of VOC and other air contaminants. Hence, it is found that nanotechnology played an important role in maintaining and controlling air quality.

REFERENCES

Abu-Daabes, M. A., & Pinto, N. G. (2005). Synthesis and characterization of a nano-structured sorbent for the direct removal of mercury vapor from flue gases by chelation. *Chemical Engineering Science*, *60*(7), 1901–1910. doi:10.1016/j.ces.2004.11.028

Albo, J., Luis, P., & Irabien, A. (2010). Carbon Dioxide Capture from Flue Gases Using a Cross-Flow Membrane Contactor and 1-Ethyl-3-methylimidazolium Ionic Liquid. *Industrial & Engineering Chemistry Research*, *49*(21), 11045–11051. doi:10.1021/ie1014266

Anenberg, S. C., Horowitz, L. W., Tong, D. Q., & West, J. J. (2010). An estimate of the global burden of anthropogenic ozone and fine particulate matter on premature human mortality using atmospheric modeling. *Environmental Health Perspectives*, *118*(9), 1189–1195. doi:10.1289/ehp.0901220 PMID:20382579

Azam, M. A., Alias, F. M., Tack, L. W., Seman, R. N. A. R., & Taib, M. F. M. (2017). Electronic properties and gas adsorption behaviour of pristine, silicon-, and boron-doped (8, 0) single-walled carbon nanotube: A first principles study. *Journal of Molecular Graphics & Modelling*, *75*, 85–93. doi:10.1016/j.jmgm.2017.05.003 PMID:28531817

Balogh, L., Swanson, D. R., Tomalia, D. A., Hagnauer, G. L., & McManus, A. T. (2001). Dendrimer–Silver Complexes and Nanocomposites as Antimicrobial Agents. *Nano Letters*, *1*(1), 18–21. doi:10.1021/nl005502p

Balogh, L., Valluzzi, R., Laverdure, K. S., Gido, S. P., Hagnauer, G. L., & Tomalia, D. A. (1999).. . *Journal of Nanoparticle Research*, *1*(3), 353–368. doi:10.1023/A:1010060404024

Balogh, L., Valluzzi, R., Laverdure, K. S., Gido, S. P., Hagnauer Bergmann, C. P., & Machado, F. M. (Eds.). (2015). *Carbon nanomaterials as adsorbents for environmental and biological applications*. New York City: Springer International Publishing.

Betha, R., Behera, S. N., & Balasubramanian, R. (2014). 2013 Southeast Asian smoke haze: Fractionation of particulate-bound elements and associated health risk. *Environmental Science & Technology*, *48*(8), 4327–4335. doi:10.1021/es405533d PMID:24646334

Bhawana, P., & Fulekar, M. H. (2012). Nanotechnology: Remediation technologies to clean up the environmental pollutants. *Research Journal of Chemical Sciences*, *2*(2), 90–96.

Bosman, A. W., Janssen, H. M., & Meijer, E. W. (1999). About dendrimers: Structure, physical properties, and applications. *Chemical Reviews*, *99*(7), 1665–1688. doi:10.1021/cr970069y PMID:11849007

Brook, R. D., Rajagopalan, S., Pope, C. A., Brook, J. R., Bhatnagar, A., Diez-Roux, A. V., & Peters, A. (2010). Particulate matter air pollution and cardiovascular disease. *Circulation*, *121*(21), 2331–2378. doi:10.1161/CIR.0b013e3181dbece1 PMID:20458016

Chen, F., Liu, S., & Yu, J. (2016). Efficient removal of gaseous formaldehyde in air using hierarchical titanate nanospheres with in situ amine functionalization. *Physical Chemistry Chemical Physics*, *18*(27), 18161–18168. doi:10.1039/C6CP03037H PMID:27331932

Chirag, M. P. N. (2015). Nanotechnology: future of environmental pollution control. *International Journal on Recent Innovation Trends in Computing and Communication, 3*(2), 79-81.

Clean Air Act, NAAQS (National Ambient Air Quality Standards (NAAQS), Air and Radiation US EPA.

De Beyer, J., Kollars, N., Edwards, N., & Cheung, H. (2004). *Research on Tobacco in China: An Annotated Bibilography of Research on Tobacco Use*. Health Effects, Policies, Farming and Industry.

Dong, H., Lin, B., Gilmore, K., Hou, T., Lee, S. T., & Li, Y. (2015). B 40 fullerene: An efficient material for CO_2 capture, storage and separation. *Current Applied Physics*, *15*(9), 1084–1089. doi:10.1016/j.cap.2015.06.008

Esrafili, M. D. (2017). N_2O reduction over a fullerene-like boron nitride nanocage: A DFT study. *Physics Letters. [Part A]*, *381*(25), 2085–2091. doi:10.1016/j.physleta.2017.04.009

Gao, L. X. (2004). Current Situation and Improved Measures of Iindoor Air Cleaner. *Period of Haerbin Industry University*, *36*(2), 199–201.

Geng, Q. J., Guo, Q. J., Cao, C. Q., & Wang, H. Q. (2008). Investigation into photocatalytic degradation of gaseous benzene in a circulated photocatalytic reactor (CPCR). *Chemical Engineering & Technology*, *31*(7), 1023–1030. doi:10.1002/ceat.200700510

Gholami, M., Nassehinia, H. R., Jonidi-Jafari, A., Nasseri, S., & Esrafili, A. (2014). Comparison of Benzene & Toluene removal from synthetic polluted air with use of Nano photocatalytic TiO_2/ZNO process. *Journal of Environmental Health Science & Engineering*, *12*(1), 45. doi:10.1186/2052-336X-12-45 PMID:24499601

Gupta, V. K., & Saleh, T. A. (2013). Sorption of pollutants by porous carbon,carbon nanotubes and fullerene- An overview. *Environmental Science and Pollution Research International*, *20*(5), 2828–2843. doi:10.100711356-013-1524-1 PMID:23430732

Hinds, W. C. (1998). *Aerosol Technology* (2nd ed.). New York: John. Wiley & Sons.

Hsu, S., & Lu, C. (2007). Modification of Single-walled Carbon Nanotubes for Enhancing Isopropyl Alcohol Vapor Adsorption from Air Streams. *Separation Science and Technology*, *42*(12), 2751–2766. doi:10.1080/01496390701515060

Hu, Y., Feng, G., Li, G., Cao, G., & Chen, X. (2005). Application of Active Carbon and Nano-titanium Dioxide photocatalysis web in air-conditioning system. *Indoor Air*, (Supplement, 15), 161–162.

Ian, S. (2012). Nanotechnology in water and air pollution treatment. *Journal Environmental Technology Reviews*, *1*(1), 136–148. doi:10.1080/21622515.2012.733966

Kim, J., Choi, S. W., Lee, J. H., Chung, Y., & Byun, Y. T. (2016). Gas sensing properties of defect-induced single-walled carbon nanotubes. *Sensors and Actuators. B, Chemical*, *228*, 688–692. doi:10.1016/j.snb.2016.01.094

Krasteva, N., Besnard, I., Guse, B., Bauer, R. E., Müllen, K., Yasuda, A., & Vossmeyer, T. (2002). Self-assembled gold nanoparticle/dendrimer composite films for vapor sensing applications. *Nano Letters*, *2*(5), 551–555. doi:10.1021/nl020242s

Lee, C. H., Shie, J. L., Tsai, C. Y., Yang, Y. T., & Chang, C. Y. (2013). Photocatalytic decomposition of indoor air pollution using dye-sensitized TiO_2 induced by anthocyanin and ru complexes. *Journal of Clean Energy Technologies*, *1*(2), 115–119. doi:10.7763/JOCET.2013.V1.27

Li, Q., Xie, R., Li, Y. W., Mintz, E. A., & Shang, J. K. (2007). Enhanced visible-light-induced photocatalytic disinfection of E. coli by carbon-sensitized nitrogen-doped titanium oxide. *Environmental Science & Technology*, *41*(14), 5050–5056. doi:10.1021/es062753c PMID:17711222

Limmongkon, Y., Johns, J., & Chaerntanyarak, L. (2013). Preparation of a TiO2-coated photocatalytic air filter for use with an electrostatic air filter pack for xylene removal. *ScienceAsia*, *39*(3), 284–293. do i:10.2306cienceasia1513-1874.2013.39.284

Limmongkon, Y., Johns, J., & Charerntanyarak, L. (2013). Preparation of a TiO_2-coated photocatalytic air filter for use with an electrostatic air filter pack for xylene removal. *Science. Asia*, *39*(1), 284–293. doi:10.2306cienceasia1513-1874.2013.39.284

Linda, Z., Yong, G. L., Martin, H., & Eric, H. (2006). Removal of VOCs by photo catalysis process using adsorption enhanced TiO_2–SiO_2 catalyst. *Chemical Engineering and Processing*, *45*(11), 959–964. doi:10.1016/j.cep.2006.01.014

Liu, C., Hsu, P. C., Lee, H. W., Ye, M., Zheng, G., Liu, N., & Cui, Y. (2015). Transparent air filter for high-efficiency PM 2.5 capture. *Nature Communications*, *6*(1), 6205. doi:10.1038/ncomms7205 PMID:25683688

Long, R. Q., & Yang, R. T. (2001). Carbon nanotubes as a superior sorbent for nitrogen oxides. *Industrial & Engineering Chemistry Research*, *40*(20), 4288–4291. doi:10.1021/ie000976k

Mayo, J. T., Yavuz, C., Yean, S., Cong, L., Shipley, H., Yu, W., & Colvin, V. L. (2007). The effect of nanocrystalline magnetite size on arsenic removal. *Science and Technology of Advanced Materials*, *8*(1-2), 71–75. doi:10.1016/j.stam.2006.10.005

McMurray, T. A., Byrne, J. A., Dunlop, P. S. M., & McAdams, E. T. (2005). Photocatalytic and electrochemically assisted photocatalytic oxidation of formic acid on TiO_2 films under UVA and UVB irradiation. *Journal of Applied Electrochemistry*, *35*(7-8), 723–731. doi:10.100710800-005-1397-1

Mohamed, E. F. (2017). Nanotechnology: Future of Environmental Air Pollution Control. *Environmental Management and Sustainable Development*, *6*(2), 429–454. doi:10.5296/emsd.v6i2.12047

Murali, K. R., Swarna, L. M., & Nakkeeran, E. (2014). Nanoparticle-Membrane Filtration of Vehicular Exhaust to Reduce Air Pollution-A review. *International Research Journal of Environment Sciences*, *3*(4), 82–86.

Muralikisrishnan, R., Swarnalakshmi, M., & Nakkeeran, E. (2014). Nanoparticle-Membrane Filtration of Vehicular Exhaust to Reduce Air Polution- A Review. *International Research Journal of Environment Sciences, 3*(4), 82-86. Retrieved from http://www.isca.in/IJENS/Archive/v3/i4/12.ISCA-IRJEvS-2014-45.pdf

Nadejda, K., Isabelle, B., Berit, G., Roland, E., Bauer, K. M., Akio, Y., & Tobias, V. (2002). Self-Assembled Gold Nanoparticle/Dendrimer Composite Film for Vapor Sensing Application. *Nanotechology Letter, 2*(5), 551–555.

Nassehinia, H. R., Gholami, M., Jafari, A. J., & Esrafily, A. (2013). Nano Photocatalytic Process Application of ZnO Nanoparticle and UV on Benzene Removal from Synthetic Air. *Asian Journal of Chemistry, 25*(6), 3427–3430.

Newkome, G. R., Moorefield, C. N., & Vögtle, F. (1996). Nomenclature. *Dendritic Molecules: Concepts, Syntheses. Perspectives - Gerontological Nursing Association*, 37–47.

Pan, B., & Xing, B. (2008). Adsorption mechanisms of organic chemicals on carbon nano-tubes. *Environmental Science & Technology, 42*(24), 9005–9013. doi:10.1021/es801777n PMID:19174865

Pandey, B., & Fulekar, M. H. (2012). Nanotechnology: Remediation Technologies to clean up the Environmental pollutants, *Research. Journal of Chemical. The Sciences, 2*(2), 90–96.

Pummakarnchana, O., Tripathi, N., & Dutta, J. (2005). Air pollution monitoring and GIS modeling: A new use of nanotechnology based solid state gas sensors. *Science and Technology of Advanced Materials, 6*(3), 251–255. doi:10.1016/j.stam.2005.02.003

Qi, L., Xie, R., Li, Y. W., Mintz, E. A., & Shang, J. K. (2007). Enhanced visible-light-induced photocatalytic disinfection of E. coli by carbon-sensitized nitrogen-doped titanium oxide. *Environmental Science & Technology, 41*(14), 5050–5056. doi:10.1021/es062753c PMID:17711222

Qijin, G., Qingjie, G., & Changing, C. (2008). Investigation into photo-catalytic degradation of gaseous benzene in a circulated photocatalytic reactor (CPCR). *Chemical Engineering & Technology, 31*(7), 1–9.

Schlupp, M., Weil, T., Berresheim, A. J., Wiesler, U. M., Bargon, J., & Müllen, K. (2001). Polyphenylene dendrimers as sensitive and selective sensor layers. *Angewandte Chemie International Edition, 40*(21), 4011–4015. doi:10.1002/1521-3773(20011105)40:21<4011::AID-ANIE4011>3.0.CO;2-C PMID:12404477

Seredych, M., & Bandosz, T. J. (2012). Manganese oxide and graphite oxide/MnO_2 composites as reactive adsorbents of ammonia at ambient conditions. *Microporous and Mesoporous Materials, 150*, 55–63. doi:10.1016/j.micromeso.2011.09.010

Simon, I., Bârsan, N., Bauer, M., & Weimar, U. (2001). Micromachined metal oxide gas sensors: Opportunities to improve sensor performance. *Sensors and Actuators. B, Chemical, 73*(1), 1–26. doi:10.1016/S0925-4005(00)00639-0

Skubal, L. R., McArthur, A. L., & Newville, M. (2008). TiO_2 films for self-detection and decontamination. *International Journal of Photoenergy, 2008*, 1–6. doi:10.1155/2008/343714

Sopyan, I. (2007). Kinetic analysis on photocatalytic degradation of gaseous acetaldehyde, ammonia and hydrogen sulfide on nanosized porous TiO_2 films. *Science and Technology of Advanced Materials*, *8*(1), 33–39. doi:10.1016/j.stam.2006.10.004

Su, F., Lu, C., Cnen, W., Bai, H., & Hwang, J. F. (2009). Capture of CO_2 from flue gas via multi-walled carbon nanotubes. *The Science of the Total Environment*, *407*(8), 3017–3023. doi:10.1016/j.scitotenv.2009.01.007 PMID:19201012

Sugg, T., Tholstrup, C., Taylor, R., & Staggers, J. (2012). *The development and application of nanoporous materials*. Texas A&M University.

Taka, A. L., Pillay, K., & Mbianda, X. Y. (2017). Nanosponge cyclodextrin polyurethanes and their modification with nanomaterials for the removal of pollutants from waste water: A review. *Carbohydrate Polymers*, *159*, 94–107. doi:10.1016/j.carbpol.2016.12.027 PMID:28038758

Timonen, K. L., Vanninen, E., De Hartog, J., Ibald-Mulli, A., Brunekreef, B., Gold, D. R., & Tarkiainen, T. (2006). Effects of ultrafine and fine particulate and gaseous air pollution on cardiac autonomic control in subjects with coronary artery disease: The ULTRA study. *Journal of Exposure Science & Environmental Epidemiology*, *16*(4), 332–341. doi:10.1038j.jea.7500460 PMID:16205787

Uzunova-Bujnova, M., Todorovska, R., Milanova, M., Kralchevska, R., & Todorovsky, D. (2009). On the spray-drying deposition of TiO_2 photocatalytic films. *Applied Surface Science*, *256*(3), 830–837. doi:10.1016/j.apsusc.2009.08.069

Vani Padmaja, G. (2016). Nanomaterials: A Measure for Air Pollutant Remedial Technologies. *International Journal of Science and Research*, *5*(10), 1307–1310.

Vossmeyer, T. B., Guse, I., Besnard, R. E., Bauer, K., Müllen, A., & Yasuda, A. (2002). Gold nanoparticle/polyphenylene dendrimer composite films: Preparation and vapor sensing properties. *Advanced Materials*, *14*(3), 238–242. doi:10.1002/1521-4095(20020205)14:3<238::AID-ADMA238>3.0.CO;2-#

Wang, S., Sun, H., Ang, H. M., & Tadé, M. O. (2013). Adsorptive remediation of environmental pollutants using novel graphene-based nanomaterials. *Chemical Engineering Journal*, *226*, 336–347. doi:10.1016/j.cej.2013.04.070

Wang, S. H., Chou, T. C., & Liu, C. C. (2003). Nano-crystalline tungsten oxide NO_2 sensor. *Sensors and Actuators. B, Chemical*, *94*(3), 343–351. doi:10.1016/S0925-4005(03)00383-6

Wu, S., Deng, F., Wei, H., Huang, J., Wang, X., Hao, Y., & Guo, X. (2014). Association of cardiopulmonary health effects with source-appointed ambient fine particulate in Beijing, China: A combined analysis from the Healthy Volunteer Natural Relocation (HVNR) study. *Environmental Science & Technology*, *48*(6), 3438–3448. doi:10.1021/es404778w PMID:24521469

Xu, M. F., Lin, S., Chen, X. M., & Peng, Y. Z. (2006). Studies on characteristics of nanostructure of N-TiO_2 thin films and photo-bactericidal action. *Journal of Zhejiang University. Science. B.*, *7*(7), 586–590. doi:10.1631/jzus.2006.B0586 PMID:16773734

Yang, X., Cao, C., Erickson, L., Hohn, K., Maghirang, R., & Klabunde, K. (2008). Synthesis of visible-light-active TiO_2-based photocatalysts by carbon and nitrogen doping. *Journal of Catalysis*, *260*(1), 128–133. doi:10.1016/j.jcat.2008.09.016

Yao, N., & Yeung, K. L. (2011). Investigation of the performance of TiO_2 photocatalytic coatings. *Chemical Engineering Journal*, *167*(1), 13–21. doi:10.1016/j.cej.2010.11.061

Zaporotskova, I. V., Boroznina, N. P., Parkhomenko, Y. N., & Kozhitov, L. V. (2016). Carbon nanotubes: Sensor properties. A review. *Modern Electronic Materials*, *2*(4), 95–105. doi:10.1016/j.moem.2017.02.002

Zhang, R., Jing, J., Tao, J., Hsu, S. C., Wang, G., Cao, J., & Shen, Z. (2013). Chemical characterization and source apportionment of PM 2.5 in Beijing: Seasonal perspective. *Atmospheric Chemistry and Physics*, *13*(14), 7053–7074. doi:10.5194/acp-13-7053-2013

Zhang, X., Yang, B., Dai, Z., & Luo, C. (2012). The gas response of hydroxyl modified SWCNTs and carboxyl modified SWCNTs to H_2S and SO_2. *Przeglad Elektrotechniczny*, *88*(7B), 311–314.

Zhang, Y., Yuan, S., Feng, X., Li, H., Zhou, J., & Wang, B. (2016). Preparation of Nanofibrous Metal–Organic Framework Filters for Efficient Air Pollution Control. *Journal of the American Chemical Society*, *138*(18), 5785–5788. doi:10.1021/jacs.6b02553 PMID:27090776

Zhao, Y. L., & Stoddart, J. F. (2009). Noncovalent functionalization of single-walled carbon nanotubes. *Accounts of Chemical Research*, *42*(8), 1161–1171. doi:10.1021/ar900056z PMID:19462997

Zhou, N., Hao, Z., Zhao, X., Maharjan, S., Zhu, S., Song, Y., & Lu, L. (2015). A novel fluorescent retrograde neural tracer: Cholera toxin B conjugated carbon dots. *Nanoscale*, *7*(38), 15635–15642. doi:10.1039/C5NR04361A PMID:26285001

Zou, L., Luo, Y., Hooper, M., & Hu, E. (2006). Removal of VOCs by photocatalysis process using adsorption enhanced TiO_2–SiO_2 catalyst. *Chemical Engineering and Processing: Process Intensification*, *45*(11), 959–964. doi:10.1016/j.cep.2006.01.014

Chapter 7
Nanomaterial Surface Modifications for Enhancement of the Pollutant Adsorption From Wastewater:
Adsorption of Nanomaterials

Hamidreza Sadegh
West Pomeranian University of Technology Szczecin, Poland

Gomaa A. M. Ali
Al-Azhar University, Egypt

Hamid Jafari Nia
Islamic Azad University, Khorramabad Branch, Iran

Zahra Mahmoodi
Islamic Azad University, Science and Research Branch, Iran

ABSTRACT

With the development of dyeing, textile, leather, paper, and other chemical industries, an increasing amount of dye wastewater containing refractory organic dyes is discharged. Undoubtedly, much high content dye wastewater will lead to serious environmental issues such as color pollution, light penetration interference, and virulence to aquatic organisms, even endanger human health. Therefore, it is an imminent problem and has become a global concern to degrade dye wastewater efficiently. So far, many techniques have been used to degrade dyeing wastewater, such as chemical degradation, biological degradation, photochemical degradation, coagulation, membrane filtration, and combined methods. These methods have certain impacts on the degradation of dye wastewater, but usually with slow degradation rate, complex and high operation costs, as well as easily causing secondary pollution. The adsorption process is a simple, effective, and low-cost way to remove dyes.

DOI: 10.4018/978-1-5225-5745-6.ch007

INTRODUCTION

Urbanization of fast human industrial civilization has resulted in the destruction of various and dangerous contaminations in water currents. This wastewater includes toxic waste materials such as metals, organic impurities, etc., which straightly produce harmful and toxic effects on biological resources. Zn(II), Cu(II), Ni(II), Cd(II), Pb(II) and Cr(VI) are the metal ions required to be eliminated from the environment and need immediate concern (Al-Degs et al., 2006; Molinari et al., 2004; Moradi et al., 2010). Heavy metal ions, azo dyes, etc. were the most general toxic ions in sewage responsible for certain problems (Babel & Kurniawan, 2003). Notwithstanding the fact that the human body needs small doses such as Zn(II), it may be too prominent health problems such as depression, lethargy, increased thirst and cause neurological symptoms. Exposure to such toxic ions can cause health problems such as liver or kidney damage, Wilson's disease, insomnia, cancer, diarrhea, nausea, vomiting, rash, asthma, cough and headache (Ernhart, 1992).

Removing these poisonous ions from wastewater is essential for protecting the health and the environment. Traditional treatments have been used to remove the ions such as reduction, precipitation, adsorption, oxidation and ion exchange. Nonetheless, due to its high efficiency and economic considerations the process of attracting is the most appropriate method (Habeeb et al., 2017a; Hagen & Hertel, 2003). Nevertheless, a part from the low adsorption efficiency of some adsorbents including activated carbon (AC), zeolites, biomaterials, nanoparticles, polymers, these adsorbents have been widely used for wastewater treatment (Arias et al., 2002; Lingamdinne et al., 2016). In order to find effective and efficient adsorbents, it has been the focus of attention of various research groups.

Significant potential to improve environmental practices has been indicated by progressing of nanoscience and nanotechnology (Brumfiel, 2003; Rickerby & Morrison, 2007). Nanostructured adsorbents have represented efficiencies and faster rates of water treatment in comparison with traditional materials (Sadegh et al., 2017b; Theron et al., 2008). Eco-efficient, affordable and environmentally friendly nanomaterials have been expanded with unique features for potential applications in detoxification of industrial waste water, groundwater, surface water and drinking water (Cloete, 2010; Machida et al., 2006).

Nanomaterials like AC, carbon nanotubes (CNTs) and graphene (Feng et al., 2010; Singh et al., 2013) are widely used in many scientific and industrial fields, including sewage treatment, catalytic treatment and have depicted a substantial role in this regard (Chen & Wang, 2006; Moradi et al., 2015).

According to the purpose of this chapter, clarify a surface modification of nanomaterials for utilizing in water purification. In particular, the important applications of nanomaterials will be focused and distinguished in the field of wastewater treatment.

SURFACE MODIFICATION TECHNIQUES

Surface modification includes a number of chemical and physical methods (Figure 1). The chemical methods are involved protonation, saturation of metal oxide bonding amine groups and organic modification of aluminosilicates. The physical methods deal with the thermal treatment of the adsorbent, thereby increasing its surface area and porosity and where impurities are removed to expose surface functional groups that were not accessible to the adsorbate earlier. Enhanced nitrate removal by surface chemical

Figure 1. Adsorbents surface modification techniques

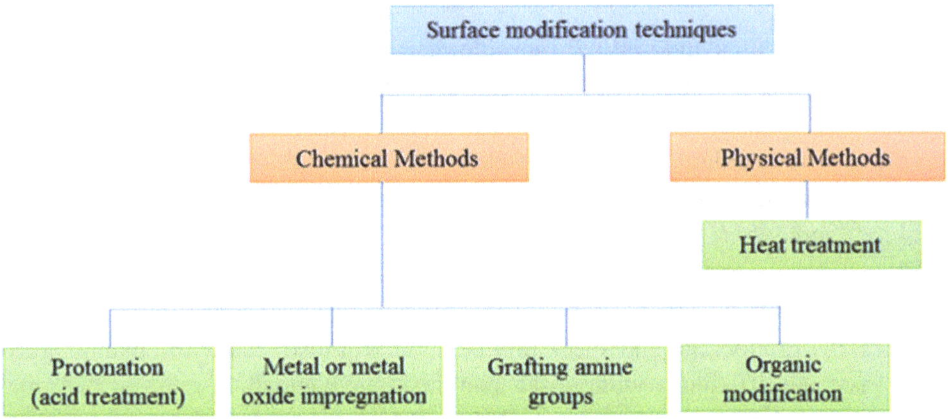

modification of adsorbents is primarily due to firstly, an increase in surface positive charges; or secondly, providing new surface functional groups having a higher affinity for nitrate. The latter method also includes the recently developed technique of coating nitrate-selective ion exchange materials on carbon electrodes and applying electric potentials to the electrodes to create positive charges on the electrodes for nitrate adsorption (Biesheuvel et al., 2011; Kim & Choi, 2012).

ADSORPTION PHENOMENON

A surface phenomenon that in which the adsorbate is accumulated on the adsorbent surface is known as the adsorption process. For containing a solution, absorbable solute comes into contact with a solid with a highly porous surface structure, liquid-solid intermolecular forces of attraction cause some of the solute molecules from the solution to be concentrated or deposited on the solid surface (Singh et al., 2011). All the bonding requirements (ionic, covalent, or metallic) of the material constituent atoms- In case of bulk materials- are filled by other atoms in the material. Although they can attract adsorbates, the atoms on the surface of the adsorbent are not wholly surrounded by other adsorbent atoms (Apul et al., 2013). However, the adsorption process is generally classified as physisorption (an adsorbate bound to the surface by weak van der Waals forces), chemisorption (an adsorbate tethered through covalent bonding), The exact nature of the bonding depends on the details of the species involved (Sadegh et al., 2016) or due to electrostatic attraction (Ekmekyapar et al., 2006). The equilibrium of adsorption between the solution and adsorbent is attained (where the adsorption of solute from the bulk onto the adsorbent is minimum). Adsorption defined as the mass transfer process where a substance becomes bound by physical and/or chemical interactions when it is transferred from the liquid phase to the surface of a solid. The main point is that the large surface area of adsorbent allows achieving a high adsorption capacity (Q_e) and surface reactivity (Kurniawan & Lo, 2009). Q_e could be obtained as $Q_e = (C_o - C_e)V / m$, where, C_o and C_e are the initial and final concentrations of the pollutants, V is the solution volume (L) and m is the adsorbent amount (g) (Habeeb et al., 2017b).

ADVANTAGES OF NANOMATERIALS AS ADSORBENTS

Due to their large surface area and optimal morphlogical properties which help to rapid removal and recovery of metal ions and dyes, the nanoadsorbents are effective and economical (Gupta et al., 2001). Nanoadsorbents by the attachment of inorganic shells and organic molecules, its surface modification stabilizes and prevents the oxidation of nanoparticles (Bhaumik et al., 2016). Resulting in enhancement of removal efficiency occurs when these surface functionalities provided sites for the uptake of specific/ selective metal ions and dyes (Gupta et al., 2001).

For organic or inorganic materials that have a high affinity to adsorb substances, Nanoadsorbents are nanoscale particles. Nanoadsorbents not only are capable of sequestering contaminants with varying the molecular size, hydrophobicity and speciation behavior but also enable the manufacturing process to consume raw materials efficiently without releasing its toxic payload, because of their high porosity, small size and an active surface (Pacheco et al., 2006). While at the nanoscale, materials show unique characteristics in order to their small size, they possess a large surface area and 'surface area to volume' ratio (Hristovski et al., 2007). These characteristics improve Q_e of the nanoparticles. These large surface area and high reactivity nanoparticles were known as better absorbing materials than conventional ones. Nanoparticles are becoming new alternatives for the treatment of wastewater for the removal of pollutants from wastewater to gain the best result of it (Natale et al., 2008).

Some metal oxides have been reported in photocatalytic degradation of organic contaminants and the adsorption of heavy metals (Wang et al., 2013b). By minimizing the distance between the sites of photon absorption, nanoparticles with a high surface area and porosity exhibit become more photocatalytic activity than their bulk counterparts (Rao et al., 2006). Numerous oxide-based nanocomposite/hybrid materials have been developed for water purification (Abdel Ghafar et al., 2015). These are composed of two or more components and can exhibit the properties of multicomponent systems (Zhang et al., 2013).

In the large-scale applications, there are considerable challenges that remaining in environmental remediation and also, the researches has revealed that the morphology, active sites, structural features, curvature of sidewalls, π-conjugative structures are some of the physical properties possessed by graphene and CNTs that have enabled them to exhibit enhanced adsorptive properties (Stafiej & Pyrzynska, 2007). On the other hand, their length ranges from nanometers to millimeters with diameters up to 100 nm and a large specific surface area, high porosity, hollow as well as layered structures. These unique properties are the cause of strong interactions which allow CNTs to interact with other molecules or atoms through π-π electronic and hydrophobic interactions thereby making them promising adsorbent material (Stafiej & Pyrzynska, 2007).

NANOADSORBENTS

Recently, nanomaterials have been widely studied as adsorbents. The smaller size increases the surface area (Gubin et al., 2005) which enhances the chemical activity and Q_e (Kalfa et al., 2009). Moreover, for persistent inorganic pollutants, a redox reaction is favored to start the ionic structure transformation (Gupta et al., 2015). However, changes in redox conditions influence the toxicity of these pollutants (Ray, 2010; Sadegh et al., 2017a). Nanoadsorbents possess two main properties: innate surface and external functionalization. Their physical, chemical and mechanical properties are also related to their extrinsic surface structure, apparent size and intrinsic composition (Mirkin et al., 1996).

In the aqueous media, the factors affecting the adsorption process are the surface area, adsorption activity, chemical activity, location of atoms on the surface, missing resistance to internal penetration and connection of energy level (Khajeh et al., 2013). Nanoparticles for removal of heavy metals should be non-toxic as adsorbent, likewise, nanoparticles have the potential to adsorb contaminations at lower concentrations, pollutants adsorbed can easily be removed and repeatedly recycled several times (Agarwal et al., 2016).

Classification of Nanoadsorbents

Nanoadsorbents are broadly classified into various groups based on their composition which mainly include metal oxide-, carbon- and polymer-based nanomaterials. In addition, recent advancements in carbon nanomaterials forms such as CNTs, carbon nanoparticles, graphene and carbon nanostructures were reported.

Moreover, various types of silicon nanomaterials are also used as nanoadsorbents such as silicon nanotube, silicon nanoparticles and silicon nano-plaiting. Nanoclays, polymer-based nanomaterials nanofibres and aerogels, are some of the nanomaterials used so as to adsorb heavy metals from sewage. The nanoparticle properties control factors include size, surface chemistry, agglomerating mode, fractal shape and dimension, chemical composition, crystalline structure and solubility. Chemical activity and fine grain size particles that it highlights two key properties compared to other materials such as titanium dioxide and alumina (Kalfa et al., 2009). Moreover, modification of nanoparticles can be carried out by some reagent to enhance its properties for metals ions pre-concentration (Khajeh & Sanchooli, 2011).

Metal Oxide-Based Nanomaterials

Oxide-based nanoparticles are inorganic nanoparticles which are usually prepared by non-metals and metals. These nanoparticles are extensively used for hazardous pollutants removal from wastewater. They include TiO_2 (Gao et al., 2008), TiO_2/dendrimers composites (Barakat et al., 2013) ZnO (Tuzen & Soylak, 2007), MgO (Gupta et al., 2011), Co_3O_4/SiO_2 (Abdel Ghafar et al., 2015), MnO_2 and Fe_3O_4 (Feng et al., 2012). Oxide-based nanoparticles are characterized by high surface area, minimum environmental impacts, less solubility and no secondary pollutants (Gupta et al., 2015). Co_3O_4/SiO_2 nanocomposite in which a single and well-crystalline Co_3O_4 phase imbedded in an amorphous SiO_2 with a crystallite size of 13.5 nm showed a great ability to absorb methelene blue (MB) dye. It showed a high removal percentage of 95.7% and high adsorption capacity of 53.87 mg g^{-1}, which is larger than the adsorption capacity of MB on other materials (Abdel Ghafar et al., 2015). In addition, manganese substituted hydroxyapatite (MnHAP) showed high adsorption capacity toward Fe(III) removal (0.389 mg g^{-1}) (Elnsar et al., 2017). The morphology of the metal oxides plays a crucial role in the adsorption process, where, MgO mesoporous nanoflakes and flowerlike microspheres exhibited excellent absorption performance for common toxic heavy metal ions and organic pollutants and are expected to be a potential absorbent in wastewater treatment (Gao et al., 2008). The hazard cadmium could be recovered at high levels of 96% using B_2O_3/TiO_2 nanocomposite (Kalfa et al., 2009). Hristovski and his co-worker reported that, TiO_2, Fe_2O_3, ZrO_2 and NiO nanopowders exhibited the highest arsenate removal in all water matrices (Hristovski et al., 2007).

Carbon-Based Nanomaterials

Carbon Nanotubes

Carbon nanotubes (CNTs) which can be divided into multi-walled carbon nanotubes (MWCNTs) and single-walled carbon nanotubes (SWCNTs) are extensively studied material having the ability in order to eliminate all kind of organic pollutants and heavy metals of wastewater (Ren et al., 2011). However, poor dispersion ability, difficulty in separation and small particles size are few problems exist for using CNTs as an adsorbent. To overcome these problems researcher modified the ordinary CNTs (Sadegh et al., 2017a; Tang et al., 2012). Modified magnetic CNTs have high dispersion ability and can be easily removed from wastewater using a magnet (Li et al., 2003a). Several studies have reported the removal of Pb(II) and Mn(II), Cu(II) (Tang et al., 2012) using Multi-walled carbon nanotubes (MWCNTs). The behavior of alumina adsorption on CNTs has been investigated for the treatment of aqueous solutions (Gupta et al., 2011). They reported that the coated CNTs exhibited better removal ability than the un-coated CNTs as the surface modification increases its overall adsorption activity.

Various kinds of surface modification techniques are reported which include acid treatment (Ren et al., 2011), metals impregnation (Tawabini et al., 2011) and functional molecules/group grafting (Ali et al., 2017; Chen et al., 2012). The characteristics of carbon nanotubes like surface area, charge, dispersion and hydrophobicity can be changed by these methods. Acid treatment of CNTs was carried out using HNO_3, H_2SO_4 and HCl (Fu & Wang, 2011; Ren et al., 2011). Treatment of acid removal of impurities is currently on the surface of carbon nanotubes. In addition, new functional groups on the surface, which leads to an increase in its activity is presented (Gupta et al., 2015). In addition, oxygen-containing groups can also be introduced with the excited microwave wave surface plasma (Chen et al., 2009).

Grafting functional molecules/functional groups on the surface of CNTs are another way of improving their surface features. This can be carried out through various methods such as plasma method, chemical modification and microwave (Chen et al., 2012). However, the plasma method, amid these techniques, is one of the best reasons for lower energy demand and an environmentally friendly process.

The adsorption capacity is highly dependent on the structural and morphological characteristics of the adsorbent. Figure 2 shows the surface area-dependence adsorption capacities of carbon nanotubes toward Hg(II) removal (Alomar et al., 2017a; 2017b; Gupta et al., 2014; Hadavifar et al., 2014). It is clear that, the adsorption capacity increased with increasing the surface area as a result of increasing the exposed surface. The linear relationship between surface area and adsorption capacity was proven and reported for H_2S removal by carbon-based materials (Habeeb et al., 2017a).

Many organic dyes can cause mutagenic and carcinogenic influences in humans with high levels of biotechnology, in fact organic dyes, one of the most hazardous substances in Industrial wastewater, are evacuated from different industries. (e.g., textiles, leather, cosmetics and paper). However, Sudan Red I, II, III and IV as a result of their toxicity or cancer is forbidden even low concentrations, they are extremely applied in other industries. More sophisticated dyes combinations including aromatic structures (Figure 3), make them highly resistant to biological degradation and resistant to oxidation. Hence, more consideration has been attracted by eliminating such compounds (Yang et al., 2013).

The prevalent priority is to develop a novel high Q_e-adsorbing material for the realization of effective control of this pollutant namely a wide variety of materials are used to remove organic dyes including AC, zeolite, clay and polymer.

Figure 2. Surface area-dependence adsorption capacities of carbon nanotubes

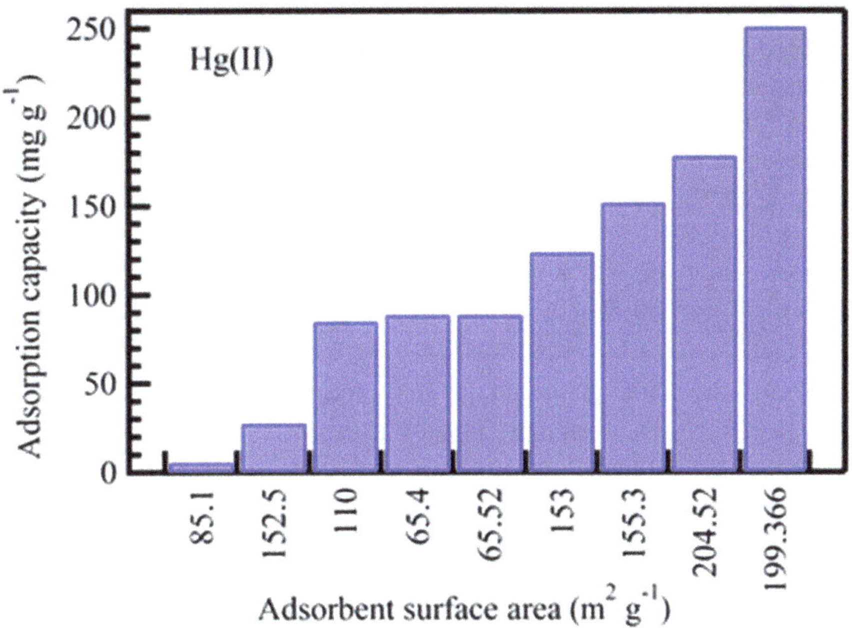

Figure 3. Chemical structures of some organic dyes

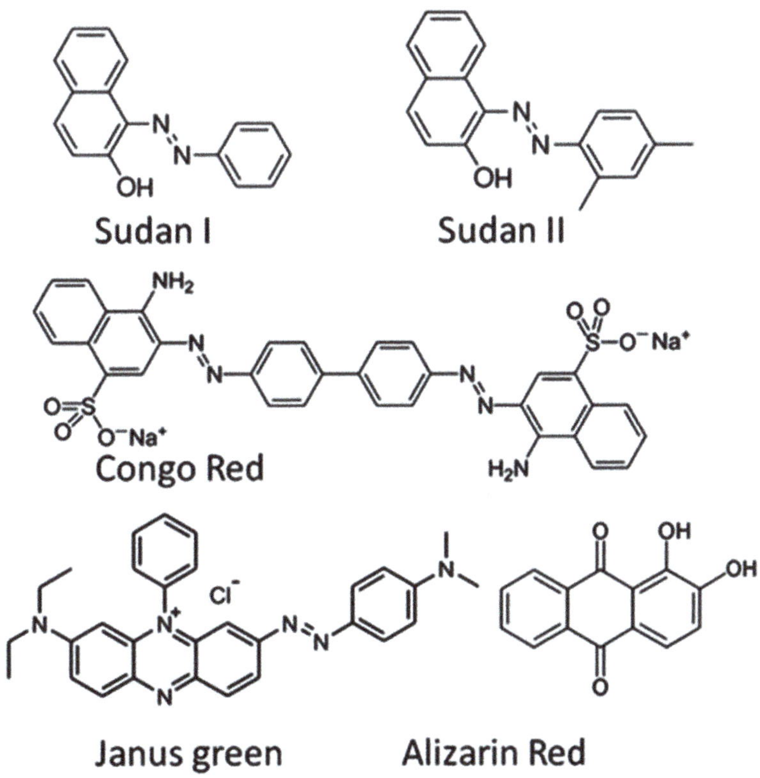

MWCNTs have been shown to outperform $Cd(OH)_2$ nanowire loaded AC with respect to their efficient removal of safranin O (Ghaedi et al., 2012a).

Though, it has been reported few reports in carbon nanotubes to remove dye from aqueous solution (Lu & Liu, 2006) and carbon nanotubes were commonly used without more treatment and the CNTs were typically directly used without further treatment (Machado et al., 2011) (Table 1).

MWCNTs have been shown to be effective in the removal of methyl red (Ghaedi & Kokhdan, 2012) and MB (Ghaedi et al., 2012b). Another work has focused on the development of CNT-impregnated chitosan hydrogel beads (CSBs) for the removal of Congo red (CR). CSBs demonstrated a higher maximum adsorption capacity (Q_m) than normal chitosan CSBs (450.4 vs. 200.0 mg g^{-1} (Chatterjee et al., 2010). Q_m could be obtained from the linear relation of Langmuir model as: $\left(C_e / Q_e\right) = \left(1 / Q_{max} K_L\right) + \left(C_e / Q_{max}\right)$, where, K_L is Langmuir constant (Abdel Ghafar et al., 2015). A new generation of CSBs prepared by using sodium dodecyl sulfate and MWCNTs to improve upon their mechanical properties has also demonstrated a high Q_m for CR (375.94 mg g^{-1}) (Chatterjee et al., 2011). Compared to MWCNTs, SWCNTs can demonstrate better adsorption properties for organic contaminants

Table 1. Removal of various organic dyes by using CNTs

CNTs Type	Modification Form	Dye Type Eliminated	Q_m (mg g^{-1})	Ref.
MWCNTs	Untreated	ECR	73.18	(Ghaedi et al., 2011b)
	Untreated	Alizarin red S Morin	161.29 26.25	(Ghaedi et al., 2011a)
	Untreated	Reactive red M-2BE	335.7	(Machado et al., 2011)
	Alkali-activated	MB MO	149 399	(Mickelson et al., 1999)
	Untreated	MB Acid red 183	59.7 5.2	(Wang et al., 2012a)
	Untreated	Acid blue 161	91.68%	(Geyikçi, 2013)
	Untreated	Reactive blue 4 Acid red 183	69 45	(Wang et al., 2012b)
	Fabricated magnetic MWCNTs by Fenton's reagent method (M-MWCNTs)	MO	28	(Yu et al., 2012)
	Oxidized	5-(4-Dimethyl benzylidene amino) rhodanine	15.52	(Ghaedi et al., 2013)
	Catalyzed pyrolysis of methane in a H_2 and N_2 flow	MB	188.68	(Li et al., 2013)
	Untreated	Triclosan	153.1	(Zhou et al., 2013)
	Alkali-activated	MB MO	149 399	(Mickelson et al., 1999)
SWCNTs	Pristine oxidized	Basic red 46	38.35 49.45	(Moradi, 2013)
	Untreated	Reactive blue 29 (RB29)	496	(Nadafi et al., 2011)
	Untreated	Reactive red 120	426.49	(Bazrafshan et al., 2012)

because of their higher special surface area. Indeed, SWCNTs are more efficient for removing benzene (9.98 mg g^{-1}) and toluene (9.96 mg g^{-1}) (Bina et al., 2012). Q_m of 496 mg g^{-1} was achieved when RB29 was removed using SWCNTs (Nadafi et al., 2011).

Recently a novel self-assembled cylindrical graphene-CNTs hybrid was developed and it achieved a Q_m of 81.97 mg g^{-1} for MB removal and the removal efficiency reached 97% for low (10 mg L^{-1}) (Ai & Jiang, 2012). Finally, Zhou et al. (2013) proposed a new concept of using entangled CNTs as porous frameworks to enhance the adsorption of organic dyes. The composites obtained through polymerization with polyaniline (PANI) possessed large surface areas. CNTs/PANI composites exhibited a 15% higher equilibrium Q_e of 13.95 mg g^{-1} compared to PANI. Some findings of organic dyes removal using CNTs are summarized in Table 1 (Zhou et al., 2013).

Zinc is one of the most common heavy metals occurring in wastewater. In water, Zn(II) ion may associate or react with neutral or ionic compounds to form inorganic salts, stable organic complexes or inorganic or organic colloids. Zn(II) ions exhibit high toxicity to aquatic organisms and it may cause high chronic toxicity in some cases. Adsorption of Zn(II) onto the CNTs, increased as e pH increased and achieved maximum between pH 8 and 11 (Shirmardi et al., 2012). SWCNTs and MWCNTs are preferably used for Zn(II) removal. Shirmardi et al. (2012) reported that the amount of Zn(II) sorbed onto the CNTs increased with a rise in temperature. Under the same conditions, Zn(II) sorption capacity of the CNTs was much greater than that of the commercially PAC, reacting that SWCNTs and MWCNTs are effective sorbents. The thermodynamic analysis revealed that the sorption of Zn(II) is endothermic and spontaneous. Zn(II) ions could be easily removed from the surface site of SWCNTs and MWCNTs by a 0.1 mol L^{-1} HNO_3. Thus both CNTs can be reused through many cycles of water treatment and regeneration.

Other trace elements such as Ni(II), Pb(II), Cd(II), Cr(VI) and Cu(II) can also be adsorbed by this powerful CNT agent. Chromium at low-level exposure can irritate the skin and cause ulceration. Long-term exposure can cause kidney and liver damages as well as damage the circulatory and nerve tissues. Once again, it is important to eliminate such traces in our wastewater. CNTs supported by AC was used to remove Cr(VI) ions from polluted water. The highest Q_m obtained from batch adsorption experiments was 9.0 mg g^{-1} (Salem Attia et al., 2014). Therefore, it seems that AC coated CNTs is most effective for the removal of Cr(VI). Natale et al. (2008) reported the removal of Cr(VI) from drinking water using CeO_2/CNTs. Q_m of CeO_2/CNTs reaches 30.2 mg g^{-1} at an equilibrium Cr(VI) concentration of 35.3 mg L^{-1} and pH 7.0. Q_m is 1.5 times higher than that of the AC, 2.0 times higher than that of Al_2O_3 and 1.8 times higher than that of ball-milled CNTs. Hu et al. (2010) studied the removal of Cr(VI) from aqueous solution using oxidized MWCNTs. The maximum removal of Cr(VI) was found at low pH and the adsorption kinetics of Cr(VI) was suitable for pseudo-second-order models. Cr(VI) removal mainly depends on the occurrence of the redox reaction on the surface of oxidized MWCNTs to the formation of Cr(III) and subsequent sorption of Cr(III) on MWCNTs. This Cr(III) sorption appears to be the leading mechanism for chromium uptake to MWCNTs to remove the anionic chromate (CrO_4^{2-}) from wastewater. Functionalized CNTs (f-CNTs) are the most suitable compared to unmodified CNTs (Richard et al., 2003). The reason for the CNTs showing excellent adsorption capability of anionic chromate is the interaction of CrO_4^{2-} with the surface oxygen-containing functional groups on the modified CNTs. Hence, f-CNTs once again can be potentially employed in adsorbing heavy metal from wastewater.

Besides Zn(II) and Cr(VI), Pb(II) is also one of the famous heavy metals occurring in wastewater. Pb(II) compounds are generally soluble in soft and slightly acidic water. Pb(II) from pipes may partially dissolve in the water owing through. It occurs in almost all water resources as well as wastewater. Despite its toxicity, the presence of lead in the water may affect the human and marine lives' health. Wang et al. (2013a) reported that the role of functional groups in the adsorption of Pb(II) is to create a chemical complex. Pb(II) (PbO and $Pb(OH)_2$ and $PbCO_3$) is adsorbed on the surface of the acidized MWCNTs. Pb(II) species adsorbed on acidized MWCNTs at the ends and defects sites. Wang et al. (2012a) reported that Mn-oxide-coated CNTs (MnO_2/CNTs) were used to remove Pb(II). Pb(II) removal of MnO_2/CNTs decreased with pH decreasing. Q_m was 78.74 mg g^{-1}, compared with CNTs, significant improvement of Pb(II) adsorption shows MnO_2/CNTs can be good Pb(II) adsorber. The adsorption followed pseudo-second-order rate and Langmuir model (Wang et al., 2013b). Results demonstrated a high Q_m of acidized MWCNTs for Pb(II) due to formation of oxygen functional groups on the surface of the CNTs. Salt formation or complex deposited occurred on the surface of CNTs when acidized MWCNTs react with Pb(II). The adsorption contact time to reach equilibrium is about 20 min, which proved to be shorter than AC (120 min). Pb(II) can be easily regenerated from the acidized MWCNTs at pH 2.

The interaction between some heavy metal ions such as Pb(II), Cd(II) and Cu(II) ions from aqueous solution adsorbed by SWCNTs and carboxylate group functionalized single-walled carbon nanotube (SWCNT–COOH) surfaces was explained. Q_m for Pb(II), Cu(II) and Cd(II) ions onto the SWCNT–COOH are 96.02, 77.00 and 55.89 mg g^{-1}, respectively, and by SWCNTs are 33.55, 24.29 and 24.07 mg g^{-1}, respectively (Moradi & Zare, 2011). The thermodynamic parameters showed that the adsorption on the SWCNT–COOH and SWCNTs is spontaneous and endothermic.

One of the most hazardous heavy metals is cadmium (Cd). In humans, long-term exposure to Cd is associated with severe renal dysfunction. Li et al. (2003a) discussed the adsorption of Cd(II) by CNTs. The experiments of the CNT dosage effect on Cd(II) adsorption react that Q_e for $KMnO_4^-$ oxidized CNTs has a sharper increase at CNT dosage from 0.03 to 0.08 g/100 mL than the as-grown H_2O_2 and HNO_3-oxidized CNTs. The removal efficiency almost reaches 100% at the CNT dosage of 0.08 g/100 mL. This shows that CNTs are able to adsorb Cd(II) with high capacity.

To further prove the efficiency of CNTs in adsorbing heavy metals, Gao et al. (2008) studied the adsorption of Ni(II), Cu(II), Zn(II) and Cd(II) on CNTs oxidized with concentrated HNO_3. The effect of process parameters such as the pH value, initial concentration of ions, the surface chemistry of the adsorbent, the kind and number of components in the adsorption system and the ratio of the metal ion species in solution was studied. For the single and binary systems, it turned out that the amount adsorbed on oxidized CNTs followed the order of $Cu(II)_{(aq)} > Ni(II)_{(aq)} > Cd(II)_{(aq)} > Zn(II)_{(aq)}$. The good correlation between the amount adsorbed and standard electrode potential indicated that the redox process may serve a role in the mechanism of adsorption. Meanwhile, for the ternary and quaternary systems, the adsorption was more complex. The order of amount adsorbed at the same concentration was $Cu(II)_{(aq)} > Cd(II)_{(aq)} > Zn(II)_{(aq)} > Ni(II)_{(aq)}$. Li et al. (2003a) studied different heavy metals (Pb(II), Cu(II) and Cd(II)) adsorption using oxidized CNTs. Q_m of Pb(II), Cu(II) and Cd(II) ions was 63.29, 23.89 and 11.01 mg g^{-1}, respectively. Pompeo and Resasco (2002) reported that the sorption capacities of metal ions to different CNTs follow roughly the order: Pb(II) > Ni(II) > Zn(II) > Cu(II) > Cd(II).

The sorption capacities of metal ions by raw CNTs are very low but significantly increased after oxidation of CNTs surfaces by HNO_3, NaOCl or $KMnO_4$ solutions. The surface-oxidized CNTs show a great potential as superior sorbents for environmental protection applications. The sorption mechanism appears mainly attributable to chemical interaction between the metal ions and the surface functional groups. The order of heavy metal ions shows that their adsorption does not depend clearly on the ionic radius of metal ions, but depends on properties of CNTs such as ionic strength, pH, CNTs mass, contact time, initial metal ion concentration and temperature.

Copper is an essential substance to human life, but in high doses, it can cause anemia, liver and kidney damages as well as stomach and intestinal irritation. This is the reason why copper should not be present in wastewater. Li et al. (2003b) reported that environmental friendly adsorbents that were CNTs immobilized by calcium alginate (CNTs/CA) were designed for copper adsorption. CNTs/CA copper adsorption properties were investigated via equilibrium studies. Experimental results showed that copper removal efficiency of the CNTs/CA is high and reaches 69.9% even at a low pH of 2.1. Q_m can attain 67.9 mg g^{-1} at the copper equilibrium concentration of 5 mg L^{-1}.

Many CNTs applications require handling in solution phase; however, CNTs have proven difficult to disperse in solvents (Rao et al. (2007). Chemical modification of SWCNTs is often required for more versatile suspension capabilities and enablement of certain applications (Chen et al., 1998). This has encouraged greater exploitation of their intrinsic properties, as well as the capability to modify these properties. In particular, the functionalization of CNTs is required for their aqueous suspensions and to allow for molecular interactions with biological systems. Native CNTs adsorb surfactants that have been found to associate with CNTs via van der Waals interactions through their hydrophobic chains, rendering CNTs hydrophilic and able to disperse in aqueous environments. SWCNTs can be isolated from aggregated bundles, allowing for spectroscopic probing of individual SWCNTs. Surfactant additives prevented nonspecific interactions between SWCNTs and proteins. Also, surfactants with modified head groups have also been used to link SWCNTs to specific molecules (Herrero & Prato, 2008). Kim and Choi (2012) reported that lipids comprise a class of molecules that interact with CNTs similarly to surfactants. This class of molecule offers control of its interactions with CNTs through modification of their hydrophobic chains, while also providing the versatility of the functionality that they impart through the modification of their head groups. Among these lipids, chemical interaction between the heavy metal ions and the surface functional groups of CNTs is the major adsorption mechanism.

Graphene

Graphene is one of allotropy of carbon having special features that make it highly favorable for several environmental applications. Graphene oxide (GO) has two-dimensional structure produced by the oxidation of graphite layer via the chemical method. The most common method used for the synthesis of GO is Hummers method (Ali et al., 2015; Marina et al., 2016). The hydrophilic groups were induced in GO which required special oxidation process (Gopalakrishnan et al., 2015; Gupta et al., 2017). The presence of these hydroxyl and carboxyl groups as functional groups in GO increases the adsorption of heavy metals (Lingamdinne et al., 2016). GO as an adsorbent for the removal of heavy metals is getting more attention due to its high surface area, mechanical strength, light weight, flexibility and chemical stability (Gopalakrishnan et al., 2015). Moreover, the presence of functional group on the surface of GO also effects the adsorption process (Zare-Dorabei et al., 2016). GO has two main features as compared to other nanomaterials such as CNTs. First, a single layer GO has two-dimensional basal planes available for the maximum adsorption of heavy metals. Secondly, it has simple synthesis process, which can be carried out by chemical exfoliation of graphite without any metallic catalyst and complicated instrument (Santhosh et al., 2016).

Moreover, GO did not require any further acid treatment to enhance its adsorption capacity as it already contains the hydrophilic functional group (Zhao et al., 2011a). Various works used graphene-based nanomaterials for the adsorption of heavy metals from wastewater. Azamat et al. (2015) studied the removal of heavy metals using GO enabled sand filter in column reactor as well. Lee and Yang

(2012) modified GO with TiO_2 and used for Pb(II), Cd(II) and Zn(II) ions adsorption. The adsorption capacity of hybrid composite reaches 65.6 mg g^{-1}, 72.8 mg g^{-1} and 88.9 mg g^{-1} for Pb(II), Cd(II) and Zn(II), respectively. Graphene and its composite show a very high efficiency for the removal of heavy metals from wastewater. Nevertheless, successful reduction in GO to pristine graphene material is still a big challenge because this reduction may decrease its mechanical and electronic properties (Santhosh et al., 2016).

Using van der Waals' forces and π-π stacking interactions, the adsorption of dyes on few-layered graphene oxide nanosheets (GONSs) could be realized. To modify the chemical and physical properties and improve the processability of graphene nanosheets (GNSs) could first be incorporated into composite materials. Whether, GNSs layer or a single GONSs layer, the high aspect ratio and large ð-electronic surface provide strong intermolecular forces among these layers and adsorbates (Denis & Iribarne, 2012). Due to the opened-up layer structure, GNSs would exhibit markedly faster adsorption kinetics than CNTs (Yu et al., 2014).

GNSs exhibited better adsorption capacities for two synthetic organic compounds (phenanthrene and biphenyl) in aqueous solutions (Apul et al., 2013). More importantly, GNSs are much cheaper than SWCNTs (Kotov, 2006). GNSs have already been used as adsorbents for the removal of cationic redX-GRL (Apul et al., 2013), MB ((Li et al., 2011a), methyl orange (Li et al., 2011b), Congo red (CR) (Du et al., 2014) and other organic materials. Q_m of ptoluenesulfonic acid, 1-naphthalene sulfonic acid and MB on GNSs reach up to 1430, 1460 and 1520 mg g^{-1}, respectively, which is the highest among all of the current nanomaterials (Wu et al., 2011). The observed adsorption capacities for the adsorption of three types of pesticides, chlorpyrifos, endosulfan and malathion, onto GONSs and GNSs from water, are as high as 1200, 1100 and 800 mg g^{-1}, respectively and GONSs and GNSs were unprecedented substrates for these adsorption technologies (Maliyekkal et al., 2013).

GNSs can be used as sorbent for Cd(II) and Co(II) ions removal from aqueous solution. It was reported that heavy metal ions sorption on nanosheets depends on pH and ionic strength. The abundant oxygen-containing functional groups on the surfaces of GONSs play an important role (Zhao et al., 2011a). Magnetite-graphene composite with a particle size of ~10 nm gave a high binding capacity for As(III) and (V) due to the increased adsorption sites in the graphene composite (Chandra et al., 2010). The strong functional groups on GO surface makes it a potential adsorbent for metal ion complexation through both electrostatic and coordinate approaches. Generally, GO showed high adsorption capacity on cationic metals. GNSs can be used for adsorption of both cationic and anionic metals. After modification of GO with organics or metal oxides, its composites can also be used for anionic metal removal due to functionalization. Cu-GO interaction in aqueous solution showed that Cu(II) causes GO sheets to be folded and form large aggregates. The coordination between Cu(II) and oxygen atoms on GO was the primary driving force. GO has a Cu(II) adsorption capacity of 46.6 mg g^{-1}, higher than that of CNTs (28.5 mg g^{-1}) and AC (4-5 mg g^{-1}) (Yang et al., 2010).

The removal of Cd(II) (Zhao et al., 2011a), Co(II) (Chen et al., 2012), Pb(II) (Zhao et al., 2011b) and U(VI) (Zhao et al., 2012) ions have been studied using few-layered GONSs. It was found that the abundant oxygen-containing functional groups on GONSs surfaces play an important role in metal sorption which was in agreement with the results of (Zhao et al., 2011a). Q_m of Cd(II) and Co(II) on GONSs at pH 6.0 and 303 K were about 106.3 and 68.2 mg g^{-1}, respectively. For Pb(II), Q_m were about 842, 1150 and 1850 mg g^{-1} at 293, 313 and 333 K, respectively (Zhao et al., 2011b), but sorption capacity of U(VI) at pH 5.0, 293 K was 97.5 mg g^{-1} (Zhao et al., 2012). Modification of GO with organic materials can change the surface functional groups for better adsorption of various metal ion species. A modified GO with thiol groups by diazonium chemistry was reported to adsorb 6-fold higher concentration of Hg(II) ions than GO and AC (Gao et al., 2011). When N- (trimethoxysilylpropyl) ethylenediamine triacetic acid was used to obtain a chelating GO for Pb(II) removal, the adsorption was fast and completed within 20 min, with an adsorption capacity of 479 mg g^{-1} at pH 6.8 (Madadrang et al., 2012). Compared to AC and CNTs, GO and GNSs present more effective adsorption for many water pollutants.

Activated Carbon

AC typically has high porosity, high surface area and prepared from readily available carbonaceous precursors such as coal, wood, coconut shells and agricultural wastes (Habeeb et al., 2017b, 2017c; Sadegh & Shahryari-ghoshekandi, 2015). Generally, the physical activation requires high temperature and longer activation time as compared to chemical activation, however, in chemical activation the AC needs a thorough washing due to the use of chemical agents (Habeeb et al., 2017d; Sadegh et al., 2016). The product formed has a very porous structure with a large surface area ranging from 500 to 2000 m^2 g^{-1} (Carrott et al., 1991; Habeeb et al., 2017d). The applicability of AC for water treatment has been demonstrated by various workers (Stenzel, 1993). Also, AC are extensively used for the removal of inorganic and organic pollutants from effluent streams and in water treatment. It was found that AC possesses a significant weak acidic ion exchange character enabling them to remove trace metal contaminants and to adsorb pollutants from wastewater. The sorption of pentavalent arsenic on a granular AC was experimentally studied (Natale et al., 2008). AC prepared from coconut tree sawdust was used as an adsorbent for the removal of Cr(VI) and mercury from aqueous solution (Gao et al., 2008; Selvi et al., 2001).

Powder AC prepared from Eucalyptus camaldulensis Patnukao et al were studied for its Q_m at 60 °C, where Q_m for Cu(II) and Pb(II) were 0.85 and 0.89 mmol g^{-1}, respectively (Patnukao et al., 2008). A novel sodium polyacrylate grafted AC was produced using gamma radiation to increase the number of functional groups on the surface. This grafting technique can be potentially applied to other adsorbents to increase the efficiency of metal ion sorption by AC (Ewecharoen et al., 2009). Their high adsorption ability and lower price make AC suitable materials for removal of heavy metal and dye (Gupta et al., 2015). The concise data in form of summary will be well elucidated from Table 2.

Polymer-Based Nanomaterials

The polymer materials are considered as alternative candidates to AC for adsorption process due to their vast surface area, perfect mechanical rigidity, adjustable surface chemistry and pore size distribution,

Table 2. Data of maximum adsorption capacity of heavy metal ions with AC

Metal ion	Q_m (mg g^{-1})	Surface Area (m^2 g^{-1})	Removal (%)	Ref.
As(V)	2.5	950	---	(Gupta et al., 2015)
Cr(VI)	3.46	486	99	(Selvi et al., 2001)
Hg(II)	110.8	1026	---	(Graydon et al., 2009)
Ni(II)	44.1 55.7	899 892	82.7 93.9	(Ewecharoen et al., 2009)
Ni(II)	44.1	899	---	(Pyrzyńska & Bystrzejewski, 2010)

and feasible regeneration under mild conditions. Polystyrene showed a desorption capacity of 1.5029 and 2.2702 mmol g^{-1} for phenol and p-chlorophenol, respectively (Abburi, 2003). In addition, polystyrene showed a high ability to absorb many other pollutants as displayed in Figure 4 (Abburi, 2003; Bilgili, 2006; Garcia et al., 1999; Pan et al., 2009; Pan et al., 2002). On the other hand, the hybrid polymer composites such as magnetic natural zeolite-polypyrrole was investigated for removal of vanadium from aqueous solution and showed a capacity of 57.803 mg g^{-1} at 25 °C (Mthombeni et al., 2015). Co(II) removal was studied using poly(acrylic acid) grafted MWCNTs from aqueous solution and showed adsorption capacity 4-times higher than pure MWCNTs (Chen et al., 2012).

Factors Affecting Adsorption Processes

Adsorptions of heavy metals from wastewater using nanoadsorbents are affected by many factors such as temperature, pH, adsorbent dose and incubation (contact) time. (Srivastava et al., 2015) reported that pH plays a vital role on adsorption of heavy metals from wastewater. Maximum adsorption for Zn(II) on magnetic nanoadsorbent was observed at pH 5.5; however, it decreases with further pH increase. Lingamdinne et al. (2016) reported the maximum removal i.e. 93% and 99.6% for Pb(II) and Cr(III) at the pH of 6.0 and 4.0 respectively. Moreover, increase in contact time increases the adsorption of heavy metals. Lingamdinne et al. (2016) reported that 120 min is the equilibrium time for maximum adsorption of Pb(II) and Cr(III) from wastewater. Nanoadsorbent characteristics also affect the adsorption, which

Figure 4. Adsorption capacities of polystyrene adsorbent toward some organic pollutants

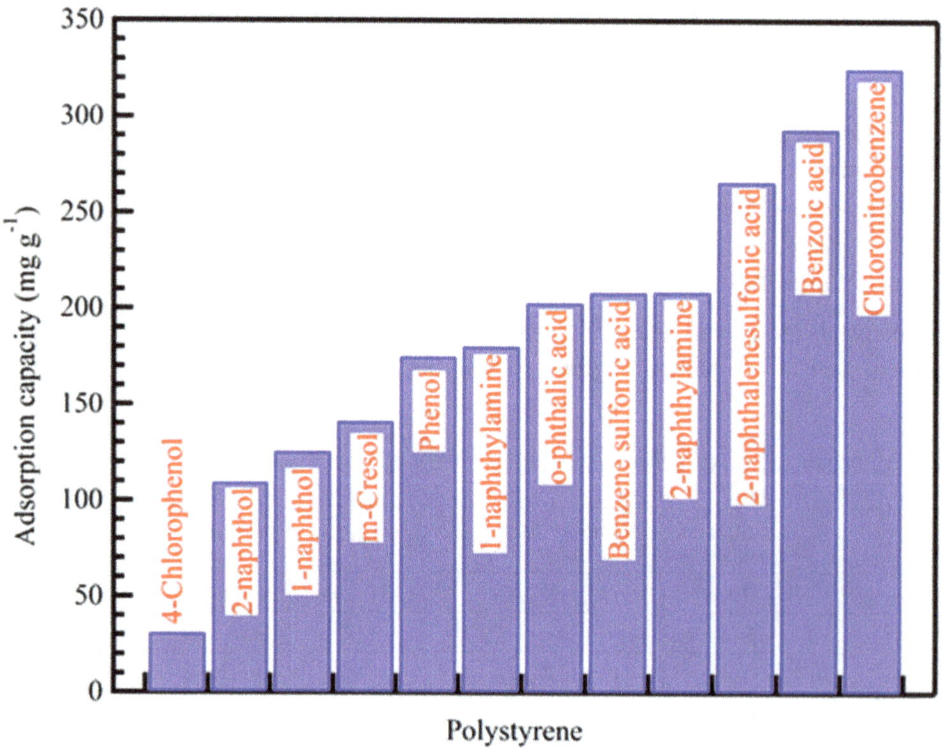

include BET surface area, surface charge, hydrophobicity and addition of new functional groups. The efficiency of various nanoadsorbents for removal of heavy metal has been reported in various studies as described in Table 3.

EVALUATION OF NANOADSORBENTS FOR WASTEWATER TREATMENT

Nanoadsorbents are widely used adsorbents for the removal of heavy metals from wastewater. CNTs and metal oxides are the most commonly used for the removal of heavy metals from aqueous solution (Lingamdinne et al., 2016; Sadegh et al., 2018). The nanoparticles have some key features such as high BET surface area, microporous structure, high dispersion ability, economically and environmentally friendly (Li et al., 2003a). However, the smaller size of particles and difficulties in separation from aqueous solution will lead to secondary pollution (Lingamdinne et al., 2016). This further affects the bioavailability and mobility of the heavy metals and causes toxicity in the environment (Wang et al., 2013a). In addition, economical reuse and regeneration are other challenges for these nanoparticles (Pan et al., 2009). However, the modification was carried out (Bahgat et al., 2011; Tarigh & Shemirani, 2013), to overcome these challenges. Nevertheless, new trends in nanoadsorbents such as organic-inorganic hybrids are a sustainable option to overcome the associated limitations of the nanoadsorption process (Lingamdinne et al., 2016).

Table 3. The efficiency of various nanoadsorbents for heavy metals removal

Nanoadsorbent	Heavy Metals	Experimental Conditions			Removal (%)	Ref.
		pH	Contact Time (min.)	Adsorbent Dose ($g L^{-1}$)		
Magnetic MWCNTs	Cr(VI)	3	600	0.1	100	(Huang et al., 2015)
Magnetic zeolite-polymer composite	V(V)	4-5	1440	0.15	73	(Mthombeni et al., 2015)
ZIF-8 nanoparticles	As(V)	7	240-420	0.20, 0.05	60.03	(Jian et al., 2015)
ZnS nanocrystals Modified magnetite	Hg(III)	1-6	5	10	99.99	(Qu et al., 2014)
Nanocomposite	Cu(II)	6.5	15	0.19	99	(Neyaz & Siddiqui, 2015)
GNSs/MnO_2	Ni(II)	---	20	5	77.04	(Varma et al., 2013)
Nanocrystalline TiO_2	As(II)	9.5	---	0.2	>98	(Pena et al., 2006)
NZVI	As(III)	7	10	1	99.9	(Rahmani et al., 2011)
Magnetic nanoparticles coated zolite	As(III)	2.5	15	0.5	95.6	(Salem Attia et al., 2014)
Zeolite materials obtained from fly ash	Pb(II)	6-7.5	90	6	>80	(Khani et al., 2016)
Magnetic nanoadsorbent	Pb(II)	6	10	20	80	(Selvi et al., 2001)
PMDA/TMSPEDA	Pb(II) Cu(II) Zn(II)	7	1440	0.01	79.6 72.36 66.8	(Alsohaimi et al., 2015)
Magnetic nanoadsorbent	Zn(II)	5.5	90	2.5	95	(Srivastava et al., 2015)

CONCLUSION

Advances in nanosciences and engineering are providing new opportunities to develop more low-cost effective adsorbents and environmentally friendly water treatment techniques. Nanomaterials have specific physicochemical properties that make them particularly attractive for wastewater purification. As a result of the remarkable structural and chemical properties of nanomaterials, their application for wastewater treatment has rapidly grown in the past several years. Previous studies obviously indicate that significant and exciting progress has been made in the aforementioned field, so that using of nanomaterials as adsorbents is a very useful and powerful tool for the removal of heavy metals and dyes due to their unique structure and specific surface characteristics. These nanomaterials are capable to remove noxious metal ions even at low concentrations i.e., up to ppb level also, with a very high selectivity and adsorption capacity. These unique and specific properties of nanosorbents make them ideal materials for wastewater treatment technology. Although carbon-based nanomaterials, such as; activated carbon, carbon nanotubes, graphene are used very efficiently and successfully in the removal of noxious metal ions from wastewater, several problems still remain; large-scale wastewater treatment is the essential one.

REFERENCES

Abburi, K. (2003). Adsorption of phenol and p-chlorophenol from their single and bisolute aqueous solutions on Amberlite XAD-16 resin. *Journal of Hazardous Materials*, *105*(1), 143–156. doi:10.1016/j.jhazmat.2003.08.004 PMID:14623424

Abdel Ghafar, H. H., Ali, G. A. M., Fouad, O. A., & Makhlouf, S. A. (2015). Enhancement of adsorption efficiency of methylene blue on $Co_3O_{4/}SiO_2$ nanocomposite. *Desalination and Water Treatment*, *53*(11), 2980–2989. doi:10.1080/19443994.2013.871343

Agarwal, S., Tyagi, I., Gupta, V. K., Ghaedi, M., Masoomzade, M., Ghaedi, A., & Mirtamizdoust, B. (2016). Kinetics and thermodynamics of methyl orange adsorption from aqueous solutions-artificial neural network-particle swarm optimization modeling. *Journal of Molecular Liquids*, *218*, 354–362. doi:10.1016/j.molliq.2016.02.048

Ai, L., & Jiang, J. (2012). Removal of methylene blue from aqueous solution with self-assembled cylindrical graphene–carbon nanotube hybrid. *Chemical Engineering Journal*, *192*, 156–163. doi:10.1016/j.cej.2012.03.056

Al-Degs, Y. S., El-Barghouthi, M. I., Issa, A. A., Khraisheh, M. A., & Walker, G. M. (2006). Sorption of Zn (II), Pb (II), and Co (II) using natural sorbents: Equilibrium and kinetic studies. *Water Research*, *40*(14), 2645–2658. doi:10.1016/j.watres.2006.05.018 PMID:16839582

Ali, G. A. M., Lih Teo, E. Y., Aboelazm, E. A. A., Sadegh, H., Memar, A. O. H., Shahryari-Ghoshekandi, R., & Chong, K. F. (2017). Capacitive performance of cysteamine functionalized carbon nanotubes. *Materials Chemistry and Physics*, *197*, 100–104. doi:10.1016/j.matchemphys.2017.05.019

Ali, G. A. M., Makhlouf, S. A., Yusoff, M. M., & Chong, K. F. (2015). Structural and electrochemical characteristics of graphene nanosheets as supercapacitor electrodes. *Reviews on Advanced Materials Science*, *40*(1), 35–43.

AlOmar, M. K., Alsaadi, M. A., Hayyan, M., Akib, S., Ibrahim, M., & Hashim, M. A. (2017a). Allyl triphenyl phosphonium bromide based DES-functionalized carbon nanotubes for the removal of mercury from water. *Chemosphere*, *167*(Supplement C), 44–52. doi:10.1016/j.chemosphere.2016.09.133 PMID:27710842

AlOmar, M. K., Alsaadi, M. A., Jassam, T. M., Akib, S., & Ali Hashim, M. (2017b). Novel deep eutectic solvent-functionalized carbon nanotubes adsorbent for mercury removal from water. *Journal of Colloid and Interface Science*, *497*(Supplement C), 413–421. doi:10.1016/j.jcis.2017.03.014 PMID:28314146

Alsohaimi, I. H., Wabaidur, S. M., Kumar, M., Khan, M. A., Alothman, Z. A., & Abdalla, M. A. (2015). Synthesis, characterization of PMDA/TMSPEDA hybrid nano-composite and its applications as an adsorbent for the removal of bivalent heavy metals ions. *Chemical Engineering Journal*, *270*, 9–21. doi:10.1016/j.cej.2015.02.008

Apul, O. G., Wang, Q., Zhou, Y., & Karanfil, T. (2013). Adsorption of aromatic organic contaminants by graphene nanosheets: Comparison with carbon nanotubes and activated carbon. *Water Research*, *47*(4), 1648–1654. doi:10.1016/j.watres.2012.12.031 PMID:23313232

Arias, M., Barral, M., & Mejuto, J. (2002). Enhancement of copper and cadmium adsorption on kaolin by the presence of humic acids. *Chemosphere*, *48*(10), 1081–1088. doi:10.1016/S0045-6535(02)00169-8 PMID:12227513

Azamat, J., Sattary, B. S., Khataee, A., & Joo, S. W. (2015). Removal of a hazardous heavy metal from aqueous solution using functionalized graphene and boron nitride nanosheets: Insights from simulations. *Journal of Molecular Graphics & Modelling*, *61*, 13–20. doi:10.1016/j.jmgm.2015.06.012 PMID:26186492

Babel, S., & Kurniawan, T. A. (2003). Low-cost adsorbents for heavy metals uptake from contaminated water: A review. *Journal of Hazardous Materials*, *97*(1), 219–243. doi:10.1016/S0304-3894(02)00263-7 PMID:12573840

Bahgat, M., Farghali, A., El Rouby, W., & Khedr, M. (2011). Synthesis and modification of multi-walled carbon nano-tubes (MWCNTs) for water treatment applications. *Journal of Analytical and Applied Pyrolysis*, *92*(2), 307–313. doi:10.1016/j.jaap.2011.07.002

Barakat, M., Al-Hutailah, R., Hashim, M., Qayyum, E., & Kuhn, J. (2013). Titania-supported silver-based bimetallic nanoparticles as photocatalysts. *Environmental Science and Pollution Research International*, *20*(6), 3751–3759. doi:10.100711356-012-1306-1 PMID:23161501

Bazrafshan, E., Mostafapour, F. K., Hosseini, A. R., Raksh Khorshid, A., & Mahvi, A. H. (2012). Decolorisation of reactive red 120 dye by using single-walled carbon nanotubes in aqueous solutions. *Journal of Chemistry*, 2013, 1-8. doi:10.1155/2013/938374

Bhaumik, M., Agarwal, S., Gupta, V. K., & Maity, A. (2016). Enhanced removal of Cr (VI) from aqueous solutions using polypyrrole wrapped oxidized MWCNTs nanocomposites adsorbent. *Journal of Colloid and Interface Science*, *470*, 257–267. doi:10.1016/j.jcis.2016.02.054 PMID:26962976

Biesheuvel, P., Zhao, R., Porada, S., & Van der Wal, A. (2011). Theory of membrane capacitive deionization including the effect of the electrode pore space. *Journal of Colloid and Interface Science*, *360*(1), 239–248. doi:10.1016/j.jcis.2011.04.049 PMID:21592485

Bilgili, M. S. (2006). Adsorption of 4-chlorophenol from aqueous solutions by xad-4 resin: Isotherm, kinetic, and thermodynamic analysis. *Journal of Hazardous Materials*, *137*(1), 157–164. doi:10.1016/j.jhazmat.2006.01.005 PMID:16487655

Bina, B., Amin, M., Rashidi, A., & Pourzamani, H. (2012). Benzene and toluene removal by carbon nanotubes from aqueous solution. *Archives of Environmental Protection*, *38*(1), 3–25. doi:10.2478/v10265-012-0001-0

Brumfiel, G. (2003). Nanotechnology: A little knowledge. *Nature*, *424*(6946), 246–248. doi:10.1038/424246a PMID:12867950

Carrott, P., Carrott, M. R., & Roberts, R. (1991). Physical adsorption of gases by microporous carbons. *Colloids and Surfaces*, *58*(4), 385–400. doi:10.1016/0166-6622(91)80217-C

Chandra, V., Park, J., Chun, Y., Lee, J. W., Hwang, I.-C., & Kim, K. S. (2010). Water-dispersible magnetite-reduced graphene oxide composites for arsenic removal. *ACS Nano*, *4*(7), 3979–3986. doi:10.1021/nn1008897 PMID:20552997

Chatterjee, S., Chatterjee, T., Lim, S.-R., & Woo, S. H. (2011). Effect of the addition mode of carbon nanotubes for the production of chitosan hydrogel core–shell beads on adsorption of Congo red from aqueous solution. *Bioresource Technology*, *102*(6), 4402–4409. doi:10.1016/j.biortech.2010.12.117 PMID:21277770

Chatterjee, S., Lee, M. W., & Woo, S. H. (2010). Adsorption of congo red by chitosan hydrogel beads impregnated with carbon nanotubes. *Bioresource Technology*, *101*(6), 1800–1806. doi:10.1016/j.biortech.2009.10.051 PMID:19962883

Chen, C., Liang, B., Ogino, A., Wang, X., & Nagatsu, M. (2009). Oxygen functionalization of multiwall carbon nanotubes by microwave-excited surface-wave plasma treatment. *The Journal of Physical Chemistry. C, Nanomaterials and Interfaces*, *113*(18), 7659.

Chen, C., & Wang, X. (2006). Adsorption of Ni (II) from aqueous solution using oxidized multiwall carbon nanotubes. *Industrial & Engineering Chemistry Research*, *45*(26), 9144–9149. doi:10.1021/ie060791z

Chen, H., Li, J., Shao, D., Ren, X., & Wang, X. (2012). Poly (acrylic acid) grafted multiwall carbon nanotubes by plasma techniques for Co (II) removal from aqueous solution. *Chemical Engineering Journal*, *210*, 475–481. doi:10.1016/j.cej.2012.08.082

Chen, J., Hamon, M. A., Hu, H., Chen, Y., Rao, A. M., Eklund, P. C., & Haddon, R. C. (1998). Solution properties of single-walled carbon nanotubes. *Science*, *282*(5386), 95–98. doi:10.1126cience.282.5386.95 PMID:9756485

Cloete, T. E. (2010). *Nanotechnology in water treatment applications*. Horizon Scientific Press.

Denis, P. A., & Iribarne, F. (2012). A first-principles study on the interaction between alkyl radicals and graphene. *Chemistry (Weinheim an der Bergstrasse, Germany)*, *18*(24), 7568–7574. doi:10.1002/chem.201103711 PMID:22532481

Du, Q., Sun, J., Li, Y., Yang, X., Wang, X., Wang, Z., & Xia, L. (2014). Highly enhanced adsorption of congo red onto graphene oxide/chitosan fibers by wet-chemical etching off silica nanoparticles. *Chemical Engineering Journal*, *245*, 99–106. doi:10.1016/j.cej.2014.02.006

Ekmekyapar, F., Aslan, A., Bayhan, Y. K., & Cakici, A. (2006). Biosorption of copper (II) by nonliving lichen biomass of Cladonia rangiformis hoffm. *Journal of Hazardous Materials*, *137*(1), 293–298. doi:10.1016/j.jhazmat.2006.02.003 PMID:16530938

Elnsar, T. A. S., Soliman, M. H., & Ayash, M. A.-E.-A. A. (2017). Modified Hydroxyapatite Adsorbent for Removal of Iron Dissolved in Water Wells in Sohag, Egypt. *Chemistry of Advanced Materials, 2*(1).

Ernhart, C. B. (1992). A critical review of low-level prenatal lead exposure in the human: 1. Effects on the fetus and newborn. *Reproductive Toxicology (Elmsford, N.Y.)*, *6*(1), 9–19. doi:10.1016/0890-6238(92)90017-N PMID:1562804

Ewecharoen, A., Thiravetyan, P., Wendel, E., & Bertagnolli, H. (2009). Nickel adsorption by sodium polyacrylate-grafted activated carbon. *Journal of Hazardous Materials*, *171*(1), 335–339. doi:10.1016/j.jhazmat.2009.06.008 PMID:19576692

Feng, J., Zhang, D.-D., Liu, Y.-F., Bai, Y., Chen, Q.-D., Liu, S.-Y., & Sun, H.-B. (2010). Magnetic nanofilm of Fe_3O_4 for highly efficient organic light-emitting devices. *The Journal of Physical Chemistry C*, *114*(14), 6718–6721. doi:10.1021/jp9122503

Feng, L., Cao, M., Ma, X., Zhu, Y., & Hu, C. (2012). Superparamagnetic high-surface-area Fe_3O_4 nanoparticles as adsorbents for arsenic removal. *Journal of Hazardous Materials*, *217*, 439–446. doi:10.1016/j.jhazmat.2012.03.073 PMID:22494901

Fu, F., & Wang, Q. (2011). Removal of heavy metal ions from wastewaters: A review. *Journal of Environmental Management*, *92*(3), 407–418. doi:10.1016/j.jenvman.2010.11.011 PMID:21138785

Gao, C., Zhang, W., Li, H., Lang, L., & Xu, Z. (2008). Controllable fabrication of mesoporous MgO with various morphologies and their absorption performance for toxic pollutants in water. *Crystal Growth & Design*, *8*(10), 3785–3790. doi:10.1021/cg8004147

Gao, W., Majumder, M., Alemany, L. B., Narayanan, T. N., Ibarra, M. A., Pradhan, B. K., & Ajayan, P. M. (2011). Engineered graphite oxide materials for application in water purification. *ACS Applied Materials & Interfaces*, *3*(6), 1821–1826. doi:10.1021/am200300u PMID:21568266

Garcia, A., Ferreira, L., Leitão, A., & Rodrigues, A. (1999). Binary Adsorption of Phenol and m-Cresol Mixtures onto a Polymeric Adsorbent. *Adsorption*, *5*(4), 359–368. doi:10.1023/A:1008904631147

Geyikçi, F. (2013). Adsorption of acid blue 161 (AB 161) dye from water by multi-walled carbon nanotubes. *Fullerenes, Nanotubes, and Carbon Nanostructures*, *21*(7), 579–593. doi:10.1080/1536383X.2011.643428

Ghaedi, M., Ghobadzadeh, P., Nasiri Kokhdan, S., & Soylak, M. (2013). Oxidized multiwalled carbon nanotubes as adsorbents for kinetic and equilibrium study of removal of 5-(4-dimethyl amino benzylidene) rhodanine. *Arabian Journal for Science & Engineering, 38*(7).

Ghaedi, M., Haghdoust, S., Kokhdan, S. N., Mihandoost, A., Sahraie, R., & Daneshfar, A. (2012a). Comparison of activated carbon, multiwalled carbon nanotubes, and cadmium hydroxide nanowire loaded on activated carbon as adsorbents for kinetic and equilibrium study of removal of Safranine O. *Spectroscopy Letters*, *45*(7), 500–510. doi:10.1080/00387010.2011.641058

Ghaedi, M., Hassanzadeh, A., & Kokhdan, S. N. (2011a). Multiwalled carbon nanotubes as adsorbents for the kinetic and equilibrium study of the removal of alizarin red S and morin. *Journal of Chemical & Engineering Data*, *56*(5), 2511–2520. doi:10.1021/je2000414

Ghaedi, M., Khajehsharifi, H., Yadkuri, A. H., Roosta, M., & Asghari, A. (2012b). Oxidized multiwalled carbon nanotubes as efficient adsorbent for bromothymol blue. *Toxicological and Environmental Chemistry*, *94*(5), 873–883. doi:10.1080/02772248.2012.678999

Ghaedi, M., & Kokhdan, S. N. (2012). Oxidized multiwalled carbon nanotubes for the removal of methyl red (MR): Kinetics and equilibrium study. *Desalination and Water Treatment*, *49*(1-3), 317–325. doi:10.1080/19443994.2012.719355

Ghaedi, M., Shokrollahi, A., Hossainian, H., & Kokhdan, S. N. (2011b). Comparison of activated carbon and multiwalled carbon nanotubes for efficient removal of eriochrome cyanine R (ECR): Kinetic, isotherm, and thermodynamic study of the removal process. *Journal of Chemical & Engineering Data*, *56*(7), 3227–3235. doi:10.1021/je200331u

Gopalakrishnan, A., Krishnan, R., Thangavel, S., Venugopal, G., & Kim, S.-J. (2015). Removal of heavy metal ions from pharma-effluents using graphene-oxide nanosorbents and study of their adsorption kinetics. *Journal of Industrial and Engineering Chemistry*, *30*, 14–19. doi:10.1016/j.jiec.2015.06.005

Graydon, J. W., Zhang, X., Kirk, D. W., & Jia, C. Q. (2009). Sorption and stability of mercury on activated carbon for emission control. *Journal of Hazardous Materials*, *168*(2), 978–982. doi:10.1016/j.jhazmat.2009.02.118 PMID:19327890

Gubin, S. P., Koksharov, Y. A., Khomutov, G., & Yurkov, G. Y. e. (2005). Magnetic nanoparticles: Preparation, structure and properties. *Russian Chemical Reviews*, *74*(6), 489–520. doi:10.1070/RC2005v074n06ABEH000897

Gupta, A., Vidyarthi, S. R., & Sankararamakrishnan, N. (2014). Enhanced sorption of mercury from compact fluorescent bulbs and contaminated water streams using functionalized multiwalled carbon nanotubes. *Journal of Hazardous Materials*, *274*(Supplement C), 132–144. doi:10.1016/j.jhazmat.2014.03.020 PMID:24780855

Gupta, V. K., Agarwal, S., Sadegh, H., Ali, G. A. M., Bharti, A. K., & Hamdy, A. S. (2017). Facile route synthesis of novel graphene oxide-β-cyclodextrin nanocomposite and its application as adsorbent for removal of toxic bisphenol A from the aqueous phase. *Journal of Molecular Liquids*, *237*, 466–472. doi:10.1016/j.molliq.2017.04.113

Gupta, V. K., Agarwal, S., & Saleh, T. A. (2011). Synthesis and characterization of alumina-coated carbon nanotubes and their application for lead removal. *Journal of Hazardous Materials*, *185*(1), 17–23. doi:10.1016/j.jhazmat.2010.08.053 PMID:20888691

Gupta, V. K., Gupta, M., & Sharma, S. (2001). Process development for the removal of lead and chromium from aqueous solutions using red mud-an aluminium industry waste. *Water Research*, *35*(5), 1125–1134. doi:10.1016/S0043-1354(00)00389-4 PMID:11268832

Gupta, V. K., Tyagi, I., Sadegh, H., Shahryari-Ghoshekandi, R., Makhlouf, A. S. H., & Maazinejad, B. (2015). Nanoparticles as adsorbent; a positive approach for removal of noxious metal ions: A review. *Science. Technology and Development*, *34*(3), 195–214. doi:10.3923td.2015.195.214

Habeeb, O. A., Kanthasamy, R., Ali, G. A. M., Sethupathi, S., & Yunus, R. B. M. (2017a). Hydrogen sulfide emission sources, regulations, and removal techniques: A review. *Reviews in Chemical Engineering*, *0*(0). doi:10.1515/revce-2017-0004

Habeeb, O. A., Ramesh, K., Ali, G. A. M., & Yunus, R. M. (2017b). Experimental design technique on removal of hydrogen sulfide using CaO-eggshells dispersed onto palm kernel shell activated carbon: Experiment, optimization, equilibrium and kinetic studies. *Journal of Wuhan University of Technology-Mater. Sci. Ed.*, *32*(2), 305–320. doi:10.100711595-017-1597-7

Habeeb, O. A., Ramesh, K., Ali, G. A. M., & Yunus, R. M. (2017c). Isothermal modelling based experimental study of dissolved hydrogen sulfide adsorption from waste water using eggshell based activated carbon. *The Malaysian Journal of Analytical Sciences*, *21*(2), 334–345. doi:10.17576/mjas-2017-2102-08

Habeeb, O. A., Ramesh, K., Ali, G. A. M., & Yunus, R. M. (2017d). Low-cost and eco-friendly activated carbon from modified palm kernel shell for hydrogen sulfide removal from wastewater: Adsorption and kinetic studies. *Desalination and Water Treatment*, *84*, 205–214. doi:10.5004/dwt.2017.21196

Hadavifar, M., Bahramifar, N., Younesi, H., & Li, Q. (2014). Adsorption of mercury ions from synthetic and real wastewater aqueous solution by functionalized multi-walled carbon nanotube with both amino and thiolated groups. *Chemical Engineering Journal*, *237*(Supplement C), 217–228. doi:10.1016/j.cej.2013.10.014

Hagen, A., & Hertel, T. (2003). Quantitative analysis of optical spectra from individual single-wall carbon nanotubes. *AIP Conference Proceedings*. 10.1063/1.1628016

Herrero, M. A., & Prato, M. (2008). Recent advances in the covalent functionalization of carbon nanotubes. *Molecular Crystals and Liquid Crystals (Philadelphia, Pa.)*, *483*(1), 21–32. doi:10.1080/15421400801898033

Hristovski, K., Baumgardner, A., & Westerhoff, P. (2007). Selecting metal oxide nanomaterials for arsenic removal in fixed bed columns: From nanopowders to aggregated nanoparticle media. *Journal of Hazardous Materials*, *147*(1), 265–274. doi:10.1016/j.jhazmat.2007.01.017 PMID:17254707

Hu, H., Wang, Z., & Pan, L. (2010). Synthesis of monodisperse Fe_3O_4@ silica core–shell microspheres and their application for removal of heavy metal ions from water. *Journal of Alloys and Compounds*, *492*(1), 656–661. doi:10.1016/j.jallcom.2009.11.204

Huang, Z., Wang, X., & Yang, D. (2015). Adsorption of Cr (VI) in wastewater using magnetic multiwall carbon nanotubes. *Water Science and Engineering*, *8*(3), 226–232. doi:10.1016/j.wse.2015.01.009

Jian, M., Liu, B., Zhang, G., Liu, R., & Zhang, X. (2015). Adsorptive removal of arsenic from aqueous solution by zeolitic imidazolate framework-8 (ZIF-8) nanoparticles. *Colloids and Surfaces. A, Physicochemical and Engineering Aspects*, *465*, 67–76. doi:10.1016/j.colsurfa.2014.10.023

Kalfa, O. M., Yalçınkaya, Ö., & Türker, A. R. (2009). Synthesis of nano B_2O_3/TiO_2 composite material as a new solid phase extractor and its application to preconcentration and separation of cadmium. *Journal of Hazardous Materials*, *166*(1), 455–461. doi:10.1016/j.jhazmat.2008.11.112 PMID:19135793

Khajeh, M., Laurent, S., & Dastafkan, K. (2013). Nanoadsorbents: Classification, preparation, and applications (with emphasis on aqueous media). *Chemical Reviews*, *113*(10), 7728–7768. doi:10.1021/cr400086v PMID:23869773

Khajeh, M., & Sanchooli, E. (2011). Synthesis and evaluation of silver nanoparticles material for solid phase extraction of cobalt from water samples. *Applied Nanoscience*, *1*(4), 205–209. doi:10.100713204-011-0028-x

Khani, R., Sobhani, S., & Beyki, M. H. (2016). Highly selective and efficient removal of lead with magnetic nano-adsorbent: Multivariate optimization, isotherm and thermodynamic studies. *Journal of Colloid and Interface Science*, *466*, 198–205. doi:10.1016/j.jcis.2015.12.027 PMID:26724702

Kim, Y.-J., & Choi, J.-H. (2012). Selective removal of nitrate ion using a novel composite carbon electrode in capacitive deionization. *Water Research*, *46*(18), 6033–6039. doi:10.1016/j.watres.2012.08.031 PMID:22980574

Kotov, N. A. (2006). Materials science: Carbon sheet solutions. *Nature*, *442*(7100), 254–255. doi:10.1038/442254a PMID:16855576

Kurniawan, T. A., & Lo, W. (2009). Removal of refractory compounds from stabilized landfill leachate using an integrated H_2O_2 oxidation and granular activated carbon (GAC) adsorption treatment. *Water Research*, *43*(16), 4079–4091. doi:10.1016/j.watres.2009.06.060 PMID:19695663

Lee, Y.-C., & Yang, J.-W. (2012). Self-assembled flower-like TiO_2 on exfoliated graphite oxide for heavy metal removal. *Journal of Industrial and Engineering Chemistry*, *18*(3), 1178–1185. doi:10.1016/j.jiec.2012.01.005

Li, B., Cao, H., & Yin, G. (2011a). $Mg(OH)_2$ @ reduced graphene oxide composite for removal of dyes from water. *Journal of Materials Chemistry*, *21*(36), 13765–13768. doi:10.1039/c1jm13368c

Li, N., Zheng, M., Chang, X., Ji, G., Lu, H., Xue, L., ... Cao, J. (2011b). Preparation of magnetic $CoFe_2O_4$-functionalized graphene sheets via a facile hydrothermal method and their adsorption properties. *Journal of Solid State Chemistry*, *184*(4), 953–958. doi:10.1016/j.jssc.2011.01.014

Li, Y., Du, Q., Liu, T., Peng, X., Wang, J., Sun, J., ... Xia, Y. (2013). Comparative study of methylene blue dye adsorption onto activated carbon, graphene oxide, and carbon nanotubes. *Chemical Engineering Research & Design*, *91*(2), 361–368. doi:10.1016/j.cherd.2012.07.007

Li, Y.-H., Ding, J., Luan, Z., Di, Z., Zhu, Y., Xu, C., ... Wei, B. (2003a). Competitive adsorption of Pb^{2+}, Cu^{2+} and Cd^{2+} ions from aqueous solutions by multiwalled carbon nanotubes. *Carbon*, *41*(14), 2787–2792. doi:10.1016/S0008-6223(03)00392-0

Li, Y.-H., Wang, S., Luan, Z., Ding, J., Xu, C., & Wu, D. (2003b). Adsorption of cadmium (II) from aqueous solution by surface oxidized carbon nanotubes. *Carbon*, *41*(5), 1057–1062. doi:10.1016/S0008-6223(02)00440-2

Lingamdinne, L. P., Koduru, J. R., Choi, Y.-L., Chang, Y.-Y., & Yang, J.-K. (2016). Studies on removal of Pb (II) and Cr (III) using graphene oxide based inverse spinel nickel ferrite nano-composite as sorbent. *Hydrometallurgy*, *165*, 64–72. doi:10.1016/j.hydromet.2015.11.005

Lu, C., & Liu, C. (2006). Removal of nickel (II) from aqueous solution by carbon nanotubes. *Journal of Chemical Technology and Biotechnology (Oxford, Oxfordshire)*, *81*(12), 1932–1940. doi:10.1002/jctb.1626

Machado, F. M., Bergmann, C. P., Fernandes, T. H., Lima, E. C., Royer, B., Calvete, T., & Fagan, S. B. (2011). Adsorption of Reactive Red M-2BE dye from water solutions by multi-walled carbon nanotubes and activated carbon. *Journal of Hazardous Materials*, *192*(3), 1122–1131. doi:10.1016/j.jhazmat.2011.06.020 PMID:21724329

Machida, M., Mochimaru, T., & Tatsumoto, H. (2006). Lead (II) adsorption onto the graphene layer of carbonaceous materials in aqueous solution. *Carbon*, *44*(13), 2681–2688. doi:10.1016/j.carbon.2006.04.003

Madadrang, C. J., Kim, H. Y., Gao, G., Wang, N., Zhu, J., Feng, H., ... Hou, S. (2012). Adsorption behavior of EDTA-graphene oxide for Pb (II) removal. *ACS Applied Materials & Interfaces*, *4*(3), 1186–1193. doi:10.1021/am201645g PMID:22304446

Maliyekkal, S. M., Sreeprasad, T., Krishnan, D., Kouser, S., Mishra, A. K., Waghmare, U. V., & Pradeep, T. (2013). Graphene: A reusable substrate for unprecedented adsorption of pesticides. *Small*, *9*(2), 273–283. doi:10.1002mll.201201125 PMID:23001848

Marina, P. E., Ali, G. A. M., See, L. M., Teo, E. Y. L., Ng, E.-P., & Chong, K. F. (2016). In situ growth of redox-active iron-centered nanoparticles on graphene sheets for specific capacitance enhancement. *Arabian Journal of Chemistry*. doi:10.1016/j.arabjc.2016.02.006

Mickelson, E., Chiang, I., Zimmerman, J., Boul, P., Lozano, J., Liu, J., ... Margrave, J. (1999). Solvation of fluorinated single-wall carbon nanotubes in alcohol solvents. *The Journal of Physical Chemistry B*, *103*(21), 4318–4322. doi:10.1021/jp9845524

Mirkin, C. A., Letsinger, R. L., Mucic, R. C., & Storhoff, J. J. (1996). A DNA-based method for rationally assembling nanoparticles into macroscopic materials. *Nature*, *382*(6592), 607–609. doi:10.1038/382607a0 PMID:8757129

Molinari, R., Poerio, T., Cassano, R., Picci, N., & Argurio, P. (2004). Copper (II) removal from wastewaters by a new synthesized selective extractant and SLM viability. *Industrial & Engineering Chemistry Research*, *43*(2), 623–628. doi:10.1021/ie030392t

Moradi, O. (2013). Adsorption behavior of basic red 46 by single-walled carbon nanotubes surfaces. *Fullerenes, Nanotubes, and Carbon Nanostructures*, *21*(4), 286–301. doi:10.1080/1536383X.2011.572317

Moradi, O., Sadegh, H., Shahryari-Ghoshekandi, R., & Norouzi, M. (2015). Application of carbon nanotubes in nanomedicine: new medical approach for tomorrow. In Handbook of Research on Diverse Applications of Nanotechnology in Biomedicine, Chemistry, and Engineering (pp. 90–128). Hershey, PA: IGI Global.

Moradi, O., & Zare, K. (2011). Adsorption of Pb (II), Cd (II) and Cu (II) ions in aqueous solution on SWCNTs and SWCNT–COOH surfaces: Kinetics studies. *Fullerenes, Nanotubes, and Carbon Nanostructures, 19*(7), 628–652. doi:10.1080/1536383X.2010.504955

Moradi, O., Zare, K., Monajjemi, M., Yari, M., & Aghaie, H. (2010). The studies of equilibrium and thermodynamic adsorption of Pb (II), Cd (II) and Cu (II) Ions from aqueous solution onto SWCNTs and SWCNT–COOH Surfaces. *Fullerenes, Nanotubes, and Carbon Nanostructures, 18*(3), 285–302. doi:10.1080/15363831003783005

Mthombeni, N. H., Mbakop, S., & Onyango, M. S. (2015). Magnetic zeolite-polymer composite as an adsorbent for the remediation of wastewaters containing vanadium. *International Journal of Environmental Sciences and Development, 6*(8), 602–605. doi:10.7763/IJESD.2015.V6.665

Nadafi, K., Mesdaghinia, A., Nabizadeh, R., Younesian, M., & Rad, M. J. (2011). The combination and optimization study on RB29 dye removal from water by peroxy acid and single-wall carbon nanotubes. *Desalination and Water Treatment, 27*(1-3), 237–242. doi:10.5004/dwt.2011.1980

Natale, F., Erto, A., Lancia, A., & Musmarra, D. (2008). Experimental and modelling analysis of As (V) ions adsorption on granular activated carbon. *Water Research, 42*(8), 2007–2016. doi:10.1016/j.watres.2007.12.008 PMID:18222519

Neyaz, N., & Siddiqui, W. (2015). Removal of Cu (II) by modified magnetite nanocomposite as a nanosorbent. *International Journal of Science and Research, 4*, 1868–1873.

Pacheco, S., Medina, M., Valencia, F., & Tapia, J. (2006). Removal of inorganic mercury from polluted water using structured nanoparticles. *Journal of Environmental Engineering, 132*(3), 342–349. doi:10.1061/(ASCE)0733-9372(2006)132:3(342)

Pan, B., Pan, B., Zhang, W., Lv, L., Zhang, Q., & Zheng, S. (2009). Development of polymeric and polymer-based hybrid adsorbents for pollutants removal from waters. *Chemical Engineering Journal, 151*(1), 19–29. doi:10.1016/j.cej.2009.02.036

Pan, B. C., Xiong, Y., Li, A. M., Chen, J. L., Zhang, Q. X., & Jin, X. Y. (2002). Adsorption of aromatic acids on an aminated hypercrosslinked macroporous polymer. *Reactive & Functional Polymers, 53*(2), 63–72. doi:10.1016/S1381-5148(02)00123-2

Patnukao, P., Kongsuwan, A., & Pavasant, P. (2008). Batch studies of adsorption of copper and lead on activated carbon from Eucalyptus camaldulensis Dehn. bark. *Journal of Environmental Sciences (China), 20*(9), 1028–1034. doi:10.1016/S1001-0742(08)62145-2 PMID:19143307

Pena, M., Meng, X., Korfiatis, G. P., & Jing, C. (2006). Adsorption mechanism of arsenic on nanocrystalline titanium dioxide. *Environmental Science & Technology, 40*(4), 1257–1262. doi:10.1021/es052040e PMID:16572784

Pompeo, F., & Resasco, D. E. (2002). Water solubilization of single-walled carbon nanotubes by functionalization with glucosamine. *Nano Letters*, *2*(4), 369–373. doi:10.1021/nl015680y

Pyrzyńska, K., & Bystrzejewski, M. (2010). Comparative study of heavy metal ions sorption onto activated carbon, carbon nanotubes, and carbon-encapsulated magnetic nanoparticles. *Colloids and Surfaces. A, Physicochemical and Engineering Aspects*, *362*(1), 102–109. doi:10.1016/j.colsurfa.2010.03.047

Qu, Z., Yan, L., Li, L., Xu, J., Liu, M., Li, Z., & Yan, N. (2014). Ultraeffective ZnS nanocrystals sorbent for mercury (II) removal based on size-dependent cation exchange. *ACS Applied Materials & Interfaces*, *6*(20), 18026–18032. doi:10.1021/am504896w PMID:25299972

Rahmani, A., Ghaffari, H., & Samadi, M. (2011). A comparative study on arsenic (III) removal from aqueous solution using nano and micro sized zero-valent iron. *Journal of Environmental Health Science & Engineering*, *8*(2), 157–166.

Rao, C. N. R., Müller, A., & Cheetham, A. K. (2006). *The chemistry of nanomaterials: synthesis, properties and applications*. John Wiley & Sons.

Rao, G. P., Lu, C., & Su, F. (2007). Sorption of divalent metal ions from aqueous solution by carbon nanotubes: A review. *Separation and Purification Technology*, *58*(1), 224–231. doi:10.1016/j.seppur.2006.12.006

Ray, P. C. (2010). Size and shape dependent second order nonlinear optical properties of nanomaterials and its application in biological and chemical sensing. *Chemical Reviews*, *110*(9), 5332–5365. doi:10.1021/cr900335q PMID:20469927

Ren, X., Chen, C., Nagatsu, M., & Wang, X. (2011). Carbon nanotubes as adsorbents in environmental pollution management: A review. *Chemical Engineering Journal*, *170*(2), 395–410. doi:10.1016/j.cej.2010.08.045

Richard, C., Balavoine, F., Schultz, P., Ebbesen, T. W., & Mioskowski, C. (2003). Supramolecular self-assembly of lipid derivatives on carbon nanotubes. *Science*, *300*(5620), 775–778. doi:10.1126cience.1080848 PMID:12730595

Rickerby, D., & Morrison, M. (2007). Nanotechnology and the environment: A European perspective. *Science and Technology of Advanced Materials*, *8*(1), 19–24. doi:10.1016/j.stam.2006.10.002

Sadegh, H., Ali, G. A. M., Abbasi, Z., & Nadagoud, M. N. (2017a). Adsorption of Ammonium Ions onto Multi-Walled Carbon Nanotubes. *Studia Universitatis Babes-Bolyai. Chemia*, *62*(2), 233–245. doi:10.24193ubbchem.2017.2.18

Sadegh, H., Ali, G. A. M., Gupta, V. K., Makhlouf, A. S. H., Shahryari-Ghoshekandi, R., Nadagouda, M. N., ... Megiel, E. (2017b). The role of nanomaterials as effective adsorbents and their applications in wastewater treatment. *Journal of Nanostructure in Chemistry*, *7*(1), 1–14. doi:10.100740097-017-0219-4

Sadegh, H., Ali, G. A. M., Makhlouf, A. S. H., Chong, K. F., Alharbi, N. S., Agarwal, S., & Gupta, V. K. (2018). MWCNTs-Fe3O4 nanocomposite for Hg(II) high adsorption efficiency. *Journal of Molecular Liquids*, *258*, 345–353. doi:10.1016/j.molliq.2018.03.012

Sadegh, H., & Shahryari-ghoshekandi, R. (2015). Functionalization of carbon nanotubes and its application in nanomedicine: A review. *Nanomedicine Journal*, *2*(4), 231–248.

Sadegh, H., Zare, K., Maazinejad, B., Shahryari-Ghoshekandi, R., Tyagi, I., Agarwal, S., & Gupta, V. K. (2016). Synthesis of MWCNT-COOH-Cysteamine composite and its application for dye removal. *Journal of Molecular Liquids*, *215*, 221–228. doi:10.1016/j.molliq.2015.12.042

Salem Attia, T. M., Hu, X. L., & Yin, D. Q. (2014). Synthesised magnetic nanoparticles coated zeolite (MNCZ) for the removal of arsenic (As) from aqueous solution. *Journal of Experimental Nanoscience*, *9*(6), 551–560. doi:10.1080/17458080.2012.677549

Santhosh, C., Velmurugan, V., Jacob, G., Jeong, S. K., Grace, A. N., & Bhatnagar, A. (2016). Role of nanomaterials in water treatment applications: A review. *Chemical Engineering Journal*, *306*, 1116–1137. doi:10.1016/j.cej.2016.08.053

Selvi, K., Pattabhi, S., & Kadirvelu, K. (2001). Removal of Cr (VI) from aqueous solution by adsorption onto activated carbon. *Bioresource Technology*, *80*(1), 87–89. doi:10.1016/S0960-8524(01)00068-2 PMID:11554606

Shirmardi, M., Mesdaghinia, A., Mahvi, A. H., Nasseri, S., & Nabizadeh, R. (2012). Kinetics and equilibrium studies on adsorption of acid red 18 (Azo-Dye) using multiwall carbon nanotubes (MWCNTs) from aqueous solution. *Journal of Chemistry*, *9*(4), 2371–2383.

Singh, S., Barick, K., & Bahadur, D. (2011). Novel and efficient three dimensional mesoporous ZnO nanoassemblies for envirnomental remediation. *International Journal of Nanoscience*, *10*(4-5), 1001-1005.

Singh, S., Barick, K., & Bahadur, D. (2013). Functional oxide nanomaterials and nanocomposites for the removal of heavy metals and dyes. *Nanomaterials and Nanotechnology*, *3*, 20. doi:10.5772/57237

Srivastava, R. R., Kim, M., Lee, J., & Ilyas, S. (2015). Liquid–liquid extraction of rhenium (VII) from an acidic chloride solution using Cyanex 923. *Hydrometallurgy*, *157*, 33–38. doi:10.1016/j.hydromet.2015.07.011

Stafiej, A., & Pyrzynska, K. (2007). Adsorption of heavy metal ions with carbon nanotubes. *Separation and Purification Technology*, *58*(1), 49–52. doi:10.1016/j.seppur.2007.07.008

Stenzel, M. H. (1993). Remove organics by activated carbon adsorption. *Chemical Engineering Progress*, *89*(4).

Tang, W.-W., Zeng, G.-M., Gong, J.-L., Liu, Y., Wang, X.-Y., Liu, Y.-Y., ... Tu, D.-Z. (2012). Simultaneous adsorption of atrazine and Cu (II) from wastewater by magnetic multi-walled carbon nanotube. *Chemical Engineering Journal*, *211*, 470–478. doi:10.1016/j.cej.2012.09.102

Tarigh, G. D., & Shemirani, F. (2013). Magnetic multi-wall carbon nanotube nanocomposite as an adsorbent for preconcentration and determination of lead (II) and manganese (II) in various matrices. *Talanta*, *115*, 744–750. doi:10.1016/j.talanta.2013.06.018 PMID:24054657

Tawabini, B. S., Al-Khaldi, S. F., Khaled, M. M., & Atieh, M. A. (2011). Removal of arsenic from water by iron oxide nanoparticles impregnated on carbon nanotubes. *Journal of Environmental Science and Health. Part A*, *46*(3), 215–223. PMID:21279891

Theron, J., Walker, J., & Cloete, T. (2008). Nanotechnology and water treatment: Applications and emerging opportunities. *Critical Reviews in Microbiology*, *34*(1), 43–69. doi:10.1080/10408410701710442 PMID:18259980

Tuzen, M., & Soylak, M. (2007). Multiwalled carbon nanotubes for speciation of chromium in environmental samples. *Journal of Hazardous Materials*, *147*(1), 219–225. doi:10.1016/j.jhazmat.2006.12.069 PMID:17258392

Varma, S., Sarode, D., Wakale, S., Bhanvase, B., & Deosarkar, M. (2013). Removal of nickel from waste water using graphene nanocomposite. *International Journal of Chemical and Physical Sciences*, *2*, 132–139.

Wang, H., Dong, Z., & Na, C. (2013a). Hierarchical carbon nanotube membrane-supported gold nanoparticles for rapid catalytic reduction of p-nitrophenol. *ACS Sustainable Chemistry & Engineering*, *1*(7), 746–752. doi:10.1021c400048m

Wang, S., Ng, C. W., Wang, W., Li, Q., & Hao, Z. (2012a). Synergistic and competitive adsorption of organic dyes on multiwalled carbon nanotubes. *Chemical Engineering Journal*, *197*, 34–40. doi:10.1016/j.cej.2012.05.008

Wang, S., Ng, C. W., Wang, W., Li, Q., & Li, L. (2012b). A comparative study on the adsorption of acid and reactive dyes on multiwall carbon nanotubes in single and binary dye systems. *Journal of Chemical & Engineering Data*, *57*(5), 1563–1569. doi:10.1021/je3001552

Wang, X., Cai, W., Liu, S., Wang, G., Wu, Z., & Zhao, H. (2013b). ZnO hollow microspheres with exposed porous nanosheets surface: Structurally enhanced adsorption towards heavy metal ions. *Colloids and Surfaces. A, Physicochemical and Engineering Aspects*, *422*, 199–205. doi:10.1016/j.colsurfa.2013.01.031

Wu, T., Cai, X., Tan, S., Li, H., Liu, J., & Yang, W. (2011). Adsorption characteristics of acrylonitrile, p-toluenesulfonic acid, 1-naphthalenesulfonic acid and methyl blue on graphene in aqueous solutions. *Chemical Engineering Journal*, *173*(1), 144–149. doi:10.1016/j.cej.2011.07.050

Yang, S.-T., Chang, Y., Wang, H., Liu, G., Chen, S., Wang, Y., ... Cao, A. (2010). Folding/aggregation of graphene oxide and its application in Cu^{2+} removal. *Journal of Colloid and Interface Science*, *351*(1), 122–127. doi:10.1016/j.jcis.2010.07.042 PMID:20705296

Yang, W., Ding, P., Zhou, L., Yu, J., Chen, X., & Jiao, F. (2013). Preparation of diamine modified mesoporous silica on multi-walled carbon nanotubes for the adsorption of heavy metals in aqueous solution. *Applied Surface Science*, *282*, 38–45. doi:10.1016/j.apsusc.2013.05.028

Yu, F., Chen, J., Chen, L., Huai, J., Gong, W., Yuan, Z., ... Ma, J. (2012). Magnetic carbon nanotubes synthesis by Fenton's reagent method and their potential application for removal of azo dye from aqueous solution. *Journal of Colloid and Interface Science*, *378*(1), 175–183. doi:10.1016/j.jcis.2012.04.024 PMID:22564767

Yu, J.-G., Zhao, X.-H., Yang, H., Chen, X.-H., Yang, Q., Yu, L.-Y., ... Chen, X.-Q. (2014). Aqueous adsorption and removal of organic contaminants by carbon nanotubes. *The Science of the Total Environment*, *482*, 241–251. doi:10.1016/j.scitotenv.2014.02.129 PMID:24657369

Zare-Dorabei, R., Ferdowsi, S. M., Barzin, A., & Tadjarodi, A. (2016). Highly efficient simultaneous ultrasonic-assisted adsorption of Pb (II), Cd (II), Ni (II) and Cu (II) ions from aqueous solutions by graphene oxide modified with 2, 2′-dipyridylamine: Central composite design optimization. *Ultrasonics Sonochemistry*, *32*, 265–276. doi:10.1016/j.ultsonch.2016.03.020 PMID:27150770

Zhang, J., Zhai, S., Li, S., Xiao, Z., Song, Y., An, Q., & Tian, G. (2013). Pb (II) removal of Fe_3O_4@ SiO_2–NH_2 core–shell nanomaterials prepared via a controllable sol–gel process. *Chemical Engineering Journal*, *215*, 461–471. doi:10.1016/j.cej.2012.11.043

Zhao, G., Li, J., Ren, X., Chen, C., & Wang, X. (2011a). Few-layered graphene oxide nanosheets as superior sorbents for heavy metal ion pollution management. *Environmental Science & Technology*, *45*(24), 10454–10462. doi:10.1021/es203439v PMID:22070750

Zhao, G., Ren, X., Gao, X., Tan, X., Li, J., Chen, C., ... Wang, X. (2011b). Removal of Pb (II) ions from aqueous solutions on few-layered graphene oxide nanosheets. *Dalton Transactions (Cambridge, England)*, *40*(41), 10945–10952. doi:10.1039/c1dt11005e PMID:21918761

Zhao, G., Wen, T., Yang, X., Yang, S., Liao, J., Hu, J., ... Wang, X. (2012). Preconcentration of U (VI) ions on few-layered graphene oxide nanosheets from aqueous solutions. *Dalton Transactions (Cambridge, England)*, *41*(20), 6182–6188. doi:10.1039/c2dt00054g PMID:22473651

Zhou, S., Shao, Y., Gao, N., Deng, J., & Tan, C. (2013). Equilibrium, kinetic, and thermodynamic studies on the adsorption of triclosan onto multi-walled carbon nanotubes. *CLEAN–Soil, Air. Water (Basel)*, *41*(6), 539–547.

Chapter 8
Nanotechnology for Water Environmental Application:
Functionalized Silica Hybrids as Nano-Sorbents

Saima Nasreen
The Women University Multan, Pakistan

Uzaira Rafique
Fatima Jinnah Women University, Pakistan

ABSTRACT

In developing countries like Pakistan, industries do not allocate budget for control and treatment of pollution due to limited resources. Most of the pollutants emitted from these industries are usually reusable but their capture from waste is highly uneconomical. This goal can be attained by synthesizing the advanced materials such as silica hybrids, the class of silica hybrids consisting of inorganic silica linked with organic ligands to form network structure. Their high porosity, surface area, and crystallinity provide a way for removing large amount of pollutants with small quantity of silica hybrids. These applications include the use of powder hybrids as adsorbent for removal of different pollutants such as heavy metals, persistent organic pollutants, and toxic organic chemicals. These functionalized hybrids can be further fabricated into thin films. This new form will be further used for catalytic degradation of various pollutants. Besides all these applications, these hybrids will have the potential to be used as sensors for detecting various pollutants in industries.

INTRODUCTION

Water is one of the most important natural resource and commodity on Earth, required for performing all vital activities of life. As water is indication of life. It is indispensable for domestic, agriculture and industrial activities as well as sustaining the earth's ecosystems. But this essential resource is under the threat. Two third portion of earth is covered with water. It is widely distributed into freshwater and

DOI: 10.4018/978-1-5225-5745-6.ch008

saline water. About 97% of water is saline (oceans and seas), not usable and only 3% water is available as freshwater. Of this total freshwater, 2% is locked in glaciers and polar icecaps while only 1% is present as ground and surface water (UNESCO, 2003; U. S. Geological Survey, 2014) readily accessible for human consumption. Globally freshwater consumption has been increased due to rapid growth in population, industrial development, agricultural activities, geological and environmental changes (Ali & Aboul-Enein., 2004; Steduto et al., 2012).

According to fact sheet of FAO (2009), agriculture sector accounts 70%, industries 20% and domestic 10% of available 1% freshwater. When it is once used and discharged back into the environment, water quality is going to be changed. According to WHO (2012) and UNICEF (2008, 2010) reports, 780-900 million people rely on unimproved drinking water supplies to meet their basic needs and almost all of them are in developing regions. United Nation Development Program (2006) reported that about 50%of all people in developing countries suffer from health problem caused by water. Pakistan was considered as water surplus country, is now become water stress country (Ensink et al., 2004) due to depletion of natural water resources.

According to the report of Pakistan Strategic Country Environmental Assessment report, available water has been decreased from 1,299 m^3 /capita (1996-1997) to 1,100 m^3/ capita (2006). It is estimated that it become less than 700 m^3 / capita in 2025, that is against the international standard of 1500 m^3/ capita. Wastewater generated from domestic, municipal, industries and agriculture runoff dump directly into water bodies without any treatment causing water pollution. Hence, it has become core issue of today and its consequences are reflected back especially into the public health, biodiversity, aquatic condition and ecosystems, by putting them in danger.

Pollutants in Aqueous Media

Scientists gave serious attention on this current issue. Pollutant (inorganic, organic) in water in different form (soluble, suspended and colloidal), having accumulative and mobile nature, once enters into food chain affecting the human health because human being is on top of food chain. Human body cannot process and disposed of and deposited in various internal organs and cause adverse effect on body function. They showed toxic and carcinogenic nature even at low concentrations (Ali & Aboul-Enein, 2006, 2009). Cadmium causes renal disorder, itai-itai disease, bone fragility and destruction of erythrocytes (Hlihor et al., 2009), Copper causes gastrointestinal catarrh and haemochromatosis (Gündoğan et al., 2004), Nitroaniline causing methemoglobinemia and so on.

Polyaromatic hydrocarbons (PAHs) belong to class of organic pollutant are basically by-products of incomplete combustion or hydrolysis of organic materials containing one or two fused benzene rings (Gao et al., 2006; Alcántara et al., 2008; Muff & Søgaard, 2010). These are hydrophobic, volatile, low water solubility and non-biodegradable having carcinogenic and mutagenic effects. Due to its persistent nature once they enter into the water systems; it is difficult to be removed. PAHs are listed as US-EPA and EU priority pollutants (Falciola et al., 2012) and their concentrations need to be controlled. The most common sources of these pollutants are paint, rubber, cement, petrochemical, and solvent manufacturing industries, disposed of their large volumes of untreated wastewater into water reservoir. Few studies regarding removal of Naphthalene and pyrene using Industrial waste (Jonker et al., 2002), organoclay composite (Ake et al., 2003), Biopolymers derived chars (Wang et al., 2007), Immature coal (Zeledon et al., 2007), mineral naphthenic oil (Luna et al., 2008), plant residue material (Li et al., 2010), rice husk (Agarry et al., 2013) have been reported due to environmental importance.

The considerable contaminants in aqueous environment, requires the development of simple and quick methods for its determination and separation. Efficient and effective techniques for the removal of such pollutants have drawn significant interest. Last few decade, different treatment techniques have been used. Most commonly used techniques are precipitation, Oxidation, floatation, gravity separation, coagulation, distillation, solvent extraction, micro and ultra filtration, electrolysis. It seems to be uneconomical due to having multi-step processes, high cost, addition of chemicals and chemical residues/ sludge, whose disposal is another issue.

Environmental Application of Nanomaterials for Wastewater Treatment

In order to release the pressure on industry to decrease emission of various pollutants into the environment, a broad range of methods with varying degrees of success have been developed to manage water pollution. These include coagulation (Balasubramanian & Madhavan, 2001), foam flotation (Lazaridis et al., 2001), filtration, ion exchange (Korngold et al., 2001), sedimentation, solvent extraction (Takeshita, 2003), electrolysis, chemical oxidation (Gonze et al., 2003), disinfection, chemical precipitation (Aguinaldo, 2007), membrane process (Li et al., 2004) and so on.

However, these methods have their own shortcomings and limitations. For example, the method based on chemical/biological oxidation, ion exchange and solvent extraction have shown low efficiency for the removal of trace levels of pollutants. Further, coagulation requires pH control and causes further problems of sludge disposal, whereas, ozonation while removing color effectively does not minimize chemical oxygen demand (COD).

The quest for simple and efficient technique ends up after the development of Adsorption processes. Adsorption is surface phenomenon in which pollutants are adhering on the surface through physical and weak chemical forces. It is simple and robust technique because it does not required additional chemicals during pre and post treatment and due to availability of a wide range of adsorbents. It is recognized as an effective, efficient and economical method.

The uniform pore size distribution along with interconnected pore structure is ideal properties of an adsorbent (Al-Dalama et al., 2005; Hong et al., 2004). Various adsorbents were reported such as organic clay, silica gel, montmorillonite and bentonite clay (Pal & Vanjara, 2001), iron oxides (Clausen & Fabricius, 2001), resins (Gurten et al., 2005), polymers (Jianguo et al., 2005), activated carbons (Duman, 2005), alumina (Villacanas et al., 2006), ion exchange resins (Cavaco et al., 2007), agricultural and industrial waste material with and without modification (Dave et al., 2011; Khosla et al., 2012) and many others have been reported for wastewater treatment.

Recently research attention is paid more towards to the nanoscience and nanotechnology due to its wide application in every sector of science and technology. It has become a versatile and promising subject for creating new materials with enhanced properties controlling size, shape, composition, fabrication and potential applications (Eggeman et al., 2006). Application of nanotechnology in order to improve the water quality using nano-adsorbent is one of the raising fields of interest. Literature have cited numerous nanoparticles as efficient adsorbent like titanium oxide (Pena et al., 2005), gold (Huang et al ., 2007), Zeolites (Shukla et al., 2007; Chen et al., 2008), multi-walled carbon nanotubes (Pillay et al., 2009), alumina (Sharma, et al., 2010), silver (Khan et al., 2011), Zinc oxide (Chen et al., 2011), silica (Shekaret al., 2012) Nickle oxide (Renu et al., 2013) and many others has been evaluated for removal of different organic and inorganic pollutant from aqueous medium.

Due to set up of stringent environmental regulation there is needed to gain greater harmony between environment and human activities. The quest for more efficient and improved material introduces hybrid materials with its unique properties. Hybrid (inorganic and organic) materials are one of emerging field in material science. Composition of inorganic and organic moiety (hybrid materials) at molecular or nano level with novel features that are different from individual one (Yokoyama et al. 2006). Inorganic imparts thermal, mechanical structural stability while organic impart the functionalities into hybrid material (Stein et al., 2000).

History of hybrid materials dates back to 1950s with successful synthesized commercial hybrid polymer (Arkles et al., 2001). It have been intensively studied due to their multi-functional properties for different applications such as electro-analytical applications (Walcarious et al., 2001), nano and ultra-membranes (Kim et al., 2012; Paun et al., 2011) transparent films (Nagappan et al., 2013) pH sensitive composites, solar cells (He et al., 2013) electrolyte (Sauza et al., 2004), separation of gases (Collins & Zhou, 2007), catalysts (Raj et al., 2013), biosensors (Fu et al., 2011), corrosion preventing coating (Chou et al., 2003), adsorbents of toxic compound (Zewail et al., 2010; Wang et al., 2012; Kim et al., 2012) and many others.

Inorganic-organic hybrids are synthesized through different route like sol-gel process, self-assembly process, dispersion of nano-building blocks, hierarchical structures and interpenetrating networks. Every route has its own characteristics exhibiting unique structure and functionalities (Sanchez et al., 2005). It has been exploited for the remediation purposes showing better results. Mostly metal salt or metal alkoxide are used as inorganic precursor and different functional group (COOH, OH, NH and many others) containing organic compound or polymers used as organic precursor.

Network modifier helped to introduce the organic moiety into the inorganic matrix (Rozes et al., 2006). Network modifier is introduced through two ways. Adding of grafted monomers within inorganic matrix formed in solvent is the first way while in situ treatment is the second way. The functional groups are responsible for interaction with pollutants.

Nano-Sorbents for Environmental Remediation

Removal of pollutants from aqueous media with novel materials with special reference to their application as adsorbent from aqueous samples including industrial effluents, municipal and drinking water actually becomes a basic requirement today. Characteristics of these novel nano-sorbents play an important role due to structure property relationship that enables the proficient degradation or chemical transformation of the target pollutant.

Each nano-sorbent has specific structure property relationship with respect to specific type of pollutants like heavy metals, PAHs, PCBs, pesticides and other organic and inorganic contaminant present in wastewater. Based on these peculiar characteristics these types of nano-materials are classified under various groups.

FUNCTIONALIZED SILICA HYBRIDS AS NANO-SORBENTS

The preference of silica over other materials for preparation of Hybrid materials is based on the fact that silicates reveal many advantages.

- Silica is transparent and does not scatter light. Silica show low optical loss in comparison to zirconia or titanium in its rutile phase.
- Silica has very high thermal resistance
- Organic-inorganic hybrids are preferred as attachment on silica surface is easier due to high number of cross-linking bonds (Arakaki et al., 2000).
- Immobilization of organic functional groups in the inorganic framework of silica renders more stability.

Another developmental aspect of Mesoporous silica materials is the incorporation of organic components either, on the silicate surface, inside the silicate wall, or trapped within the channels to permit the tuning and modification of bulk properties and at the same time stabilizing the materials towards hydrolysis.

The resultant hybrids exhibited high catalytic activity and increased binding to guest molecules. This is due to dual contribution in a hybrid; where mechanical, thermal and optical properties are due to inorganic species and organic moiety provides flexibility, toughness and hydrophobicity. Thus, organic-inorganic hybrid materials have desired combinations of both components.

In recent years, design and preparation of organic–inorganic hybrid materials have attracted considerable attention in the chemical synthesis (Wight & Davis, 2002; Park et al., 2007; Stein et al., 2000; Inagaki et al., 2000). The post-synthesis grafting method involves modification of the pre-fabricated Mesoporous material with organosilane compound after surfactant removal. In another method, tetraalkoxysilane and organo-alkoxysilane are directly co-condensed with Si-C bonds through sol-gel process.

Co-condensation has been demonstrated using ionic neutral surfactant and non-surfactant templates (Wei et al., 2000; Feng et al., 2000). The synthesis of monodisperse surface-modified silica spheres often include different alkoxysilanes (Si-OR), where R varies from methyl to vinyl (Uricanu et al., 2004; Wu et al., 2007; Lee et al., 2007).

The direct method offers advantage of providing uniform surface coverage in a single step and better control over the amount of organic groups incorporated in the structure. This route also enables high reactivity, better purity, avoidance of corrosive by products, improved control of product structure and is an easy, cost-effective and excellent way to incorporate inorganic compounds into an organic one.

Organic-inorganic hybrids represent an important class of synthetic engineering materials. The combination of properties of individual components and development of new properties on hybridization opened a fast expanding area of research having diverse applications (Wang et al., 2003a; Wang et al., 2003b; Wang et al., 2005; Wang et al., 2007). These Hybrids have progressively found their place as economical adsorbents with significant synthetic advances and technological development.

Surface area, particle size and surface morphology are actually basic features of nano-sorbents that make them unique for specific sorbates (pollutants) due to their structure property relationship. Following Table 1 list these basic features for two nano silica (SBA-15 and MSU-H) and of five functionalized silica hybrids (AM, GM, MM, PM and VM) with different organic moieties. All these nanomaterials have specific adsorbent properties for heavy metals, PAHs and phenols.

Application for the Removal of Metal Ions

Metals are problematic environmental pollutants, with well-known toxic effects on living systems. Metals cannot be degraded or destroyed (Davydova et al., 2005). To a small extent they enter our bodies

Table 1.Description of Nano Silica and functionalized mesoporous Silica hybrids (Nasreen et al., 2015)

Sample Code	Details	Surface Morphology SEM	TEM	Surface Area m^2/g	Particle Size
SBA-15	TMOS Precursor withP104 as Structure directing agent	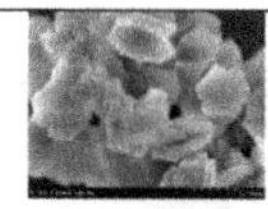	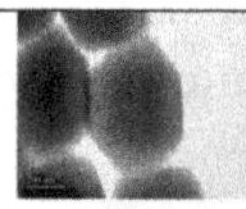	521.13	11.51 nm
MSU-H	Na_2SiO_3 Precursor withP104 as Structure directing agent			580.27	10.34 nm
AM	Functionalized MSU-H with 3-Aminopropyltrimethoxysilane (APTMS)	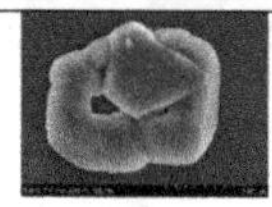	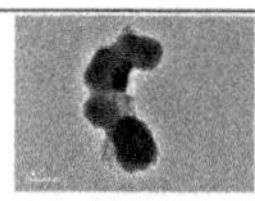	44.28	135.48 nm
GM	Functionalized MSU-H with 3-Glucidoxypropyltrimethoxysilane (GPTMS)		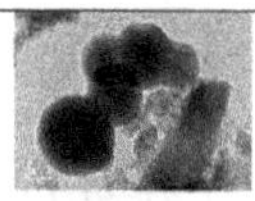	4.77	1.25 μm
MM	Functionalized MSU-H with 3-Methacryloxypropyltrimethoxysilane (MPTMS)	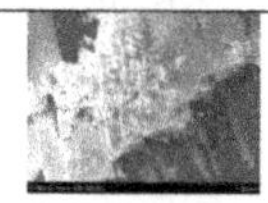	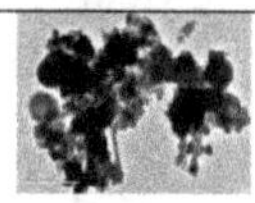	86.08	69.69 nm
VM	Functionalized MSU-H with Vinyltrimethoxysilane (VTMS)		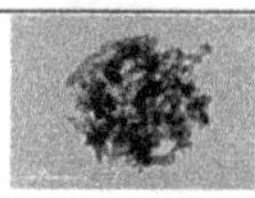	36.86	162.75 nm
PM	Functionalized MSU-H with Phenyltrimethoxysilane (PTMS)		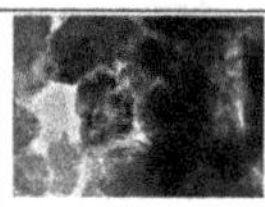	114.47	52.41 nm

via food, drinking water and air. Their presence in compartments of environment is posing threats. The discharge from industries ending up into the water bodies is adversely affecting the life of humans as well as aquatic organisms. Some metals such as lead and mercury easily cross the placenta and damage the brain (Levine et al., 2006). Scientists are faced with a number of challenges directly related to metal pollution.

Metals such as Cu, Cd, Pb, Hg, Ni, Co and Zn are introduced in the environment as a result of natural process of weathering and erosion of rocks. These and few other metals are added to the ecosystems through wastewater polluted by industrial process such as chemical manufacturing, metal finishing, tannery and use of fertilizers and pesticides (Bradl, 2005).

The identification of metals in the different environmental compartments is a great challenge. Ground water sources are continuously being contaminated by the release of toxic pollutants. The most disastrous threat is posed by mercury, arsenic, lead and chromium.

Mercury is a known and potential human carcinogen and neuro poison. The refused and discarded consumer products like fluorescent light bulbs, electrical fixtures, auto switches, thermostats, and medical equipment releases mercury to the environment. The organic and inorganic ligands make strong complexes, making mercury soluble in oxidized aquatic systems.

The chemistry of arsenic is complex due to its presence in several oxidation states. Arsenic compounds are used to make special glass, semi-conductors, dyes, and drugs. Seafood and drinking water also contain arsenic. Many arsenic compounds sorb strongly to soils and are therefore transported in groundwater and surface water.

Lead and lead containing compounds are very toxic for human health and the environment. It causes inhibition of some enzymes activity and disruption in the nervous system resulting in severe health problems, ending up by death. Beside that it causes lower IQs, behavioral changes and concentration disorder in children (Gupta et al., 2001; Lenntech, 2011). Lead is listed as No. 2 out of 783 hazardous substances Comprehensive Environmental Response, Compensation, and Liability Act (CERCLA) (EPA, 1992).

Chromium is widely used in industrial applications as corrosion inhibitor (Owlad, 2009) in cooling towers, metal plating, leather tanning, and dye manufacture (Kotas & Stasicka, 2000). Chromium can be found in the natural environment, however, large doses of Cr (VI) are linked to cancer, skin ulcers, and other maladies (Agency for Toxic Substances and Disease Registry, 1994). Evidence of human health impacts have also been shown in several DNA interaction studies with chromium (Plaper et al. 2002). The removal of metal ions from wastewaters is a much needed subject for research. Different approaches and methods are attempted by scientists, but most of these processes are unacceptable, owing to the disposal of sludge, their high cost, low efficiency and inapplicability to a wide range of pollutants (Barakat, 2011).

Synthesized Mesoporous Silica hybrid materials are applied in batch experiment to assess the feasibility of these materials as adsorbents for the remediation of toxic metals.

Each of the functionalized silica hybrids are applied as adsorbents for the removal of chromium, lead, mercury and arsenic from the aqueous media in a batch protocol. The uptake of each metal was analyzed on ICP-MS. The adsorption results on Mesoporous Silica indicate a general increase in removal potential with increasing time. The maximum adsorption potential secured by SBA-15 and MSU-H follows the same sequence i.e, Hg < As < Pb < Cr. However, MSU-H is found relatively better adsorbent. This preference can be assigned to BET and TG characterization revealing smaller particle size, larger surface area and increased kinetic energy at higher operating temperatures. The well-defined porous structure of the synthesized adsorbents further defines their contribution to high loading capacities for metal ions from aqueous solutions. The significantly high percentage removal of toxic metals ranging from 71%-99% by the Mesoporous Silica is a good contribution of present research towards environmental remediation. Further, adsorption results reiterate the use of sodium silicate for synthesis of Mesoporous Silica.

Functionalized silica hybrids are observed to have maximum adsorption potential for chromium (98%) and least for mercury (78%) as noted earlier for Mesoporous Silica. However, there is a shift over in the mid-order sequence for lead and arsenic.

Each of the functionalized silica hybrids with five different organic moieties reveals interesting adsorption trends for the uptake of metals. The graphical presentations given in following Figure 1 show striking unanimously adhered decreasing adsorption sequence of Pb $\geq$ Cr > As > Hg. On further probe,

Figure 1. Adsorption trends (%R) as a function of (a) silica based hybrids and (b) metal ions

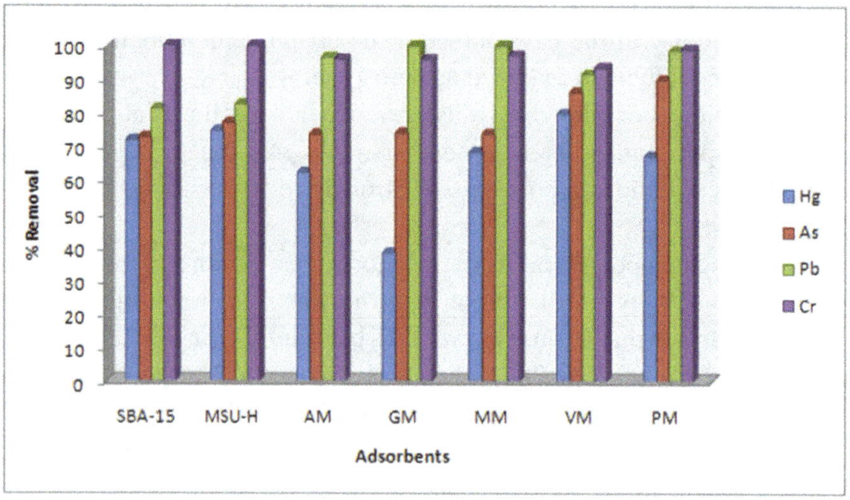

A

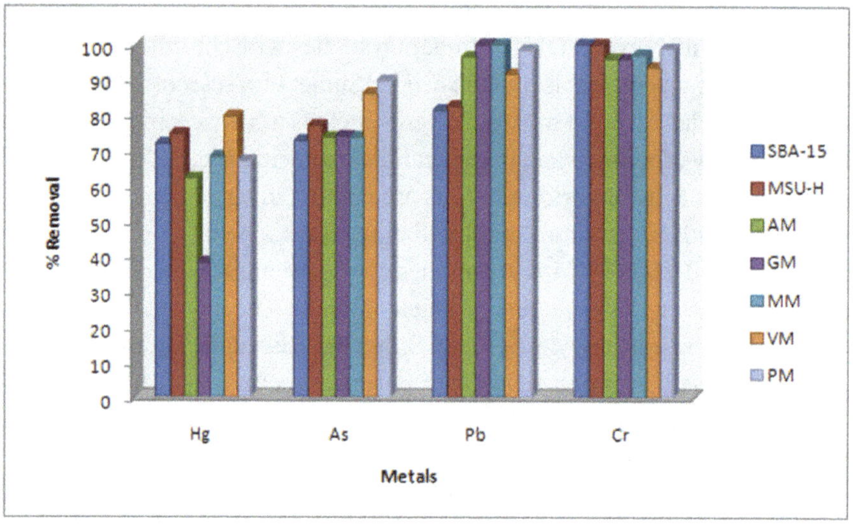

b

it appears that GM and PM have almost superimposed adsorption for Pb and Cr with significantly lower removal percentage of Hg by the earlier than later. On the contrary, a regular incremental increase from mercury to chromium is demonstrated by AM, VM, and MM. It is also noted that effect of organic moiety is manifested on the percentage removal. The minimum and maximum adsorption for the metals depicts range of 62-96%, 38-99%, 68-99%, 79-93% and 67-98% on AM, GM, MM, VM, and PM, respectively.

The contribution of functionalization in the adsorptive removal of metals is also considered. The results indicate that non-functional Mesoporous Silica is relatively less efficient than hybrids. Improved adsorption capacity for metals is also attributed to functionalization by (Antochshuk & Jaroniec, 2002; Antochshuk et al., 2003; Olkhovyk et al., 2004).

Application for the Removal of Phenols

Phenols are important aromatic compounds having antioxidant properties. It can inhibit the oxidative degradation of organic materials. Phenols are natural constituent in a number of biological aerobic organisms such as human blood plasma, mammalian urine, pine needles and oil from tobacco leaves. Phenol derivatives such as α-tocopherol is a component of vitamin E, thymol and carvacrol are component of lignin, from which phenol is liberated by hydrolysis.

Commercially, phenol is used in the production of phenolic resins phenol–formaldehyde resin called Bakelite phenolphthalein used as an indicator. Dilute solutions of Phenols are useful antiseptics, since Phenols are more acidic than aliphatic alcohols. However, its toxic fumes cause kidney damage. It is likely that Phenol solution contains dioxins. Consequently, phenol has only limited use in pharmaceuticals today because of its toxicity. Phenols penetrate deep into the tissue, leading to gangrene through damage to blood vessels (Nair et al., 2008). Ingestion of phenols in concentration from 10 to 240 mg/L for a long time causes mouth irritation, diarrhea, and excretion of dark urine and vision problems (Navarro et al., 2008).

Most of the phenol absorbed by the body is excreted in the urine as phenol and/or its metabolites. Only smaller quantities are excreted with feces or exhaled. The refractory Phenols form stable free radicals. Such a property is undesirable (Nguyen et al., 2003) that makes it an important pollutant in wastewater. Phenols are released by industries that produce chlorophenols for use as fungicides and insecticides in agriculture sector.

A number of phenolic compounds like chlorinated, nitrated, methylated or alkylated are prevalent in the environment due to chemical processing industries and use of numerous pesticides. For instance, N-acetylated aminophenol is a component of Paracetamol (Nagaraja et al., 2003). 3-aminophenol, 2-aminophenol and 2.4-diaminophenol is used as biomarker for analysis of drugs, precursor for indol synthesis, preparation of hair dyes (Chen et al., 2004), respectively. Nitrophenols result due to environmental reaction of phenol with nitrite ions.

Toxicity of Phenolic compounds is due to hydrophobic nature and property to generate reactive radicals. Phenolic compounds in potable water emit an unpleasant odor and flavor in concentration as low as 5μg/L and are poisonous to aquatic life, plants and humans. Furthermore, the position of substitution in phenol molecule also affects the toxic action. p-nitrophenol is classified as a priority pollutant and potential environmental toxicant due to its rapid breakdown in water. The Maximum Contaminant Level of 1 μg/L in drinking water (USEPA, 1986) is defined for phenols.

The significant environmental risks urge for the rapid removal and detoxification of phenols. Different physical, chemical and biological treatment processes are frequently employed (Zylstra et al., 2000). For instance, adsorption, ultrasonic irradiation and microwave assisted oxidation are preferred for removal of p-nitrophenol. Further, different materials for adsorption are investigated by a number of researchers. Roostaei et al., (2004) applied activated carbons, activated alumina, filtrasorb-400, silica gel and zeolites for adsorption of phenol from water. Cardenas et al., (2005) attempted porous clay heterostructure as adsorbents for removal of phenol and dichlorophenols.

The application of synthesized Mesoporous Silica and functionalized hybrids is extended for the decontamination of Phenolic compounds using protocol of batch adsorption process. The results are demonstrated graphically to exhibit different relations as a function of time. The first observation is the dormant or insignificant role of varying contact time on percentage adsorption. It is encouraging to report that appreciable quantities of induced phenols on each synthesized material are adsorbed. This is evident by the removal percentage of more than 90% acquired on Mesoporous Silica and functionalized hybrids. It is understood from the literature that addition of surfactant sodium dodecyl sulfate (SDS) reduces the adsorption for phenols (Shawabkeh & Abu-Nameh, 2007). However, the enhanced adsorption in the present study may be explained on the basis that non-ionic surfactant (P104) facilitates adsorption in comparison to ionic (SDS).

Comparative analysis of Mesoporous Silicas revealed the analogous adsorption potential for 4-aminophenol and p-nitrophenol on SBA-15 and MSU-H as shown in following figure 2. It suggests that

Figure 2. Adsorption trends (%R) as a function of (a) silica based hybrids and (b) phenols

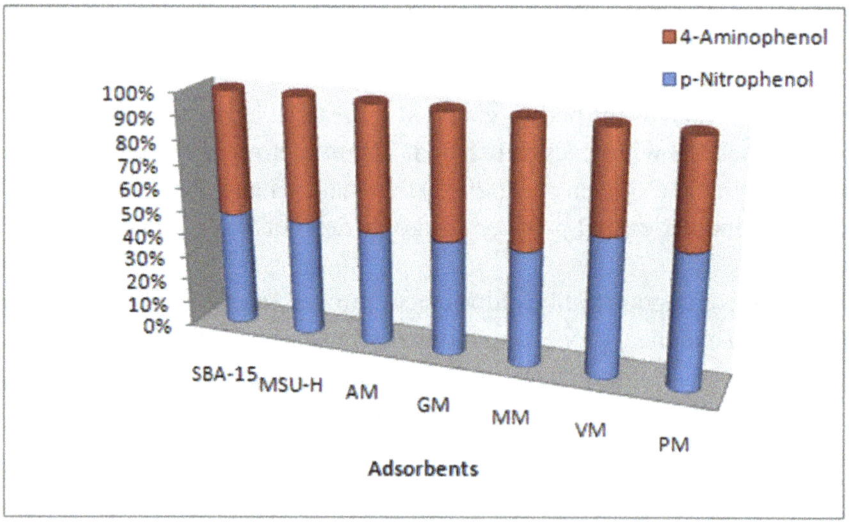

A

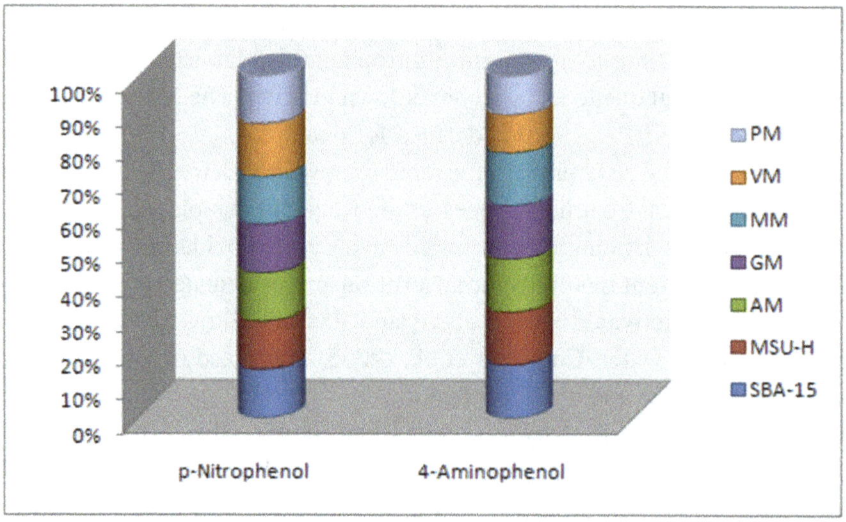

b

silica framework provides major retention irrespective of the substituent on phenol. Other studies suggest that silica retains its properties even after adsorbent regeneration process (Khalid et al., 2004) and high Si content in zeolites is beneficial for making good adsorption potential (Khalid et al., 2004). This can also be proposed that Si-N ($-NH_2$ or $-NO_3$) linkage is significant for better adsorption, and this is available in both phenols.

On the bases of results of adsorption trends (see Figure 2), Functionalized hybrids can conveniently be categorized as potential adsorbents for (a) 4-aminophenol (b) p-nitrophenol. It is clearly distinguished that AM, GM MM and VM, PM belong to class (a) and (b), respectively. The postulate again stands true that lower adsorption of 4-aminophenol owns to unsaturation of vinyl and phenyl.

Figure 2 clearly draws the comparative analysis of Mesoporous Silica and functionalized hybrids as adsorbents for the removal of phenols with nitro and amino groups. The Mesoporous Silica develop comparable adsorption efficiency to functionalized hybrids (AM, GM and MM). This reflects that organic moiety of –amino, -glucidoxy and -methacrylate is not contributing significantly in providing binding sites for phenol in the silica framework.

On the extreme, the polarization of functionalized hybrids having organic moiety of –phenyl (PM) and –vinyl (VM) is clearly visible. This polarization from normalization of Mesoporous Silica extends more disparity for the removal of p-nitrophenols in comparison to 4-aminophenols. Thus, it can be concluded that p-nitrophenols develop relatively comparable and better retention than Mesoporous Silica on adsorbent class (a) and (b), respectively.

Application for the Removal of Polycyclic Aromatic Hydrocarbons

An important class of gas-phased organic compounds grouped as Polycyclic Aromatic Hydrocarbons (PAH) is characterized by having low vapor pressure and hydrophobic property. PAHs are formed during incomplete combustion of coal, gasoline, wood (Schauer et al., 2001; Shen et al. 2011), forest fires and volcanic eruptions (Ohura et al., 2003; Sinha et al., 2005). The residential and industrial heating process is estimated to accounts for 80% of total PAH emissions (Manoli et al., 2000).

Polycyclic aromatic hydrocarbon (PAHs) compounds are included on the priority pollutant lists of the European Union (EU) and United States Environmental Protection Agency (USEPA) because of their mutagenic, carcinogenic, and endocrine-disrupting properties (Vidal et al., 2011). PAHs are also included in a class of persistent organic pollutants (POPs) due to low water solubility, electro-chemical stability and resistant to biodegradation. PAHs also accumulate in air, water bodies, soil and food as contaminants. PAHs contamination in the ground (Nadim et al., 2000) and carcinogenic potential (IARC, 2009) in refinery workers has also been recognized. Naphthalene is associated with significant carcinogenic effects in humans (National Toxicology Program, 2000; Deutsche Forschungsgemeinschaft, 2001) likely due to exposure to Cigarette smoke. Mainstream versus side stream contains 0.3 - 4.0 μg and 7.8-46 μg naphthalene (Hoffmann et al., 2001) in one cigarette smoke. According to (Chang et al., 2004), naphthalene is a frequently recorded pollutant and can pose a threat to ground-water resources (Williams et al., 2003).

The removal of these pollutants is of great concern among researchers and environmentalists. A number of international protocols demand the need to control the emission or release of these pollutants from the industrial sector of the world. Compliance to the protocols led to the development of different materials that present the ability to rectify the problem of environmental pollution.

It is understood that PAHs are not easy to degrade by biological or chemical methods. Therefore, it seems imperative to develop effective adsorbents for removing PAHs from the environment. There are very few studies on the adsorption of PAHs. Meso-structured inorganic silicas are evaluated to have high adsorption capacity (Lee et al., 2004; Sanz et al., 2012; Shahbazi et al., 2012). Further, organic/inorganic hybrids have been successfully used as adsorbents (Wang et al., 2006) due to high potential charge association for adsorption.

This section is directed towards the application of synthesized silica based hybrids as adsorbent for the remediation of organic pollutants. The environmental contaminants selected as representatives of organic substances for the present investigation are naphthalene and phenanthrene in a batch mode was the selected experimental design for control of organic pollutants. The organic compounds with fused aromaticity are selected as environmental pollutants for the present study. The synthesized materials are attempted as candidates for the remediation of naphthalene and phenanthrene as representative of planar and non-planar PAHs, respectively.

The results depicted in the following figure 3 show the adsorption capacity of 66-72% on Mesoporous Silica. This and amine-functionalized SBA-15 is reported to be good adsorbent for naphthalene. The presence of free pairs of electrons on the amine group of amine-functionalized Mesoporous Silicas should increase their sorptive properties for organic compounds in aqueous solutions. NH_2-SBA-15 is better at adsorbing halophenols than SBA-15 (Anbia & Amir Mahmoodi, 2011). A deeper insight shows relatively lower retention of phenanthrene than naphthalene. Lower adsorption is likely inclined to 3-fused rings in non-planar arrangement finding its way difficult to diffuse into mesopores of silica framework due to steric hindrance of larger molecule.

Another aspect of these organic moieties in functionalized hybrids is identified as shown in Figure 3. Here, adsorption of phenanthrene is reduced specifically by AM and GM down to 10 orders of magnitude. This also indicates that other organic moieties do not have a discriminatory look for Naphthalene and phenanthrene.

The broad conclusion of the present adsorption trends is the development of two adsorbent categories (see Figure 3). Mesoporous silica and functionalized hybrids showing comparable, relatively lower and appreciably lower adsorption for phenanthrene in comparison to Naphthalene.

A general comparison of silica based hybrids as adsorbents for the removal of PAHs and Phenols highlight the different efficacy extent. Phenols are observed to take the lead by showing more adsorption than PAHs. Larger molecular weight and fused aromaticity likely appear as hindrances to better retention of selected PAHs.

Application for Nitrogen Sorption

The success of any adsorbent material is tested on two important parameters. The scaling up of in-silico experiments to the industrial application. Another aspect is the life time of the proposed adsorbents. This can conveniently be judged by the regeneration ability defining number of cyclic operations. These two approaches add to the significance of 3Rs principle and sustainable development. Keeping in view, the present study envisages the effective utilization of synthesized functionalized silica based hybrids. For this purpose, adsorption experiments are designed to define the adsorption-desorption Hysteresis loop for Nitrogen gas.

Figure 3. Adsorption trends (%R) as a function of (a) silica based hybrids and (b) PAHs

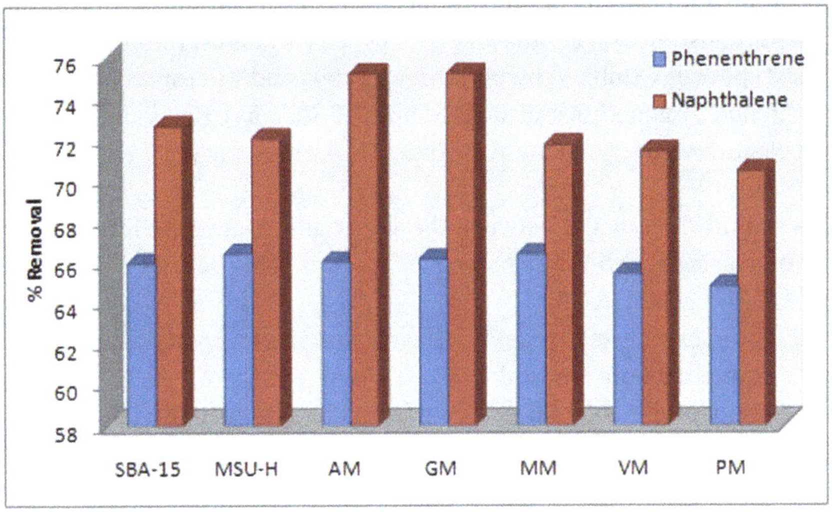

a

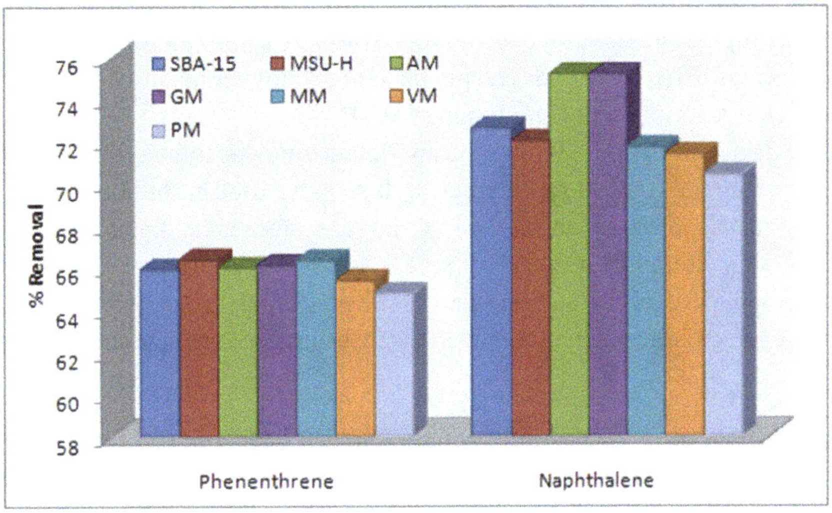

b

Gas adsorption process on Mesoporous silica is understood as an active interaction of nitrogen molecule with the solid surface. The porous structure and large surface area of adsorbent facilitates in temporary or permanent linkage of incoming molecule. The extent of linkage defines the interactive forces operative.

The amount of pollutant on each functionalized hybrid as a function of pressure results in adsorption. The possible structure and mechanism involved can be predicted on the basis of shapes of curves. Six types of adsorption- desorption curves are reported in the literature. A brief overview is sketched below.

Type I is characteristic of microporous materials showing slope to be sharp, horizontal and tail-curve when relative pressure is low, high and near to 1, respectively. The macroporous or nonporous materials follow Type II curves related to monolayer to multilayer adsorption. The type III defines weak to strong interaction between adsorbent-adsorbate and adsorbate-adsorbate, in respective order demonstrated by a rapid increase of quantity adsorbed on multilayers. The Type IV hysteresis loop is peculiar to Mesoporous materials. Whereas, Type V is exhibited by both Mesoporous and Microporous adsorbents. The rarest curve is Type VI exhibited by nonporous surface. Shapes of Six curve types are given below in figure 4:

The possible mechanisms governing the Adsorption Hysteresis curves are;

1. The transitory equilibrium that is between the vapor and condensed form of induced sample is characteristic of adsorbent defined as Independent Pores. This develops H1 type Hysteresis loop. For example, MCM-41 or SBA-15.
2. The contribution of interconnected pores of varying pore size and pore shapes, leads to H_2 hysteresis that represents disordered Pore Network
3. The interconnectivity of pores on surface with irregular symmetry and pore size are classified as Disordered Pores.

The adsorption of N_2 as representative of gaseous pollutant is attempted on presently synthesized materials. The breakthrough curves are presented in Figures 5 and 6. It is observed that quantity of N_2 adsorbed increases at higher pressure on Mesoporous silica. A saturation of 360cm^3/g is attained linearly by MSU-H, whereas SBA-15 extends steeply beyond and higher quantity of 660cm^3/g is adsorbed reflecting twofold better adsorption potential than MSU-H.

The better performance of SBA-15 for gaseous pollutants may postulate that reduced pore size and enhanced pore volume are the controlling parameters. It is understood earlier that MSU-H better retains the liquid pollutants (metals, phenols) attributed to its larger surface area. It also proposes that 'Independent Pore diffusion' is the dominant mechanism for SBA-15. The desorption curve at higher and lower pressure for SBA-15 and MSU-H can be designated as Type IV and Type I hysteresis, respectively. The comparison of Mesoporous silica suggests that SBA-15 has more regeneration capacity attributed to

Figure 4. Types of sorption isotherms and hysteresis loops (IUPAC, 1985)

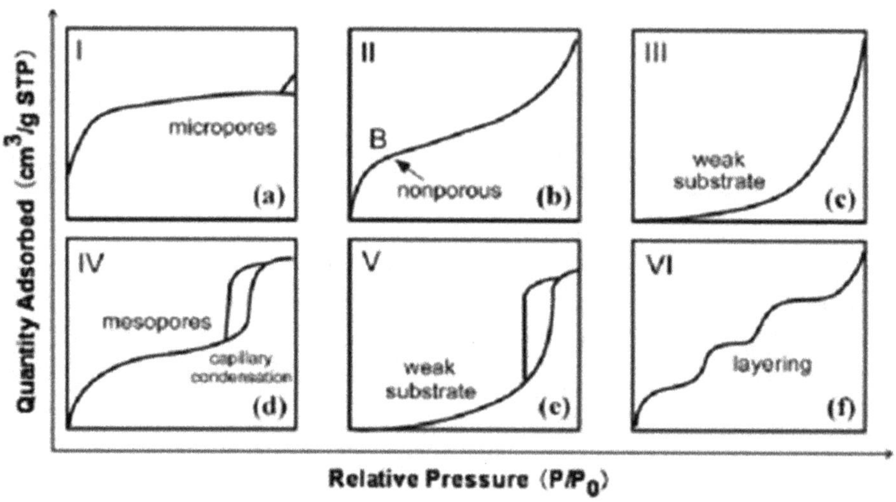

Figure 5. N_2 breakthrough curves for mesoporous silica (a) SBA-5 and (b) MSU-H

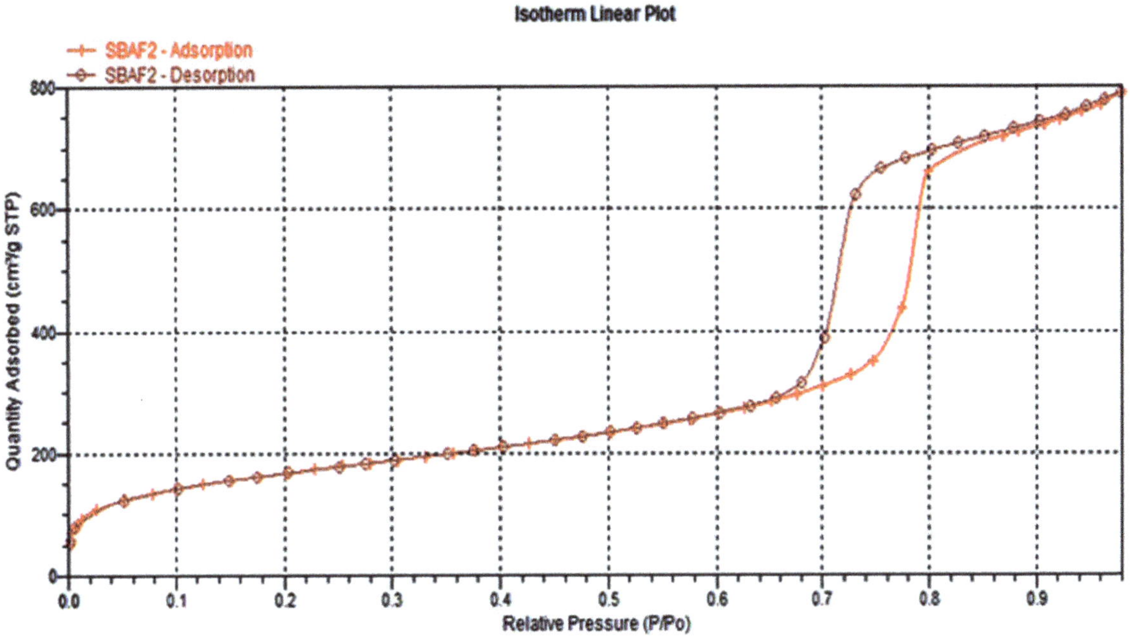

5a

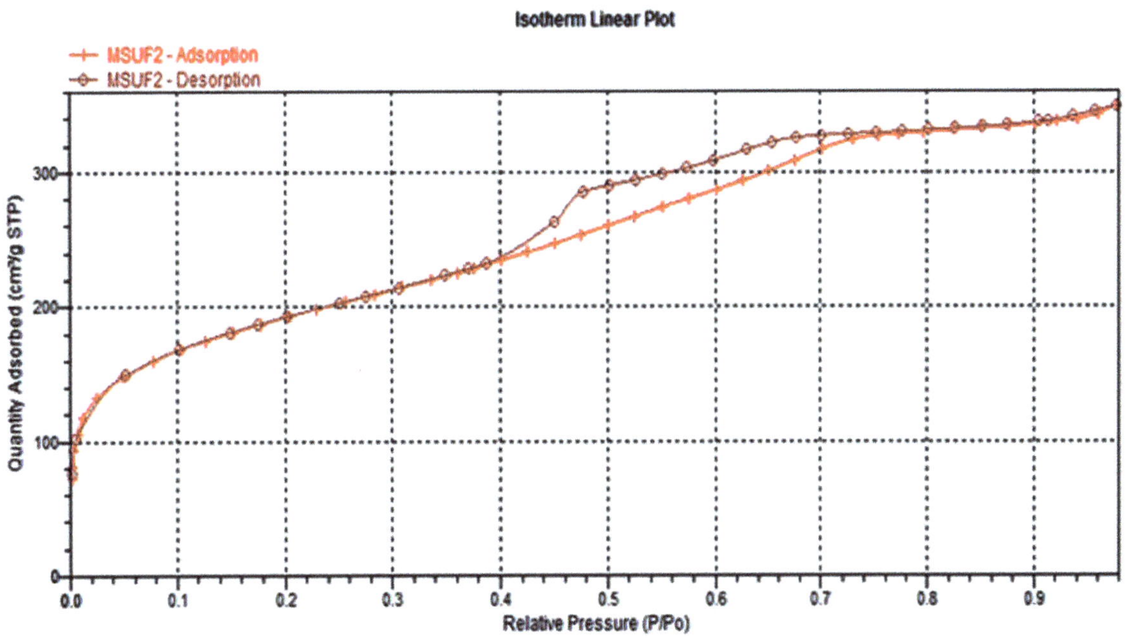

5b

hexagonal Mesoporous than homogeneously distributed lamellar structure of MSU-H as presented in following figure 5 a-b.

The breakthrough curves can conveniently be designated as Type I and Type II shown in following figure 6 a-e. The –amino, -methacrylate, -vinyl and –phenyl moieties follow a monotonic approach homogeneously distributing the induced nitrogen gas from the pores and getting it condensed onto the

Figure 6a. N_2 breakthrough curves for functionalized silica hybrids (AM)

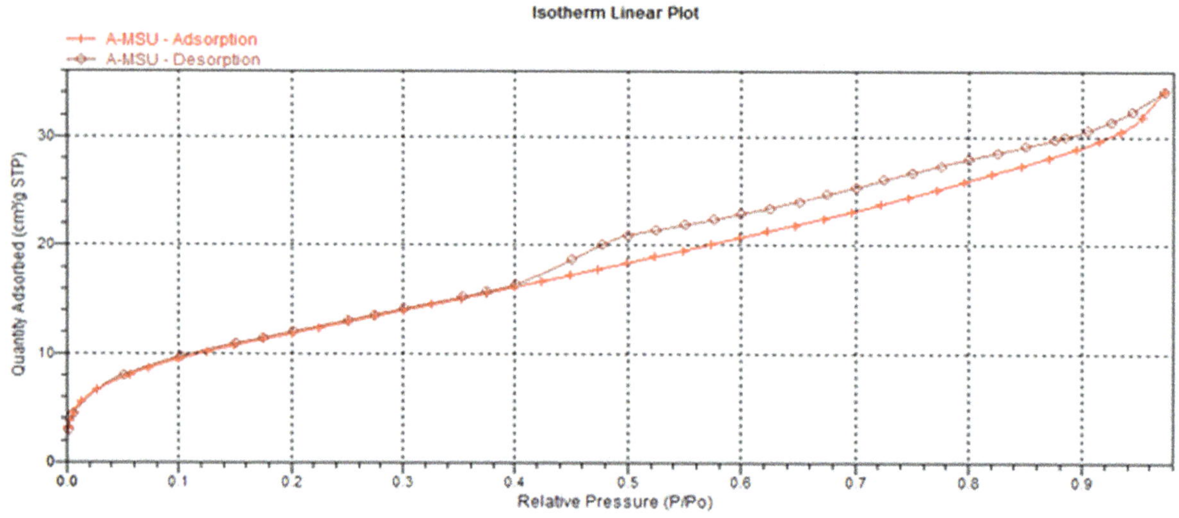

Figure 6b. N_2 breakthrough curves for functionalized silica hybrids (GM)

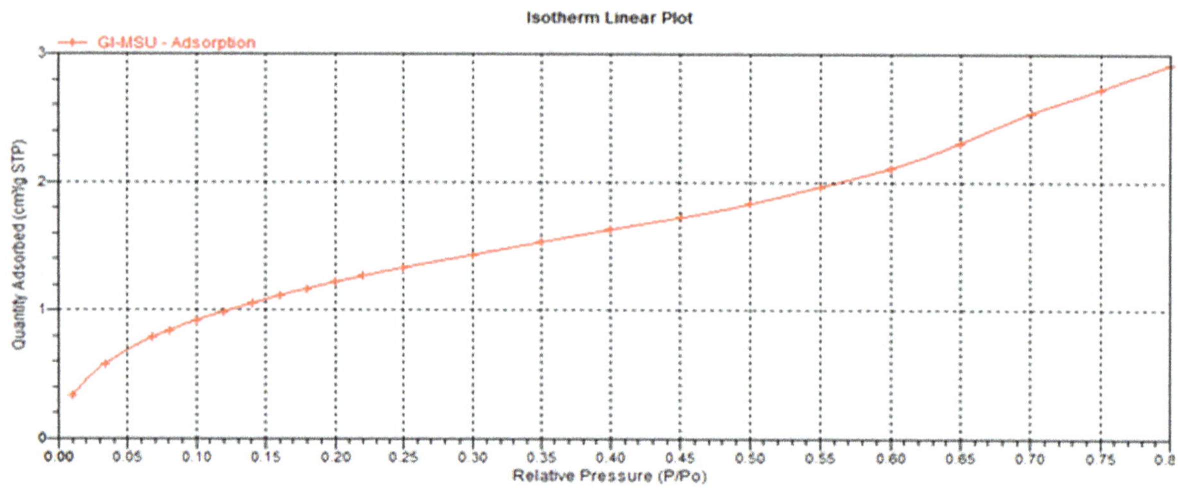

Figure 6c. N_2 breakthrough curves for functionalized silica hybrids (MM)

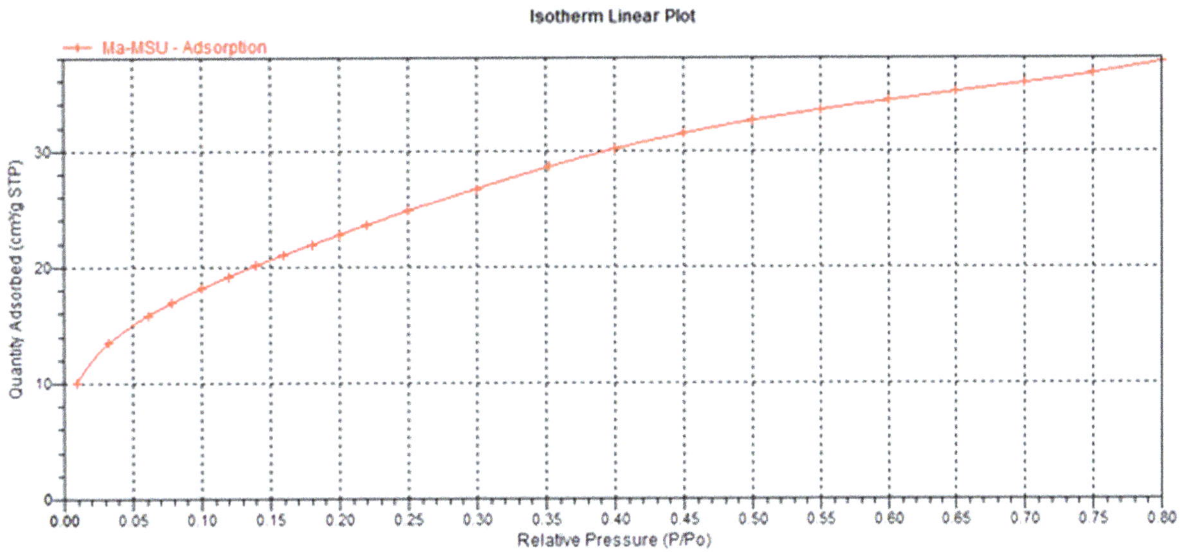

Figure 6d. N_2 breakthrough curves for functionalized silica hybrids (VM)

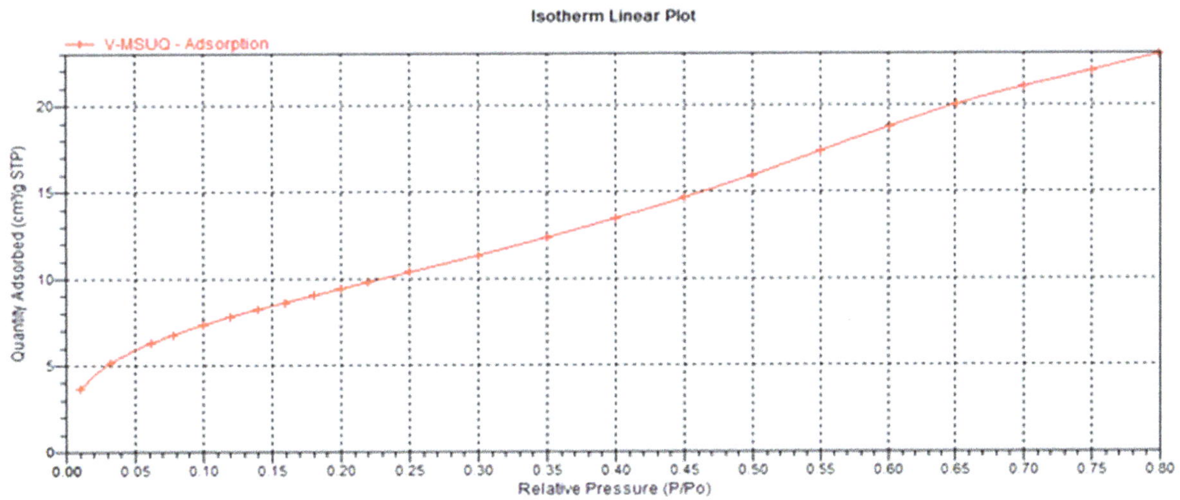

surface. On the extreme, Type II is exhibited by GM proposing its surface to be nonporous. On average lower N_2 adsorption is noted for functionalized silica hybrids in comparison to Mesoporous silica. The adsorption potential for Nitrogen follows the sequence:

PM ($48cm^3/g$) > MM ($38cm^3/g$) > AM ($34cm^3/g$) > VM ($26cm^3/g$) > GM ($3cm^3/g$).

Figure 6e. N_2 breakthrough curves for functionalized silica hybrids (PM)

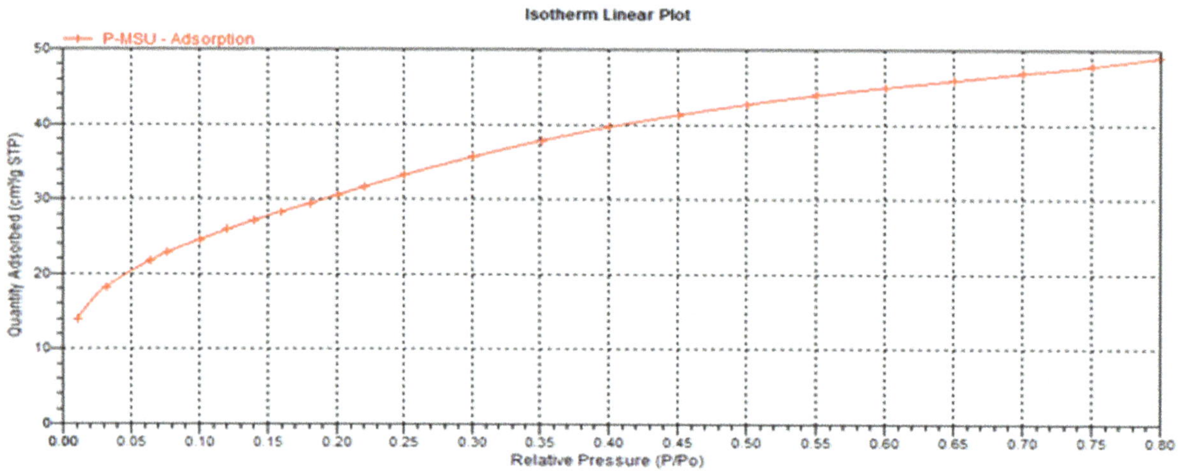

ACKNOWLEDGMENT

The authors thank NISP Lab and FabLab at University of Maryland, College Park USA, Chemistry Labs in University of Malaya, Malaysia and Fatima Jinnah Women University, Rawalpindi, Pakistan, for providing the necessary techniques for characterization of synthesized materials. Authors also gratefully acknowledge HEC (Higher Education Commission, Pakistan) for providing IRSIP fellowship to Saima Nasreen for pursuing her doctoral research at UMD, USA.

REFERENCES

Agarry, S. E., Ogunleye, O. O., & Aworanti, O. A. (2013). Biosorption equilibrium, kinetic and thermodynamic modelling of naphthalene removal from aqueous solution onto modified spent tea leaves. *Environmental Technology*, *34*(7), 825–839. doi:10.1080/09593330.2012.720616 PMID:23837334

Aguado, J., Arsuaga, J. M., & Arencibia, A. (2005). Adsorption of aqueous mercury (II) on propylthiolfunctionalized mesoporous silica obtained by cocondensation. *Industrial & Engineering Chemistry Research*, *44*(10), 3665–3671. doi:10.1021/ie0487585

Aguado, J., Arsuaga, J. M., & Arencibia, A. (2008). Influence of synthesis conditions on mercury adsorption capacity of propylthiol functionalized SBA-15 obtained by co-condensation. *Microporous and Mesoporous Materials*, *109*(1-3), 513–524. doi:10.1016/j.micromeso.2007.05.061

Aguinaldo, J. T. (2007). Integrated Precipitative Softening and Immersed Spiral Wound. *Conference Proceedings*.

Ake, C. L., Wiles, M. C., Huebner, H. J., McDonald, T. J., Cosgriff, D., Richardson, M. B., & Phillips, T. D. (2003). Porous organoclay composite for the sorption of polycyclic aromatic hydrocarbons and pentachlorophenol from groundwater. *Chemosphere*, *51*(9), 835–844. doi:10.1016/S0045-6535(03)00040-7 PMID:12697173

Alcántara, M. T., Gómez, J., & Sanromán, M. A. (2008). Combined treatment of PAHs contaminated soils using the sequence extraction with surfactant- electrochemical degradation. *Chemosphere*, *70*(8), 1438–1444. doi:10.1016/j.chemosphere.2007.08.070 PMID:17936331

Ali, I., & Aboul-Enein, H. Y. (2004). *Chiral Pollutants: Distribution, Toxicity and Analysis by Chromatography and Capillary Electrophoresis*. Chichester, UK: John Wiley & Sons.

Ali, I., & Aboul-Enein, H. Y. (2006). *Instrumental Methods in Metal Ions Speciation: Chromatography, Capillary Electrophoresis and Electrochemistry*. New York: Taylor & Francis Ltd. doi:10.1201/9781420019407

Ali, I., Aboul-Enein, H. Y., & Gupta, V. K. (2009). *Nano Chromatography and Capillary Electrophoresis: Pharmaceutical and Environmental Analyses*. Hoboken, NJ: John Wiley & Sons.

Anbia, M., & Amirmahmoodi, S. (2011). Adsorption of phenolic compoundsfrom aqueous solutions using functionalized SBA-15 asa nano-sorbent. *Scientia Iranica*, *18*(3), 446–452. doi:10.1016/j.scient.2011.05.007

Antochshuk, V., Olkhovyk, O., Jaroniec, M., Park, I.-S., & Ryoo, R. (2003). Benzoylthiourea modified mesoporous silica for mercury(II) removal. *Langmuir*, *19*(7), 3031–3034. doi:10.1021/la026739z

Arakaki, L. N. H., Nunes, L. M., Simoni, J. A., & Airoldi, C. (2000). Ethyleneimine anchored on thiol-modified silica gel surface - Adsorption of divalent cations and calorimetric data. *Journal of Colloid and Interface Science*, *228*(1), 46–51. doi:10.1006/jcis.2000.6842 PMID:10882492

Arakaki, L. N. H., Nunes, L. M., Simoni, J. A., & Airoldi, C. (2000). Ethyleneimine anchored on thiol-modified silica gel surface - Adsorption of divalent cations and calorimetric data. *Journal of Colloid and Interface Science*, *228*(1), 46–51. doi:10.1006/jcis.2000.6842 PMID:10882492

Balasubramanian, N., & Madhavan, K. (2001). Arsenic removal from Industrial Effluents through electrocoagulation. *Journal of Chemical Engineering & Tecnology*, *24*(5), 519–521. doi:10.1002/1521-4125(200105)24:5<519::AID-CEAT519>3.0.CO;2-P

Barakat, M. A. (2011). New trends in removing heavy metals from industrial wastewater. *Arabian Journal of Chemistry*, *4*(4), 361–377. doi:10.1016/j.arabjc.2010.07.019

Biskup, B., & Subotic, B. (2004). Kinetic analysis of the exchange processes between sodium ions from zeolite A and cadmium, copper and nickel ions from solutions. *Separation and Purification Technology*, *37*(1), 17–31. doi:10.1016/S1383-5866(03)00220-X

Boissiere, C., Larbot, A., & Prouzet, E. (2000). Synthesis of mesoporous MSU-X materials using inexpensive silica sources. *Chemistry of Materials*, *12*(7), 1937–1940. doi:10.1021/cm001012m

Bradl, H. B. (2005). *Heavy Metals in the Environment: Origin, Interaction and Remediation*. Amsterdam, The Netherlands: Elsevier.

Burleigh, M. C., & Dai, S. (2004). Functionalized Nanoporous Adsorbents for Environmental Remediation. *Series on Chemical Engineering*, *4*, 756–771. doi:10.1142/9781860946561_0024

Cao, R., Brakenhielm, E., Pawliuk, R., Wariaro, D., Post, M. J., Wahlberg, E., ... Cao, Y. (2003). Angiogenic synergism, vascular stability and improvement of hind-limb ischemia by a combination of PDGF-BB and FGF-2. *Nature Medicine*, *9*(5), 604–613. doi:10.1038/nm848 PMID:12669032

Cardenas, S. A., Velazquez, T. G., Revilla, G. O., Cortez, M., & Perera, B. G. (2005). Adsorption of phenol and dichlorophenol from aqueous solutions by porous clay heterostructure. *Journal of the Mexican Chemical Society*, *49*(3), 287–291.

Chang, C. F., Chang, C. Y., Chen, K. H., Tsai, W. T., Shie, J. L., & Chen, Y. H. (2004a). Adsorption of Naphthalene on Zeolite from Aqueous Solution. *Journal of Colloid and Interface Science*, *277*(1), 29–34. doi:10.1016/j.jcis.2004.04.022 PMID:15276034

Che, S., Garcia-Bennett, A. E., Yokoi, T., Sakamoto, K., Kunieda, H., Terasaki, O., & Tatsumi, T. (2003). Mesoporous Silica of Novel Structures with Periodic Modulations Synthesized by Anionic Surfactant Templating Route. *Nature Materials*, *2*, 801–805. doi:10.1038/nmat1022 PMID:14634644

Chen, J., Jiang, J., Zhang, F., Yu, H., & Zhang, J. (2004). Cytotoxic environmentally relevant chlorophenols on L929 cells and their mechanisms. *Cell Biology and Toxicology*, *20*(3), 183–196. doi:10.1023/B:CBTO.0000029468.89746.64 PMID:15250542

Cincotti, A., Mameli, A., Locci, A. M., Orru, R., & Cao, G. (2006). Heavy metals uptake by Sardinian natural zeolites: Experiment and modeling. *Industrial & Engineering Chemistry Research*, *45*(3), 1074–1084. doi:10.1021/ie050375z

Corriu, R. J. P., Mehdi, A., & Reye, C. (2005). Molecular chemistry and nanosciences: On the way to interactive materials. *Journal of Materials Chemistry*, *15*(40), 4285–4294. doi:10.1039/b504536n

Da'na, E., & Sayari, A. (2010). Optimization of copper removal efficiency by adsorption on amine modified SBA-15: Experimental design methodology. *Chemical Engineering Journal*, *167*(1), 91–98. doi:10.1016/j.cej.2010.12.005

Da'na, E., & Sayari, A. (2011). Adsorption of copper on amine-functionalized SBA-15 prepared by co-condensation: Equilibrium properties. *Chemical Engineering Journal*, *166*(1), 445–453. doi:10.1016/j.cej.2010.11.016

Da'na, E., Silva, N. D., & Sayari, A. (2011). Adsorption of copper on amine-functionalized SBA-15 prepared by co-condensation: Kinetics properties. *Chemical Engineering Journal*, *166*(1), 454–459. doi:10.1016/j.cej.2010.11.017

Dang, V. B. H., Doan, H. D., Dang-Vuc, T., & Lohi, A. (2009). Equilibrium and kinetics of biosorption of cadmium(II) and copper(II) ions by wheat straw. *Bioresource Technology*, *100*(1), 211–219. doi:10.1016/j.biortech.2008.05.031 PMID:18599289

Davydova, S. (2005). Heavy metals as toxicants in big cities. *Microchemical Journal, 79*(1-2), 133–136. doi:10.1016/j.microc.2004.06.010

Deutsche, F. G. (2001). *Commission for the investigation of health hazards of chemical compounds in the work area MAK and BAT values 2001*. Report no. 37. Wiley-VCH, Weinheim.

Ensink, J. H., Mahmood, T., van der Hoek, W., Raschid-Sally, L., & Amerasinghe, F. P. (2004). A nationwide assessment of wastewater use in Pakistan: An obscure activity or a vitally important one? *Water Policy, 6*(3), 197–206.

Feng, Q., Xu, J., Dong, H., Li, S., & Wei, Y. (2000). Synthesis of Polystyrene-Silica Hybrid Mesoporous Materials via the Nonsurfactant-template Sol-gel Process. *Journal of Materials Chemistry, 10*(11), 2490–2494. doi:10.1039/b003170o

Fryxell, G. E. (2006). The synthesis of functional mesoporous materials. *Inorganic Chemistry Communications, 9*(11), 1141–1150. doi:10.1016/j.inoche.2006.06.012

Fryxell, G. E., Lin, Y., Fiskum, S., Birnbaum, J. C., Wu, H., Kemner, K., & Kelly, S. (2005). Actinide Sequestration Using Self-Assembled Monolayers on Mesoporous Supports (SAMMS). *Environmental Science & Technology, 39*(5), 1324–1331. doi:10.1021/es049201j PMID:15787373

Fryxell, G. E., Mattigod, S. V., Lin, Y., Wu, H., Fiskum, S., Parker, K., ... Feng, X. (2007). Design and synthesis of self-assembled monolayers on mesoporous supports (SAMMS): The importance of ligand posture in functional nanomaterials. *Journal of Materials Chemistry, 17*(28), 2863–2874. doi:10.1039/b702422c

Gao, B., Gao, Y., & Li, Y. (2010). Preparation and chelation adsorption property of composite chelating material poly(amidoxime)/SiO_2 towards heavy metal ions. *Chemical Engineering Journal, 158*(3), 542–549. doi:10.1016/j.cej.2010.01.046

Gao, Z., Pereira, T., Choi, O., Wang, Y., & Hahn, H. T. (2006). Surface functionalized alumina nanoparticle filled polymeric nanocomposites with enhanced mechanical properties. *Journal of Materials Chemistry, 16*(27), 2800–2808. doi:10.1039/b603020c

Ge, P., Li, F., & Zhang, B. (2010). Synthesis of modified mesoporous materials and comparative studies of removal of heavy metal from aqueous solutions. *Polish Journal of Environmental Studies, 19*, 301–308.

Gier, S., & Johns, W. D. (2000). Heavy metal-adsorption on micas and clay minerals studied by X–ray photoelectron spectroscopy. *Applied Clay Science, 16*(5-6), 289–299. doi:10.1016/S0169-1317(00)00004-1

Gonze, E., Commenges, N., Gonthier, Y., & Bernis, A. (2003). High frequency ultrasound as a pre- or a post-oxidation for paper mill wastewaters and landfill leachate treatment. *Chemical Engineering Journal, 92*(1–3), 215–225. doi:10.1016/S1385-8947(02)00258-9

Gündoğan, R., Acemioğlu, B., & Alma, M. H. (2004). Copper (II) adsorption from aqueous solution by herbaceous peat. *Journal of Colloid and Interface Science, 269*(2), 303–309. doi:10.1016/S0021-9797(03)00762-8 PMID:14654388

Gupta, V., Gupta, M., & Sharma, S. (2001). Process development for the removal of lead and chromium from aqueous solutions using red mud an aluminum industry waste. *Water Research*, *35*(5), 1125–1134. doi:10.1016/S0043-1354(00)00389-4 PMID:11268832

Hatton, B., Landskron, K., Whitnall, W., Perovic, D., & Ozin, G. A. (2005). Past, Present, and Future of Periodic Mesoporous Organosilicas The PMOs. *Accounts of Chemical Research*, *38*(4), 305–312. doi:10.1021/ar040164a PMID:15835877

Hlihor, R. M., & Gavrilescu, M. (2009). Removal of some environmentally relevant heavy Metals using low-cost natural sorbents. *Environmental Engineering and Management Journal*, *8*(2), 353–372.

Hoffmann, D., Hoffmann, I., & El-Bayoumy, K. (2001). The less harmful cigarette: A controversial issue. A tribute to Ernst L. Wynder. *Chemical Research in Toxicology*, *14*(7), 767–790. doi:10.1021/tx000260u PMID:11453723

Hoffmann, F., Cornelius, M., Morell, J., & Froeba, M. (2006). Silica-based Mesoporous organic-inorganic hybrid materials. *Angewandte Chemie International Edition*, *45*(20), 3216–3251. doi:10.1002/anie.200503075 PMID:16676373

Hudson, S. P., Padera, R. F., Langer, R., & Kohane, D. S. (2008). The biocompatibility of mesoporous silicates. *Biomaterials*, *29*(30), 4045–4055. doi:10.1016/j.biomaterials.2008.07.007 PMID:18675454

Im, H.-J., Yost, T. L., Yang, Y., Bramlett, J. M., Yu, X., Fagan, B. C., ... Xue, Z.-L. (2006). ACS Symp. Ser.: Vol. 943. *Organofunctional Sol-Gel Materials for Toxic Metal Separation*. Nuclear Waste Management, Accomplishments of the Environmental Management Science Program. doi:10.1021/bk-2006-0943.ch012

Inagaki, S., Guan, S., Fukushima, Y., Ohsuna, T., & Terasaki, O. (2000). Nanoporous Materials II, Proceedings of the Conference on Access in Nanoporous Materials, 2000. *Studies in Surface Science and Catalysis*, *129*, 155. doi:10.1016/S0167-2991(00)80209-8

International Agency for Research on Cancer (IARC). (2009). IARC monographs on the evaluation of carcinogenic risks to humans, Vols 1-97 (until 2009). IARC Publications.

IUPAC Recommendations,. (1985). *Pure and Applied Chemistry*, *57*, 603. doi:10.1351/pac198557040603

Jal, P. K., Patel, S., & Mishra, B. K. (2004). Chemical modification of silica surface by immobilization of functional groups for extractive concentration of metal ions. *Talanta*, *62*(5), 1005–1028. doi:10.1016/j.talanta.2003.10.028 PMID:18969392

Jaroniec, M., & Olkhovyk, O. (2006). Synthesis and Characterization of Ordered Mesoporous Materials for Removal of Heavy Metal Ions. *NATO Science Series II.*, *228*, 325–336.

Jonker, M. T. O., & Koelmans, A. A. (2002). Sorption of polycyclic aromatic hydrocarbons. *Journal of Molecular Catalysis. B, Enzymatic*, *66*, 302–310.

Katiyar, A., Yadav, S., Smirniotis, P. G., & Pinto, N. G. (2006). Synthesis of ordered large pore SBA-15 spherical particles for adsorption of biomolecules. *Journal of Chromatography. A*, *1122*(1-2), 13–20. doi:10.1016/j.chroma.2006.04.055 PMID:16716334

Khalid, M., Joly, G., Renaud, A., & Magnoux, P. (2004). Removal of Phenol from Water by Adsorption Using Zeolites. *Journal of Industrial and Engineering Chemistry Research, 43*, 5275-5280.

Kickelbick, G. (2004). Hybrid inorganic-organic mesoporous materials. *Angewandte Chemie International Edition, 43*(24), 3102–3104. doi:10.1002/anie.200301751 PMID:15199556

Kim, S. S., Pauly, T. R., & Pinnavaia, T. J. (2000). Non-ionic surfactant assembly of wormhole silica molecular sieves from water soluble silicates. *Chemical Communications*, (10), 835–836. doi:10.1039/b001012j

Kim, S. W., Son, S. U., Lee, S. I., Hyeon, T., & Chung, Y. K. (2000). Cobalt on mesoporous silica: The first heterogeneous Pauson-Khand catalyst. *Journal of the American Chemical Society, 122*(7), 1550–1551. doi:10.1021/ja9939237

Kim, Y., & Yi, J. (2004). Advances in environmental technologies via the application of mesoporous materials. *Journal of Industrial and Engineering Chemistry, 10*(1), 41–51.

Korngold, E., Belayer, N., & Aronvol, L. (2001). Removal of arsenic from drinking water by anion exchangers. *Desalination, 141*(1), 81–84. doi:10.1016/S0011-9164(01)00391-5

Kotas, J., & Stasicka, Z. (2000). Chromium occurrence in the environment and methods of its speciation. *Environmental Pollution, 107*(3), 263–283. doi:10.1016/S0269-7491(99)00168-2 PMID:15092973

Kruk, M., Jaroniec, M., & Sayari, A. (2000). New insights into pore-size expansion of mesoporous silicates using long-chain amines. *Microporous and Mesoporous Materials, 35*, 545–553. doi:10.1016/S1387-1811(99)00249-8

Lazaridis, N. K., Matis, K. A., & Webb, M. (2001). Flotation of metal-loaded clay anion exchangers. Part I: The case of chromates. *Chemosphere, 42*(4), 373–378. doi:10.1016/S0045-6535(00)00143-0 PMID:11100788

Lebeau, B., Gaslain, F., Fernandez-Martin, C., & Babonneau, F. (2009). In V. Valtchev, S. Mintova, & M. Tsapatsis (Eds.), *Ordered Porous Solids* (pp. 283–308). Oxford, UK: Elsevier Ltd. doi:10.1016/B978-0-444-53189-6.00011-1

Lebeau, B., Patarin, J., & Sanchez, C. (2004). Design and Properties of Hierarchically Organized Hybrid Organic-Inorganic Nanocomposites. *Advances in Technology of Materials and Materials Processing, 6*, 298–307.

Lee, J. W., Shim, W. G., & Moon, H. (2004). Adsorption equilibrium and kinetics for capillary condensation of trichloroethylene on MCM-41 and MCM-48. *Microporous and Mesoporous Materials, 73*(3), 109–119. doi:10.1016/j.micromeso.2004.04.020

Lee, Y. G., Park, J. H., Oh, C., Oh, S. G., & Kim, Y. C. (2007). Preparation of highly monodispersed hybrid silica spheres using a one-step sol-gel reaction in aqueous solution. *Langmuir, 23*(22), 10875–10878. doi:10.1021/la702462b PMID:17902722

Lenntech, B. V. (2011). *Lead and water*. Water Treatment Solutions. Available at: http://www.lenntech.com/periodic/water/lead/lead-and-water.htm

Levine, T. R., Kim, R. K., Park, H. S., & Hughes, M. (2006). Deception detection accuracy is a predictable linear function of message veracity base-rate: A formal test of Park and Levine's probability model. *Communication Monographs*, *73*(3), 243–260. doi:10.1080/03637750600873736

Li, J., Wang, L., Qi, T., Zhou, Y., Liu, C., Chu, J., & Zhang, Y. (2008). Different N-containing functional groups modified mesoporous adsorbents for Cr(VI) sequestration: Synthesis, characterization and comparison. *Microporous and Mesoporous Materials*, *110*(2-3), 442–450. doi:10.1016/j.micromeso.2007.06.033

Li, S. Z., Li, X. Y., Cui, Z. F., & Wang, D. Z. (2004). Application of ultrafiltration to improve the extraction of antibiotics. *Separation and Purification Technology*, *34*(1-3), 115–123. doi:10.1016/S1383-5866(03)00185-0

Li, Y., Chen, B., & Zhu, L. (2010). Enhanced sorption of polycyclic aromatic hydrocarbons from aqueous solution by modified pine bark. *Bioresource Technology*, *101*(19), 7307–7313. doi:10.1016/j.biortech.2010.04.088

Liu, A. M., Hidajat, K., Kawi, S., & Zhao, D. Y. (2000a). A New Class of Hybrid Mesoporous Materials with functionalized Organic Monolayers for selective Adsorption of Heavy Metal ions. *Chemical Communications*, *13*(13), 1145–1146. doi:10.1039/b0026611

Liu, J., Shin, Y., Nie, Z., Chang, J. H., Wang, L.-Q., Fryxell, G. E., ... Exarhos, G. J. (2000b). Molecular Assembly in Ordered Mesoporosity: A New Class of Highly Functional Nanoscale Materials. *The Journal of Physical Chemistry A*, *104*(36), 8328–8339. doi:10.1021/jp0009812

Liu, J., Shin, Y., Wang, L.-Q., Exarhos, G. J., Chang, J. H., Fryxell, G. E., ... Samuels, W. D. (2003). In S. L. Scott, C. M. Crudden, & C. W. Jones (Eds.), *Nanostructured Catalysts* (pp. 297–310). New York: Kluwer Academic/Plenum Publishers. doi:10.1007/978-0-387-30641-4_13

Luna, F. M. T., Pontes-Filho, A. A., Trindade, E. D., Silva, I. J., Azevedo, D. C., & Cavalcante, C. L. (2008). Removal of aromatic compounds from mineral naphthenic oil by adsorption. *Industrial & Engineering Chemistry Research*, *47*(9), 3207–3212. doi:10.1021/ie071476v

Manoli, E., Samara, C., Konstantinou, I., & Albanis, T. (2000). Polycyclic aromatic hydrocarbons in the bulk precipitation and surface waters of Northern Greece. *Chemosphere*, *41*(12), 1845–1855. doi:10.1016/S0045-6535(00)00134-X PMID:11061306

Mercier, L., & Bell, V. (2000). Chapter. In H. Mostaghaci (Ed.), Environment Conscious Materials: Ecomaterials (pp. 599–606). Montreal, Canada: Canadian Institute of Mining, Metallurgy and Petroleum.

Muff, J., & Søgaard, E. G. (2010). Electrochemical degradation of PAH compounds in process water: A kinetic study on model solutions and a proof of concept study on runoff water from harbour sediment purification. *Water Science and Technology*, *61*(8), 2043–2051. doi:10.2166/wst.2010.129 PMID:20389002

Nadim, F., Hoag, G. E., Liu, S., Carley, R. J., & Zack, P. (2000). Detection and remediation of soil aquifer systems contaminated with petroleum products: An overview. *Journal of Petroleum Science Engineering*, *26*(1-4), 169–178. doi:10.1016/S0920-4105(00)00031-0

Nagaraja, P., Yathirajan, H., Raju, C., Vasantha, R., Nagendra, P., & Kumar, M. (2003). 3-aminophenol as a novel coupling agent for the spectrophotometric determination of sulfonamide derivatives. *Il Farmaco*, *58*(12), 1295–1300. doi:10.1016/S0014-827X(03)00093-4 PMID:14630242

Nagarale, R. K., Gohil, G. S., & Shahi, V. K. (2006). Recent developments on ion-exchange membranes and electro-membrane processes. *Advances in Colloid and Interface Science*, *119*(2-3), 97–130. doi:10.1016/j.cis.2005.09.005 PMID:16325751

Nair, C.I., & Jayachandran, K., & Shashidhar. (2008). Biodegradation of Phenol. *African Journal of Biotechnology*, *7*(25), 4951–4958.

National Toxicology Program (NTP). (2000). *Toxicology and carcinogenesis studies of naphthalene (CAS no. 91-20-3) in F344/N rats (inhalation studies)*. Technical report series no. 91-20-3. NIH Publication no. 01-4434. US Department of Health and Human Services, Public Health Service, National Institutes of Health.

Navarro Abel, E., Portales Rosario, F., Sun-Kou María, R., & Llanos Bertha, P. (2008). Effect of pH on phenol biosorption by marine seaweeds. *Journal of Hazardous Materials*, *156*(1–3), 405–411. doi:10.1016/j.jhazmat.2007.12.039 PMID:18249490

Newalkar, B. L., Komarneni, S., & Katsuki, H. (2000). Rapid synthesis of mesoporous SBA-15 molecular sieve by a microwave-hydrothermal process. *Chemical Communications*, (23): 2389–2390. doi:10.1039/b0074411

Newalkar, B. L., Olanrewaju, J., & Komarneni, S. (2001). Direct synthesis of titanium-substituted mesoporous SBA-15 molecular sieveunder microwave-hydrothermal conditions. *Chemistry ofMaterials*, *13*(2), 552–557. doi:10.1021/cm000748g

Nguyen, M. T., Kryachko, E. S., & Vanquickenborne, L. G. (2003-d), The Chemistry of Phenols. In Z. Rappoport (Ed.), *General and theoretical aspects of phenols*. John Wiley and Sons, Ltd.

O'Connell, D. W., Birkinshaw, C., & O'Dwyer, T. F. (2008). Heavy metal adsorbents prepared from the modification of cellulose: A review. *Bioresource Technology*, *99*(15), 6709–6724. doi:10.1016/j.biortech.2008.01.036 PMID:18334292

Ohura, T., Amagai, T., Fusaya, M., & Matsushita, H. (2003). Polycyclic aromatic hydrocarbons in indoor and outdoor environments and factors affecting their concentrations. *Environmental Science & Technology*, *38*(1), 77–83. doi:10.1021/es030512o PMID:14740720

Olkhovyk, O., Antochshuk, V., & Jaroniec, M. (2004). Benzoylthiourea-modified MCM-48 mesoporous silica for mercury(II) adsorption from aqueous solutions. *Colloids and Surfaces. A, Physicochemical and Engineering Aspects*, *236*(1-3), 69–72. doi:10.1016/j.colsurfa.2003.12.032

Owlad, Aroua, M. K., Daud, W. A. W., & Baroutian, S. (2009). Removal of Hexavalent Chromium-Contaminated Water and Wastewater: A Review. *Water, Air, and Soil Pollution*, *200*(1), 59–77. doi:10.100711270-008-9893-7

Pan, B. J., Pan, B. C., Zhang, M. W., Lv, L., Zhang, Q. X., & Zheng, S. R. (2009). Development of polymeric and polymer-based hybrid adsorbents for pollutants removal from waters. *Chemical Engineering Journal, 151*(1-3), 19–29. doi:10.1016/j.cej.2009.02.036

Pang, Y., Zeng, G., Tang, L., Zhang, Y., Liu, Y., Lei, X., ... Xie, G. (2011). PEI-grafted magnetic porous powder for highly effective adsorption of heavy metal ions. *Desalination, 281*, 278–284. doi:10.1016/j.desal.2011.08.001

Park, J., Joo, J. C., Kwon, S.G., Jang, Y., & Hyeon, T. (2007). Synthesis of Monodisperse Spherical Nanocrystals. Anegwandte Chemie International Edition, 46, 4630- 4660.

Pérez-Quintanilla, D., Del Hierro, I., Fajardo, M., & Sierra, I. (2006). Preparation of 2-mercaptobenzothiazole-derivatized mesoporous silica and removal of Hg(II) from aqueous solution. *Journal of Environmental Monitoring C, 8*(1), 214–222. doi:10.1039/B507983G PMID:16395482

Plaper, A., Jenko-Brinovec, S., Premzl, A., Kos, J., & Raspor, P. (2002). Genotoxicity of trivalent chromium in bacterial cells. Possible effects on DNA topology. *Chemical Research in Toxicology, 15*(7), 943–949. doi:10.1021/tx010096q PMID:12119005

Rahman, A. K. M. R., Hossain, S. M., & Akramuzzaman, M. M. (2010). Distribution of heavy metals in rice plant cultivated in industrial effluent receiving soil. *Environment Asia, 3*(2), 15–19.

Rahmani, H. R., & Rezaei, M. (2007). Effluent of industrial effluents of Zob- Ahan on soil, water and vegetable plants. *Journal of Applied Sciences (Faisalabad)*, 72519–72522.

Repo, E., Warchoł, J. K., Bhatnagar, A., & Sillanpää, M. (2011). Heavy metals adsorption by novel EDTA-modified chitosan-silica hybrid materials. *Journal of Colloid and Interface Science, 358*(1), 261–267. doi:10.1016/j.jcis.2011.02.059 PMID:21440904

Roostaei, N., & Tezel, F. H. (2004). Removal of phenol from aqueous solutions by adsorption. *Journal of Environmental Management, 70*(2), 157–164. doi:10.1016/j.jenvman.2003.11.004 PMID:15160741

Sanchez, C., Boissiere, C., Grosso, D., Laberty, C., & Nicole, L. (2008). Design, Synthesis, and Properties of Inorganic and Hybrid Thin Films Having Periodically Organized Nanoporosity. *Chemistry of Materials, 20*(3), 682–737. doi:10.1021/cm702100t

Sanz, R., Calleja, G., Arencibia, A., & Sanz-Pérez, E. S. (2012). Amino functionalized mesostructured SBA-15 silica for CO_2 capture: Exploring the relation between the adsorption capacity and the distribution of amino groups by TEM. *Microporous and Mesoporous Materials, 158*, 309–317. doi:10.1016/j.micromeso.2012.03.053

Sayari, A., & Hamoudi, S. (2001). Periodic mesoporous silica-based organic-inorganic nano-composite materials. *Chemistry of Materials, 13*(10), 3151–3168. doi:10.1021/cm0110391

Schauer, J. J., Kleeman, M. J., Cass, G. R., & Simoneit, B. R. T. (2001). Measurement of emissions from air pollution sources, 3. C1–C29 organic compounds from fireplace combustion of wood. *Environmental Science & Technology, 35*(9), 1716–1728. doi:10.1021/es001331e PMID:11355184

Shahbazi, A., Younesi, H., & Badiei, A. (2012). Batch and fixed-bed column adsorption of Cu(II), Pb(II) and Cd(II) from aqueous solution onto functionalised SBA-15 mesoporous silica. *Canadian Journal of Chemical Engineering*, *91*(4), 739–750. doi:10.1002/cjce.21691

Shawabkeh, R. A., & Abu-Nameh, E. S. M. (2007). Absorption of Phenol and Methylene Blue by Activated Carbon from Pecan Shells. *Colloid Journal*, *69*(3), 355–359. doi:10.1134/S1061933X07030143

Shen, H., Tao, S., Wang, R., Wang, B., Shen, G., Li, W., ... Sun, K. (2011). Global time trends in PAH emissions from motor vehicles. *Atmospheric Environment*, *45*(12), 2067–2073. doi:10.1016/j.atmosenv.2011.01.054 PMID:24198716

Simsek, E. B., Duranoglu, D., & Beker, U. (2012). Heavy metal adsorption by magnetic hybrid-sorbent: An experimental and theoretical approach. *Separation Science and Technology*, *47*(9), 1334–1340. doi:10.1080/01496395.2012.672845

Sinha, R., Peters, U., Cross, A. J., Kulldorff, M., Weissfeld, J. L., Pinsky, P. F., ... Colorectal, T. O. C. P. (2005). Meat, meat cooking methods and preservation, and risk for colorectal adenoma. *Cancer Research*, *65*(17), 8034–8041. doi:10.1158/0008-5472.CAN-04-3429 PMID:16140978

Slowing, I. I., Trewyn, B. G., & Lin, V. S. Y. (2007). Mesoporous silica nanoparticles for intracellular delivery of membrane-impermeable proteins. *Journal of the American Chemical Society*, *129*(28), 8845–8849. doi:10.1021/ja0719780 PMID:17589996

Soler-Illia, G. J. de A.A., Sanchez, C., Lebeau, B., & Patarin, J. (2002). Chemical strategies to design textured materials: From microporous and mesoporous oxides to nanonetworks and hierarchical structures. *Chemical Reviews*, *102*(11), 4093–4138. doi:10.1021/cr0200062 PMID:12428985

Steduto, P., Faurès, J.-M., Hoogeveen, J., Winpenny, J. T., & Burke, J. J. (2012). Coping with Water Scarcity: An Action Framework for Agricultural and Food Security. Food and Agricultural Organization of the United Nations.

Stein, A., Melde, B. J., & Schroden, R. C. (2000). Hybrid Inorganic-Organic Mesoporous Silicates-Nanoscopic Reactors Coming of Age. *Advanced Materials*, *12*(19), 1403–1419. doi:10.1002/1521-4095(200010)12:19<1403::AID-ADMA1403>3.0.CO;2-X

Suchithra, P. S., Vazhayal, L., Mohamed, A. P., & Ananthakumar, S. (2012). Mesoporous organic-inorganic hybrid aerogels through ultrasonic assisted sol-gel intercalation of silica-PEG in bentonite for effective removal of dyes, volatile organic pollutants and petroleum products from aqueous solution. *Chemical Engineering Journal*, *200–202*, 589–600. doi:10.1016/j.cej.2012.06.083

Takeshita, K., Watanabe, K., Nakano, Y., & Watanabe, M. (2003). Solvent extraction of Cd(II) and Zn(II) with the organophosphorus extractant D2EHPA and the aqueous nitrogen-donor ligand TPEN. *Hydrometallurgy*, *70*(1-3), 63–71. doi:10.1016/S0304-386X(03)00046-X

Tavlarides, L. L., & Lee, J. S. (2001). Synthesis and application of functionalized organo-ceramic selective adsorbents. *Solvent Extraction and Ion Exchange*, *14*, 169–227.

UNESCO. (2003). *Water for People, Water for Life*. United Nations World Water Development Report. Retrieved from www.unesdoc.unesco.org

UNICEF/WHO. (2010). *Progress on Sanitation and Drinking Water: 2010 Update*. WHO/UNICEF Joint Monitoring Programme for Water Supply and Sanitation.

Uricanu, V., Donescu, D., Banu, A. G., Serban, S., Olteanu, M., & Dudau, M. (2004). Organic-inorganic hybrids made from polymerizable precursors. *Materials Chemistry and Physics*, *85*(1), 120–130. doi:10.1016/j.matchemphys.2003.12.024

U.S. Department of the Interior. (n.d.). *U.S. Geological Survey*. Retrieved from http://water.usgs.gov/edu/earthwherewater.html

Uzun, I., & Güzel, F. (2000). Adsorption of some heavy metal ions from aqueous solution by activated carbon and comparison of percent adsorption results of activated carbon with those of some other adsorbents. *Turkish Journal of Chemistry*, *24*, 291–297.

Valderrama, C., Cortina, J. L., Farran, A., Gamisans, X., & Lao, C. (2007). Kinetics of Sorption of polyaromatic Hydrocarbons onto Granular Activated Carbon and Macronet Hyper-Cross-Linked Polymers (MN200). *Journal of Colloid and Interface Science*, *310*(1), 35–46. doi:10.1016/j.jcis.2007.01.039 PMID:17367802

Vidal, M., Domínguez, J., & Luís, A. (2011). Spatial and temporal patterns of polycyclic aromatic hydrocarbons (PAHs) in eggs of a coastal bird from northwestern Iberia after a major oil spill. *The Science of the Total Environment*, *409*(13), 2668–2673. doi:10.1016/j.scitotenv.2011.03.025 PMID:21482431

Vinu, A., Hossain, K. Z., & Ariga, K. (2005). Recent advances in functionalization of mesoporous silica. *Journal of Nanoscience and Nanotechnology*, *5*(3), 347–371. doi:10.1166/jnn.2005.089 PMID:15913241

Walcarius, A., Etienne, M., Sayen, S., & Lebeau, B. (2003). Grafted silicas in electroanalysis: Amorphous versus ordered mesoporous materials. *Electroanalysis*, *15*(5-6), 414–421. doi:10.1002/elan.200390048

Wan, Y., & Zhao, D. Y. (2007). On the Controllable Soft-Templating Approach to Mesoporous Silicates. *Chemical Reviews*, *107*(7), 2821–2860. doi:10.1021/cr068020s PMID:17580976

Wang, L., Wu, X. L., Xu, W. H., Huang, X. J., Liu, J. H., & Xu, A. W. (2012). Stable organic-inorganic hybrid of polyaniline/α-zirconium phosphate for efficient removal of organic pollutants in water environment. *ACS Applied Materials & Interfaces*, *4*(5), 2686–2692. doi:10.1021/am300335e PMID:22545781

Wang, L., Zhang, J., & Wang, A. (2011). Fast removal of methylene blue from aqueous solution by adsorption onto chitosan-g-poly (acrylic acid)/attapulgite composite. *Desalination*, *266*(1-3), 33–39. doi:10.1016/j.desal.2010.07.065

Wang, W., & Gu, B. (2005). Self-assembly of Two- and Three-Dimensional Particle Arrays by Manipulating Hydrophobicity of Silica Nanoparticles. *The Journal of Physical Chemistry B*, *109*(47), 22175–22180. doi:10.1021/jp053692s PMID:16853885

Wang, W., Gu, B., & Liang, L. (2007). Effect of anionic surfactants on synthesis and self-assembling of silica colloidal nanoparticles. *Journal of Colloid and Interface Science*, *313*(1), 169–173. doi:10.1016/j.jcis.2007.04.042 PMID:17512000

Wang, W., Gu, B., Liang, L. Y., & Hamilton, W. (2003, November). (2003-a), Fabrication of Near Infrared Photonic Crystals using Highly Monodispersed Submicrometer SiO_2 Spheres. *The Journal of Physical Chemistry B*, *107*(44), 12113–12117. doi:10.1021/jp034161b

Wang, W., Gu, B., Liang, L. Y., & Hamilton, W. (2003, April). (2003-b), Fabrication of Two and Three-Dimensional Silica Nanocolloidal Particle Arrays. *The Journal of Physical Chemistry B*, *107*(15), 3400–3404. doi:10.1021/jp0221800

Wang, X., Chan, J. C. C., Tseng, Y. H., & Cheng, S. (2006). Synthesis, characterization and catalytic activity of ordered SBA-15 materials containing high loading of diamine functional groups. *Microporous and Mesoporous Materials*, *95*(1-3), 57–65. doi:10.1016/j.micromeso.2006.05.003

Wang, X., Chan, J. C. C., Tseng, Y. H., & Cheng, S. (2006). Synthesis, characterization and catalytic activity of ordered SBA-15 materials containing high loading of diamine functional groups. *Microporous and Mesoporous Materials*, *95*(1-3), 57–65. doi:10.1016/j.micromeso.2006.05.003

Wang, X., & Xing, B. (2007). Sorption of organic contaminants by biopolymer-derived chars. *Environmental Science & Technology*, *41*(24), 8342–8348. doi:10.1021/es071290n PMID:18200861

Washmon-Kriel, L., Jimenez, V. L., & Balkus, K. J. Jr. (2000). Cytochrome c immobilization into meoporous molecular sieves. *Journal of Molecular Catalysis. B, Enzymatic*, *10*(5), 453–469. doi:10.1016/S1381-1177(99)00123-X

Wei, Y., Feng, Q., Xu, J., Dong, H., Qiu, K.-Y., Jansen, S. A., ... Ong, K. (2000). Polymethacrylate-Silica Hybrid Nanoporous Materials: A Bridge between Inorganic and Polymeric Molecular Sieves. *Advanced Materials*, *12*(19), 1448–1450. doi:10.1002/1521-4095(200010)12:19<1448::AID-ADMA1448>3.0.CO;2-W

WHO/UNICEF. (2012). Joint Monitoring Programme for Water Supply and Sanitation. Progress on Sanitation and Drinking-water. (NLM classification: WA 670)

Wight, A. P., & Davis, M. E. (2002). Design and Preparation of Organic−Inorganic Hybrid Catalysts. *Chemical Reviews*, *102*(10), 3589–3614. doi:10.1021/cr010334m PMID:12371895

Wight, A. P., & Davis, M. E. (2002). Design and Preparation of Organic−Inorganic Hybrid Catalysts. *Chemical Reviews*, *102*(10), 3589–3614. doi:10.1021/cr010334m PMID:12371895

Williams, T. D., Gensberg, K., Minchin, S. D., & Chipman, J. K. (2003). A DNA expression array to detect toxic stress response in European flounder (*Platichthys flesus*). *Aquatic Toxicology (Amsterdam, Netherlands)*, *65*(2), 141–157. doi:10.1016/S0166-445X(03)00119-X PMID:12946615

Wu, Z. J., Han, H., Han, W., Kim, B., Ahn, K. H., & Lee, K. (2007). Controlling the hydrophobicity of submicrometer silica spheres via surface modification for nanocomposite applications. *Langmuir*, *23*(14), 7799–7803. doi:10.1021/la700386z PMID:17559243

Yoshitake, H. (2005). Highly-controlled synthesis of organic layers on mesoporous silica: Their structure and application to toxic ion adsorptions. *New Journal of Chemistry*, *29*(9), 1107–1117. doi:10.1039/b504957a

Yu, C., Yu, Y., & Zhao, D. (2000). Highly ordered large caged cubic mesoporous silica structures template by triblock PEO–PBO–PEO copolymer. *Journal of Chemical Communications*, (7), 575-576.

Zaitseva, N., Zaitsev, V., & Walcarius, A. (2013). Chromium(VI) removal via reduction–sorption on bi-functional silica adsorbents. *Journal of Hazardous Materials*, *250–251*, 454–461. doi:10.1016/j.jhazmat.2013.02.019 PMID:23500426

Zeledón-Toruño, Z. C., Lao-Luque, C., de las Heras, F. X. C., & Sole-Sardans, M. (2007). Removal of PAHs from water using an immature coal (leonardite). *Chemosphere*, *67*(3), 505–512. doi:10.1016/j.chemosphere.2006.09.047 PMID:17109923

Zeleňák, V., Badaničová, M., Halamová, D., Čejka, J., Zukal, A., Murafa, N., & Goerigk, G. (2008). Amine-modified ordered mesoporous silica: Effect of pore size on carbon dioxide capture. *Chemical Engineering Journal*, *144*(2), 336–342. doi:10.1016/j.cej.2008.07.025

Zewail, T. M., & El-Garf, S. A. M. (2010). Preparation of agriculture residue based adsorbents for heavy metal removal. *Desalination and Water Treatment*, *22*(1-3), 363–370. doi:10.5004/dwt.2010.1245

Zylstra, G. J., Bang, S. W., Newman, L. M., & Perry, L. L. (2000). *Biodegradation of Nitroaromatic Compounds and Explosives. In Microbial Degradation of Mononitrophenols and Mononitrobenzoates* (pp. 145–160). Boca Raton, FL: Lewis Publishers.

Chapter 9
Application of Magnetic Nanomaterials for Water Treatment

Ambreen Lateef
University of the Punjab, Pakistan

Rabia Nazir
Pakistan Council of Scientific and Industrial Research, Pakistan

ABSTRACT

The availability of clean drinking water becomes a critical issue for all the people of the world due to a rapid increase in population and industrialization. The water bodies get contaminated due to the discharge of wastewater, that will not only disturb the aquatic life but also badly affect human health. Therefore, different methods are adopted to treat the contaminated water to make it clean and safe for people. In last few years, the nanomaterials have gained much attention for water treatment because of their unique properties. Among all nanomaterials, magnetic nanomaterials are considered more efficient and attractive because of their easy separation and reusable property. In this chapter, a brief review related to synthesis and characterization of MNM was studied along with their application in removal of dyes, heavy metals, and microbes from wastewater through simple adsorption processes.

INTRODUCTION

The access to clean water has become one of most serious issue facing people all over the world. As the world's population is expected to hype from 6.5 billion and to 9 billion by the year 2025, the demand for water is also expected to increase. The water quality is affected very badly due to anthropogenic activities (Xu et al., 2012). Both organic and inorganic pollutants are present in water because of agriculture, domestic and industrial activities, which effect the human health and in turn deteriorate the ecosystem (Gupta, Kumar, Nayak, Saleh, & Barakat, 2013).

DOI: 10.4018/978-1-5225-5745-6.ch009

Water is a basic necessity of all life available on earth; only 3% of the total water comprises of freshwater and a small proportion i.e. 0.01% of this freshwater is accessible for human use. Exploitation of water resources due to lack of proper management for sustainable use, deterioration of ecosystems coupled with the environmental pollution resulted in scarcity and depletion of water reservoirs. Use of contaminated water affects all the biotic components of ecosystem (Azizullah, Khattak, Richter, & Häder, 2011). Polluted water can cause mutagenic, teratogenic, carcinogenic, reproductive etc. impairments in living organisms due to the presence of toxic substances like heavy metals, dyes, pesticide, phosphates, nitrates etc (Nabeela et al., 2014).

Major sources of these toxic compounds are industrial effluents, disposal of sewage sludge and run-off from agricultural fields. Once pollutants enter the water bodies, they either get dissolved into the water or may remain suspended, and then they enter the food chain which may result into bioaccumulation or biomagnification. The presence of toxic substances poses threat to overall ecosystems. Trace amounts of some elements are essential for water quality, but when their concentration exceeds permissible levels, they become toxic compounds affecting the environment (Brabec, Schulte, & Richards, 2002).

Currently different methods are developed and practiced to treat and remove pollutants from the water to make it reusable. For water and wastewater treatment commonly physical, chemical and biological methods are applied/ used. In these methods, different techniques like adsorption, membrane filtration and catalytic degradation are employed to make them more efficient (Burkhard, Deletic, & Craig, 2000). Although these methods are effective, but are expensive, time consuming and require more land and man power for their operation (Anjum, Miandad, Waqas, Gehany, & Barakat, 2016). To combat with this above scenario, there is a dire need to develop and adopt new technology to remove contaminants and to ensure availability of safe and clean water to people.

Nanotechnology and Nanomaterials

Nanotechnology is emerging as an interdisciplinary research and provides innovative solutions for the recent problems. Nanotechnology is "the design, fabrication and application of nanostructures or nanomaterials and the fundamental understanding of the relationships between physical properties or phenomena and material dimensions". Nanomaterials are the important factors in the revolution of nanotechnology to elucidate the water issues globally (Qu, Brame, Li, & Alvarez, 2013).

Nanomaterials have unique physicochemical properties due to their small size and large surface area, which make them attractive for their application in water treatment(Amin, 2009). Different kind of nanomaterials like metal, magnetic, carbon based and clay based nanomaterials are used for the treatment of water. These nanomaterials can be synthesized by top down and bottom up approach. Two different approaches are used for synthesis of nanocomposites (Figure 1) i.e. top-down approach which includes physical methods and bottom up approach includes wet methods (Ajayan, 2003; Oliveira & Machado, 2013).

The benefit of physical methods is production of large amount of nanocomposites, while synthesis of equal sized nanocomposites is difficult to attain through top-down approach. In comparison, wet chemical methods give uniformity in size of nanocomposites where controlled particle size can be achieved. Although by varying conditions of reaction, different shapes (nano-rods, nanowires, nanotubes etc.) of nanocomposites can be synthesized.

Figure 1. Schematic representations of top down and bottom up approaches for synthesis of nanomaterials

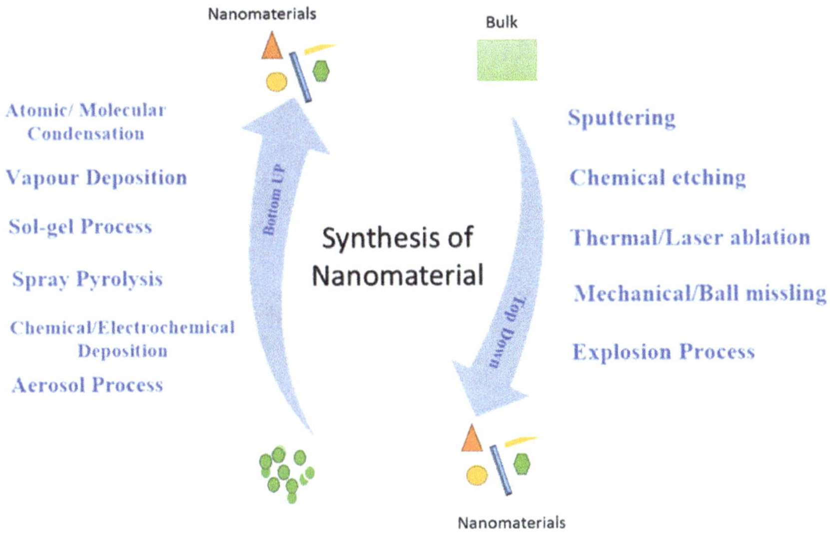

Nanomaterials have become an emerging solution for providing potential, efficient, cost-effective and ecofriendly solution for water treatment to improve the quality of water (Santhosh et al., 2016). There are many different kinds of nanomaterials researched that have potential applications in the field of water and wastewater treatment but the focus of this chapter is on the magnetic nanomaterials (MNM), because they are more efficient and have potential to treat both organic and inorganic contaminants and play a significant role in the treatment of water.

Magnetic Nanomaterials

Magnetic nanomaterials (MNM) are considered as supplementary and complementary sources for water treatment technologies because of their advantage over non-magnetic materials, as these can be easily separated using magnetic field from water treatment (Ambashta & Sillanpää, 2010). Magnetism is a unique physical property that independently helps in water purification by influencing the physical properties of contaminants in water. In addition, its combination with other processes enables an improvised efficient purification technology. This technique allows one to design processes where the particles not only remove compounds from water but also can easily be removed again and then be recycled or regenerated as represented in Figure 2. (Ditsch, Lindenmann, Laibinis, Wang, & Hatton, 2005).

MNM includes various kind of iron based nanomaterials and are grouped as iron oxide nanoparticle, nanocomposites and filtration membranes. These are synthesized by different methods and are discussed presented in Table 1.

Water Treatment / Water Pollutants

MNM are very efficient materials that have been used effectively for removal of different pollutants, which are discussed here briefly:

Figure 2. Schematic representation of use of MNM and their removal from wastewater

Dyes

Dyes are the most important pollutant discharged from the industry wastewater and contaminate the water bodies globally (Khan & Malik, 2014). These dyes are organic compounds and used in large quantity for the dying of different substrates like paper, leather, fur, hair, waxes, plastics and textile materials. Dyes are carcinogenic in nature and cause serious threats to human's health and ecosystem (Chequer et al., 2013). These are discharged into water bodies and reduces the oxygen concentration because of hydrosulfides and interrupts the light penetration which badly impact aquatic ecosystem (Carmen & Daniela, 2012) Different methods and adsorbent are use d for the removal but now a days magnetic but a now magnetic nanomaterials are getting more attention because of their easy separation.

Keyhanian et al. (2016), synthesized the magnetic nanoparticles (Fe_3O_4) by chemical precipitation method for the removal of methyl violet dye from aqueous solution by adsorption (Keyhanian, Shariati, Faraji, & Hesabi, 2016). Similar modified Fe_3O_4 magnetic nanoparticles was employed to eradicate the reactive red-120 dye from wastewater using ionic liquid (Absalan, Asadi, Kamran, Sheikhian, & Goltz, 2011). While the Fe_3O_4 magnetic nanoparticles synthesized by Muthukumaran et al. (2016), using chemical precipitation method and studied the adsorption models which represent the adsorption capacity of 166.6mg/g for crystal violet dye (Muthukumaran, Sivakumar, & Thirumarimurugan, 2016). In the same way, magnetic nanoparticles (Fe_3O_4) coated with activated maize cob by co-precipitation method to prepare Fe_3O_4-MCP nanocomposite for the elimination of cation methylene blue dye. The results indicated that adsorption capacity of dye enhances due to coating of maize cob (Tan, Morad, Teng, Norli, & Panneerselvam, 2012).

The anions dyes can't be easily adsorbed directly on the surface of magnetic nanoparticles. Thus, to enhance adsorption capacity these MNM were modified as illustrated in Figure 3. Magnetic ferrite nanoparticles and modified with surfactant, were used for the adsorption of direct green 6, direct red 31, and direct 23. The results of adsorption studies revealed that at different concentrations of dye (90, 120, 150 and 200 mg/L) the percentage of removal of 63, 45, 30 and 23% for DR23, 97, 90, 78 and 45% for DR31 and 51, 48, 42 and 37% for DG6, respectively (Mahmoodi, Abdi, & Bastani, 2014). Similarly Zargar et al. (2009), used the Cetyltrimethylammonium bromide (CTAB) surfactant for the functionalization of magnetic Fe_3O_4 NPs and conducted their batch studies for the adsorption and elimination of acid red 27 dye he optimized the pH, temperature and adsorbent dose and find the adsorption capacity

Table 1. Synthesis and characterization of different Magnetic Nanomaterials (MNM)

MNM	Synthesis method	Particle size	Reaction conditions	Precursor	Characterization	References
hematite (α-Fe_2O_3), magnetite (Fe_3O_4), and maghemite (γ-Fe_2O_3) particles	Sol gel method		140oC for 24h, 330 °C for 6 h, 240 °C for 2 h, acidic condition	$Fe(OH)_3$	FT-IR, XRD, SEM, magnetic properties, Coercivity	(Itoh & Sugimoto, 2003)
Maghemite nanoparticles	Sol gel method		In presence of N_2, pH 8	NH_4OH	XRD,TEM,VSM,BET	(Hu, Chen, & Lo, 2005)
Magnetite nanoparticles	Sol gel method	10nm	In presence of N_2, pH 8	NH_4OH	XRD, TEM, BET, Zeta potential	(Hu, Lo, & Chen, 2004)
Cobalt ferrite	Wet chemical	15-48nm	pH 11-12, temp 80°C, and centrifuge at 3000 rpm speed	oleic acid as surfactant and coating material	XRD, TEM and SEM	(Maaz, Mumtaz, Hasanain, & Ceylan, 2007)
Chitosan-iron nanoparticles	Microemulsion	4.5nm	Temp 100°C, pH 10 with constant stirring	toluene as organic phase, sodium bis(2-ethylhexyl) sulfosuccinate (AOT) and sodium dodecyl sulfate (SDS) (2/1, w/w) as surfactant	FT-IR, XRD, HRTEM, Magnetic properties	(Pineda et al., 2014)
Silica iron nanocomposite	Co-precipitation		Temp 60°C, pH 6	NH_4OH, silica	,FTIR, XSP, XRD, TEM and particle size	(X. Liu, Guan, Ma, & Liu, 2004)
Humic acid/ Fe_3O_4 nanoparticles (Fe_3O_4/HA)	Co-precipitation	10nm	90°C, constant stirring	$FeCl_3 \cdot 6H2O$, $FeSO_4$.7H2O, NH_4OH and humic acid	FT-IR, XRD, TEM, BET particle size	(J.-f. Liu, Zhao, & Jiang, 2008)
Fe_3O_4/chitosan	Co-precipitation	10-11nm	50°C temp with three different chitosan concentrations 0.125, 0.25, and 0.5%	$FeCl_3 \cdot 6H_2O$ (0.32 M) and 50 mL of $FeCl_2 \cdot 4H_2O$ (0.2 M)	XRD, STEM,VSM,TGA	(Gregorio-Jauregui et al., 2012)
Montmorillonite-supported magnetite nanoparticles	Co-precipitation	15nm	In presence of N_2,pH.8, 60°C, 30min	$FeCl_3 \cdot 6H_2O$, $FeCl_2 \cdot 4H_2O$, NH_4OH	XRD,TEM,EDX,XPS,BET,DCS	(Yuan et al., 2009)
Montmorillonite-supported magnetite nanoparticles	Hydro sol method		pH.4, centrifuge at 4000rpm for 5min,	$FeCl_3 \cdot 6H_2O$, $FeCl_2 \cdot 4H_2O$, NH_4OH, HCl	XRD,TEM,EDX,XPS,BET,DCS	(Yuan et al., 2009)
Fe_3O_4- diatomite	Co-precipitation	15nm	pH.7, temp 200°C	$FeCl_3 \cdot 6H_2O$, $FeCl_2 \cdot 4H_2O$, NH_4OH, HCl	XRD,TEM,EDX,XPS,BET,DCS	(Yuan et al., 2010)
Fe_3O_4- Zeolite	Co-precipitation		pH 8, temp 343 K	$FeCl_3 \cdot 6H_2O$, $FeCl_2 \cdot 4H_2O$, NaOH	XRD,magnetization measurements, chemical analyses, N2 adsorption isotherms and Mo¨ssbauer spectroscopy	Luiz C.A. Oliveira
Fe_3O_4–carbon nanotube	Two step deposition		Ph 11-12, temp 50°C for 30min	Polyaniline as surfactant	FT-IR, XRD, SEM AND TGA	(Anjum et al., 2016)
Magnetite- corncob nanocomposite	Chemical method	50nm	100°C,	$FeCl_3 \cdot 6H2O$, NaOH	XRD,SEM,TEM.VSM	(Nethaji, Sivasamy, & Mandal, 2013)
PPy/Fe_3O_4 magnetic nanocomposite	chemical oxidative polymerization		Shaken for 6h, temp 100°C	Polypyrrole coating material	ATR-FTIR, XRD,FE-SEM,HR-TEM,ESR	(Bhaumik, Maity, Srinivasu, & Onyango, 2011)
Magnetic hydrogel nanocomposite	Chemical photopolymerization			2-acrylamido-2-methyl-1-propansulfonic acid (AMPS)	XRD, FR-IT,SEM	(Ozay, Ekici, Baran, Aktas, & Sahiner, 2009)

of 1.05 mg/g (Zargar, Parham, & Hatamie, 2009). Dalali et al. (2011), applied the same CTAB surfactant for modification of magnetite nanoparticles and used these MNPs for the decay or adsorption of Nyloset Yellow E-RK (NY) dye using Taguchi Method. The adsorption equilibrium data were fitted to Langmuir isotherm rather than Freundlich isotherm. The maximum adsorption capacity, qmax, obtained from Langmuir's model was 136 mg g^{-1}(Dalali, Khoramnezhad, Habibizadeh, & Faraji, 2011). While the polyacrylic acid applied as surfactant to modify the surface of magnetic nanoparticle (Fe_3O_4) and were used as adsorbent for the removal of methylene blue and crystal violet. The polyacrylic acid- Fe_3O_4 nanoparticles showed higher adsorption capacity for methylene blue (199mg g^{-1}) and lower for crystal violet (116mg g^{-1})(Liao, Wu, & Chen, 2005; Mak & Chen, 2004).

The silica coated magnetic nanoparticles were employed for the removal of methyl orange dye using batch scale experiment. The adsorption studied were conducted at optimized parameters and the results demonstrated that adsorption capacity of 34.29mg g_{-1} was found(Shariati-Rad, Irandoust, Amri, Feyzi, & Ja'fari, 2014). Chang et al. (2005), use the Chitosan-Fe_3O_4 nanocomposite for removal of acid orange 12 and acid green 25 found the adsorption capacity of 1.883 and 1.471mg g^{-1}, respectively (Chang & Chen, 2005). The Fe_3O_4–wheat straw nanocomposite prepared by simple precipitation method, employed for the removal of methylene blue dye with adsorption capacity of 1374.6 mg g^{-1} (Ebrahimian Pirbazari, Saberikhah, & Habibzadeh Kozani, 2014). The magnetic Fe_3O_4 nanoparticles were incorporated into the multi walled carbon nanotubes (MWCNT) through wet chemical method and employed as nano-adsorbent for the adsorption of neutral red and methylene blue. The results demonstrate that maximum adsorption capacity of neutral red and methylene blue were 77.5 and 42.3 mg/g, respectively obtain at 60min time interval with 0.05g of adsorbent dose (Qu, Huang, Yu, Chen, & Kong, 2008).

Similar study was conducted by Gong et al. (2009), who used magnetic multi walled carbon nanotube (MMWCNT) nanocomposite for removal of cationic dyes (methylene blue, neutral red and brilliant cresyl blue) and studied their kinetics and adsorption models to evaluate the efficiency of nanocomposite (Gong et al., 2009). Haham et al.(2015), synthesized magnetic nanocomposite by impregnation

Figure 3. schematic representation of surface modification of MNM for the removal of dyes from water

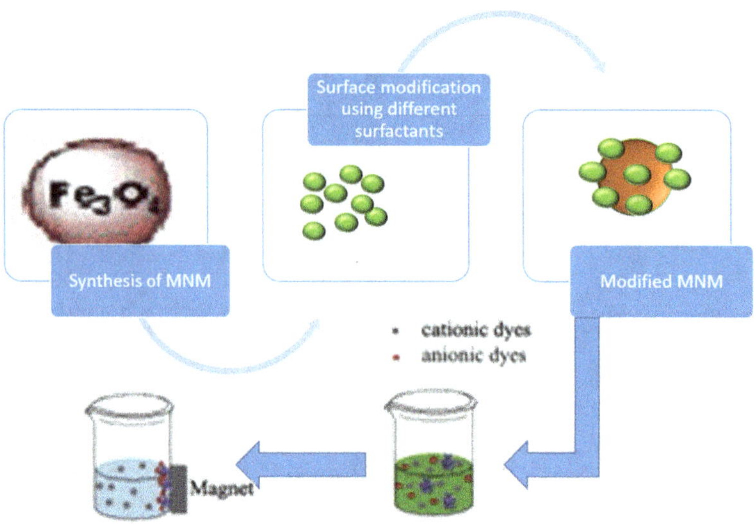

of magnetic Fe_3O_4 and Fe/Pd nanoparticles into activated carbon fabrics (ACF) and investigated their comparative adsorptive effects on azo dye (orange II) in the presence of H_2O_2(Fenton catalysis). The adsorption capacity of dyes follow this order Fe/Pd/ACF > Fe/ACF > ACF, represents that magnetic nanoparticles-activated carbon fabric nanocomposite as more efficient adsorbent (Haham, Grinblat, Sougrati, Stievano, & Margel, 2015).

Heavy Metal Removal

Different kinds of contaminants like heavy metal ions, bacteria, organic pollutants are discharged from many industries as wastewater that contaminates the water bodies. The amount of these pollutants in wastewater increases due to rapid growth of industrial sector(Fu & Wang, 2011). Among all these water contaminants, heavy metals are considered as more harmful and toxic, because these are non-biodegradable and have the ability of deposition. These can accumulate in organisms and become a part of food chain which will affect the both aquatic and terrestrial ecosystem (Kobya, Demirbas, Senturk, & Ince, 2005). The heavy metals include Pb^{2+}, Ar+,Cr^{2+}, Cd^{2+}, Zn^{2+}, Ni^{2+} and Hg^{2+} ion which are discharged from different industries like including chemical manufacturing, battery manufacturing industries, metallurgical, leather tanning, and mining (Wang et al., 2012). These are carcinogenic in nature that's why permissible limits are defined for water (Kumar & Puri, 2012). Hence, the removal of these toxic metal ions from wastewater is become very a crucial issue globally.

Different kinds of nanomaterials are used for removal of the heavy metals due to their unique properties. MNM gained much importance due to their easy separation after the treatment from the water (Hua et al., 2012). This section details the removal of heavy metals is discussed one by one.

Chromium

Hu et al. (2004), studied the removal of hexavalent chromium Cr (VI) using magnetite nanoparticles by conducting batch scale adsorption studies (Hu et al., 2004). While in his second experiment maghematite nanoparticles were employed for Cr (VI) that resulted in 90% of adsorption in just 15min. The regeneration studies of maghemite nanoparticles confirmed that these materials can be reused, even up to six successive adsorption–desorption processes with same metal removal capacity (Hu et al., 2005).

Similarly, Yuan and his researchers synthesized magnetite nanocomposites using different support materials for removal of Cr (VI). In the first experiment, he synthesized the magnetite nanoparticles doped montmorillonite clay nanocomposite by co-precipitation and hydrosol methods for treatment of Cr (VI). The comparative results showed that nanocomposites prepared by hydrosol method showed better results for adsorption of Cr (VI) than the other nanocomposites prepared by co-precipitation method (Yuan et al., 2009). In second experiment, magnetite nanoparticles were impregnated with diatomite support material to synthesized nanocomposite and employed to remove Cr (VI) through adsorption process (Yuan et al., 2010).

Magnetite nanoparticles incorporated into corncob based activated charcoal support material were als employed for the removal the Cr (VI) by adsorption process (Nethaji et al., 2013). All these MNM have low adsorption capacity of removal of Cr (VI), so the new polymer based nanocomposite (polypyrrole/Fe_3O_4 magnetic nanocomposite) was developed by Bhaumaik and his coworkers for the elimination of Cr (VI). This adsorbent is highly efficient and showed 100% removal efficiency for Cr (VI) at pH 2 with 200mg/L of Cr (VI) concentration (Bhaumik et al., 2011). MNM were incorporated into AMPS

(2-acrylamido-2-methyl-1-propansulfonic acid) hydrogels to synthesize nanocomposites for elimination of Cr (III) from aqueous solutions. These magnetic hydrogel nanocomposites proved to be more effective for treatment of Cr(III) than conventional materials (Ozay et al., 2009).

Copper

The second toxic metal is copper (Cu (II)) which is discharged from many industries like paint and pigment, electroplating, fertilizer and electrical industries. Various kind of MNM were used for treatment of Cu(ll) but for more efficient removal of Cu(II), magnetite nanoparticles were functionalized with various materials. Hao et al. (2010), used 1, 6-hexadiamine ligand for surface of magnetite nanoparticles to improve the removal capacity of MNPs for Cu(II). The results revealed that 98% adsorption of Cu(ll) occurred due to attachment of amine group on the surface (Hao, Man, & Hu, 2010). In the same way, magnetic nanoparticles modified with the amine group, showed maximum adsorption capacity of 22.4 mg/g at pH 6 and temperature of 298K (Hui, Xiao, Hua, Rui, & Zuo, 2013).

The adsorption of magnetic nanoparticles functionalized with amine group enhances with increases in pH because at higher pH amine group make complex with Cu^{+2} ion. However, at lower pH amine group undergoes protonation (Dave & Chopda, 2014). Maghemite nanoparticles coated with chitosan (CCMNPs) were functionalized using a ecofriendly reagent α-ketoglutaric acid employed for eradication of Cu(ll) from water (Zhou, Nie, Branford-White, He, & Zhu, 2009). In another study, chitosan carboxymethylated was attached to Fe_3O_4 nanoparticles surface by covalent bond and used for removal of Cu(ll) and maximum adsorption of 21.5mg/g was found (Chang & D.-H. Chen, 2005). Magnetic composite microspheres were synthesized using co-precipitation method by Fe_3O_4 nanoparticles and polyacrylic acid-chitosan, as an efficient adsorbent for elimination of Cu(II) (Yan et al., 2012). The CS/PPA-Fe_3O_4 microspheres have exhibited higher adsorption capacity than CS-Fe_3O_4 microspheres. The Fe_3O_4-polyvinyl acetate-iminodiacetic acid containing EDTA was also found useful adsorbent for removal of Cu(II) (Tseng et al., 2009). Khandanlou et al. (2015), prepared Fe_3O_4 - rice straw nanocomposite for the adsorption of Cu(ll) that led to adsorption capacity of 75.54% (Khandanlou et al., 2015)

Lead

Lead Pb(II) is a major heavy metal pollutant and used in various manufacture process of industries like battery manufacturing, lead smelter, paper and pulp industries. Lead is very dangerous for the health of adults and children's (Araki, Sato, Yokoyama, & Murata, 2000). Thus, there are different methods and treatments employed for their removal and MNM are considered best for their removal and recovery. Fe_3O_4 nanoadsorbents were employed for the removal of Pb(II) ions from aqueous solution through batch-adsorption process. The effect of optimum parameters were studied on adsorption of of Pb(II) in detail. Maximum adsorption capacity of 36mg/g was attained in just 30min and amount of adsorption increase with increase in temperature which indicates endothermic adsorption reaction(Nassar, 2010).

Chang et al. (2012), synthesized maghemite (γ-Fe_2O_3) nanoparticles by co-precipitation for eradication of Pb(ll) from electroplating wastewater. The maximum adsorption of Pb(II) conquered within 15 min (Cheng et al., 2012). MNPs-Ca-alginate coated with Phanerochaete chrysosporium adsorbed about 90% of Pb(II) after repeated five cycles (Xu et al., 2012). Similarly, Wang et al. (2012), synthesized water soluble magnetite nanoparticles using hydrothermal approach. Their high stability and solubility enhances removal of Pb(II) than insoluble magnetite nanoparticles (Wang, Li, Jiang, & Zhao, 2012). Recently,

Zhang et al. (2013) reported the application of monodisperse amine-terminated (Fe_3O_4@SiO_2–NH_2) nanocomposite for the removal of Pb(ll). These MNM showed effective Pb (II) removal at pH range of 2-6 and these amine groups provide effective chelating sites for adsorption of Pb (II) (Zhang et al., 2013).

Cadmium

Contamination of water by Cd(ll) due to different anthropogenic activities is one of the biggest environmental concerns. Cadmium is released to the environment by a number of industries such as metal finishing, metallurgical, alloying, electroplating, pigments, batteries, fertilizers, textile operations, tanneries, zinc smelting and by acid mine drainage. Cadmium acts as an enzyme inhibitor and also affects the liver, kidney and bones and causes nausea and vomiting (Ehrampoush, Miria, Salmani, & Mahvi, 2015).

Magnetic hydroxyapatite nanoparticles (MNHAP) was synthesized by hydrothermal method and used for the treatment of Cd(ll)(Xu et al., 2012). Similarly, Chen et al. (2017), synthesized the Fe_3O_4 sulfonated magnetic (Fe_3O_4-SO_3H MNP) nanoparticle for Cd(ll) removal by surface modification with sulfo group to improve the adsorption capacity. Batch scale adsorption studies were conducted and maximum adsorption of Cd(ll) was observed at 25°C (Chen et al., 2017). Gong et al. (2012), used shellac coated iron oxide magnetic nanoparticles for removal of Cd(ll). The adsorption capacity of Cd(ll) enhanced upto 18.80 mg g^{-1} by the use of Shellac coating, shellac is natural, biodegradable and renewable resin (Gong et al., 2012).

Magnetic alumina nanocomposite (MANC) was synthesized by the simple co-precipitation method for the eradication of Cd (ll). The maximum adsorption of 626mg g^{-1} of Cd(ll) was attained at pH 6 (El-Latif, Ibrahim, Showman, & Hamide, 2013). Ehrampoush et al. (2015), synthesized the iron oxide nanoparticles using tangerine peel extract for removal of Cd (ll). The adsorption studies were conducted to optimize the parameters, 90% of Cd (ll) removal was achieved at pH 4 with adsorbent dose of 0.4g (Ehrampoush et al., 2015). The magnetic nanoparticles (MNPs) modified with sawdust (SD) to improve the efficiency of Cd (ll) adsorption. The results of adsorption studies demonstrated that the adsorption capacity of SD-MNP (1000mg g^{-1}) was almost double than the MNP (588.8mg g^{-1})(Shah, Jan, Khan, & Amir, 2016).

Arsenic

Arsenic (As) contamination in water is a widespread problem globally. Millions of people depend on arsenic-contaminated groundwater. Arsenic poisoning leads to fatal diseases such as skin and internal cancers. Hence, to remove As different methods were practiced but now a day magnetic nanomaterials gain much importance due to their easy separation (Parida, Lolage, Angal, & Rautaray, 2017). The magnetic γ-Fe_2O_3 nanoparticles were simply synthesized using co-precipitation method for removal of As(III) and As(V) by adsorption process. The results revealed that adsorption of As(III) and As(V) was not affected by the pH but temperature had marked influence on the adsorption, at 50°C maximum adsorption occurred (Lin, Lu, & Liu, 2012).

In addition to that, thermodynamic and spectroscopic studies were also conducted to observe the adsorption behavior of magnetic nanoparticles on As(lll) and As(V)(Lin et al., 2012; C.-H. Liu et al., 2015). Mayo et al. (2007), has studied the effect of magnetic nanoparticle (magnetite) on the adsorption of both As(III) and As(V). He reported that the adsorption capacity of nanoparticle improves 200 times when size of nanoparticles reduces from 300 nm to 12nm(Mayo et al., 2007). The Fe_3O_4 nanoparticles coated with ascorbic acid were synthesized by hydrothermal method to eliminate As(III) and As(V) from water by adsorption. The results demonstrated that adsorption data fitted well into Langmuir model and maximum adsorption capacity of 46.06mgg^{-1} for As(lll) and 16.56 mg g^{-1} for As(V)(Feng, Cao, Ma, Zhu, & Hu, 2012). Zhang and his co-workers synthesized two bimetallic nanomaterials i.e; $MnFe_2O_4$ and $CoFe_2O_4$ by chemical precipitation method for removal of both arsenite (As (III)) and arsenate (As(V)). The adsorption were conducted as a function of temperature, contact time, pH and dose concentration. The results demonstrated that the maximum adsorption capacities of arsenite and arsenate on $MnFe_2O_4$ were 94 and 90 mg g^{-1}, and on $CoFe_2O_4$ were 100 and 74 mg g^{-1}, respectively. The bimetallic nanocomposites were more efficient than Fe_3O_4 for arsenic removal (Zhang, Niu, Cai, Zhao, & Shi, 2010). Different kind of surfactants was used in literature to enhance the adsorption capacity of MNM by reducing the oxidation of magnetic particles. As Jin et al. (2012), has reported that magnetic nanoparticles were modified by Cetyltrimethylammonium bromide (CTAB), which improved the adsorption capacity of MNPs for arsenate (Jin, Liu, Tong, & Hou, 2012).

Microbes

Now a days, people are facing problems of infectious diseases globally and also microorganism's resistance increases towards antibiotics (Organization, 2012). Most of bacteria, that cause infection have strong resistant for the antibiotics which are mostly used to eliminate the infection (Bhutta, Sommerfeld, Lassi, Salam, & Das, 2014). The researchers are working to find alternative antibiotics to fight against bacteria and microorganisms (Kebede et al., 2013). Both gram-positive and gram- negative bacterial strains are harmful and source of human diseases. A number of synthesized nanoparticles of different metal oxides are considered as good inhibitors for many bacterial strains but MNM havve gained much importance due to their size and stable behavior (Iwamoto & Ishigaki, 2013). These magnetic iron oxide nanoparticles penetrate on or into the membrane of cells, that may hinder or block the protein synthesis and transcription mechanism which ultimate led to death of cell as presented in Figure 4(Kashmiri & Mankar, 2014). The size of MNM directly affect the zone of inhibition because small size of nanoparticles have large surface area for attachment with bacteria (Panáček et al., 2006).

Figure 4. Magnetic nanoparticles penetration into the cell and damage the cell by inhibiting the metabolic activity of bacterium

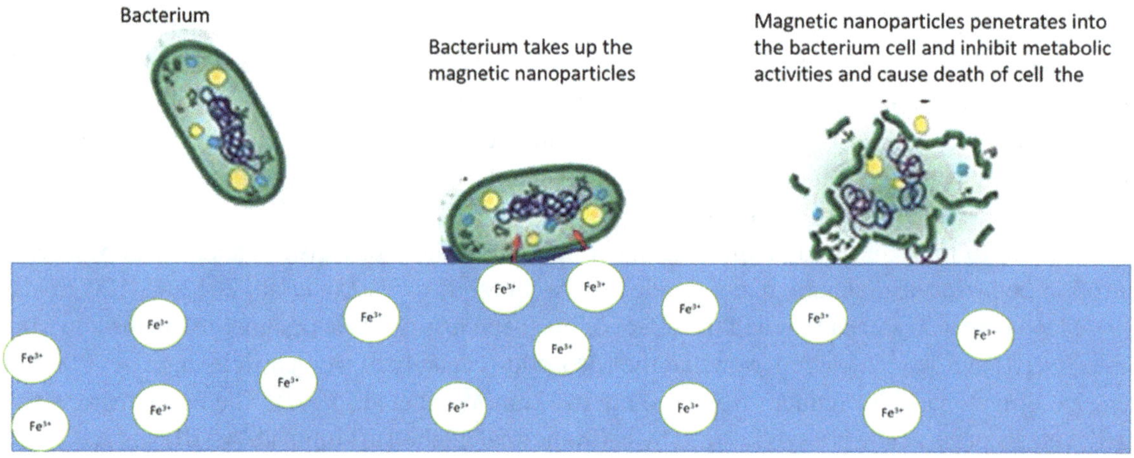

Ismail and his coworkers were synthesized iron oxide nanoparticles (α-Fe_2O_3) by laser ablation method and studied their antimicrobial activity against gram-positive; *Staphylococcus aureus* and gram-negative; *Escherichia coli*, *Pseudomonus aeruginosa* and *Serratia marcescens* by the well diffusion method. The results showed that magnetic iron oxide nanoparticles inhibit the *Staphylococcus aureus* growth more rapidly than the others bacteria(Ismail, Sulaiman, Abdulrahman, & Marzoog, 2015). To stabilize and enhance their efficiency of MNM, coated with different materials like polymers. As reported by the Zhang et al. (2015), that poly-allylamine-hydrochloride (PAAH) used as surfactant to improve the removal efficiency of bacteria from the water. These are applied for the removal of bacteria as well as a tool to control the single bacterial cell (Zhang et al., 2015). Because non-modified MNPs were considered as cytotoxic and genotoxic for bacterial and mammalian cells (Brunner et al., 2006). However, it will raised the issues related to toxic effects of MNM for pathogenic removal from water (Boxall, Tiede, & Chaudhry, 2007). The functionalization of MNM were done with different materials like silica and poly-allylamine-hydrochloride (PAAH) to solve this issue related to toxicity of MNM(Zhang et al., 2015). Xu has synthesized the MNM functionalized with PAAH by simple co-precipitation approach for safe removal of *Escherichia coli*, *Acinetobacter baumanii*, *Pseudomonas aeruginosa* and *Bacillus cereus* from water (Xu, Li, Zhu, Huang, & Zhang, 2014). Santosh et al. (2014), also prepared the graphene based Fe_3O_4 nanocomposites (G- Fe_3O_4) and studied their antibacterial behavior against *E coli* at different concentrations of 0, 25, 50, 100, 200 mg L^{-1} of adsorbent, the results represents that G- Fe_3O_4 nanocomposite inhibit the bacterial growth by rupturing the cell wall after 60min as presented in figure (Santhosh et al., 2014). Similarly He in his next studies, investigate the influence of $MnFe_2O_4$-G on *E. coli* and findings of this study demonstrated that at 100mg L^{-1} concentration of $MnFe_2O_4$-G nanocomposite, about 82% bacterial population become stagnant which shows their bacteriostatic activity (Chella et al., 2015).

Figure 5. SEM image represents the E.coli bacteria control (a) and rupturing effect of graphene–Fe_3O_4 nanocomposite on cell (B); source:(Santhosh et al., 2014)

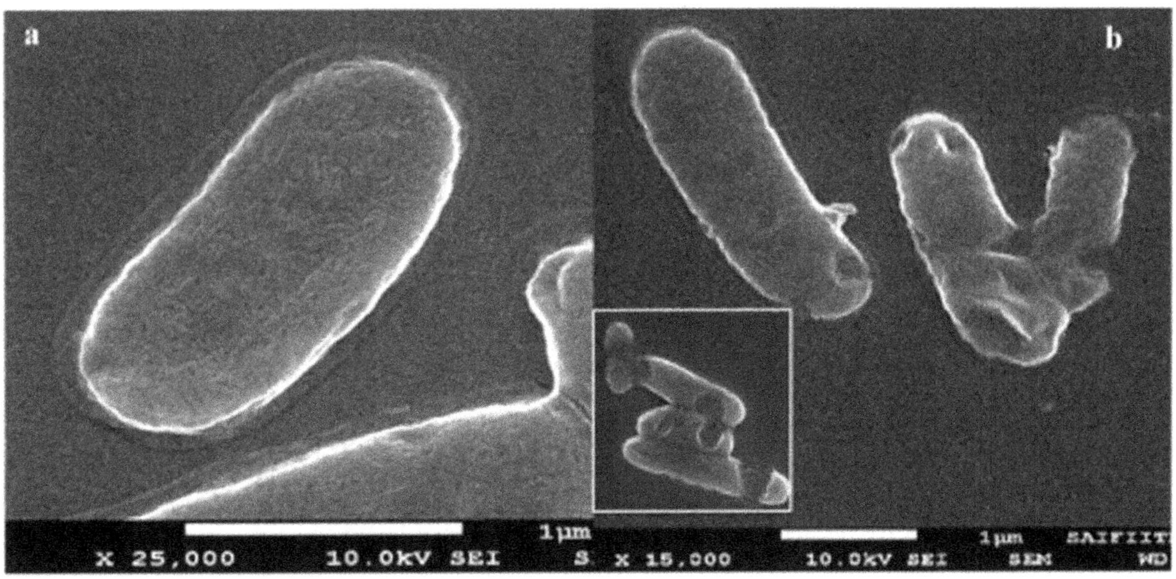

CONCLUSION

Water and wastewater treatment technologies have been much rearmed with the advent of MNM which are considered very effective adsorbents for the removal of organic, inorganic and biological contaminants. These materials offer various advantages due to their easy recovery and reuse by simply applying magnetic field making MNM more cost effective. In this chapter, use of different kind of MNM were discussed briefly which were cited in literature for removal of different pollutants. The use of different magnetic materials that are either pure magnetite or are in form of composites or modified by using of surfactants have seen to impact the adsorption capacities to high level. In addition to that various other parameters like synthesis methods, pH, and temperature affect the effectiveness of the treatment process. In short, the MNM proved to be viable materials for the water remediation and treatment.

REFERENCES

Absalan, G., Asadi, M., Kamran, S., Sheikhian, L., & Goltz, D. M. (2011). Removal of reactive red-120 and 4-(2-pyridylazo) resorcinol from aqueous samples by Fe 3 O 4 magnetic nanoparticles using ionic liquid as modifier. *Journal of Hazardous Materials*, *192*(2), 476–484. doi:10.1016/j.jhazmat.2011.05.046 PMID:21645968

Ajayan, P. M. (2003). Bulk metal and ceramics nanocomposites. *Nanocomposite Science and Technology*, 1-76.

Ambashta, R. D., & Sillanpää, M. (2010). Water purification using magnetic assistance: A review. *Journal of Hazardous Materials*, *180*(1), 38–49. doi:10.1016/j.jhazmat.2010.04.105 PMID:20488616

Amin, N. K. (2009). Removal of direct blue-106 dye from aqueous solution using new activated carbons developed from pomegranate peel: Adsorption equilibrium and kinetics. *Journal of Hazardous Materials*, *165*(1), 52–62. doi:10.1016/j.jhazmat.2008.09.067 PMID:18986765

Anjum, M., Miandad, R., Waqas, M., Gehany, F., & Barakat, M. A. (2016). Remediation of wastewater using various nano-materials. *Arabian Journal of Chemistry*. doi:10.1016/j.arabjc.2016.10.004

Araki, S., Sato, H., Yokoyama, K., & Murata, K. (2000). Subclinical neurophysiological effects of lead: A review on peripheral, central, and autonomic nervous system effects in lead workers. *American Journal of Industrial Medicine*, *37*(2), 193–204. doi:10.1002/(SICI)1097-0274(200002)37:2<193::AID-AJIM5>3.0.CO;2-J PMID:10615100

Azizullah, A., Khattak, M. N. K., Richter, P., & Häder, D.-P. (2011). Water pollution in Pakistan and its impact on public health—a review. *Environment International*, *37*(2), 479–497. doi:10.1016/j.envint.2010.10.007 PMID:21087795

Bhaumik, M., Maity, A., Srinivasu, V. V., & Onyango, M. S. (2011). Enhanced removal of Cr(VI) from aqueous solution using polypyrrole/Fe3O4 magnetic nanocomposite. *Journal of Hazardous Materials*, *190*(1), 381–390. doi:10.1016/j.jhazmat.2011.03.062 PMID:21497438

Bhutta, Z. A., Sommerfeld, J., Lassi, Z. S., Salam, R. A., & Das, J. K. (2014). Global burden, distribution, and interventions for infectious diseases of poverty. *Infectious Diseases of Poverty*, *3*(1), 21. doi:10.1186/2049-9957-3-21 PMID:25110585

Boxall, A. B., Tiede, K., & Chaudhry, Q. (2007). *Engineered nanomaterials in soils and water: how do they behave and could they pose a risk to human health?* Academic Press.

Brabec, E., Schulte, S., & Richards, P. L. (2002). Impervious surfaces and water quality: A review of current literature and its implications for watershed planning. *CPL Bibliography*, *16*(4), 499–514.

Brunner, T. J., Wick, P., Manser, P., Spohn, P., Grass, R. N., Limbach, L. K., ... Stark, W. J. (2006). In vitro cytotoxicity of oxide nanoparticles: Comparison to asbestos, silica, and the effect of particle solubility. *Environmental Science & Technology*, *40*(14), 4374–4381. doi:10.1021/es052069i PMID:16903273

Burkhard, R., Deletic, A., & Craig, A. (2000). Techniques for water and wastewater management: A review of techniques and their integration in planning. *Urban Water*, *2*(3), 197–221. doi:10.1016/S1462-0758(00)00056-X

Carmen, Z., & Daniela, S. (2012). *Textile organic dyes–characteristics, polluting effects and separation/elimination procedures from industrial effluents–a critical overview.* Paper presented at the Organic Pollutants Ten Years After the Stockholm Convention-Environmental and Analytical Update. 10.5772/32373

Chang, Y.-C., & Chen, D.-H. (2005). Preparation and adsorption properties of monodisperse chitosan-bound Fe 3 O 4 magnetic nanoparticles for removal of Cu (II) ions. *Journal of Colloid and Interface Science*, *283*(2), 446–451. doi:10.1016/j.jcis.2004.09.010 PMID:15721917

Chang, Y. C., & Chen, D. H. (2005). Adsorption Kinetics and Thermodynamics of Acid Dyes on a Carboxymethylated Chitosan-Conjugated Magnetic Nano-Adsorbent. *Macromolecular Bioscience*, *5*(3), 254–261. doi:10.1002/mabi.200400153 PMID:15768445

Chella, S., Kollu, P., Komarala, E. V. P., Doshi, S., Saranya, M., Felix, S., ... Venugopal, V. (2015). Solvothermal synthesis of MnFe 2 O 4-graphene composite—Investigation of its adsorption and antimicrobial properties. *Applied Surface Science*, *327*, 27–36. doi:10.1016/j.apsusc.2014.11.096

Chen, K., He, J., Li, Y., Cai, X., Zhang, K., Liu, T., ... Liu, J. (2017). Removal of cadmium and lead ions from water by sulfonated magnetic nanoparticle adsorbents. *Journal of Colloid and Interface Science*, *494*(Supplement C), 307–316. doi:10.1016/j.jcis.2017.01.082 PMID:28161502

Cheng, Z., Tan, A. L. K., Tao, Y., Shan, D., Ting, K. E., & Yin, X. J. (2012). Synthesis and characterization of iron oxide nanoparticles and applications in the removal of heavy metals from industrial wastewater. *International Journal of Photoenergy*.

Chequer, F. M. D., Oliveira, G. A. R. d., Ferraz, E. R. A. c., Cardoso, J. C., Zanoni, M. V. B., & Oliveira, D. P. d. (2013). Textile Dyes: Dyeing Process and Environmental Impact. In M. Günay (Ed.), Eco-Friendly Textile Dyeing and Finishing. Rijeka: InTech.

Dalali, N., Khoramnezhad, M., Habibizadeh, M., & Faraji, M. (2011). Magnetic removal of acidic dyes from waste waters using surfactant-coated magnetite nanoparticles: optimization of process by Taguchi method. *Proceedings of International Conference on Environmental and Agriculture Engineering IPCBEE*.

Dave, P. N., & Chopda, L. V. (2014). Application of iron oxide nanomaterials for the removal of heavy metals. *Journal of Nanotechnology*.

Ditsch, A., Lindenmann, S., Laibinis, P. E., Wang, D. I., & Hatton, T. A. (2005). High-gradient magnetic separation of magnetic nanoclusters. *Industrial & Engineering Chemistry Research*, *44*(17), 6824–6836. doi:10.1021/ie048841s

Ebrahimian Pirbazari, A., Saberikhah, E., & Habibzadeh Kozani, S. S. (2014). Fe3O4–wheat straw: Preparation, characterization and its application for methylene blue adsorption. *Water Resources and Industry*, *7-8*(Supplement C), 23–37. doi:10.1016/j.wri.2014.09.001

Ehrampoush, M. H., Miria, M., Salmani, M. H., & Mahvi, A. H. (2015). Cadmium removal from aqueous solution by green synthesis iron oxide nanoparticles with tangerine peel extract. *Journal of Environmental Health Science & Engineering*, *13*(1), 84. doi:10.118640201-015-0237-4 PMID:26682059

El-Latif, M. M. A., Ibrahim, A. M., Showman, M. S., & Hamide, R. R. A. (2013). Alumina/Iron Oxide nano composite for cadmium ions removal from aqueous solutions. *International Journal of Nonferrous Metallurgy*, *2*(02), 47–62. doi:10.4236/ijnm.2013.22007

Feng, L., Cao, M., Ma, X., Zhu, Y., & Hu, C. (2012). Superparamagnetic high-surface-area Fe3O4 nanoparticles as adsorbents for arsenic removal. *Journal of Hazardous Materials*, *217-218*(Supplement C), 439–446. doi:10.1016/j.jhazmat.2012.03.073 PMID:22494901

Fu, F., & Wang, Q. (2011). Removal of heavy metal ions from wastewaters: A review. *Journal of Environmental Management*, *92*(3), 407–418. doi:10.1016/j.jenvman.2010.11.011 PMID:21138785

Gong, J.-L., Wang, B., Zeng, G.-M., Yang, C.-P., Niu, C.-G., Niu, Q.-Y., ... Liang, Y. (2009). Removal of cationic dyes from aqueous solution using magnetic multi-wall carbon nanotube nanocomposite as adsorbent. *Journal of Hazardous Materials*, *164*(2), 1517–1522. doi:10.1016/j.jhazmat.2008.09.072 PMID:18977077

Gong, J., Chen, L., Zeng, G., Long, F., Deng, J., Niu, Q., & He, X. (2012). Shellac-coated iron oxide nanoparticles for removal of cadmium(II) ions from aqueous solution. *Journal of Environmental Sciences (China)*, *24*(7), 1165–1173. doi:10.1016/S1001-0742(11)60934-0 PMID:23513435

Gregorio-Jauregui, K. M., Pineda, M. G., Rivera-Salinas, J. E., Hurtado, G., Saade, H., Martinez, J. L., ... López, R. G. (2012). One-step method for preparation of magnetic nanoparticles coated with chitosan. *Journal of Nanomaterials*, *2012*, 4. doi:10.1155/2012/813958

Gupta, V. K., Kumar, R., Nayak, A., Saleh, T. A., & Barakat, M. (2013). Adsorptive removal of dyes from aqueous solution onto carbon nanotubes: A review. *Advances in Colloid and Interface Science*, *193*, 24–34. doi:10.1016/j.cis.2013.03.003 PMID:23579224

Haham, H., Grinblat, J., Sougrati, M.-T., Stievano, L., & Margel, S. (2015). Engineering of iron-based magnetic activated carbon fabrics for environmental remediation. *Materials (Basel)*, *8*(7), 4593–4607. doi:10.3390/ma8074593 PMID:28793459

Hao, Y.-M., Man, C., & Hu, Z.-B. (2010). Effective removal of Cu (II) ions from aqueous solution by amino-functionalized magnetic nanoparticles. *Journal of Hazardous Materials*, *184*(1), 392–399. doi:10.1016/j.jhazmat.2010.08.048 PMID:20837378

Hu, J., Chen, G., & Lo, I. M. C. (2005). Removal and recovery of Cr(VI) from wastewater by maghemite nanoparticles. *Water Research*, *39*(18), 4528–4536. doi:10.1016/j.watres.2005.05.051 PMID:16146639

Hu, J., Lo, I., & Chen, G. (2004). Removal of Cr (VI) by magnetite. *Water Science and Technology*, *50*(12), 139–146. PMID:15686014

Hua, M., Zhang, S., Pan, B., Zhang, W., Lv, L., & Zhang, Q. (2012). Heavy metal removal from water/wastewater by nanosized metal oxides: A review. *Journal of Hazardous Materials*, *211*, 317–331. doi:10.1016/j.jhazmat.2011.10.016 PMID:22018872

Hui, L., Xiao, D., Hua, H., Rui, L., & Zuo, P. (2013). Adsorption behavior and adsorption mechanism of Cu (II) ions on amino-functionalized magnetic nanoparticles. *Transactions of Nonferrous Metals Society of China*, *23*(9), 2657–2665. doi:10.1016/S1003-6326(13)62782-X

Ismail, R. A., Sulaiman, G. M., Abdulrahman, S. A., & Marzoog, T. R. (2015). Antibacterial activity of magnetic iron oxide nanoparticles synthesized by laser ablation in liquid. *Materials Science and Engineering C*, *53*, 286–297. doi:10.1016/j.msec.2015.04.047 PMID:26042717

Itoh, H., & Sugimoto, T. (2003). Systematic control of size, shape, structure, and magnetic properties of uniform magnetite and maghemite particles. *Journal of Colloid and Interface Science*, *265*(2), 283–295. doi:10.1016/S0021-9797(03)00511-3 PMID:12962662

Iwamoto, T., & Ishigaki, T. (2013). *Fabrication of iron oxide nanoparticles using laser ablation in liquids*. Paper presented at the Journal of Physics: Conference Series. 10.1088/1742-6596/441/1/012034

Jin, Y., Liu, F., Tong, M., & Hou, Y. (2012). Removal of arsenate by cetyltrimethylammonium bromide modified magnetic nanoparticles. *Journal of Hazardous Materials*, *227-228*(Supplement C), 461–468. doi:10.1016/j.jhazmat.2012.05.004 PMID:22703733

Kashmiri, Z., & Mankar, S. (2014). Free radicals and oxidative stress in bacteria. *International Journal of Current Microbiology and Applied Sciences*, *3*, 34–40.

Kebede, A., Singh, A. K., Rai, P. K., Giri, N. K., Rai, A. K., Watal, G., & Gholap, A. V. (2013). Controlled synthesis, characterization, and application of iron oxide nanoparticles for oral delivery of insulin. *Lasers in Medical Science*, *28*(2), 579–587. doi:10.100710103-012-1106-3 PMID:22581389

Keyhanian, F., Shariati, S., Faraji, M., & Hesabi, M. (2016). Magnetite nanoparticles with surface modification for removal of methyl violet from aqueous solutions. *Arabian Journal of Chemistry*, *9*(Supplement 1), S348–S354. doi:10.1016/j.arabjc.2011.04.012

Khan, S., & Malik, A. (2014). Environmental and Health Effects of Textile Industry Wastewater. In A. Malik, E. Grohmann, & R. Akhtar (Eds.), *Environmental Deterioration and Human Health: Natural and anthropogenic determinants* (pp. 55–71). Dordrecht: Springer Netherlands. doi:10.1007/978-94-007-7890-0_4

Khandanlou, R., Ahmad, M. B., Masoumi, H. R. F., Shameli, K., Basri, M., & Kalantari, K. (2015). Rapid adsorption of copper (II) and lead (II) by rice straw/Fe3O4 nanocomposite: Optimization, equilibrium isotherms, and adsorption kinetics study. *PLoS One*, *10*(3), e0120264. doi:10.1371/journal.pone.0120264 PMID:25815470

Kobya, M., Demirbas, E., Senturk, E., & Ince, M. (2005). Adsorption of heavy metal ions from aqueous solutions by activated carbon prepared from apricot stone. *Bioresource Technology*, *96*(13), 1518–1521. doi:10.1016/j.biortech.2004.12.005 PMID:15939281

Kumar, M., & Puri, A. (2012). A review of permissible limits of drinking water. *Indian Journal of Occupational and Environmental Medicine*, *16*(1), 40. doi:10.4103/0019-5278.99696 PMID:23112507

Liao, M. H., Wu, K. Y., & Chen, D. H. (2005). Fast Adsorption of Crystal Violet on Polyacrylic Acid-Bound Magnetic Nanoparticles. *Separation Science and Technology*, *39*(7), 1563–1575. doi:10.1081/SS-120030802

Lin, S., Lu, D., & Liu, Z. (2012). Removal of arsenic contaminants with magnetic γ-Fe2O3 nanoparticles. *Chemical Engineering Journal*, *211-212*(Supplement C), 46–52. doi:10.1016/j.cej.2012.09.018

Liu, C.-H., Chuang, Y.-H., Chen, T.-Y., Tian, Y., Li, H., Wang, M.-K., & Zhang, W. (2015). Mechanism of arsenic adsorption on magnetite nanoparticles from water: Thermodynamic and spectroscopic studies. *Environmental Science & Technology*, *49*(13), 7726–7734. doi:10.1021/acs.est.5b00381 PMID:26055623

Liu, J., Zhao, Z., & Jiang, G. (2008). Coating Fe3O4 magnetic nanoparticles with humic acid for high efficient removal of heavy metals in water. *Environmental Science & Technology*, *42*(18), 6949–6954. doi:10.1021/es800924c PMID:18853814

Liu, X., Guan, Y., Ma, Z., & Liu, H. (2004). Surface modification and characterization of magnetic polymer nanospheres prepared by miniemulsion polymerization. *Langmuir*, *20*(23), 10278–10282. doi:10.1021/la0491908 PMID:15518525

Maaz, K., Mumtaz, A., Hasanain, S. K., & Ceylan, A. (2007). Synthesis and magnetic properties of cobalt ferrite (CoFe2O4) nanoparticles prepared by wet chemical route. *Journal of Magnetism and Magnetic Materials*, *308*(2), 289–295. doi:10.1016/j.jmmm.2006.06.003

Mahmoodi, N. M., Abdi, J., & Bastani, D. (2014). Direct dyes removal using modified magnetic ferrite nanoparticle. *Journal of Environmental Health Science & Engineering*, *12*(1), 96–96. doi:10.1186/2052-336X-12-96 PMID:24991427

Mak, S.-Y., & Chen, D.-H. (2004). Fast adsorption of methylene blue on polyacrylic acid-bound iron oxide magnetic nanoparticles. *Dyes and Pigments*, *61*(1), 93–98. doi:10.1016/j.dyepig.2003.10.008

Mayo, J., Yavuz, C., Yean, S., Cong, L., Shipley, H., Yu, W., ... Colvin, V. (2007). The effect of nanocrystalline magnetite size on arsenic removal. *Science and Technology of Advanced Materials*, *8*(1-2), 71–75. doi:10.1016/j.stam.2006.10.005

Muthukumaran, C., Sivakumar, V. M., & Thirumarimurugan, M. (2016). Adsorption isotherms and kinetic studies of crystal violet dye removal from aqueous solution using surfactant modified magnetic nanoadsorbent. *Journal of the Taiwan Institute of Chemical Engineers*, *63*, 354–362. doi:10.1016/j.jtice.2016.03.034

Nabeela, F., Azizullah, A., Bibi, R., Uzma, S., Murad, W., Shakir, S. K., ... Häder, D.-P. (2014). Microbial contamination of drinking water in Pakistan—a review. *Environmental Science and Pollution Research International*, *21*(24), 13929–13942. doi:10.100711356-014-3348-z PMID:25056753

Nassar, N. N. (2010). Rapid removal and recovery of Pb (II) from wastewater by magnetic nanoadsorbents. *Journal of Hazardous Materials*, *184*(1), 538–546. doi:10.1016/j.jhazmat.2010.08.069 PMID:20837379

Nethaji, S., Sivasamy, A., & Mandal, A. B. (2013). Preparation and characterization of corn cob activated carbon coated with nano-sized magnetite particles for the removal of Cr(VI). *Bioresource Technology*, *134*(Supplement C), 94–100. doi:10.1016/j.biortech.2013.02.012 PMID:23500565

Oliveira, M., & Machado, A. (2013). Preparation of polymer-based nanocomposites by different routes. *Nanocomposites: synthesis, characterization and applications*, 1-22.

Organization, W. H. (2012). *The Global Report for Research on Infectious Diseases of Poverty*. World Health Organization.

Ozay, O., Ekici, S., Baran, Y., Aktas, N., & Sahiner, N. (2009). Removal of toxic metal ions with magnetic hydrogels. *Water Research*, *43*(17), 4403–4411. doi:10.1016/j.watres.2009.06.058 PMID:19625066

Panáček, A., Kvítek, L., Prucek, R., Kolář, M., Večeřová, R., Pizúrová, N., ... Zbořil, R. (2006). Silver colloid nanoparticles: Synthesis, characterization, and their antibacterial activity. *The Journal of Physical Chemistry B*, *110*(33), 16248–16253. doi:10.1021/jp063826h PMID:16913750

Parida, P., Lolage, M., Angal, A., & Rautaray, D. (2017). Iron Oxide Nanoparticles to Remove Arsenic from Water. In S. Ranjan, N. Dasgupta, & E. Lichtfouse (Eds.), *Nanoscience in Food and Agriculture 4* (pp. 279–299). Cham: Springer International Publishing. doi:10.1007/978-3-319-53112-0_10

Pineda, M. G., Torres, S., López, L. V., Enríquez-Medrano, F. J., De León, R. D., Fernández, S., ... López, R. G. (2014). Chitosan-coated magnetic nanoparticles prepared in one-step by precipitation in a high-aqueous phase content reverse microemulsion. *Molecules (Basel, Switzerland)*, *19*(7), 9273–9287. doi:10.3390/molecules19079273 PMID:24991759

Qu, S., Huang, F., Yu, S., Chen, G., & Kong, J. (2008). Magnetic removal of dyes from aqueous solution using multi-walled carbon nanotubes filled with Fe 2 O 3 particles. *Journal of Hazardous Materials*, *160*(2), 643–647. doi:10.1016/j.jhazmat.2008.03.037 PMID:18430510

Qu, X., Brame, J., Li, Q., & Alvarez, P. J. (2013). Nanotechnology for a safe and sustainable water supply: Enabling integrated water treatment and reuse. *Accounts of Chemical Research*, *46*(3), 834–843. doi:10.1021/ar300029v PMID:22738389

Santhosh, C., Kollu, P., Doshi, S., Sharma, M., Bahadur, D., Vanchinathan, M. T., ... Grace, A. N. (2014). Adsorption, photodegradation and antibacterial study of graphene–Fe 3 O 4 nanocomposite for multipurpose water purification application. *RSC Advances*, *4*(54), 28300–28308. doi:10.1039/C4RA02913E

Santhosh, C., Velmurugan, V., Jacob, G., Jeong, S. K., Grace, A. N., & Bhatnagar, A. (2016). Role of nanomaterials in water treatment applications: A review. *Chemical Engineering Journal*, *306*(Supplement C), 1116–1137. doi:10.1016/j.cej.2016.08.053

Shah, J., Jan, M. R., Khan, M., & Amir, S. (2016). Removal and recovery of cadmium from aqueous solutions using magnetic nanoparticle-modified sawdust: Kinetics and adsorption isotherm studies. *Desalination and Water Treatment*, *57*(21), 9736–9744. doi:10.1080/19443994.2015.1030777

Shariati-Rad, M., Irandoust, M., Amri, S., Feyzi, M., & Ja'fari, F. (2014). Magnetic solid phase adsorption, preconcentration and determination of methyl orange in water samples using silica coated magnetic nanoparticles and central composite design. *International Nano Letters*, *4*(4), 91–101. doi:10.100740089-014-0124-5

Tan, K. A., Morad, N., Teng, T. T., Norli, I., & Panneerselvam, P. (2012). Removal of Cationic Dye by Magnetic Nanoparticle (Fe3O4) Impregnated onto Activated Maize Cob Powder and Kinetic Study of Dye Waste Adsorption. *APCBEE Procedia*, *1*(Supplement C), 83–89. doi:10.1016/j.apcbee.2012.03.015

Tseng, J.-Y., Chang, C.-Y., Chang, C.-F., Chen, Y.-H., Chang, C.-C., Ji, D.-R., ... Chiang, P.-C. (2009). Kinetics and equilibrium of desorption removal of copper from magnetic polymer adsorbent. *Journal of Hazardous Materials*, *171*(1), 370–377. doi:10.1016/j.jhazmat.2009.06.030 PMID:19595507

Wang, L., Li, J., Jiang, Q., & Zhao, L. (2012). Water-soluble Fe 3 O 4 nanoparticles with high solubility for removal of heavy-metal ions from waste water. *Dalton Transactions (Cambridge, England)*, *41*(15), 4544–4551. doi:10.1039/c2dt11827k PMID:22358186

Wang, X., Guo, Y., Yang, L., Han, M., Zhao, J., & Cheng, X. (2012). Nanomaterials as sorbents to remove heavy metal ions in wastewater treatment. *Journal of Environmental & Analytical Toxicology*, *2*(7), 1000154. doi:10.4172/2161-0525.1000154

Xu, P., Zeng, G. M., Huang, D. L., Feng, C. L., Hu, S., Zhao, M. H., ... Xie, G. X. (2012). Use of iron oxide nanomaterials in wastewater treatment: A review. *The Science of the Total Environment*, *424*, 1–10. doi:10.1016/j.scitotenv.2012.02.023 PMID:22391097

Xu, Y., Li, C., Zhu, X., Huang, W. E., & Zhang, D. (2014). Application of magnetic nanoparticles in drinking water purification. *Environmental Engineering and Management Journal*, *13*(8).

Yan, H., Yang, L., Yang, Z., Yang, H., Li, A., & Cheng, R. (2012). Preparation of chitosan/poly (acrylic acid) magnetic composite microspheres and applications in the removal of copper (II) ions from aqueous solutions. *Journal of Hazardous Materials*, *229*, 371–380. doi:10.1016/j.jhazmat.2012.06.014 PMID:22749139

Yuan, P., Fan, M., Yang, D., He, H., Liu, D., Yuan, A., ... Chen, T. (2009). Montmorillonite-supported magnetite nanoparticles for the removal of hexavalent chromium [Cr(VI)] from aqueous solutions. *Journal of Hazardous Materials*, *166*(2), 821–829. doi:10.1016/j.jhazmat.2008.11.083 PMID:19135796

Yuan, P., Liu, D., Fan, M., Yang, D., Zhu, R., Ge, F., ... He, H. (2010). Removal of hexavalent chromium [Cr(VI)] from aqueous solutions by the diatomite-supported/unsupported magnetite nanoparticles. *Journal of Hazardous Materials*, *173*(1), 614–621. doi:10.1016/j.jhazmat.2009.08.129 PMID:19748178

Zargar, B., Parham, H., & Hatamie, A. (2009). Fast removal and recovery of amaranth by modified iron oxide magnetic nanoparticles. *Chemosphere*, *76*(4), 554–557. doi:10.1016/j.chemosphere.2009.02.065 PMID:19345980

Zhang, D., Berry, J. P., Zhu, D., Wang, Y., Chen, Y., Jiang, B., ... Davison, P. A. (2015). Magnetic nanoparticle-mediated isolation of functional bacteria in a complex microbial community. *The ISME Journal*, *9*(3), 603–614. doi:10.1038/ismej.2014.161 PMID:25191996

Zhang, J., Zhai, S., Li, S., Xiao, Z., Song, Y., An, Q., & Tian, G. (2013). Pb (II) removal of Fe 3 O 4@ SiO 2–NH 2 core–shell nanomaterials prepared via a controllable sol–gel process. *Chemical Engineering Journal*, *215*, 461–471. doi:10.1016/j.cej.2012.11.043

Zhang, S., Niu, H., Cai, Y., Zhao, X., & Shi, Y. (2010). Arsenite and arsenate adsorption on coprecipitated bimetal oxide magnetic nanomaterials: MnFe2O4 and CoFe2O4. *Chemical Engineering Journal*, *158*(3), 599–607. doi:10.1016/j.cej.2010.02.013

Zhou, Y.-T., Nie, H.-L., Branford-White, C., He, Z.-Y., & Zhu, L.-M. (2009). Removal of Cu 2+ from aqueous solution by chitosan-coated magnetic nanoparticles modified with α-ketoglutaric acid. *Journal of Colloid and Interface Science*, *330*(1), 29–37. doi:10.1016/j.jcis.2008.10.026 PMID:18990406

Chapter 10
Wastewater Treatment in Removal of Heavy Metals:
Nanotechnology Applications in Environmental Engineering

Tehseen Yaseen
Pakistan Council of Scientific and Industrial Research, Pakistan

Anum Yaseen
University of Punjab, Pakistan

ABSTRACT

Nanotechnology is the area of nano science that shows great potential to establish a new process for wastewater treatment. It has been applied on a nanometer scale level. Currently, limited water resources and real treatment of wastewater is a chief requirement for the growing economy. It is in great demand to introduce the progressive wastewater treatment technologies. Therefore, the modern innovative processes in nanomaterial sciences have been appealing the target of scientists. The chapter addresses the developments in nanotechnology with respect to wastewater treatment, especially the removal of heavy metals and to the environmental applications. It will discuss the application of different classes of nanomaterials for wastewater treatment in removal of heavy metals and its possible effects to the environment. Therefore, the scope is to offer an overview of how nanomaterials are causing concerns related to heavy metal removal for water and in the surrounding environment.

INTRODUCTION

Advancement in science and technology are always a source of great attention for scientists and researchers to all over the world, specifically when the emphasis is on some particular area. In recent years, as a growing field, the Nanotechnology has been used in various applications. Similarly, new nano structures and materials are in great demand (Baruah & Dutta, 2009). Among the natural resources, water is the most abundant and important, but only about 1% of that resource is available for human consumption

DOI: 10.4018/978-1-5225-5745-6.ch010

(Grey et al., 2013; Adeleye et al., 2016). It is calculated that about 1.1 billion people faces deficiency of adequate drinking water (WHO, 2015), due to the increasing cost of potable water, rising populations, and diversity in climatic and environmental concerns (Adeleye et al., 2016). The obvious test in the water supply chain is regular contamination of freshwater resources by a number of inorganic and organic pollutants (Schwarzenbach et al., 2006). However, the previously used methods of treatment are less reliable to completely eliminate the evolving contaminants like heavy metals and fulfill the need of strict water quality standards (Qu et al., 2012). This can be obtained either by the establishment of new methods or by refining the existing methods through some intercessions. Among these emerging technologies, developments in nanotechnology has proved an incredible potential for the remediation of water contamination and various other environmental issues (Zare, Najafi, & Sadegh, et al., 2013; Gupta et al., 2015). Nanotechnology and nano-science have been shown the significant development in different fields and producing new and advance materials. Nanotechnology has been mentioned, in other different works, as one of the reliable method for water management (Stone et al., 2010; Zhang, & Fang, 2010). It could be categorized on the basis of nano-materials characteristics into different classes: nano-catalysts, nano-adsorbents and nano-membranes (Zhang et al., 2014; Mohmood et al., 2013) Furthermore, there is a significant need for advanced water technologies to ensure a high quality of water, eliminate chemical and biological pollutants, and intensify industrial production processes of wastewater. In this regard, nanotechnology is one of the ideal options for advance wastewater treatment processes. Furthermore, these nano-particles can be joined with biological processes (anaerobic digestion, algal membrane, microbial fuel cell) to enhance the water purification process. Each process has its own merits and specific pollutant removal efficiency (Theron, Zussman, & Yarin, 2004).

Heavy metals are considered to be the main contaminants in wastewater especially at high concentrations causing health problems in human beings and raising serious environmental issues (Chowdhury et al., 2016). These metal ions have toxicity potential, and might cause carcinogenicity in living organisms. The rapid development of industries leads to the production of huge amount of toxic heavy metals ion. They are directly or indirectly added into the water resources specifically in developing countries. As an effect of their non-biodegradability nature and the potential to aggregate in living organisms, treatment of industrial wastewaters is a particular concern for removing toxic heavy metals such as PbII, CdII, NiII, CoII, MnII.etc. At high and the low level concentrations (Fu & Wang, 2011). There are many different techniques used for eliminating heavy metal ions as revealed by literature. Such as activated carbon (Rao et al., 2002; Kadirvelu et al., 2001) was used for the elimination of contaminants in the water that has been replaced by less expensive adsorbing materials. There are several other adsorbents, such as fly ash (Rao et al., 2002), zeolite (Sheta et al., 2003), seaweeds (Vijayaraghavan et al., 2005), have been employed by different researchers and informed for the elimination of metal based pollutants in the wastewaters. Unfortunately, all these techniques have drawbacks including cost and sludge issues. Therefore, nanotechnology has been widely used to overcome the drawbacks of traditional methods and to meet the needs of current wastewater treatment. There are different research and development policies regarding nanotechnology perceptions are in development to increase the accuracy and efficiency of traditional methods (Contreras et al., 2017). It is clear with recent findings that nanotechnology can change the applications and properties of research and industrial materials, which are achieved because of the size of nano particles.

Currently, a great consideration has been dedicated to the application and creation of nano-materials to remove toxic and heavy metal ions from wastewater. The need to produce and control the nano-particles discovers use in enlightening the quality of water in the environment. For the importance of water quality and growing practices in the field of nanotechnology, endeavors have been create to discuss regarding removal of various heavy metals of wastewater treatment by employing nano-materials. Several studies have been published on the involvement of nano-materials for water treatment, current work specifically point out management of wastewater with emphasis on elimination of certain heavy metals from wastewater as presented in figure 1.

In this chapter we will discuss the previous researches in order to make the connection that will help to pay for effective use in the future. The chapter will conclude at certain points to figure out future direction.

Waste Water

The disposal of polluted wastewater has been dramatically increased during the past decades because of certain growing activities such as industrialization, agricultural and urbanization practices (Renuka, Sood, Ratha, Prasanna, & Ahluwalia, 2013). The main problem related to a lacking management system for wastewater treatment leads to environ-mental problems and the human illness. In addition, government and researchers have been implemented and focused on encouraging possible technologies for eliminating pollutants from wastewater. Commonly, the wastewater is categorized as industrial wastewater and domestic wastewater. The domestic waste- water contains liquid and solid liberations from a non-industrial practice that includes garbage, bacteria, toxic, and non-toxic organisms, viruses, sewage, detergents (Johnson, Manjreker, & Halligan, 1973). But the disposal of unprocessed wastewater from the industries is considered as the chief origin of water pollution. The waste initiating from the industrial side contains a varying amount of pollutants are disposed-off into the rivers or other water resources. At early disposal, the wastewater may include a high amount of inorganic and organic pollutants. Wastewater produced from industries is due to chemicals, treatment of paper and pulp, textile, the fabrication process, and from different sources such as the boiler, production line and cooling tower, etc. and on the basis of already mentioned reasons, the properties and the composition of pollutant in water discharge varies (Carstea, Bridgeman, Baker, & Reynolds, 2016). The pollutants from industries

Figure 1. Overview of chapter

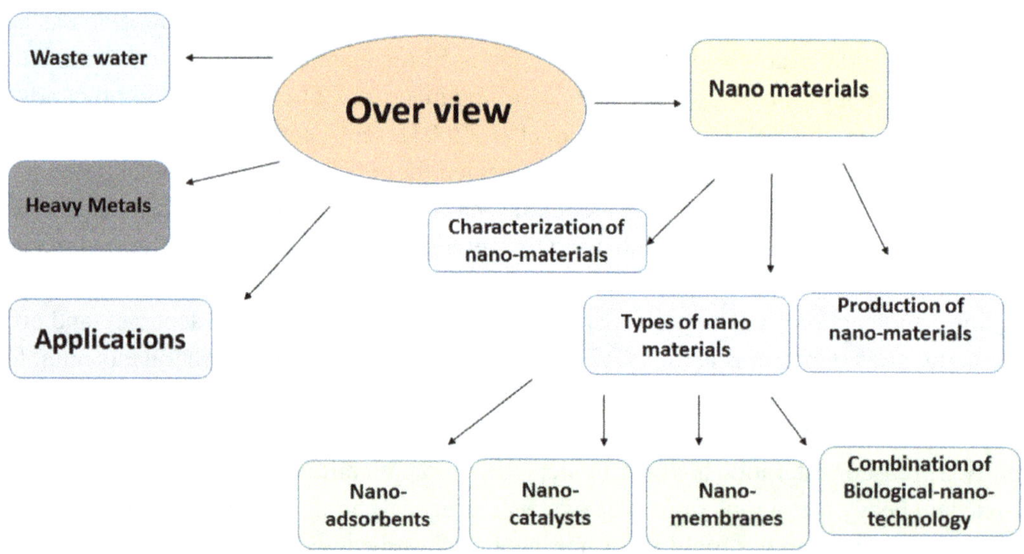

that come in the environment through wastewater are dyes, oils, pesticides, heavy metals and aromatic hydrocarbons (Ngah, Teong, & Hanafiah, 2011) and these pollutants are specified as main the hazard to the individual's health and environment (Xu et al., 2012).

Heavy Metals

Contaminants produced in the discharge is categorized as inorganic and organic contaminants that have a diverse range of toxicity in it. Physical, Chemical and Biological treatments are commonly used in the treatment of organic contaminants. These treatments are not useful for the inorganic contaminants such as heavy metals. Due to their properties like oxidation-reduction characteristics, complex formation and solubility, the heavy metal disintegration plays an important role (Lee & Pandey, 2012). The element having the atomic weight in ranging 63.5t o 200.6 with a specific gravity higher than 5.0 shows that the element is heavy metal (Srivastava & Majumder, 2008). They seem as a natural element in the environment. The word heavy metal denotes to the element that is toxic and having has a higher density even at low concentration. Currently, heavy metals are referred as the main issue for the treatment of wastewater in the environment, because the high danger is related to human health and ecosystems even at the low concentration. Due to its accumulation, non-biodegradable, flexibility, and endurance, the heavy metal contamination is a main environmental problem (Peligro, Pavlovic, Rojas, & Barriga, 2016). Industries like paper industries, pesticides, tanneries, metal plating industries, mining operations, etc., releases the heavy metal discharge into the environment which is found to be toxic and non- biodegradable or harmful to human health and other biological environments. There is a possibility to convert them into less harmful materials. Toxic metals can endure, either in or mixed or chemical form, they are difficult to eliminate from the wastewater Han, Fu, Cheng, Tang, & Wu, 2016). Heavy metals in exposed waters causes the oxygen insufficiency, algal blooms and to the end of aquatic life. When they released into the rivers, the heavy metals were transformed into hydrated ions which are in high toxic form than the metal atoms. These hydrated ions disrupt the enzymatic process as well as the absorption is faster in it. Therefore the elimination of heavy metals is necessary to lesser the public health risks. To control water pollution, Environmental Protection Agency (EPA) and World Health Organization (WHO) have set the most permissible discharge level of heavy metal into the environment. Permissible limits of heavy metals in drinking water is mentioned in table 1 (Kumar & Puri, 2012).

However, the discharged waste comprises a high level of heavy metals than the acceptable limits which lead to the problems for human health and environmental. The concentrations of heavy metals (mg L^{-1}) in the river Ganga water at different study sites is mentioned in table 2.

Types of Heavy Metals and Their Effects

The toxic heavy metals include such as nickel, zinc, arsenic, cobalt, chromium copper, antimony, and cadmium, etc. which produces the toxic and dangerous effects to the living environment. Ionic forms of metals like Hg^{2+}, Pb^{2+}, As^{3+} Cd^{2+}, Ag^{+} and reacts with different bio particles present in the body to form toxic substances which are essential to separate. The toxic properties are depending on the raised bio amplification and concentration to such an extent. Oxidation and ligand state performs a major part in bioavailability of heavy metals. When the amount of heavy metals is more than the allowed limit, then the heavy metal is act as toxic metal if it will disturb the metabolism of the cell (Singh & Pant, 2004). The toxicity of metal effects the health by lowering the functions of the cerebral and nervous

Table 1. Permissible limits of heavy metals in drinking water (Kumar & Puri, 2012)

Heavy Metals	Permissible Limits				
	WHO	USEPA	ISI	CPCB	ICMR
Iron (mg/l)	0.1	-	0.3	1.0	1.0
Cupper(mg/l)	1.0	1.3	0.05	1.5	1.5
Mercury (mg/l)	0.001	0.002	0.001	No relaxation	0.001
Cadmium (mg/l)	0.005	0.005	0.01	No relaxation	0.01
Arsenic (mg/l)	0.05	0.05	0.05	No relaxation	0.05
Lead (mg/l)	0.05	-	0.10	No relaxation	0.05
Zinc (mg/l)	5.0	-	5.0	15.0	0.10
Chromium (mg/l)	0.1	-	0.05	No relaxation	-

WHO: World Health Organization, USEPA: United States Environmental Protection Agency, ISI: Indian Standard Institution, ICMR: Indian Council of Medical Research, CPCB: Central Pollution Control Board.

Table 2. The concentrations of heavy metals (mg L^{-1}) in the river Ganga water at different study sites

Study Area	Cd	Cr	Cu	Fe	Hg	Mn	Ni	Pb	Zn	References
Allabad	ND-18	ND-30	-	-	-	-	-	18-86	26-122	Gupta et al., 2009
Berhampore	1-2	10-18	3-7	365-1744	-	181-712	41-84	8-21	65-95	Kar et al., 2008
Diamond Harbour	-	-	5-90	30-560	150	90-350	-	12-62	150-710	Sarkar et al., 2007
Haridwar		43-196	101-178			28.7-16		108-690	113-219	Rai et al., 2012
Kaushambi	-		ND=1000	ND-600	-	-	-	ND-9	ND-980	Chaudhary et al., 2017
Mirzapur	-	13.37	-	38.0	19.75-72.77	34.25	67.25	34.25	94.25	Sharma et al., 1992
Rishra-Konnagar	0.043	0.281	0.155	-	-	-	-	0.041-0.058	0.545-0.691	Bhattacharya et al., 2008

BDL- Below detection limit; ND- Not detected

system, harm the kidneys, blood content, lungs and other organs, draw out memory loss, increase the blood pressure, an increase of allergies, weakness, in the human body. Cell death also takes place due to the formation of free radicals and these radicals are responsible for the oxidative stress (Krystofova et al., 2009). Principal and mechanisms of oxidative stresses by heavy metals in terms of damage to plant cells is elaborated in figure 2 (Sharma & Dietz, 2009).

Due to these special effects, health authorities are growing worldwide and different regulatory bodies have implemented allowed limits to the heavy metal release effluent. Researchers also focusing on establishing the treatment methods due to the presence of dangerous effects in wastewaters (Dhir, Sharmila, & Saradhi, 2009)

Figure 2. Principal and mechanism of oxidative stress by heavy metals in term of damage to plant cell (Sharma & Dietz, 2009)

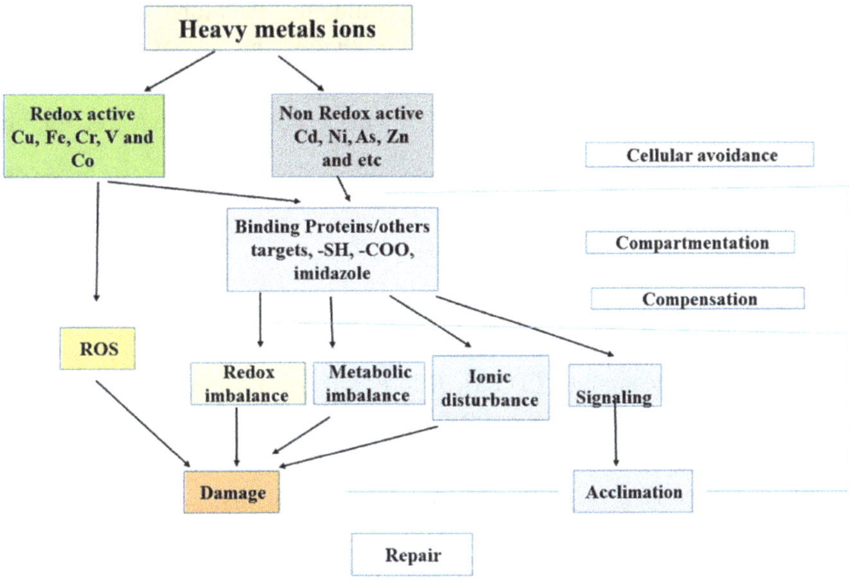

NANO MATERIALS

Nano-materials are the tiniest structures that have been developed and widely used, having the size of a few nanometers (Chaturvedi, Dave, & Shah, 2012. More accurately, the nano-particles are one dimensioned structure components at least less than 100 nm (Amin, Alazba, & Manzoor, 2014). Nano-materials have been settled in different forms such as nanotubes, nanowires, films, particles, colloids, quantum and dots (Lubick & Betts, 2008). In case of water treatment application, various efficient, cost-effective and eco-friendly nano-materials have been produced possessing specific characteristics for possible removal of contaminants from industrial effluents, to the drinking and ground water (Gupta et al., 2015).

Preparation of Nano-Particle

Preparation of nano-particles is the foremost vital phase in the method of wastewater treatment by using engineered nano-particle. In the synthesis of nano-particles, there are two two major steps used such as, the bottom-up technique (which includes the organization of each smaller molecules, molecule-by-molecule or atom-by-atom for extra complicated associations by use of specific chemical or physical forces through chemical synthesis, self-assembling, and positional assembling), while the top-down technique (which includes the production of nano-particles of large size, such as micro-scale or granular sizes, materials through chemical or mechanical steps involving milling, etching etc.). There are many other different techniques employed for the production of nano-particles based on the required properties, morphology, applications, that involves precipitation (Phiwdang, Suphankij, Mekprasart, & Pecharapa, 2013), catalytic growth (Li et al., 2003), co-precipitation (Petcharoen & Sirivat, 2012), mechanical alloying/milling (Arbain, Othman, & Palaniandy, 2011), impregnation (Kukovecz et al., 2002), mechanochemical synthesis (Tsuzuki & McCormick, 2004). The molecular structure and configuration of nano particles can be changed by using these methods. Among different techniques established to produce nano-particles so far, sol-gel technique is mostly commonly and widely employed because it offers better control over morphology, density, surface area, and molecular structure, of the particles synthesized for water treatment procedures. Moreover, on the basis of shapes and size of nano-materials, it could be categorized into nano-tubes, nano-fibres, nano-polymers, nano- wires, nano-belts, quantum dots, nano-capsules, nano-springs, nano-particles. Each of the nano-materials has its own specific properties which are widely explored (Dubey et al., 2017).

Characterization of Nano-Particles

There are different techniques involved in the characterization of engineered nanoparticles like, scanning electron microscopy (SEM), Fourier transform IR, differential scanning calorimetry (DSC), (FTIR) spectroscopy, thermo-gravimetric analysis (TGA), differential thermo-gravimetric analysis (DTA), transmission electron microscopy (TEM), triple-point-nitrogen-Brunuaer-Emmett-Teller (BET) adsorption, X-ray diffraction (XRD). Generally, stated procedures are employed for the production of various kind of nano-particles by little alteration for different applications like environmental sciences, electronics, optical, materials, biomedical, biotechnology, etc. The assortment of the technique relays on the properties, size, starting materials, and the specified uses of the nano-particles (Dubey et al., 2017)

Methodologies Adopted for Waste Water Treatment

Nano-Adsorbents

Now a days, nano-particle have been considered as their ability for adsorbents. The small size of the nano -particles enhance the area of surface (Gubin, Koksharov, Khomutov, & Yurkov, 2005) that increases the chemical reactivity and capacity to adsorb the nano-particles of metals through the surface (Khallaf, Oladeji, & Chow, 2008). This mechanisim relay on adsorption coefficient K d and recitation partitioning of contaminant such as organic contaminants or heavy metals at equilibrium conditions (Hu, Wang, & Pan, 2010). Furthermore, for the insistent inorganic contaminants redox reaction was chosen to begin the ionic configuration alteration (Gupta et al., 2015). Therefore, many scientists intensely agree that variations in the redox reaction effect the toxicity of contaminants (Chen & Mao, 2007). The commonly employed nano-particles for the removal or adsorption of these heavy metals are nominated as ferric oxides, grapheme, zinc oxide, magnesium oxide, manganese oxide, titanium oxide, activated carbon, and carbon nanotubes (Gupta et al., 2015). Nano-adsorbents retain two major characteristics, external functionalization, and innate surface. Their chemical, physical, and material characteristic are associated to their extrinsic surface structure, obvious intrinsic composition and size (Mirkin, Letsinger, Mucic, & Storhoff, 1996). In the aquatic system, the factors infusing the adsorption mechanism are increased surface area, chemical reaction, adsorption specificity, distribution of atoms to the surface, deficient of diffusion resistance and extraordinary surface covering potential (Khajeh, Laurent, & Dastafkan, 2013). Nano-particles employed as adsorbent for the elimination of heavy metals must be nontoxic and should have the tendency to adsorb pollutants of even at low concentration (ppb), high adsorption capacity, adsorbed contaminants which can be simply eliminated on the adsorbent surface and could be reprocessed for several times (Dubey et al., 2017).

The methodology beside the elimination of a metallic particle by the adsorption process includes three common steps that is the diffusion of film which involves in carrying of adsorbed particles through a surface film in the outside of the adsorbent, in the pore diffusion which considers the adsorbate diffusion through the pores of the adsorbent, while intra-particle diffusion involves the solute adsorption at the inside of the surfaces of the adsorbent. Graphic description of different methods of adsorption of metal ions by nanoparticles is elaborated in figure 3 (Dubey et al., 2017).

In order to capture the original mechanism involved in the adsorption of toxic metallic species by nanoparticles, the effect of the adsorption mechanisms by employing different analytical techniques, such as X-Ray Diffraction (XRD), Extended X-ray Absorption Fine Structure (EXAFS), infrared (IR) spectroscopy, X-ray Photoelectron Spectroscopy (XPS) and spectroscopy have been widely inspected. The drive of this functional and structural investigations is to offer understanding of the possible mechanism related to the elimination which may comprise either one or the mixture of two or more methods such as physical adsorption (Nassar, 2010), surface-complexation (Xu et al., 2013), and electrostatic interaction (Jordan et al., 2013).

In wastewater treatment the heavy metals adsorption by employing the nano-adsorbents are influenced by different parameters such as pH, temperature, incubation/contact time, and adsorbent dose. Srivastava et al. (2015) stated that pH shows a critical role on heavy metals adsorption from wastewater. Highest adsorption of Zn (II) on magnetic nano-adsorbent was reported at pH 5.5; although, it declines with more rise in pH. Furthermore, Lingamdinne et al. (2016) observed the highest adsorption such as 99.6% and 93% for Cr(III) and Pb(II) at the pH of 4.0 and 6.0 respectively. Furthermore, it was observed that the adsorption of heavy metals is increased by increasing the contact time from solution for the reason that it offers long time for adsorption process. In start, the adsorption of metals is very high because levels

Figure 3. Graphic description of different methods of adsorption of metal ions by nano particles (Dubey et al., 2017)

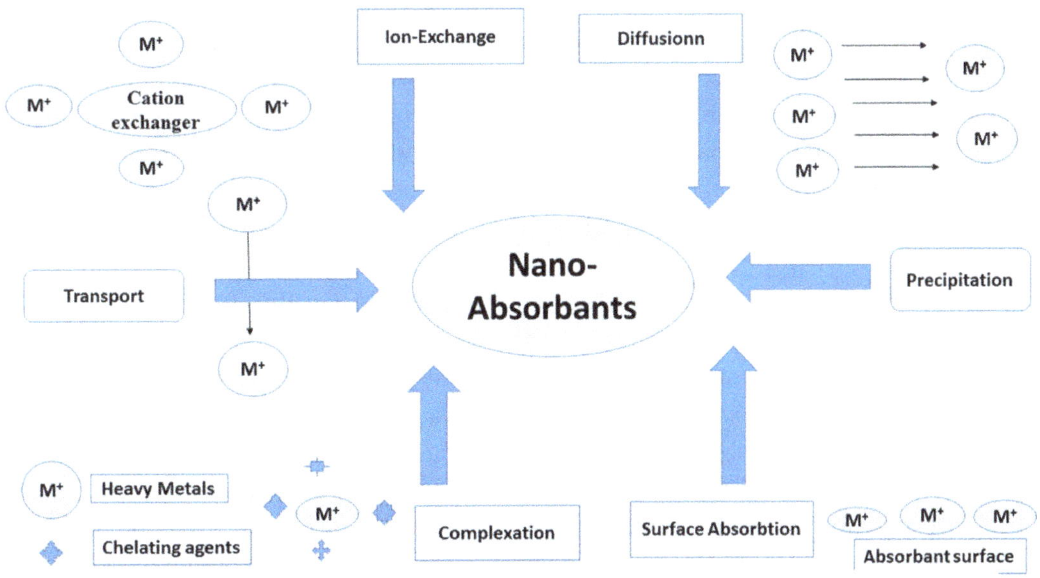

of metals are high, further it drops as time increases because of the coverage of active sites (Shirsath & Shirivastava, 2015). Nano-adsorbent features also influence the heavy metals adsorption that involved surface charge, surface area, and accumulation of new functional groups. These amendment strategies upgrade the adsorption tendency of nano-adsorbents (Wang et al., 2015; Tarigh & Shemirani, 2013).

Nano-Catalysts

The nano-catalysts, particularly from inorganic materials like metal oxides and semiconductors, are attaining substantial consideration of the researchers in application of wastewater management. There are many different kinds of nano-catalysts are used for wastewater treatment like electrocatalysts lphotocatalysts (Dutta, Maji, & Adhikary, 2014), and Fenton based catalysts (Kurian & Nair, 2015) for enhancing chemical oxidation of organic contaminants (Ma, Wang, & Na, 2015) and antimicrobial movements (Chaturvedi, Dave, & Shah, 2012).

Nano-Membranes

In currently used more modernized wastewater treatment processes, nano-materials fabricated by membrane filtration technology is the most operative approach (Ho, Chan, Blondy, Yeung, & Schrotter, 2012). Nanotechnology concerns go more than state-of-the-art enactment of water treatment mem- branes and permit new functionality, like catalytic reactivity, fouling resistance, and high permeability (Pendergast & Hoek, 2011). The basic aim for the application of this technique is their advantages in terms of quality of treated water, low space requisite for plant and effective dis- infection (Jang et al., 2015). In addition, it is very efficient, cost effective, and easy to design as matched to other treatment procedures (Guo, Zhang, Cai, & Zhao, 2016). Alongside the separation of heavy metals from wastewater, nano-materials reside in the novel membrane exhibit a vital role in the chemical decomposition of organic contaminants (Yang, Geng, & Chen, et al., 2015). The alignments of these kinds of membrane are one dimensional nano- materials (consisting of inorganic and organic materials) like nanotubes, nanofibers and nanoribbons (Liu, Li, Hao, & Yao, 2014). For discriminate filtration and nano-particle elimination, a membrane which is fabricated with carbonaceous nanofibers (CNFs) exhibited extraordinary selective removal/filtration ability under high pressure (Liang et al., 2010). Moreover, the catching ability of different nano-particles and other several small molecules can be effectively increased through interconnecting nano-particles and negative charged atoms on the titanate-nanoribbon membrane which is also known as macroscopic disk (Liu, Li, Hao, & Yao, 2014).

Combination of Biological-Nano Technology Processes

Nano scale engineering and science technologies proposed that the current issues related to water quality could be fixed by employing nano absorbent, nano-catalysts, nanotubes, micro- molecules nano powder and nanostructured catalytic membranes, (S. K. Gupta, Behari, & Kesari, 2006). All these nano- colloids and particles had a significant effect on the water treatment process (Diallo & Savage, 2005). Literature showed that incorporation of the biological wastewater treatment mechanism with modernized nanotechnology resulted in the proficient water purification scheme (Yin, Zhu, & Deng, 2013).

APPLICATION OF NANOPARTICLES TO THE REMOVAL OF HEAVY METAL POLLUTANTS FROM WASTEWATER

Being particular to inorganic contaminants, several heavy metals are very famous and specifically linked as water contaminants with the elevated level of carcinogenicity and danger. Different application of using nano-materials for heavy metal removal is presented in table 3.

Arsenic is considered as a well-known alarming metal due to its lethality and symptoms. Moreover, there are several other metals have been reported, such as Chromium (Cr), Selenium (Se), Mercury (Hg), Lead (Pb), Cadmium (Cd), which exhibit severe toxicities. A dose more the permissible limit, several metals lead to injurious effects on health like, Zinc is needed for the normal growth of a person, but elevated level higher than the reasonable limit of imprisonment, lead to mental fever and poor development, Cobalt creates queasiness, carcinoma asthma, and regurgitating. Furthermore, cobalt is liable for gastrointestinal, liver and thyroid complications. Due to the adulteration, the ground- waters and surface water at specific areas of different countries are infected and not consider to suitable for drinking purposes. Therefore, it is crucially needed to remove the elevated level of heavy metal from water source to make it safe for drinking purpose. There are several studies have been reported such as elimination of arsenic (As V) at 16 nano-particles of metal oxides has been described by Hristovski,

Table. 3. Applications of nano-materials for the removal of heavy metals in wastewater

Nano-Materials	Heavy Metals	References
, Fe_2O_3, NiO, ZrO_2, and TiO_2,	Arsenic	(Hristovski, Baumgardner, & Westerhoff, 2007)
Nano zero-valent iron		(Zhu, Jia, Wu, & Wang, 2009)
Activated carbon with nano zero valent iron	Arsenate	(Zhu, Jia, Yao, Wu, & Wang, 2009)
Magnetite Maghemite	arsenic (V), arsenic (III),	(Chowdhury & Yanful, 2010)
Magnetic hydroxyapatite (MNHAP)	Zn^{2+}, Cd^{2+}	(Feng et al., 2010)
Magnetic adsorbing material (MNP-OPP) with Fe_3O_4	Cd^{2+}	(Gupta & Nayak, 2012)
Organo-montmorillonite with iron	Cr (VI)	(Venkatesham, Madhu, Satyanarayana, & Preetham, 2013)
Chitosan-coated magnetic (CCMNPs) with biodegradable,α-ketoglutaric acid α-KA-CCMNPs	Cu^{2+}	(Zhou, Nie, Branford-White, He, & Zhu, 2009)
Nano-sized ZnO with via gel combustion	Pb (II)	(Venkatesham, Madhu, Satyanarayana, & Preetham, 2013)
Silica with thiol coated onto magnetic nano-particles (TF-SCMNPs)	Hg (II)	(Hakami, Zhang, & Banks, 2012)
Cellulose modified nano-scale manganese oxide composite (C–NMOC)	Pb(II)	(Maliyekkal, Lisha, & Pradeep, 2010)
Nano-Maghemite particles	Se (VI)	(Jordan et al., 2013)
MCM-41 with nano-scale zero-valent iron (n-ZVI)	Cr (VI)	(Petala et al., 2013)
Silver nanoparticles impregnated with carbon nanotubes (Ag-MWCNTs)	Cd (II), Cu(II)	(Ramana, Yu, & Seshaiah, 2013)
MnO_2 coated multiwalled carbon nanotubes (MWCNTs/MnO_2)	Pb (II)	(Salam, 2013)

Baumgardner, & Westerhoff, (2007), out of which, Fe_2O_3, NiO, ZrO_2, and TiO_2, were considered as the most important. Each type of nano-particles were reported as highly significant with elimination potential of 98% excluding ZrO_2. Correspondingly, started carbon encumbered with nano zero-valent iron was employed for the arsenic adsorption reported by Zhu, Jia, Wu, & Wang, (2009). The highest adsorption potential was about 1.997 mg/g in 2.0 mg/L. The analysis of the influence of the other ions exposed that the occurrence of silicate and phosphate ions reduced the elimination of arsenic, besides the influence of other ions like carbonate, oxalate and sulfate was observed to be irrelevant. The researcher stated that this type of adsorbent is considered as most suitable and also suggested it for the management of arsenic polluted drinking water.

Furthermore, Zhu, Jia, Yao, Wu, and Wang, (2009) manufactured another nano material like activated carbon infused with nano zero valent iron, which further applied in the reduction of arsenate. In addition, the adsorption tendency was estimated as 15.4 mg/g at equilibrium level of 1.0 mg/L and at 6.5 ph. While the elimination of arsenate was reduced with the existence of silicate and phosphate, whereas the influence of different anions and cations on the elimination of arsenic were found negligible. Because of its reasonable eliminating capacity, this process was recommended by the scientists for treatment of drinking water. In the same way, Chowdhury & Yanful, (2010) reported removal of from aqueous solutions by using magnetite maghemite nano-particles, At pH 2.0, the highest sorption of arsenic was reported at Ph. 2, with the range of 3.71 mg/g and 3.69 for arsenic (V) and arsenic (III), respectively, at 1.5 mg/L of starting concentration. Moreover, researcher described and analyzed the drawbacks of arsenic intake in the occurrence of a phosphate anion. Therefore, a mutual relation was noticed with reduction in Arsenic removal on increasing level of phosphate. Results revealed that 60% intake of arsenic was acquired in the groundwater, comprising 1.13 mg/L of arsenic and N5.0 mg/L phosphate. In addition, Feng et al., (2010) produced magnetic hydroxyapatite nano-particles (MNHAP) to analyse its capability elimination of Zn^{2+} and Cd^{2+} ions through aqueous solutions. The effect of different parameters that influence this process have been evaluated such as adsorbent level, initial metal ion level, and pH and contact time. The highest sorption potential of Zn^{2+} and Cd^{2+} was estimated at 2.151 and 1.964 mmol/g, respectively with ideal adsorbent concentration of 0.1 g/L. on the other hand the desorption analysis, the most capable eluent employed for desorption of metal ions was observed to be EDTA together with 67% and 66.2% of Zn^{2+} and Cd^{2+} ions liberated, respectively. However, it was concluded that the most obvious benefit of the synthesized MNHAP nano materials at large specific surface area (142.5m 2 /g) and paramagnetic behavior was the precision of parting from aqueous solutions after the finishing of process. Moreover, V. Gupta & Nayak, (2012) synthesized a unique adsorbent by altering agricultural waste, orange- peel, into a magnetic adsorbing material (MNP-OPP) through precipitating it with Fe_3O_4. Adsorption of cadmium into magnetic orange-peel powder (OPP), Fe_3O_4 nano-particles (MNP) and MNP–OPP, was analyzed at various investigational situations, like the strength of ionic solution, pH, natural organic matter, reaction time between absorbent and adsorbate, temperature conditions and adsorbate concentration,. The findings revealed that a rapid kinetics and proved the prepared adsorbent MNP–OPP as more proficient as compare those of OPP and MNP. Furthermore, organo-montmorillonite with iron nano-particles were assimilated to create elimination of Cr VI ions as reported by Wu et al., (2012). These findings showed that a level at 0.47g/L, maximum elimination of 106 mg Cr/g Fe was attained. While the adsorption potentials of unsupported and supported iron nano-particles toward elimination of Cr VI ions were relatively analyzed, while the outcomes of comparison revealed that the removal of Cr VI was very high with organo-montmorillonite with iron nano-particles. In another study, Chitosan-coated magnetic nano-particles (CCMNPs) together with a eco-friendly and biodegradable biologic reagent, α-ketoglutaric acid(α-KA) has been prepared by Zhou, Nie, Branford-White, He, and Zhu, (2009) and employed for elimination of Cu^{2+} ions from water solution. The findings revealed that no major desorption hysteresis has been observed. Furthermore, the great recovery and stability capability of the prepared nano-particles such as (α-KA-CCMNPs) confirmed that the senovel magnetic nano-particles have high ability as for, eliminating of Cu^{2+} from wastewater. Similarly another, unique nano-material based on cellulose modified nano-scale manganese oxide composite (C–NMOC) was prepared by Maliyekkal, Lisha, & Pradeep, (2010) and analysed their capability for Pb II ions elimination from solutions. In addition, venkatesham, Madhu, Satyanarayana, & Preetham, (2013) reported the removal of Pb II ions on nano-sized ZnO nanoparticles produced by gel combustion process. In addition, the average of BET particular surface area of the prepared zinc oxide and particle size was measured as 80.425 m^2/g and 40 nm and respectively. While the effect of metalic ion concentration, on the adsorbent level of the adsorption method and pH was analyzed and it has been observed that adsorption of Pb was highest at the reduced level of Pb and acidic solution at pH 4.0, while it remains persistent in basic solution. Moreover, it also shows an increased with increment in the adsorbent level such as highest at 1.0 g. However, the values of correlation coefficient (R^2) were acquired Freundlich and Langmuir adsorption isotherm as, 0.9999 and 0.9987 respectively, have been observed approximately the same. However, both the isotherms precisely suited to Pb II adsorption on nano-ZnO particles.

Furthermore, application of silica with thiol coated onto magnetic nano-particles (TF-SCMNPs) has been reported by Hakami, Zhang, and Banks, (2012) employing condensation mechanism. The maximum adsorption for Hg II onto TF-SCMNPs of 50 nm has been observed as 207.7 mg/g when 4 mg/L of TF SCMNPs were applied. It was also noticed that the adsorption mechanism was not influenced by the existence of anions and light cautions during experimental conditions. Similarly, work on the removal of Se VI ions by using nano-Maghemite particles have been analyzed both at microscopic and macroscopic levels by Jordan et al. (2013). The highest level of elimination was obtained low pH such as in acidic condition and reduced with increase in pH and with ionic strength of the solution. The process involve in the elimination was elucidated on the basis of electro-phoretic mobility measurement which recommended the development of outer-sphere multifaceted layer in the removal of Se (VI) on to nano-maghemite.

Moreover, Recillas et al. (2011) discovered the capability of TiO_2 and CeO_2, Fe_3O_4 NMs for the elimination of Pb(II), recording adsorption abilities of 83 mg Pb g^{-1}, 189 mg Pb g^{-1}, and 159 mg Pb g^{-1}, respectively. Similarly, Pena et al. (2005) used nanocrystalline TiO_2 as adsorbent to eliminate arsenate As(V) and arsenite As(III) along with complete conversion of As(III) in to As(V), by photocatalytic oxidation. Recently, Hua et al. (2012) mentioned different studies to elaborate the capacity of nanosized metal oxides for heavy metals removal, as well as other approaches including the functionalization of NMs, such as supporting with zeolites or coating with poly(3,4-ethylenedioxythiophene), surface modification with amino groups to increase the efficiency of metal removal.

On the other hand, Petala et al., (2013) stated 100 percent Cr VI elimination onto MCM-41 with nano-scale zero-valent iron (n-ZVI) prepared through saturating the silica matrix (mesoporous) with ferric chloride, monitored at pH 3.0 by chemical reduction with $NaHB_4$ and 6 mg/L of Cr VI concentration. However, the findings revealed that a maximum pH-dependent decline in the potential of the material, while the kinetics was well described by a pseudo-first-order kinetic model.

Ramana, Yu, and Seshaiah, (2013) prepared silver nanoparticles impregnated with multi walled carbon nanotubes (Ag-MWCNTs) by using chemical reduction process for elimination of Cd II and Cu II ions from water sample. The results concluded that the highest adsorption of Cd (II) ions obtained was

98.36% and for Cu II was 99.54% at pH 6.0 and 7.0, respectively. Furthermore, it has been noticed that this adsorption mechanism was controlled by using chemical methods either by exchanging or sharing of electrons in between the adsorbent and the adsorbate in the solution.

In the similar way, 97.3% elimination of Pb II ions from solutions at pH 9.0 onto MnO_2 coated multiwalled carbon nanotubes ($MWCNTs/MnO_2$) was stated by Salam, (2013). Similarly, the equilibrium level achieved after 5 min of the reaction time and the adsorption mechanism was observed to be described well by pseudo-second-order with Elovich model with R^2 N 0.92 and R^2 N 0.99 at temperatures determined for the analysis.

CONCLUSION

In a current situation, there is a major requirement for more advanced water technologies to confirm a quality of water, remove biological and chemical contaminants, and strengthen industrial level production processes of wastewater. In this concern, nanotechnology is considered to be a one of the perfect options for more advance wastewater management processes. Therefore, the critical perception into several struggles have been taken by other researchers for environmental elimination, regarding the exclusion of many different metallic species from water illustrate that the involvement of nano-technology in wastewater and water treatment methods verified to be a blessing for environmental protection. Nano-particles, in relationship to other traditional cost effective materials provide many benefits, like fast removal, low regeneration, dose requirement, ease of separation, large surface area to volume ratio and reusability. Though, their requisite in less amount and their reuse and regeneration potential creates their practical application economical. They could also able to eliminate inorganic contaminants at much low level of metal solutions in different experimental situation of temperature and pH. As the foundation of several results of different nano materials involved processes deliberated herein, so, it can be concluded that the nano-particles could be a stable, potential and efficient option for elimination of heavy metals from waste-waters and water.

REFERENCES

Adeleye, A. S., Conway, J. R., Garner, K., Huang, Y., Su, Y., & Keller, A. A. (2016). Engineered nanomaterials for water treatment and remediation: Costs, benefits, and applicability. *Chemical Engineering Journal, 286*, 640–662. doi:10.1016/j.cej.2015.10.105

Amin, M., Alazba, A., & Manzoor, U. (2014). A review of removal of pollutants from water/wastewater using different types of nanomaterials. *Advances in Materials Science and Engineering*.

Arbain, R., Othman, M., & Palaniandy, S. (2011). Preparation of iron oxide nanoparticles by mechanical milling. *Minerals Engineering, 24*(1), 1–9. doi:10.1016/j.mineng.2010.08.025

Badruddoza, A. Z. M., Shawon, Z. B. Z., Tay, D. W. J., Hidajat, K., & Uddin, M. S. (2013). Endocrine disrupters and toxic metal ions removal by carboxymethyl-β-cyclodextrin polymer grafted onto magnetic nanoadsorbents. *Chemical Engineering Journal, 27*(1), 69–73.

Baruah, S., & Dutta, J. (2009). Nanotechnology applications in pollution sensing and degradation in agriculture: A review. *Environmental Chemistry Letters*, *7*(3), 191–204. doi:10.100710311-009-0228-8

Bhattacharya, A. K., Mandal, S. N., & Das, S. K. (2008). Heavy metals accumulation in water, sediment and tissues of different edible fishes in upper stretch of Gangetic West Bengal. *Trends in Applied Sciences Research*, *3*(1), 61–68. doi:10.3923/tasr.2008.61.68

Brumfiel, G. (2003). Nanotechnology: A little knowledge. *Nature, 424*(6946), 246-248.

Carstea, E. M., Bridgeman, J., Baker, A., & Reynolds, D. M. (2016). Fluorescence spectroscopy for wastewater monitoring: A review. *Water Research*, *95*, 205–219. doi:10.1016/j.watres.2016.03.021 PMID:26999254

Chaturvedi, S., Dave, P. N., & Shah, N. (2012). Applications of nano-catalyst in new era. *Journal of Saudi Chemical Society*, *16*(3), 307–325. doi:10.1016/j.jscs.2011.01.015

Chaudhary, M., Mishra, S., & Kumar, A. (2017). Estimation of water pollution and probability of health risk due to imbalanced nutrients in River Ganga, India. *International Journal of River Basin Management*, *15*(1), 53–60. doi:10.1080/15715124.2016.1205078

Chen, X., & Mao, S. S. (2007). Titanium dioxide nanomaterials: Synthesis, properties, modifications, and applications. *Chemical Reviews*, *107*(7), 2891–2959. doi:10.1021/cr0500535 PMID:17590053

Chowdhury, S. R., & Yanful, E. K. (2010). Arsenic and chromium removal by mixed magnetite–maghemite nanoparticles and the effect of phosphate on removal. *Journal of Environmental Management*, *91*(11), 2238–2247. doi:10.1016/j.jenvman.2010.06.003 PMID:20598797

Dhir, B., Sharmila, P., & Saradhi, P. P. (2009). Potential of aquatic macrophytes for removing contaminants from the environment. *Critical Reviews in Environmental Science and Technology*, *39*(9), 754–781. doi:10.1080/10643380801977776

Diallo, M. S., & Savage, N. (2005). Nanoparticles and water quality. *Journal of Nanoparticle Research*, *7*(4), 325–330. doi:10.100711051-005-8543-x

Dubey, S., Banerjee, S., Upadhyay, S. N., & Sharma, Y. C. (2017). Application of common nano-materials for removal of selected metallic species from water and wastewaters: A critical review. *Journal of Molecular Liquids*, *240*, 656–677. doi:10.1016/j.molliq.2017.05.107

Dutta, A. K., Maji, S. K., & Adhikary, B. (2014). γ-Fe 2 O 3 nanoparticles: An easily recoverable effective photo-catalyst for the degradation of rose bengal and methylene blue dyes in the waste-water treatment plant. *Materials Research Bulletin*, *49*, 28–34. doi:10.1016/j.materresbull.2013.08.024

Feng, Y., Gong, J.-L., Zeng, G.-M., Niu, Q.-Y., Zhang, H.-Y., Niu, C.-G., ... Yan, M. (2010). Adsorption of Cd (II) and Zn (II) from aqueous solutions using magnetic hydroxyapatite nanoparticles as adsorbents. *Chemical Engineering Journal*, *162*(2), 487–494. doi:10.1016/j.cej.2010.05.049

Grey, D., Garrick, D., Blackmore, D., Kelman, J., Muller, M., & Sadoff, C. (2013). Water security in one blue planet: twenty-first century policy challenges for science. *Philosophical Transactions of the Royal Society A, 371*(2002), 20120406.

Gubin, S. P., Koksharov, Y. A., Khomutov, G., & Yurkov, G. Y. e. (2005). Magnetic nanoparticles: Preparation, structure and properties. *Russian Chemical Reviews*, *74*(6), 489–520. doi:10.1070/RC2005v-074n06ABEH000897

Guo, J., Zhang, Q., Cai, Z., & Zhao, K. (2016). Preparation and dye filtration property of electrospun polyhydroxybutyrate–calcium alginate/carbon nanotubes composite nanofibrous filtration membrane. *Separation and Purification Technology*, *161*, 69–79. doi:10.1016/j.seppur.2016.01.036

Gupta, A., Rai, D. K., Pandey, R. S., & Sharma, B. (2009). Analysis of some heavy metals in the riverine water, sediments and fish from river Ganges at Allahabad. *Environmental Monitoring and Assessment*, *157*(1), 449–458. doi:10.100710661-008-0547-4 PMID:18850290

Gupta, S. K., Behari, J., & Kesari, K. K. (2006). Low Frequencies Ultrasonic Treatment of Sludge. *Asian Journal of Water, Environment and Pollution*, *3*(2), 101–105.

Gupta, V., & Nayak, A. (2012). Cadmium removal and recovery from aqueous solutions by novel adsorbents prepared from orange peel and Fe 2 O 3 nanoparticles. *Chemical Engineering Journal*, *180*, 81–90. doi:10.1016/j.cej.2011.11.006

Gupta, V., Tyagi, I., Sadegh, H., Shahryari-Ghoshekandi, R., Makhlouf, A., & Maazinejad, B. (2015). Nanoparticles as adsorbent; a positive approach for removal of noxious metal ions: A review. *Sci. Technol. Dev*, *34*(3), 195–214. doi:10.3923td.2015.195.214

Hakami, O., Zhang, Y., & Banks, C. J. (2012). Thiol-functionalised mesoporous silica-coated magnetite nanoparticles for high efficiency removal and recovery of Hg from water. *Water Research*, *46*(12), 3913–3922. doi:10.1016/j.watres.2012.04.032 PMID:22608609

Han, W., Fu, F., Cheng, Z., Tang, B., & Wu, S. (2016). Studies on the optimum conditions using acid-washed zero-valent iron/aluminum mixtures in permeable reactive barriers for the removal of different heavy metal ions from wastewater. *Journal of Hazardous Materials*, *302*, 437–446. doi:10.1016/j.jhazmat.2015.09.041 PMID:26521089

Ho, H. L., Chan, W. K., Blondy, A., Yeung, K. L., & Schrotter, J.-C. (2012). Experiment and modeling of advanced ozone membrane reactor for treatment of organic endocrine disrupting pollutants in water. *Catalysis Today*, *193*(1), 120–127. doi:10.1016/j.cattod.2012.03.059

Hristovski, K., Baumgardner, A., & Westerhoff, P. (2007). Selecting metal oxide nanomaterials for arsenic removal in fixed bed columns: From nanopowders to aggregated nanoparticle media. *Journal of Hazardous Materials*, *147*(1), 265–274. doi:10.1016/j.jhazmat.2007.01.017 PMID:17254707

Hu, H., Wang, Z., & Pan, L. (2010). Synthesis of monodisperse Fe 3 O 4@ silica core–shell microspheres and their application for removal of heavy metal ions from water. *Journal of Alloys and Compounds*, *492*(1), 656–661. doi:10.1016/j.jallcom.2009.11.204

Hua, M., Zhang, S., Pan, B., Zhang, W., Lv, L., & Zhang, Q. (2012). Heavy metal removal from water/wastewater by nanosized metal oxides: A review. *Journal of Hazardous Materials*, *211*, 317–331. doi:10.1016/j.jhazmat.2011.10.016 PMID:22018872

Jang, J. H., Lee, J., Jung, S.-Y., Choi, D.-C., Won, Y.-J., Ahn, K. H., ... Lee, C.-H. (2015). Correlation between particle deposition and the size ratio of particles to patterns in nano-and micro-patterned membrane filtration systems. *Separation and Purification Technology*, *156*, 608–616. doi:10.1016/j.seppur.2015.10.056

Johnson, R. F., Manjreker, T. G., & Halligan, J. E. (1973). Removal of oil from water surfaces by sorption on unstructured fibers. *Environmental Science & Technology*, *7*(5), 439–443. doi:10.1021/es60077a003 PMID:22283537

Jordan, N., Ritter, A., Foerstendorf, H., Scheinost, A., Weiß, S., Heim, K., ... Reuther, H. (2013). Adsorption mechanism of selenium (VI) onto maghemite. *Geochimica et Cosmochimica Acta*, *103*, 63–75. doi:10.1016/j.gca.2012.09.048

Kadirvelu, K., Thamaraiselvi, K., & Namasivayam, C. (2001). Adsorption of nickel (II) from aqueous solution onto activated carbon prepared from coirpith. *Separation and Purification Technology*, *24*(3), 497–505. doi:10.1016/S1383-5866(01)00149-6

Kar, D., Sur, P., Mandai, S. K., Saha, T., & Kole, R. K. (2008). Assessment of heavy metal pollution in surface water. *International Journal of Environmental Science and Technology*, *5*(1), 119–124. doi:10.1007/BF03326004

Khajeh, M., Laurent, S., & Dastafkan, K. (2013). Nanoadsorbents: Classification, preparation, and applications (with emphasis on aqueous media). *Chemical Reviews*, *113*(10), 7728–7768. doi:10.1021/cr400086v PMID:23869773

Khallaf, H., Oladeji, I. O., & Chow, L. (2008). Optimization of chemical bath deposited CdS thin films using nitrilotriacetic acid as a complexing agent. *Thin Solid Films*, *516*(18), 5967–5973. doi:10.1016/j.tsf.2007.10.079

Krystofova, O., Shestivska, V., Galiova, M. V., Novotny, K., Kaiser, J., Zehnalek, J., ... Kizek, R. (2009). Sunflower Plants as Bioindicators of Environmental Pollution with Lead (II) Ions. *Sensors (Basel)*, *9*(7), 5040–5505. doi:10.339090705040 PMID:22346686

Kukovecz, A., Balogi, Z., Konya, Z., Toba, M., Lentz, P., Niwa, S.-I., ... Kiricsi, I. (2002). Synthesis, characterisation and catalytic applications of sol–gel derived silica–phosphotungstic acid composites. *Applied Catalysis A, General*, *228*(1), 83–94. doi:10.1016/S0926-860X(01)00969-3

Kumar, M., & Puri, A. (2012). A review of permissible limits of drinking water. *Indian Journal of Occupational and Environmental Medicine*, *16*(1). PMID:23112507

Kurian, M., & Nair, D. S. (2015). Heterogeneous Fenton behavior of nano nickel zinc ferrite catalysts in the degradation of 4-chlorophenol from water under neutral conditions. *Journal of Water Process Engineering*, *8*, 37–49. doi:10.1016/j.jwpe.2014.10.011

Lee, J., & Pandey, B. D. (2012). Bio-processing of solid wastes and secondary resources for metal extraction–a review. *Waste Management (New York, N.Y.)*, *32*(1), 3–18. doi:10.1016/j.wasman.2011.08.010 PMID:21925857

Li, Y.-H., Ding, J., Luan, Z., Di, Z., Zhu, Y., Xu, C., ... Wei, B. (2003). Competitive adsorption of Pb 2+, Cu 2+ and Cd 2+ ions from aqueous solutions by multiwalled carbon nanotubes. *Carbon*, *41*(14), 2787–2792. doi:10.1016/S0008-6223(03)00392-0

Liang, H. W., Wang, L., Chen, P. Y., Lin, H. T., Chen, L. F., He, D., & Yu, S. H. (2010). Carbonaceous nanofiber membranes for selective filtration and separation of nanoparticles. *Advanced Materials*, *22*(42), 4691–4695. doi:10.1002/adma.201001863 PMID:20859940

Lingamdinne, L. P., Koduru, J. R., Choi, Y. L., Chang, Y. Y., & Yang, J. K. (2016). Studies on removal of Pb (II) and Cr (III) using graphene oxide based inverse spinel nickel ferrite nano-composite as sorbent. *Hydrometallurgy*, *165*, 64–72. doi:10.1016/j.hydromet.2015.11.005

Liu, T., Li, B., Hao, Y., & Yao, Z. (2014). MoO 3-nanowire membrane and Bi 2 Mo 3 O 12/MoO 3 nano-heterostructural photocatalyst for wastewater treatment. *Chemical Engineering Journal*, *244*, 382–390. doi:10.1016/j.cej.2014.01.070

Lubick, N., & Betts, K. (2008). *Silver socks have cloudy lining| Court bans widely used flame retardant*. ACS Publications.

Ma, H., Wang, H., & Na, C. (2015). Microwave-assisted optimization of platinum-nickel nanoalloys for catalytic water treatment. *Applied Catalysis B: Environmental*, *163*, 198–204. doi:10.1016/j.apcatb.2014.07.062

Maliyekkal, S. M., Lisha, K. P., & Pradeep, T. (2010). A novel cellulose–manganese oxide hybrid material by in situ soft chemical synthesis and its application for the removal of Pb (II) from water. *Journal of Hazardous Materials*, *181*(1), 986–995. doi:10.1016/j.jhazmat.2010.05.112 PMID:20638968

Mirkin, C. A., Letsinger, R. L., Mucic, R. C., & Storhoff, J. J. (1996). A DNA-based method for rationally assembling nanoparticles into macroscopic materials. *Nature, 382*(6592), 607-609.

Mohmood, I., Lopes, C. B., Lopes, I., Ahmad, I., Duarte, A. C., & Pereira, E. (2013). Nanoscale materials and their use in water contaminants removal—a review. *Environmental Science and Pollution Research International*, *20*(3), 1239–1260. doi:10.100711356-012-1415-x PMID:23292223

Ngah, W. W., Teong, L., & Hanafiah, M. (2011). Adsorption of dyes and heavy metal ions by chitosan composites: A review. *Carbohydrate Polymers*, *83*(4), 1446–1456. doi:10.1016/j.carbpol.2010.11.004

Pena, M. E., Korfiatis, G. P., Patel, M., Lippincott, L., & Meng, X. (2005). Adsorption of As (V) and As (III) by nanocrystalline titanium dioxide. *Water Research*, *39*(11), 2327–2337. doi:10.1016/j.watres.2005.04.006 PMID:15896821

Pendergast, M. M., & Hoek, E. M. (2011). A review of water treatment membrane nanotechnologies. *Energy & Environmental Science*, *4*(6), 1946–1971. doi:10.1039/c0ee00541j

Petala, E., Dimos, K., Douvalis, A., Bakas, T., Tucek, J., Zbořil, R., & Karakassides, M. A. (2013). Nanoscale zero-valent iron supported on mesoporous silica: Characterization and reactivity for Cr (VI) removal from aqueous solution. *Journal of Hazardous Materials*, *261*, 295–306. doi:10.1016/j.jhazmat.2013.07.046 PMID:23959249

Petcharoen, K., & Sirivat, A. (2012). Synthesis and characterization of magnetite nanoparticles via the chemical co-precipitation method. *Materials Science and Engineering B*, *177*(5), 421–427. doi:10.1016/j.mseb.2012.01.003

Phiwdang, K., Suphankij, S., Mekprasart, W., & Pecharapa, W. (2013). Synthesis of CuO nanoparticles by precipitation method using different precursors. *Energy Procedia*, *34*, 740–745. doi:10.1016/j.egypro.2013.06.808

Qiu, L., Peng, Y., Liu, B., Lin, B., Peng, Y., Malik, M. J., & Yan, F. (2012). Polypyrrole nanotube-supported gold nanoparticles: An efficient electrocatalyst for oxygen reduction and catalytic reduction of 4-nitrophenol. *Applied Catalysis A, General*, *413*, 230–237. doi:10.1016/j.apcata.2011.11.013

Rai, U. N., Prasad, D., Verma, S., Upadhyay, A. K., & Singh, N. K. (2012). Biomonitoring of metals in Ganga water at different ghats of Haridwar: Implications of constructed wetland for sewage detoxification. *Bulletin of Environmental Contamination and Toxicology*, *89*(4), 805–810. doi:10.100700128-012-0766-9 PMID:22869393

Ramana, D. V., Yu, J. S., & Seshaiah, K. (2013). Silver nanoparticles deposited multiwalled carbon nanotubes for removal of Cu (II) and Cd (II) from water: Surface, kinetic, equilibrium, and thermal adsorption properties. *Chemical Engineering Journal*, *223*, 806–815. doi:10.1016/j.cej.2013.03.001

Rao, M., Parwate, A. V., & Bhole, A. G. (2002). Removal of Cr 6^+ and Ni 2^+ from aqueous solution using bagasse and fly ash. *Waste Management (New York, N.Y.)*, *22*(7), 821–830. doi:10.1016/S0956-053X(02)00011-9 PMID:12365785

Rao, M. M., Ramesh, A., Rao, G. P. C., & Seshaiah, K. (2006). Removal of copper and cadmium from the aqueous solutions by activated carbon derived from Ceiba pentandra hulls. *Journal of Hazardous Materials*, *129*(1), 123–129. PMID:16191464

Raval, N. P., Shah, P. U., & Shah, N. K. (2016). Adsorptive removal of nickel (II) ions from aqueous environment: A review. *Journal of Environmental Management*, *179*, 1–20. doi:10.1016/j.jenvman.2016.04.045 PMID:27149285

Recillas, S., García, A., González, E., Casals, E., Puntes, V., Sanchez, A., & Font, X. (2011). Use of CeO_2, TiO_2 and Fe_3O_4 nanoparticles for the removal of lead from water: Toxicity of nanoparticles and derived compounds. *Desalination*, *277*(1), 213–220. doi:10.1016/j.desal.2011.04.036

Renuka, N., Sood, A., Ratha, S. K., Prasanna, R., & Ahluwalia, A. S. (2013). Evaluation of microalgal consortia for treatment of primary treated sewage effluent and biomass production. *Journal of Applied Phycology*, *25*(5), 1529–1537. doi:10.100710811-013-9982-x

Salam, M. A. (2013). Coating carbon nanotubes with crystalline manganese dioxide nanoparticles and their application for lead ions removal from model and real water. *Colloids and Surfaces. A, Physicochemical and Engineering Aspects*, *419*, 69–79. doi:10.1016/j.colsurfa.2012.11.064

Sarkar, S. K., Saha, M., Takada, H., Bhattacharya, A., Mishra, P., & Bhattacharya, B. (2007). Water quality management in the lower stretch of the river Ganges, east coast of India: An approach through environmental education. *Journal of Cleaner Production*, *15*(16), 1559–1567. doi:10.1016/j.jclepro.2006.07.030

Schwarzenbach, R. P., Escher, B. I., Fenner, K., Hofstetter, T. B., Johnson, C. A., Von Gunten, U., & Wehrli, B. (2006). The challenge of micropollutants in aquatic systems. *Science*, *313*(5790), 1072–1077. doi:10.1126cience.1127291 PMID:16931750

Sharma, S. S., & Dietz, K. (2009). The relationship between metal toxicity and cellular redox imbalance. *Trends in Plant Science*, *14*(1), 43–45. doi:10.1016/j.tplants.2008.10.007 PMID:19070530

Sharma, Y. C., Prasad, G., & Rupainwar, D. C. (1992). Heavy metal pollution of river Ganga in Mirzapur, India. *The International Journal of Environmental Studies*, *40*(1), 41–53. doi:10.1080/00207239208710712

Sheta, A. S., Falatah, A. M., Al-Sewailem, M. S., Khaled, E. M., & Sallam, A. S. H. (2003). Sorption characteristics of zinc and iron by natural zeolite and bentonite. *Microporous and Mesoporous Materials*, *61*(1), 127–136. doi:10.1016/S1387-1811(03)00360-3

Shirsath, D. S., & Shirivastava, V. S. (2015). Adsorptive removal of heavy metals by magnetic nanoadsorbent: An equilibrium and thermodynamic study. *Applied Nanoscience*, *5*(8), 927–935. doi:10.100713204-014-0390-6

Singh, T. S., & Pant, K. (2004). Equilibrium, kinetics and thermodynamic studies for adsorption of As (III) on activated alumina. *Separation and Purification Technology*, *36*(2), 139–147. doi:10.1016/S1383-5866(03)00209-0

Srivastava, N., & Majumder, C. (2008). Novel biofiltration methods for the treatment of heavy metals from industrial wastewater. *Journal of Hazardous Materials*, *151*(1), 1–8. doi:10.1016/j.jhazmat.2007.09.101 PMID:17997034

Srivastava, R. R., Kim, M. S., Lee, J. C., & Ilyas, S. (2015). Liquid–liquid extraction of rhenium (VII) from an acidic chloride solution using Cyanex 923. *Hydrometallurgy*, *157*, 33–38. doi:10.1016/j.hydromet.2015.07.011

Stone, V., Nowack, B., Baun, A., van den Brink, N., von der Kammer, F., Dusinska, M., ... Fernandes, T. F. (2010). Nanomaterials for environmental studies: Classification, reference material issues, and strategies for physico-chemical characterisation. *The Science of the Total Environment*, *408*(7), 1745–1754. doi:10.1016/j.scitotenv.2009.10.035 PMID:19903569

Tarabara, V. V. (2009). Multifunctional nanomaterial-enabled membranes for water treatment. *Nanotechnology Applications for Clean Water, 5*, 59-75.

Tarigh, G. D., & Shemirani, F. (2013). Magnetic multi-wall carbon nanotube nanocomposite as an adsorbent for preconcentration and determination of lead (II) and manganese (II) in various matrices. *Talanta*, *115*, 744–750. doi:10.1016/j.talanta.2013.06.018 PMID:24054657

Theron, J., Cloete, T. E., & de Kwaadsteniet, M. (2010). Current Molecular and Emerging Nanobiotechnology Approaches for. *Nanotechnology in Water Treatment Applications*, 39.

Theron, S., Zussman, E., & Yarin, A. (2004). Experimental investigation of the governing parameters in the electrospinning of polymer solutions. *Polymer*, *45*(6), 2017–2030. doi:10.1016/j.polymer.2004.01.024

Tsuzuki, T., & McCormick, P. G. (2004). Mechanochemical synthesis of nanoparticles. *Journal of Materials Science*, *39*(16), 5143–5146. doi:10.1023/B:JMSC.0000039199.56155.f9

Venkatesham, V., Madhu, G., Satyanarayana, S., & Preetham, H. (2013). Adsorption of lead on gel combustion derived nano ZnO. *Procedia Engineering*, *51*, 308–313. doi:10.1016/j.proeng.2013.01.041

Vijayaraghavan, K., Jegan, J., Palanivelu, K., & Velan, M. (2005). Biosorption of cobalt (II) and nickel (II) by seaweeds: Batch and column studies. *Separation and Purification Technology*, *44*(1), 53–59. doi:10.1016/j.seppur.2004.12.003

Wang, H., Yuan, X., Wu, Y., Zeng, G., Chen, X., Leng, L., ... Li, H. (2015). Facile synthesis of amino-functionalized titanium metal-organic frameworks and their superior visible-light photocatalytic activity for Cr (VI) reduction. *Journal of Hazardous Materials*, *286*, 187–194. doi:10.1016/j.jhazmat.2014.11.039 PMID:25585267

World Health Organization. (2015). *Drinking-Water-Fact Sheet No. 391*. WHO.

Wu, P., Li, S., Ju, L., Zhu, N., Wu, J., Li, P., & Dang, Z. (2012). Mechanism of the reduction of hexavalent chromium by organo-montmorillonite supported iron nanoparticles. *Journal of Hazardous Materials*, *219*, 283–288. doi:10.1016/j.jhazmat.2012.04.008 PMID:22521796

Xu, M., Wang, H., Lei, D., Qu, D., Zhai, Y., & Wang, Y. (2013). Removal of Pb (II) from aqueous solution by hydrous manganese dioxide: Adsorption behavior and mechanism. *Journal of Environmental Sciences (China)*, *25*(3), 479–486. doi:10.1016/S1001-0742(12)60100-4 PMID:23923420

Xu, P., Zeng, G. M., Huang, D. L., Feng, C. L., Hu, S., Zhao, M. H., ... Xie, G. X. (2012). Use of iron oxide nanomaterials in wastewater treatment: A review. *The Science of the Total Environment*, *424*, 1–10. doi:10.1016/j.scitotenv.2012.02.023 PMID:22391097

Yang, B., Geng, P., & Chen, G. (2015). One-dimensional structured IrO 2 nanorods modified membrane for electrochemical anti-fouling in filtration of oily wastewater. *Separation and Purification Technology*, *156*, 931–941. doi:10.1016/j.seppur.2015.10.040

Yin, J., Zhu, G., & Deng, B. (2013). Multi-walled carbon nanotubes (MWNTs)/polysulfone (PSU) mixed matrix hollow fiber membranes for enhanced water treatment. *Journal of Membrane Science*, *437*, 237–248. doi:10.1016/j.memsci.2013.03.021

Zare, K., Najafi, F., Sadegh, H., & ghoshekandi, R. (2013). Studies of ab initio and Monte Carlo simulation on interaction of fluorouracil anticancer drug with carbon nanotube. *Journal of Nanostructure in Chemistry*, *3*(1), 71. doi:10.1186/2193-8865-3-71

Zhang, L., & Fang, M. (2010). Nanomaterials in pollution trace detection and environmental improvement. *Nano Today*, *5*(2), 128–142. doi:10.1016/j.nantod.2010.03.002

Zhang, Q., Xu, R., Xu, P., Chen, R., He, Q., Zhong, J., & Gu, X. (2014). Performance study of ZrO 2 ceramic micro-filtration membranes used in pretreatment of DMF wastewater. *Desalination*, *346*, 1–8. doi:10.1016/j.desal.2014.05.006

Zhou, Y.-T., Nie, H.-L., Branford-White, C., He, Z.-Y., & Zhu, L.-M. (2009). Removal of Cu 2+ from aqueous solution by chitosan-coated magnetic nanoparticles modified with α-ketoglutaric acid. *Journal of Colloid and Interface Science*, *330*(1), 29–37. doi:10.1016/j.jcis.2008.10.026 PMID:18990406

Zhu, H., Jia, Y., Wu, X., & Wang, H. (2009). Removal of arsenite from drinking water by activated carbon supported nano zero-valent iron. *Huan jing ke xue=. Huan Jing Ke Xue, 30*(6), 1644–1648. PMID:19662844

Zhu, H., Jia, Y., Yao, S., Wu, X., & Wang, S. (2009). Removal of arsenate from drinking water by activated carbon supported nano zero-valent iron. *Huan jing ke xue=. Huan Jing Ke Xue, 30*(12), 3562–3567. PMID:20187387

Chapter 11
Nanoparticles for Degradation of Organic Pollutants

Sana Aslam
Government College Women University Faisalabad, Pakistan

Matloob Ahmad
Government College University Faisalabad, Pakistan

ABSTRACT

Water pollution is becoming a serious threat to the environment. The access to safe drinking water, especially for the poor, is reaching an alarmingly low level, and the number of people suffering from kidney, stomach, and liver diseases, etc. is increasing every day. Among various pollutants, organic pollutants have major contribution in the contamination of water. These pollutants include toxic dyes, pharmaceutical compounds, industrial chemicals, and organic solvents, etc. Various routes are being employed for the removal/degradation of these pollutants. The use of nanoparticles (i.e., metal nanoparticles, nanocomposites, carbon nanotubes, and graphene oxide-based nanocomposite materials) possesses significant position as photocatalysts. This chapter has a focus on the environment remediation using these nanomaterials and contains the recent work in this field.

INTRODUCTION

Fresh water is essential for the existence of life on the globe. The supply of fresh water has health, environmental, economical and social impacts. It is estimated that 10-20 million people die every year due to water borne diseases (Leonard *et al.*, 2003). Moreover, about 0.78 billion population do not have access to safe drinking water (WHO and UNICEF, 2013). Diarrhea, a disease caused by polluted water is responsible for the deaths of five to six thousand children every day (Ashbol, 2004; Hutton *et al.*, 2007). The situation is expected to become worse in coming years as there are no expectations to get increased the supply of safe and fresh drinking water. Moreover, according to UN population projection study, the world population is expected to increase by 2.5 billion from now to 2050. The increased population would have increased demands for fresh water and obviously, the health and quality of life would suffer even more (Rockstrom, 2003). The world has understood that the intensity of worse situation could be reduced or controlled by proper use of fresh water resources.

DOI: 10.4018/978-1-5225-5745-6.ch011

The 20th century is recognized as a century of industrial revolution. The development of industrial products has huge impacts on the global life; both positive and negative impacts. So far the supply of fresh water is concerned; it has badly affected the access to safe water for aquatic life, plants and humans. Industrial waste water is accompanied by a number of organic and inorganic pollutants which are ultimately polluting the world's water resources. The major industries having major contribution in water contamination include textile, dye, leather, rubber, petrochemical, coal, plastics, pharmaceutical, paper and food industries *etc*. Moreover, intensified usage of pesticides in agriculture and aquaculture are another big source of water contamination (Njoku et al., 2014).

The main objectives of this chapter are;

- To introduce the role of organic pollutants in the contamination of water.
- To discuss the recent work on the synthesis of nanomaterials for water treatment.
- To discuss the recent developments in the titled field.

ORGANIC POLLUTANTS AND THEIR REMEDIATION

Organic pollutants are hazardous substances which have adverse effects on human health like dysfunction of liver, adverse effects on brain, reproductive system, kidney and central nervous system (Ali *et al*., 2012; Costa *et al*., 2012). Currently, effective and advance treatment techniques are used for the purification of wastewater like biological oxidation, chemical oxidation via oxidants (e.g. O_3 or H_2O_2), coagulation, flocculation, and sedimentation (Duran *et al*., 2011), advanced oxidation processes photo-Fenton treatment (Salman *et al*., 2011), membrane processes (Altınışık *et al*., 2010), adsorption, electrochemical (Ahmed *et al*., 2014) and photocatalytic oxidation/degradation and combined methods (Ahmaruzzaman & Gayatri, 2010).

The right choice of process for removal of organic pollutants depends upon the nature of the chemical moiety or functional groups to be removed like cationic and anionic species, carboxylic/sulfonic acids, condensed aromatic rings, zwitter ions and their concentration in waste water sample.

Despite the availability of variety of methodologies for hazardous organic substances removal, the adsorption technique is still considered as the most effective and best methodology. Moreover, both soluble and insoluble organic contaminants are removed without production of dangerous by-products.

Waste water treatment has become a promising field of research in recent days. Among various techniques mentioned above, the use of nanomaterials emerged as a worthwhile method.

Nanoparticles are generally classified into two major groups of inorganic and organic nanoparticles.

Organic nanoparticles include

1. Carbon nanoparticles (fullerenes).
2. Graphene oxide (GO) based on Nanomaterials.
3. Polymer based on Nanomaterials.
4. Photocatalyst based on Nanomaterials etc.

Whereas, inorganic nanoparticles include:

Table 1. The detailed examples of various adsorbents reported in literature for removal of organic pollutants

Adsorbent Class	Examples
Nanometer materials	Nano-graphite/Fe_3O_4 composite (Li *et al.*, 2014), alkali activated MCNT (Salam *et al.*, 2014), Fe_3C-multiwalled carbon nanotubes (MCNT) (Konicki *et al.*, 2012), CNT-Al_2O_3 (Ihsanullah *et al.*, 2015), Pt/poly(2-methoxyaniline)-MCNT (Raoof *et al.*, 2014), MCNT-COOH (Sadegh *et al.*, 2015) etc.
Industrial wastes	coal fly ash (Zhang *et al.*, 2011), silica gel (Nie *et al.*, 2015), waste tires (Saleh *et al.*, 2014), dried activated sludge and fly ash (Aksu and Yener, 2001), polyurethane foam (Vinhal *et al.*, 2015) etc
Sea materials	Chitosan/surfactant (Zhang *et al.*, 2014), crosslinked chitosan beads (Zhang *et al.*, 2011), peat moss (Aroguz, 2006), shrimp shell (Kousha *et al.*, 2015), Spirogyra sp. (Khataee *et al.*, 2013), brown seaweed biomass (Rathinam *et al.*, 2011) etc.
Metal–organic frameworks (MOF)	iron terephthalate, Cr-based MOFs, Cu-BTC, MIL- 101-Cr and Cu-BTC *etc.*
Metal oxides and hydroxides	Fe_3O_4/ZnCr-layered double hydroxide composite (Chen *et al.*, 2012), Mg–Fe layer double hydroxides (Abdelkader *et al.*, 2011), TiO_2 (Feng *et al.*, 2015), MgAl-layer and ZnCr- double hydroxide (Liu *et al..*, 2013), magnesium hydroxide (Kamaraj *et al.*, 2014), Fe_3O_4/MgAl layer double hydroxides (Shan *et al.*, 2014) *etc.*
Soil and ore materials	Montmorillonite clay (Lin *et al.*, 2006), hematite (Saikia *et al.*, 2013), magnetite, quartz (El-Zahhar *et al.*, 2014), acrisol and phaeozem (Wang *et al.*, 2013), Moroccan clays (Elmoubarki *et al.*, 2015), zeolite (Guo *et al.*, 2015), halloysite–magnetite-based composite, calcined Lapindo volcanic mud (Jalil *et al.*, 2010), red mud (Zhang *et al.*, 2014), river sediment (Gao *et al.*, 2011), kaolinite (De Sales *et al.*, 2015; Bonetto *et al.*, 2015) *etc.*
Agricultural wastes	Tomato (Güzel *et al.*, 2014), rice husk (Sharma, 2011), peanut husk (Noreen *et al.*, 2013), sesame hull (Feng *et al.*, 2011), sugarcane bagasse (Saad *et al.*, 2010) *etc.*

1. Magnetic nanoparticles.
2. Noble metal nanoparticles (e.g. Ag and Au).
3. Semiconductor nanoparticles (e.g. TiO_2 and ZnO).

Nanoparticles were classified into three types by Ruffini-Castiglione and Cremonini (Ruffini-Castiglione and Cremonini, 2009; Costa et al., 2012).

1. Natural (e.g. Lunar dust or volcanic, mineral composites).
2. Incidental (consequential from anthropogenic activity, e.g., coal combustion, diesel exhaust welding fumes).
3. Engineered (e.g. nanozinc, nanoaluminium, nanogold, ZnO, TiO_2 and Al_2O_3 (Lin & Xing, 2007; Duran *et al.*, 2011).

Nanoparticles are considerably gaining the importance in environmental remediation processes. They are being used in detoxification of pollutants, as adsorbents for phosphate removal (Wen *et al.*, 2014), for fluoride removal (Ali *et al.*, 2015), for Congo dye removal (Ali *et al.*, 2016), in adsorption of ametryn from water (Ali *et al.*, 2016), for removal of ibuprofen drug residue from water (Ali *et al.*, 2016), in removal of pantoprazole drug residue from water (Ali *et al.*, 2016) and for adsorption of propranolol (Ali *et al.*, 2017).

In short, the formation of various types of nanomaterials like nanocatalysts, nanosorbents, nanostructured catalytic membranes, dendrimers, and zeolites, were promising for removing toxic metals, to disinfect infectious microbes, and eliminate inorganic and organic solutes from wastewater.

CARBON NANOTUBES (CNTS) FOR REMOVAL OF ORGANIC POLLUTANTS

Activated carbon (AC) has excellent adsorption potential for various organic contaminants and is therefore used as a commercial adsorbent (Demirbas, 2009; Hosseini *et al.*, 2011). Among various drawbacks, activated carbon adsorbents have limited availability, off-site transport, difficult recovery, low adsorption capability, thermal treatments and material lost during regeneration processes. Keeping in view these certain shortcomings, the development of novel adsorbent materials is under focus. Nano-structured material is one of the excellent option for the removal of harmful and toxic organic pollutants from wastewater (Mohmood *et al.*, 2013). Carbon nanotubes (CNTs) are excellent adsorbents for toxic material on nanoscale (Iijima, 1991; Latorre *et al.*, 2012), with layered and hollow structure having large surface area. CNT adsorbents are of three types:

1. Single-walled CNTs (SWCNTs).
2. Multi-walled CNTs (MWCNTs).
3. Functionalized CNTs (f-CNTs) (Ruelle *et al.*, 2012; Ma *et al.*, 2011b).

CNTs are considered as efficient adsorbents for pharmaceuticals removal due to their large micropore volume along with high specific surface area. Endocrine disrupting compounds (EDCs), a group from pharmaceutical pollutants, which interfere with human reproductive systems via mimicking or blocking the action of natural hormones is efficiently removed from water (Joseph *et al.*, 2011a). 17α-Ethinyl estradiol is removed efficiently using SWCNTs incorporated with ultrafiltration membrane systems.

Norfloxacin and ofloxacin removal were reported by different types of CNTs from aqueous solutions such as three p-MWCNTs with diameters range (15-50 nm), with carboxylated, hydroxylated and graphitized functional MWCNTs and with carboxylated, hydroxylated and purified functionalized

Figure 1. Chemical Structures of Pharmaceutical Compounds Investigated for Removal by CNTs

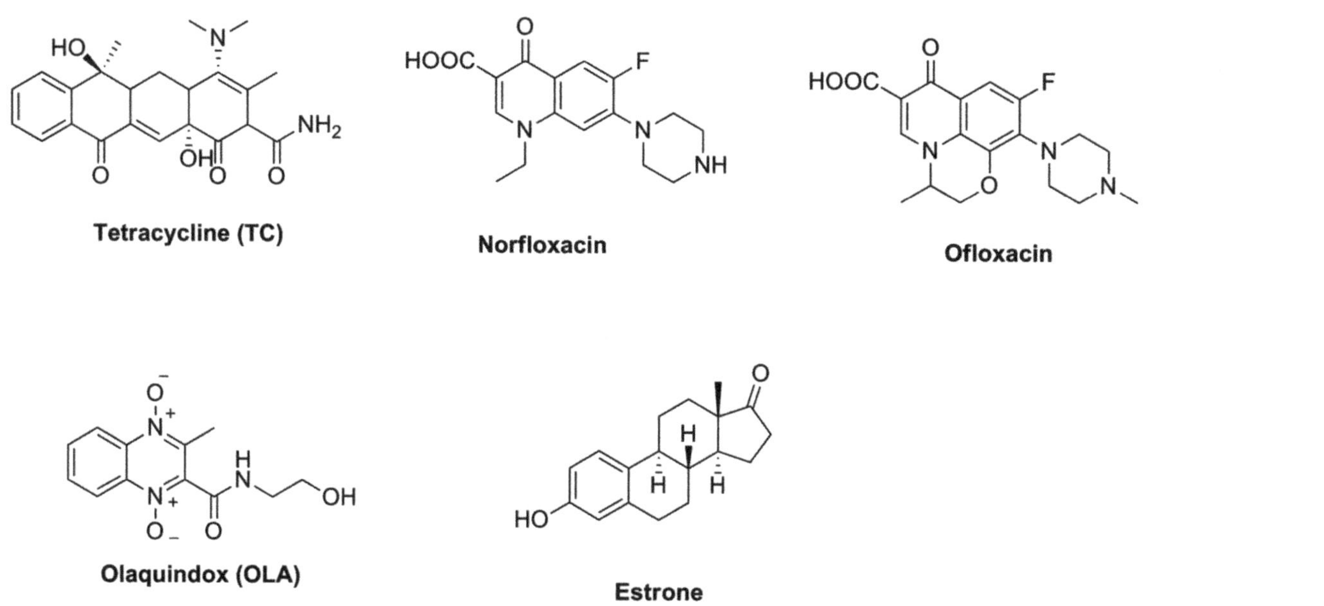

SWCNTs. The hydrophobic nature of CNTs contributes more to their adsorption potential (Peng *et al.*, 2012). Adsorption of antibiotics by CNTs was reported like sulfapyridine and sulfamethoxazole were effectively adsorbed on CNT/sand porous system (Tian *et al.*, 2013). Ciprofloxacin was removed by magnetic meso-porous carbon nano-composite (Fe_3O_4/C) (Shi *et al.*, 2013), While carbamazepine and diclofenac sodium was removed by hybrid type of adsorbent *i.e.* CNTs/Al_2O_3 (Wei *et al.*, 2013).

Multi-walled CNTs (MWCNTs) potently removed arsenazo(III), methyl red (MR) (Ghaedi *et al.*, 2011c), Eriodirome Cyanine R (ECR) (Ghaedi *et al.*, 2011b) and methyl orange (MO) (Hosseini *et al.*, 2011) from wastewater as compared to simple activated carbon. MWCNTs also reported for the removal of herbicides like dichlobenil and diuron from soiled water (Chen *et al.*, 2011). While diuron removal was also studied by using oxidized MWCNTs, as pore volume and surface area increases by oxidation treatment (Deng *et al.*, 2012a). A phenoxy acid herbicide i.e. 4-chloro-2-methylphenoxyacetic acid (MCPA) was selectively adsorbed by SWCNTs as compared to metal oxide nanoparticles (De Martino *et al.*, 2012).

GRAPHENE OXIDE (GO) FOR REMOVAL OF ORGANIC POLLUTANTS

Graphene oxide (GO) is one of the good choice as an adsorbent due to its high surface area, low production costs, good dispersibility (Perreault *et al.*, 2015; Ramesha *et al.*, 2011). The mechanism of adsorbent action includes electrostatic interactions, π-π interactions and ion exchange (Yu *et al.*, 2017).

Currently, graphene oxide nanosheets are used for the removal of aromatic contaminants (Apul *et al.*, 2013) *e.g.* biphenyl while graphene oxide based foams effectively used for the removal of organic dyes (Wu *et al.*, 2015), motor oil gasoline, petroleum and diesel (Guo *et al.*, 2015) from wastewater.

Polysulfone (PS) is a hydrophobic class of thermally stable and biocompatible polymers. PS mechanically processed into strong porous ultrafiltration membranes which can be widely used for water microbiologic depuration and haemodialysis on industrial scale. PS membranes doped with various kinds of nanomaterials like modified graphene oxide (Kochameshki *et al.*, 2017), carbon nanotubes (Maphutha *et al.*, 2013) and silica nanoparticles (Sen *et al.*, 2014), in order to enhance their applications like removal of oil residues from wastewater via filtration (Goh *et al.*, 2015).

Arsenate was successfully removed from water through filtration by using polysulfone membranes doped with fabricated graphene oxide (Rezaee *et al.*, 2015). Polysulfone composite membranes doped with graphene oxide and were fabricated by using phase inversion technique. It was used as adsorbent for the removal of various types of hazardous organic pollutants from water (Zambianchi *et al.*, 2017).

METALLIC NANOPARTICLES

Metal Nanoparticles are simply prepared from the metals precursors. These NPs have potent optical and electrical properties due to distinguished localized surface plasmon resonance (LSPR), consequently having applications in various research domains. Noble and alkali metals nanoparticles like Ag, Au and Cu have a broad and intense absorption band in the visible region of electromagnetic spectrum. The metals nanoparticles currently used as a cutting-edge material, controlled by their specific size, shape, facet and synthetic pathway.

Silver (Ag) is extensively reported substance because of its microbial inactivation and less toxicity in water (Spadaro *et al.*, 1974; Zhao & Stevens Jr, 1998; Inoue *et al.*, 2002; Kumar *et al.*, 2004). So keeping in view as effective biocides (Sondi & Salopek-Sondi, 2004; Baker *et al.*, 2005; Pan´aˇcek *et al.*, 2006; Shrivastava *et al.*, 2007; Feng *et al.*, 2000) is well-reported with antibacterial mechanism (Yamanaka *et al.*, 2005; Pal *et al.*, 2007). Silver nanoparticles are prepared from its salts e.g. $AgNO_3$ and AgCl. Truncated triangular Ag NPs (Morones *et al.*, 2005) with small size (8 nm) (Makhluf *et al.*, 2005) exhibited effective antibacterial activity as compared to rod-shaped and spherical nanoparticles. The mechanisms of bactericidal action involved interactions with DNA, damaging the bacterial membranes via generation of free radicals (Xiu *et al.*, 2011; Xiu *et al.*, 2012), and enzyme damage (Kim *et al.*, 2007; Liau *et al.*, 1997; Danilczuk *et al.*, 2006).

Nanocomposites and antimicrobial nanofibers were developed by incorporation of various types of polymers with Ag nanoparticles (Balogh *et al.*, 2001; Chen *et al.*, 2003; Botes & Cloete, 2010).

Nanofiber mats prepared by incorporation of Ag nanoparticles with in poly (ε-caprolactone-) based polyurethane as antimicrobial nanofilters (Jeon e*t al.*, 2008). Ag nanoparticles based nanofibers were exhibited antimicrobial applications (Lala e*t al.*, 2007; Chen & Chiang, 2008; Vimala e*t al.*, 2009).

Gold nanoparticles coating enhances the electronic stream during sampling of SEM consequently image quality of SEM increases.

PHOTOCATALYST BASED ON NANOMATERIALS

Photocatalysts is known as advanced oxidation technology, for their potent action to remove the organic contaminants from wastewater (Pan *et al.*, 2010). Photocatalysis ideally leads to green chemistry technology due to their simple end-products like H_2O, CO_2, and other minerals (Weon *et al.*, 2016). The mechanism of action is the generation of electron (e^-) and positive hole (h^+) pairs upon light irradiation with photon energy equal to or greater than the energy of the photocatalyst band gap. These pairs drift to the surface followed by their reaction with species adsorbed on photocatalyst surface as result active radicals like $\cdot OH^-$, $.O_2^-$, h^+, *etc.* produced. These radicals having a potential to degrade or mineralized the various organic contaminants *e.g.* phenol (Yu *et al.*, 2012), dinitrophenol (Ma *et al.*, 2014), trichlorophenol (Carević *et al.*, 2016), bisphenol A (Chiang & Doong, 2014), 6-hydroxymethyl uracil (Zhao *et al.*, 2014), aromatic organoarsenic compounds (Zheng *et al.*, 2014), spiramycin (Vaiano *et al.*, 2015), cyanotoxins (Fotiou *et al.*, 2016), flame retardant (Antonopoulou *et al.*, 2016), levofloxacin drug (Kaur & Kansal, 2016) etc.

For example praseodymium zirconate ($Pr_2Zr_2O_7$) nanomaterials as a highly active photocatalyst was prepared by reaction of praseodymium nitrate with zirconyl nitrate in the presence of benzene tricarboxylic acid as capping and stabilization agent followed by fabrication in the presence of connecting agent i.e. propylene glycol. These nanostructures were utilized for the organic contaminants degradation e.g. rhodamine B, methyl orange, methylene blue, and eosin Y.

Biself-doped Bi_2MoO_6 nanocrystals support the separation as well as transfer of photogenerated electron-hole pairs (Ding *et al.*, 2016). Moreover, it also changes the position of conduction and valence bands. Rare earth Gd^{3+} doped on Bi_2MoO_6 not only contract the lattice but also decreases the crystal size (Yu *et al.*, 2016). Bi_2MoO_6 is also doped with B (Zhao *et al.*, 2009), Zn^{2+} (Zhang *et al.*, 2016) and rare earth Ce (Dai *et al.*, 2016) has also been reported.

Fluorine doped Bi_2MoO_6 nanocrystals were prepared by solvothermal calcination method. The doping with F^- anions causes the lattice shrinkage for the host O_2^- anions, consequently crystal size decreases as well as increases the crystallinity. In addition, the oxygen species in F-Bi_2MoO_6 nanocrystals and F^- anions adsorbed on catalyst surface withdraws the photoexcited electrons as a result potentially enhancing the photoexcited electron-hole pairs separation. Fluorinated Bi_2MoO_6 nanocrystals have a great potential for the removal of targeted organic pollutants such as Rhodamine B dye, phenol, 4-chlorophenol and bisphenol A (Zhen *et al.*, 2017).

Organic-inorganic hybrids like bismuth terephthalate doped with combination of $Bi(NO_3)_3$ and halide anion e.g. NaI, NaBr, NaCl, and NaF show excellent photocatalytic adsorption performance for the pollutants removal due to their large porous structures and surface areas as well as regularly lamellar morphology. Visible photodegradation of rhodamine B was significantly enhanced by hybrid doped with NaI and NaF (Zhao *et al.*, 2017).

The two step synthetic route for Fe_3O_4/PDA hybrid material was reported, including the preparation of Fe_3O_4 nanoparticles firstly by the reaction of $FeCl_3$ in DEG and anhydrous CH_3COONa followed by the dispersion of these nanoparticles in dopamine within tris buffer solution. These hybrid were chelated metal ions and have used for the removal of organic dyes *e.g.* tartrazine and methylene blue as well as heavy metals like Cu^{2+}, Hg^{2+}, Ag^+ (Zhang *et al.*, 2014).

NANOPARTICLES/NANOCOMPOSITES FOR THE DEGRADATION OF ORGANIC DYES/PHENOLS

Among the organic pollutants, dyes are the major ones which are found in waste water in large quantity. Most of the dyes, especially azo-dyes are toxic in nature which turns the water unsafe for aquatic life. These hazardous pollutants possess stable color and complicated chemical structure. Their removal from waste water requires high costs. More than 100,000 dye industries are working across the globe which is producing more than 700,000 metric tons of dyes every year (Sen *et al.*, 2011). It is assumed that about 20% of total dye productions are finally lost in industrial waste water.

Photocatalysis is considered one of the promising candidates for complete destruction of organic pollutants without generation of secondary pollutants. Nanoparticles play effective role in the degradation of organic dyes especially by photocatalytic degradation. A photocatalyst is a substance which accelerates the photoreaction without being consumed (Fox, 1988). These catalysts possess unusual characteristics due to high surface to volume ratio. Pure titanium and zinc oxides are well known for their affectivity as photocatalysts for degradation of organic pollutants (Gupta *et al.*, 2012; Pomies *et al.*, 2013; Verma *et al.*, 2012; Brillas and Martinez-Huitle, 2015).

Ag/SnO_2 nanoparticles prepared by sol–gel method were recently studied for photocatalytic degradation of methylene blue. It was observed that the photocatalytic activity of 84% was achieved when a 5wt% of silver nano particles were used in Ag/SnO_2 in UV light.

The photocatalytic degradation of Amaranth dye (anionic dye) was also carried out in the presence of Ag/SnO2 nanoparticles under the UV light. The removal was studied at different pH levels. At pH=4, the dye removal upto 53% was observed. It is worthwhile to mention here that amaranth dye is banned in USA due to its toxic nature (Ahmad *et al.*, 2017).

The phenomenon of degradation of organic dye is a combined effect of adsorption and photocatalytic reactivity under UV light. The adsorption depends upon the surface area and pore structure of the

Figure 2. Chemical structures of Methylene Blue (Cationic Dye) and Amaranth (anionic dye) respectively

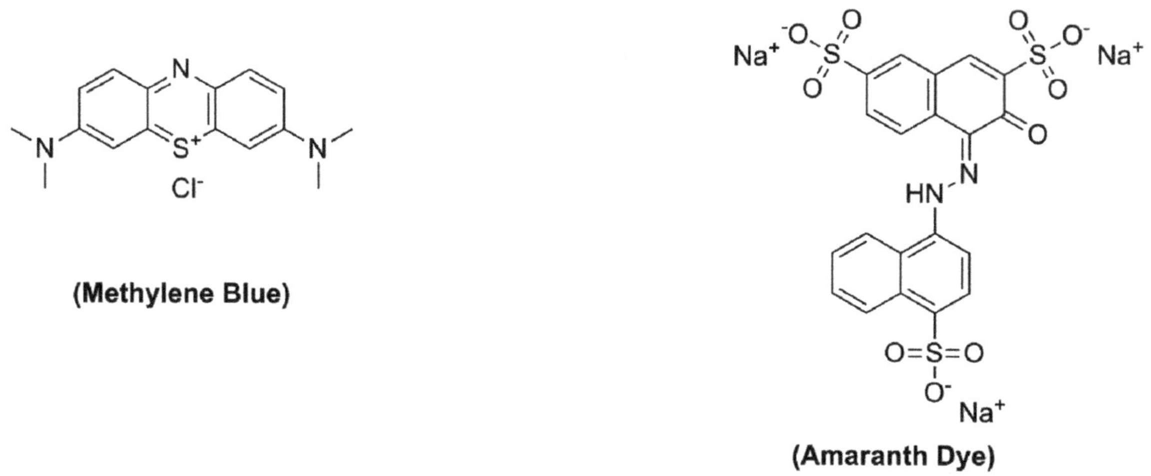

oxide. It is now understood that the adsorption is a prime factor of degradation of dye. First of all, the dye molecules are adsorbed on the surface of photocatalyst which are followed by degradation under the influence of UV light. The UV light when irradiates the semiconductor surface results in the formation of positive hole and electron. The electron in the conduction band is responsible for the formation of hydroxyl radical, hydrogen peroxideradical and superoxide anion radical, which are reactive species. These reactive radicals combine with organic molecules that undergo the formation of various intermediates which decompose in UV light to ultimately form carbon dioxide (Ngah & Fatinathan, 2008; Chen *et al.*, 2008; Ghorai *et al.*, 2008; Ishibashi *et al.*, 2000).

A detailed mechanism of photocatlytic degradation was published by Royal Society of Chemistry as shown in Figure 3.

Figure 3. Proposed mechanism of photocatalytic degradation of methylene blue (Xia et al., 2015)

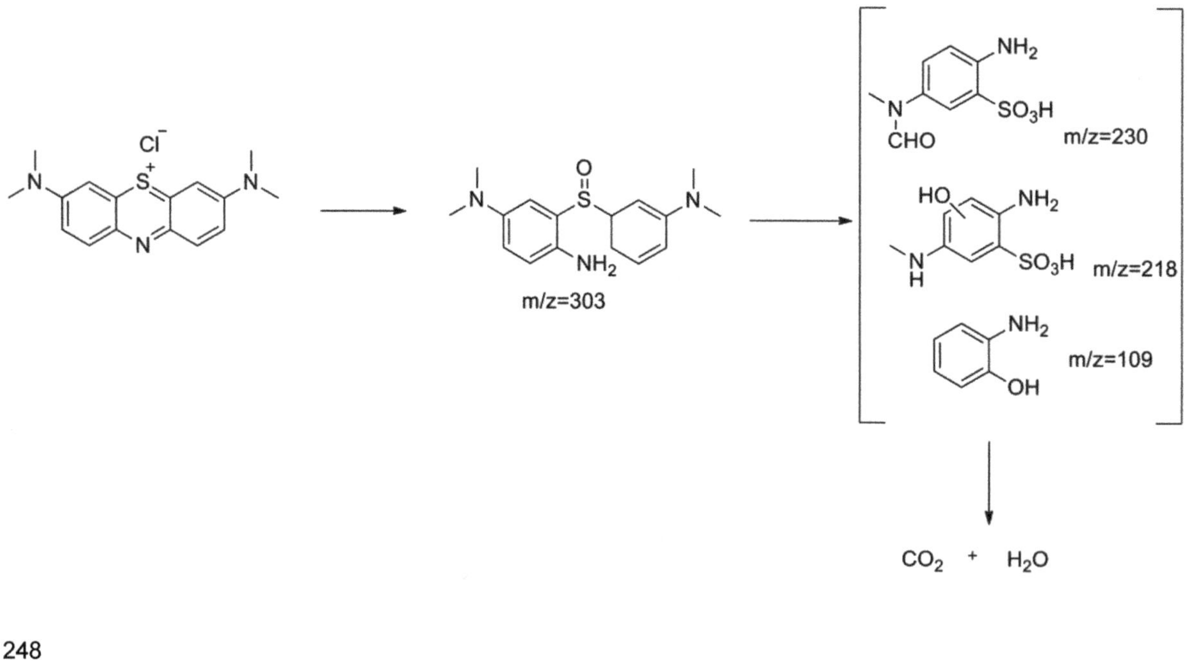

Xu *et al* reported recently the synthesis of ZIF-derived N-doped porous carbon adsorbents. For convenience, Xu marked ZIF/Sucrose and ZIF/Dicyandiamide composites-derived materials Carbon-ZS and Carbon-ZD, respectively. The ZIF-8 derived material was also obtained and marked as Carbon-Z. These substances are used for their application to remove methylene blue (MB) from wastewater. Carbon-ZD emerged as an excellent MB adsorbent. Moreover, when the excessive amounts of carbon-ZD were used, it was observed that carbon-ZD have the ability to remove MB as much as 100% which indicates that Carbon-ZD is an excellent candidate for MB removal from wastewater. Carbon-ZD displayed an excellent regeneration and reusability profile, as well as anti-interference ability when MO and MB pollutants coexist in wastewater (Xu *et al.*, 2017).

Rhodamine B (RhB), another organic pollutant was removed from its aqueous solution by using Cu_2O/rGO (reduced graphene oxide) nanocomposites. The nanocomposites were employed as photocatalysts in the presence of UV light and the RhB removal ability was observed as more than 95% (Huang *et al.*, 2017).

Another photocatalyst, F-Bi_2MoO_6 displayed good photocatalytic degradation ability for the removal of 4-chlorophenol, phenol and dyes. The photocatalyst was further subjected to F-doping to enhance its photocatalytic activity. The catalyst surface possesses fluoride ions and oxygen vacancies in Bi_2MoO_6 to trap electrons which boost the photocatalytic reaction by generating more reactive radicals (h+, •OH, $•O_2^{2-}$) and positive holes that finally participate in the degradation of organic pollutants. The F-doping leads an improvement in the oxidation ability of the catalyst and thus, enabled the catalyst for its promising application in water treatment (Yu *et al.*, 2017).

WO_3 supported on zeolite-Y (WO_3-ZY) also appeared as excellent photocatalyst for the removal of RhB. The proposed degradation mechanism is shown in Figure 4 (Jothivenkatachalam *et al.*, 2014).

Metal-Organic Framework (MOF)

Metal-organic framework (MOF) is another class of materials based on nanometals used for organic pollutant degradation. Stable and hydrophilic $MgFe_2O_4$@MOF is reported to possess large surface area (519.86 m^2g^{-1}) and rapid removal (within 5 min). Rhodamine B (RB) and Rhodamine 6G (Rh6G) were successfully removed by using the MOF from water samples. $MgFe_2O_4$@MOF possesses significant adsoption capacity towards RhB and Rh6G. Moreover, the material could be recycled for reuse by washing with acetonitrile (Tian *et al.*, 2018).

Biosynthesis of metal nanoparticles emerged as an interesting route of synthesis. Bello *et al.* reported the biosynthesis of silver nanoparticles (AgNps) from the aqueous extract of H. thebaica fruit. The biosynthesized nanoparticles were applied for removal of congo red dye. The experimental results showed that the particles degraded congo red dye efficiently. Moreover, these nanoparticles have potential for both treatments of cancer as well as purification of drinking water (Bello *et al.*, 2017).

The mechanism of photocatalytic degradation of congo red is reported in literature as given Figure 7 (Murica *et al.*, 2011).

ZnO/oligoaniline based nanocomposite was studied for photodegradation of MB, RhB and CR. Photocatalytic activity of ZnO/oligoaniline nanocomposites for degradation of MB, RhB and CR dyes was determined and the results indicate their excellent ability of degradation (Chen *et al.*, 2017).

Figure 4. Proposed mechanism of photocatalytic degradation of Rhodamine B (RhB)

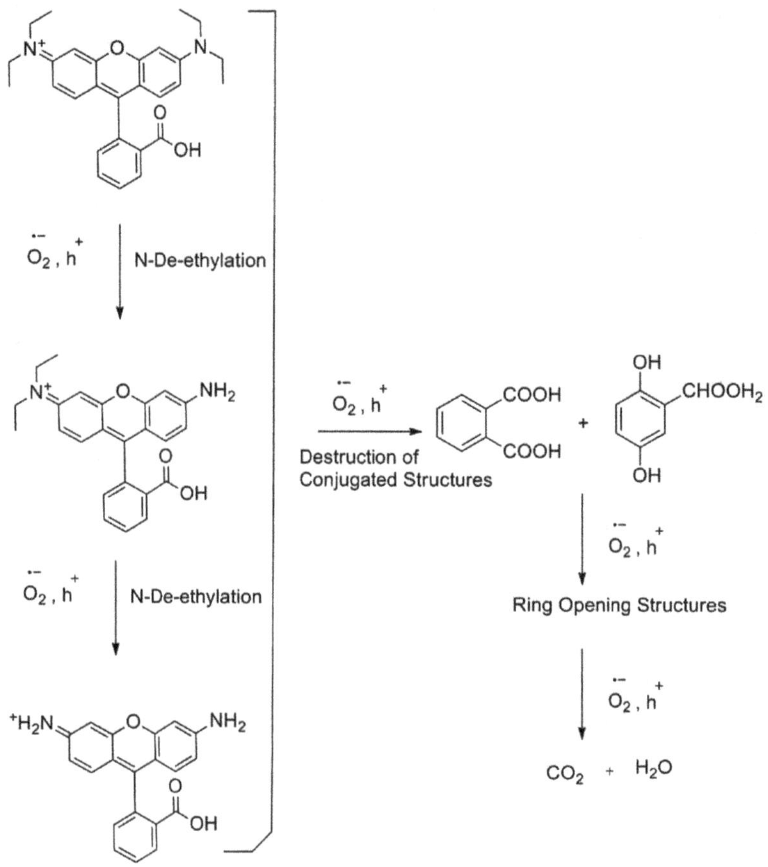

Figure 5. Structure of Rhodamine B and Rhodamine 6G

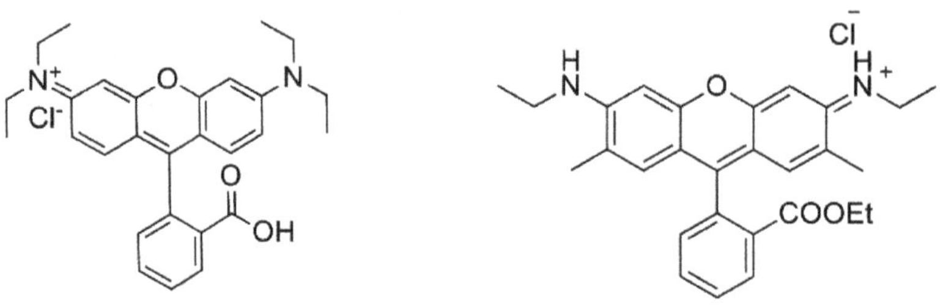

Gold Nanoparticles

Recently, gold nanoparticles (AuNPs) were synthesized by using green chemistry approach. Lagerstroemia speciosa leaf extract (LSE) [L. speciosa] was employed for green synthesis of AuNPs which was complete within 30 min at 25°C. The dyes; MB, MO, bromocresol green and the phenols, bromophenol blue (BPB) & 4-nitrophenol (4-NP) was tested for removal in the presence of sodium borohydride. The nanoparticles showed efficient results as the characteristic color of dyes disappeared after few minutes of experiment (Choudhary *et al.*, 2017).

Figure 6. Structure of Congo red dye

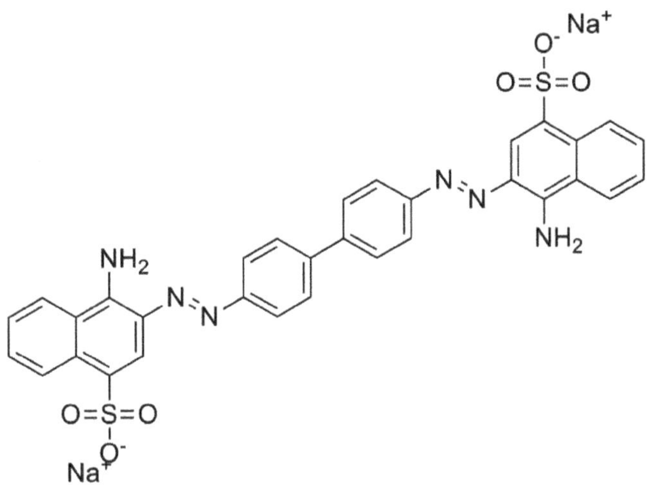

Figure 7. Mechanism of photocatalytic degradation of congo red dye

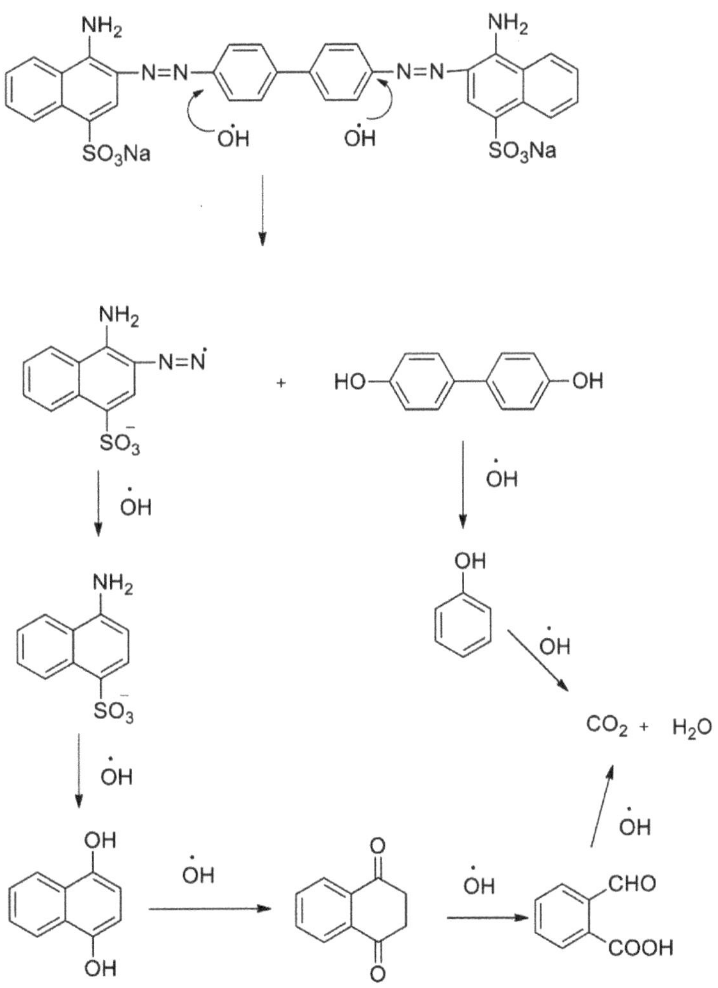

Iron Based Metal Organic Frame Works

The synthesis of iron based metal organic framework loaded on iron oxide nanoparticles (Fe3O4@ MIL-100(Fe)) and the study of its capability for the removal of methyl red was reported. The results indicate that the adsorption capacity of Fe3O4@MIL-100(Fe) for MR is significant. Therefore, it should be concluded that Fe3O4@MIL-100(Fe) nano adsorbent can be effectively utilized for the removal of methyl red from the aqueous media (Dadfarnia *et al.*, 2015).

Nanoparticle Cu (I) oxide and titania in an inorganic polymer (geopolymer) matrix modified by CTAB was studied for MB removal. These composites displayed significant potential for water treatment (Falah *et al.*, 2016).

NANOPARTICLE SYNTHESIS

There are different potent synthetic approaches for vast variety of nanoparticles and continuous research efforts carried on for the discovery of novel, effective and cheap menology. One the potent and frequently used method is liquid phase routes. Citrate route is the most convenient and generally used synthetic route for gold nanoparticles, developed by Turkevich *et al.* (1951) The reaction involved AuCl, sodium citrate as a stabilizing and reducing agent in water to obtained aqueous gold nanoparticle sols, exhibited spherical morphology with comparatively narrow size distribution i.e. 20±1.5 nm (Turkevich *et al.*, 1951). These nanoparticles showed potent optical properties like gold nanorods have absorption spectral band at longer wavelengths (Murphy *et al.*, 2005).

Semiconducting nanoparticles with well defined particles size and distribution pattern were prepared by hot injection synthetic method, which represented significant milestone (Donega *et al.*, 2005). Organometallic precursor injected rapidly into hot coordinating solvent, as a result produces instantaneous nucleation which grows further into the nanocrystals at lower temperature (Park *et al.*, 2007). Other important synthetic routes for metal oxides includes polyol-mediated syntheses (Feldmann, 2003), heat-up methods (Kwon & Hyeon, 2011), nonaqueous/nonhydrolytic, (Mutin & Vioux, 2009) and hydrolytic (Hench & West, 1990) sol–gel methods. Nonaqueous sol–gel approaches for nanoscale materials were more successful due to high crystallinity, a wide range of shapes and sizes, compositions as well as electronically conductive, magnetic, or ferroelectric materials properties as compared to aqueous processes (Niederberger & Pinna, 2009). The synthetic route involved the reaction of different metallic precursors like metal alkoxides, acetates, halides and acetylacetonates, with specific organic solvent or a mixture of definite organic solvents in the presence of surfactants followed by heating in the of range 50–250$_0$C. Different modes of heating include in an oil bath, in a microwave reactor or in an autoclave (Niederberger, 2007).

CONCLUSION

The work described above indicates the effectiveness of nanomaterials for the degradation of organic dyes *i.e.,* methylene blue, methyl orange, congo red, rhodamine B etc. as well as phenols and pharmaceutical residues. Carbon nanotubes and graphene oxide based nanocomposites are observed as excellent materials for removal/degradation of these pollutants. Moreover, the role of these nanocomposites and their photocatalytic activity as recent field of research is discussed. The recent work on nanometals for water treatment is also discussed. The green synthesis of metal nanoparticles using various plant extracts is a valuable addition in recent work. These particles have effectively been applied for degradation of pollutants. These materials could pose a solution to water pollution. So, there is a continued need for development of novel and effective nanocomposites for water treatment.

REFERENCES

Abdelkader, N. B., Bentouami, A., Derriche, Z., Bettahar, N., & de Ménorval, L. C. (2011). Synthesis and characterization of Mg–Fe layer double hydroxides and its application on adsorption of orange G from aqueous solution. *Chemical Engineering Journal, 169*(1-3), 231–238. doi:10.1016/j.cej.2011.03.019

Aditya, D., Rohan, P., & Suresh, G. (n.d.). Nano-adsorbents for wastewater treatment: a review. *Research Journal of Chemical & Environmental Sciences, 15,* 1033–40.

Ahmaruzzaman, M., & Gayatri, S. L. (2010). Activated tea waste as a potential low-cost adsorbent for the removal of p-nitrophenol from waste water. *Journal of Chemical & Engineering Data, 55*(11), 4614–4623. doi:10.1021/je100117s

Ahmed, M. A., & Abdel Messih, M. F. (2017). Synthesis of metallic silver nanoparticles decorated mesoporous SnO2 for removal of methylene blue dye by coupling adsorption and photocatalytic processes. *Journal of Photochemistry and Photobiology A Chemistry, 346*, 77–88. doi:10.1016/j.jphotochem.2017.05.048

Ahmed, M. J. K., Ahmaruzzaman, M., & Reza, R. A. (2014). Lignocellulosic-derived modified agricultural waste: Development, characterization and implementation in sequestering pyridine from aqueous solutions. *Journal of Colloid and Interface Science, 428*, 222–234. doi:10.1016/j.jcis.2014.04.049 PMID:24910057

Aksu, Z., & Yener, J. (2001). A comparative adsorption/biosorption study of mono-chlorinated phenols onto various sorbents. *Waste Management (New York, N.Y.), 21*(8), 695–702. doi:10.1016/S0956-053X(01)00006-X PMID:11699627

Ali, I., Al-Othman, Z. A., & Alharbi, O. M. (2016). Uptake of pantoprazole drug residue from water using novel synthesized composite iron nano adsorbent. *Journal of Molecular Liquids, 218*, 465–472. doi:10.1016/j.molliq.2016.02.088

Ali, I., AL-Othman, Z. A., & Alwarthan, A. (2016). Green synthesis of functionalized iron nano particles and molecular liquid phase adsorption of ametryn from water. *Journal of Molecular Liquids, 221*, 1168–1174. doi:10.1016/j.molliq.2016.06.089

Ali, I., AL-Othman, Z. A., & Alwarthan, A. (2016). Molecular uptake of congo red dye from water on iron composite nano particles. *Journal of Molecular Liquids, 224*, 171–176. doi:10.1016/j.molliq.2016.09.108

Ali, I., AL-Othman, Z. A., & Alwarthan, A. (2016). Synthesis of composite iron nano adsorbent and removal of ibuprofen drug residue from water. *Journal of Molecular Liquids, 219*, 858–864. doi:10.1016/j.molliq.2016.04.031

Ali, I., Alothman, Z. A., & Alwarthan, A. (2017). Uptake of propranolol on ionic liquid iron nanocomposite adsorbent: Kinetic, thermodynamics and mechanism of adsorption. *Journal of Molecular Liquids*, *236*, 205–213. doi:10.1016/j.molliq.2017.04.028

Ali, I., Alothman, Z. A., & Sanagi, M. M. (2015). Green synthesis of iron nano-impregnated adsorbent for fast removal of fluoride from water. *Journal of Molecular Liquids*, *211*, 457–465. doi:10.1016/j.molliq.2015.07.034

Ali, I., Asim, M., & Khan, T. A. (2012). Low cost adsorbents for the removal of organic pollutants from waste water. *Journal of Environmental Management*, *113*, 170–183. doi:10.1016/j.jenvman.2012.08.028 PMID:23023039

Altınışık, A., Gür, E., & Seki, Y. (2010). A natural sorbent, Luffa cylindrica for the removal of a model basic dye. *Journal of Hazardous Materials*, *179*(1-3), 658–664. doi:10.1016/j.jhazmat.2010.03.053 PMID:20378245

Antonopoulou, M., Karagianni, P., & Konstantinou, I. K. (2016). Kinetic and mechanistic study of photocatalytic degradation of flame retardant Tris (1-chloro-2-propyl) phosphate (TCPP). *Applied Catalysis B: Environmental*, *192*, 152–160. doi:10.1016/j.apcatb.2016.03.039

Apul, O. G., Wang, O., Zhou, Y., & Karanfil, Y. (2013). Adsorption of aromatic organic contaminants by graphene nanosheets: Comparison with carbon nanotubes and activated carbon. *Water Research*, *47*(4), 4648–1654. doi:10.1016/j.watres.2012.12.031 PMID:23313232

Aroguz, A. Z. (2006). Kinetics and thermodynamics of adsorption of azinphosmethyl from aqueous solution onto pyrolyzed (at 600 °C) ocean peat moss (Sphagnum sp.). *Journal of Hazardous Materials*, *135*(1-3), 100–105. doi:10.1016/j.jhazmat.2005.11.027 PMID:16442228

Ashbolt, N. J. (2004). Microbial contamination of drinking water and disease outcomes in developing regions. *Toxicology*, *198*(1–3), 229–238. doi:10.1016/j.tox.2004.01.030 PMID:15138046

Baker, C., Pradhan, A., Pakstis, L., Pochan, D. J., & Shah, S. I. (2005). Synthesis and antibacterial properties of silver nanoparticles. *Journal of Nanoscience and Nanotechnology*, *5*(2), 244–249. doi:10.1166/jnn.2005.034 PMID:15853142

Balogh, L., Swanson, D. R., Tomalia, D. A., Hagnauer, G. L., & McManus, A. T. (2001). Dendrimer-Silver Complexes and Nanocomposites as Antimicrobial Agents. *Nano Letters*, *1*(1), 18–21. doi:10.1021/nl005502p

Bello, B. A., Khan, S. A., Khan, J. A., Syed, F. Q., Mirza, M. B., Shah, L., & Khan, S. B. (2017). Anticancer, antibacterial and pollutant degradation potential of silver nanoparticles from *Hyphaene thebaica*. *Biochemical and Biophysical Research Communications*, *490*(3), 889–894. doi:10.1016/j.bbrc.2017.06.136 PMID:28648600

Bonetto, L. R., Ferrarini, F., deMarco, C., Crespo, J. S., Guégan, R., & Giovanela, M. (2015). Removal of methyl violet 2B dye from aqueous solution using a magnetic composite as an adsorbent. *Journal of Water Process Engineering*, *6*, 11–20. doi:10.1016/j.jwpe.2015.02.006

Botes, M., & Cloete, T. E. (2010). The potential of nanofibers and nanobiocides in water purification. *Critical Reviews in Microbiology*, *36*(1), 68–81. doi:10.3109/10408410903397332 PMID:20088684

Brillas, E., & Martínez-Huitle, C. A. (2015). Decontamination of wastewaters containing synthetic organic dyes by electrochemical methods. An updated review. *Applied Catalysis B: Environmental*, *166*, 603–643. doi:10.1016/j.apcatb.2014.11.016

Carević, M. V., Abazović, N. D., Novaković, T. B., Pavlović, V. B., & Čomor, M. I. (2016). Zirconium dioxide nanopowders with incorporated Si4+ ions as efficient photocatalyst for degradation of trichlorophenol using simulated solar light. *Applied Catalysis B: Environmental*, *195*, 112–120. doi:10.1016/j.apcatb.2016.05.005

Chen, A. H., Liu, S. C., Chen, C. Y., & Chen, C.-Y. (2008). Comparative adsorption of Cu(II), Zn(II), and Pb(II) ions in aqueous solution on the crosslinked chitosan with epichlorohydrin. *Journal of Hazardous Materials*, *154*(1-3), 184–191. doi:10.1016/j.jhazmat.2007.10.009 PMID:18031930

Chen, C. Y., & Chiang, C. L. (2008). Preparation of cotton fibers with antibacterial silver nanoparticles. *Materials Letters*, *62*(21-22), 3607–3609. doi:10.1016/j.matlet.2008.04.008

Chen, D., Li, Y., Zhang, J., Zhou, J. Z., Guo, Y., & Liu, H. (2012). Magnetic Fe3O4/ZnCr-layered double hydroxide composite with enhanced adsorption and photocatalytic activity. *Chemical Engineering Journal*, *185–186*, 120–126. doi:10.1016/j.cej.2012.01.059

Chen, G. C., Shan, X. Q., Pei, Z. G., Wang, H. H., Zheng, L. R., Zhang, J., & Xie, Y. (2011). Adsorption of diuron and dichlobenil on multiwalled carbon nanotubes as affected by lead. *Journal of Hazardous Materials*, *188*(1-3), 156–163. doi:10.1016/j.jhazmat.2011.01.095 PMID:21324587

Chen, M., Bao, C., Cun, T., & Huang, Q. (2017). One-pot synthesis of ZnO/oligoaniline nanocomposites with improved removal of organic dyes in water: Effect of adsorption on photocatalytic degradation. *Materials Research Bulletin*, *95*, 459–467. doi:10.1016/j.materresbull.2017.08.017

Chen, Y., Wang, L., Jiang, S., & Yu, H. (2003). Study on Novel Antibacterial Polymer Materials (I) Preparation of Zeolite Antibacterial Agents and Antibacterial Polymer Composite and Their Antibacterial Properties. *Journal of Polymer Materials*, *20*(3), 279–284.

Chiang, L. F., & Doong, R. A. (2014). Cu–TiO_2 nanorods with enhanced ultraviolet- and visible-light photoactivity for bisphenol degradation. *Journal of Hazardous Materials*, *277*, 84–92. doi:10.1016/j.jhazmat.2014.01.047 PMID:24556011

Choudhary, B. C., Paul, D., Gupta, T., Tetgure, S. R., Garole, V. J., Borse, A. U., & Garole, D. J. (2017). Photocatalytic reduction of organic pollutant under visible light by green route synthesized gold nanoparticles. *Journal of Environmental Sciences (China)*, *55*, 236–246. doi:10.1016/j.jes.2016.05.044 PMID:28477818

Costa, A. S., Romão, L. P. C., Araújo, B. R., Lucas, S. C. O., Maciel, S. T. A., Wisniewski, A. Jr, & Alexandre, M. R. (2012). Environmental strategies to remove volatile aromatic fractions (BTEX) from petroleum industry wastewater using biomass. *Bioresource Technology*, *105*, 31–39. doi:10.1016/j.biortech.2011.11.096 PMID:22178487

Dadfarnia, S., Haji Shabani, A. M., Moradi, S. E., & Emami, S. (2015). Methyl red removal from water by iron based metal-organic frameworks loaded onto iron oxide nanoparticle adsorbent. *Applied Surface Science*, *330*, 85–93. doi:10.1016/j.apsusc.2014.12.196

Dai, Z., Qin, F., Zhao, H. P., Ding, J., Liu, Y. L., & Chen, R. (2016). Crystal Defect Engineering of Aurivillius Bi_2MoO_6 by Ce Doping for Increased Reactive Species Production in Photocatalysis. *ACS Catalysis*, *6*(5), 3180–3192. doi:10.1021/acscatal.6b00490

Danilczuk, M., Lund, A., Sadlo, J., Yamada, H., & Michalik, J. (2006). Conduction electron spin resonance of small silver particles, *Spectrochimica Acta-art A. Molecular and Biomolecular Spectroscopy*, *63*(1), 189–191. doi:10.1016/j.saa.2005.05.002 PMID:15978868

De Martino, A., Iorio, M., Xing, B. S., & Capasso, R. (2012). Removal of 4-chloro-2-methylphenoxyacetic acid from water by sorption on carbon nanotubes and metal oxide nanoparticles. *RSC Advances*, *2*(13), 5693–5700. doi:10.1039/c2ra00038e

De Sales, P. F., Magriotis, Z. M., Rossi, M. A. L. S., Resende, R. F., & Nunes, C. A. (2015). Comparative analysis of tropaeolin adsorption onto raw and acid-treated kaolinite: Optimization by response surface methodology. *Journal of Environmental Management*, *151*, 144–152. doi:10.1016/j.jenvman.2014.12.008 PMID:25559496

Demirbas, A. (2009). Agricultural based activated carbons for the removal of dyes from aqueous solutions: A review. *Journal of Hazardous Materials*, *167*(1-3), 1–9. doi:10.1016/j.jhazmat.2008.12.114 PMID:19181447

Deng, J., Shao, Y. S., Gao, N. Y., Deng, Y., Tan, C. Q., Zhou, S. Q., & Hu, X. (2012a). Multiwalled carbon nanotubes as adsorbents for removal of herbicide diuron from aqueous solution. *Chemical Engineering Journal*, *193*, 339–347. doi:10.1016/j.cej.2012.04.051

Ding, X., Ho, W., Shang, J., & Zhang, L. (2016). Self-doping promoted photocatalytic removal of no under visible light with bi2moo6: Indispensable role of superoxide ions. *Applied Catalysis B: Environmental*, *182*, 316–325. doi:10.1016/j.apcatb.2015.09.046

Donega, M. C., Liljeroth, P., & Vanmaekelbergh, D. (2005). Physicochemical evaluation of the hot-injection method, a synthesis route for monodisperse nanocrystals. *Small*, *1*(12), 1152–1162. doi:10.1002mll.200500239 PMID:17193409

Duran, C., Ozdes, D., Gundogdu, A., & Senturk, H. B. (2011). Kinetics and isotherm analysis of basic dyes adsorption onto almond shell (Prunus dulcis) as a low cost adsorbent. *Journal of Chemical & Engineering Data*, *56*(5), 2136–2147. doi:10.1021/je101204j

El-Zahhar, A. A., Awwad, N. S., & El-Katori, E. E. (2014). Removal of bromophenol blue dyes from industrial waste water by synthesizing polymer-clay composite. *Journal of Molecular Liquids*, *199*, 454–461. doi:10.1016/j.molliq.2014.07.034

Elmoubarki, R., Mahjoubi, F. Z., Tounsadi, H., Moustadraf, J., Abdennouri, M., Zouhri, A., ... Barka, N. (2015). Adsorption of textile dyes on raw and decanted Moroccan clays: Kinetics, equilibrium and thermodynamics. *Water Resources and Industry*, *9*, 16–29. doi:10.1016/j.wri.2014.11.001

Falah, M., MacKenzie, K. J. D., Knibbe, R., Page, S. J., & Hanna, J. V. (2016). New composites of nanoparticle Cu (I) oxide and titania in a novelinorganic polymer (geopolymer) matrix for destruction of dyes andhazardous organic pollutants. *Journal of Hazardous Materials*, *318*, 772–782. doi:10.1016/j.jhazmat.2016.06.016 PMID:27329791

Feldmann, C. (2003). Polyol-Mediated Synthesis of Nanoscale Functional Materials. *Advanced Functional Materials*, *13*(2), 101–107. doi:10.1002/adfm.200390014

Feng, J. T., Zhu, J. W., Lv, W., Li, J. J., & Yan, W. (2015). Effect of hydroxyl group of carboxylic acids on the adsorption of acid red G and methylene blue on TiO2. *Chemical Engineering Journal*, *269*, 316–322. doi:10.1016/j.cej.2015.01.109

Feng, Q. L., Wu, J., Chen, G. Q., Cui, F. Z., Kim, T. N., & Kim, J. O. (2000). A mechanistic study of the antibacterial effect of silver ions on *Escherichia coli* and Staphylococcus aureus. *Journal of Biomedical Materials Research*, *52*(4), 662–668. doi:10.1002/1097-4636(20001215)52:4<662::AID-JBM10>3.0.CO;2-3 PMID:11033548

Feng, Y. F., Yang, F., Wang, Y. Q., Ma, L., Wu, Y. H., Kerr, P. G., & Yang, L. Z. (2011). Basic dye adsorption onto an agro-based waste material-sesame hull (Sesamum indicum L.). *Bioresource Technology*, *102*(22), 10280–10285. doi:10.1016/j.biortech.2011.08.090 PMID:21962534

Fotiou, T., Triantis, T. M., Kaloudis, T., O'Shea, K. E., Dionysiou, D. D., & Hiskia, A. (2016). Assessment of the roles of reactive oxygen species in the UV and visible light photocatalytic degradation of cyanotoxins and water taste and odor compounds using $CeTiO_2$. *Water Research*, *90*, 52–61. doi:10.1016/j.watres.2015.12.006 PMID:26724439

Fox, M. (1988). Photocatalytic Oxidation of Organic Substances. In Photocatalysis and Environment: Trends and Applications. New York: Academic Publishers.

Gao, P., Feng, Y. J., Zhang, Z. H., Liu, J. F., & Ren, N. Q. (2011). Comparison of competitive and synergetic adsorption of three phenolic compounds on river sediment. *Environmental Pollution*, *159*(10), 2876–2881. doi:10.1016/j.envpol.2011.04.047 PMID:21636192

Ghaedi, M., Shokrollahi, A., Hossainian, H., & Kokhdan, S. N. (2011b). Comparison of activated carbon and multiwalled carbon nanotubes for efficient removal of eriochrome cyanine R (ECR): Kinetic, isotherm, and thermodynamic study of the removal process. *Journal of Chemical & Engineering Data*, *56*(7), 3227–3235. doi:10.1021/je200331u

Ghaedi, M., Shokrollahi, A., Tavallali, H., Shojaiepoor, F., Keshavarz, B., Hossainian, H., ... Purkait, M. K. (2011c). Activated carbon and multiwalled carbon nanotubes as efficient adsorbents for removal of arsenazo(III) and methyl red from waste water. *Toxicological and Environmental Chemistry*, *93*(3), 438–449. doi:10.1080/02772248.2010.540244

Ghorai, T. K., Biswas, S. K., & Pramanik, P. (2008). Surface modification with both phosphorylcholine and stearyl groups to adjust hydrophilicity and hydrophobicity. *Applied Surface Science*, *254*, 498–7504.

Goh, K., Setiawan, L., Wei, L., Si, R., Fane, A. G., Wang, R., & Chen, Y. (2015). Preparation of graphene oxide-based hydrogels as efficient dye adsorbents for wastewater treatment. *Nanoscale Research Letters*, *10*(1), 272–281. doi:10.118611671-015-0931-2 PMID:26123269

Guo, H., Jiao, T., Qingrui, Z., Wenfeng, G., Qiuming, P., & Xuehai, Y. (2015). Preparation of graphene oxide-based hydrogels as efficient dye adsorbents for wastewater treatment. *Nanoscale Research Letters, 10*(1), 272–281. doi:10.118611671-015-0931-2 PMID:26123269

Guo, Z. B., Liu, J., & Liu, F. L. (2015). Selective adsorption of p-CNB and o-CNB in silica-coating HZSM-5 zeolite. *Microporous and Mesoporous Materials, 213*, 8–13. doi:10.1016/j.micromeso.2015.04.006

Gupta, V. K., Ali, I., Saleh, T. A., Nayak, A., & Agarwal, S. (2012). Chemical treatment technologies for waste-water recycling—an overview. *RSC Advances, 2*(16), 6380–6388. doi:10.1039/c2ra20340e

Güzel, F., Sayğılı, H., Sayğılı, G. A., & Koyuncu, F. (2014). Elimination of anionic dye by using nanoporous carbon prepared from an industrial biowaste. *Journal of Molecular Liquids, 194*, 130–140. doi:10.1016/j.molliq.2014.01.018

Hench, L. L., & West, J. K. (1990). The sol-gel process. *Chemical Reviews, 90*(1), 33–72. doi:10.1021/cr00099a003

Hosseini, S. J., Kokhdan, S. N., Ghaedi, A. M., & Moosavian, S. S. (2011). Comparison of multiwalled carbon nanotubes and activated carbon for efficient removal of methyl orange: Kinetic and thermodynamic investigation. *Fresenius Environmental Bulletin, 20*, 219–234.

Hosseini, S. J., Kokhdan, S. N., Ghaedi, A. M., & Moosavian, S. S. (2011). Comparison of multiwalled carbon nanotubes and activated carbon for efficient removal of methyl orange: Kinetic and thermodynamic investigation. *Fresenius Environmental Bulletin, 20*, 219–234.

Huang, H., Zhang, J., Jiang, L., & Zang, Z. (2017). Preparation of cubic Cu2O nanoparticles wrapped by reduced graphene oxide for the efficient removal of rhodamine B. *Journal of Alloys and Compounds, 718*, 112–115. doi:10.1016/j.jallcom.2017.05.132

Hutton, G., Haller, L., & Bartram, J. (2007). *Economic and Health Effects of Increasing Coverage of Low Cost Household Drinking Water Supply and Sanitation Interventions.* World Health Organization.

Ihsanullah, H. A., Asmaly, T. A., Saleh, T., Laoui, V. K., Gupta, V. K., & Atieh, M. A. (2015). Enhanced adsorption of phenols from liquids by aluminum oxide/carbon nanotubes: Comprehensive study from synthesis to surface properties. *Journal of Molecular Liquids, 206*, 176–182. doi:10.1016/j.molliq.2015.02.028

Iijima, S. (1991). Helical microtubules of graphitic carbon. *Nature, 354*(6348), 56–58. doi:10.1038/354056a0

Inoue, Y., Hoshino, M., Takahashi, H., Noguchi, T., Murata, T., Kanzaki, Y., ... Sasatsu, M. (2002). Bactericidal activity of Ag-zeolite mediated by reactive oxygen species under aerated conditions. *Journal of Inorganic Biochemistry, 92*(1), 37–42. doi:10.1016/S0162-0134(02)00489-0 PMID:12230986

Ishibashi, K., Fujishima, A., Watanabe, T., & Hashimoto, K. (2000). Quantum yields of active oxidative species formed on TiO2 photocatalyst. *Journal of Photochemistry and Photobiology A Chemistry, 134*(1-2), 139–142. doi:10.1016/S1010-6030(00)00264-1

Jalil, A. A., Triwahyono, S., Adam, S. H., Rahim, N. D., Aziz, M. A. A., Hairom, N. H. H., ... Mohamadiah, M. K. A. (2010). Adsorption of methyl orange from aqueous solution onto calcined Lapindo volcanic mud. *Journal of Hazardous Materials*, *181*(1-3), 755–762. doi:10.1016/j.jhazmat.2010.05.078 PMID:20538408

Jeon, H. J., Kim, J. S., Kim, T. G., Kim, J. H., Yu, W. R., & Youk, J. H. (2008). Preparation of poly(ε-caprolactone)-based polyurethane nanofibers containing silver nanoparticles. *Applied Surface Science*, *254(*18), 5886–5890. doi:10.1016/j.apsusc.2008.03.141

Joseph, L., Heo, J., Park, Y. G., Flora, J. R. V., & Yoon, Y. (2011a). Adsorption of bisphenol A and 17 alphaethinyl estradiol on single walled carbon nanotubes from seawater and brackish water. *Desalination*, *281*, 68–74. doi:10.1016/j.desal.2011.07.044

Jothivenkatachalam, K., Prabhu, S., Nithya, A., & Jeganathan, K. (2014). Facile synthesis of WO_3 with reduced particle size on zeolite and enhanced photocatalytic activity. *RSC Advances*, *4*(41), 21221–21229. doi:10.1039/C4RA01376J

Kamaraj, R., Davidson, D. J., Sozhan, G., & Vasudevan, S. (2014). An in situ electrosynthesis of metal hydroxides and their application for adsorption of 4-chloro-2-methylphenoxyacetic acid (MCPA) from aqueous solution. *Journal of Environmental Chemical Engineering*, *2*(4), 2068–2077. doi:10.1016/j.jece.2014.08.027

Kaur, A., & Kansal, S. K. (2016). Bi_2WO_6 nanocuboids: An efficient visible light active photocatalyst for the degradation of levofloxacin drug in aqueous phase. *Chemical Engineering Journal*, *302*, 194–203. doi:10.1016/j.cej.2016.05.010

Khataee, A. R., Vafaei, F., & Jannatkhah, M. (2013). Biosorption of three textile dyes from contaminated water by filamentous green algal Spirogyra sp.: Kinetic, isotherm and thermodynamic studies. *International Biodeterioration & Biodegradation*, *83*, 33–40. doi:10.1016/j.ibiod.2013.04.004

Kim, J. S., Kuk, E., & Yu, K. N. (2007). Antimicrobial effects of silver nanoparticles Nanomedicine. *Nanotechnology. Biology, and Medicine*, *3*(1), 95–101.

Kochameshki, M. G., Marjani, A., Mahmoudian, M., & Farhadi, K. (2017). Grafting of diallyldimethylammonium chloride on graphene oxide by RAFT polymerization for modification of nanocomposite polysulfone membranes using in water treatment. *Chemical Engineering Journal*, *309*, 206–221. doi:10.1016/j.cej.2016.10.008

Konicki, W., Pełech, I., Mijowska, E., & Jasińska, I. (2012). Adsorption of anionic dye direct red 23 onto magnetic multi-walled carbon nanotubes-Fe3C nanocomposite: Kinetics, equilibrium and thermodynamics. *Chemical Engineering Journal*, *210*, 87–95. doi:10.1016/j.cej.2012.08.025

Kousha, M., Tavakoli, S., Daneshvar, E., Vazirzadeh, A., & Bhatnagar, A. (2015). Central composite design optimization of acid blue 25 dye biosorption using shrimp shell biomass. *Journal of Molecular Liquids*, *207*, 266–273. doi:10.1016/j.molliq.2015.03.046

Kumar, V. S., Nagaraja, B. M., Shashikala, V., Padmasri, A. H., Madhavendra, S. S., Raju, B. D., & Rao, K. S. R. (2004). Highly efficient Ag/C catalyst prepared by electro-chemical deposition method in controlling microorganisms in water. *Journal of Molecular Catalysis A Chemical*, *223*(1-2), 313–319. doi:10.1016/j.molcata.2003.09.047

Kwon, S. G., & Hyeon, T. (2011). Formation Mechanisms of Uniform Nanocrystals via Hot-Injection and Heat-Up Methods. *Small*, *7*(19), 2685–2702. doi:10.1002mll.201002022 PMID:21805623

Lala, N. L., Ramaseshan, R., Bojun, L., Sundarrajan, S., Barhate, R. S., Ying-jun, L., & Ramakrishna, S. (2007). Fabrication of nanofibers with antimicrobial functionality used as filters protection against bacterial contaminants. *Biotechnology and Bioengineering*, *97*(6), 1357–1365. doi:10.1002/bit.21351 PMID:17274060

Latorre, C. H., Méndez, J. A., García, J. B., Martín, S. G., & Crecente, R. M. P. (2012). Carbon nanotubes as solid-phase extraction sorbents prior to atomic spectrometric determination of metal species: A review. *Analytica Chimica Acta*, *749*, 16–35. doi:10.1016/j.aca.2012.09.001 PMID:23036463

Leonard, P., Hearty, S., Brennan, J., Dunne, L., Quinn, J., Chakraborty, T., & O'Kennedy, R. (2003). Advances in biosensors for detection of pathogens in food and water. *Enzyme and Microbial Technology*, *32*(1), 3–13. doi:10.1016/S0141-0229(02)00232-6

Li, C., Dong, Y., Yang, J., Li, Y., & Huang, C. (2014). Modified nano-graphite/Fe_3O_4 composite as efficient adsorbent for the removal of methyl violet from aqueous solution. *Journal of Molecular Liquids*, *196*, 348–356. doi:10.1016/j.molliq.2014.04.010

Liau, S. Y., Read, D. C., Pugh, W. J., Furr, J. R., & Russell, A. D. (1997). Interaction of silver nitrate with readily identifiable groups: Relationship to the antibacterial action of silver ions. *Letters in Applied Microbiology*, *25*(4), 279–283. doi:10.1046/j.1472-765X.1997.00219.x PMID:9351278

Lin, D. H., & Xing, B. S. (2007). Phytotoxicity of nanoparticles: Inhibition of seed germination and root growth. *Environ. Poll.*, *150*(2), 243–250. doi:10.1016/j.envpol.2007.01.016

Lin, S., Hsiao, R., & Juang, R. (2006). Removal of soluble organics from water by a hybrid process of clay adsorption and membrane filtration. *Journal of Hazardous Materials*, *135*(1-3), 134–140. doi:10.1016/j.jhazmat.2005.11.030 PMID:16359788

Liu, X., Zhao, X. F., Zhu, Y., & Zhang, F. Z. (2013). Experimental and theoretical investigation into the elimination of organic pollutants from solution by layered double hydroxides. *Applied Catalysis B: Environmental*, *140–141*, 241–248. doi:10.1016/j.apcatb.2013.04.008

Ma, J., Zhang, L. Z., Wang, Y. H., Lei, S. L., Luo, X. B., Chen, S. H., ... Au, C. T. (2014). Mechanism of 2,4-dinitrophenol photocatalytic degradation by ζ-Bi2O3/Bi2MoO6 composites under solar and visible light irradiation. *Chemical Engineering Journal*, *251*, 371–380. doi:10.1016/j.cej.2014.04.085

Ma, X. M., Tsige, M., Uddin, S., & Talapatra, S. (2011b). Application of carbon nanotubes for removing organic contaminants from water. *Materials Express*, *1*(3), 183–200. doi:10.1166/mex.2011.1023

Makhluf, S., Dror, R., Nitzan, Y., Abramovich, Y., Jelinek, R., & Gedanken, A. (2005). Microwave-assisted synthesis of nanocrystalline MgO and its use as a bacteriocide. *Advanced Functional Materials*, *15*(10), 1708–1715. doi:10.1002/adfm.200500029

Maphutha, S., Moothi, K., Meyyappan, M., & Iyuke, S. E. (2013). A carbon nanotube-infused polysulfone membrane with polyvinyl alcohol layer for treating oil-containing waste water. *Scientific Reports*, *3*(1), 1509. doi:10.1038rep01509 PMID:23518875

Mohmood, I., Lopes, C. B., Lopes, I., Ahmad, I., Duarte, A. C., & Pereira, E. (2013). Nanoscale materials and their use in water contaminants removal—a review. *Environmental Science and Pollution Research International*, *20*(3), 1239–1260. doi:10.100711356-012-1415-x PMID:23292223

Morones, J. R., Elechiguerra, J. L., Camacho, A., Holt, K., Kouri, J. B., Ramírez, J. T., & Yacaman, M. J. (2005). T-bactericidal effect of silver nanoparticles. *Nanotechnology*, *16*(10), 2346–2353. doi:10.1088/0957-4484/16/10/059 PMID:20818017

Murcia, M. D., Gómez, M., Gómez, E., Gómez, J. L., & Christofi, N. (2011). Photodegradation of congo red using XeBr, KrCl and Cl_2 barrier discharge excilamps: A kinetics study. *Desalination*, *281*, 364–371. doi:10.1016/j.desal.2011.08.011

Murphy, C. J., Sau, T. K., Gole, A. M., Orendorff, C. J., Gao, J., Gou, L., ... Li, T. (2005, July). Anisotropic Metal Nanoparticles: Synthesis, Assembly, and Optical Applications. *The Journal of Physical Chemistry B*, *109*(29), 13857–13870. doi:10.1021/jp0516846 PMID:16852739

Mutin, P. H., & Vioux, A. (2009). Nonhydrolytic Processing of Oxide-Based Materials: Simple Routes to Control Homogeneity, Morphology, and Nanostructure. *Chemistry of Materials*, *21*(4), 582–596. doi:10.1021/cm802348c

Ngah, W. S., & Fatinathan, S. (2008). Adsorption of Cu(II) ions in aqueous solution using chitosan beads, chitosan-GLA beads and chitosan-alginate beads. *Chemical Engineering Journal*, *143*(1-3), 62–72. doi:10.1016/j.cej.2007.12.006

Nie, L. R., Lu, J., Zhang, W., He, A., & Yao, S. (2015). Ionic liquid-modified silica gel as adsorbents for adsorption and separation of water-soluble phenolic acids from Salvia militiorrhiza Bunge. *Seperation and Purification Technology*. .2015.01.03710.1016/j.seppur

Niederberger, M. (2007). Nonaqueous Sol–Gel Routes to Metal Oxide Nanoparticles. *Accounts of Chemical Research*, *40*(9), 793–800. doi:10.1021/ar600035e PMID:17461544

Niederberger, M., & Pinna, N. (2009). *Metal Oxide Nanoparticles in Organic Solvents: Synthesis, Formation, Assembly and Application*. London: Springer-Verlag. doi:10.1007/978-1-84882-671-7

Njoku, V. O., Islam, M. A., Asif, M., & Hameed, B. H. (2014). Utilization of sky fruit husk agricultural waste to produce high quality activated carbon for the herbicide bentazon adsorption. *Chemical Engineering Journal*, *251*, 183–191. doi:10.1016/j.cej.2014.04.015

Noreen, S., Bhatti, H. N., Nausheen, S., Sadaf, S., & Ashfaq, M. (2013). Batch and fixed bed adsorption study for the removal of drimarine black CL-B dye from aqueous solution using a lignocellulosic waste: A cost affective adsorbent. *Industrial Crops and Products*, *50*, 568–579. doi:10.1016/j.indcrop.2013.07.065

Pal, S., Tak, Y. K., & Song, J. M. (2007). Does the antibacterial activity of silver nanoparticles depend on the shape of the nanoparticle A study of the gram-negative bacterium Escherichia coli. *Applied and Environmental Microbiology, 73*(6), 1712–1720. doi:10.1128/AEM.02218-06 PMID:17261510

Pan, J. H., Dou, H., Xiong, Z., Xu, C., Ma, J., & Zhao, X. S. (2010). Porous photocatalysts for advanced water purifications. *Journal of Materials Chemistry, 20*(22), 4512–4528. doi:10.1039/b925523k

Panáček, A., Kvítek, L., Prucek, R., Kolář, M., Večeřová, R., Pizúrová, N., ... Zbořil, R. (2006). Silver colloid nanoparticles: Synthesis, characterization, and their antibacterial activity. *The Journal of Physical Chemistry B, 110*(33), 16248–16253. doi:10.1021/jp063826h PMID:16913750

Park, J., Joo, J., Kwon, S. G., Jang, Y., & Hyeon, T. (2007). Synthesis of monodisperse spherical nanocrystals. *Angewandte Chemie International Edition, 46*(25), 4630–4660. doi:10.1002/anie.200603148 PMID:17525914

Peng, H. B., Pan, B., Wu, M., Liu, Y., Zhang, D., & Xing, B. S. (2012). Adsorption of ofloxacin and norfloxacin on carbon nanotubes: Hydrophobicity- and structure-controlled process. *Journal of Hazardous Materials, 233*, 89–96. doi:10.1016/j.jhazmat.2012.06.058 PMID:22819959

Perreault, F., Fonseca de Faria, A., & Elimelech, M. (2015). Environmental applications of graphene-based nanomaterials. *Chemical Society Reviews, 44*(16), 5861–5896. doi:10.1039/C5CS00021A PMID:25812036

Pomiès, M., Choubert, J. M., Wisniewski, C., & Coquery, M. (2013). Modelling of micropollutant removal in biological wastewater treatments: A review. *The Science of the Total Environment, 443*, 733–748. doi:10.1016/j.scitotenv.2012.11.037 PMID:23228719

Ramesha, G. K., Kumara, A. V., Muralidhara, H. B., & Sampath, S. S. (2011). Graphene and graphene oxide as effective adsorbents toward anionic and cationic dyes. *Journal of Colloid and Interface Science, 361*(1), 270–277. doi:10.1016/j.jcis.2011.05.050 PMID:21679961

Raoof, J., Hosseini, S. R., & Rezaee, S. (2014). Preparation of Pt/poly(2-methoxyaniline)/multiwalled carbon nanotube nanocomposite and its application for electrocatalytic oxidation of methanol. *Journal of Molecular Liquids, 200*, 196–204. doi:10.1016/j.molliq.2014.09.043

Rathinam, A., Rao, J. R., & Nair, B. U. (2011). Adsorption of phenol onto activated carbon from seaweed: Determination of the optimal experimental parameters using factorial design. *Journal of Taiwan Institute of Chemical Engineers, 42*(6), 952–956. doi:10.1016/j.jtice.2011.04.003

Rezaee, R., Nasseri, S., Mahvi, A. H., Nabizadeh, R., Mousavi, S. A., Rashidi, A., ... Nazmara, S. (2015). Fabrication and characterization of a polysulfone-graphene oxide nanocomposite membrane for arsenate rejection from water. *Journal of Environmental Health Science & Engineering, 13*(1), 61–71. doi:10.118640201-015-0217-8 PMID:26301096

Rockstr¨om, J. (2003). Water for food and nature in drought-prone tropics: Vapour shift in rain-fed agriculture. *Philosophical Transactions of the Royal Society of London. Series B, Biological Sciences, 358*(1440), 1997–2009. doi:10.1098/rstb.2003.1400 PMID:14728794

Ruelle, B., Peeterbroeck, S., Godfroid, T., Bittencourt, C., Hecq, M., Snyders, R., & Dubois, P. (2012). Selective grafting of primary amines onto carbon nanotubes via free-radical treatment in microwave plasma post-discharge. *Polymers*, *4*(1), 296–315. doi:10.3390/polym4010296

Ruffini-Castiglione, M., & Cremonini, R. (2009). Article. *Caryologia*, *62*, 161–165. doi:10.1080/00087114.2004.10589681

Saad, S. A., Isa, K. M., & Bahari, R. (2010). Chemically modified sugarcane bagasse as a potentially low-cost biosorbent for dye removal. *Desalination*, *264*(1-2), 123–128. doi:10.1016/j.desal.2010.07.015

Sadegh, H., Shahryari-ghoshekandi, R., Agarwal, S., Tyagi, I., Asif, M., & Gupta, V. K. (2015). Microwave-assisted removal of malachite green by carboxylate functionalized multi-walled carbon nanotubes: Kinetics and equilibrium study. *Journal of Molecular Liquids*, *206*, 151–158. doi:10.1016/j.molliq.2015.02.007

Saikia, N., Sarma, J., Borah, J. M., & Mahiuddin, S. (2013). Adsorption of 3,4-dihydroxybenzoic acid onto hematite surface in aqueous medium: Importance of position of phenolic –OH groups and understanding of the same using catechol as an auxiliarymodel. *Journal of Colloid and Interface Science*, *398*, 227–233. doi:10.1016/j.jcis.2013.02.028 PMID:23518302

Salam, M. A., El-Shishtawy, R. M., & Obaid, A. Y. (2014). Synthesis of magnetic multi-walled carbon nanotubes/magnetite/chitin magnetic nanocomposite for the removal of Rose Bengal from real and model solution. *Journal of Industrial and Engineering Chemistry*, *20*(5), 3559–3567. doi:10.1016/j.jiec.2013.12.049

Saleh, T. A., Al-Saadi, A. A., & Gupta, V. K. (2014). Carbonaceous adsorbent prepared from waste tires: Experimental and computational evaluations of organic dyemethyl orange. *Journal of Molecular Liquids*, *191*, 85–91. doi:10.1016/j.molliq.2013.11.028

Salman, J. M., Njoku, V. O., & Hameed, B. H. (2011). Adsorption of pesticides from aqueous solution onto banana stalk activated carbon. *Chemical Engineering Journal*, *174*(1), 41–48. doi:10.1016/j.cej.2011.08.026

Sen, D., Ghosh, A. K., Majumder, S., Bindal, R. C., & Tiwari, P. K. (2014). Novel polysulfone-spray-dried silica composite membrane for water purification: Preparation, characterization and performance evaluation. *Separation and Purification Technology*, *123*, 79–86. doi:10.1016/j.seppur.2013.12.019

Sen, T. K., Afroze, S., & Ang, H. (2011). Equilibrium, kinetics and mechanism of removal of methylene blue from aqueous solution by adsorption onto pine cone biomass of Pinus radiate. *Water, Air, and Soil Pollution*, *218*(1-4), 499–515. doi:10.100711270-010-0663-y

Shan, R. R., Yan, L. G., Yang, K. Y., Yu, S. J., Hao, Y. F., Yu, H. Q., & Du, B. (2014). Magnetic Fe3O4/MgAl- LDH composite for effective removal of three red dyes from aqueous solution. *Chemical Engineering Journal*, *252*, 38–46. doi:10.1016/j.cej.2014.04.105

Sharma, Y. C. (2011). Adsorption characteristics of a low-cost activated carbon for the reclamation of colored effluents containing malachite green. *Journal of Chemical & Engineering Data*, *56*(3), 478–484. doi:10.1021/je1008922

Shi, S., Fan, Y. W., & Huang, Y. M. (2013). Facile low temperature hydrothermal synthesis of magnetic mesoporous carbon nanocomposite for adsorption removal of ciprofloxacin antibiotics. *Industrial & Engineering Chemistry Research, 52*(7), 2604–2612. doi:10.1021/ie303036e

Shrivastava, S., Bera, T., Roy, A., Singh, G., Ramachand, P., & Dash, D. (2007). Characterization of enhanced antibacterial effect of novel silver nanoparticles. *Nanotechnology, 18*(22), 225103. doi:10.1088/0957-4484/18/22/225103

Sondi, I., & Salopek-Sondi, B. (2004). Silver nanoparticles as antimicrobial agent: A case study on E. coli as a model for Gramnegative bacteria. *Journal of Colloid and Interface Science, 275*(1), 177–182. doi:10.1016/j.jcis.2004.02.012 PMID:15158396

Spadaro, J. A., Berger, T. J., Barranco, S. D., Chapin, S. E., & Becker, R. O. (1974). Antibacterial effects of silver electrodes with weak direct current. *Antimicrobial Agents and Chemotherapy, 6*(5), 637–642. doi:10.1128/AAC.6.5.637 PMID:15825319

Sui, Z. Y., Meng, Q. H., Zhang, X. T., Ma, R., & Cao, B. (2012b). Green synthesis of carbon nanotube-graphene hybrid aerogels and their use as versatile agents for water purification. *Journal of Materials Chemistry, 22*(18), 8767–8771. doi:10.1039/c2jm00055e

Sweetman, L. J., Nghiem, L., Chironi, I., Triani, G., Panhuis, M. I. H., & Ralph, S. F. (2012). Synthesis, properties and water permeability of SWNT buckypapers. *Journal of Materials Chemistry, 22*(27), 13800–13810. doi:10.1039/c2jm31382k

Tian, H., Peng, J., Lv, T., Sun, C., & He, H. (2018). Preparation and performance study of MgFe2O4/metal–organic framework composite for rapid removal of organic dyes from water. *Journal of Solid State Chemistry, 257*, 40–48. doi:10.1016/j.jssc.2017.09.017

Tian, Y., Gao, B., Morales, V. L., Chen, H., Wang, Y., & Li, H. (2013). Removal of sulfamethoxazole and sulfapyridine by carbon nanotubes in fixed-bed columns. *Chemosphere, 90*(10), 2597–2605. doi:10.1016/j.chemosphere.2012.11.010 PMID:23232047

Turkevich, J., Stevenson, P. C., & Hillier, J. (1951). A study of the nucleation and growth processes in the synthesis of colloidal gold. *Discussions of the Faraday Society, 11*, 55–75. doi:10.1039/df9511100055

Vaiano, V., Sacco, O., Sannino, D., & Ciambelli, P. (2015). Photocatalytic removal of spiramycin from wastewater under visible light with N-doped TiO2 photocatalysts. *Chemical Engineering Journal, 261*, 3–8. doi:10.1016/j.cej.2014.02.071

Verma, A. K., Dash, R. R., & Bhunia, P. (2012). A review on chemical coagulation/flocculation technologies for removal of colour from textile wastewaters. *Journal of Environmental Management, 93*(1), 154–168. doi:10.1016/j.jenvman.2011.09.012 PMID:22054582

Vimala, K., Samba Sivudu, K., Murali Mohan, Y., Sreedhar, B., & Mohana Raju, K. (2009). Controlled silver nanoparticles synthesis in semi-hydrogel networks of poly(acrylamide) and carbohydrates: A rational methodology for antibacterial application. *Carbohydrate Polymers, 75*(3), 463–471. doi:10.1016/j.carbpol.2008.08.009

Vinhal, J. O., Lage, M. R., Carneiro, J. W. M., Lima, C. F., & Cassella, R. J. (2015). Modeling, kinetic, and equilibrium characterization of paraquat adsorption onto polyurethane foamusing the ion-pairing technique. *Journal of Environmental Management*, *156*, 200–208. doi:10.1016/j.jenvman.2015.03.022 PMID:25846000

Wang, A. A., Li, S. X., Teng, Y., Liu, W. X., Wu, L. H., Zhang, H. B., ... Christie, P. (2013). Adsorption and desorption characteristics of diphenylarsenicals in two contrasting soils. *Journal of Environmental Sciences (China)*, *25*(6), 1172–1179. doi:10.1016/S1001-0742(12)60148-X PMID:24191607

Wei, H., Deng, S., Huang, Q., Nie, Y., Wang, B., Huang, J., & Yu, G. (2013). Regenerable granular carbon nanotubes/alumina hybrid adsorbents for diclofenac sodium and carbamazepine removal from aqueous solution. *Water Research*, *47*(12), 4139–4147. doi:10.1016/j.watres.2012.11.062 PMID:23579087

Wen, Z., Zhang, Y., & Dai, C. (2014). Removal of phosphate from aqueous solution using nanoscale zerovalent iron (nZVI). *Colloids and Surfaces. A, Physicochemical and Engineering Aspects*, *457*, 433–440. doi:10.1016/j.colsurfa.2014.06.017

Weon, S., & Choi, W. (2016). TiO_2 Nanotubes with Open Channels as Deactivation-Resistant Photocatalyst for the Degradation of Volatile Organic Compounds. *Environmental Science & Technology*, *50*(5), 2556–2563. doi:10.1021/acs.est.5b05418 PMID:26854616

World Health Organization (WHO) and UNICEF. (2013). *Progress on Sanitation and Drinking-Water*. Geneva, Switzerland: World Health Organization.

Wu, Y., Yi, N., Huang, L., Zhang, T., Fang, S., Chang, H., ... Chen, Y. (2015). Three-dimensionally bonded spongy graphene material with super compressive elasticity and near-zero Poisson's ratio. *Nature Communications*, *6*(1), 6141–6149. doi:10.1038/ncomms7141 PMID:25601131

Wu, Yu, Liu, Dionysiou, Yang, Wang, & Liu. (2017). *Novel Fluorinated Bi_2MoO_6 Nanocrystals for Efficient Photocatalytic Removal of Water Organic Pollutants under Different Light Source Illumination*. Academic Press.

Xia, S., Zhang, L., Pan, G., Qian, P., & Ni, Z. (2015). Photocatalytic degradation of methylene blue with a nanocomposite system: Synthesis, photocatalysis and degradation pathways. *Physical Chemistry Chemical Physics*, *17*(7), 5345–5351. doi:10.1039/C4CP03877K PMID:25611297

Xiu, Z. M., Ma, J., & Alvarez, P. J. J. (2011). Differential effect of common ligands and molecular oxygen on antimicrobial activity of silver nanoparticles versus silver ions. *Environmental Science & Technology*, *45*(20), 9003–9008. doi:10.1021/es201918f PMID:21950450

Xiu, Z. M., Zhang, Q. B., Puppala, H. L., Colvin, V. L., & Alvarez, P. J. J. (2012). Negligible particle-specific antibacterial activity of silver nanoparticles. *Nano Letters*, *12*(8), 4271–4275. doi:10.1021/nl301934w PMID:22765771

Xu, S., Lv, Y., Zeng, X., & Cao, D. (2017). ZIF-Derived Nitrogen-doped porous carbons as highly efficient adsorbents for removal of organic compounds from wastewater. *Chemical Engineering Journal*, *323*, 502–511. doi:10.1016/j.cej.2017.04.093

Yamanaka, M., Hara, K., & Kudo, J. (2005). Bactericidal actions of a silver ion solution on *Escherichia coli*, studied by energy filtering transmission electron microscopy and proteomic analysis. *Applied and Environmental Microbiology*, *71*(11), 7589–7593. doi:10.1128/AEM.71.11.7589-7593.2005 PMID:16269810

Yu, C., Wu, Z., Liu, R., Dionysiou, D. D., Yang, K., Wang, C., & Liu, H. (2017). Novel Fluorinated Bi_2MoO_6 nanocrystals for efficient photocatalytic removal of water organic pollutants under different light source illumination. *Applied Catalysis B: Environmental*, *209*, 1–11. doi:10.1016/j.apcatb.2017.02.057

Yu, C. L., Fan, Q. Z., Xie, Y., Chen, J. C., Shu, Q., & Yu, J. C. (2012). Sonochemical fabrication of novel square-shaped F doped TiO2 nanocrystals with enhanced performance in photocatalytic degradation of phenol. *Journal of Hazardous Materials*, *237–238*, 38–45. doi:10.1016/j.jhazmat.2012.07.072 PMID:22947181

Yu, C. L., Wu, Z., Liu, R. Y., He, H. B., Fan, W. H., & Xue, S. S. (2016). The effects of Gd3þ doping on the physical structure and photocatalytic performance of Bi2MoO6 nanoplate crystals. *Journal of Physics and Chemistry of Solids*, *93*, 7–13. doi:10.1016/j.jpcs.2016.02.008

Yu, W., Zhan, S., Shen, Z., Zhou, Q., & Yang, D. (2017). Efficient removal mechanism for antibiotic resistance genes from aquatic environments by graphene oxide nanosheet. *Chemical Engineering Journal*, *313*, 836–846. doi:10.1016/j.cej.2016.10.107

Zambianchi, M., Durso, M., Liscio, A., Treossi, E., Bettini, C., Capobianco, M. L., … Melucci, M. (2017). *Graphene oxide doped polysulfone membrane adsorbers for the removal of organic contaminants from water*. DOI: 10.1016/j.cej.2017.05.143

Zhang, C. L., Qiao, G. L., Zhao, F., & Wang, Y. (2011). Thermodynamic and kinetic parameters of ciprofloxacin adsorption onto modified coal fly ash from aqueous solution. *Journal of Molecular Liquids*, *163*(1), 53–56. doi:10.1016/j.molliq.2011.07.005

Zhang, L., Cheng, Z., Guo, X., Jiang, X., & Liu, R. (2014). Process optimization, kinetics and equilibrium of orange G and acid orange 7 adsorptions onto chitosan/surfactant. *Journal of Molecular Liquids*, *197*, 353–367. doi:10.1016/j.molliq.2014.06.007

Zhang, L. Y., Zhang, H. Y., Guo, W., & Tian, Y. L. (2014). Removal of malachite green and crystal violet cationic dyes from aqueous solution using activated sintering process red mud. *Applied Clay Science*, *93–94*, 85–93. doi:10.1016/j.clay.2014.03.004

Zhang, Q. Y., Deng, S. D., Yu, G., & Huang, J. (2011). Removal of perfluorooctane sulfonate from aqueous solution by crosslinked chitosan beads: Sorption kinetics and uptake mechanism. *Bioresource Technology*, *102*(3), 2265–2271. doi:10.1016/j.biortech.2010.10.040 PMID:21044835

Zhang, S., Zhang, Y., Bi, G., Liu, J., Wang, Z., Xu, Q., ... Li, X. (2014). Mussel-inspired polydopamine biopolymer decorated with magneticnanoparticles for multiple pollutants removal. *Journal of Hazardous Materials*, *270*, 27–34. doi:10.1016/j.jhazmat.2014.01.039 PMID:24525161

Zhang, X. B., Zhang, L., Hu, J. S., & Huang, X. H. (2016). Facile hydrothermal synthesis and improved photocatalytic activities of Zn^{2+} doped Bi_2MoO_6 nanosheets. *RSC Advances*, *6*(38), 32349–32357. doi:10.1039/C6RA06972J

Zhao, C., Pelaez, M., Dionysiou, D. D., Pillai, S. C., Byrne, J. A., & O'Shea, K. E. (2014). UV and visible light activated TiO2 photocatalysis of 6-hydroxymethyl uracil, a model compound for the potent cyanotoxin cylindrospermopsin. *Catalysis Today*, *224*, 70–76. doi:10.1016/j.cattod.2013.09.042

Zhao, G., & Stevens, S. E. Jr. (1998). Multiple parameters for the comprehensive evaluation of the susceptibility of *Escherichia coli* to the silver ion. *Biometals*, *11*(1), 27–32. doi:10.1023/A:1009253223055 PMID:9450315

Zhao, X., Chen, X., & Hu, J. (2017). Composition-dependent dual halide anion-doped bismuth terephthalate hybrids for enhanced pollutants removal. *Microporous and Mesoporous Materials*, *244*, 284–290. doi:10.1016/j.micromeso.2016.11.003

Zhao, X., Qu, J. H., Liu, H. J., Qiang, Z. M., Liu, R. P., & Hu, C. Z. (2009). Photoelectrochemical degradation of anti-inflammatory pharmaceuticals at Bi2MoO6–boron-doped diamond hybrid electrode under visible light irradiation. *Applied Catalysis B: Environmental*, *91*(1-2), 539–545. doi:10.1016/j.apcatb.2009.06.025

Zheng, S., Jiang, W. J., Cai, Y., Dionysiou, D. D., & O'Shea, K. E. (2014). Adsorption and photocatalytic degradation of aromatic organoarsenic compounds in TiO2 suspension. *Catalysis Today*, *224*, 83–88. doi:10.1016/j.cattod.2013.09.040

Chapter 12
Wastewater Treatment and Role of Green Synthesized Metal Oxide Nanocomposites

Durr-e-Shahwar Noman
University of South Australia, Australia

Gary Owens
University of South Australia, Australia

ABSTRACT

Water pollution by metalloids is a global environmental concern. Owing to their propensity for bioaccumulation, water solubility, and interaction with environment, they are threatening both human and ecosystem health. Inherent limitations like low efficiency, sensitive operating conditions, and high capital and operating costs are associated with conventional removal methods which restricts adoption of these technologies on large scale. While adsorption is commonly recognized as both an effective and affordable remediation technology, many common adsorbents often have inherited limitations including non-renewability and high operating costs. Thus, limitations in conventional remediation technologies have headed to the rapid progression of new avenues for advanced treatment technologies for metalloid pollutant removal such as green nanotechnology. In contrast to many of the currently available adsorbents, nanoparticles often have unique properties such as tiny size, more active sites and big surface area, easy separation, and high reactivity that enhance removal efficiencies.

INTRODUCTION

Access to high quality water is indispensable for life. Worldwide, the excessive consumption of water (Vajnhandl & Valh, 2014), compounded with a general deterioration in the quality of both surface and ground waters due to anthropogenic activities such as agriculture, is a major threat to water security and ecosystem health. Seventy percent of fresh water is currently used for agriculture (Steinfeld et al., 2006) of which irrigation accounts for 66% of all water extractions (Scanlon, Jolly, Sophocleous, & Zhang, 2007; Tingey-Holyoak, 2014). Dairy farming, geothermal power utilization, contemporary use

DOI: 10.4018/978-1-5225-5745-6.ch012

of arsenic-based pesticides and industrial activities are contributing anionic pollutants (phosphate, nitrate, nitrite, arsenate and chromium (VI)) to water bodies (Bundschuh & Maity, 2015; Kuppusamy et al., 2016). Elevated levels of these pollutants in turn are governing the production of algae and posing a threat to fresh water resources (Arshadi, Foroughifard, Etemad Gholtash, & Abbaspourrad, 2015). Annually exposure to poor water quality results in the death of more than five million people because of water-borne diseases (FAO, 2010).

In the previous few decades, the removal of anionic pollutants from water and wastewater has been extensively examined in order to safeguard standing water sources such as estuaries and lakes from eutrophication. However, these often costly techniques, are only effective for treating point-source pollution, and may involve synthetic toxic chemicals with potential to detrimentally impact the environment. There is therefore a need to explore cost-effective, efficient, and eco-friendly alternative treatment methods which can be deployed *in situ* to tackle more persistent diffuse pollution sources.

Of all the emerging techniques, green nanotechnology is perhaps the most promising; being, efficient, eco-friendly and cost-effective. Recently, large-scale bio-synthesis (green synthesis) of energy-efficient, low-cost and nontoxic metallic nanoparticles (MNPs) from plant extracts has generated considerable interest as a substitute for more traditional chemical and physical methods (Mittal, Chisti, & Banerjee, 2013; Qu, Alvarez, & Li, 2013). Plant extracts performs as both capping (stabilizing) and reducing agents. In plant extracts, the biomolecules, such as amino acids, proteins, enzymes, citrates, vitamins, flavonoids, organic acids, and polysaccharides are environmentally benign and mediate the metal ion bio-reduction (Iravani, 2011).

Many different types of plants (e.g. aloevera, azadirachta indica, eucalyptus, lemon grass, tea) are currently being widely used to prepare MNPs (Mittal et al., 2013). The source of the plant extracts greatly affects the final characteristics of the formed MNPs (Kumar & Yadav, 2009). Usually, bio-reduction via plant extracts, takes place at room temperature within a few minutes and simply involves mixing a liquid phase plant extract with the liquid phase solution mixture of the associated salt of metal. Unlike chemical procedures, which often require expensive instruments and end in the discharge of toxic chemicals to the environment, biological methods are more simple, eco-friendly and result in more monodispersed MNPs (Bindhu & Umadevi, 2013; Ibrahim, 2015). Some researchers have also reported the use of green NPs for wastewater treatment. For example, Wang et al. (2014) utilized *n*Fe, synthesized through eucalyptus leaf extracts for eutrophic wastewater treatment, obtaining removal efficiencies of 30.4%, 71.7%, and 84.5% for total P, total N and COD, respectively.

For practical removal of pollutants via NPs many different approaches are currently being investigated. Many researchers have simply used NPs in their raw form as nano adsorbents (Qu et al., 2013). However, to overcome the problems associated with NP regeneration and agglomeration many researchers have also prepared different composite structures; such as nanocomposite membranes and core-shell microspheres; with MNPs as nanofillers (Yin & Deng, 2015). The improved porous structure and mechanical strength makes these nanocomposites a more effective for pollutant removal from wastewater. Currently, the use of synthetic polymer hosts in nanocomposites are not preferred because of the associated high costs and toxic environmental effects. However as an alternative, biopolymer hosts are becoming more popular because of their renewability, low cost and environmentally friendly nature (Gehrke, Geiser, & Somborn-Schulz, 2015).

Recent literature (Lofrano et al., 2016) advocates that most of the common complications related with the use of NPs and polymer nanocomposites (i.e effect of other chemical species present in composite wastewater feeds, *in situ* wastewater treatment and complex operational conditions) have still not been adequately addressed. Over the past five years many researchers have synthesized and used different nano adsorbents (i.e. silver, gold and titanium) which all impart toxic effects to environmental entities when directly released (Lofrano et al., 2016). Approaches for better dispersion of nanomaterials and compatibility of the nanofillers with biopolymers hosts are thus needed to tackle the problem of NPs toxicity upon direct release. To date, there are limited reports of using *n*ZnO and *n*MgO as nano adsorbents and nanofillers in biopolymer hosts for anionic pollutant removal. There are also lab-based studies on the use of NPs and nanocomposites for pollutant removal, but limited studies report the large scale fabrication for practical *in situ* wastewater treatment.

Stress on Fresh Water Resources

Globally, stress on limited freshwater resources has dramatically increased due to population growth, urbanization, mounting food demand and the cumulative impacts from climate change, socioeconomic, industrial and agricultural sectors (Ridoutt & Pfister, 2010). Since almost forty percent of the world's inhabitants reside in zones where water bodies are over loaded owing to competition and scarcity, a recent World Economic Forum report stated that the greatest risk facing many societies is the impacts of water scarcity (World Economic Forum, 2015). Even wherever water supplies are relatively plenteous, they are gradually becoming at risk due to increasing water demand. According to the United Nations Organization, two thirds of the world's populace, by 2025, will experience moderate to severe water dearth (United Nations Development Programme, 2006). Many African, Middle Eastern and South Asian countries will face the threat of serious water shortages in the next two decades. Whereas, in developing countries the issue is further worsened by a lack of trained professionals, proper management and financial constraints (Azizullah, Khattak, Richter, & Hader, 2011).

Despite worldwide attempts by irrigators to implement improved technologies and practices to decrease water-use, since 70% of global water usage occurs in the agricultural sector (Wheeler, Zuo, & Loch, 2015) any degradation in water resources dramatically increases agricultural demand. Worldwide increases in intensified farming and livestock production are also triggering an overexploitation of surface and groundwater resources and may endanger the freshwater abundance for future generations (Ridoutt & Pfister, 2013). These practices are consequently contributing to greater inputs of nutrients to water; in particularly; nitrogen (N) and phosphate fertilizers (Doole, Marsh, & Ramilan, 2013). Other common sources of nutrients (i.e. phosphate and nitrate) come from natural geological sources, soil erosion, community sewerage schemes, urban runoff and septic tanks (Abell, Hamilton, & Paterson, 2011). Elevated concentrations of anionic pollutants like phosphate can reduce water quality over time resulting in aquatic and algal blooms, which deplete dissolved oxygen (Arshadi et al., 2015). Such eutrophication of surface water results in severe environmental costs (Pretty et al., 2003) and requires a multidisciplinary approach to deal with it (Sevcenco et al., 2015).

NANOTECHNOLOGY

A nanoparticle is defined as having at least one dimension < 100 nm (Fendler, 2007; Narayanan & Sakthivel, 2011). Unlike larger bulk materials, nanomaterials exhibit very unique properties and consequently have found numerous applications because of their surface energy, high surface-to-volume ratio and morphology. In fact a new "nanotechnology" industry has evolved around the application of nanomaterials to real word problems (Herlekar, Barve, & Kumar, 2014)

Nanoparticles Synthesis

Nanoparticles (NPs) are usually synthesized by common top-down and bottom-up schemes (Narayanan & Sakthivel, 2011). In the top-down method (size reduction) the large loose material is slowly shattered down to nano-sized particles. While in the bottom-up technique, molecules or atoms are agglomerated to nanometer sized molecular structures. Common top-down or bottom-up strategies are summarized below (Figure 1) (Mittal et al., 2013).

Figure 1. Top-down or bottom-up strategies used in nanoparticle synthesis

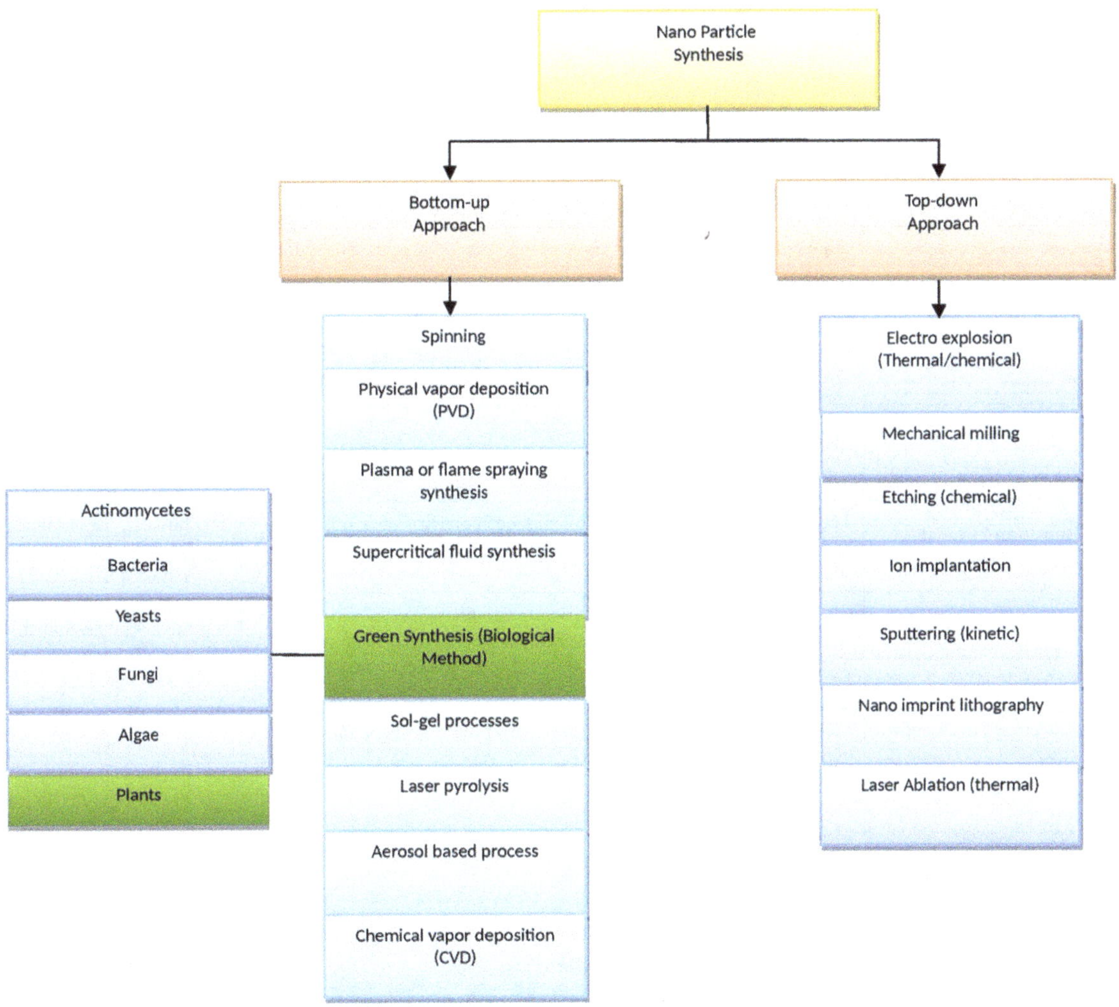

Nanoparticles Characterization

Accurate NP characterization in terms of shape, size, homogeneity and surface area is important to determine their efficiency and effectiveness for adsorption process. Numerous methods are commonly used for characterization of NPs, nanostructures, nanomaterials and nanocomposite thin films. Some methods for structural characterization include EDX (Electron Dispersive X-ray), SEM (Scanning Electron Microscopy), FT-IR, XRD (X-Ray Diffraction) and TEM (Transmission Electron Microscopy) (Mahendra Rai & Duran, 2011). During nanomaterial fabrication, while obtaining a small size is an important consideration, matching the specific NPs characteristics consistently to the application is also important, i.e. NPs should have the same size, morphology, shape and crystal structure and chemical composition (Tiwari, Behari, & Sen, 2008). Figure 2 illustrates the size, morphology and shape of two distinct NPs, a sponge-like magnesium oxide NP and rod-like zinc oxide NP (Rafiq, Nazir, Durr-e-Shahwar, Shah, & Ali, 2014).

As summarized in Table 1., NPs characterization techniques generally involve the characterization of size, surface area, shape, and dispersity which reveal important information such as optical properties, surface charges, topographic features, morphologies, purity and porosity nature (Kuppusamy, Yusoff, Maniam, & Govindan, 2014; Mittal et al., 2013).

Table 1. Nanoparticle characterization techniques

Technique	Characteristics Determined	Reference
Atomic force Microscope (AFM)	Topological appearance and range of the size and height distribution with the nature of the material in terms of porosity.	(Suresh et al., 2015)
Dynamic Light Scattering (DLS)	Size distribution and surface charges of the particles suspended in a liquid.	(Elumalai, Velmurugan, Ravi, Kathiravan, & Adaikala Raj, 2015)
Energy Dispersive Spectroscopy (EDS)	Elemental composition of metal NPs.	(Elumalai, Velmurugan, Ravi, Kathiravan, & Ashokkumar, 2015b; Strasser et al., 2010)
Fourier Transform Infrared Spectroscopy (FTIR)	The organic functional groups and surface chemistry such as hydroxyls and carbonyls attached to the surface of NPs and the other surface chemical deposits.	(Elumalai, Velmurugan, Ravi, Kathiravan, & Ashokkumar, 2015c)
Powder X-ray diffraction (XRD)	Characterization of the crystal structure and phase identification, size purity of the NPs. X-rays enter into the NPs and subsequent diffraction pattern is matched with specific standards to find structural information.	(Yuvakkumar, Suresh, Nathanael, Sundrarajan, & Hong, 2014)
Scanning Electron Microscopy (SEM)	Morphological characterization, crystalline size and shape (at the nanometer to micrometer scale).	(Vanathi et al., 2014)
Transmission Electron Microscopy (TEM)	Morphological characterization (at the nanometer to micrometer scale). It has a 1000 times higher resolution as equated with the SEM.	(Elumalai, Velmurugan, Ravi, Kathiravan, & Ashokkumar, 2015a; Schaffer, Hohenester, Trügler, & Hofer, 2009)
UV–visible Spectrophotometry	Determines optical properties of MNPs.	(Elumalai & Velmurugan, 2015; Shahverdi, Shakibaie, & Nazari, 2011)
Zeta potential	Potential effective electric charge.	(S. Kuppusamy et al., 2016)

Figure 2. (a) Overview of agglomerated hexagonal wurtzite structure of nZnO (53.7 nm) synthesized using sedum alfredii Hance leaf extract (Jiao Qu, Luo, & Hou, 2011), (b) Rod-like nZnO (24.7 nm) taken at 50 micro meter resolution (Rafiq et al., 2014)

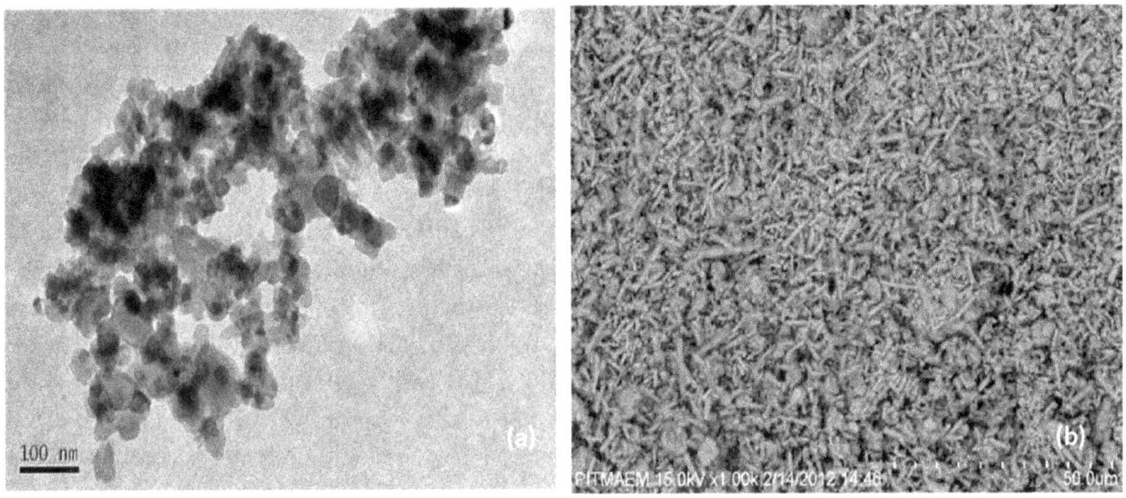

Applications of Nanotechnology in Water and Wastewater Treatment

Interest in nanotechnology-empowered wastewater treatments are increasing because such treatments have the potential to overcome some of the major deficiencies of existing treatments while simultaneously providing greater and more economic use of scarce water resources (Tiwari et al., 2008). Adsorption is the most commonly applied nano-technology to remove both inorganic and organic pollutants from water/wastewater. Compared to NPs, effectiveness of other adsorbents such as biomass and charcoal is normally very limited as the number of active sites and surface area are much less owing to their larger particle size. Due to NPs enhanced selectivity and adsorption kinetics nano-adsorbents can play a vital role in pollutant removal. In particular, nano-adsorbents have high surface area and large numbers of sorption sites, short tunable pore size and intra-particle diffusion distance and novel surface chemistry (Prasse, Stalter, Schulte-Oehlmann, Oehlmann, & Ternes, 2015). Different types of nano materials are used depending upon the particular application, where for adsorption applications, carbon nano tubes, nano scale metal oxide, nano fibers with core shell structure are all commonly used. Nano-zeolites, *n*Ag, carbon nano tubes, and nano-magnetite are also all commonly used in nano-composite membranes synthesis. However, for photo catalysis, disinfection and microbial control, $n\mathrm{TiO_2}$, fullerene derivatives, *n*Ag and carbon nano tubes are frequently used (Ali, 2012).

Surface Chemistry of Metal Oxide Adsorbents

Depending on the type of adsorbent, surface chemistry may affect the adsorbate/adsorbent interaction. This is especially true for the adsorption of ions onto oxidic adsorbents. Oxidic adsorbents such as metal oxides are characterized by crystalline structures where positively charged metal ions and negatively charged oxygen or hydroxide ions are arranged in such a manner that the different charges compensate each other. In aqueous solutions, the negative charges of the surface oxygen ions are neutralized by protons, whereas the positive charges of the surface metal ions are neutralized by hydroxide ions. As a result, the surface of oxidic adsorbents is covered with surface OH groups. These groups are subject to protonation or deprotonation depending on the solution pH and consequently the surface is positively charged at low pH and negatively charged at high pH.. Between these extremes, a pH value exists at which the sum of negative charges equals the sum of positive charges and the net charge of the surface is zero. This point is referred to as the point of zero charge (pzc). The pH_{pzc} is an important adsorbent parameter that aids in understanding the adsorption of charged species and the influence of pH on the adsorption process (Bhatnagar, 2013; Ruthven, 1984).

Generally, the adsorption of charged species onto charged surfaces can be expected to be strongly influenced by electrostatic attraction or repulsion forces. Dissolved ions can be bound to the surface OH groups by different mechanisms: specific adsorption (i.e. surface complex formation) and nonspecific adsorption. These mechanisms are strongly related to the structure of the electric double layer that surrounds the charged solid particle. The term surface complex formation includes two different types of reactions, the formation of inner-sphere complexes and the formation of outer-sphere complexes. The model of outer-sphere complex formation presumes that ions can also be bound to the surface sites by chemical bonds without losing their hydration water. That means that a water molecule is located between the ion and the adsorption site. Therefore, the distance to the surface is larger and the binding strength is weaker in comparison to inner-sphere complex formation (Bhatnagar, 2013; Worch, 2012).

The very fast adsorption of anionic pollutants via nanoadsorbents is usually attributed to either simple electrostatic attractions, or due to the driving force of a significant concentration gradient at higher initial pollutant concentrations which both results in the transportation of anionic pollutants to the adsorbents surface. Kinetics studies have shown that here the adsorption of anionic pollutants such as phosphate onto ZnO (Luo et al., 2015), hydrous zirconium oxide (Lin et al., 2017), $Fe_3O_4@SiO_2$ core/shell magnetic NPs (Li Lai, Qiang Xie, Lina Chi, Wei Gu, & Deyi Wu, 2016) cerium–zirconium binary oxide NPs (Yu Su, Yang, Sun, Li, & Shang, 2015), iron (Fe^{+3}) oxide/hydroxide NPs (Zelmanov & Semiat, 2015) is primarily considered to be electrostatic attractions between adsorbent active sites and adsorbate (Bhatnagar, 2013).

Nano-Adsorbents

A large number of different kinds of NPs, like metals based NPs (metal oxides i.e. iron oxide, titanium dioxide, zeolites, magnetic NPs and alumina) (Abid, Kanematsu, Young, & Kennedy, 2013), carbonaceous NPs (carbon nano tubes) (Bo Pan & Xing, 2008), polymeric NPs (dendrimers) (Huang & Keller, 2013) are being used as nano adsorbents for a range of different pollutants including heavy metals, radionuclides and organic contaminants (Hornyak, Moore, Tibbals, & Dutta, 2008).

Metal Oxide Nanoparticles as Powerful Nano-Adsorbents

Application of metal oxide nano particles (MONPs) is widespread because of their novel properties (Kharissova et al., 2013) such as monodisperse nature, purity, shape, size and crystalline structure. MONPs can be easily synthesized and chemically modified (Makarov et al., 2014). While both chemical and physical processes are commonly used to synthesis MONPs with predefined (tunable) characteristics (Diaz et al., 2015), these techniques are normally expensive and hazardous to living organisms and the environment (Gan, Ng, Huang, & Li, 2012). Indeed many of the old wet chemical methods of

MONPs synthesis involved harmful materials (Joo & Zhao, 2016). Furthermore, dry methods like UV irradiation, lithography and aerosol are also not considered eco-friendly. Therefore, green synthesis of biocompatible, clean, non-toxic and eco-friendly MONPs has been proposed both intracellularly and extracellularly (Sharma, Yngard, & Lin, 2009). These green synthesized MO nano-adsorbents be able to be effortlessly regenerated by altering solution pH (Y. C. Sharma, Srivastava, Singh, Kaul, & Weng, 2009) and although some authors have reported reduced adsorption capacity after regeneration in some circumstances (Deliyanni, Bakoyannakis, Zouboulis, & Matis, 2003), their overall adsorption capability was generally preserved following numerous regeneration cycles (Hu, Chen, & Lo, 2006). Thus, in general, the great adsorption capacity, easy separation, low cost and renewal capacity have made metal based nano adsorbents and attractive and economically viable technology for water purification (Qu et al., 2013).

Green Synthesis of Metal Oxide Nanoparticles

For some time only prokaryotes were tested for the ability to bio-reduce or bio-sorb metal ions and convert toxic metals to non-toxic metal salts. Other biosynthetic approaches, including microbial synthesis of NPs, have been conducted at the interface between nanotechnology and microbiology. For example, biosynthesis of energy-efficient, low-cost, and non-toxic metallic NPs by viruses, fungi, bacteria, yeasts and actinomycetes have all been reported by researchers (Iravani, 2011). However, because of some limitation in growth and safety amongst microorganisms, plants are considered to be the safest organism for the biosynthesis of MONPs (Herlekar et al., 2014; Mittal et al., 2013). Recently, different eukaryotic plants have been recognized as possessing reducing and stabilizing potential enabling them to transform inorganic metal ions into metal NPs. Such mediated plants extract synthesis is now more feasible than ever before and is gaining more popularity due to perceived benefits of lower costs, shorter production times and volumes and being a sustainable source of phototrophs.

Plant Extracts Mediated Synthesis of Metal Nanoparticles

Plants contain a complex variety of different metabolites like proteins, enzymes and other reducing agents (Narayanan & Sakthivel, 2010, 2011). Thus, the production of numerous MONPs via plant extracts is an easy process, where NP shape and size can be regulated through the judicious use of different pH and temperature (Makarov et al., 2014). This approach uses extracts of various plant parts and multiple salts/acids of metals (like copper, gold, iron, platinum, silver, zinc and many others) for MONPs synthesis (Sougata Ghosh et al., 2012; Park, 2014; Rai & Yadav, 2013).

FTIR has shown that plant metabolites all reduce metal ions to NPs (Shankar, Rai, Ahmad, & Sastry, 2004, 2005). However, different plants have different concentrations of active compounds, which influences NPs morphology since hexagons, triangles, pentagons, spheres, cubes, nanowires, ellipsoids and nanorods can all be formed by a variety of plants extracts (Makarov et al., 2014; Rauwel, Küünal, Ferdov, & Rauwel, 2015). Table 2 summarizes the MNPs size and morphology synthesized by different plants extracts.

Metal Ion Reduction and Role of Plant Compounds (Metabolites)

There are three main phases in the synthesis of MNPs by plant extracts 1) Reduction, 2) Growth and 3) Stabilization (Figure 3). During the reduction phase, metal ions reduction occurs with consequential

Table 2. Morphology and size of MNPs synthesized by different plants extracts

Plant Name	Botanical Name	Metal	Morphology	Size (nm)	Reference
Rose geranium	*Elargonium Graveolens*	Gold	decahedral icosahedral shaped	20–40	(Shankar, Ahmad, & Sastry, 2003)
Lemon grass	*Cymbopogon Flexuosus*	Gold	Nano spheres & Nano triangles	0.05–18	(Shankar et al., 2005)
Neem (Indian lilac)	*Azadirachta Indica*	Gold	triangles and hexagons	50–100	(Shankar et al., 2004)
Neem (Indian lilac)	*Azadirachta indica*	Silver	polydispersity spherical	5–25	(Shankar et al., 2004)
Aloe vera	*Aloe barbadensis*	In_2O_3	cubic	5–50	(Maensiri et al., 2008)

Figure 3. Insides into nanoparticle synthesis (M+ metal ion)

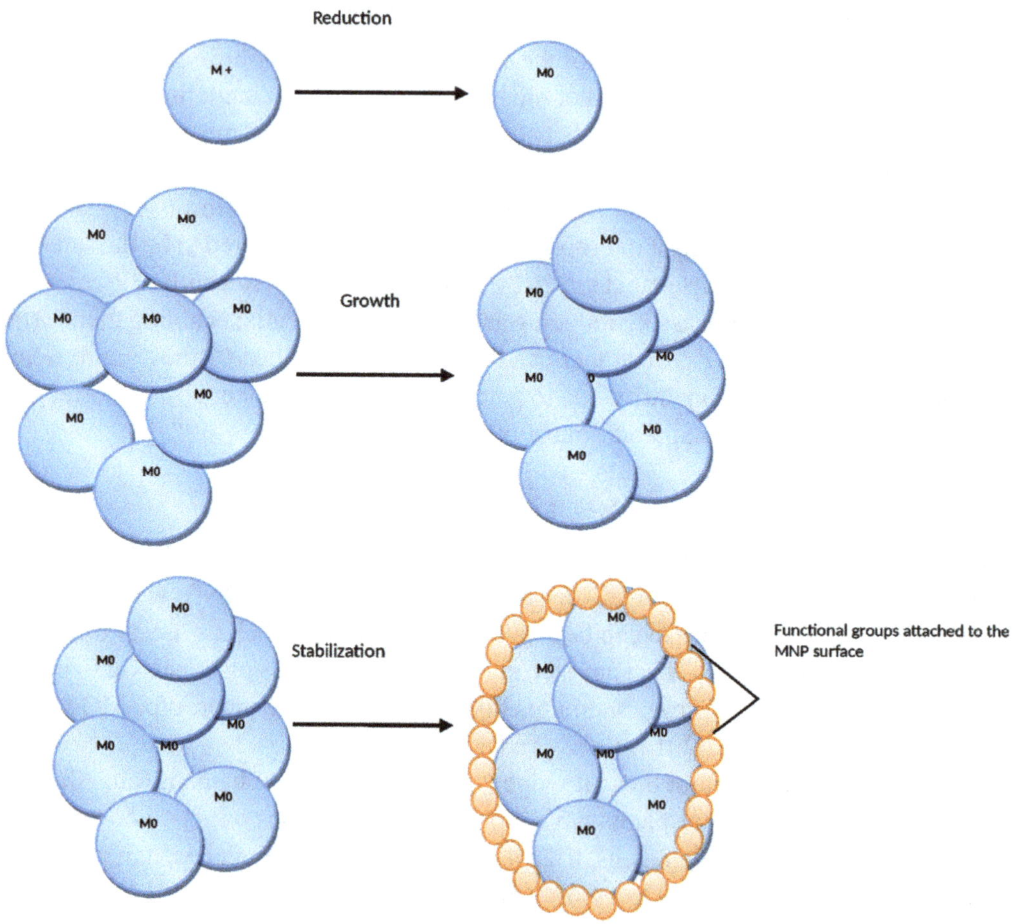

nucleation of reduced metal atoms. In the second growth stage, NPs in close proximity agglomerate and overall NP size increases. This is the direct creation of NPs by a heterogeneous nucleation and growth process leading to metal ion reduction (aka Ostwald ripening), supplemented by boom in the thermodynamic stability of NPs. Finally, in the stabilization phase, reactions slow and an equilibrium is established which determines the final NPs form (Si & Mandal, 2007). The duration of the growth phase in part determines the final aggregate form (i.e. nanohexahedrons, nanoprisms, nanotubes and various other irregularly shaped nanoparticles) (Glusker, Katz, & Bock, 1999). Whereas, the termination phase is dependent upon the plant extract type because here NPs acquire the most energetically favorable conformation (Kim et al., 2010).

Among various metabolites, the terpenoid content of plants is most commonly linked with NPs production because these are the most typical organic polymers manufactured by plants from 5-carbon isoprene units acting as strong antioxidants (Shankar et al., 2003). Thus terpeniods have been cited as being responsible for the transformation of Zn^{+} to *nZn* using Aloe vera extract (Ali et al., 2016). It has been reported that the main terpenoid, Eugenol, present in cinnamon extracts (*Cinnamomum zeylanisum)* played a key part in the bioreduction of both silver and gold ions (Sathishkumar et al., 2009). FTIR spectroscopy revealed that proton dissociation from Eugenol's hydroxyl group resulted in the formation of resonance structures which facilitated oxidation and as an outcome the active reduction of metal ions resulted in NPs synthesis (Sathishkumar et al., 2009; Singh, Talat, Singh, & Srivastava, 2010).

Flavonoids are another big cluster of polyphenolic compounds ubiquitous in many plants. NPs are formed by the flavonoids via a combination of chelation (by carbonyl groups π-electrons) and metal ion reduction (by flavonoid functional groups). Transformation of flavonoids from enol-form transformed to keto-form also discharge reactive hydrogen atoms that may reduce the metal ions (Singh et al., 2010). For instance, *Ocimum basilicum* (sweet basil) extracts containing rosmarinic and luteolin acid were transformed from the enol- to the keto- form for reduction of Ag ions (Ahmad et al., 2010) (Figure 4). Reactive hydrogen is released from luteolin when the enol form is formed which further converts of

Figure 4. Reduction of Ag ions by transformation of the enol- to the keto- form

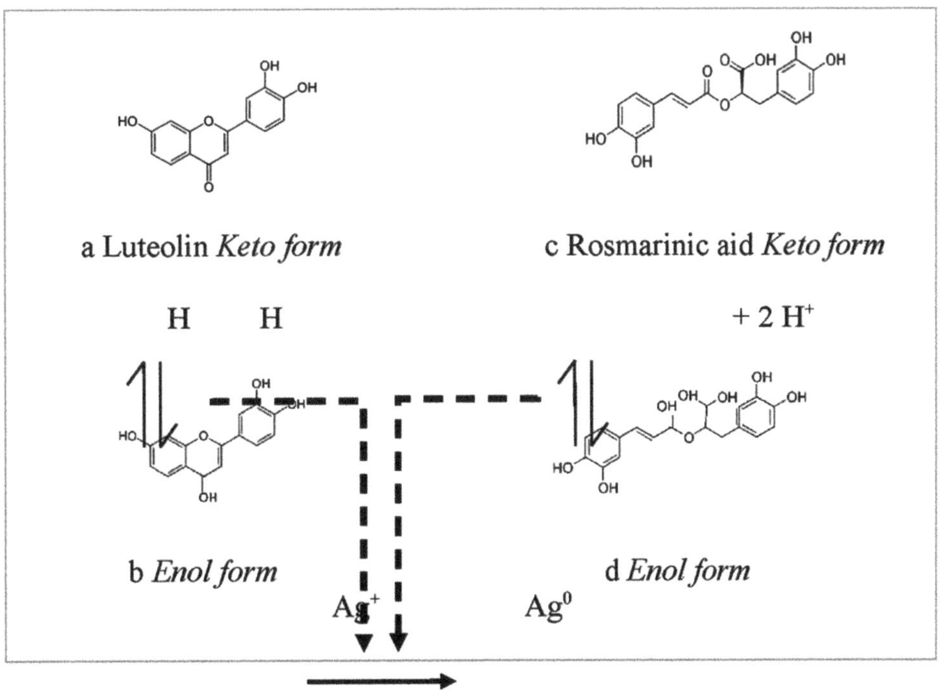

Ag+ ion to Ag metal (Figure 4 a and b). The keto form of the rosemarinic acid again liberates the reactive hydrogen to make an unstable intermediate enol form of the same acid. However, this structure is unstable because it contains two hydroxyl group on a single carbon atom and returns back to its keto form (Figure 4 c and d) (Ahmad et al., 2010).

Sugars including monosaccharides like glucose (which contain aldehyde groups) can also perform as reducing agents. They comprise a keto-group (fructose), which transforms ketone to an aldehyde and exhibits antioxidant ability. The reducing ability of both di- and poly-saccharides is reliant on the monosaccharide components. Like the disaccharides, lactose and maltose also have also some reducing ability, because their monomers have an open chain form.

However, there are also some sugars like sucrose, which have no reducing ability because the fructose and glucose monomers are joined in such a manner that an open chain is not accessible. Glucose is a strong reducing agent than the fructose, due to the inherit antioxidant prospective of the fructose. When the aldehyde group in sugar is oxidized in to a carboxylic acid group this leads to the reduction of metal ions (Panigrahi, Kundu, Ghosh, Nath, & Pal, 2004).

Moreover, FTIR spectroscopy has shown that amino acids and carbonyl groups of the aspartic and glutamic acid's core chain or side chains from proteins are associated with the NPs (Glusker et al., 1999; Zayed, Eisa, & Shabaka, 2012) as they bind with metal ions to reduce them further. Different studies have revealed that amino acids like arginine, cysteine and lysine all bind strongly with the silver ions (Tan, Lee, & Wang, 2010). Furthermore, natural α-amino acids such as tryptophan are also a very strong reducing agent for gold ions (Glusker et al., 1999).

Physical Factors Affecting Metal Nanoparticles Formation Through Plant Extracts

Several researches revealed that shape, yield and size of MNPs is mainly dependent upon the pH of the reaction mixture since pH changes result in the alteration of natural phytochemical charges which are present in an extract thereby affecting their capability to bind and subsequently reduce both metal anions and cations (Gan & Li, 2012). Thus, at highly acidic pH, MNP aggregation is more dominant than either reduction or nucleation of reduced atoms, and this is why most gold NPs are synthesized at pH values which are alkaline, whereas NPs do not synthesize at acidic pHs (Ghodake, Deshpande, Lee, & Jin, 2010). Elevated temperatures result in increased reaction rates, increased NP synthesis (Das, Gogoi, & Bora, 2011; Lukman, Gong, Marjo, Roessner, & Harris, 2011) and increased nucleation rates (Das et al., 2011). For example, *n*Ag derived from Alfalfa plants (*M. sativa*) and lemon verbena (*Aloysia citrodora*) only formed at temperatures above 30 °C (Lukman et al., 2011). The electrochemical potential of the ion of interest also plays a critical role in the reduction process, since if the electrochemical potential for metal ion reduction is too high, plants extracts may not effectively reduce metal ions (Haverkamp & Marshall, 2008). Agglomeration of reduced metal atoms to NPs is also dependent upon the reaction time so that the faster NPs formation is the more uniform and smaller the NPs formed are (Makarov et al., 2014).

1. Zinc Oxide (ZnO) Nanoparticle

Zinc oxide nanoparticles (*n*ZnO) are commonly used in textiles, industrial coatings, personal care products (sunscreens and cosmetics), antibacterial agents, dye-sensitized solar cells and electronic and optical materials. They are also effective as fertilizers to improve Zn insufficiency in soils (Amutha et al., 2016; Wang et al., 2013). The relative exposure pathways, commercial applications and use in the water/wastewater treatment process are implausible to lead to direct release in to environment, although other uses such as fertilizer applications may lead to their direct discharge to the soil (Milani et al., 2012). Around 550 tons or *n*ZnO is being used annually in cosmetics, paints and coatings, plastics/polymers (Mitrano, Motellier, Clavaguera, & Nowack, 2015). The amount of *n*ZnO released to the environment (Lide, 2005; Majedi, Kelly, & Lee, 2014) is dependent upon the temperature and organic acid content (Joo & Zhao, 2016). ZnO is a very stable non-toxic material. It effectively removes As in the pH range 5.8-6.8 (Singh et al., 2013). *n*ZnOs improve hydrophilicity and compaction of nanocomposite membrane and decrease membrane fouling when used as nano fillers in MMMs fabrication (Alhoshan, Alam, Dass, & Al-Homaidi, 2013). Moreover, Zn-Cl_2-activated carbon (Namasivayam & Sangeetha, 2004), Zn-Al hydroxides (Cheng et al., 2009) and zinc oxide (Luo et al., 2015) showed good phosphate adsorption abilities. However, up to now, anionic pollutant adsorption by *n*ZnO has not been reported.

2. Magnesium Oxide (MgO) Nanoparticles

Magnesium oxide (MgO) is a multipurpose oxide material with a low dielectric constant and refractive index, large band gap and good thermodynamic stability with various applications in ceramics, catalysis, toxic waste remediation, as an additive in paint, refractory, antibacterial properties, and also as a superconducting material (Shah & Qurashi, 2009; W. Wang, Qiao, Chen, Tan, & Li, 2009). To date, removal of anionic pollutants using *n*MgO has scarcely been reported (Sathyamoorthy, Mageshwari, Mali, Priyadharshini, & Patil, 2013). However, use of both nanoscale and ultrafine MgO particles is a favorable technique for the adsorption of toxic chemical agents. Hence, it is significant to focus the current research effort on the preparation of *n*MgO for the removal of anionic pollutants (Ramanujam & Sundrarajan, 2014).

Until now, *n*MgO were almost exclusively produced by dehydration of $Mg(OH)_2$ or by decomposition of various Mg salts using hydrothermal (Kumari, Li, Vannoy, Leblanc, & Wang, 2009), flame spray pyrolysis (Yi, Wenzhong, Yitai, Li, & Zhiwen, 1996), thermal evaporation (Yang, Sha, Wang, Wang, & Yang, 2006), chemical vapor deposition (Hao, Meng, Ye, Zhang, & Zhang, 2005), sol–gel (Phuoc, Howard, Martello, Soong, & Chyu, 2008) and also surfactant techniques (Ouraipryvan, Sreethawong, & Chavadej, 2009). However, these synthesis methods are costly and require complex processes, sophisticated instruments, and the use of environmentally toxic chemicals and hazardous organic solvents which do not simply biodegrade in the environment. Therefore, evolving a bio-synthetic process for synthesizing *n*MgO under slight reaction conditions is a stimulating research goal (Mageshwari, Mali, Sathyamoorthy, & Patil, 2013; Moorthy, Ashok, Rao, & Viswanathan, 2015).

a. Green Materials Used for ZnO and MgO Nanoparticles Synthesis

While a variety of physicochemical methods are engaged in NPs fabrication, "green" synthesis using eco-friendly and non-toxic biological methods is attaining popularity (Mittal et al., 2013). These, safer alternate biological synthetic approaches can result in NPs with distinctive properties and allow for precise shape and controlled structure formation (Geetha, Kavitha, & Ramesh, 2015).

Several studies have been conducted for *n*ZnO synthesis through plant extracts. Crystalline *n*ZnO (72.5 nm) were produced by *Physalis alkekengi L.* and transmission electron microscope images confirmed that the produced NPs were poly dispersed (Qu, Yuan, Wang, & Shao, 2011). *n*ZnO are similar to *n*TiO_2 in that they show high UV absorption efficiency and also photocatalytic efficiency through the release of hydrogen peroxide (Dhakras, 2011). Another study showed that hexagonal *n*ZnO from a *Sedum alfredii* Hance plant extract (Zn-hyper-accumulator) were agglomerated, but single *n*ZnO were pseudospherical with an average size (53.7 nm) (Jiao Qu et al., 2011). Highly spherical and stable *n*ZnO (25 to 40 nm) were also synthesized by zinc nitrate and *aloe barbadensis miller* leaf extracts (Kharissova et al., 2013). At aloe leaf extract concentrations greater than 25% (w/v) more than 95% NP transformation occurred yielding poly dispersed NPs. However, by changing the concentrations of the leaf extract in solution the particle size could be varied (Sangeetha, Rajeshwari, & Venckatesh, 2011).

Some plants have already been successfully used for the green synthesis of both *n*ZnO and *n*MgO. Table 3 lists green synthesized *n*ZnO and *n*MgO according to their respective size distribution. The table indicates few studies report biosynthesis of *n*MgO probably because of its less redox potential compared to other metals, however these studies shows the potential of biosynthesis of *n*MgO using plant extracts.

Plant extracts having elevated concentrations of proteins, alkaloids, amino acids, enzymes, polysaccharides and alcoholic compounds (Kesharwani, Yoon, Hwang, & Rai, 2009) are more likely to facilitate the rapid bio- metal ions reduction to MNPs and promote their successive growth and stabilization. Thus a successful green synthesis needs plants containing high phenol levels. For instance, banana and acacia plants are renowned for their elevated secondary metabolites content which makes them promising candidates for green synthesis of MONPs (Bankar, Joshi, Kumar, & Zinjarde, 2010).

b. Acacia

Acacia is a monophyletic genus of flowering plants in the *Fabaceae* family, commonly known as thorn trees. *Acacia* occurs in a variety of open, tropical to subtropical parts of the world. Out of the 1350 existing species worldwide, in terms of geographical species distribution, 163 are considered to be common or widespread species) of which 9 are native to Australasia (Maslin & McDonald, 2004). Secondary metabolites reported in various *Acacia* species include amines, simple alkaloids, terpenes, flavonoids, phenolic substances and polysaccharides (gums) (Seigler, 2003). Gum acacia (GA), is a biopolymer polysaccharide derived from the branches and stems of the Acacia senegal tree (Boland et al., 2006). Acacia has been extensively used in the preparation of metallic NPs were it acts as stabilizing and reducing agent and is popular because of its biocompatibility and non-toxicity (Bajpai & Kumari, 2015; Dong, Zhang, Cai, & Cao, 2014). Two species of *Acacia*, *Acacia pycnantha* (Golden Wattle) and *Acacia dealbata* (Silver Wattle) are a favorable plant candidate for NP green synthesis due to their widespread availability, abundance and high bioreduction potential as inferred from their high phenolic content.

CURRENT WASTEWATER TREATMENT METHODS

In order to control anionic derived eutrophication and conserve the natural aquatic environment, efficient techniques for the elimination of anionic nutrient pollutants from water bodies is of significant global importance. Current methodologies for water treatment fall in to many different categories (Figure 5) (Gupta, Ali, Saleh, Nayak, & Agarwal, 2012). Despite widespread adoption all of these current methods have some limitations, including poor efficacy, sensitive operational conditions and release of a secondary sludge which requires additional costly final disposal (Bhatnagar & Sillanpää, 2011). These drawbacks, combined with the need to develop further cost effective approaches for recuperating

Table 3. Green materials previously used for nZnO and nMgO synthesis

Green Materials	Morphology	Size (nm)	Reference
nZnO			
Ocimum tenuiflorum leaf extract	Hexagonal	11-25	(Raut, Thorat, & Thakre, 2015)
Murraya koenigii	Hexagonal and spherical	12	(Elumalai, Velmurugan, et al., 2015a)
Moringa oleifera	Spherical and granular	16–20	(Elumalai, Velmurugan, et al., 2015c)
Azadirachta indica leaf extract	Spherical shape	20 to 45	(Elumalai & Velmurugan, 2015)
Hibiscus subdariffa leaf extract	Spherical	12-46	(Bala et al., 2015)
Olea europea leaf extract	Crystalline	18-30	(Awwad, Albiss, & Ahmad, 2014)
Tamarindus indica leaf extract	Spherical	19	(Elumalai, Velmurugan, et al., 2015b)
	Hexagonal	37	
Solanum nigrum	Hexagonal (wurtzite)	20 and 30	(Ramesh, Anbuvannan, & Viruthagiri, 2015)
Aloe Barbadensis Miller leaf extract	Poly dispersed	25 to 40	(Sangeetha et al., 2011)
Parthenium hysterophorus L.	Spherical	27 ± 5	(Rajiv, Rajeshwari, & Venckatesh, 2013)
	Hexagonal	84 ± 2	
Vitex trifolia leaf extract	Spherical shape	28	(Elumalai, Velmurugan, Ravi, Kathiravan, & Adaikala Raj, 2015)
Hibiscus rosa-sinensis	Crystalline	30-35	(Devi & Gayathri, 2014)
Zingiber officinale (ginger)	Spherical	30-50	(Raj & E, 2015)
Sargassum Muticum (s.muticum) aqueous extract (Brown marine macro algae)	Hexagonal structures	30-57	(Azizi, Ahmad, Namvar, & Mohamad, 2014)
Eichhornia crassipes leaf extract	Spherical shape	32	(Vanathi et al., 2014)
Aloe extract	Spherical	40	(Gunalan, Sivaraj, & Rajendran, 2012)
Ocimum basilicum l. Var. Purpurascens benth.-lamiaceae leaf extract	Hexagonal (wurtzite)	50	(Abdul Salam, Sivaraj, & Venckatesh, 2014)
Nephelium Lappaceum leaf & peel extract	Hexagonal	50	(Yuvakkumar et al., 2014)
Citrus aurantifolia extracts	Spherical shape	50-200	(Ain Samat & Md Nor, 2013)
Cassiafistula plant extract	Sponge	5–15	(Suresh et al., 2015)
Sedum alfredii	Hexagonal (wurtzite) & Pseudo-spherical	53.7	(Jiao Qu et al., 2011)
Sedum alfredii hance leaf extract	Hexagonal (wurtzite)	53.7	(Jiao Qu et al., 2011)
Physalis alkekengi	Crystalline	72.5	(J. Qu et al., 2011)
Physalis alkekengi leaf extract	Polydispersity	72.5	(J. Qu et al., 2011)
Camellia sinesis	Crystalline	85.3	(R. K. Shah, Boruah, & Parween, 2015)
Cassia Auriculata(tanners cassia)	Spherical	100	(P.Ramesh, A.Rajendran, & M.Meenakshisundaram, 2014)
nMgO			
Aloe vera gel MgO (Fe^{3+} decorated)	Spherical	12–22	(Anilkumar et al., 2015)
Azadirachta indica	Crystalline	25 and 27	(Sankara Rao et al., 2015)
Emblica officinalis	Cubic structure	27	(Ramanujam & Sundrarajan, 2014)
Orange Fruit Waste	Cubic	29	(Rao, Ashok, Rao, Chakra, & A, 2015)

Figure 5. Classification of wastewater treatment technologies

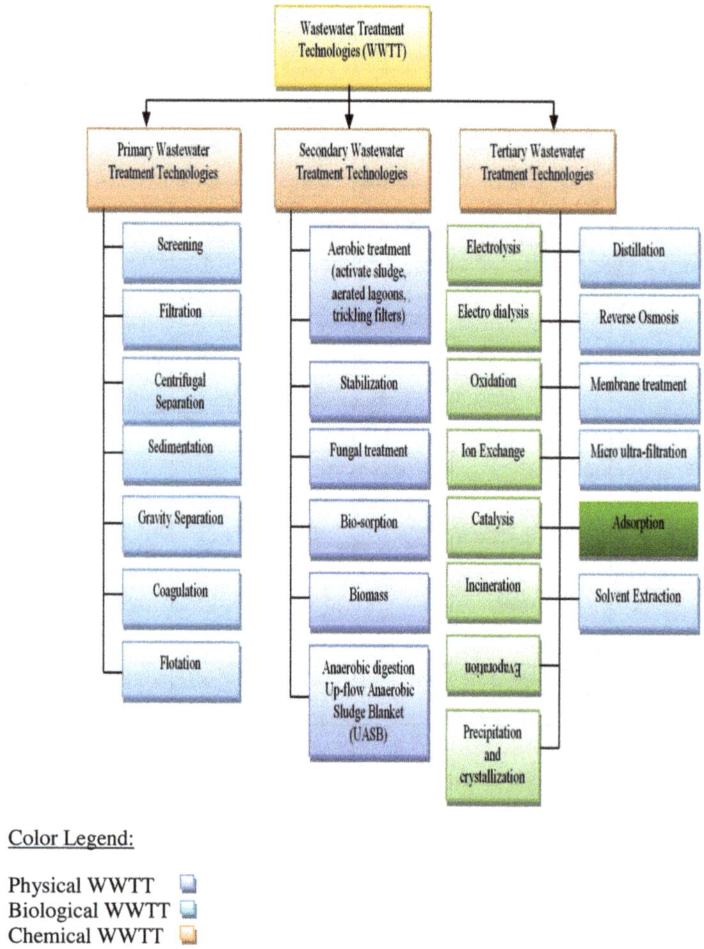

contaminants, have led to the improvement of diverse new separation technologies together with nano adsorption (Gupta et al., 2012).

Removal of Anionic Pollutants From Wastewater

Removal of anionic pollutants from agro-industrial and domestic wastewaters prior to environmental discharge is increasingly important and often enforced legislatively (Naghash & Nezamzadeh-Ejhieh, 2015; Xu, Deng, Liu, & Peng, 2012). There are numerous conventional and advanced technologies for anion removal from wastewater (Ayyasamy et al., 2007; Sullivan, Tyrer, Cheeseman, & Graham, 2010; Ungureanu et al., 2015; Jie Xie, Wang, Wu, & Kong, 2014). Among these, the adsorption method is favored practically owing to its flexibility, simplicity, cost-effectiveness and easy operation (Ogata et al., 2016; Xing Xu, Gao, Jin, & Yue, 2016). Several other approaches have successfully been employed for the removal of phosphate, nitrate, chromium and arsenic (Lee, 2011). Some of these methods and their removal efficiencies are listed in Table 4.

Table 4. Approaches employed for the removal of phosphate, nitrate, nitrite and arsenic

Pollutant removal methods from different adsorbents	*Initial concentration of pollutant (mg L^{-1})*	*Adsorbent dose (g)*	*pH*	*Temperature (K)*	*Contact Time (hours)*	*Maximum adsorption capacity Q_m*		*Reference*
						(mg g^{-1})	*Percentage (%)*	
Phosphate								
La(III)-modified bentonite	5	-	5.5	298.15	1	-	95	(Kuroki et al., 2014)
Red mud/polyaluminum chloride composite coagulant	5	0.1475	6.5	298.15	0.16	-	95	(Ni, He, Wang, & Luan, 2015)
Fe_3O_4@SiO_2 Core/shell magnetic nanoparticles functionalized with hydrous lanthanum oxide	2	0.2	7.18	298.15	24	27.8	90	(Lai, Xie, Chi, Gu, & D. Wu, 2016)
Nanoscale zerovalent iron particles supported on Spondias Purpurea seed waste	150	0.05	6	298.15	5	-	86	(Arshadi et al., 2015)
Natural and surface modified coir pith	50	2	3	303.15	12	50	82	(Krishnan & Haridas, 2008)
Coal fly ash	500	0.2	7	298.15	-	-	77	(Jie Xie et al., 2014)
Fly ash was modified by sulfuric acid	60	1	7	298.15	1	-	75	(Xu, Deng, Liu, & Peng, 2010)
Hybrid sorbent zeolite and lanthanum Hydroxide	100	0.1	8	298.15	24	-	72	(Xie, Wang, Fang, Li, & Wu, 2014)
Fly ash	100	0.5	10.9	298.15	6	-	68	(Lu, Bai, Zhu, & Shan, 2009)
Steel slag through magnetic separation	22.8	2	5.2	298.15	2	-	62	(Xiong et al., 2008)
Hybrid fibrous exchanger containing hydrated ferric oxide nanoparticles	5	0.05	8.5	294.15	24	162	-	(You et al., 2016)
Amorphous Zirconium oxide nanoparticles	10	0.1	6.2	298.15	8	99	-	(Su, Cui, Li, Gao, & Shang, 2013)
Activated aluminum Oxide and lanthanum oxide	5	1	5	298.15	-	86.8	-	(Jie Xie, Lin, Li, Wu, & Kong, 2015)
Chitosan-polymer composites	10	0.1	5	298.15	0.6	74.85	-	(Rajeswari, Amalraj, & Pius, 2015)
Green synthesized novel hybrid sorbent of zeolite/lanthanum Hydroxide	100	0.1	6-7	298.15	5	71.94	-	(Xie et al., 2014)
Modified chitosan beads	20	0.2	4	288.15	0.83	60.6	-	(Liu & Zhang, 2015)
Nanostructured iron(III)–copper(II) binary oxides	5.2	0.2	5	298.15	2	39.8	-	(Li, Gao, Zhang, & Zhang, 2014)
Magnetic polymer particles (PAM/oleic acid/Fe_3O_4)	900	0.5	3	298.0	-	37.82	-	(Lin, Chen, Chen, & Chiou, 2013)
Mesoporous ZrO_2	30	0.6	2.82	298.15	5	29.7	-	(Liu, Sun, Yin, & Hu, 2008)
Zirconium oxide-based nanocomposite	10	0.5	7	298.15	6	21.3	-	(Chen et al., 2015)
Arsenic								
Nanocomposites of graphene oxide hydrated zirconium oxide	50	0.1	6.2	298.15	8	99.01	-	(Su et al., 2013)
ZVI reduced graphite-oxide composites	3-8	0.4	7	298.15	4	As III 35.83 As V 29.04	-	(Wang et al., 2014)
Cupric oxide nanoparticles	100	2	9.3	298.15	0.5	As III 26.9 As V 22.9	-	(Martinson & Reddy, 2009)
Forager Sponge-loaded superparamagnetic iron oxide nanoparticles	100	0.3 m mol g^{-1}	3.8	296.15	1	As III 0.43* As V 0.91*	-	(Morillo, Perez, & Valiente, 2015)
Nano-iron/oyster shell composites	1.8	0.3	6.8	293.15	24	-	As III 97	(Fan et al., 2015)

continued on following page

Table 4. Continued

Pollutant removal methods from different adsorbents	Initial concentration of pollutant (mg L^{-1})	Adsorbent dose (g)	pH	Temperature (K)	Contact Time (hours)	Maximum adsorption capacity Q_m		Reference
						(mg g^{-1})	Percentage (%)	
Arsenate								
Zinc oxide nanoparticles acetate functionalized	2	0.025	6.8	298.15	1	-	99	(Singh et al., 2013)
Biochar supported zero-valent iron nanocomposite	100	0.04	6.8	293.15	24	-	97	(Wang, Gao, Li, Creamer, & He, 2016)
Superporous p(3-acrylamidopropyl) trimethyl ammonium Chloride cryogels	400	0.3	9.3	298.15	0.5	-	95	(Sahiner, Demirci, Sahiner, Yilmaz, & Al-Lohedan, 2015)
Amorphous zirconium oxide nanoparticles	100	0.01	7	298.15	1	124.5	-	(Cui, Li, Gao, & Shang, 2012)
Nitrate								
Iron nanoparticles produced by arc discharge	5	1.33	2	298.15	24	-	96	(Kassaee, Motamedi, Mikhak, & Rahnemaie, 2011)
Zero-valent iron NPs supported on high surface area nanographenes	50	0.005	7	298.15	5	-	91	(Salam, Fageeh, Al-Thabaiti, & Obaid, 2015)
Sugar beet pulp loaded with Zr(IV)	-	-	6	298.15	24	63	-	(Hassan, Kassem, & Abd El-Kader, 2010)
Activated carbon with impregnated almond shell	10–50	-	6.2	293.15	2	16–17	-	(Rezaee, Godini, Dehestani, & Khavanin, 2008)
Iron NPs by green tea & eucalyptus leaves extracts	20	1	8.45	298.15	2	13.06 & 9.6	-	(Wang, Lin, Chen, Megharaj, & Naidu, 2014)
$ZnCl_2$ treated coconut granular activated carbon	5–200	-	5.5	298.15	2	10.2	-	(Bhatnagar et al., 2008)
Wheat straw charcoal	0–25	-	-	288.15	0.16	1.10	-	(Mishra & Patel, 2009)
Carbon nanotubes	-	-	<5	298.15	1	25 *	-	(Khani & M. Mirzaei, 2008)
Bamboo charcoal	-	-	2-4	303	-	1.04×10^{-1} *	-	(Ohe, Nagae, Nakamura, & Baba, 2003)
Chitosan coated zeolite	10–3100	-	-	277.15-293.15	72	0.6–0.74 *	-	(Arora et al., 2010)
Chromium (VI)								
Diatomic supported magnetite NPs	-	0.5	2-2.5	298.15	2	-	99	(Yuan et al., 2010)
Guar Gum-nano ZnO composite	25	1	7	298.15	0.83	-	96	(Khan, Nazir, Ali, & Kumar, 2013)
$MnFe_2O_4$ NPs	50-100	5	2	298.0	0.16	-	90	(Hu, Lo, & Chen, 2007)
ZnO NPs under UV light irradiation	20	1	7	298.15	2	-	80	(Siboni, Samadi, Yang, & Lee, 2011)
Hematite NPs	1	0.6	7	298.15	8	-	47	(Xiao, Sun, Zhang, & Li, 2015)
Magnetite NPs	20	0.1	3.8	303.15	3	-	42	(Wang, Jin, Chen, Megharaj, & Naidu, 2014)
CuO NPs	25	0.001	4	298.15	24	245	-	(Nogueira, Giroto, Neto, & Ribeiro, 2016)

continued on following page

Table 4. Continued

Pollutant removal methods from different adsorbents	*Initial concentration of pollutant (mg L^{1-})*	*Adsorbent dose (g)*	*pH*	*Temperature (K)*	*Contact Time (hours)*	*Maximum adsorption capacity Q_m*		*Reference*
						(mg g^{-1})	*Percentage (%)*	
Fe NPs decorated on graphene	-	-	4.25	293.15	4	162	-	(Jabeen et al., 2011)
Nano hydro-talcite/SiO2 composite	4	1	-	298.15	24	85.4	-	(Pérez et al., 2015)
Carbon nanotubes supported ceria NPs	-	0.1	3.7.4	298.15	24	30.2	-	(Di et al., 2006)
Maghemite NPs	50-150	5	2.5	298.15	0.25	19.2	-	(Hu, Chen, & Lo, 2005)

- not reported
* mmol g^{-1}

Color Legend:

- Physical WWTT
- Biological WWTT
- Chemical WWTT

CONCLUSION

A review of the current relevant literature shows that more research is required to overcome the drawbacks of existing water treatment technologies and to introduce more environmental friendly, lower cost, efficient and green nano-adsorbents. Furthermore, synthetic polymers have been employed by researches around the world to fabricate nanocomposite membrane for heavy metals removal. However, there are few investigations on using biopolymers for membrane fabrication and its subsequent use for anionic pollutant adsorption. Moreover, characterization of these nano adsorbents before and after application will also reveal further in depth adsorption mechanisms which will support further enhancement of removal efficiencies and lowering manufacturing costs.

REFERENCES

Abdul Salam, H., Sivaraj, R., & Venckatesh, R. (2014). Green synthesis and characterization of zinc oxide nanoparticles from Ocimum basilicum L. var. purpurascens Benth.-Lamiaceae leaf extract. *Materials Letters*, *131*, 16–18. doi:10.1016/j.matlet.2014.05.033

Abell, J. M., Hamilton, D. P., & Paterson, J. (2011). Reducing the external environmental costs of pastoral farming in New Zealand: Experiences from the Te Arawa lakes, Rotorua. *Australasian Journal of Environmental Management*, *18*(3), 139–154. doi:10.1080/14486563.2011.591520

Abid, A. D., Kanematsu, M., Young, T. M., & Kennedy, I. M. (2013). Arsenic removal from water using flame-synthesized iron oxide nanoparticles with variable oxidation states. *Aerosol Science and Technology*, *47*(2), 169–176. doi:10.1080/02786826.2012.735380 PMID:23645964

Ahmad, N., Sharma, S., Alam, M. K., Singh, V. N., Shamsi, S. F., Mehta, B. R., & Fatma, A. (2010). Rapid synthesis of silver nanoparticles using dried medicinal plant of basil. *Colloids and Surfaces. B, Biointerfaces*, *81*(1), 81–86. doi:10.1016/j.colsurfb.2010.06.029 PMID:20656463

Ahmad, R., Nelson, P. N., & Kookana, R. S. (2006). The molecular composition of soil organic matter as determined by 13C NMR and elemental analyses and correlation with pesticide sorption. *European Journal of Soil Science*, *57*(6), 883–893. doi:10.1111/j.1365-2389.2005.00784.x

Ahn, J., Chung, W.-J., Pinnau, I., & Guiver, M. D. (2008). Polysulfone/silica nanoparticle mixed-matrix membranes for gas separation. *Journal of Membrane Science*, *314*(1-2), 123–133. doi:10.1016/j.memsci.2008.01.031

Ain Samat, N., & Md Nor, R. (2013). Sol–gel synthesis of zinc oxide nanoparticles using Citrus aurantifolia extracts. *Ceramics International*, *39*, S545–S548. doi:10.1016/j.ceramint.2012.10.132

Alhoshan, M., Alam, J., Dass, L. A., & Al-Homaidi, N. (2013). Fabrication of Polysulfone/ZnO Membrane: Influence of ZnO Nanoparticles on Membrane Characteristics. *Advances in Polymer Technology*, *32*(4), 1–7. doi:10.1002/adv.21369

Ali, I. (2010). The Quest for Active Carbon Adsorbent Substitutes: Inexpensive Adsorbents for Toxic Metal Ions Removal from Wastewater. *Separation and Purification Reviews*, *39*(3-4), 95–171. doi:10.1080/15422119.2010.527802

Ali, I. (2012). New generation adsorbents for water treatment. *Chemical Reviews*, *112*(10), 5073–5091. doi:10.1021/cr300133d PMID:22731247

Ali, I., Asim, M., & Khan, T. A. (2012). Low cost adsorbents for the removal of organic pollutants from wastewater. *Journal of Environmental Management*, *113*, 170–183. doi:10.1016/j.jenvman.2012.08.028 PMID:23023039

Ali, K., Dwivedi, S., Azam, A., Saquib, Q., Al-Said, M. S., Alkhedhairy, A. A., & Musarrat, J. (2016). Aloe vera extract functionalized zinc oxide nanoparticles as nanoantibiotics against multi-drug resistant clinical bacterial isolates. *Journal of Colloid and Interface Science*, *472*, 145–156. doi:10.1016/j.jcis.2016.03.021 PMID:27031596

Amutha, C., Thanikaikarasan, S., Ramadas, V., Bahadur, S. A., Natarajan, B., & Kalyani, R. (2016). Synthesis, characterization and antibacterial efficiency of ZnO nanoparticles using rice as soft biotemplate. *Optik - International Journal for Light and Electron Optics, 127*(10), 4281-4286. doi:10.1016/j.ijleo.2016.01.124

Anilkumar, M. R., Nagaswarupa, H. P., Nagabhushana, H., Sharma, S. C., Vidya, Y. S., Anantharaju, K. S., ... Gurushantha, K. (2015). Bio-inspired route for the synthesis of spherical shaped MgO:Fe(3+) nanoparticles: Structural, photoluminescence and photocatalytic investigation. *Spectrochimica Acta. Part A: Molecular and Biomolecular Spectroscopy*, *149*, 703–713. doi:10.1016/j.saa.2015.05.003 PMID:25988816

Arora, M., Eddy, N. K., Mumford, K. A., Baba, Y., Perera, J. M., & Stevens, G. W. (2010). Surface modification of natural zeolite by chitosan and its use for nitrate removal in cold regions. *Cold Regions Science and Technology*, *62*(2–3), 92–97. doi:10.1016/j.coldregions.2010.03.002

Arshadi, M., Foroughifard, S., Etemad Gholtash, J., & Abbaspourrad, A. (2015). Preparation of iron nanoparticles-loaded Spondias purpurea seed waste as an excellent adsorbent for removal of phosphate from synthetic and natural waters. *Journal of Colloid and Interface Science*, *452*, 69–77. doi:10.1016/j.jcis.2015.04.019 PMID:25919431

Asmatullah, Q., Qureshi, S. N., & Shakoori, A. R. (1998). Hexavalent Chromium-induced Congenital Abnormalities in Chick Embryos. *Journal of Applied Toxicology*, *18*(3), 167–171. doi:10.1002/(SICI)1099-1263(199805/06)18:3<167::AID-JAT492>3.0.CO;2-8 PMID:9685044

Awwad, A. M., Albiss, B., & Ahmad, A. L. (2014). Green synthesis, characterization and optical properties of zinc oxide nanosheets using Olea europea leaf extract. *Advanced Materials Letters*, *5*(9), 520–524. doi:10.5185/amlett.2014.5575

Ayyasamy, P. M., Shanthi, K., Lakshmanaperumalsamy, P., Lee, S.-J., Choi, N.-C., & Kim, D.-J. (2007). Two-stage removal of nitrate from groundwater using biological and chemical treatments. *Journal of Bioscience and Bioengineering*, *104*(2), 129–134. doi:10.1263/jbb.104.129 PMID:17884658

Azizi, S., Ahmad, M. B., Namvar, F., & Mohamad, R. (2014). Green biosynthesis and characterization of zinc oxide nanoparticles using brown marine macroalga Sargassum muticum aqueous extract. *Materials Letters*, *116*, 275–277. doi:10.1016/j.matlet.2013.11.038

Azizullah, A., Khattak, M. N., Richter, P., & Hader, D. P. (2011). Water pollution in Pakistan and its impact on public health--a review. *Environment International*, *37*(2), 479–497. doi:10.1016/j.envint.2010.10.007 PMID:21087795

Baijal, M., Singh, S., Shukla, R. N., & Sanwal, G. G. (1972). Enzymes Of The Banana Plant: Optimum Conditions For Extraction. *Phytochemistry*, *11*(3), 929–936. doi:10.1016/S0031-9422(00)88435-6

Bajpai, S. K., & Kumari, M. (2015). A green approach to prepare silver nanoparticles loaded gum acacia/poly(acrylate) hydrogels. *International Journal of Biological Macromolecules*, *80*, 177–188. doi:10.1016/j.ijbiomac.2015.06.048 PMID:26123815

Bala, N., Saha, S., Chakraborty, M., Maiti, M., Das, S., Basu, R., & Nandy, P. (2015). Green synthesis of zinc oxide nanoparticles using Hibiscus subdariffa leaf extract: Effect of temperature on synthesis, antibacterial activity and anti-diabetic activity. *RSC Advances*, *5*(7), 4993–5003. doi:10.1039/C4RA12784F

Bandara, N., Zeng, H., & Wu, J. (2013). Marine mussel adhesion: Biochemistry, mechanisms, and biomimetics. *Journal of Adhesion Science and Technology*, *27*(18-19), 2139–2162. doi:10.1080/01694243.2012.697703

Banerjee, S. S., & Chen, D.-H. (2007). Fast removal of copper ions by gum arabic modified magnetic nano-adsorbent. *Journal of Hazardous Materials*, *147*(3), 792–799. doi:10.1016/j.jhazmat.2007.01.079 PMID:17321674

Bankar, A., Joshi, B., Kumar, A. R., & Zinjarde, S. (2010). Banana peel extract mediated synthesis of gold nanoparticles. *Colloids and Surfaces. B, Biointerfaces*, *80*(1), 45–50. doi:10.1016/j.colsurfb.2010.05.029 PMID:20620890

Bhatnagar, A. (2013). *Application of Adsorbents for Water Pollution Control*. Sharjah, UAE: Bentham Science Publishers.

Bhatnagar, A., Ji, M., Choi, Y. H., Jung, W., Lee, S. H., Kim, S. J., ... Kang, J. W. (2008). Removal of Nitrate from Water by Adsorption onto Zinc Chloride Treated Activated Carbon. *Separation Science and Technology*, *43*(4), 886–907. doi:10.1080/01496390701787461

Bhatnagar, A., & Sillanpää, M. (2011). A review of emerging adsorbents for nitrate removal from water. *Chemical Engineering Journal*, *168*(2), 493–504. doi:10.1016/j.cej.2011.01.103

Bibi, M., Hashmi, M. Z., & Malik, R. N. (2015). Human exposure to arsenic in groundwater from Lahore district, Pakistan. *Environmental Toxicology and Pharmacology*, *39*(1), 42–52. doi:10.1016/j.etap.2014.10.020 PMID:25434761

Bindhu, M. R., & Umadevi, M. (2013). Synthesis of monodispersed silver nanoparticles using Hibiscus cannabinus leaf extract and its antimicrobial activity. *Spectrochimica Acta. Part A: Molecular and Biomolecular Spectroscopy*, *101*, 184–190. doi:10.1016/j.saa.2012.09.031 PMID:23103459

Boland, D. J., Brooker, M. I. H., Chippendale, G. M., Hall, N., Hyland, B. P. M., Johnston, R. D., . . . Turner, J. D. (2006). Forest Trees Of Australia (5th ed.). CSIRO Publishing.

Brett, R., Brent, C., Nanthi, S.,B., Santiago, M., Marc, G., Christopher, M., . . . Georgina, M. (2004). *Arsenic in the New Zealand environment*. Paper presented at the SuperSoil 2004: 3rd Australian New Zealand Soils Conference, Sydney, Australia.

Bundschuh, J., & Maity, J. P. (2015). Geothermal arsenic: Occurrence, mobility and environmental implications. *Renewable & Sustainable Energy Reviews*, *42*, 1214–1222. doi:10.1016/j.rser.2014.10.092

Camargo, J. A., Alonso, A., & Salamanca, A. (2005). Nitrate toxicity to aquatic animals: A review with new data for freshwater invertebrates. *Chemosphere*, *58*(9), 1255–1267. doi:10.1016/j.chemosphere.2004.10.044 PMID:15667845

Cao, G., & Wang, Y. (2004). *Nanostructures and nanomaterials: synthesis, properties and applications*. London: Imperial College Press. doi:10.1142/p305

Chen, L., Zhao, X., Pan, B., Zhang, W., Hua, M., Lv, L., & Zhang, W. (2015). Preferable removal of phosphate from water using hydrous zirconium oxide-based nanocomposite of high stability. *Journal of Hazardous Materials*, *284*, 35–42. doi:10.1016/j.jhazmat.2014.10.048 PMID:25463215

Cheng, L., Liu, S., & Dixon, K. (1998). Analysis of repair and mutagenesis of chromium-induced DNA damage in yeast, mammalian cells, and transgenic mice. *Environmental Health Perspectives*, *106*(Suppl 4), 1027–1032. doi:10.1289/ehp.98106s41027 PMID:9703488

Cheng, X., Huang, X., Wang, X., Zhao, B., Chen, A., & Sun, D. (2009). Phosphate adsorption from sewage sludge filtrate using zinc–aluminum layered double hydroxides. *Journal of Hazardous Materials*, *169*(1–3), 958–964. doi:10.1016/j.jhazmat.2009.04.052 PMID:19443104

Chithrani, B. D., Ghazani, A. A., & Chan, W. C. W. (2006). Determining the Size and Shape Dependence of Gold Nanoparticle Uptake into Mammalian Cells. *Nano Letters*, *6*(4), 662–668. doi:10.1021/nl052396o PMID:16608261

Choong, T. S. Y., Chuah, T. G., Robiah, Y., Gregory Koay, F. L., & Azni, I. (2007). Arsenic toxicity, health hazards and removal techniques from water: An overview. *Desalination*, *217*(1-3), 139–166. doi:10.1016/j.desal.2007.01.015

Cui, H., Li, Q., Gao, S., & Shang, J. K. (2012). Strong adsorption of arsenic species by amorphous zirconium oxide nanoparticles. *Journal of Industrial and Engineering Chemistry*, *18*(4), 1418–1427. doi:10.1016/j.jiec.2012.01.045

Daraei, P., Madaeni, S. S., Ghaemi, N., Salehi, E., Khadivi, M. A., Moradian, R., & Astinchap, B. (2012). Novel polyethersulfone nanocomposite membrane prepared by PANI/Fe3O4 nanoparticles with enhanced performance for Cu(II) removal from water. *Journal of Membrane Science*, *415-416*, 250–259. doi:10.1016/j.memsci.2012.05.007

Das, R. K., Gogoi, N., & Bora, U. (2011). Green synthesis of gold nanoparticles using Nyctanthes arbortristis flower extract. *Bioprocess and Biosystems Engineering*, *34*(5), 615–619. doi:10.100700449-010-0510-y PMID:21229266

Deliyanni, E. A., Bakoyannakis, D. N., Zouboulis, A. I., & Matis, K. A. (2003). Sorption of As(V) ions by akaganeite-type nanocrystals. *Chemosphere*, *50*(1), 155–163. doi:10.1016/S0045-6535(02)00351-X PMID:12656241

Devi, R. S., & Gayathri, R. (2014). Green Synthesis of Zinc Oxide Nanoparticles by using Hibiscus rosa-sinensis. *International Journal of Current Engineering and Technology*, *4*(4).

Dhakras, P. A. (2011). Nanotechnology Applications in Water Purification and Waste Water Treatment: A Review. The Institute of Electrical and Electronics Engineers (professional association).

Di, Z.-C., Ding, J., Peng, X.-J., Li, Y.-H., Luan, Z.-K., & Liang, J. (2006). Chromium adsorption by aligned carbon nanotubes supported ceria nanoparticles. *Chemosphere*, *62*(5), 861–865. doi:10.1016/j.chemosphere.2004.06.044 PMID:16403560

Diaz, C., Barrera, G., Segovia, M., Valenzuela, M. L., & Osiak, M. (2015). #x, & Dwyer, C. (2015). Crystallizing Vanadium Pentoxide Nanostructures in the Solid-State Using Modified Block Copolymer and Chitosan Complexes. *Journal of Nanomaterials*, *13*. doi:10.1155/2015/105157

Dong, C., Zhang, X., Cai, H., & Cao, C. (2014). Facile and one-step synthesis of monodisperse silver nanoparticles using gum acacia in aqueous solution. *Journal of Molecular Liquids*, *196*, 135–141. doi:10.1016/j.molliq.2014.03.009

Doole, G. J., Marsh, D., & Ramilan, T. (2013). Evaluation of agri-environmental policies for reducing nitrate pollution from New Zealand dairy farms accounting for firm heterogeneity. *Land Use Policy*, *30*(1), 57–66. doi:10.1016/j.landusepol.2012.02.007

Dubey, S. P., Dwivedi, A. D., Lahtinen, M., Lee, C., Kwon, Y. N., & Sillanpaa, M. (2013). Protocol for development of various plants leaves extract in single-pot synthesis of metal nanoparticles. *Spectrochimica Acta. Part A: Molecular and Biomolecular Spectroscopy*, *103*, 134–142. doi:10.1016/j.saa.2012.11.021 PMID:23257341

Ebert, K., Fritsch, D., Koll, J., & Tjahjawiguna, C. (2004). Influence of inorganic fillers on the compaction behaviour of porous polymer based membranes. *Journal of Membrane Science*, *233*(1–2), 71–78. doi:10.1016/j.memsci.2003.12.012

Ehiowemwenguan, G., Emoghene, A. O., & Inetianbor, J. E. (2014). Antibacterial and phytochemical analysis of Banana fruit peel. *IOSR Journal Of Pharmacy*, *4*(8), 18–25. doi:10.9790/3013-0408018025

Elumalai, K., & Velmurugan, S. (2015). Green synthesis, characterization and antimicrobial activities of zinc oxide nanoparticles from the leaf extract of Azadirachta indica (L.). *Applied Surface Science*, *345*, 329–336. doi:10.1016/j.apsusc.2015.03.176

Elumalai, K., Velmurugan, S., Ravi, S., Kathiravan, V., & Adaikala Raj, G. (2015). Bio-approach: Plant mediated synthesis of ZnO nanoparticles and their catalytic reduction of methylene blue and antimicrobial activity. *Advanced Powder Technology*, *26*(6), 1639–1651. doi:10.1016/j.apt.2015.09.008

Elumalai, K., Velmurugan, S., Ravi, S., Kathiravan, V., & Ashokkumar, S. (2015a). Bio-fabrication of zinc oxide nanoparticles using leaf extract of curry leaf (Murraya koenigii) and its antimicrobial activities. *Materials Science in Semiconductor Processing*, *34*, 365–372. doi:10.1016/j.mssp.2015.01.048

Elumalai, K., Velmurugan, S., Ravi, S., Kathiravan, V., & Ashokkumar, S. (2015b). Facile, eco-friendly and template free photosynthesis of cauliflower like ZnO nanoparticles using leaf extract of Tamarindus indica (L.) and its biological evolution of antibacterial and antifungal activities. *Spectrochim Acta A Mol Biomol Spectrosc, 136 Pt B*, 1052-1057. doi:10.1016/j.saa.2014.09.129

Elumalai, K., Velmurugan, S., Ravi, S., Kathiravan, V., & Ashokkumar, S. (2015c). Green synthesis of zinc oxide nanoparticles using Moringa oleifera leaf extract and evaluation of its antimicrobial activity. *Spectrochimica Acta. Part A: Molecular and Biomolecular Spectroscopy*, *143*, 158–164. doi:10.1016/j.saa.2015.02.011 PMID:25725211

Eppler, A. S., Rupprechter, G., Anderson, E. A., & Somorjai, G. A. (2000). Thermal and Chemical Stability and Adhesion Strength of Pt Nanoparticle Arrays Supported on Silica Studied by Transmission Electron Microscopy and Atomic Force Microscopy. *The Journal of Physical Chemistry B*, *104*(31), 7286–7292. doi:10.1021/jp0006429

Fan, L., Zhang, S., Zhang, X., Zhou, H., Lu, Z., & Wang, S. (2015). Removal of arsenic from simulation wastewater using nano-iron/oyster shell composites. *Journal of Environmental Management*, *156*, 109–114. doi:10.1016/j.jenvman.2015.03.044 PMID:25839745

FAO. (2010). *Information System on Water and Agriculture*. Retrieved from http://www.fao.org/nr/water/aquastat/water_use/index6.stm

Fei, B., Qian, B., Yang, Z., Wang, R., Liu, W. C., Mak, C. L., & Xin, J. H. (2008). Coating carbon nanotubes by spontaneous oxidative polymerization of dopamine. *Carbon*, *46*(13), 1795–1797. doi:10.1016/j.carbon.2008.06.049

Feldheim, D., & Foss, C. (2002). *Metal nanoparticles: synthesis, characterization, and applications*. Boca Raton, FL: CRC Press.

Fendler, J. H. (2007). *Nanoparticles and Nanostructured Films: Preparation, Characterization, and Applications*. Wiley-VCH Verlag GmbH.

Gaballah, I., & Kilbertus, G. (1998). Recovery of heavy metal ions through decontamination of synthetic solutions and industrial effluents using modified barks. *Journal of Geochemical Exploration*, *62*(1–3), 241–286. doi:10.1016/S0375-6742(97)00068-X

Gan, P. P., & Li, S. F. Y. (2012). Potential of plant as a biological factory to synthesize gold and silver nanoparticles and their applications. *Reviews in Environmental Science and Biotechnology*, *11*(2), 169–206. doi:10.100711157-012-9278-7

Gan, P. P., Ng, S. H., Huang, Y., & Li, S. F. Y. (2012). Green synthesis of gold nanoparticles using palm oil mill effluent (POME): A low-cost and eco-friendly viable approach. *Bioresource Technology*, *113*, 132–135. doi:10.1016/j.biortech.2012.01.015 PMID:22297042

Gao, H., Sun, Y., Zhou, J., Xu, R., & Duan, H. (2013). Mussel-inspired synthesis of polydopamine-functionalized graphene hydrogel as reusable adsorbents for water purification. *ACS Applied Materials & Interfaces*, *5*(2), 425–432. doi:10.1021/am302500v PMID:23265565

Geetha, D., Kavitha, S., & Ramesh, P. S. (2015). A novel bio-degradable polymer stabilized Ag/TiO2 nanocomposites and their catalytic activity on reduction of methylene blue under natural sun light. *Ecotoxicology and Environmental Safety*, *121*, 126–134. doi:10.1016/j.ecoenv.2015.04.042 PMID:25976106

Gehrke, I., Geiser, A., & Somborn-Schulz, A. (2015). Innovations in nanotechnology for water treatment. *Nanotechnology, Science and Applications*, *8*, 1–17. doi:10.2147/NSA.S43773 PMID:25609931

Ghodake, G. S., Deshpande, N. G., Lee, Y. P., & Jin, E. S. (2010). Pear fruit extract-assisted room-temperature biosynthesis of gold nanoplates. *Colloids and Surfaces. B, Biointerfaces*, *75*(2), 584–589. doi:10.1016/j.colsurfb.2009.09.040 PMID:19879738

Ghosh, S., Patil, S., Ahire, M., Kitture, R., Gurav, D. D., Jabgunde, A. M., . . . Chopade, B. A. (2012). Gnidia glauca flower extract mediated synthesis of gold nanoparticles and evaluation of its chemocatalytic potential. *Journal of Nanobiotechnology, 10*(17), 1-9.

Ghosh, S., Patil, S., Ahire, M., Kitture, R., Kale, S., Pardesi, K., ... Chopade, B. A. (2012). Synthesis of silver nanoparticles using Dioscorea bulbifera tuber extract and evaluation of its synergistic potential in combination with antimicrobial agents. *International Journal of Nanomedicine*, *7*, 483–496. doi:10.2147/IJN.S24793 PMID:22334779

Glusker, J. P., Katz, A. K., & Bock, C. W. (1999). Metal Ions In Biological Systems. *The Rigaku Journal*, *16*(2), 8–17.

Gunalan, S., Sivaraj, R., & Rajendran, V. (2012). Green synthesized ZnO nanoparticles against bacterial and fungal pathogens. *Progress in Natural Science: Materials International*, *22*(6), 693–700. doi:10.1016/j.pnsc.2012.11.015

Guo, L., Liu, Q., Li, G., Shi, J., Liu, J., Wang, T., & Jiang, G. (2012). A mussel-inspired polydopamine coating as a versatile platform for the in situ synthesis of graphene-based nanocomposites. *Nanoscale*, *4*(19), 5864–5867. doi:10.1039/c2nr31547e PMID:22945453

Gupta, V. K., Ali, I., Saleh, T. A., Nayak, A., & Agarwal, S. (2012). Chemical treatment technologies for waste-water recycling—an overview. *RSC Advances*, *2*(16), 6380. doi:10.1039/c2ra20340e

Halek, F. S., Farahani, S. K., & Hosseini, S. M. (2015). Fabrication of poly(ether sulfone) based mixed matrix membranes modified by TiO2 nanoparticles for purification of biodiesel produced from waste cooking oils. *Korean Journal of Chemical Engineering*, *33*(2), 629–637. doi:10.100711814-015-0161-4

Hanley, N., Shogren, J. F., & White, B. (2007). *Environmental Economics* (2nd ed.). Basingstoke, UK: Palgrave Macmillan.

Hao, Y., Meng, G., Ye, C., Zhang, X., & Zhang, L. (2005). Kinetics-Driven Growth of Orthogonally Branched Single-Crystalline Magnesium Oxide Nanostructures. *The Journal of Physical Chemistry B*, *109*(22), 11204–11208. doi:10.1021/jp0505451 PMID:16852367

Hassan, M. L., Kassem, N. F., & Abd El-Kader, A. H. (2010). Novel Zr(IV)/sugar beet pulp composite for removal of sulfate and nitrate anions. *Journal of Applied Polymer Science*, *117*(4), 2205–2212. doi:10.1002/app.32063

Haverkamp, R. G., & Marshall, A. T. (2008). The mechanism of metal nanoparticle formation in plants: Limits on accumulation. *Journal of Nanoparticle Research*, *11*(6), 1453–1463. doi:10.100711051-008-9533-6

Herlekar, M., Barve, S., & Kumar, R. (2014). Plant-Mediated Green Synthesis of Iron Nanoparticles. *Journal of Nanoparticles*, *2014*, 1–9. doi:10.1155/2014/140614

Hornyak, G. L., Moore, J. J., Tibbals, H. F., & Dutta, J. (2008). *Fundamentals of Nanotechnology*. CRC Press Taylor & Francis Group.

Hu, J., Chen, G., & Lo, I. M. C. (2005). Removal and recovery of Cr(VI) from wastewater by maghemite nanoparticles. *Water Research*, *39*(18), 4528–4536. doi:10.1016/j.watres.2005.05.051 PMID:16146639

Hu, J., Chen, G., & Lo, I. M. C. (2006). Selective Removal of Heavy Metals from Industrial Wastewater Using Maghemite Nanoparticle: Performance and Mechanisms. *Journal of Environmental Engineering*, *132*(7), 709–715. doi:10.1061/(ASCE)0733-9372(2006)132:7(709)

Hu, J., Lo, I. M. C., & Chen, G. (2007). Comparative study of various magnetic nanoparticles for Cr(VI) removal. *Separation and Purification Technology*, *56*(3), 249–256. doi:10.1016/j.seppur.2007.02.009

Huang, Y., & Keller, A. A. (2013). Magnetic Nanoparticle Adsorbents for Emerging Organic Contaminants. *ACS Sustainable Chemistry & Engineering*. doi:10.1021c400047q

Hussain, F., Hojjati, M., Okamoto, M., & Gorga, R. E. (2006). Review article: Polymer-matrix Nanocomposites, Processing, Manufacturing, and Application: An Overview. *Journal of Composite Materials*, *40*(17), 1511–1575. doi:10.1177/0021998306067321

Ibrahim, H. M. M. (2015). Green synthesis and characterization of silver nanoparticles using banana peel extract and their antimicrobial activity against representative microorganisms. *Journal of Radiation Research and Applied Sciences*, *8*(3), 265–275. doi:10.1016/j.jrras.2015.01.007

Iravani, S. (2011). Green synthesis of metal nanoparticles using plants. *Green Chemistry*, *13*(10), 2638. doi:10.1039/c1gc15386b

Jabeen, H., Chandra, V., Jung, S., Lee, J. W., Kim, K. S., & Kim, S. B. (2011). Enhanced Cr(vi) removal using iron nanoparticle decorated graphene. *Nanoscale*, *3*(9), 3583–3585. doi:10.1039/c1nr10549c PMID:21814702

Jeon, E. K., Seo, E., Lee, E., Lee, W., Um, M. K., & Kim, B. S. (2013). Mussel-inspired green synthesis of silver nanoparticles on graphene oxide nanosheets for enhanced catalytic applications. *Chemical Communications (Cambridge)*, *49*(33), 3392–3394. doi:10.1039/c3cc00115f PMID:23435315

Jiang, J., Oberdörster, G., & Biswas, P. (2009). Characterization of size, surface charge, and agglomeration state of nanoparticle dispersions for toxicological studies. *Journal of Nanoparticle Research*, *11*(1), 77–89. doi:10.100711051-008-9446-4

Joo, S. H., & Zhao, D. (2016). Environmental dynamics of metal oxide nanoparticles in heterogeneous systems: A review. *Journal of Hazardous Materials*. doi:10.1016/j.jhazmat.2016.02.068 PMID:26961405

Kassaee, M. Z., Motamedi, E., Mikhak, A., & Rahnemaie, R. (2011). Nitrate removal from water using iron nanoparticles produced by arc discharge vs. reduction. *Chemical Engineering Journal*, *166*(2), 490–495. doi:10.1016/j.cej.2010.10.077

Kawasaki, N., Matsushige, K., Komatsu, K., Kohzu, A., Nara, F. W., Ogishi, F., ... Imai, A. (2011). Fast and precise method for HPLC-size exclusion chromatography with UV and TOC (NDIR) detection: Importance of multiple detectors to evaluate the characteristics of dissolved organic matter. *Water Research*, *45*(18), 6240–6248. doi:10.1016/j.watres.2011.09.021 PMID:21959091

Kazi, T. G., Arain, M. B., Baig, J. A., Jamali, M. K., Afridi, H. I., Jalbani, N., ... Niaz, A. (2009). The correlation of arsenic levels in drinking water with the biological samples of skin disorders. *The Science of the Total Environment*, *407*(3), 1019–1026. doi:10.1016/j.scitotenv.2008.10.013 PMID:19027142

Kesharwani, J., Yoon, K. Y., Hwang, J., & Rai, M. (2009). Phytofabrication of Silver Nanoparticles by Leaf Extract of Datura metel: Hypothetical Mechanism Involvedin Synthesis. *Journal of Bionanoscience*, *3*(1), 1–6. doi:10.1166/jbns.2009.1008

Khan, T. A., Nazir, M., Ali, I., & Kumar, A. (2013). Removal of Chromium(VI) from aqueous solution using guar gum–nano zinc oxide biocomposite adsorbent. *Arabian Journal of Chemistry*. doi:10.1016/j.arabjc.2013.08.019

Khan, T. A., Nazir, M., Ali, I., & Kumar, A. Removal of Chromium(VI) from aqueous solution using guar gum–nano zinc oxide biocomposite adsorbent. *Arabian Journal of Chemistry*. doi:10.1016/j.arabjc.2013.08.019

Khani, A., & Mirzaei, M. (2008). *Comparative study of nitrate removal from aqueous solution using powder activated carbon and carbon nanotubes*. Paper presented at the 2nd International IUPAC Conference on Green Chemistry, Russia.

Kharissova, O. V., Dias, H. V., Kharisov, B. I., Perez, B. O., & Perez, V. M. (2013). The greener synthesis of nanoparticles. *Trends in Biotechnology*, *31*(4), 240–248. doi:10.1016/j.tibtech.2013.01.003 PMID:23434153

Kilpimaa, S., Runtti, H., Kangas, T., Lassi, U., & Kuokkanen, T. (2014). Removal of phosphate and nitrate over a modified carbon residue from biomass gasification. *Chemical Engineering Research & Design*, *92*(10), 1923–1933. doi:10.1016/j.cherd.2014.03.019

Kilpimaa, S., Runtti, H., Kangas, T., Lassi, U., & Kuokkanen, T. (2015). Physical activation of carbon residue from biomass gasification: Novel sorbent for the removal of phosphates and nitrates from aqueous solution. *Journal of Industrial and Engineering Chemistry*, *21*, 1354–1364. doi:10.1016/j.jiec.2014.06.006

Kim, J., Rheem, Y., Yoo, B., Chong, Y., Bozhilov, K. N., Kim, D., ... Myung, N. V. (2010). Peptide-mediated shape- and size-tunable synthesis of gold nanostructures. *Acta Biomaterialia*, *6*(7), 2681–2689. doi:10.1016/j.actbio.2010.01.019 PMID:20083240

Krishnan, K. A., & Haridas, A. (2008). Removal of phosphate from aqueous solutions and sewage using natural and surface modified coir pith. *Journal of Hazardous Materials*, *152*(2), 527–535. doi:10.1016/j.jhazmat.2007.07.015 PMID:17706344

Kumar, V., & Yadav, S. K. (2009). Plant-mediated synthesis of silver and gold nanoparticles and their applications. *Journal of Chemical Technology and Biotechnology (Oxford, Oxfordshire)*, *84*(2), 151–157. doi:10.1002/jctb.2023

Kumari, L., Li, W. Z., Vannoy, C. H., Leblanc, R. M., & Wang, D. Z. (2009). Synthesis, characterization and optical properties of Mg(OH)2 micro-/nanostructure and its conversion to MgO. *Ceramics International*, *35*(8), 3355–3364. doi:10.1016/j.ceramint.2009.05.035

Kuppusamy, P., Yusoff, M. M., Maniam, G. P., & Govindan, N. (2014). Biosynthesis of metallic nanoparticles using plant derivatives and their new avenues in pharmacological applications – An updated report. *Saudi Pharmaceutical Journal*. doi:10.1016/j.jsps.2014.11.013 PMID:27330378

Kuppusamy, S., Thavamani, P., Megharaj, M., Venkateswarlu, K., Lee, Y. B., & Naidu, R. (2016). Potential of Melaleuca diosmifolia leaf as a low-cost adsorbent for hexavalent chromium removal from contaminated water bodies. *Process Safety and Environmental Protection*, *100*, 173–182. doi:10.1016/j.psep.2016.01.009

Kuroki, V., Bosco, G. E., Fadini, P. S., Mozeto, A. A., Cestari, A. R., & Carvalho, W. A. (2014). Use of a La(III)-modified bentonite for effective phosphate removal from aqueous media. *Journal of Hazardous Materials*, *274*, 124–131. doi:10.1016/j.jhazmat.2014.03.023 PMID:24769849

Lai, L., Xie, Q., Chi, L., Gu, W., & Wu, D. (2016). Adsorption of phosphate from water by easily separable Fe3O4@SiO2 core/shell magnetic nanoparticles functionalized with hydrous lanthanum oxide. *Journal of Colloid and Interface Science*, *465*, 76–82. doi:10.1016/j.jcis.2015.11.043 PMID:26641568

Lai, L., Xie, Q., Chi, L., Gu, W., & Wu, D. (2016). Adsorption of phosphate from water by easily separable Fe3O4@SiO2 core/shell magnetic nanoparticles functionalized with hydrous lanthanum oxide. *Journal of Colloid and Interface Science*, *465*, 76–82. doi:10.1016/j.jcis.2015.11.043 PMID:26641568

Lata, S., & Samadder, S. R. (2016). Removal of arsenic from water using nano adsorbents and challenges: A review. *Journal of Environmental Management*, *166*, 387–406. doi:10.1016/j.jenvman.2015.10.039 PMID:26546885

Lee, H., Dellatore, S. M., Miller, W. M., & Messersmith, P. B. (2007). Mussel-Inspired Surface Chemistry for Multifunctional Coatings. *Science*, *318*(5849), 426–430. doi:10.1126cience.1147241 PMID:17947576

Lee, K. S. (2011). Three ways of linking laboratory endeavours to the realm of policies. *European Journal of the History of Economic Thought*, *18*(5), 755–776. doi:10.1080/09672567.2011.616593

Li, G., Gao, S., Zhang, G., & Zhang, X. (2014). Enhanced adsorption of phosphate from aqueous solution by nanostructured iron(III)–copper(II) binary oxides. *Chemical Engineering Journal*, *235*, 124–131. doi:10.1016/j.cej.2013.09.021

Lide, D. R. (2005). *Handbook of Chemistry and Physics* (86th ed.). Cleveland, OH: Chemical Rubber Company.

Lin, J., Zhan, Y., Wang, H., Chu, M., Wang, C., He, Y., & Wang, X. (2017). Effect of calcium ion on phosphate adsorption onto hydrous zirconium oxide. *Chemical Engineering Journal*, *309*, 118–129. doi:10.1016/j.cej.2016.10.001

Lin, Y.-F., Chen, H.-W., Chen, Y.-C., & Chiou, C.-S. (2013). Application of magnetite modified with polyacrylamide to adsorb phosphate in aqueous solution. *Journal of the Taiwan Institute of Chemical Engineers*, *44*(1), 45–51. doi:10.1016/j.jtice.2012.09.005

Liu, H., Sun, X., Yin, C., & Hu, C. (2008). Removal of phosphate by mesoporous ZrO2. *Journal of Hazardous Materials*, *151*(2-3), 616–622. doi:10.1016/j.jhazmat.2007.06.033 PMID:17658689

Liu, R., Guo, Y., Odusote, G., Qu, F., & Priestley, R. D. (2013). Core-shell Fe3O4 polydopamine nanoparticles serve multipurpose as drug carrier, catalyst support and carbon adsorbent. *ACS Applied Materials & Interfaces*, *5*(18), 9167–9171. doi:10.1021/am402585y PMID:24010676

Liu, X., & Zhang, L. (2015). Removal of phosphate anions using the modified chitosan beads: Adsorption kinetic, isotherm and mechanism studies. *Powder Technology*, *277*, 112–119. doi:10.1016/j.powtec.2015.02.055

Lofrano, G., Carotenuto, M., Libralato, G., Domingos, R. F., Markus, A., Dini, L., ... Meric, S. (2016). Polymer functionalized nanocomposites for metals removal from water and wastewater: An overview. *Water Research*, *92*, 22–37. doi:10.1016/j.watres.2016.01.033 PMID:26827255

Lu, S. G., Bai, S. Q., Zhu, L., & Shan, H. D. (2009). Removal mechanism of phosphate from aqueous solution by fly ash. *Journal of Hazardous Materials*, *161*(1), 95–101. doi:10.1016/j.jhazmat.2008.02.123 PMID:18434007

Lukman, A. I., Gong, B., Marjo, C. E., Roessner, U., & Harris, A. T. (2011). Facile synthesis, stabilization, and anti-bacterial performance of discrete Ag nanoparticles using Medicago sativa seed exudates. *Journal of Colloid and Interface Science*, *353*(2), 433–444. doi:10.1016/j.jcis.2010.09.088 PMID:20974473

Luo, Z., Zhu, S., Liu, Z., Liu, J., Huo, M., & Yang, W. (2015). Study of phosphate removal from aqueous solution by zinc oxide. *Journal of Water and Health*, *13*(3), 704–713. doi:10.2166/wh.2015.210 PMID:26322756

Luximon-Ramma, A., Bahorun, T., Soobrattee, M. A., & Aruoma, O. I. (2002). Antioxidant Activities of Phenolic, Proanthocyanidin, and Flavonoid Components in Extracts of Cassia fistula. *Journal of Agricultural and Food Chemistry*, *50*(18), 5042–5047. doi:10.1021/jf0201172 PMID:12188605

Lynge, M. E., Westen, R., Postmab, A., & Stadler, B. (2011). Polydopamine—a nature-inspired polymer coating for biomedical science. *Nanoscale*, *3*(12), 4916–4928. doi:10.1039/c1nr10969c PMID:22024699

Maensiri, S., Laokul, P., Klinkaewnarong, J., Phokha, S., Promarak, V., & Seraphin, S. (2008). Indium oxide (In2O3) nanoparticles using Aloe vera plant extract: Synthesis and optical properties. *Optoelectronics And Advanced Materials*, *2*(3), 161–165.

Mageshwari, K., Mali, S. S., Sathyamoorthy, R., & Patil, P. S. (2013). Template-free synthesis of MgO nanoparticles for effective photocatalytic applications. *Powder Technology*, *249*, 456–462. doi:10.1016/j.powtec.2013.09.016

Majedi, S. M., Kelly, B. C., & Lee, H. K. (2014). Combined effects of water temperature and chemistry on the environmental fate and behavior of nanosized zinc oxide. *The Science of the Total Environment*, *496*, 585–593. doi:10.1016/j.scitotenv.2014.07.082 PMID:25108799

Makarov, V. V., Love, A. J., Sinitsyna, O. V., Makarova, S. S., Yaminsky, I. V., Taliansky, M. E., & Kalinina, N. O. (2014). "Green" Nanotechnologies: Synthesis of Metal Nanoparticles Using Plant. *Acta naturae*, *6*(20), 35-44.

Martinson, C. A., & Reddy, K. J. (2009). Adsorption of arsenic(III) and arsenic(V) by cupric oxide nanoparticles. *Journal of Colloid and Interface Science*, *336*(2), 406–411. doi:10.1016/j.jcis.2009.04.075 PMID:19477461

Maslin, B. R., & McDonald, M. W. (2004). AcaciaSearch- Evaluation of Acacia as a woody crop option for southern Australia. Rural Industries Research and Development Corporation; Dept. Conservation and Land Management.

Milani, N., McLaughlin, M. J., Stacey, S. P., Kirby, J. K., Hettiarachchi, G. M., Beak, D. G., & Cornelis, G. (2012). Dissolution kinetics of macronutrient fertilizers coated with manufactured zinc oxide nanoparticles. *Journal of Agricultural and Food Chemistry*, *60*(16), 3991–3998. doi:10.1021/jf205191y PMID:22480134

Mishra, P. C., & Patel, R. K. (2009). Use of agricultural waste for the removal of nitrate-nitrogen from aqueous medium. *Journal of Environmental Management*, *90*(1), 519–522. doi:10.1016/j.jenvman.2007.12.003 PMID:18294755

Mitrano, D. M., Motellier, S., Clavaguera, S., & Nowack, B. (2015). Review of nanomaterial aging and transformations through the life cycle of nano-enhanced products. *Environment International*, *77*, 132–147. doi:10.1016/j.envint.2015.01.013 PMID:25705000

Mittal, A. K., Chisti, Y., & Banerjee, U. C. (2013). Synthesis of metallic nanoparticles using plant extracts. *Biotechnology Advances*, *31*(2), 346–356. doi:10.1016/j.biotechadv.2013.01.003 PMID:23318667

Mohapatra, D., Mishra, S., & Sutar, N. (2010). Banana And Its By Product Utilisation: An Overview. *Journal of Scientific and Industrial Research*, *69*, 323–329.

Mook, W. T., Chakrabarti, M. H., Aroua, M. K., Khan, G. M. A., Ali, B. S., Islam, M. S., & Abu Hassan, M. A. (2012). Removal of total ammonia nitrogen (TAN), nitrate and total organic carbon (TOC) from aquaculture wastewater using electrochemical technology: A review. *Desalination*, *285*, 1–13. doi:10.1016/j.desal.2011.09.029

Moorthy, S. K., Ashok, C. H., Rao, K. V., & Viswanathan, C. (2015). Synthesis and Characterization of Mgo Nanoparticles by Neem Leaves through Green Method. *Materials Today: Proceedings*, *2*(9), 4360–4368. doi:10.1016/j.matpr.2015.10.027

Morillo, D., Perez, G., & Valiente, M. (2015). Efficient arsenic(V) and arsenic(III) removal from acidic solutions with Novel Forager Sponge-loaded superparamagnetic iron oxide nanoparticles. *Journal of Colloid and Interface Science*, *453*, 132–141. doi:10.1016/j.jcis.2015.04.048 PMID:25982936

Mrówczyński, R., Turcu, R., Leostean, C., Scheidt, H. A., & Liebscher, J. (2013). New versatile polydopamine coated functionalized magnetic nanoparticles. *Materials Chemistry and Physics*, *138*(1), 295–302. doi:10.1016/j.matchemphys.2012.11.059

Murugan, R., & Parimelazhagan, T. (2014). Comparative evaluation of different extraction methods for antioxidant and anti-inflammatory properties from Osbeckia parvifolia Arn. – An in vitro approach. *Journal of King Saud University - Science, 26*(4), 267-275. doi:10.1016/j.jksus.2013.09.006

Nadagouda, M. N., Desai, I., Cruz, C., & Yang, D. J. (2012). Novel Pd based catalyst for the removal of organic and emerging contaminants. *RSC Advances*, *2*(19), 7540. doi:10.1039/c2ra20562a

Naghash, A., & Nezamzadeh-Ejhieh, A. (2015). Comparison of the efficiency of modified clinoptilolite with HDTMA and HDP surfactants for the removal of phosphate in aqueous solutions. *Journal of Industrial and Engineering Chemistry*, *31*, 185–191. doi:10.1016/j.jiec.2015.06.022

Namasivayam, C., & Sangeetha, D. (2004). Equilibrium and kinetic studies of adsorption of phosphate onto ZnCl2 activated coir pith carbon. *Journal of Colloid and Interface Science*, *280*(2), 359–365. doi:10.1016/j.jcis.2004.08.015 PMID:15533408

Narayanan, K. B., & Sakthivel, N. (2010). Biological synthesis of metal nanoparticles by microbes. *Advances in Colloid and Interface Science*, *156*(1-2), 1–13. doi:10.1016/j.cis.2010.02.001 PMID:20181326

Narayanan, K. B., & Sakthivel, N. (2011). Green synthesis of biogenic metal nanoparticles by terrestrial and aquatic phototrophic and heterotrophic eukaryotes and biocompatible agents. *Advances in Colloid and Interface Science*, *169*(2), 59–79. doi:10.1016/j.cis.2011.08.004 PMID:21981929

Ni, F., He, J., Wang, Y., & Luan, Z. (2015). Preparation and characterization of a cost-effective red mud/polyaluminum chloride composite coagulant for enhanced phosphate removal from aqueous solutions. *Journal of Water Process Engineering*, *6*, 158–165. doi:10.1016/j.jwpe.2015.04.003

Nogueira, A. E., Giroto, A. S., Neto, A. B. S., & Ribeiro, C. (2016). CuO synthesized by solvothermal method as a high capacity adsorbent for hexavalent chromium. *Colloids and Surfaces. A, Physicochemical and Engineering Aspects*, *498*, 161–167. doi:10.1016/j.colsurfa.2016.03.022

Nora'aini, A., Wahab Mohammad, A., Jusoh, A., Hasan, M. R., Ghazali, N., & Kamaruzaman, K. (2005). Treatment of aquaculture wastewater using ultra-low pressure asymmetric polyethersulfone (PES) membrane. *Desalination*, *185*(1-3), 317–326. doi:10.1016/j.desal.2005.03.084

Ogata, F., Ueda, A., Tanei, S., Imai, D., & Kawasaki, N. (2016). Simultaneous removal of phosphate and nitrite ions from aqueous solutions using modified soybean waste. *Journal of Industrial and Engineering Chemistry*, *35*, 287–294. doi:10.1016/j.jiec.2016.01.007

Ohe, K., Nagae, Y., Nakamura, S., & Baba, Y. (2003). Removal of nitrate anion by carbonaceous materials prepared from bamboo and coconut shell. *Journal of Chemical Engineering of Japan*, *36*(4), 511–515. doi:10.1252/jcej.36.511

Orphanides, A., Goulas, V., & Gekas, V. (2013). Effect of Drying Method on the Phenolic Content and Antioxidant Capacity of Spearmint. *Czech Journal of Food Sciences*, *31*(5), 509–513. doi:10.17221/526/2012-CJFS

Ouraipryvan, P., Sreethawong, T., & Chavadej, S. (2009). Synthesis of crystalline MgO nanoparticle with mesoporous-assembled structure via a surfactant-modified sol–gel process. *Materials Letters*, *63*(21), 1862–1865. doi:10.1016/j.matlet.2009.05.068

Pal, S., Tak, Y. K., & Song, J. M. (2007). Does the antibacterial activity of silver nanoparticles depend on the shape of the nanoparticle? A study of the Gram-negative bacterium Escherichia coli. *Applied and Environmental Microbiology*, *73*(6), 1712–1720. doi:10.1128/AEM.02218-06 PMID:17261510

Pan, B., Han, F., Nie, G., Wu, B., He, K., & Lu, L. (2014). New Strategy To Enhance Phosphate Removal from Water by Hydrous Manganese Oxide. *Environmental Science & Technology*, *48*(9), 5101–5107. doi:10.1021/es5004044 PMID:24730751

Pan, B., & Xing, B. (2008). Adsorption Mechanisms of Organic Chemicals on Carbon Nanotubes. *Environmental Science & Technology*, *42*(24), 9005–9013. doi:10.1021/es801777n PMID:19174865

Panigrahi, S., Kundu, S., Ghosh, S. K., Nath, S., & Pal, T. (2004). General method of synthesis for metal nanoparticles. *Journal of Nanoparticle Research*, *6*(4), 411–414. doi:10.100711051-004-6575-2

Park, Y. (2014). New paradigm shift for the green synthesis of antibacterial silver nanoparticles utilizing plant extracts. *Toxicological Research*, *30*(3), 169–178. doi:10.5487/TR.2014.30.3.169 PMID:25343010

Pendergast, M. M., & Hoek, E. M. V. (2011). A review of water treatment membrane nanotechnologies. *Energy & Environmental Science*, *4*(6), 1946. doi:10.1039/c0ee00541j

Pendergast, M. T. M., Nygaard, J. M., Ghosh, A. K., & Hoek, E. M. V. (2010). Using nanocomposite materials technology to understand and control reverse osmosis membrane compaction. *Desalination*, *261*(3), 255–263. doi:10.1016/j.desal.2010.06.008

Pérez, E., Ayele, L., Getachew, G., Fetter, G., Bosch, P., Mayoral, A., & Díaz, I. (2015). Removal of chromium(VI) using nano-hydrotalcite/SiO2 composite. *Journal of Environmental Chemical Engineering*, *3*(3), 1555–1561. doi:10.1016/j.jece.2015.05.009

Phuoc, T. X., Howard, B. H., Martello, D. V., Soong, Y., & Chyu, M. K. (2008). Synthesis of Mg(OH)2, MgO, and Mg nanoparticles using laser ablation of magnesium in water and solvents. *Optics and Lasers in Engineering*, *46*(11), 829–834. doi:10.1016/j.optlaseng.2008.05.018

Prasse, C., Stalter, D., Schulte-Oehlmann, U., Oehlmann, J., & Ternes, T. A. (2015). Spoilt for choice: A critical review on the chemical and biological assessment of current wastewater treatment technologies. *Water Research*, *87*, 237–270. doi:10.1016/j.watres.2015.09.023 PMID:26431616

Pretty, J. N., Mason, C. F., Nedwell, D. B., Hine, R. E., Leaf, S., & Dils, R. (2003). Environmental costs of freshwater eutrophication in England and Wales. *Environmental Science & Technology*, *37*(2), 201–208. doi:10.1021/es020793k PMID:12564888

Qu, J., Luo, C., & Hou, J. (2011). Synthesis of ZnO nanoparticles from Zn-hyperaccumulator (Sedum alfredii Hance) plants. *Micro & Nano Letters*, *6*(3), 174. doi:10.1049/mnl.2011.0004

Qu, J., Yuan, X., Wang, X., & Shao, P. (2011). Zinc accumulation and synthesis of ZnO nanoparticles using Physalis alkekengi L. *Environmental Pollution*, *159*(7), 1783–1788. doi:10.1016/j.envpol.2011.04.016 PMID:21549461

Qu, X., Alvarez, P. J., & Li, Q. (2013). Applications of nanotechnology in water and wastewater treatment. *Water Research*, *47*(12), 3931–3946. doi:10.1016/j.watres.2012.09.058 PMID:23571110

Rafiq, Z., Nazir, R., Durr-e-Shahwar, Shah, M. R., & Ali, S. (2014). Utilization of magnesium and zinc oxide nano-adsorbents as potential materials for treatment of copper electroplating industry wastewater. *Journal of Environmental Chemical Engineering*, *2*(1), 642–651. doi:10.1016/j.jece.2013.11.004

Rai, M., & Duran, N. (2011). *Metal Nanoparticles in Microbiology*. Springer. doi:10.1007/978-3-642-18312-6

Rai, M., & Yadav, A. (2013). Plants as potential synthesiser of precious metal nanoparticles: Progress and prospects. *IET Nanobiotechnology / IET*, *7*(3), 117–124. doi:10.1049/iet-nbt.2012.0031 PMID:24028810

Raj, L. F. A., & e, J. (2015). Biosynthesis and Characterization of Zinc Oxide Nanoparticles Using Root Extract of Zingiber Officinale. *Oriental Journal of Chemistry*, *31*(1), 51–56. doi:10.13005/ojc/310105

Rajabi, H., Ghaemi, N., Madaeni, S. S., Daraei, P., Astinchap, B., Zinadini, S., & Razavizadeh, S. H. (2015). Nano-ZnO embedded mixed matrix polyethersulfone (PES) membrane: Influence of nanofiller shape on characterization and fouling resistance. *Applied Surface Science*, *349*, 66–77. doi:10.1016/j.apsusc.2015.04.214

Rajeswari, A., Amalraj, A., & Pius, A. (2015). Removal of phosphate using chitosan-polymer composites. *Journal of Environmental Chemical Engineering*, *3*(4), 2331–2341. doi:10.1016/j.jece.2015.08.022

Rajiv, P., Rajeshwari, S., & Venckatesh, R. (2013). Bio-fabrication of zinc oxide nanoparticles using leaf extract of Parthenium hysterophorus L. and its size-dependent antifungal activity against plant fungal pathogens. *Spectrochimica Acta. Part A: Molecular and Biomolecular Spectroscopy, 112*, 384–387. doi:10.1016/j.saa.2013.04.072 PMID:23686093

Ramanujam, K., & Sundrarajan, M. (2014). Antibacterial effects of biosynthesized MgO nanoparticles using ethanolic fruit extract of Emblica officinalis. *Journal of Photochemistry and Photobiology. B, Biology*, *141*, 296–300. doi:10.1016/j.jphotobiol.2014.09.011 PMID:25463681

Ramesh, M., Anbuvannan, M., & Viruthagiri, G. (2015). Green synthesis of ZnO nanoparticles using Solanum nigrum leaf extract and their antibacterial activity. *Spectrochim Acta A Mol Biomol Spectrosc, 136 Pt B*, 864-870. doi:10.1016/j.saa.2014.09.105

Ramesh, P., Rajendran, A., & Meenakshisundaram, M. (2014). Green Syntheis of Zinc Oxide Nanoparticles Using Flower Extract Cassia Auriculata. *Journal of Nanoscience and Nanotechnology*, *2*(1), 41–45.

Rao, K. G., Ashok, C., Rao, K. V., & Chakra, C. S., & A, A. (2015). Eco-Friendly Synthesis of MgO Nanoparticles from Orange Fruit Waste. *International Journal of Advanced Research in Physical Science*, *2*(3), 1–6.

Raut, S., Thorat, P. V., & Thakre, R. (2015). Green Synthesis of Zinc Oxide (ZnO) Nanoparticles Using Ocimum Tenuiflorum Leaves. *International Journal of Science and Research*, *4*(5), 2319–7064.

Rauwel, P., Küünal, S., Ferdov, S., & Rauwel, E. (2015). A Review on the Green Synthesis of Silver Nanoparticles and Their Morphologies Studied via TEM. *Advances in Materials Science and Engineering*, *2015*, 1–9. doi:10.1155/2015/682749

Rezaee, A., Godini, H., Dehestani, S., & Khavanin, A. (2008). Application of impregnated almond shell activated carbon by zinc and zinc sulfate for nitrate removal from water. *Iranian Journal of Environmental Health Sciences & Engineering*, *5*, 125–130.

Ridoutt, B. G., & Pfister, S. (2010). A revised approach to water footprinting to make transparent the impacts of consumption and production on global freshwater scarcity. *Global Environmental Change*, *20*(1), 113–120. doi:10.1016/j.gloenvcha.2009.08.003

Ridoutt, B. G., & Pfister, S. (2013). A new water footprint calculation method integrating consumptive and degradative water use into a single stand-alone weighted indicator. *The International Journal of Life Cycle Assessment*, *18*(1), 204–207. doi:10.100711367-012-0458-z

Robinson, B., Clothier, B., Bolan, N. S., Mahimairaja, S., Greven, M., Moni, C., . . . Milne, G. (2004). *Arsenic in the New Zealand environment*. Paper presented at the SuperSoil 2004: 3rd Australian New Zealand Soils Conference, Sydney, Australia.

Ruthven, D. M. (1984). *Principles of Adsorption and Adsorption Processes* (1st ed.). ohn Wiley and Sons, Inc.

Sahaa, R. K., Acharyaa, S., Shovon, S. S. H., & Royb, P. (2013). Medicinal activities of the leaves of Musa sapientum var. sylvesteris in vitro. *Asian Pacific Journal of Tropical Biomedicine*, *3*(6), 476–482. doi:10.1016/S2221-1691(13)60099-4 PMID:23730561

Sahiner, N., Demirci, S., Sahiner, M., Yilmaz, S., & Al-Lohedan, H. (2015). The use of superporous p(3-acrylamidopropyl)trimethyl ammonium chloride cryogels for removal of toxic arsenate anions. *Journal of Environmental Management*, *152*, 66–74. doi:10.1016/j.jenvman.2015.01.023 PMID:25617870

Salam, M. A., Fageeh, O., Al-Thabaiti, S. A., & Obaid, A. Y. (2015). Removal of nitrate ions from aqueous solution using zero-valent iron nanoparticles supported on high surface area nanographenes. *Journal of Molecular Liquids*, *212*, 708–715. doi:10.1016/j.molliq.2015.09.029

Sangeetha, G., Rajeshwari, S., & Venckatesh, R. (2011). Green synthesis of zinc oxide nanoparticles by aloe barbadensis miller leaf extract: Structure and optical properties. *Materials Research Bulletin*, *46*(12), 2560–2566. doi:10.1016/j.materresbull.2011.07.046

Sankara Rao, B., Sunandana, C. S., Srikanth, V. V. S. S., Rao, T. N., Jain, P. K., Moorthy, S. K., . . . Viswanathan, C. (2015). International Conference on Nano Science & Engineering Application (ICONSEA-2014) Centre For Nano Science and Technology Synthesis and Characterization of Mgo Nanoparticles by Neem Leaves through Green Method. *Materials Today: Proceedings, 2*(9), 4360-4368. doi:10.1016/j.matpr.2015.10.027

Sathishkumar, M., Sneha, K., Won, S. W., Cho, C. W., Kim, S., & Yun, Y. S. (2009). Cinnamon zeylanicum bark extract and powder mediated green synthesis of nano-crystalline silver particles and its bactericidal activity. *Colloids and Surfaces. B, Biointerfaces*, *73*(2), 332–338. doi:10.1016/j.colsurfb.2009.06.005 PMID:19576733

Sathyamoorthy, R., Mageshwari, K., Mali, S. S., Priyadharshini, S., & Patil, P. S. (2013). Effect of organic capping agent on the photocatalytic activity of MgO nanoflakes obtained by thermal decomposition route. *Ceramics International*, *39*(1), 323–330. doi:10.1016/j.ceramint.2012.06.028

Scanlon, B. R., Jolly, I., Sophocleous, M., & Zhang, L. (2007). Global impacts of conversions from natural to agricultural ecosystems on water resources: Quantity versus quality. *Water Resources Research, 43*(3). doi:10.1029/2006wr005486

Schaffer, B., Hohenester, U., Trügler, A., & Hofer, F. (2009). High-resolution surface plasmon imaging of gold nanoparticles by energy-filtered transmission electron microscopy. *Physical Review B: Condensed Matter and Materials Physics*, *79*(4), 041401. doi:10.1103/PhysRevB.79.041401

Seigler, D. S. (2003). Phytochemistry of Acacia—sensu lato. *Biochemical Systematics and Ecology*, *31*(8), 845–873. doi:10.1016/S0305-1978(03)00082-6

Sepeur, S. (2008). *Nanotechnology: technical basics and applications*. Hannover: Vincentz.

Sevcenco, A. M., Paravidino, M., Vrouwenvelder, J. S., Wolterbeek, H. T., van Loosdrecht, M. C., & Hagen, W. R. (2015). Phosphate and arsenate removal efficiency by thermostable ferritin enzyme from Pyrococcus furiosus using radioisotopes. *Water Research*, *76*, 181–186. doi:10.1016/j.watres.2015.02.054 PMID:25817554

Shah, M. A., & Qurashi, A. (2009). Novel surfactant-free synthesis of MgO nanoflakes. *Journal of Alloys and Compounds*, *482*(1-2), 548–551. doi:10.1016/j.jallcom.2009.04.129

Shah, R. K., Boruah, F., & Parween, N. (2015). Synthesis and Characterization of ZnO Nanoparticles using Leaf Extract of Camellia sinesis and Evaluation of their Antimicrobial Efficacy. *International Journal of Current Microbiology and Applied Sciences*, *4*(8), 444–450.

Shahverdi, A., Shakibaie, M., & Nazari, P. (2011). Basic and practical procedures for microbial synthesis of nanoparticles. In M. Rai & N. Duran (Eds.), *Metal nanoparticles in microbiology*. Berlin: Springer. doi:10.1007/978-3-642-18312-6_8

Shankar, S. S., Ahmad, A., & Sastry, M. (2003). Geranium Leaf Assisted Biosynthesis of Silver Nanoparticles. *Biotechnology Progress*, *19*(6), 1627–1631. doi:10.1021/bp034070w PMID:14656132

Shankar, S. S., Rai, A., Ahmad, A., & Sastry, M. (2004). Rapid synthesis of Au, Ag, and bimetallic Au core–Ag shell nanoparticles using Neem (Azadirachta indica) leaf broth. *Journal of Colloid and Interface Science*, *275*(2), 496–502. doi:10.1016/j.jcis.2004.03.003 PMID:15178278

Shankar, S. S., Rai, A., Ahmad, A., & Sastry, M. (2005). Controlling the Optical Properties of Lemongrass Extract Synthesized Gold Nanotriangles and Potential Application in Infrared-Absorbing Optical Coatings. *Chemistry of Materials*, *17*(3), 566–572. doi:10.1021/cm048292g

Sharma, K., Ko, E. Y., Assefa, A. D., Ha, S., Nile, S. H., Lee, E. T., & Park, S. W. (2015). Temperature-dependent studies on the total phenolics, flavonoids, antioxidant activities, and sugar content in six onion varieties. *Journal of Food and Drug Analysis*, *23*(2), 243–252. doi:10.1016/j.jfda.2014.10.005 PMID:28911379

Sharma, V. K., Yngard, R. A., & Lin, Y. (2009). Silver nanoparticles: Green synthesis and their antimicrobial activities. *Advances in Colloid and Interface Science*, *145*(1–2), 83–96. doi:10.1016/j.cis.2008.09.002 PMID:18945421

Sharma, Y. C., Srivastava, V., Singh, V. K., Kaul, S. N., & Weng, C. H. (2009). Nano-adsorbents for the removal of metallic pollutants from water and wastewater. *Environmental Technology*, *30*(6), 583–609. doi:10.1080/09593330902838080 PMID:19603705

Shumilla, J. A., Broderick, R. J., Wang, Y., & Barchowsky, A. (1999). Chromium(VI) Inhibits the Transcriptional Activity of Nuclear Factor-B by Decreasing the Interaction of p65 with cAMP-responsive Element-binding Protein-binding Protein. *The Journal of Biological Chemistry*, *274*(51), 36207–36212. doi:10.1074/jbc.274.51.36207 PMID:10593907

Si, J., & Yang, H. (2011). Preparation and characterization of bio-compatible Fe3O4@Polydopamine spheres with core/shell nanostructure. *Materials Chemistry and Physics*, *128*(3), 519–524. doi:10.1016/j.matchemphys.2011.03.039

Si, S., & Mandal, T. K. (2007). Tryptophan-based peptides to synthesize gold and silver nanoparticles: A mechanistic and kinetic study. *Chemistry*, *13*(11), 3160–3168. doi:10.1002/chem.200601492 PMID:17245786

Siboni, M. S., Samadi, M. T., Yang, J. K., & Lee, S. M. (2011). Photocatalytic reduction of Cr(VI) and Ni(II) in aqueous solution by synthesized nanoparticle ZnO under ultraviolet light irradiation: A kinetic study. *Environmental Technology*, *32*(13-14), 1573–1579. doi:10.1080/09593330.2010.543933 PMID:22329148

Singh, A. K., Talat, M., Singh, D. P., & Srivastava, O. N. (2010). Biosynthesis of gold and silver nanoparticles by natural precursor clove and their functionalization with amine group. *Journal of Nanoparticle Research*, *12*(5), 1667–1675. doi:10.100711051-009-9835-3

Singh, N., Singh, S. P., Gupta, V., Yadav, H. K., Ahuja, T., Tripathy, S. S., & Rashmi. (2013). A process for the selective removal of arsenic from contaminated water using acetate functionalized zinc oxide nanomaterials. *Environmental Progress & Sustainable Energy*, *32*(4), 1023–1029. doi:10.1002/ep.11698

Steinfeld, H., Gerber, P., Wassenaar, T., Castel, V., Rosales, M., & Haan, C. d. (2006). *Livestock's Long Shadow environmental issues and options*. Retrieved from Room, Italy: Stott, K. J., & Gourley, C. J. P. (2016). Intensification, nitrogen use and recovery in grazing-based dairy systems. *Agricultural Systems*, *144*, 101–112. doi:10.1016/j.agsy.2016.01.003

Strasser, P., Koh, S., Anniyev, T., Greeley, J., More, K., Yu, C., . . . Nilsson, A. (2010). Lattice-strain control of the activity in dealloyed core–shell fuel cell catalysts. *Nat Chem*, *2*(6), 454-460. Retrieved from http://www.nature.com/nchem/journal/v2/n6/suppinfo/nchem.623_S1.html

Su, Y., Cui, H., Li, Q., Gao, S., & Shang, J. K. (2013). Strong adsorption of phosphate by amorphous zirconium oxide nanoparticles. *Water Research*, *47*(14), 5018–5026. doi:10.1016/j.watres.2013.05.044 PMID:23850213

Su, Y., Yang, W., Sun, W., Li, Q., & Shang, J. K. (2015). Synthesis of mesoporous cerium–zirconium binary oxide nanoadsorbents by a solvothermal process and their effective adsorption of phosphate from water. *Chemical Engineering Journal*, *268*, 270–279. doi:10.1016/j.cej.2015.01.070

Sullivan, C., Tyrer, M., Cheeseman, C. R., & Graham, N. J. D. (2010). Disposal of water treatment wastes containing arsenic — A review. *The Science of the Total Environment*, *408*(8), 1770–1778. doi:10.1016/j.scitotenv.2010.01.010 PMID:20153878

Sun, S., Murray, C. B., Weller, D., Folks, L., & Moser, A. (2000). Monodisperse FePt Nanoparticles and Ferromagnetic FePt Nanocrystal Superlattices. *Science*, *287*(5460), 1989–1992. doi:10.1126cience.287.5460.1989 PMID:10720318

Suresh, D., Nethravathi, P. C., Udayabhanu, Rajanaika, H., Nagabhushana, H., & Sharma, S. C. (2015). Green synthesis of multifunctional zinc oxide (ZnO) nanoparticles using Cassia fistula plant extract and their photodegradative, antioxidant and antibacterial activities. *Materials Science in Semiconductor Processing*, *31*, 446–454. doi:10.1016/j.mssp.2014.12.023

Sureshkumar, M., Lee, P.-N., & Lee, C.-K. (2011). Stepwise assembly of multimetallic nanoparticles via self-polymerized polydopamine. *Journal of Materials Chemistry*, *21*(33), 12316. doi:10.1039/c1jm11914a

Tan, Y. N., Lee, J. Y., & Wang, D. I. C. (2010). Uncovering the Design Rules for Peptide Synthesis of Metal Nanoparticles. *Journal of the American Chemical Society*, *132*(16), 5677–5686. doi:10.1021/ja907454f PMID:20355728

Thayumanavan, B., & Sadasivam, S. (1984). Physicochemical basis for the preferential uses of certain rice varieties. *Qual Plant Plant Foods Hum Nutr*, *34*(4), 253–259. doi:10.1007/BF01126554

Tingey-Holyoak, J. L. (2014). Water sharing risk in agriculture: Perceptions of farm dam management accountability in Australia. *Agricultural Water Management*, *145*, 123–133. doi:10.1016/j.agwat.2014.02.011

Tiwari, D. K., Behari, J., & Sen, P. (2008). Application of Nanoparticles in Waste Water Treatment. *World Applied Sciences Journal*, *3*(3), 417–433.

Tran, H. V., Tran, L. D., & Nguyen, T. N. (2010). Preparation of chitosan/magnetite composite beads and their application for removal of Pb(II) and Ni(II) from aqueous solution. *Materials Science and Engineering C*, *30*(2), 304–310. doi:10.1016/j.msec.2009.11.008

Ünal, M. Ü. (2007). Properties of polyphenol oxidase from Anamur banana (Musa cavendishii). *Food Chemistry*, *100*(3), 909–913. doi:10.1016/j.foodchem.2005.10.048

Ungureanu, G., Santos, S., Boaventura, R., & Botelho, C. (2015). Arsenic and antimony in water and wastewater: Overview of removal techniques with special reference to latest advances in adsorption. *Journal of Environmental Management*, *151*, 326–342. doi:10.1016/j.jenvman.2014.12.051 PMID:25585146

United Nations Development Programme. (2006). *Coping with water scarcity; Challenge of the twenty-first century* [Press release]. UNDP.

Vajnhandl, S., & Valh, J. V. (2014). The status of water reuse in European textile sector. *Journal of Environmental Management*, *141*, 29–35. doi:10.1016/j.jenvman.2014.03.014 PMID:24768832

Vanathi, P., Rajiv, P., Narendhran, S., Rajeshwari, S., Rahman, P. K. S. M., & Venckatesh, R. (2014). Biosynthesis and characterization of phyto mediated zinc oxide nanoparticles: A green chemistry approach. *Materials Letters*, *134*, 13–15. doi:10.1016/j.matlet.2014.07.029

Wang, C., Luo, H., Zhang, Z., Wu, Y., Zhang, J., & Chen, S. (2014). Removal of As(III) and As(V) from aqueous solutions using nanoscale zero valent iron-reduced graphite oxide modified composites. *Journal of Hazardous Materials*, *268*, 124–131. doi:10.1016/j.jhazmat.2014.01.009 PMID:24480523

Wang, P., Menzies, N. W., Lombi, E., McKenna, B. A., Johannessen, B., Glover, C. J., ... Kopittke, P. M. (2013). Fate of ZnO nanoparticles in soils and cowpea (Vigna unguiculata). *Environmental Science & Technology*, *47*(23), 13822–13830. doi:10.1021/es403466p PMID:24195448

Wang, S., Gao, B., Li, Y., Creamer, A. E., & He, F. (2016). Adsorptive removal of arsenate from aqueous solutions by biochar supported zero-valent iron nanocomposite: Batch and continuous flow tests. *Journal of Hazardous Materials*. doi:10.1016/j.jhazmat.2016.01.052 PMID:26852252

Wang, T., Jin, X., Chen, Z., Megharaj, M., & Naidu, R. (2014). Simultaneous removal of Pb(II) and Cr(III) by magnetite nanoparticles using various synthesis conditions. *Journal of Industrial and Engineering Chemistry*, *20*(5), 3543–3549. doi:10.1016/j.jiec.2013.12.047

Wang, T., Lin, J., Chen, Z., Megharaj, M., & Naidu, R. (2014). Green synthesized iron nanoparticles by green tea and eucalyptus leaves extracts used for removal of nitrate in aqueous solution. *Journal of Cleaner Production*, *83*, 413–419. doi:10.1016/j.jclepro.2014.07.006

Wang, W., Qiao, X., Chen, J., Tan, F., & Li, H. (2009). Influence of titanium doping on the structure and morphology of MgO prepared by coprecipitation method. *Materials Characterization*, *60*(8), 858–862. doi:10.1016/j.matchar.2009.02.002

Wheeler, S. A., Zuo, A., & Loch, A. (2015). Watering the farm: Comparing organic and conventional irrigation water use in the Murray–Darling Basin, Australia. *Ecological Economics*, *112*, 78–85. doi:10.1016/j.ecolecon.2015.02.019

Worch, E. (2012). *Adsorption Technology in Water Treatment, Fundamentals, Processes, and Modeling*. Gottingen: GmbH & Co. KG. doi:10.1515/9783110240238

World Economic Forum. (2015). *Global Risks. 10*. Retrieved from http://www3.weforum.org/docs/WEF_Global_Risks_2015_Report15.pdf

Wuyts, N., De Waele, D., & Swennen, R. (2006). Extraction and partial characterization of polyphenol oxidase from banana (Musa acuminata Grande naine) roots. *Plant Physiology and Biochemistry*, *44*(5-6), 308–314. doi:10.1016/j.plaphy.2006.06.005 PMID:16814556

Xiao, Q., Sun, Y., Zhang, J., & Li, Q. (2015). Size-dependent of chromium (VI) adsorption on nano α-Fe2O3 surface. *Applied Surface Science*, *356*, 18–23. doi:10.1016/j.apsusc.2015.08.005

Xie, J., Lin, Y., Li, C., Wu, D., & Kong, H. (2015). Removal and recovery of phosphate from water by activated aluminum oxide and lanthanum oxide. *Powder Technology*, *269*, 351–357. doi:10.1016/j.powtec.2014.09.024

Xie, J., Wang, Z., Fang, D., Li, C., & Wu, D. (2014). Green synthesis of a novel hybrid sorbent of zeolite/lanthanum hydroxide and its application in the removal and recovery of phosphate from water. *Journal of Colloid and Interface Science*, *423*, 13–19. doi:10.1016/j.jcis.2014.02.020 PMID:24703662

Xie, J., Wang, Z., Wu, D., & Kong, H. (2014). Synthesis and properties of zeolite/hydrated iron oxide composite from coal fly ash as efficient adsorbent to simultaneously retain cationic and anionic pollutants from water. *Fuel*, *116*, 71–76. doi:10.1016/j.fuel.2013.07.126

Xie, Y., Yan, B., Xu, H., Chen, J., Liu, Q., Deng, Y., & Zeng, H. (2014). Highly regenerable mussel-inspired Fe(3)O(4)@polydopamine-Ag core-shell microspheres as catalyst and adsorbent for methylene blue removal. *ACS Applied Materials & Interfaces*, *6*(11), 8845–8852. doi:10.1021/am501632f PMID:24787615

Xiong, J., He, Z., Mahmood, Q., Liu, D., Yang, X., & Islam, E. (2008). Phosphate removal from solution using steel slag through magnetic separation. *Journal of Hazardous Materials*, *152*(1), 211–215. doi:10.1016/j.jhazmat.2007.06.103 PMID:17703877

Xu, K., Deng, T., Liu, J., & Peng, W. (2010). Study on the phosphate removal from aqueous solution using modified fly ash. *Fuel*, *89*(12), 3668–3674. doi:10.1016/j.fuel.2010.07.034

Xu, K., Deng, T., Liu, J., & Peng, W. (2012). Phosphate Removal from Digested Sludge Supernatant Using Modified Fly Ash. *Water Environment Research*, *84*(5), 411–416. doi:10.2175/106143012X13347678384323 PMID:22852426

Xu, X., Gao, B., Jin, B., & Yue, Q. (2016). Removal of anionic pollutants from liquids by biomass materials: A review. *Journal of Molecular Liquids*, *215*, 565–595. doi:10.1016/j.molliq.2015.12.101

Xu, X., Lin, L., Papelis, C., Myint, M., Cath, T. Y., & Xu, P. (2015). Use of drinking water treatment solids for arsenate removal from desalination concentrate. *Journal of Colloid and Interface Science*, *445*, 252–261. doi:10.1016/j.jcis.2014.12.090 PMID:25622050

Yang, Q., Sha, J., Wang, L., Wang, J., & Yang, D. (2006). MgO nanostructures synthesized by thermal evaporation. *Materials Science and Engineering C*, *26*(5-7), 1097–1101. doi:10.1016/j.msec.2005.09.082

Ye, Q., Zhou, F., & Liu, W. (2011). Bioinspired catecholic chemistry for surface modification. *Chemical Society Reviews*, *40*(7), 4244–4258. doi:10.1039/c1cs15026j PMID:21603689

Yi, X., Wenzhong, W., Yitai, Q., Li, Y., & Zhiwen, C. (1996). Deposition and microstructural characterization of MgO thin films by a spray pyrolysis method. *Surface and Coatings Technology*, *82*(3), 291–293. doi:10.1016/0257-8972(95)02762-9

Yin, J., & Deng, B. (2015). Polymer-matrix nanocomposite membranes for water treatment. *Journal of Membrane Science*, *479*, 256–275. doi:10.1016/j.memsci.2014.11.019

You, X., Farran, A., Guaya, D., Valderrama, C., Soldatov, V., & Cortina, J. L. (2016). Phosphate removal from aqueous solutions using a hybrid fibrous exchanger containing hydrated ferric oxide nanoparticles. *Journal of Environmental Chemical Engineering*, *4*(1), 388–397. doi:10.1016/j.jece.2015.11.032

Yuan, P., Liu, D., Fan, M., Yang, D., Zhu, R., Ge, F., ... He, H. (2010). Removal of hexavalent chromium [Cr(VI)] from aqueous solutions by the diatomite-supported/unsupported magnetite nanoparticles. *Journal of Hazardous Materials*, *173*(1–3), 614–621. doi:10.1016/j.jhazmat.2009.08.129 PMID:19748178

Yuvakkumar, R., Suresh, J., Nathanael, A. J., Sundrarajan, M., & Hong, S. I. (2014). Novel green synthetic strategy to prepare ZnO nanocrystals using rambutan (Nephelium lappaceum L.) peel extract and its antibacterial applications. *Materials Science and Engineering C*, *41*, 17–27. doi:10.1016/j.msec.2014.04.025 PMID:24907732

Zayed, M. F., Eisa, W. H., & Shabaka, A. A. (2012). Malva parviflora extract assisted green synthesis of silver nanoparticles. *Spectrochimica Acta. Part A: Molecular and Biomolecular Spectroscopy*, *98*, 423–428. doi:10.1016/j.saa.2012.08.072 PMID:23010627

Zelmanov, G., & Semiat, R. (2015). The influence of competitive inorganic ions on phosphate removal from water by adsorption on iron (Fe+3) oxide/hydroxide nanoparticles-based agglomerates. *Journal of Water Process Engineering*, *5*, 143–152. doi:10.1016/j.jwpe.2014.06.008

Zeng, H., Hwang, D. S., Israelachvili, J. N., & Waite, J. H. (2010). Strong reversible Fe3+-mediated bridging between dopa-containing protein films in water. *Proceedings of the National Academy of Sciences of the United States of America*, *107*(29), 12850–12853. doi:10.1073/pnas.1007416107 PMID:20615994

Zhang, M., He, X., Chen, L., & Zhang, Y. (2010). Preparation of IDA-Cu functionalized core–satellite Fe3O4/polydopamine/Au magnetic nanocomposites and their application for depletion of abundant protein in bovine blood. *Journal of Materials Chemistry*, *20*(47), 10696. doi:10.1039/c0jm01336f

Zheng, Y., & Wang, A. (2012). Ag nanoparticle-entrapped hydrogel as promising material for catalytic reduction of organic dyes. *Journal of Materials Chemistry*, *22*(32), 16552. doi:10.1039/c2jm32774k

Zhu, L.-P., Yu, J.-Z., Xu, Y.-Y., Xi, Z.-Y., & Zhu, B.-K. (2009). Surface modification of PVDF porous membranes via poly(DOPA) coating and heparin immobilization. *Colloids and Surfaces. B, Biointerfaces*, *69*(1), 152–155. doi:10.1016/j.colsurfb.2008.11.011 PMID:19117736

KEY TERMS AND DEFINITIONS

%: Percentage.
AFM: Atomic force microscope.
As: Arsenic.
COD: Chemical oxygen demand.
Cr: Chromium.
DLS: Dynamic light scattering.
DMAC: N,N-dimethylacetamide.
EDS: Energy dispersive spectroscopy.
EDX: Electron dispersive x-ray.
EG: Ethylene glycol.
FT-IR: Fourier transform infrared spectroscopy.
GA: Gum acacia.
HMO: Hydrous manganese oxide.
Mg: Magnesium.
MMMs: Mix matrix membranes.
MNP: Metallic nanoparticle.
MONPs: Metal oxide nanoparticles.
NPs: Nanoparticles.
NZVI: Nanoscale zerovalent iron particles.
PDA: Polydopamine.
PES: Polyethersulfone.
PVP: Polyvinylpyrrolidone.
SEM: Scanning electron microscopy.
TEM: Transmission electron microscopy.
XRD: X-ray diffraction.
Zn: Zinc.

Chapter 13
Recent Progresses in Membranes for Proton Exchange Membrane Fuel Cell (PEMFC) for Clean and Environmentally Friendly Applications

Jay Pandey
Indian Institute of Technology Delhi, India

ABSTRACT

Fuel cell has become an emerging renewable energy device with potential to meet energy demand by portable and transport applications with zero-emission, easy operation, and compact design. The chapter provides an insight into design and development of membranes for PEMFCs and recent progresses made in membranes so far. Although majority of research has focused on fluorinated and non-fluorinated membranes, these polymeric membranes have showed deteriorated properties at elevated temperature (>80oC) and lower relative humidity (30%). Considering the major issues with polymeric membranes, the authors have reviewed inorganic-organic nanocomposite membranes showing improved physical and electrochemical properties at elevated temperature and lower relative humidity. Recently, metal-organic framework (MOF), a novel and unique material, has attracted tremendous attention due to their enhanced proton conductivity, easy functionality, and stability. MOFs have also exhibited excellent compatibility with different polymeric materials that are also discussed in this chapter.

DOI: 10.4018/978-1-5225-5745-6.ch013

INTRODUCTION

Energy is among one of the necessities of human being. Heavy energy consumption worldwide, rapid depletion of available fossil fuels and their high polluting nature led to extensive research in the area of alternative sources of energy for achieving a clean and sustainable environment. At present scenario, it is hard to meet the total world energy demand only by one of the environmentally clean energy generation technologies. However, it could be possible to meet certain fraction of energy demand through rapidly developing energy conversion/storage technologies. One such class is the energy consumption by stationary, mobile and portable electronic devices, whose typical power requirement range is 0.5 W to 1000 kW, which is 10-20% of total world energy consumption (Carter et al., 2013). Addressing their power demand through new and renewable sources can reduce demand on the large stationary power generation systems. In last few decades, fuel cell, which utilizes large range of hydrocarbon fuels, has emerged as a potential candidate to fulfill the energy demand of portable applications largely. Performance of fuel cells depends on the judicious design and selection of effective and robust materials like solid proton exchange membranes (PEMs), electro-catalysts and bipolar plates.

Ion-exchange membrane (IEM), separating the electrodes, is the one of the major components of a fuel cell device. IEM plays a crucial role in enhancing the efficiency of the device based on its ion-transport properties. Typical IEM comprises of organic polymers matrix with acid functionalities such as carboxylic, sulfonic and phosphonic groups that dissociate when solvated with water which leads to transport of protons via vehicular (molecular diffusion) or Grotthuss mechanism (structure diffusion) (Strathmann et al., 2004). An ideal IEM should exhibit high proton conductivity, low electronic conductivity, less fuel crossover, cost-effective and high stability.

In past few decades, role of membrane in low temperature fuel cell such as polymer electrolyte membrane fuel cell (PEMFC) is a matter of debate among scientists due to deteriorating performance of device in terms of lower power density and its limited range of operation. To date, Nafion, developed by DuPont, USA in 1960, globally used as a PEM due to its homogeneous structure, high proton conductivity and remarkable stability. Unfortunately, Nafion is highly expensive ($ 200/m^2) and exhibits poor performance at lower humidity (30%) and high temperature (>80°C) (Robert et al., 2011). Moreover, Nafion is very prone to fuel crossover (10^{-7} cm^2/s) and demonstrated dimensional instability. To address these issues, lost of efforts are being made to design and develop several polymeric membranes to overcome the complexities associated with Nafion.

In this direction, much emphasis has been also made to focus on inorganic-organic composite ion-exchange membranes synthesized by blending inorganic ion-exchangers like heteropoly acids (HPAs), zirconium phosphates (ZrP), inorganic metal oxides, carbon nanotubes (CNTs), graphene oxides (GO) into polymer matrix (PVA, PVDF, PEEK, PS) to enhance the properties of composite membrane (Nagarale et al., 2010). Recently, metal-organic frameworks (MOFs), a unique class of inorganic-organic co-ordination polymer, have also attracted tremendous attention due to their enhanced proton conductivity, easy functionality, and excellent structural stability (Yeo et al., 2014). These MOFs have also exhibited excellent compatibility with different polymers that make them a potential candidate as fuel cell membrane material. Herein, our aim is to brief recent progresses made in room temperature fuel cell membranes suitable for portable applications and address the current challenges.

FUEL CELL: GENERAL BACKGROUND

Fuel cell, an electrochemical energy conversion device, converting the chemical energy of molecules directly into electricity, is one of the most demanding and attractive alternate sources of renewable energy due to its modular features, compactness, high efficiency (60%), easy operation and eco-friendliness (Energy et al., 2000). Figure 1 schematically shows a classification of fuel cells based on operating temperature, charged carrier, type of fuels and oxidants. Details of fuel cells and their role are in Table 1.

PEMFC, also called proton exchange membrane fuel cells, delivers high power density with advantages of light weight and volume. PEMFC, operating at 50-80°C, mainly comprises of a solid electrolyte and electro-catalysts for anodic oxidation and cathodic reduction (Kraytsberg et al., 2014). Alkaline fuel cells (AFCs) use a solution of potassium hydroxide in water as the electrolyte and non-precious metals as a catalyst for anode and cathode. AFCs are closely related to conventional PEM fuel cells, which operate in the temperature range of 80-120°C except that they use an alkaline membrane instead of a polymeric

Figure 1. Broad classification of fuel cell on the basis of their operating temperature and charge carrier

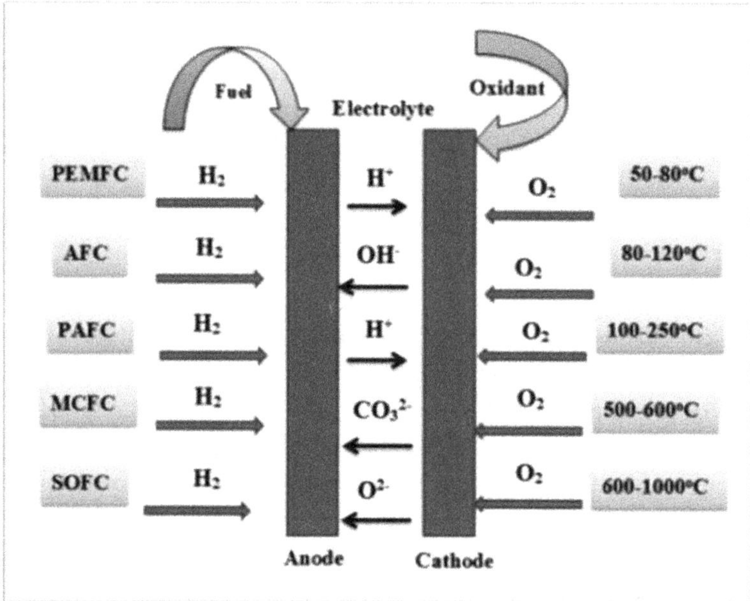

Table 1. Types of fuel cells based on their power output and electrolyte

Fuel Cell	Temperature (°C)	Power Output	Electrolyte
PEMFC	50-80	1 W- 500 kW	Polymer
AFC	80-120	10-200 KW	Aqueous alkaline solution
PAFC	100-250	<10 MW	Molten phosphoric acid
MCFC	500-650	100 MW	Molten alkaline carbonate
SOFC	800-1000	250 W-100 MW	Ceramics

(Hayre et al., 2016)

membrane (Pan et al., 2017). Phosphoric acid fuel cell (PAFC) is a type of fuel cell that uses liquid phosphoric acid as an electrolytic medium and porous carbon electrodes imbedded with nanoparticles of platinum catalyst in temperature range of 100-250°C (Eapen et al., 2016). Molten carbonate fuel cells (MCFCs) are high-temperature (500-650°C) fuel cells that use molten carbonate salt mixture, suspended in chemically inert ceramic lithium aluminum oxide matrix, as an electrolyte and non-precious metals catalysts at the anode and cathode (Cassir et al., 2016). Solid oxide fuel cell (SOFC) operates at a very high temperature in the range of 800-1000°C and utilizes ceramic materials as solid electrolyte and non-precious metal catalysts as electro-catalyst for fuel oxidation and oxidant reduction (Zhu et al., 2003).

Amongst several types of fuel cells as discussed above, PEMFC is the most researched and investigated type of fuel cell. However, PEMFC has problems associated with storage and handling of H_2 fuel which has very low energy density (4.8 Whcm^{-3}) compared to other fuels (see Table 2) (Zelenay et al., 2004). Moreover, cost and availability of the hydrogen is a major issue for the technologists. Contrary to this, direct methanol fuel cell (DMFC), a subcategory of PEMFC, uses aqueous methanol directly as fuel, has the potential to fulfill energy demand by portable applications. However, DMFC is envisioned not to be apt for powering large vehicles and for stationary applications. DMFC is clean, environmentally friendly and potentially economical energy resource with very high gravimetric energy density (1600 Whkg^{-1}) compared to conventional lithium ion batteries (200 Whkg^{-1}) and zero refueling time (Wee et al., 2007). DMFC has high power density as well as better fuel cell efficiency (45–60%) compared to IC engines and lower emission compared to PEMFCs (Kamarudin et al., 2009). A PEMFC/DMFC single cell is a two-compartment electrochemical cell (Figure 2) where the fuel hydrogen/methanol is fed to the anode side and the oxygen, as oxidant, is supplied at the cathode side of DMFC (Kraytsberg et al., 2014).

Considering the major issue related to room temperature fuel cells, it can be observed that the electro-catalyst and proton conducting membrane itself contributes about 40% to the total fuel cell cost. In addition, the PEMFC technologies also face many technological challenges under real conditions and need significant efforts in converting fuel cell technology a commercially viable technology worldwide. In the

Figure 2. Schematic of single cell DMFC

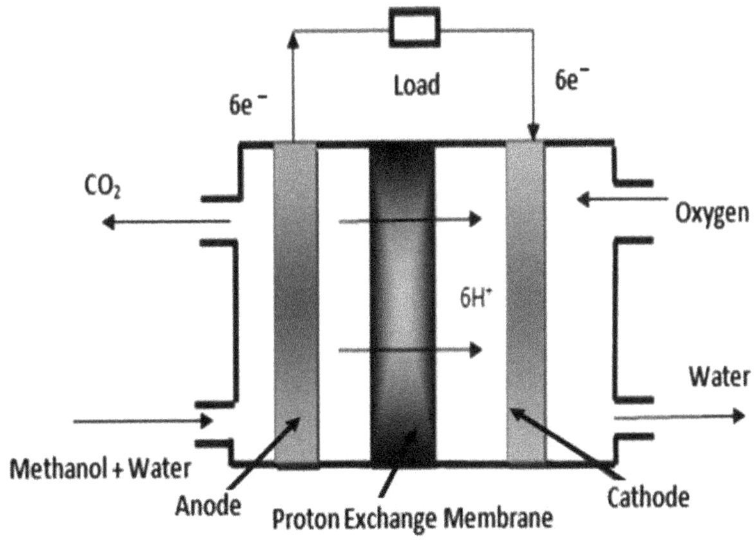

Table 2. Different types of fuels and their energy density

Fuel	Energy density ($Whcm^{-3}$)
Hydrogen, liquid	2.7
Hydrogen, gas	0.5
Methanol	4.8
Ethanol	6.3
Ethylene	5.9

(Zelenay et al., 2004)

book chapter, we have highlighted the recent advances made in membranes suitable for room temperature fuel cell for portable applications. The most advantageous aspect a fuel cell device is its compactness that can be easily integrated in various power devices for portable, stationary and transportation applications (Steel et al., 2001). In order to increase the power output the single cell device can be stacked by connecting them in series to achieve a power range from mW to MW as per requirements. Figure 3 depicts the multiple fuel cell applications for portable to stationary devices of varied power range.

FUEL CELL MEMBRANE

Proton exchange membrane (PEM), a core component of room temperature fuel cell (PEMFC/DMFC), is usually a solid electrolyte that allows selective transport of protons from anode to cathode. A broad classification of conventional PEMs based on material types is shown in Figure 4.

PEM mainly comprises of a polymeric matrix and fixed ionic functional groups. The polymeric matrix provides mechanical, chemical and thermal strength to the membrane and fixed functional groups impart the ion exchange ability to the membrane. Nafion series (Nafion-117, 112, and 115), sulfonated poly ether ether ketone (SPEEK), sulfonated poly(benzimidazole) (s-PBI), sulfonated poly(ether sul-

Figure 3. Applications of fuel cell devices

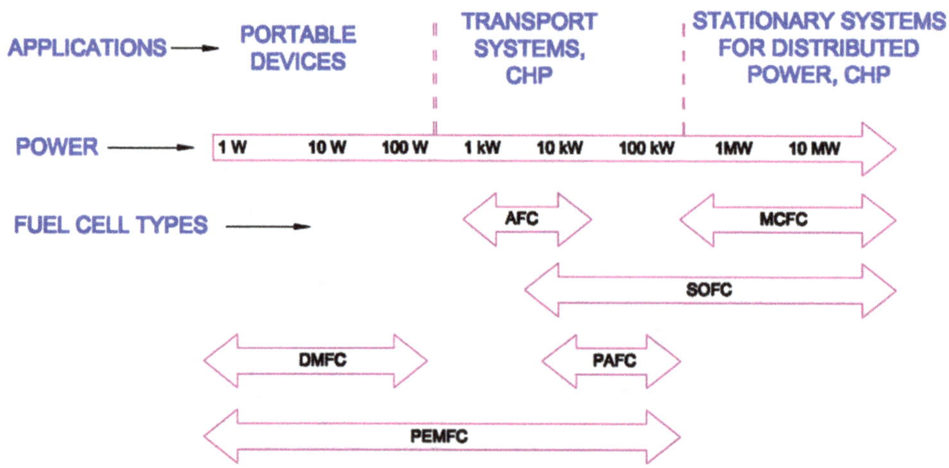

Figure 4. Classification of proton exchange membrane (PEM) for fuel cell

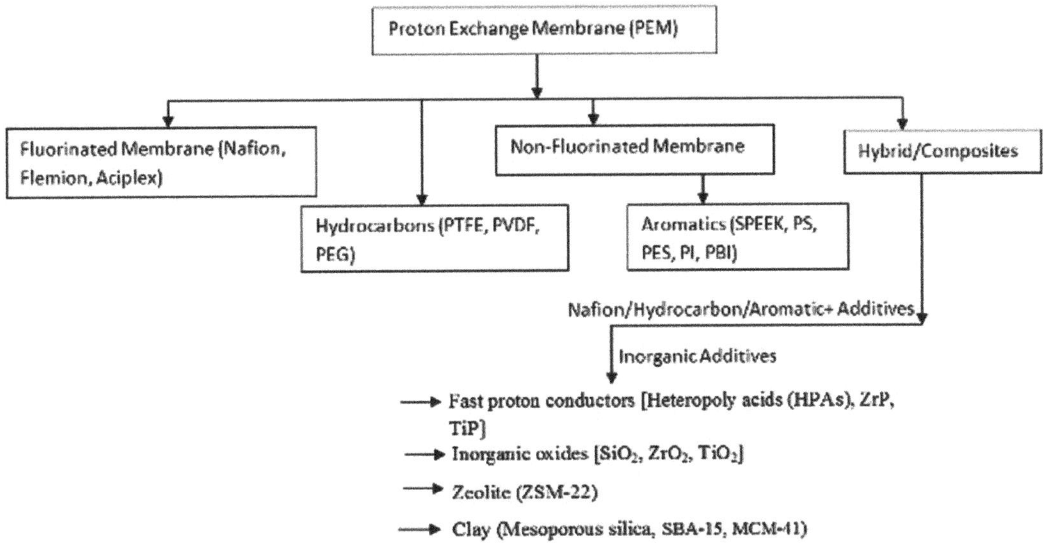

fone) (s-PES) membranes are most commonly used polymeric membrane (Lee et al., 2009). DuPont's Nafion is the most commonly used membrane for DMFC and has excellent properties, but suffers a few disadvantages such as high cost, \$200 m^{-2}, high methanol crossover (> 10^{-6} $cm^{-2}s^{-1}$) and poor proton conductivity at elevated temperatures (above 80°C) (Dimitrova et al., 2002). These disadvantages are also common problems with other widely used polymeric membranes that led to developments of composite membranes for DMFC. Therefore, synthesis of inorganic-organic composite membranes have gained major attention because it possess excellent properties of inorganic ion exchanger (high surface area, porosity, excellent thermal stability, IEC, high water holding capacity) and organic part (chemically stable, high mechanical strength, easy functionalization (Ruffmann et al., 2003).

Polymeric Membrane

Typical polymeric membrane comprises of polymers with acid functionalities such as carboxylic, sulfonic and phosphonic groups which dissociate when solvated with water which leads to transport of protons via vehicular (molecular diffusion) or Grotthuss mechanism (structure diffusion) (Kim et al., 2015). The commercially available membranes for room temperature fuel cells are Nafion, Dow, Flemion, Aciplex and Hyflon. All these membranes are having fixed sulfonic groups with varied degree of polymerization as indicated in the membrane's structure in Figure 5. Among all these membranes, Nafion is best suited for room temperature fuel cells due to its best performance at optimum fuel cell conditions.

State of the Art: Nafion

To date Nafion, developed by DuPont in 1960, being globally used in fuel cell. Nafion is a chemically stabilized copolymer of perfluorosulfonic acid (PFSA) and poly (tetrafluroethylene) (PTFE) in acidic

Figure 5. Chemical structure of perflourinated polymer electrolyte membranes for fuel cell (Rikukawa, 2000)

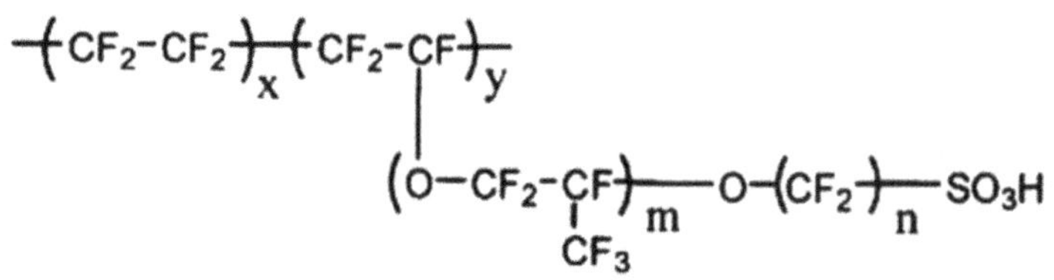

Nafion®117	m≧1, n=2, x=5-13.5, y=1000
Flemion®	m=0, 1; n=1-5
Aciplex®	m=0, 3; n=2-5, x=1.5-14
Dow membrane	m=0, n=2, x=3.6-10

(H^+) form [4]. The hydrophobic PTFE support provides mechanical strength to the membrane having fixed negatively charged sulfonic groups ($-SO_3^-H^+$), which are highly hydrophilic in nature, imparts ion-selective moieties to Nafion membrane as shown in Figure 6. The key characteristic of Nafion membrane is its homogeneous structure resulting into excellent ion-transport properties. So far, Nafion has shown the most promising performance over room temperature fuel cell conditions and it is the most studied membrane due to its unique structure, shape and morphology as shown in Figure 7 and Figure 8. Slade et al., 2002 have reported the details of many Nafion series membranes (Nafion-117, 115, 112). This series of membranes shows very high proton conductivity in the range of 0.08–0.1 Scm^{-1}. However, high methanol crossover (10^{-7} cm^2s^{-1}) and cost are main drawbacks of the Nafion series membranes for DMFC application. The comparative details of Nafion series membranes are provided in Table 3.

Similar PFSA polymers are Flemion (Asahi Glass), Aciplex (Asahi Chemicals), Dow and Solvay membranes have same structure like Nafion and their physical and electrochemical properties are listed in Table 4.

Figure 6. Struture of the state-of-the-art Nafion

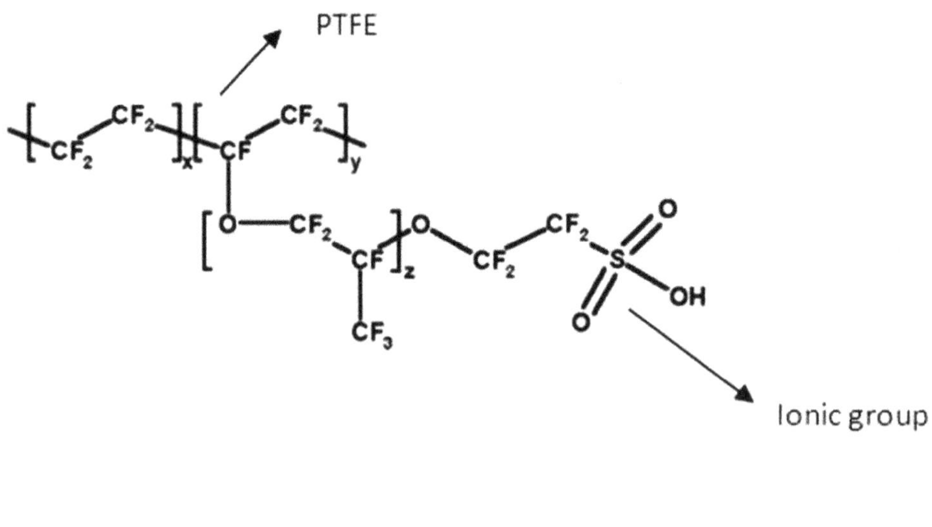

Figure 7. Microstructure and morphology of the state-of-the-art Nafion (a) close to surface and (b) in the middle (Chen et al., 2006)

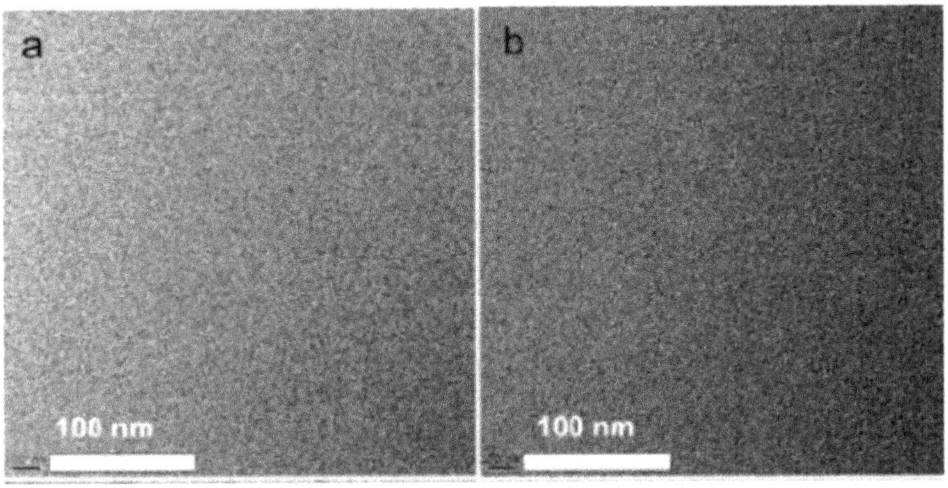

Figure 8. (a) Illustration of porous framework in Nafion and (b) view of cluster network model (Smitha et al., 2005)

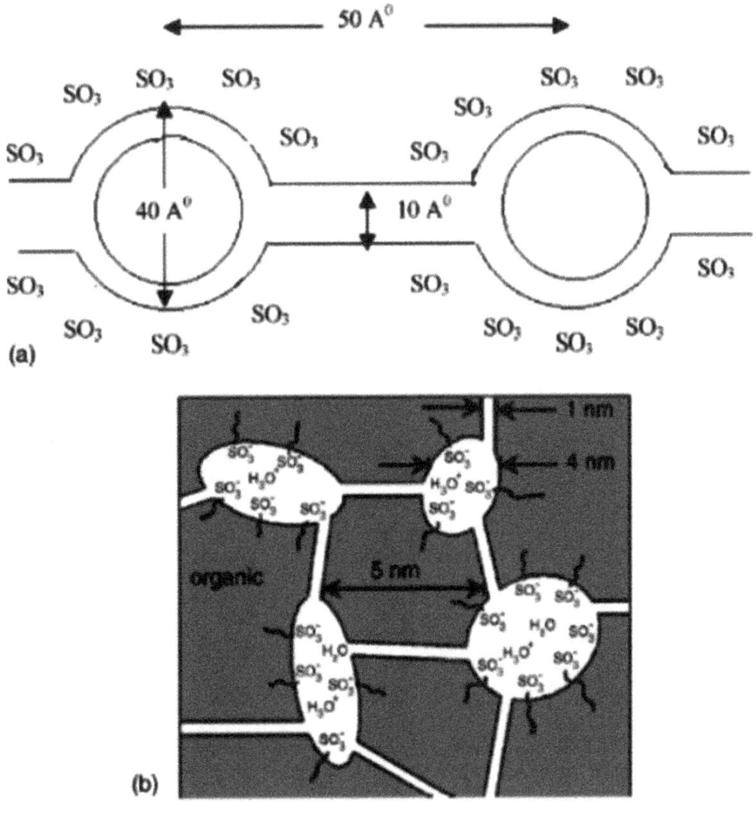

Table 3. Comparison of membrane properties with Nafion

Property	Nafion-112	Nafion-115	Nagion-117
EW (equivalent weight)	1100	1100	1100
Ion-exchange capacity (IEC), meq/g	0.91	0.91	0.91
Thickness (μm)	50	125	175
Proton conductivity (S/cm)	0.06-0.07	0.08-0.09	0.1
Diffusion coefficient (cm²/s)	7.8 ×10^{-7}	11.8×10^{-7}	12.8×10^{-7}

(**Coutanceau et al., 2006**).

Table 4. Comparison of membrane properties with commercially available perfluorosulfonic acid (PFSA) membranes

Membrane	Equivalent Weight)	Ion-Exchange Capacity (IEC), meq/g	Thickness (μm)	Proton Conductivity (S/cm)
Aciplex	1000-1200	0.83-1.00	25-100	0.1
Flemion	1000	1	50-150	0.09-0.1
DoW	800	1.25	125	0.1-0.2

(Kim et al., 2015).

Development of Alternative Polymeric Membrane

Keeping in view the problem associated with the Nafion, many efforts are being made to overcome the complexities with Nafion. So far various types of membranes are used in PEMFC but none of them has demonstrated promising performance than Nafion. Some of the well-studied polymeric membranes are SPEEK, s-PBI, s-PES, s-PSU, PEG-sPPSU etc.

Bauer et al., (2000) reported SPEEK membrane for DMFC synthesized by sulfonation of PEEK polymer. SPEEK, structure shown in Figure 9, has very high IEC (2–3 meqg^{-1}), tensile strength and thermal stability. However, it has low proton conductivity 0.057 Scm^{-1} compared to Nafion-117 and is also very expensive. Wainright et al., (1995) reported s-PBI membrane with good properties at high temperature (>100°C) and low humidity (~30%). However, poor proton conductivity (0.46 mScm^{-1} at

Figure 9. Chemical structures of SPEEK (Bauer et al., 2000)

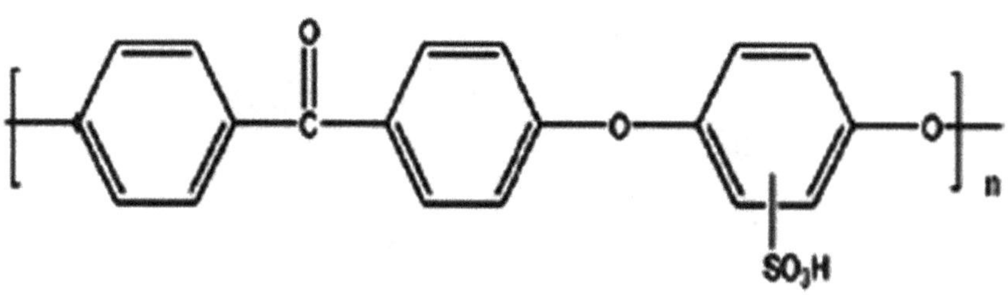

room temperature). Methanol crossover (of the order of 10^{-9} cm^2s^{-1}) was found to be much lower than Nafion-117. Figure 10 illustrates the chemical structure of s-PBI. Lufrano et al., (2000) used s-PSU membrane (see Figure 11) for DMFC applications. The synthesized membrane showed very high IEC (1.68–2.33 $meqg^{-1}$) and good proton conductivity (0.01 Scm^{-1}) at room temperature. However, methanol crossover was very high. Zhao et al., (2013) have reported the sulfonated polyethersulfone (SPES), as shown in Figure 12, was synthesized through a feasible way by introducing sulfonic groups onto amino-substituted PES. The water contact angles of the modified PES membranes decreased from 84° to 68°, and the water fluxes had a dramatic increase from 162 to 1912 mL/m^2 mmHg h when the blended amount of SPES increased from 0 to 2 wt.%. Feng et al., (2017) has studied the effect of PEG doping on the membrane formation and properties of sulfonated polyphenylenesulfone (s-PPSU) membranes (Poly-

Figure 10. Chemical structures of s-PBI (Wainright et al., 1995)

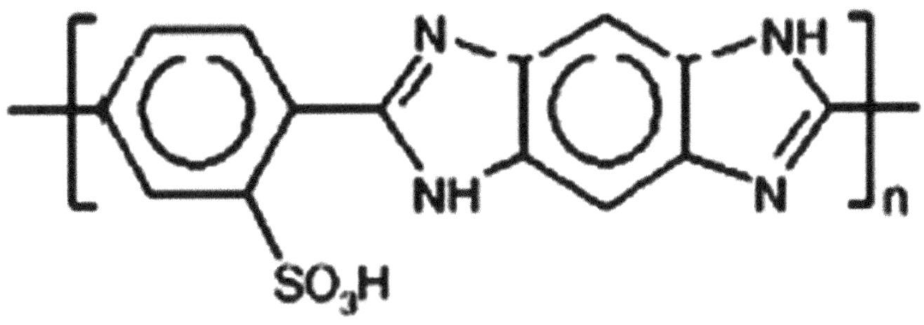

Figure 11. Chemical structures of s-PSU (Lufrano et al., 2000)

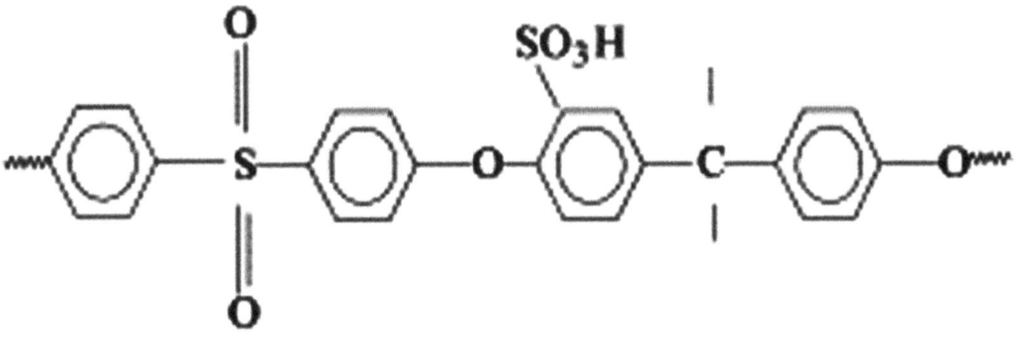

Figure 12. Chemical structures of s-PES (Zhao et al., 2013)

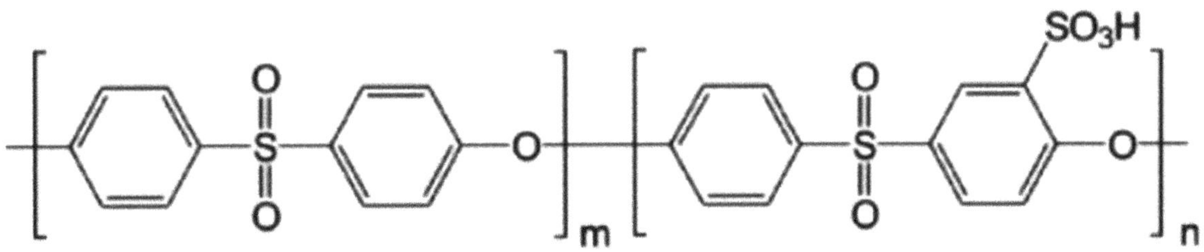

ethylene (PEG) - polyphenylenesulfone (s-PPSU)). It was observed that addition of PEG leads to more prominent change in s-PPSU than in PPSU alone and also the addition of PEG suppresses formation of macro-voids in membranes. The modified surface of the membrane with enhanced proton conductivity and good stability can be directly used in fuel cell applications. Figure 13 shows the chemical structures of PEG and s-PPSU.

Inorganic-Organic Nanocomposite Membrane

Inorganic-organic nanocomposite membranes comprising of stable polymeric support doped with sub-micron to nano-sized inorganic ion-exchangers. These inorganic ion-exchangers with high surface area, porosity and excellent thermal stability endow improved surface and ion-transport properties. To-date, many work has been reported on inorganic-organic composite membranes by doping the inorganic ion-exchangers such zirconium phosphate, hygroscopic metal oxides, zeolites, Heteropoly acids (HPAs), CNT/MWCNT/graphene, metal-organic framework (MOFs) onto polymeric matrix.

Inorganic Ion-Exchanger

Handful of literatures is available on inorganic ion-exchangers in the preparation of inorganic-organic composite membranes for room temperature fuel cell applications.

Heteropoly Acids (HPAs)

Among various types of HPAs known, phosphotungstic acid (PWA) has gained much interest in recent times due to its high proton conductivity (0.1 Scm^{-1}) and good oxygen affinity. Mioc et al., 1997 described the various types of HPAs and reported very high proton conductivity (0.02–0.1 Scm^{-1}) ascribed to high bronsted acidity and thereof higher acidic sites and good IEC (0.2–1.5 $meqg^{-1}$). Izumi et al., 1999 used tetra ethyl orthosilicate (TEOS) derived silica to immobilize the PWA. The PWA clusters, incorporated with inorganic silica framework did not dissociate and showed high proton conductivity 0.02 Scm^{-1} at 25°C. Ramani et al., 2005 used cations (Cs^+, Tl^+, NH_4^+) for effiecient immobilization of PWA. Tang et al., 2010 (Figure 14) synthesized PWA based inorganic membrane for DMFC applications and reported proton conductivity 0.01–0.08 Scm^{-1} at room temperature and 0.06–0.08 Scm^{-1} at 70–100°C. This was mainly due to easy availability of acid sites for proton conduction. Moreover, the PWA has self-humidifying properties as shown in Fig. 14 indicating its potential to be used at elevated temperatures.

Figure 13. Chemical structures of (a) PEG and (b) s-PPSU (Feng et al., 2017)

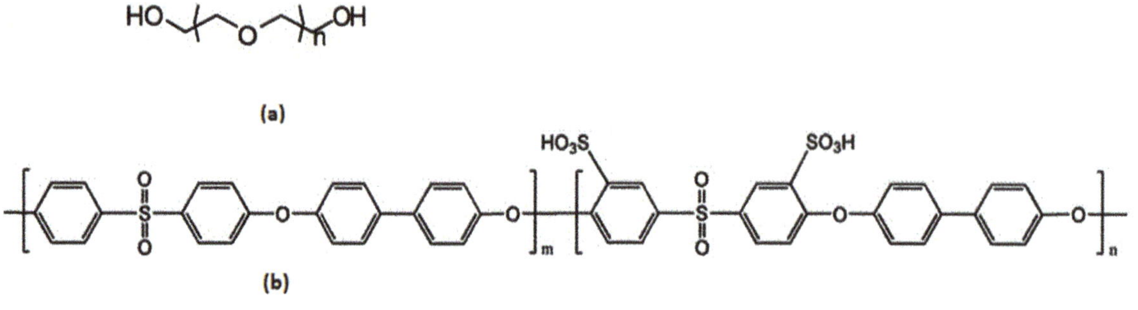

Figure 14. (a) Small angle XRD pattern (in inset HRTEM micrograph) (b) N_2 adsorption/desorption isotherm (c) and (d) proposed proton conduction pathways for self-assembled HPW/meso-silica membrane and mixed HPW/meso-silica membrane respectively. (Tang et al., 2010)

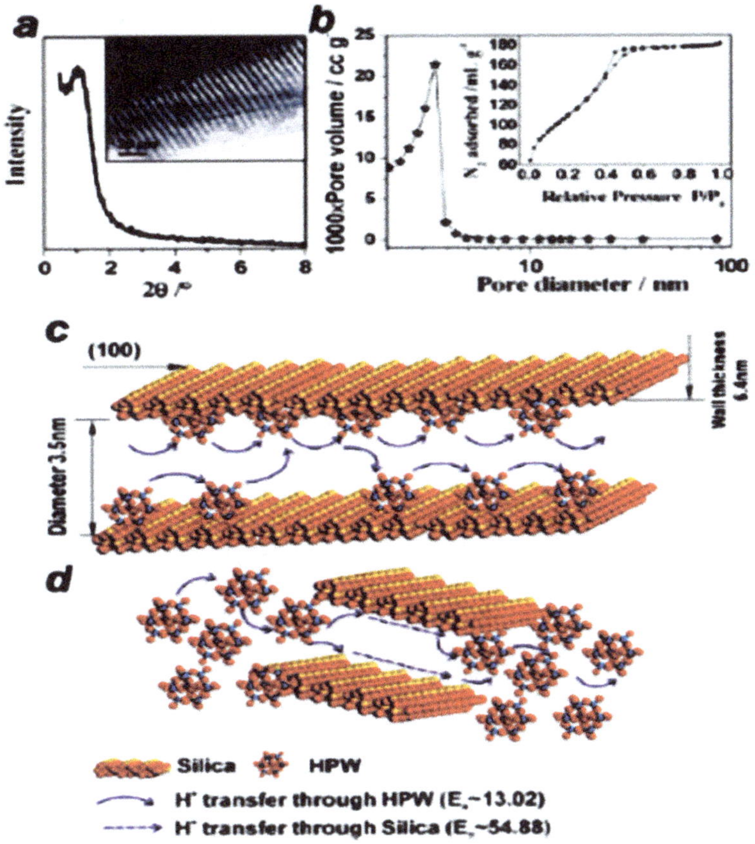

Acid Salts of Tetravalent Metals

Recently, many progresses are made on fast proton conductors (acid salts of tetravalent metals [M (IV). $HPO_4.nH_2O$) where, M is the tetravalent metal (Zr, Ti, Sn)) comprising of structurally bound hydroxyl protons which hold water molecules even at very high temperatures indicating great potential for this type of inorganic ion exchanger for solid-state proton conduction. Among these, ZrP has very high thermal stability as well as very high IEC (6.64 $meqg^{-1}$)(Alberti et al., 1983). Only a few research papers have reported on ZrP based ion-exchange membranes for DMFCs and PEMFCs. Alberti et al., (2000) and Clearfield et al., (1988) reported a very high IEC (4–8 $meqg^{-1}$) for ZrP, but low transport number (< 0.85) and proton conductivity (10^{-2}–10^{-3} Scm^{-1}).

Metal Oxides

Silica (SiO_2), titania (TiO_2), zirconia (ZrO_2) are most coomonly used metal oxides ion-exchangers for proton conduction applications. Li et al., (2002) have reported very high IEC (1–2 $meqg^{-1}$) for silica with very poor proton conductivity in the range of 10^{-4} to 10^{-5} Scm^{-1}. Relatively low proton conductivity was due to poor mobility of the protons in the lattices of silica frameworks.

Modified Carbons (CNT, Graphene)

Organic modifiers carbons such as graphene oxide (GO), carbon nanotubes (CNTs), multiwalled CNTs (MWCNT) have attracted great interest due their unique structure and surface properties with a two-dimensional (2D) layered material with epoxy, hydroxyl, and carboxyl groups on the surface and edge which provide active sites for hydrogen bonding and thereby facile proton conduction can be achieved as reported by Yang et al., (2015). It was observed that the presence of these oxygen-containing functional groups facilitate the hydration of GO and helps to hold more water molecules and thus improve the proton conductivity. Owing to this feature and also large surface area, as well as intrinsic stability and fuel crossover barrier, GO is one of the best nano-fillers to be used in PEMFCs (Chien et al., 2013; Bayer et al., 2014; Lue et al., 2015). Recently, Hatakeyama et al., (2014) found that the monolayer GO sheet exhibited a proton conductivity as high as 10^{-2} S/cm which is much higher than bulk GO (10^{-4} S/cm) or acidulated GO (10^{-5} S/cm). Cele et al., (2010) have synthesized nanofibers of CNT, as shown in Figure 15, and demonstrated enhanced proton conductivity in the range of 10^{-4} S/cm higher as compared to pristine CNT.

Figure 15. Morphology for (a) pCNTs, (b) oCNTs and (c) fCNTs (Cele et al., 2010)

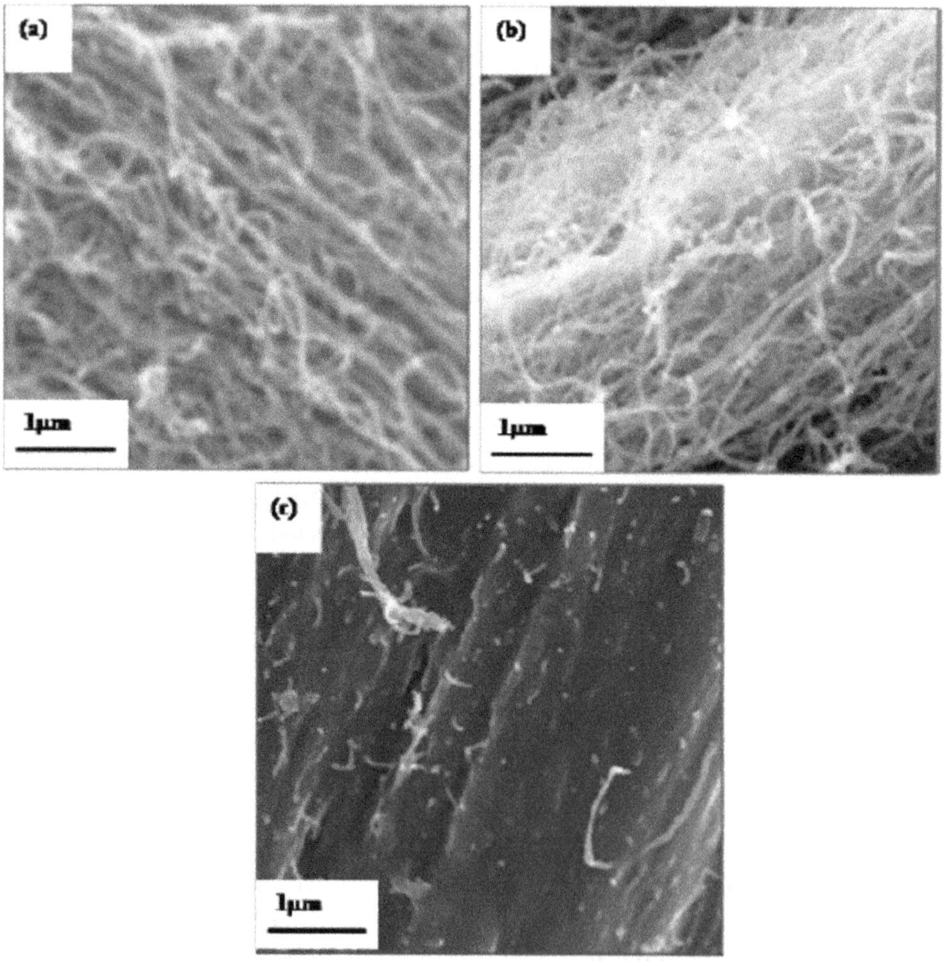

Zeolites

Zeolites, class of porous frameworks with AFI structure, are widely used ion-exchangers in fuel cell applications due to their unique structure, high surface area (200–300 m^2g^{-1}) and small pore size (1–2 nm). Mckeen et al., (2008) reported proton conductivity (10^{-5}-10^{-6} Scm^{-1}) with lower water uptake showing right potential for fuel cell applications. Relatively lower ion transport was due to the rigid network structure.

Metal-Organic Frameworks (MOFs)

Metal organic frameworks (MOFs), unique class of porous and crystalline materials, are 3D porous structures made of metal ions or clusters of ions linked together by organic molecules. MOFs are considered promising solid porous materials as they possess tunable pore size, higher pore volume and surface area having ability to accommodate guest molecules into their pores [6]. In recent advances of MOFs, the proton-conducting MOFs are targeted as novel candidates for solid proton conductors in fuel cells as they possess hydrophilic open channels responsible for proton conductivity. Recently, Ramaswamy et al., (2014) have reported an excellent review on several proton conducting MOFs developed so far having proton conductivity in the order of magnitude 10^{-3}-10^{-6} Scm^{-1} lower than conventional and highly studied Nafion-117 (0.1 Scm^{-1}). Ramaswamy et al., (2015) also reported the most performant proton conducting MOF, $Mg_2(H_2O)_4(H_6L)\bullet H_2O$ [PCMOF-10], that conducts proton under both high humidity and anhydrous conditions. The single crystal structure of PCMOF-10 has been illustared in Figure 16. The proton conductivity measured at 70°C under 95% relative humidity equals to 3.55×10^{-2} Scm^{-1}. A three dimensional proton conducting metal–organic framework (PCMOF-5) reported by Taylor et al. [49] conducts protons above 10^{-3} S/cm at 60 °C and 98% relative humidity as shown in Figure 17. PCMOF-5 with free phosphonic acidic sites shows high humidity stability under fully hydrated conditions and therefore well suitable for fuel cell applications.

Polymeric Support

Several organic polymers have been used as membrane supporting materials. Selection of the polymers is done on the basis of their cost, stability and density. Different polymeric matrices like PVDF, PTFE, poly(propylene) (PP), poly(ether sulfone) (PES), PVA have been used, however the polymeric materials of high tensile strength and good methanol resistance are preferred.

Poly(tetrafluoroethylene) (PTFE)

In an excellent review on PEMs, Peighambardoust et al., (2010) listed the various DMFC membranes (including Nafion-117) which use PTFE support and described the physical and chemical properties of PTFE. It has very high mechanical strength in the range of 20–30 MPa and low density 2.2 gcm^{-3}.

Poly(Ether Ether Ketone) (PEEK)

Ahmad el al., (2010) used PEEK as a polymeric membrane material for the synthesis of sulfonated poly(ether ether ketone) (SPEEK) membrane and reported very high mechanical strength (90–100 MPa). Additionally, the density of PEEK (1.32 gcm^{-3}) is also lower compared to PTFE polymer.

Figure 16. (a) Single crystal structure for PCMOF10 and (b) Hydrogen bonding network between lattice water molecule and intralayer oxygen atoms (Ramaswamy et al., 2015)

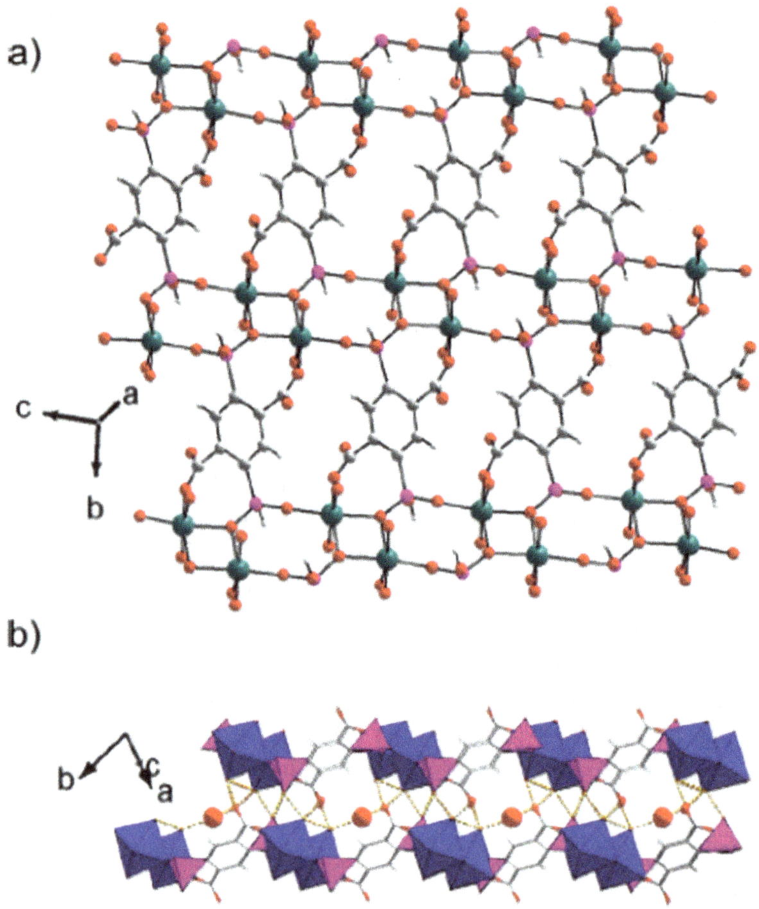

Poly(Benzimidazole) (PBI)

Wycisk et al., (2006) used PBI based membrane for DMFC application. The reported mechanical strength of 160 MPa was very high compared to PTFE and PEEK. Moreover, its density is also low (1.3 gcm^{-3}).

PVDF

Liu et al., (2011) reported a review on PVDF polymers which have mechanical strength in the range of 50–80 MPa and density of 1.78 gcm^{-3}. The PVDF membranes are also used in pervaporation processes like dehydration of ethanol. PVDF's methanol selectivity is very low (20% for methanol and 80% for water).

PVA

Maiti et al., (2011) reported a rigorous review on the PVA based membranes for DMFC showing the merits of PVA in comparison to other polymers. PVA is preferred due to its biodegradable nature, high mechanical strength (80–100 MPa) with low density of 1.19–1.31 gcm^{-3}. Additionally it is easy in film

Figure 17. Proton conductivity as a function of temperature for PCMOF-5 measured at 90% and 98% RH (Taylor et al., 2013)

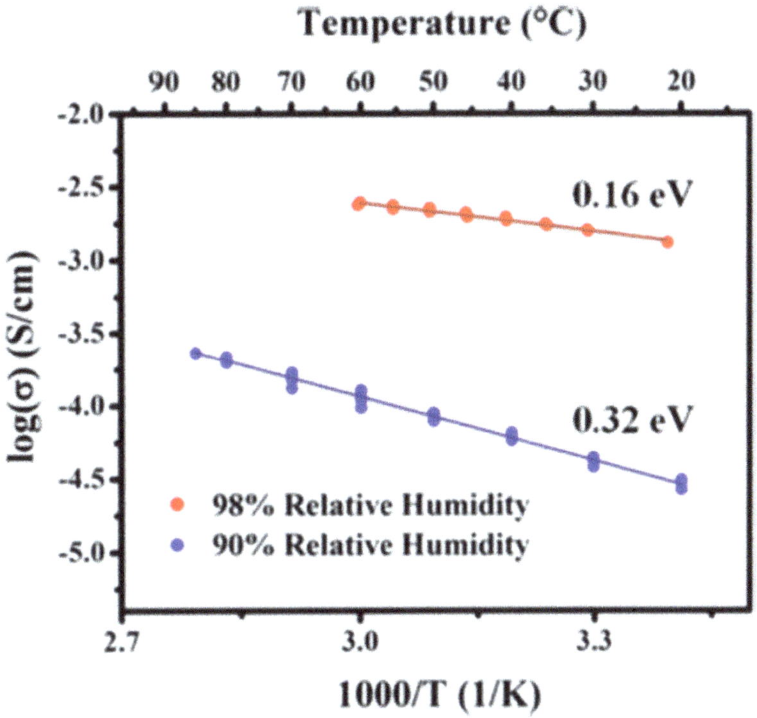

forming, very economical and highly methanol resistant. Unfortunately, PVA is highly water soluble and therefore must be crosslinked to make it water resistant before applying to fuel cell. Figure 18 below depicts various routes for PVA crosslinking. Recently the synthesis of inorganic-organic hybrid membrane has gained major attention due to the excellent combined properties of inorganic ion exchanger (high thermal stability, excellent water holding capacity at higher temperatures) and organic support (chemically stable, high mechanical strength). Amongst various inorganic-organic composite membranes, the Nafion-117 based inorganic-organic composite membranes have gained tremendous attention for DMFC application. The most widely studied Nafion-117 based membranes are as follows.

NAFION BASED NANOCOMPOSITE MEMBRANE

To improve the fuel cell performance at elevated temperature (>80°C) and lower humidity (30%), Nafion membrane was modified using submicron to nanosized inorganic ion-exchangers as dopants to improve the Nafion's properties. Mechanism for the proton conduction through pristine Nafion and modified Nafion is illustrated in Figure 19. In this direction, Cele et al., (2010) reported synthesis of carbon nanotubes (CNTs) doped Nafion composite membranes (Nafion-MWCNT) via melt-blending at 250°C using three different types of CNTs i.e. pure CNTs, (pCNTs), oxidized CNTs (oCNTs) and amine functionalized CNTs (fCNTs). It is assumed that upon doping of nano-CNTs in the nano-pores of Nafion, unique structure and high surface area of the CNT endows more proton transport channels

Figure 18. Cross linked PVA matrix (Maiti et al., 2011)

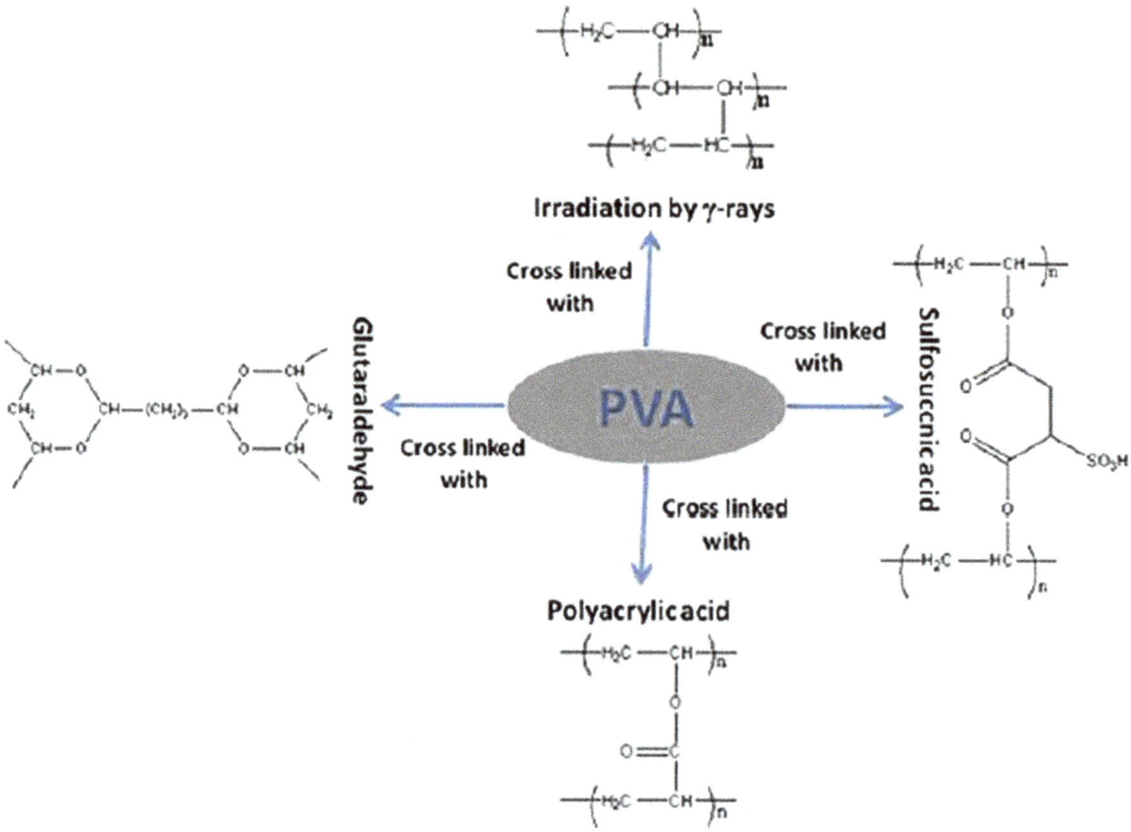

Figure 19. Illustration for vehicular and hopping mechanism for proton conduction in (a) pristine membranes and (b) composite membranes (Kim et al., 2015)

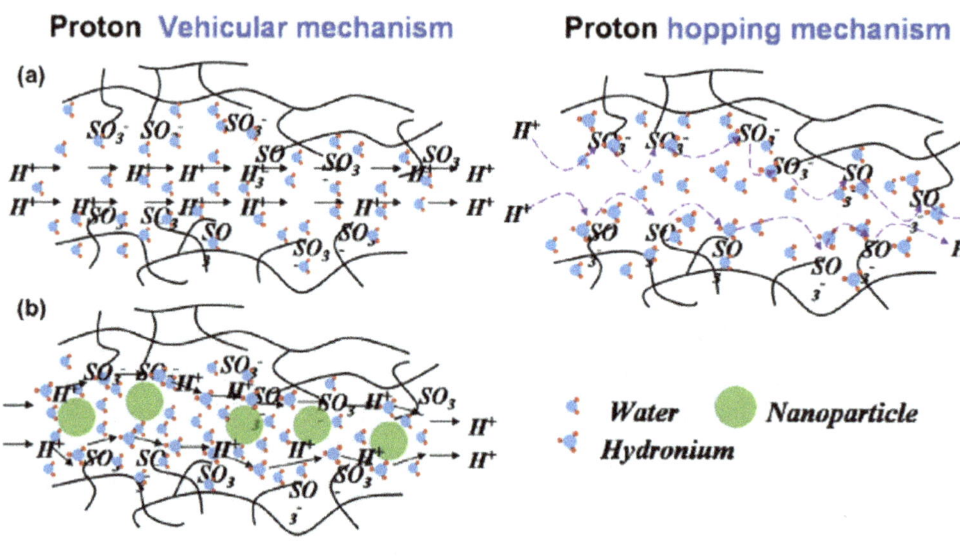

and holds water molecules quite tightly. It was also reported by Alberti et al., (2000) that doping of the colloidal dispersion of inorganic nano-particles into a polymeric solution allows the inorganic nano-particles to disperse properly in the porous polymeric membrane and facilitates active sites for proton conduction. Chen et al., (2008) showed that Nafion-ZrP composite membrane synthesized by in situ method had a better DMFC performance than composite membranes synthesized by direct impregnation method. These membranes showed reduced methanol crossover but low proton conductivity (10^{-3} Scm^{-1}). Smitha et al., (2005) synthesized Nafion-ZrP based composite membrane and reported good IEC. However, proton conductivity (2-3 $mScm^{-1}$) was very low due to doped ZrP particles. Interestingly the doped ZrP improved the thermal stability of membrane and reduced the methanol crossover but the tensile strength and the oxidative stability was a lower compared to Nafion-117. Nagarale et al., (2004) have reviewed the synthesized Nafion-silica and Nafion-titania membranes which were highly thermally stable with low methanol crossover and proton conductivity of 23–24 $mScm^{-1}$. Tricoli et al., (2003) have reported detailed synthesis and characterization of Nafion-zeolite membrane that showed very low value of proton conductivity 1.8–5.1 $mScm^{-1}$. Lufrano et al., (2013) have reviewed the PWA based Nafion membrane (Nafion-PWA) for DMFC and reported low value of IEC (0.7–0.8 $meqg^{-1}$) compared to pristine Nafion-117 (0.9 $meqg^{-1}$). Reported proton conductivity was in the range of 5–14 $mScm^{-1}$. Nafion-PWA membrane possesses better conductivity compared to other Nafion based membranes due to higher acid sites in PWA. Cho et al., (2005) and Alwin et al., (2011) doped PVDF in the pores of Nafion-117 i.e. Nafion-PVDF and reported the membrane conductivity of 2.5 $mScm^{-1}$ at room temperature and 4 $mScm^{-1}$ at 100°C. Methanol crossover of the doped membrane was lower than Nafion-117. Many works have been also reported on PVA based membranes for DMFC (Nafion-PVA). However, in most of the cases, DMFC performance was poor mainly due to lower proton conductivity of the composite membrane. A detailed review on PVA based membranes, reported by Maiti et al., (2011) who compared the properties with the Nafion-117. Methanol permeability value was in the range 10^{-7} to 10^{-8} cm^2s^{-1} which is an order of magnitude lower than the Nafion-117. Sahu et al., (2008) reported the proton conductivity 0.033 Scm^{-1} at 100°C for PVA-Nafion, was much lower compared to pristine Nafion-117 proton conductivity at 90°C (0.1 Scm^{-1}). Deluca et al., (2006) used PVA as a polymeric support to hold Nafion nanoparticles but measured low value of the proton conductivity in the range of 0.9-5 $mScm^{-1}$, however, the reported methanol crossover ($3.1–6.5 \times 10^{-7}$ cm^2s^{-1}) was much lower as shown in Figure 20.

MOF-Nafion Based Nanocomposite Membrane

In past few years, significant progresses have been made in MOF-polymer composite membranes for molecular separation and gas storage (Yeo et al., 2014). Contrary to this, very little is known on proton conducting MOF–polymer composite membranes for fuel cell applications. Moreover, there are only few structure related properties of MOFs that recommend them MOFs as suitable materials for proton conduction. MOFs possess both the lattice water as well as structurally bound water that can retain water even at high temperature. Compared to the classic inorganic proton conductors, MOFs can be easily embedded into polymers. Importantly, MOFs have better proton conductivity than zeolites, mesoporous silica, mofified carbons. This makes MOFs promising as solid proton conducting materials in proton conducting membrane based fuel cell device. Researchers have also tried to make thin films of pristine MOFs but failed mainly due to the complexities associated with the formation of highly flexible as well as mechanically and thermally stable thin films. Therefore, the best alternative was to use MOF as a

Figure 20. Selectivity (open bars) and proton conductivity (solid bars) of Nafion® 117 and Nafion®/ PVA blend membranes (Deluca et al., 2006)

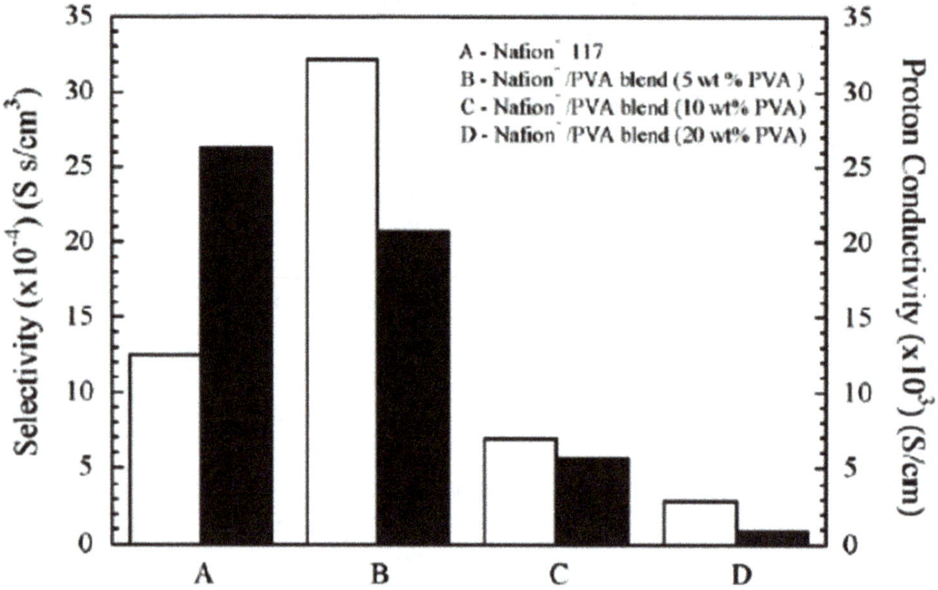

dopant for composite membrane formation. However, the MOF dispersion onto the polymer solution and the MOF–matrix compatibility are some of the critical issues working with MOF–polymer composite membranes (Zhu et al., 2014). Yang et al., (2015) have fabricated Nafion based composite membrane modified by a ZIF-8@GO types of MOF. The ZIF-8@GO/Nafion hybrid membrane exhibits proton conductivity 0.28 S/cm at 120°C and 40% RH (Figure 21), resulting from a synergetic effect of zeolitic-imidazole framework (ZIF-8) and graphene oxide (GO). Based on the literature on polymeric membranes, inorganic ion-exchanges and composite membranes, we have summarized the proton conductivities of the different proton exchange membranes (PEMs) measured at different temperatures as shown in Table 5.

Figure 21. (a) Temperature-dependent proton conductivity plots of different ZIF-8@GO loadings (b) water uptake for recast Nafion and ZIF-8@GO/Nafion-1 under 60 and 120°C at 40% RH (Yang et al., 2015)

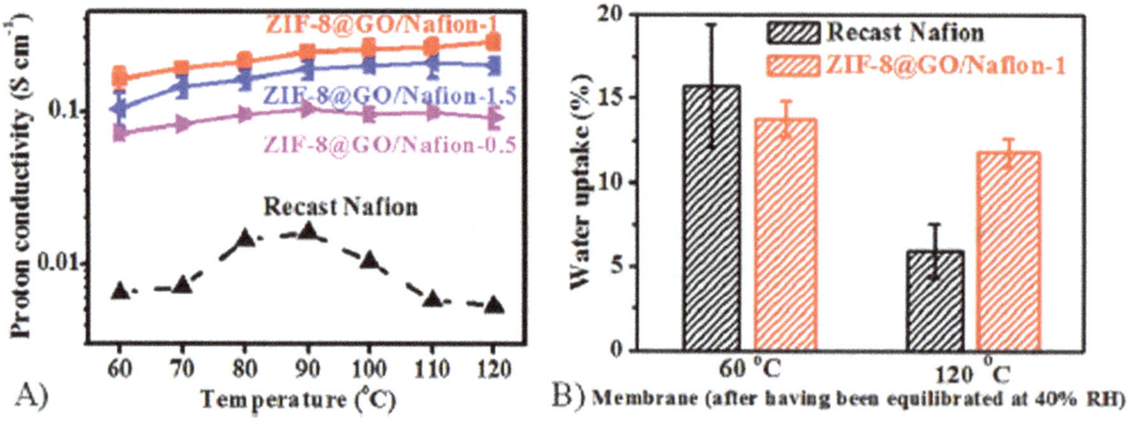

Table 5. Proton conductivity of the different proton exchange membranes (PEMs) and inorganic ion-exchangers measured at different temperatures.

Proton Exchange Membrane	Proton Conductivity, Scm^{-1}	References
Polymeric Membrane		
1. Nafion-115	0.074 (20°C), 0.0078 (80°C)	Ahmad el al., 2010
2. SPEEK	0.057 (25°C)	Bauer et al., 2000
3. Nafion-117	0.017 (70°C), 0.027 (80°C), 0.03 (90°C)	Slade et al., 2002
4. s-PBI	0.0046 (25°C)	Wainright et al., 1995
5. s-PSU	0.01 (25°C)	Lufrano et al., 2000
6. Nafion-112	0.08 (25°C)	Coutanceau et al., 2006
Nanocomposite Membrane		
1. Nafion-Silica	0.023–0.024 (40°C)	Nagarale et al., 2004
2. Nafion-ZrP	0.001 (40°C)	Chen et al., 2008
3. Nafion-PVDF	0.0025 (25°C)	Cho et al., 2005; Alwin et al., 2011
4. Nafion-PVA	0.001–0.01 (25°C)	Sahu et al., 2008; Maiti et al., 2011
5. Nafion-PWA	0.005–0.014	Lufrano et al., 2013
Inorganic Ion-Exchanger		
1. PWA	0.01–0.08 (25°C)	Tang et al., 2010
2. Silica	10^{-3}–10^{-5} (25°C)	Li et al., 2002
3. Zeolite	10^{-3}–10^{-6} (25°C)	Mckeen et al., 2008
4. ZrP	0.024–0.06 (25°C)	Clearfield et al., 1988

**Table 6* at last of chapter summarizes the abbreviations.

NON-NAFION BASED NANOCOMPOSITE MEMBRANE

Considering very high cost of Nafion based composite membrane mainly due to PTFE support, researcher moved to non-Nafion based inorganic-organic composite membranes and compared their properties with the conventional membranes. SPEEK doped with sulfonated GO based composite membranes (SPEEK-sulfonated graphene oxide (sGO)) for PEMFC was synthesized by Kumar et al., (2014). SGO/SPEEK composite membrane showed a very good conductivity of 0.055 S cm^{-1} at 80 °C and 30% RH higher than pristine SPEEK indicating their suitability for fuel cell operating under similar conditions. Chen et al., (2010) synthesized PVDF-PWA nanofiber for fuel cell applications. The PWA particles were mixed with the PVDF powder followed by electro-spinning for fibrous membrane structure. Very low value of the IEC (0.22 $meqg^{-1}$) and proton conductivity (0.4 $mScm^{-1}$) was reported as shown in Figure 22 however, methanol crossover was not reported which is one of key issues with DMFCs. Xu et al., (2004) used PWA-PVA membrane for DMFC application. Very low values of IEC (0.2–0.7 $meqg^{-1}$) and proton conductivity (10^{-4} Scm^{-1}) were reported. At the same time, very low methanol crossover, in the range of 10^{-7}-10^{-9} cm^2s^{-1}, was also measured. The reduced methanol crossover ascribed to the PVA polymeric matrix that has very poor affinity for methanol molecules. Pandey et al., (2014) have synthesized PVA-Si-PWA based composite membrane for room temperature operated DMFC as illustrated in Figure 23. Composite membrane was synthesized by dispersing sub-micron to nano sized silica immobilized

Figure 22. Proton conductivity of PVDF-PWA based nanofiber membranes (Chen et al., 2010)

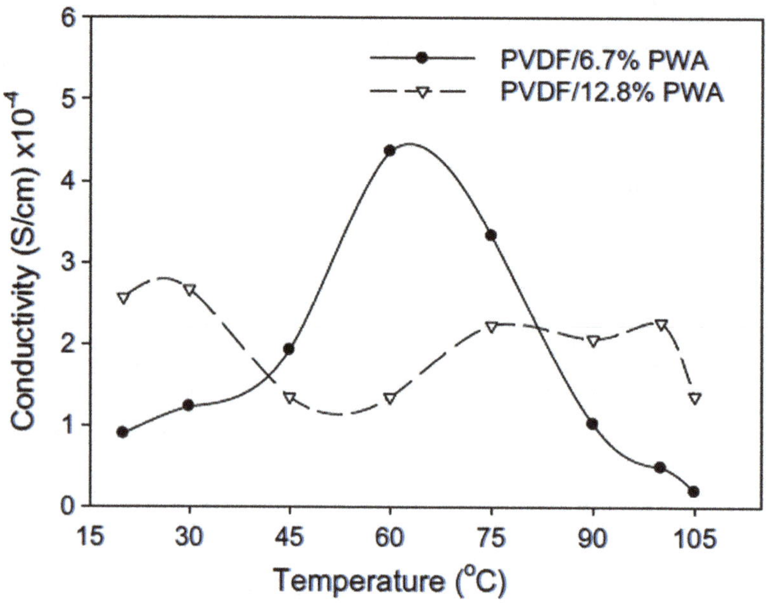

Figure 23. Synthesis process for Si-PWA doped PVA membrane (Pandey et al., 2014)

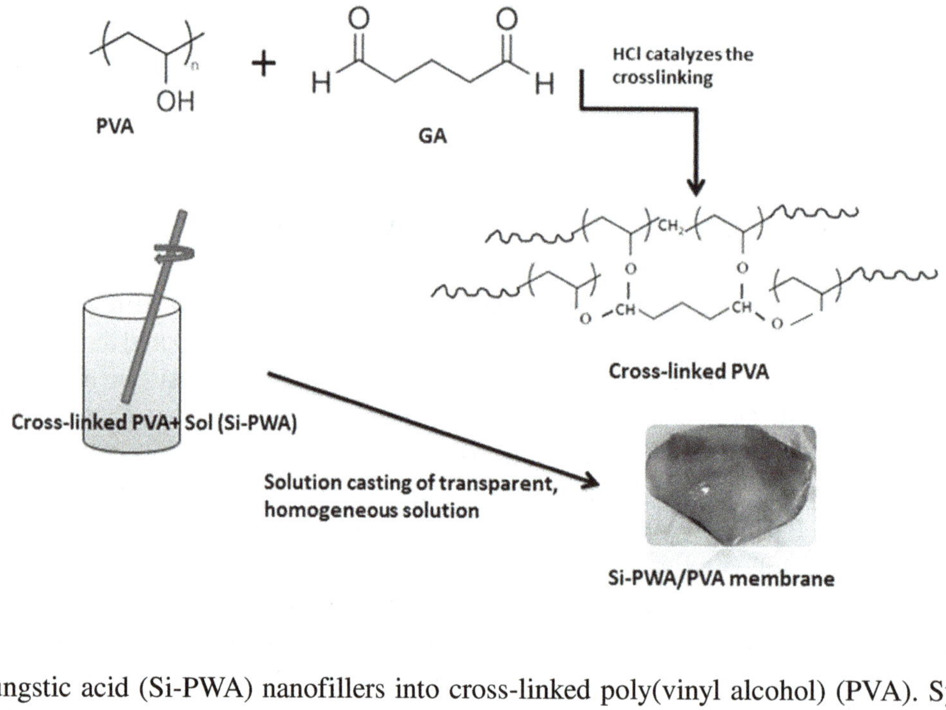

phosphotungstic acid (Si-PWA) nanofillers into cross-linked poly(vinyl alcohol) (PVA). Synthesized membrane showed good water uptake of 35% with methanol uptake (8.4%) and thereof reduced methanol permeability (1.6×10^{-7} cm^{2}s^{-1}) compared to Nafion-117. Proton conductivity increased from 7.04 mS cm^{-1} to 10.5 mS cm^{-1} with increase in temperature from 30 °C to 50 °C. The synthesized composite membrane also exhibited excellent mechanical and thermal stability compared to Nafion-117. Among various classes of available inorganic fillers, carbon nanotubes (CNTs), a cylindrical graphene tubes comprising of multiple layers of graphene, have drawn considerable attention due to their unique structure, active sites for proton conduction, high aspect ratios of 100–1000, high specific surface area, light weight, and remarkable mechanical properties. Because of these amazing properties, CNTs have been widely investigated as candidate for low to high temperature operating fuel cell materials. Recently Cui et al., (2015) reported the incorporation of silica-coated CNTs as novel dopant into the SPEEK (SPEEK/SiO_2@CNT) for DMFC applications. The SiO2@CNT loading of 5 wt.% showed decline in methanol crossover compared to pristine membrane without compromising proton conductivity which remained above 10^{-2} S cm^{-1} at room temperature. Figure 24 shows the water uptake and swelling ratio and Nyquist plots of the SPEEK/SiO_2@CNTs composite membranes with different SiO_2@CNTs contents. Auimviriyavat et al., (2011) reported ferrierite zeolite, a version of zeolitic structure, as inorganic nano-dopant in SPEEK. The composite membrane displayed improved proton conductivity with reduced methanol permeability, and high selectivity (ratio of proton conductivity and methanol crossover) of 7.148 $\times 10^{-3}$ Scm^{-1}. Lee et al., (2016) have studied the preparation of ZrP doped SPEEK composite membranes using a electron beam irradiation approach for fuel cell applications. Water uptake and IEC of the cross-linked SPEEK/ZrP hybrid membranes were higher than SPEEK and showed increasing trend with increase in ZrP particles loading content. Furthermore, under high temperature (120°C) and low humidity conditions (35%), the cross-linked SPEEK/ZrP hybrid membrane displayed a higher proton conductivity (1.41×10^{-3} S/cm) compared with the cross-linked SPEEK membrane (1.13×10^{-4} S/cm). Qian et al., (2012) reported synthesis of sulfonated polybenzimidazole/zirconium phosphate (sPBI/ZrP) composite membranes using solution-casting approach. It was observed that due to the addition of ZrP, the conductivity of sPBI-ZrP composite membrane increased to 0.08 Scm^{-1} at 180°C indicating that sPBI/ZrP composite membranes are promising polymer electrolyte for high temperature PEMFCs applications.

Figure 24. (a) Water uptake-swelling ratio and (b) Nyquist plots for SPEEK/SiO_2@CNTs membranes with different SiO_2@CNTs loadings (Cui et al., 2015)

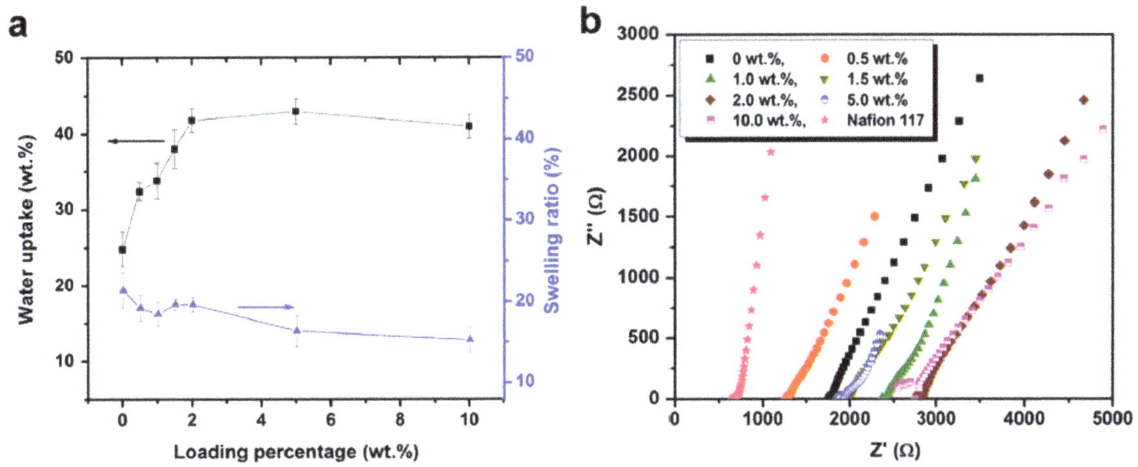

MOF Doped Nanocomposite Membrane

As stated above, the MOF-polymer composites have proven them as good proton conducting membrane materials for fuel cell applications from low to intermediate temperature operation. So far very limited work has been reported on Nafion-MOF doped composite membrane which is discussed already. Here we have summarized below the proton conducting MOF-polymer composite membrane developed so far for fuel cell applications. Liang et al., (2013) reported a MOF–polymer composite membrane (PVP-MOF) using poly(vinylpyrrolidone) (PVP) as support and $\{[Ca(_D\text{-Hpmpc})(H_2O)_2].2HO_{0.5}\}_n$ MOF as filer and studied the proton conductivity at low relative humidity (53% RH). The improved proton conductivity (2.8×10^{-5} Scm^{-1}) of the composite membrane as compared with the pristine MOF (6.9×10^{-6} Scm^{-1}) is due to the high hydrophilic nature of PVP polymer under same humidified conditions. Wu et al., (2013) synthesized a sulfonated poly(2,6–dimethyl–1,4–phenylene oxide) (SPPO) supported Fe–MIL–101–NH_2 MOF composite membrane (SPPO-MOF). The proton conductivity values measured at room temperature was 0.1 Scm^{-1}) and 0.25 Scm^{-1} at 90 °C, comparable to Nafion. The much higher proton conductivity of the composite membrane could be also attributed to the SPPO matrix which itself is highly proton conducting organic polymer causing very high cost of composite membrane. Erkartal et al., (2016) demonstrated unique ternary composite membrane consisting of poly(vinyl alcohol) (PVA), poly (2-acrylamido-2-methylpropane sulfonic acid) (PAMPS), zeolitic imidazolate framework-8 (ZIF-8), prepared by physical blending and casting methods as shown in Figure 25. The reported proton conductivity of the composite under fully hydrated state is 0.134 Scm^{-1} measured at 80°C which is attributed to the doped ZIF-8 particles in PVA and PAMPS.

Figure 25. Route for the preparation of the PVA-PAMPS-MOF based composite membranes (Erkartal et al., 2016)

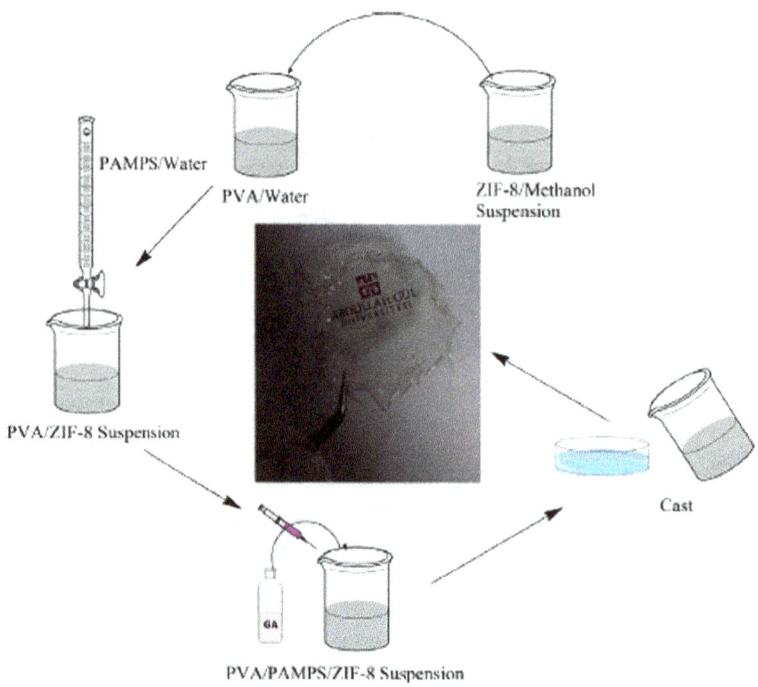

APPLICATION OF FUEL CELL MEMBRANE

Based on structure and properties of proton conducting membranes, they are utilized in many energy devices for portable and transport applications such as PEMFCs, DMFCs, redox flow batteries (RFBs), PEM electrolysis, microbial fuel cells (MFCs) and super-capacitors which are discussed below in detail.

PEMFC and DMFC

So far most of the nanocomposite membranes are being used in room temperature fuel cell (PEMFC) due to their high power output and efficiency. However the cost of hydrogen as a fuel and other components (platinum based catalysts) of PEMFCs are very high that led to development of other possible applications for fuel cell membrane. In this section, we have summarized the performance of Nafion-117 and other successfully developed membranes in single cell PEMFC and DMFC under different operating conditions.

Pulido et al., (2017) have reviewed recent developments in the titania (TiO_2) doped nanocomposite membranes for PEMFC. They have reported doping of nanosized titania onto polymeric matrix such as Nafion, SPEEK, PTFE, SPES, sulfonted polystyrene ethylene butylene polystyrene (SPSEBS). It was noticed that doping of TiO_2 reduced H_2 crossover significantly with improvement in proton conductivity. Figure 26 shows the comparative PEMFC performance for TiO_2 doped composite membranes.

Performance of DMFC with Nafion-117 has been reported by Neburchilov et al., (2007) and maximum reported current density for Nafion-117 was 50 mAcm^{-2} measured at 80°C and 1 M of methanol concentration. However, high methanol concentration deterioration the DMFC performance (i.e. OCV) due to high methanol crossover. Additionally the loss in the proton conductivity above 80°C also resulted into

Figure 26. Power density for single cell PEMFC by different TiO_2 doped polymeric nanocomposite membranes (Pulido et al., 2017)

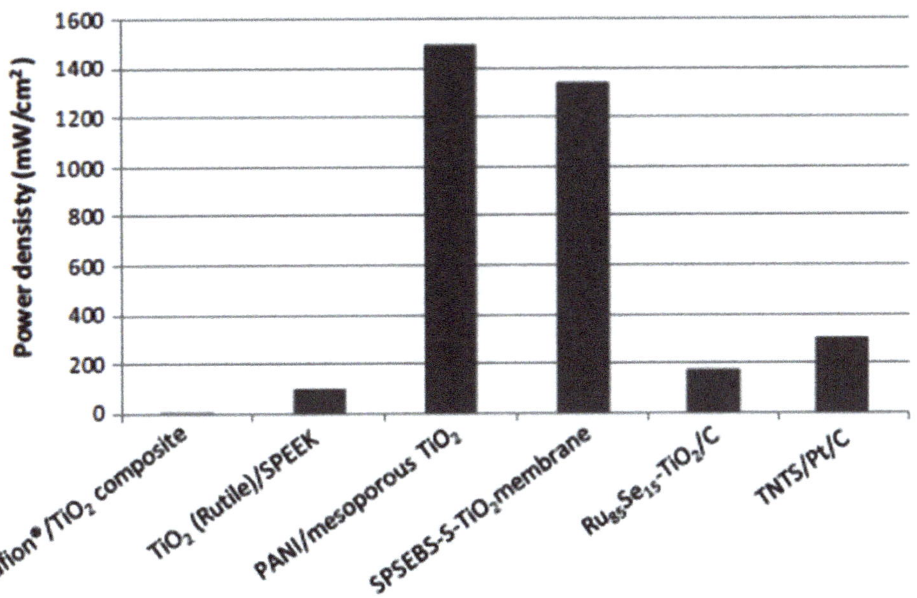

poor DMFC performance. Therefore, for good performance of the DMFC with Nafion-117 membrane, operating temperature is limited to 80°C. For Nafion-ZrP membranes in DMFC application (Arbizzani et al., 2010) the reported maximum power density was 80 mWcm^{-2} at 0.5 V and 90°C. Very low OCV (0.6 V) was measured at 0.5 M methanol feed concentration due to high methanol crossover for Nafion-ZrP composite membrane. OCV increased from 0.549 V to 0.575 V with increase in temperature from 70 to 90°C. Reported maximum current density was 650 mAcm^{-2} and 120 mAcm^{-2} respectively for the Nafion-silica membrane (Jung et al., 2002). For Nafion-PWA, very high value of OCV 0.935 V and current density 82 mAcm^{-2} has been reported by Amirinejad et al., (2011) for the Nafion-PWA membrane and the performance was better compared to pristine Nafion-117 at the similar conditions. Maximum power density of 83 mWcm^{-2} and OCV 0.8 V is reported for the PVDF-Nafion-117 composite membrane (Cho et al., 2005). The performance was found to be better than pristine Nafion-117 membrane due to lower methanol crossover. Maximum current density of 60 mAcm^{-2} was measured at room temperature for the PVA-Nafion-117 membrane as shown in Figure 27. However a very low value of power density about 8 mWcm^{-2} and OCV (0.6 V) was obtained (Deluca et al., 2006).

Vanadium Redox Flow Battery (VRFB)

VRFB is one of the promising technologies for large-scale renewable energy storage. Design of ion-exchange membrane (IEM) with desired properties like low-cost, mechanically chemically stable, low vanadium ion permeability and high proton conductivity is one of the major challenges. Considering high cost of fuel cell stacks, in recent past many efforts have been also made in flow batteries for energy storage. High energy density and good cyclic stability.

Figure 27. Power (open) and polarization (solid) curve of Nafion®/PVA (20 wt% PVA) blend membrane at 2, 4, 8, 16 M methanol concentrations. (Deluca et al., 2006)

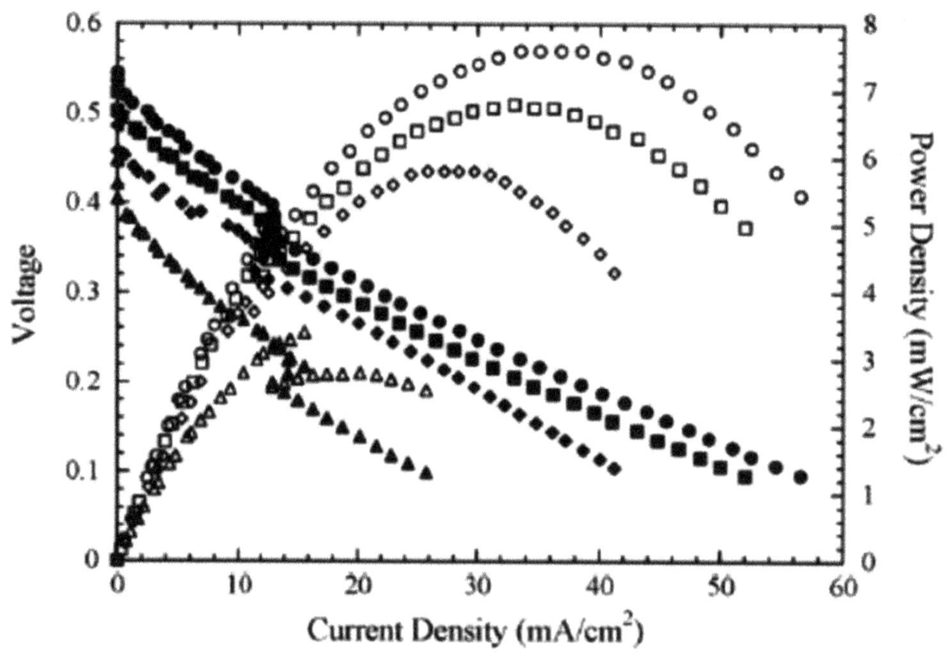

Pandey et al., (2016) reported performance of Si-PWA/PVA based nanocomposite membrane in VRFB and compared the performance with Nafion-115. Vanadium ion permeability (0.69×10^{-7} cm min^{-1}) of membrane was significantly lower than Nafion-115 with lower OCV decay. Single-cell VRB with membrane exhibited lower voltage during charge and higher during discharge with excellent cyclic durability compared Nafion-115. Figure 28 indicates the charge-discharge curve for Si-PWA/PVA composite membrane.

On the other hand, Seepana et al., (2017) synthesized novel PTFE-ZrP membrane using a simple brush coating method. Single-cell VRFB with membrane showed 80% energy efficiency below and very high columbic efficiency (95%) with little contamination of electrolyte due to cross mixing.

Microbial Fuel Cell (MFC)

MFC is a bio-electrochemical system for generation of hydrogen and electricity by action microorganisms. MFCs use PEMs for proton conduction and separating the bio-anode and cathode. Presently, MFCs are being used in wastewater treatments.

Chae et al., (2014) fabricated nanofiber-reinforced (NFR) composite PEM based on SPEEK and tested in MECs. TheNFR-PEM into MECs improved H_2 production with an overall h efficiency 86.4% higher compared to 77.2% for Nafion. The higher proton selectivity of NFR-PEM reduced anodic pH decrease harmful for anode bacteria, by facilitating selective proton transfer rather than other competing cations, which led to overall improved hydrogen gas yield in the MEC as the NFR-PEM was installed.

Recently, Park et al., (2017) have also reported sulfonated poly(arylene ether sulfone) (SPAES)/ polyimide nanofiber (PIN) based PEM for MFC. Hydrogen production efficiency of MFC increased by 32.4% compared to Nafion-211 with high hydrogen purity (90.3%) than for Nafion-211 (61.8%) which suggests that the novel membrane has good potential for MFCs, particularly when protons and other competing cations are present altogether in electrolytic solution.

Figure 28. (a) OCV decay and (b) charge–discharge curves of single cell VRFB with Si-PWA/PVA and Nafion-115 membrane (Pandey et al., 2016)

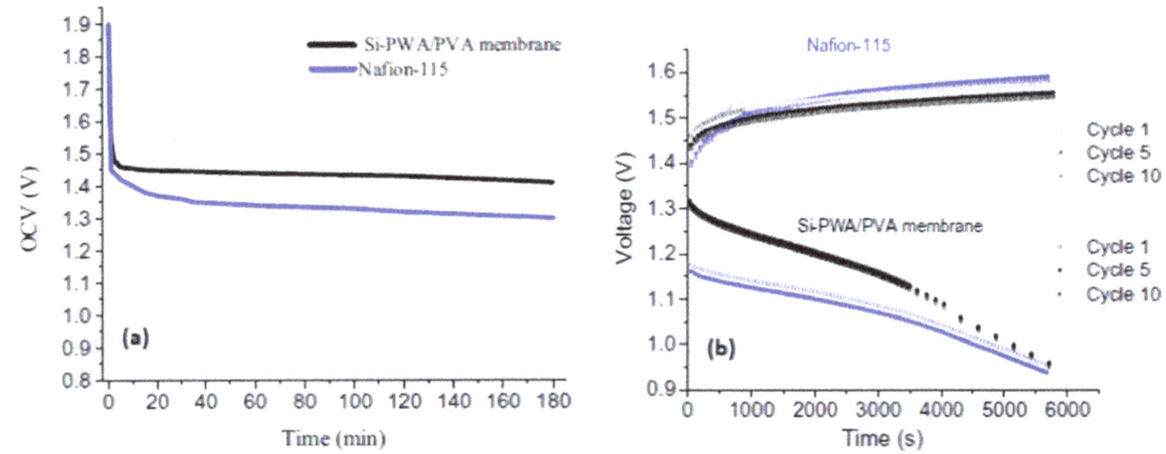

PEM Electrolysis

Electrolysis is a well know electro-chemical process for production hydrogen by electrolysis of water. In electrolysis process, the proton conducting membrane plays a key role in improving process yield. Many solid acids, novel conducting materials, as discussed above, are potential candidates as proton conducting electrolytes in PEM electrlysis. As per US patent (US 9437895 B2) by Huang et al., (2014), it is reported that H_2-Cl_2 proton exchange membrane based fuel cell electrolyzer using a PEM membrane, was successfully developed for cogeneration of electricity and HCL indicating huge potential of PEM in electrolysis process for production of useful chemicals.

Supercapacitors

Other than VRFBs, supercapacitor is also an emerging device for large energy storage due to their high specific capacitance (400-500 F/g) and excellent durability under real conditions. In supercapacitors, the PEM acts as separator as reported by Szubzda et al., (2017) who used low-energy plasma modification of PI and PP based nonwoven fabrics for supercapacitors. On using membrane as the separator the results showed improving trend in wettability of separators by the electrolyte that reduces the internal resistance of the supercapacitor and increasing the specific power of device.

CHALLENGES FOR THE NEXT GENERATION FUEL CELL MEMBRANES

Fuel cell technologies have shown tremendous potential to reduce massive load on conventional fossil fuels and boosting contribution of renewable energy to total world energy production that is far behind from fossil fuel contribution (88%). However, the cost of fuel cell has to be reduced significantly because in the year 2017 the cost of PEMFC is \$30/kW where PEM cost includes \$17/m^2 and \$14/KW for MEA as per recent report by Department of Energy (DOE), USA. This could be only possible by efficient design and developments of PEMs and electro-catalysts for fuel cells. Based on the requisite membrane properties and issues, we have identified the key challenges and targets for researchers for room temperature fuel cell membranes. Figure 29 illustrates the key properties of the state-of-the-art Nafion membrane and current challenges for the membranes under developments. In order to achieve the targets following steps should be taken into consideration.

1. Proper selection of thermo-stable base polymeric materials with low cost
2. Selection of highly proton conducting dopants.
3. Adopting novel and advanced synthesis methodology such as UV-grafting, infiltration, blending/recasting, electro-spinning, electron beam irradiation etc.
4. Precise control over the size, shape and composition of dopants.
5. Optimized thickness of the membrane with good flexibility
6. Understanding role of membrane's structure-properties relationship
7. Efficient design of MEA to reduce the interfacial resistance
8. Optimum water content per molecule of ionic group to facilitate fast proton conduction

Figure 29. Key properties of state-of-the-art Nafion-117 and current challenges for the membranes under development

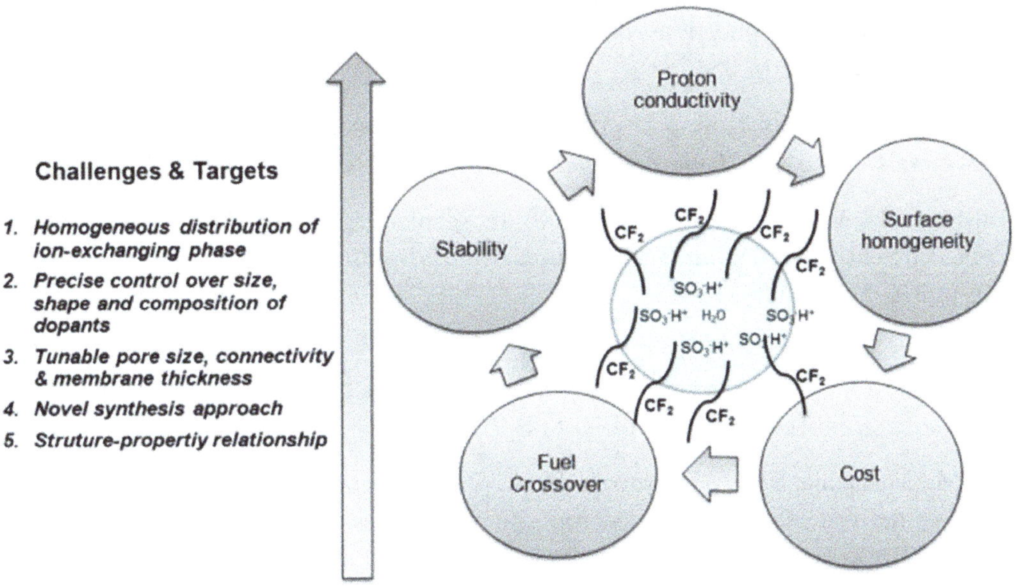

State-of-the-art Nafion-117

CONCLUSION AND RECOMMENDATIONS

Present book chapter critically reviews the recent developments and advances in fuel cell membranes for portable and transport applications. Although, state-of-the-art Nafion has demonstrated excellent proton conductivity and stability in low temperature fuel cell however, deteriorated properties at high temperature and lower humidity, fuel crossover and high cost are the technical barriers for worldwide commercialization. On the other hand, the inorganic-organic nanocomposite membranes have shown good potential as alternative to Nafion as they exhibit good stability, minimal fuel crossover and cost-effective. Unfortunately, achieving the proton conductivity of nanocomposite membrane comparable to Nafon has been a challenge among researchers. In this direction, recently MOF-polymer based nanocomposite membranes have also emerged as replacement of Nafion for room temperature fuel cells.

It is noteworthy that the novel synthetic routes have also great role in modifying the structure and morphology of the membrane that leads to improve the surface and ion-transport properties. Therefore, it is a challenge to have judicious selection of membrane materials, synthesis methodology to have best optimized membrane properties for fuel cell operation. In fact, a precise control over synthesis processing parameters also results into optimized membrane properties and thus enhanced performance. Precisely understanding the structure-property relations play a crucial role in solving the problem.

Lastly, the proton conducting membrane being the heart of fuel cell device, any quantum leap, innovation and improvements in the design and development will be huge economic reformations in fuel cell technology. Unfortunately, until date due to lack of efficient and effective design of fuel cell components, the prevalence of fuel cell technology, for clean and an environment friendly application, among us is still a long way to go.

REFERENCES

Ahmad, H., Kamarudin, S. K., Harsan, U. A., & Daud, W. R. W. (2010). Overview of hybrid membranes for direct methanol fuel cell applications. *International Journal of Hydrogen Energy*, *35*(5), 2160–2175. doi:10.1016/j.ijhydene.2009.12.054

Alberti, G., Casciola, M., & Costantino, U. (1983). Inorganic ion-exchange membranes made of acid salts of tetravalent metals. *Journal of Membrane Science*, *16*, 137–149. doi:10.1016/S0376-7388(00)81305-8

Alberti, G., Casciola, M., & Palombari, R. (2000). Inorganic-Organic Proton Conducting Membranes for Fuel Cells and Sensors at Medium Temperatures. *Journal of Membrane Science*, *172*(1-2), 233–239. doi:10.1016/S0376-7388(00)00332-X

Alwin, S., Bhat, S. D., Sahu, A. K., Jalajakshi, A., Sridgar, P., Sridgar, S., & Shukla, A. K. (2011). Modified-pore-filled-PVDF-membrane electrolytes for direct methanol fuel cells. *Journal of the Electrochemical Society*, *158*(2), B91–B98. doi:10.1149/1.3518774

Amirinejad, M., Madaenia, S. S., & Navarrab, M. A. (2011). Preparation and characterization of phosphotungstic acid-derived salt/Nafion nanocomposite membranes for proton exchange membrane fuel cells. *Journal of Power Sources*, *196*(3), 988–998. doi:10.1016/j.jpowsour.2010.08.062

Arbizzani, C., Donnadia, A., Pica, M., Saganappa, M., Varzi, A., Casciola, M., & Mastragostino, M. (2010). Methanol permeability and performance of Nafion-zirconium phosphate composite membranes in active and passive direct methanol fuel cells. *Journal of Power Sources*, *195*(23), 7751–7756. doi:10.1016/j.jpowsour.2009.07.034

Auimviriyavat, J., Changkhamchom, S., & Sirivat, A. (2011). Development of poly (ether ether ketone) (Peek) with inorganic filler for direct methanol fuel cells (DMFCS). *Industrial & Engineering Chemistry Research*, *50*(22), 12527–12533. doi:10.1021/ie2006005

Bauer, B., Jones, D. J., & Roziere, J. (2000). Electrochemical characterization of sulfonated polyetherketone membranes. *Journal of New Materials for Electrochemical Systems*, *3*, 93–98.

Bayer, T., Bishop, S. R., Nishihara, M., Sasaki, K., & Lyth, S. M. (2014). Characterization of a graphene oxide membrane fuel cell. *Journal of Power Sources*, *272*, 239–247. doi:10.1016/j.jpowsour.2014.08.071

Carter, D., & Wing, J. (2013). The fuel cell industry review 2013. *Fuel Cell Today*, *1*, 1–46.

Cassir, M., Ceballos, A. M., Ringuede, A., & Lair, V. (2016). Molten carbonate fuel cells. *Compendium of Hydrogen Energy*, *3*, 71–87. doi:10.1016/B978-1-78242-363-8.00003-7

Cele, N. P. (2010). Carbon Nanotubes Based NafionComposite Membranes for Fuel Cell Applications. *Fuel Cells (Weinheim)*, *10*, 64–71.

Chae, K. J., Kim, K. Y., Choi, M. J., Yang, E., Kim, S., Ren, X., & Lee, M. (2014). Sulfonated polyether ether ketone (SPEEK)-based composite proton exchange membrane reinforced with nanofibers for microbial electrolysis cells. *Chemical Engineering Journal*, *254*, 393–398. doi:10.1016/j.cej.2014.05.145

Chen, L., Yu, T. L., Lin, H. L., & Yeh, S. H. (2008). Nafion/PTFE and zirconium phosphate modified Nafion/PTFE composite membranes for direct methanol fuel cells. *Journal of Membrane Science*, *307*(1), 10–20. doi:10.1016/j.memsci.2007.03.008

Chen, Y., Guo, J., & Kim, H. (2010). Preparation of poly(vinylidene fluoride)/phosphotungstic acid composite nanofiber membranes by electrospinning for proton conductivity. *Reactive & Functional Polymers*, *70*(1), 69–74. doi:10.1016/j.reactfunctpolym.2009.10.006

Chien, H. C., Tsai, L. D., Huang, C. P., Kang, C. Y., Lin, J. N., & Chang, F. C. (2013). Sulfonated graphene oxide/Nafion composite membranes for high-performance direct methanol fuel cells. *International Journal of Hydrogen Energy*, *38*(31), 13792–13801. doi:10.1016/j.ijhydene.2013.08.036

Cho, K. Y., Jung, H. Y., Choi, N. S., Sung, S. J., Pak, J. K., Choi, J. H., & Sung, Y. E. (2005). A coated Nafion membrane with a PVDF copolymer/Nafion blend for direct methanol fuel cells (DMFCs). *Solid State Ionics*, *176*(39-40), 3027–3030. doi:10.1016/j.ssi.2005.09.048

Cho, K. Y., Jung, H. Y., Choi, N. S., Sung, S. J., Pak, J. K., Choi, J. H., & Sung, Y. E. (2005). A coated Nafion membrane with a PVDF copolymer/Nafion blend for direct methanol fuel cells (DMFCs). *Solid State Ionics*, *176*(39-40), 3027–3030. doi:10.1016/j.ssi.2005.09.048

Clearfield, A. (1988). Role of Ion Exchange in Solid-state Chemistry. *Chemical Reviews*, *88*(1), 125–148. doi:10.1021/cr00083a007

Coutanceau, C., Koffi, R. K., Leger, J. M., Marestin, K., Mercier, R., Nayoze, C., & Capron, P. (2006). Development of materials for mini DMFC working at room temperature for portable applications. *Journal of Power Sources*, *160*(1), 334–339. doi:10.1016/j.jpowsour.2006.01.073

Cui, L., Geng, Q., Gong, C., Liu, H., Zheng, G., Wang, G., ... Wen, S. (2015). Novel sulfonated poly (ether ether ketone)/silica coated carbon nanotubes high-performance composite membranes for direct methanol fuel cell. *Polymers for Advanced Technologies*, *26*(5), 457–464. doi:10.1002/pat.3473

Deluca, N. W., & Elabd, Y. A. (2006). Direct methanol fuel cell performance of Nafion®/poly(vinyl alcohol) blend membranes. *Journal of Power Sources*, *163*(1), 386–391. doi:10.1016/j.jpowsour.2006.09.009

Dimitrova, P., Friedrich, K. A., Stimming, U., & Vogt, B. (2002). Modified Nafion-based membranes for use in direct methanol fuel cells. *Solid State Ionics*, *150*(1-2), 115–122. doi:10.1016/S0167-2738(02)00267-9

Eapen, D. E., Suseendiran, S. R., & Rengaswamy, R. (2016). Phosphoric acid fuel cells. *Compendium of Hydrogen Energy*, *3*, 57–70. doi:10.1016/B978-1-78242-363-8.00002-5

Energy, U. S. D. (2000). *Fuel cell handbook* (5th ed.). Science Applications International Corporation E&G Services Parson Inc.

Erkartal, M., Usta, H., Citir, M., & Sen, U. (2016). Proton conducting poly(vinyl alcohol) (PVA)/ poly (2-acrylamido-2-methylpropane sulfonic acid) (PAMPS)/ zeolitic imidazolate framework (ZIF) ternary composite membrane. *Journal of Membrane Science*, *499*, 156–16. doi:10.1016/j.memsci.2015.10.032

Feng, Y., Han, G., Chung, T. S., Weber, M., Widjojo, N., & Maletzko, C. (2017). Effects of polyethylene glycol on membrane formation and properties of hydrophilic sulfonated polyphenylenesulfone (sPPSU) membranes. *Journal of Membrane Science*, *531*, 27–35. doi:10.1016/j.memsci.2017.02.040

Hatakeyama, K., Razaul, K. M., Ogata, C., Tateishi, H., Taniguchi, T., Koinuma, M., ... Matsumoto, Y. (2014). Optimization of proton conductivity in graphene oxide by filling sulfate ions. *Chemical Communications*, *50*(93), 14527–14530. doi:10.1039/C4CC07273A PMID:25302574

Holmberg, Z. B., Li, W., Wang, X., Munoz, W., & Yan, R. Y. (2006). Nafion/Zeolite Nanocomposite Membrane by in Situ Crystallization for a Direct Methanol Fuel Cell. *Chemistry of Materials*, *18*(24), 5669–5675. doi:10.1021/cm060841q

Huang, F. Y. (2014). *H_2-Cl_2 proton exchange membrane fuel cells, fuel cell assemblies including the same and systems for cogeneration of electricity and HCL.* US Patent no. US9437895 B2.

Izumi, Y., Hisano, K., Hida, T. (1999). Acid catalysis of silica-included heteropolyacid in polar reaction media. *Applied Catalysis, A, 181*, 277-282.

Jung, D. H., Chob, S. Y., Pecka, D. H., Shina, D. R., & Kim, J. S. (2002). Performance evaluation of a Nafion/silicon oxide hybrid membrane for direct methanol fuel cell. *Journal of Power Sources*, *106*(1-2), 173–177. doi:10.1016/S0378-7753(01)01053-9

Kamarudin, S. K., Ahmad, F., & Daud, W. R. W. (2009). Overview on application of direct methanol fuel cell (DMFC) for portable electronic devices. *International Journal of Hydrogen Energy*, *34*(16), 6902–6916. doi:10.1016/j.ijhydene.2009.06.013

Kim, D. J., Jo, M. J., & Nam, S. Y. (2015). A review of polymer–nanocomposite electrolyte membranes for fuel cell application. *Journal of Industrial and Engineering Chemistry*, *21*, 36–52. doi:10.1016/j.jiec.2014.04.030

Kim, D. J., Jom, M. J., & Nam, S. Y. (2015). A review of polymer–nanocomposite electrolyte membranes for fuel cell application. *Journal of Industrial and Engineering Chemistry*, *21*, 36–52. doi:10.1016/j.jiec.2014.04.030

Kraytsberg, A., & Eli, Y. E. (2014). Review of Advanced Materials for Proton Exchange Membrane Fuel Cells. *Energy & Fuels*, *28*(12), 7303–7330. doi:10.1021/ef501977k

Kumar, R., Mamlouk, M., & Scott, K. (2014). Sulfonated polyether ether ketone-sulfonated graphene oxide composite membranes for polymer electrolyte fuel cells. *RSC Advances*, *4*(2), 617–623. doi:10.1039/C3RA42390E

Lee, J. K., Li, W., & Manthiram, A. (2009). Poly(arylene ether sulfone)s containing pendant sulfonic acid groups as membrane materials for direct methanol fuel. *Journal of Membrane Science*, *330*(1-2), 73–79. doi:10.1016/j.memsci.2008.12.043

Lee, S. Y., Woo, H. S., Song, J. M., Sohn, J. Y., & Shin, J. (2016). Preparation of cross-linked SPEEK-zirconium phosphate hybrid membranes by electron beam irradiation. *Polymer Composites, 2016*. doi:10.1002/pc.24147

Li, H., & Nogami, M. (2002). Pore-controlled proton conducting silica films. *Advanced Materials*, *14*(12), 912–914. doi:10.1002/1521-4095(20020618)14:12<912::AID-ADMA912>3.0.CO;2-L

Liang, X., Zhang, F., Feng, W., Zou, X., & Zhu, G. (2013). From metal–organic framework (MOF) to MOF–polymer composite membrane: Enhancement of low-humidity proton conductivity. *Chemical Science (Cambridge)*, *4*(3), 983–992. doi:10.1039/C2SC21927A

Liu, F., Hashim, N. A., Liu, Y., Moghareh Abed, M. R., & Li, K. (2011). Review: Progress in the production and modification of PVDF membranes. *Journal of Membrane Science*, *375*(1-2), 1–27. doi:10.1016/j.memsci.2011.03.014

Lue, S. J., Pai, Y. L., Shih, C. M., Wu, M. C., & Lai, S. M. (2015). Novel bilayer well-aligned Nafion/graphene oxide composite membranes prepared using spin coating method for direct liquid fuel cells. *Journal of Membrane Science*, *493*, 212–223. doi:10.1016/j.memsci.2015.07.007

Lufrano, F., Baglio, V., Stati, P., Antonucci, V., & Arico, A. S. (2013). Performance analysis of polymer electrolyte membranes for direct methanol fuel cells. *Journal of Power Sources*, *243*, 519–534. doi:10.1016/j.jpowsour.2013.05.180

Lufrano, F., Squardrito, G., Patti, A., & Passalacqua, E. (2000). Sulfonated polysulfone as promising membranes for polymer electrolyte fuel cells. *Journal of Applied Polymer Science*, *77*(6), 1250–1257. doi:10.1002/1097-4628(20000808)77:6<1250::AID-APP9>3.0.CO;2-R

Maiti, J., Kakati, N., Lee, S. H., Jee, S. H., & Yoon, Y. S. (2011). PVA nano composite membrane for DMFC application. *Solid State Ionics*, *201*(1), 21–26. doi:10.1016/j.ssi.2011.07.013

Mckeen, J. C., Yan, Y. S., & Davis, M. E. (2008). Proton conductivity of acid-functionalized zeolite beta, MCM-41 and MCM-48: Effect of acid strength. *Chemistry of Materials*, *20*(16), 5122–5124. doi:10.1021/cm801418r

Mioc, U. B., Milonjic, S. K., Malovic, D., Stamenkovic, V., Colomband, P., Mitrovic, M. M., & Dimitrijevic, R. (1997). Structure and proton conductivity of 12-tungstophosphoric acid doped silica. *Solid State Ionics*, *97*(1-4), 239–246. doi:10.1016/S0167-2738(97)00089-1

Nagarale, R. K., Gohil, G. S., Shahi, V. K., & Rangarajan, R. (2004). Organic-inorganic hybrid membrane: Thermally stable cation-exchange membrane prepared by sol-gel method. *Macromolecules*, *37*(26), 10023–10030. doi:10.1021/ma048404p

Nagarale, R. K., Shin, W., & Singh, P. K. (2010). Progress in ionic organic-inorganic composite membranes for fuel cell applications. *Polymer Chemistry*, *1*(4), 388–408. doi:10.1039/B9PY00235A

Neburchilov, V., Martin, J., Wang, H., & Zhang, J. (2007). A review of polymer electrolyte membranes for direct methanol fuel cells. *Journal of Power Sources*, *160*(2), 221–238. doi:10.1016/j.jpowsour.2007.03.044

O'Hayre, R., Cha, S. W., Colella, W., & Prinz, F. B. (2016). *Fuel Cell fundamentals*. John Wiley and Sons. doi:10.1002/9781119191766

Pan, Z. F., Chen, R., An, L., & Li, Y. S. (2017). Alkaline anion exchange membrane fuel cells for cogeneration of electricity and valuable chemicals. *Journal of Power Sources*, *365*, 430–455. doi:10.1016/j.jpowsour.2017.09.013

Pandey, J., Mir, F. Q., & Shukla, A. (2014). Synthesis of silica immobilized phosphotungstic acid (Si-PWA)-poly(vinyl alcohol) (PVA) composite ion-exchange membrane for direct methanol fuel cell. *International Journal of Hydrogen Energy*, *39*(17), 9473–9481. doi:10.1016/j.ijhydene.2014.03.237

Pandey, J., & Tankal, B. R. (2016). Performance of the vanadium redox-flow battery (VRB) for Si PWA/PVA nanocomposite membrane. *Journal of Solid State Electrochemistry*, *8*(8), 2259–2265. doi:10.100710008-016-3244-1

Park, S. G., Chae, K. J., & Lee, M. (2017). A sulfonated poly(arylene ether sulfone)/polyimide nanofiber composite proton exchange membrane for microbial electrolysis cell application under the coexistence of diverse competitive cations and protons. *Journal of Membrane Science*, *540*, 165–173. doi:10.1016/j.memsci.2017.06.048

Peighambardoust, S. J., Rowshanzamir, S., & Amjadi, M. (2010). Review of the proton exchange membranes for fuel cell applications. *International Journal of Hydrogen Energy*, *35*(17), 9349–9384. doi:10.1016/j.ijhydene.2010.05.017

Pulido, J. M. O., Martinez, J. R. C., Stoller, M., & Ferez, A. M. (2017). *Organic/TiO2 Nanocomposite Membranes: Recent Developments, Springer International Publishing AG 2017*. Organic–Inorganic Composite Polymer Electrolyte Membranes; doi:10.1007/978-3-319-52739-02

Qian, W., Shang, Y., Fang, M., Wang, S., Xie, X., Wang, J., ... Mao, Z. (2012). Sulfonated polybenzimidazole/zirconium phosphate composite membrane for high temperature applications. *International Journal of Hydrogen Energy*, *37*(17), 12919–12924. doi:10.1016/j.ijhydene.2012.05.076

Ramani, V., Kunz, H. R., & Fenton, J. M. (2005). Stabilized heteropolyacid/Nafion composite membranes for elevated temperature/low relative humidity PEFC operation. *Electrochimica Acta*, *50*(5), 1181–1187. doi:10.1016/j.electacta.2004.08.015

Ramaswamy, P., Wong, N. E., Benjamin, S. G., & Shimizu, G. K. H. (2015). A water stable magnesium MOF that conducts proton over 10^{-2} Scm^{-1}. *Journey of the American Chemical Society.*, *137*(24), 7640–7643. doi:10.1021/jacs.5b04399 PMID:26053659

Ramaswamy, P., Wong, N. E., & Shimizu, G. K. H. (2014). MOFs as proton conductors- challenges and opportunities. *Chemical Society Reviews*, *43*(16), 5913–5932. doi:10.1039/C4CS00093E PMID:24733639

Rikukawa, M., & Sanui, K. (2000). Proton-conducting polymer electrolyte membranes based on hydrocarbon polymers. *Progress in Polymer Science*, *25*(10), 1463–1502. doi:10.1016/S0079-6700(00)00032-0

Robert, C., Valle, K., & Sanchez, C. (2011). Design and properties of functional hybrid organic–inorganic membranes for fuel cells. *Chemical Society Reviews*, *40*(2), 961–1005. doi:10.1039/c0cs00144a PMID:21218233

Ruffmann, B., Silva, H., & Schulte, B. (2003). Organic/inorganic composite membranes for application in DMFC. *Solid State Ionics*, *162*, 269–275. doi:10.1016/S0167-2738(03)00240-6

Sahu, A. K., Selvarani, S., Pitchumani, G., Sridhar, P., Shukla, A. K., Narayanan, N., ... Chandrakumar, N. (2008). PVA-PSSA Membrane with Interpenetrating Networks and its Methanol Crossover Mitigating Effect in DMFCs. *Journal of the Electrochemical Society*, *155*(7), B686–B695. doi:10.1149/1.2912040

Seepana, M. M., Pandey, J., & Shukla, A. (2017). Design and synthesis of highly stable poly(tetrafluoroethylene-zirconium phosphate (PTFE-ZrP) ion-exchange membrane for vanadium redox flow battery (VRFB). *Ionics*, *23*(6), 1471–1480. doi:10.100711581-016-1967-8

Slade, S., Campbell, S. A., Ralph, T. R., & Walsh, F. C. (2002). Ionic Conductivity of an Extruded Nafion 1100 EW Series of Membranes. *Journal of the Electrochemical Society*, *149*(12), A1556–A1564. doi:10.1149/1.1517281

Smitha, B., Sridhar, S., & Khan, A. A. (2005). Solid polymer electrolyte membrane for fuel cell applications. *Journal of Membrane Science*, *259*(1-2), 10–26. doi:10.1016/j.memsci.2005.01.035

Smitha, B., Sridhar, S., & Khan, A. A. (2005). Solid polymer electrolyte membranes for fuel cell applications-a review. *Journal of Membrane Science*, *259*(1-2), 10–26. doi:10.1016/j.memsci.2005.01.035

Steele, B. C., & Heinzel, H. A. (2001). Materials for fuel cell technologies. *Nature*, *414*(6861), 345–352. doi:10.1038/35104620 PMID:11713541

Strathmann, H. (2004). Ion-Exchange Membrane Separation Processes. *Elsevier Science Technology*, *9*, 10–30.

Szubzda, B., Szmaja, A., Ozimek, M., & Mazurkiewicz, S. (2017). Polymer membranes as separators for supercapacitors. Applied Physics A. *Materials and Processing*, *117*, 1801–1809.

Tang, H., Pan, M., & Jiang, S. P. (2010). One- step synthesized HPW/meso-silica inorganic proton exchange membranes for fuel cells. *Chemical Communications*, *46*(24), 4351–4353. doi:10.1039/c003129a PMID:20464019

Taylor, J. M., Dawson, K. W., & Shimizu, G. K. H. (2013). A Water-Stable Metal–Organic Framework with Highly Acidic Pores for Proton-Conducting Applications. *Journal of the American Chemical Society*, *135*(4), 1193–1196. doi:10.1021/ja310435e PMID:23305324

Tricoli, V., & Nannetti, F. (2003). Zeolite-Nafion composites as ion conducting membrane materials. *Electrochimica Acta*, *48*(18), 2625–2633. doi:10.1016/S0013-4686(03)00306-2

Wainright, J. S., Wang, J. T., Weng, D., Savinell, R. F., & Litt, M. (1995). Acid doped polybenzimidazoles a new polymer electrolyte. *Journal of the Electrochemical Society*, *142*(7), L121–L123. doi:10.1149/1.2044337

Wee, J. H. (2007). Feasibility study on direct methanol fuel cells for laptop, computers based on a cost comparison with lithium ion batteries. *Journal of Power Sources*, *173*(1), 424–436. doi:10.1016/j.jpowsour.2007.04.084

Wu, B., Lin, X., Ge, L., Wu, L., & Xu, T. (2013). A novel route for preparing highly proton conductive membrane materials with metal-organic frameworks. *Chemical Communications*, *49*(2), 143–145. doi:10.1039/C2CC37045J PMID:23108195

Wycisk, R., Chisholm, J., Lee, J., Lin, J., & Pintauro, P. N. (2006). Direct methanol fuel cell membranes from Nafion–polybenzimidazole blends. *Journal of Power Sources*, *163*(1), 9–17. doi:10.1016/j.jpowsour.2005.11.056

Xu, W., Liu, C., Xue, X., Su, Y., Lv, Y., Xing, W., & Lu, T. (2004). New proton exchange membranes based on poly(vinyl alcohol) for DMFCs. *Solid State Ionics*, *171*(1-2), 121–127. doi:10.1016/j.ssi.2004.04.009

Yang, L., Tang, B., & Wu, P. (2015). Metal–organic framework–graphene oxide composites: A facile method to highly improve the proton conductivity of PEMs operated under low humidity. *Journal of Materials Chemistry. A, Materials for Energy and Sustainability*, *3*(31), 15838–15842. doi:10.1039/C5TA03507D

Yeo, Z. Y., Chai, S. P., Zhu, P. W., & Mohamed, A. R. (2014). An overview: Synthesis of thin films/membranes of metal organic frameworks and its gas separation performances. *RSC Advances*, *4*(97), 54322–54334. doi:10.1039/C4RA08884K

Zelenay, P., & Piela, P. (2004). Researchers redefine the DMFC roadmap. *The Fuel Cell Review*, *1*, 17–23.

Zhao, W., Mou, Q., Zhang, X., Shi, J., Sun, S., & Zhao, C. (2013). Preparation and characterization of sulfonated polyethersulfone membranes by a facile approach. *European Polymer Journal*, *49*(3), 738–751. doi:10.1016/j.eurpolymj.2012.11.018

Zhu, Q. L., & Xu, Q. (2014). Metal–organic framework composites. *Chemical Society Reviews*, *43*(16), 5468–5512. doi:10.1039/C3CS60472A PMID:24638055

Zhu, W. Z., & Deevi, S. C. (2003). A review on the status of anode materials for solid oxide fuel cells. *Materials Science and Engineering*, *362*(1-2), 228–239. doi:10.1016/S0921-5093(03)00620-8

APPENDIX

Table 6. List of abbreviations

AFC	Alkaline membrane fuel cell
CNTs	Carbon nanotubes
DMFC	Direct methanol fuel cell
GO	Graphene oxide
HPA	Heteropoly acid
IEC	Ion exchange capacity
IEM	Ion-exchange membrane
IMCs-HPW	Imidazole microcapsules loaded with phosphotungstic acid
MCFC	Molten carbonate fuel cell
MEA	Membrane electrode assembly
MMT	Montmorillonite
OCV	Open circuit voltage
PAFC	Phosphoric acid fuel cell
PBI	Poly(benzimidazole)
PEM	Proton exchange membrane
PEMFC	Polymer electrolyte membrane fuel cell
PES	Poly(ether sulfone)
PI	Poly(imide)
PP	Poly(propylene)
PSA	Particle size analyzer
PSSA	poly(styrene sulfonic acid)
PSU	Poly(sulfone)
PTFE	Poly(tetrafluoroethylene)
PVA	Poly(vinyl alcohol)
PVDF	Poly(vinylidene fluoride)
PWA	Phosphotungstic acid
SEM	Scanning electron microscopy
sGO	sulfonated graphene oxide
Si-PWA	Silica immobilized phosphotungstic acid
SOFC	Solid oxide fuel cell
SPEEK	Sulfonated poly(ether ether ketone)
TEM	Transmission electron microscopy
TEOS	Tetraethoxyorthosilicate
TGA	Thermo gravimetric analysis
TNTs	Titanium nanotubes
XRD	X-ray diffraction
ZrP	Zirconium phosphate

Chapter 14 Toxicocinetic and Mechanisms of Action of Nanoparticles

Abderrezak Khelfi
National Center of Toxicology, Algeria

ABSTRACT

Human exposure to nanoparticles has been dramatically increased in the past 25 years as a result of the rapidly developing field of nanotechnology. Many have recognized the importance of identifying potential effects on human health associated with the manufacture and use of these important technology. Many questions remain unanswered regarding the short- and long-term effect, systemic toxicity, and carcinogenicity. Engineered nanoparticles can be taken up by the human body via inhalation, ingestion, dermal uptake, and injection. They can reach the bloodstream and ultimately affect multiple body organs such as liver and spleen or even transcend the blood-brain barrier. Because of the huge diversity of materials used and the wide range in size of nanoparticles, these effects will vary a lot. Local and systemic adverse effects consist of primarily inflammatory reactions. Other observed effects include generation of reactive oxygen species and subsequent oxidative stress, disruption of proteins, DNA, mitochondria and membrane structures, as well as changes in cell signaling pathways.

INTRODUCTION

Human being has been exposed to nanoparticles (NP) throughout its evolution via inhalation of ultrafine particles (UFP) produced by fires, volcanic action and other natural phenomena. This exposure has been dramatically increased in the past 25 years as a result of the tremendous development of nanotechnology, which has produced large numbers of synthetic nano-sized particles. With nanomaterials (NM) becoming an integral part of the economy and a relevant component of consumer products, many have recognized the importance of prospectively identifying the human health and environmental risks associated with the manufacture and use of these materials. As NM based products enter the market, there is an urgent need of advanced researches in order to prevent dramatic consequences of any health issues caused by nanotechnology-driven products. Entire populations are at high risk for NM exposure due to the wide availability of these new products and many questions remain unanswered regarding the short- and long-

DOI: 10.4018/978-1-5225-5745-6.ch014

term effects of NM on several living organisms. These synthetic particles can be taken up by different routes such as inhalation, ingestion, dermal uptake and injection (Jain, 2017).

Inhalational toxicity is to be expected, given the known effects of inhaled fine particulate matter. It has been proposed that dermal exposure will be the most relevant route of exposure, but there is considerably less literature regarding dermal effects and absorption (Curtis, Greenberg, Kester, Phillips, & Krieger, 2006). The ability of NP to penetrate beyond the epidermal and/or dermal layers is important to consider, along with their subsequent bioaccumulation and systemic effects. Data on NP toxicity often conflict when comparing in vitro and in vivo studies and less defined still are the potential effects of nanoproducts on fetal development and the environment.

There are still many unanswered questions about the fate of engineered NP in human being. They can reach the bloodstream and ultimately affect multiple organs. Biodistribution, movement through tissues, phagocytosis, opsonization and endocytosis of NP are all likely to have an impact on potential toxicity. Because of the huge diversity of materials used and the wide range in size of NP, these effects vary a lot. The main concern is about smaller particles that can easily enter cells. The interaction of NP with biological systems is affected by other factors such as shape, chemical composition, surface chemistry, surface charge and aggregation state. These parameters can modify cellular uptake, protein binding, translocation from portal of entry to the target site and the possibility of causing tissue injury.

One of the most important fact about NP is their remarkable surface reactivity, a characteristic that may result in toxicity effects. Evidence from the literature states that our understanding of NP properties is incomplete and toxicity as well as biodistribution to specific targets occur for NP with particular characteristics. Some NP exhibit the ability to translocate into different type of cells and organs such as liver, spleen, kidneys, etc or even transcend the blood-brain barrier (BBB).

Local and systemic adverse effects consist usually of inflammatory reactions. Some deleterious effects, including inflammatory reactions and pulmonary tissue injury, have been demonstrated following inhalational exposure to nano-sized TiO_2. These inflammatory reactions may result in fibrosis and granulomatous reactions.

Other observed pathological effects include generation of reactive oxygen species (ROS) and subsequent oxidative stress, disruption of proteins, DNA, mitochondria and membrane structures as well as changes in cell signaling pathways such as calcium and cytokine pathways.

To date, we have barely scratched the lessons to learn and to comprehend the effects of these tiny molecules on cells, tissues and whole organisms. It is clear that additional studies using human subjects must be undertaken in order to adequately predict short and long-term effects of NP. This chapter focuses on the current knowledge of nanotoxicity, highlights areas where new information is available and suggests directions for additional and future research. The results of early cytotoxicity studies are reviewed by mechanism pathways.

TOXICOCINETIC OF NANOPARTICLES

Absorption

NP can enter human body and penetrate some important organs such as lungs, intestine and skin. Their penetration depends mainly on two parameters; size and surface characteristics. These parameters usually modify protein binding, cellular uptake, translocation from portal of entry to the target site.

Figure 1. Toxicocinetic of nanoparticles

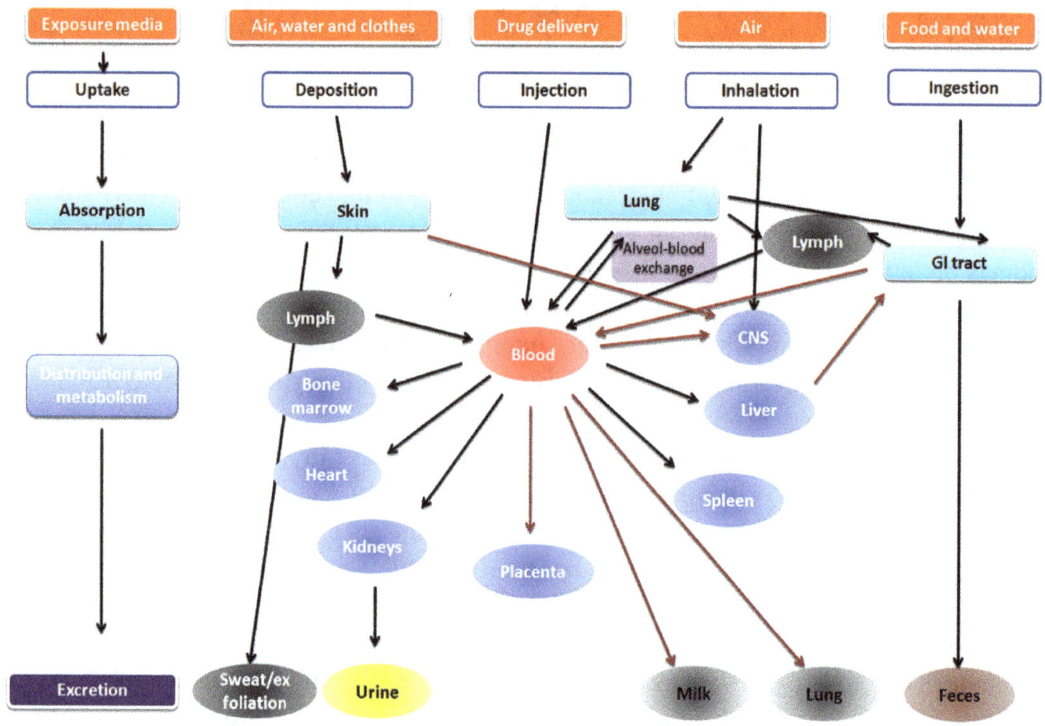

It is known that natural barriers such as skin and intestinal epithelium can be bypassed by NP through injection while other barriers such as BBB and placenta tissue are very selective. As shown in Figure 1 the conceivable paths of uptake and transport of NP are multiple. Absorption efficiency are greatly depending on how NP penetrate the body. The effectiveness of the absorption is different with uptake through the skin as the least effective and injection administration as the most effective route. Although research is still ongoing, most of research studies on dermal absorption of NP confirm that skin cannot be penetrated by particles even when they are in the nanoscale range. According to some studies, neither TiO_2 NP nor quantum dots (QD) are able to reach bloodstream, even if TiO_2 NP were found in the stratum corneum (Beer, 2016). However, NP may arrive to bloodstream if the skin is impaired by wounds, sunburn or skin diseases. In addition, other conditions such as stretching of the skin and follicular openings may influence barrier function of the skin. Overall, there is currently no consensus about the ability of NP to be absorbed through the skin.

Pulmonary system is usually the most exposed to NP. Following penetration, NP can penetrate through different tissue compartments of lungs and eventually reach the capillaries and circulating cells and constituents. Depending on size, NP are deposited more or less deep in the respiratory tract (Figure 2). Usually, particles with a diameter >10 µm are deposited in the upper respiratory tract, particles between 2 and 10 µm can reach trachea, bronchi and bronchioles, and particles <1 µm can reach the alveoli. NP that reach the alveoli are mainly cleared by alveolar macrophages. Besides the size, the charge of NP is important as positively charged particles are more efficiently absorbed than negatively charged or neutral NP (Zhao & Castranova, 2011). If the particles persist they are able to access the pulmonary interstitium either through diffusion or, more likely, transcytosis through the alveolar epithelium. Thereafter, NP can

Figure 2. Penetration of NP by respiratory route

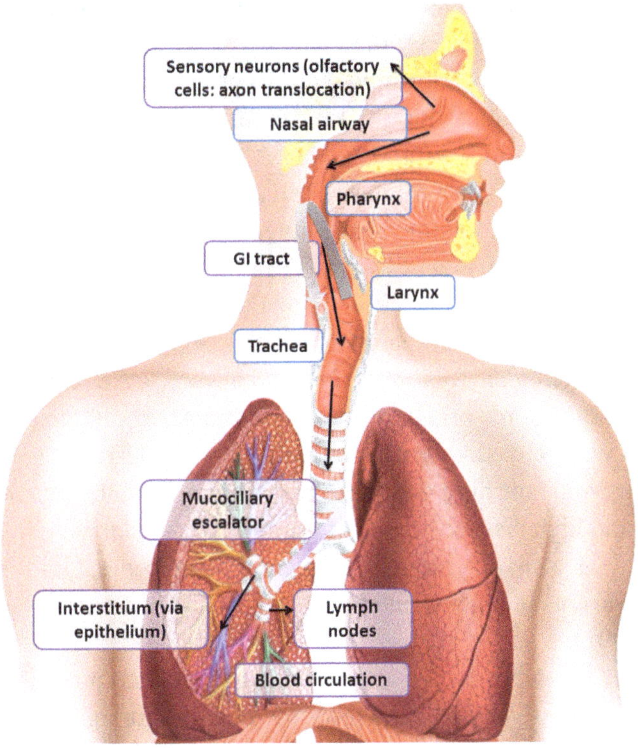

cross the endothelium of the capillary, enter the bloodstream and translocate to different organs. Another way can be done as inhaled NP may be cleared from the lungs via the mucociliary escalator and reach the gastrointestinal tract (GIT).

NP that are introduced into GIT are absorbed by intestine exchanges and more precisely through a process called persorption by paracellular translocation in the epithelial cell layer. It is thought that tight junctions in the mucosa allow NP to be transported through the epithelium and translocate into sub-epithelial regions either through lymph tracts or veins.

Distribution

Once NP has entered bloodstream they will be distributed to various tissues by partitioning and diffusion processes. They may translocate to several organs including liver, bone marrow, spleen, heart, kidneys, brain and lymph nodes, where they may be deposited. Translocation and biodistribution are very dependent on the specific properties of NP as they influence the way NP interact with cells, body fluids and proteins. A general rule is that smaller NP have a greater biodistribution compared to larger NP. Binding of some NP to proteins can influence the biokinetic of distribution and interaction with organs. For example, if these proteins promote cellular uptake of some NP in specific organs or immune cells the biodistribution of these particles might be limited. In one study, a series of NIR fluorescent NP were varied in different parameters such as chemical composition, size, shape and surface charge in rat models after lung instillation (Choi et al., 2010). The obtained data suggested that NP with hydrodynamic diameter <34 nm and non-cationic surface charge NP translocate rapidly from the lung to mediastinal lymph nodes. NP of hydrodynamic diameter <6 nm can traffic rapidly from the lungs to lymph nodes and bloodstream with subsequent clearance by the kidneys.

A biokinetic study was performed to determine the influence of particle size on the tissue distribution of spherical-shaped gold NP in rat model (De Jong et al., 2008). For all tested gold NP, the largest part was demonstrated to be present in liver and spleen. A clear difference was observed between the distribution of 10 nm particles and larger particles. The 10 nm particles were present in various organ systems including blood, liver, spleen, kidney, heart, lung, thymus, testis and brain whereas the larger particles were only detected in blood, liver and spleen. These results demonstrate that tissue distribution of gold NM depends on size with the smallest NP showing the most widespread organ distribution. They may also leak out of capillaries and get into the interstitium.

Smaller particles apparently circulate for longer time and in some cases can cross the BBB to lodge in the brain. In fact, before getting to some organs, organic and inorganic NP have to bypass some biological barriers. Rajagopalan et al. (1996) demonstrated that fullerenes derivatives have the high degree of protein binding in bloodstream and the high volume of distribution probably resulting from the bypass of several biological barriers and high distribution into tissues (Rajagopalan, Wudl, Schinazi, & Boudinot, 1996). Non-organic NP may also cross biological barriers. For example, Gold NP administered orally were widely distributed to various body areas including the brain to an extent inversely proportional to size (Hillyer & Albrecht, 2001).

At the target site, NP may trigger toxic processes leading to tissue injury. They may also enter cells of target organs and remain sequestered into cells for some time before moving to other organs or being excreted. Depending on the chemical composition, NP may remain unchanged, become modified or even metabolized.

One of the most important biokinetic behaviors of NP are cellular endocytosis and transcytosis. Intracellular uptake is affected mainly by surface charges. Cationic NP interact with negatively charged cell membrane components, contributing to increased cellular uptake of these particles compared to anionic once. However, interaction of NP with cell membrane components can disrupt its integrity, contributing to overall toxicity.

Most of the cellular uptake of NP occur via active transport mechanisms into the cell. Various mechanisms of internalization have been demonstrated including clathrin-coated pits, caveolae, receptor-mediated endocytosis, phargocytosis and pinocytosis. These cellular uptake mechanisms are greatly dependent on NP size and surface coating. Overall, large NP are mainly taken up via phagocytosis by mainly macrophages. Smaller particles are primarily processed by endocytic pathways. Caveolae are plentiful in pulmonary capillaries and type I pneumocytes but are sparse in capillaries of the brain (Figure 3).

Because of phagocytosis process, the reticuloendothelial system (e.g., liver and spleen) is the primary site of deposition for several NP by Kupffer cells within liver and macrophages within spleen sinusoids. Opsonization is the process that contributes to the uptake of NP by the liver and several other tissues.

In some cases, surface coating allow NP to bind to specific receptors on cell surface and increase their uptake. This process is used for drug delivery by targeting receptors that are exclusively expressed at the surface of specific cells.

As mentioned before, NP undergo transcytosis across respiratory epithelium into the interstitium. Albumin and lecithin, two components of alveolar lining fluid, may facilitate this process. Within the interstitium, NP may cause local effects or gain access to the circulatory system directly or via lymphatic tracts, providing an alternative systemic access. Another potential mechanism for extrapulmonary

Figure 3: Mechanisms of action of nanoparticles

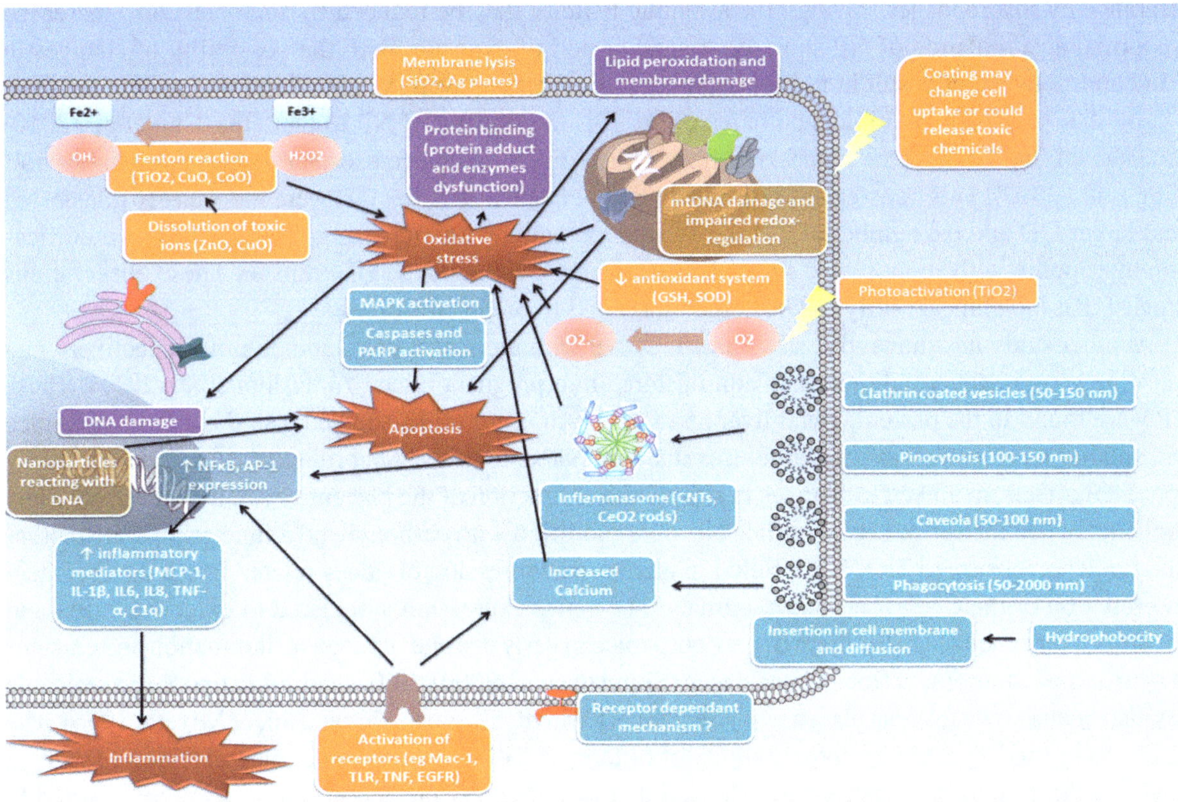

diffusion is uptake by nerves in respiratory epithelium with subsequent spread to the CNS via axonal translocation. Several data suggest that some NP like UFP, manganese containing NP can move from olfactory nerve endings in the nose to the brain (Figure 2) (Oberdorster et al., 2004; Rao et al., 2003). Thus, there are two possible pathways for NP to reach the brain. The first pathway is mediated though the uptake of NP by sensory nerve endings present in the airway epithelia. The other pathway is NP bypass of the BBB via systemic circulation.

The BBB is a specialized system that protects the brain from different xenobiotics. Owing to their special properties, some NP have the capacity to bypass the BBB either by passive diffusion or by carrier-mediated endocytosis and to accumulate in different regions of the brain. NP may produce potential toxicity effects on neural cells and influence brain physiology leading to severe side effects. They may not be easily eliminated from the brain, which results in the extension of their toxic effects. The limited self-regenerative ability of neurons represents a susceptibility factor that increases the harmful effects of some NP on the brain. All these properties suggest that more cautions are needed in industrial applications of NP containing products.

Another important barrier to NP is skin. It's a physical barrier where the epidermis, which contains mostly dead skin cells, plays an important role in the elimination of NP. However, some NP, such as QD, have shown undesirable uptake by skin cells. This nonselective uptake is an important consideration for NP as they could potentially result in cytotoxic and immunogenic effects in the skin. Intracellular uptake of NP may be a mechanism by which NP localize to the avascular epidermis and thus escape the clearance by macrophages. Though the systemic toxicity may be reduced by such behavior, increased epidermal accumulation of NP and a prolonged period of exposure with the possibility of changes in skin conditions, could result in further local adverse effects or even systemic effects.

Another possibility of biological barrier bypass is the transfer of NP to the fetus. Chu et al. (2010) have shown that QD may be transferred from female mice to their fetuses across the placental barrier (Chu et al., 2010). This transfer was size and dose dependant as smaller QD were more easily transferred than larger QD and the number of QD transferred increases with increasing dosage. Surface modification by capping with an inorganic silica shell or organic PEG reduced QD transfer. These observations suggest that the clinical utility of QD could be limited in pregnant women.

Another study has shown that silica and TiO_2 NP with diameters of 70 and 35 nm, respectively, can cause pregnancy abnormalities when administered into pregnant mice (Yamashita et al., 2011). These NP were found in the placenta, fetal liver and fetal brain. The results showed also that the treated mice had smaller uteri as well as smaller fetuses than untreated controls. According to authors, these developmental effects are linked to changes in structure and function of the placenta, which were surprisingly abolished when the surfaces of the silica NP were modified with carboxyl and amine groups. In another study, mice exposure to TiO_2 NP resulted in an altered expression of genes related to the development and function of the CNS of newborn animals. The genes expression associated to oxidative stress and apoptosis were changed in the brain of new born mice in early age and induced inflammation and changes in neurotransmitter production in later stages (Shimizu et al., 2009). C60, as a highly lipophilic molecule has also been shown to cross placenta barrier in rats (Tsuchiya, Oguri, Yamakoshi, & Miyata, 1996). The results suggested a potential fetotoxic ability of fullerenes with harmful developmental effects (Tsuchiya et al., 1996). Finally, an experimental showed that fluorescently labeled porous silicon NP injected to pregnant mice were found deep in brain tissues (80% of cortical depth) of embryos (Yuryev et al., 2016).

Metabolism and Dissolution

NP metabolism has not been very well investigated and understood. According to the available data, NP are predominantly found not to be metabolized, but this process seems to be dependent on the chemical composition of the considered NP. For example, QD seem to have a very long half-life in the body (several weeks or month). In contrast, nanoscaled liposomes are likely to be much easier degraded and metabolized, particularly if they are able to fuse with cell membrane components.

As a general rule, non-metabolized NP might be less excreted and, therefore, accumulate in the body. This might especially be a problem for chronic exposure or repeated administration of nanomedicines or when an interaction with DNA occurs which could result in carcinogenesis.

Dissolution is a key factor in NP transformation in biological systems. In some cases, it may result in toxic effects. For example, some of toxic effects of NP consisting of ZnO, CuO or Ag have been attributed to the released metal ions (Figure 3) (Beer, Foldbjerg, Hayashi, Sutherland, & Autrup, 2012; Mortimer, Kasemets, & Kahru, 2010). The dissolution of NP can occur inside cells usually by the acidic environment of endosomes and lysosomes that are thought to contribute significantly to the degradation of NP and to the release of toxic metal ions (Xia et al., 2008). The important parameters for NP dissolution are size, roughness, coating and aggregation state of NP.

Excretion and Clearance

After being metabolized or dissolved, NP derived compounds might be available for subsequent excretion. The two major routes for excretion are through feces and urine (Figure 1). Biodegradable NM are more easily excreted by the body and does no longer pose a health threat. However, excretion of non-biodegradable NM might take very long time and extend potential effects of NP on biological components.

For the excretion via urine, the blood is filtered in the kidney through the renal glomerula. NP with a size <8 nm can be filtered whereas particles that are larger accumulate in the mononuclear phagocyte system (MPS) around the liver, kidney, spleen and lymph nodes (Baeza-Squiban, Boland, Hussain, & Marano, 2009). This system is particularly well developed in the liver and the macrophages, called Kupffer cells, are responsible for clearance of the biggest part of the particles in human body.

Although there is still a large demand for investigations on the fate of non-biodegradable NP, it seems that particles are either rapidly cleared by the kidney or taken up by the mononuclear phagocyte system and persist in the body. For example, water-soluble single-walled carbon nanotubes (SWCNT) have been shown to be excreted via the renal route, whereas, TiO_2 NP accumulate in the liver and spleen for several weeks (Baeza-Squiban et al., 2009).

Surface charges seem also to play an important role in NP distribution and excretion in biological systems. Positively charged NP are rapidly cleared from the blood and their aggregates accumulate in the liver and lung. Neutral NP are less internalized by macrophages of the liver and spleen. Neutral surface charges increase, therefore, the half-life of NP in the blood and the availability for uptake by other organs.

MECHANISMS OF ACTION OF NANOPARTICLES

The major toxicological pathways activated by NP in cells include oxidative stress, inflammation and genotoxicity. It is know that some engineered NP have also the potential to inhibit cell proliferation and decrease cell viability. Also, some NP have the ability to pass the normal phagocytosis defense without being detected and thereby get access to, for example, blood and the nervous system.

All these effects are due to the unique properties displayed by materials at the nanoscale, which are profoundly dependant on their chemical composition, coating, etc which might interfere with the physiology of the target living cell, leading to unpredictable effects.

From toxicological point of view, recent researches reported that engineered NP may enter into or get in contact with living organisms at realistic doses, but do not cause acute toxic effects. However, their prolonged interaction with living organisms may ultimately disrupt normal activities and lead to malfunction or diseases. The vast majority of studies on nanotoxicity are on short-term effects while long-term effects after chronic exposures are mainly unknown.

Oxidative Stress/Generation of Reactive Oxygen Species

Oxidative stress is one of the most frequent biological events associated with NP toxicity. Among all mechanisms, it causes the major toxic effect on cells. ROS formation can be produced by the following ways:

- ROS formed directly on NP surfaces;

- Released transition metals from NP that act as catalysts for ROS formation;
- Mitochondrial damages induced by NP by disturbing the balance in the respiratory chain;
- Activation of macrophages and neutrophils by NP leading to the excessive production of ROS.

Table 1 lists outcomes of ROS generation, while Figure 3 shows types of toxicity that NP-mediated ROS generation and oxidative stress can produce.

Although the specific mechanisms of action of NP are not completely elucidated, it seems plausible that NP surface is the key factor in ROS formation. Additionally, the mechanisms by which NP generate ROS, both inside and outside the cell, seem to be specific to the type of NP.

It appears that pro-oxidative organic hydrocarbons such as polycyclic aromatic hydrocarbons and quinones as well as transition metals such as copper, vanadium, chromium, cobalt, nickel, titanium and iron are likely to participate in most of these reactions.

Several carbon nanotubes (CNT) production processes require a metal catalyst, often transition metals, resulting in significant concentrations of metal and metal oxide impurities such as nickel and iron. The so-called decorating nickel and iron in CNT products may make up as much as 25% and 40% by weight, respectively. Experiments with biological and non-biological systems demonstrated a significant potential of non-purified SWCNT to generate ROS and free radicals with subsequent accumulation of biomarkers of oxidative stress and cytotoxicity. In fact, some metals are particularly effective as catalysts of oxidative stress and inflammatory responses in different biological systems (cells, tissues and biological fluids) that can be particularly damaging. Metal ions such as Ag^{+}, Cu^{+}, Fe^{2+}, Ni^{2+}, Mn^{2+} and Cr^{5+} can be released from NP, producing intracellular ROS via a Fenton-type reaction (Figure 3).

Pumera et al. (2009) correlated the redox behavior of basic amino-acids, arginine and histidine, in the presence of nickel and iron residual catalyst impurities within the SWCNT. They showed that both

Table 1. Outcomes from nanoparticle effects

Experimental NM Effects	Possible Pathophysiological Outcomes
Oxidative stress and ROS generation	Lipid, protein and DNA injury Apoptosis **Energy Perturbation (mitochondrial injury)**
Inflammation	Oxidative stress Tissue infiltration with inflammatory cells Fibrosis Granulomas
Uptake by reticulo-endothelial system	Sequestration and storage in liver, spleen, lymph nodes Possible organ enlargement and dysfunction
Protein denaturation and degradation	Loss of enzyme activity Autoantigenicity
Nuclear uptake	DNA damage
Uptake in neuronal tissue	Brain and peripheral nervous system injury
Perturbation of phagocytic function	Chronic inflammation Fibrosis Granulomas
Altered cell cycle regulation	Cell cycle arrest
DNA damage and genotoxicity	Mutagenesis and carcinogenesis

nickel and iron participate in the oxidation of arginine, while only the iron component participates in the oxidation of histidine (Pumera, Iwai, & Miyahara, 2009). These findings suggest that metallic impurities in SWCNT can cause oxidative damage to proteins in biological systems (Martin Pumera & Iwai, 2009).

Oxidative species generated during inflammatory response can interact with transition metals and trigger redox-cycling cascades with a remarkable oxidizing potential to deplete endogenous reserves of antioxidant agents and induce oxidative damage to macromolecules. It is known that metal based NP without any protective coating have the catalytical ability to induce major damage in cells. They can generate hydroxyl radicals and hydrogen peroxide. Additionally, some semiconductors, when photoexcited, can create electron-hole pairs that can produce ROS (Michalet et al., 2005).

Large quantities of ROS and free radicals in cells impair normal cell physiological redox-regulated activities and cause cellular damage by interacting with their components (lipids, proteins and DNA) in an uncontrolled way. They may induce lipid peroxidation as well as DNA, RNA and protein oxidation. These effects may cause internal oxidative stress, which gives rise to inflammation, apoptosis and genotoxicity. For example, lipid peroxidation has been shown to be the primary cause of fullerene cytotoxicity (Sayes et al., 2005). Some studies have shown that nano-sized TiO_2 increases lipid peroxidation and production of ROS such as superoxide, hydrogen peroxide, hydroxyl radicals and nitric oxide (Gurr, Wang, Chen, & Jan, 2005; Kang et al., 2008).

Oxidative stress induction seem to be independent to cell type with some NP. For example, CNT have been shown to induce ROS generation in several cell lines including human keratinocytes (Manna et al., 2005; Shvedova et al., 2003).

Generation of oxygen radicals has been observed for both inorganic NP, CNT and lipid-based materials (Rodriguez-Yañez, Muñoz, & Albores, 2013; Sarkar, Ghosh, & Sil, 2014; Zhong et al., 2013). This was shown also for different types of NP and nanofibers from different origins (C60 fullerenes, TiO_2, diesel exhaust particulates, etc) (Krug & Wick, 2011).

As mentioned before, the mechanism for free radical production by metal oxide NP involves active redox cycling at the surface of the particle as well as various pro-oxidant functional groups present on the NP. Other NP induce oxidative stress state via indirect mechanisms such as alteration of mitochondrial function. NP that target mitochondria increase electron leakage from the respiratory chain, which results in mitochondrial membrane damage and depolarization (loss of transmembrane potential leading to mitochondria swelling and apoptosis). This action may also overwhelm antioxidant actions, resulting in apoptosis via internal pathway.

Some studies suggest that ROS generation may be due to the metabolism of NP to reactive intermediates or inflammatory responses involving macrophage release of oxidant radicals (Lowe, Hunter-Ellul, & Wilkerson, 2013). In fact, NP that are engulfed by macrophages can increase the oxidative burst by these cells resulting in enhanced local oxidative stress.

Finally, many studies are using unrealistic high particle concentrations or purely characterized NP. Therefore, definite and especially general conclusions cannot be drawn on ROS generation by NP and toxicity effects at this point.

Interaction With Endogenous Molecules

The interactions of NP with cells, body fluids and proteins play an important role in their biological effects and ability to distribute throughout the body. Understanding these interactions is important for predicting potential toxic effects. NP binding to proteins may generate complexes that are more mobile and able to move in tissue sites that are usually inaccessible. Accelerated protein denaturation or degradation on NP surface may lead to structural and/or functional changes including interference in enzyme function.

It is important to note that NP not only interact with macromolecules but also act as adsorbents for important nutrients. NP can adsorb large amounts of nutrients from cell media and cause cell starvation because of their very large surface areas. In fact, it has been demonstrated that SWCNT, based on their large surface area, decrease micronutrient concentrations of cell culture medium, which is called physisorption. Physisorption is a very weak and specific phenomena that influences a wide range of nutrients such as amino acids and vitamins (folate) (Guo et al., 2008). It has also been demonstrated that SWCNT can adsorb important cofactors of enzymes such as NAD^+ by carboxylic chemical groups. This can potentially lead to disturbances in cell metabolism and energy transfer (Scipioni, Pumera, Boero, Miyahara, & Ohno, 2010).

Additionally, CNT may exert toxicity on cells by phospholipids-NP interactions. This effect may arise from the insertion of CNT in lipid bilayer resulting in its alteration and leading to increased diffusion of low weight molecules. Simulation studies suggest that CNT may also produce membrane pores through which solvated ions can pass (Peter & Hummer, 2005). Experimental studies using liposomes and RAW 264.7 cells confirmed membrane damaging potential of MWCNT as well (Shimizu et al., 2013). The disruption of plasma membrane has been identified in electron microscopy. These NP bind to membrane receptor with collagenous structure (MARCO) and disrupt plasma membrane (Hirano, Fujitani, Furuyama, & Kanno, 2012; Hirano, Kanno, & Furuyama, 2008).

In another study, atomic force microscopy was used to show that aqueous solutions of poly(amidoamine) dendrimers cause the formation of holes of 15-40 nm diameter in intact lipid bilayers (Mecke et al., 2004). In contrast, carboxyl-terminated core-shell tectodendrimer clusters do not create holes in the lipid membrane but instead show a strong affinity to adsorb to the edges of existing bilayer defects.

Apoptosis

Apoptosis is a relevant mode of cell death in response to some NP in in vitro and in vivo models. Poly(ADP)ribose polymerase (PARP) activation was observed in response to MWCNT (Pacurari et al., 2008). Some studies demonstrated that CNT can also stimulate apoptotic cell death pathways involving mitochondria (Figure 3) (Chen et al., 2012; Wang et al., 2012).

The role of necrosis in cell death due to NP exposure has also been suggested. Cell cycle arrest has also been demonstrated after incubation with SWCNT (Cui, Tian, Ozkan, Wang, & Gao, 2005).

Complement system may also play an important role in cell death after exposure to some NP. In fact, some studies has demonstrated that CNT may bind to several proteins in plasma and cells including complement proteins leading to complement system activation (Hamad et al., 2008; Salvador-Morales, Basiuk, Basiuk, Green, & Sim, 2008; Carolina Salvador-Morales et al., 2007).

Inflammation

Inflammatory responses to NP are both common and detrimental to human body. They may be restricted to the site of exposure or may include systemic immune effects. The effects of NP on immune systems may occur in different organs such as lungs, liver, heart and brain. Inflammation is initiated by a number of pathways including cell surface receptor engagement by NP and immune cell activation following internalization (Figure 3).

If not bound by endogenous antioxidants or degraded by the action of antioxidative enzymes, free radicals produced by NP can also, via oxidative stress, trigger inflammatory reactions. Alternatively, the activation of intracellular signaling proteins like NFκB, peroxisome proliferator-activated receptor as well as extracellular signaling molecules like cytokines, PDGF and TGF could stimulate downstream signaling pathways resulting in oxidative stress, inflammatory response, chemotaxis, fibrosis and cell death at the site of NP exposure (Godugu, Singh, & Poduri, 2015). Again, the modification of size, surface charge and surface functionalization groups directs the body's immune response to NP and provides the potential to induce inflammatory reaction.

Different forms of inflammation can be triggered after NP exposure. For example, after instillation or inhalation of high doses of CNT or TiO_2, fibrosis and bronchial granulomas were observed to form in test animals and to strongly affect lung function (Krug & Wick, 2011).

An unusual property of immune response is that some type of CNT are not effectively recognized by macrophages. The lack of macrophage recognition is typical of non-functionalized CNT. Their chemical modifications are consistently associated with increased recognition by macrophages. Poor engulfment of non-functionalized SWCNT by macrophages probably results from this lack of recognition of signals on their surfaces.

Different mechanism can be triggered if CNT are internalized inside cells with the induction of several stress related genes. The activation of transcription factors like NFκB and AP-1 by these NP lead to high production of inflammatory cytokines. Incubation of SWCNT and MWCNT with mesothelial cells showed an increase in activation of ERK, JNK and p38 (Pacurari et al., 2008; Pacurari et al., 2008). Further, AP-1, NFκB and Akt activation was also observed after incubation of cells with SWCNT (Manna et al., 2005; Sato et al., 2005). The activation of NFκB and ROS generation lead usually to the production of cytokines. The activation of protein kinases may also lead to the activation of NFκB and could account for CNT toxicity (Hirano et al., 2008).

Surface charge of NP is a key a factor playing an important role in cytotoxic and inflammatory effects of NP, likely via influencing cellular localization or due to variations in tissue receptor sensitivity to particle charge leading therefore to changes in cytokine production. For example, carboxylic acid-coated QD induce the release of the pro-inflammatory cytokines IL-1β, IL-6 and IL-8. Cell viability has been shown to be reduced following exposure to these kind of NP (Ryman-Rasmussen, Riviere, & Monteiro-Riviere, 2007). Liposomal charged NP may also induce inflammation. Both negatively and positively charged liposomes enhance the release of IL-1β from THP-1 cells with no effects in case of neutral liposome exposure (Zhong et al., 2013). The charged liposomes did not increase TNF-α release, suggesting that they did not activate the NFκB pathway. The large surface with charged groups of some NP allows for more interactions with tissue receptors such as those found in the plasma membranes of brain microglia or respiratory epithelial cells. As an example, negatively charged airborne NP activate the TRPV1 vanilloid receptor found on these cells, which stimulates pro-inflammatory cytokine release (Verones & Oortgiesen, 2001).

Another important contributor to inflammation is the binding of serum proteins to NP surface. Once bound to NP, different proteins can unfold to reveal epitopes that are normally hidden within the protein structure. These NP-protein complexes are recognized by cell-surface receptors such as Mac-1, which induce through NFκB the release of pro-inflammatory cytokines (Figure 3). Alternatively, NP can activate intracellular inflammasomes following internalization, which leads to the activation of NFκB and the production of caspases that cleave pro-IL-1β and pro-IL-8 to functional inflammatory cytokines IL-1β and IL-8 (Godugu et al., 2015). Inflammasome activation has been reported for numerous NP including amino-functionalized polystyrenes, silica, Ag NP and CNT (Lunov et al., 2011; Sandberg et al., 2012; Sun et al., 2015; Yang, Kim, Kim, & Choi, 2012). Cathepsins released from the endosomal compartment seem to be a major contributor to inflammasome activation and may be central to NP-induced cytotoxicity (Mortimer & Minchin, 2017). A recent study showed that inflammasome activation is dependent on size and surface charge with cationic NP demonstrating the greatest propensity to release IL-1β in mouse bone marrow dendritic cells and human mononuclear cells (Neumann et al., 2014).

Genotoxicity

One of the most hazardous areas governing regulatory health risk assessment of NP is genotoxicity as DNA damage may initiate and promote carcinogenesis.

Genotoxicity potential of NP varies highly among different types of NP and many reports are available on the negative genotoxic potential of various types of NP (Barnes et al., 2008; He et al., 2009; Shinohara, Matsumoto, Endoh, Maru, & Nakanishi, 2009).

Mechanisms of potential genotoxic effects of NP can be diverse and their elucidation can be demanding. Overall, NP can induce genotoxic effects by either direct interaction with genetic material or by indirect action through ROS generation or transition metal release from NP core.

In some studies, engineered NP were found to cause DNA damages such as chromosomal fragmentation, single and double strand breaks, oxidative adducts, structural changes of the DNA (point mutations, deletions, duplications, inversion and translocation of chromosome segments), changes in the number of chromosomes as well as alterations in gene expression profiles.

NP get inside cells through different mechanisms including passive diffusion, receptor-mediated endocytosis and clathrinor caveolin-mediated endocytosis (Figure 3). Internalization through these mechanisms often culminates with lysosome compartmentalization. It is known that while biodegradable NM may not interfere with the functional activities of such machinery, non-biodegradable NM can dramatically perturb vesicle trafficking and lysosomal fate, resulting in severe cellular dysfunctions. Some NM escape the endosomal trafficking and continue their way into the cytoplasm in cell organelles and eventually the nucleus, affecting therefore DNA integrity. The most relevant example is SWCNT that can cross cell membranes in fibroblasts and enter the nucleus. Nuclear internalization may occur through diffusion across the nuclear membrane, transport through the nuclear pores or by accidental recruitment during mitosis and meiosis. Several studies report that NP with a diameter between 1 and 10 nm are able to enter the nuclear pores, while those presenting larger sizes (16-60 nm) can access DNA only when the nuclear envelope is disrupted for example during cell division or meiosis.

The molecular mechanisms activated in the nucleus in response to nanodeposits are modification in DNA metabolism, DNA structural damages, DNA transcription changes as well as epigenetic modification. Disruption of essential nuclear proteins may occur after high nuclear deposition of NP and may lead to genotoxic outcomes.

Though SiO_2, CNT, Ag NP, ZnO NP and TiO_2 NP can all reach the nucleus they do not induce similar genotoxic potentials. Depending on the chemical composition different effects can be observed. Size dependent entry of SiO_2 NP into the nucleus and disturbance of nuclear organization indicates the direct effect of these NP on the nucleus (Chen & von Mikecz, 2005). Strong binding of CNT to DNA with potential DNA aggregation in a dose-dependent manner was observed, which may explain the altered gene expressions upon NP exposure reported in various studies (An, Liu, Ji, & Jin, 2010). Chromosomal aberrations, mutagenesis and DNA strand breaks have been reported for Ag NP, CNT, fullerenes and most metal oxide NP both in in vitro and in vivo studies (Magdolenova et al., 2014; Patlolla, Hackett, & Tchounwou, 2015; Zhang, Xu, Zheng, Su, & Wang, 2015). TiO_2 NP form usually aggregates included in large vesicles, which are fused to the nucleus and modifying its form. Subsequent alteration of the nucleus shape can impair correct segregation of chromosomes during mitosis (Tino, Ambrosone, Marchesano, & Tortiglione, 2014). Of significant interest, cellular internalization of ultra-fine superparamagnetic iron-oxide NM (d-USPIONs) was shown to correlate with the induction of genotoxicity with large chromosomal damage. The induced chromosomal damage was primarily fragmentation, thus indicating a clastogenic mode of action of these materials. Furthermore, d-USPIONs have the potential to cause other DNA damage that could lead to the initiation and progression of cancer (Singh, 2009).

According to the phase of cell cycle, nanodeposits in the nucleus affect different processes. During the interphase, NP can interfere with replication of DNA and/or DNA transcription by binding to single-strand DNA or by incorporation into DNA duplexes. When cells undergo mitosis NP may disrupt chromosome structure by inducing double strand breaks and altering the ability to go through cell division. An et al. (2010) demonstrated a direct interaction between CNT and DNA in *E. coli* (An et al., 2010). The authors provided evidence that CNT bind to single-stranded DNA and incorporate into DNA duplex structures, probably during DNA replication, suggesting that these NP could represent a serious hazard for DNA replication processes.

Besides the structural and functional effects of NP on DNA, aneugenic effects with the impairment of the machinery responsible for chromosome segregation (mitotic spindle, centrioles and associated proteins tubulin polymerization) have also been found to be induced by some NP. In vitro studies demonstrated that TiO_2 NP can induce multipolar spindle formation and impairment of chromosome alignment and segregation (Gonzalez et al., 2010; Huang, Chueh, Lin, Shih, & Chuang, 2009).

Two of the most frequently reported NP-associated toxicities that may lead to dramatic change in biological activities are ROS generation and inflammation. NP induced oxidative stress play pivotal role in further pathophysiological effects including genotoxicity. For example, SiO_2 NP and TiO_2 NP that are able to generate ROS can introduce DNA mutations leading to purine and pyrimidine lesions, and also induce strand breaks in the DNA. QD have been reported to induce free-radical mediated DNA nicking, while QD photosensitized TiO_2 films were shown to induce free-radical mediated damage to bacteria and DNA (Green & Howman, 2005; Lu et al., 2005).

Another source of ROS production is the endogenous disruption of mitochondria that can suffer the presence of NP that results from the interruption of the respiratory chain and the release of large amounts of ROS. Overproduction of ROS and interruption of energy generation may also cause DNA damage. Depletion of endogenous antioxidant system (glutathione and superoxide dismutase) can also underline NP genotoxicity (Ambrosi & Pumera, 2010; Barillet et al., 2010; Folkmann et al., 2009; Park et al., 2008).

There are mixed results in the available studies to determine the genotoxicity of fullerenes derivatives. There seems to be low potential for genotoxicity and carcinogenicity based on a limited number of studies in lower order organisms such as *Drosophila* and *Salmonella* (Babynin, Nuretdinov, Gubskaya, & Barabanshchikov, 2002; Zakharenko et al., 1997). In vitro analysis of *Salmonella* model indicated DNA damage by a mechanism related to excessive generation of free radicals (Sera, Tokiwa, & Miyata, 1996). Studies on mouse embryonic stem cells with MWCNT demonstrated the potential of genotoxicity induction and indicated the need of careful examination of this effect (Zhu, Chang, Dai, & Hong, 2007). Both in vitro and in vivo tests indicated the positive genotoxicity of MWCNT in type II pneumocytes and rat lung epithelial cells exposed to MWCNT. Further, fluorescent in situ hybridization assay evidenced clastogenic and aneugenic effects (Muller et al., 2008).

Dhawan et al. (2006) demonstrated a positive response with strong correlation between the genotoxic response and C60 concentration (Dhawan et al., 2006). The genotoxic potential of SWCNT on lung fibroblast cells by comet assay and micronucleus test indicated also a positive genotoxic response (Kisin et al., 2007).

For inorganic NP, human epidermal cells exposed to ZnO NP demonstrated dose-dependent genotoxic effects with significant levels of DNA damage. Additional effects following ZnO NP exposure include enhanced lipid peroxidation, glutathione depletion and decreased catalase and superoxide dismutase activity (Sharma et al., 2009).

In one study treating the mice with TiO_2 NP in drinking water resulted in the significant DNA damage (Godugu et al., 2015). Another study demonstrated that exposure to nano-sized TiO_2 disturbs cell cycle progression and duplicated genome segregation, leading to chromosomal instability and cell transformation (Huang et al., 2009). These findings raise concern about potential health hazards associated with TiO_2 NP exposure. Several other studies conducted on murine models have earned TiO_2 a group 2B (possible human carcinogen) by the International Agency for Research on Cancer, although the carcinogenic risk of nano-sized TiO_2 in humans has not been exhibited yet (Baan et al., 2006; "Carbon black, titanium dioxide, and talc," 2010; Dankovic, Kuempel, & Wheeler, 2007).

In contrast, other studies reported the lack of genotoxicity of TiO_2 NP (Hackenberg et al., 2011). Epidemiological studies on workers that were exposed to TiO_2 NP were inconclusive and could not show an association between exposure to these particles and an increased cancer risk (Ng, Li, Bay, & Yung, 2010).

As reported here, discordance amongst studies exists surrounding the capability of nano-sized TiO_2 to induce DNA damage and genotoxicity. These inconsistencies may be explained by the use of different cell lines, concentrations of TiO_2 NP and/or exposure times. Different cell lines have varying thresholds for DNA repair and upregulation of antioxidant or other adaptive mechanisms. They likely become overwhelmed at differing concentrations and exposure times, which would account for the variations in observed DNA damage and genotoxicity.

Cobalt-chromium NP have been shown to damage human fibroblast cells (Bhabra et al., 2009). According to Bhabra et al. (2009) the damage is mediated by a novel mechanism involving transmission of purine nucleotides (such as ATP) and intercellular signaling through connexin gap junctions or hemichannels and pannexin channels. The outcome, which includes DNA damage without significant cell death is different from that observed in cells subjected to direct exposure to NP.

Finally, coating technology plays an important role in NP ability to induce genotoxic effects. For example, coated Ag NP exhibited more severe damage than uncoated NP (Ahamed et al., 2008). PEG-coated and carboxylic acid-coated QD exhibit nuclear localization, while other QD exhibit localization within the cytoplasm. QD coated with biotin have been shown to nick supercoiled DNA in vitro (Green & Howman, 2005).

Overall, there is limited information about NP genotoxicity, considering the large amount of different types of NP that has been developed. The results are conflicting, often due to limited information available on the physicochemical properties of the investigated NP or variations in the experimental settings hardly to compare. Due to inconsistencies in the literature, it is difficult to draw conclusions on the features of NP that promote genotoxicity. Thus, further attention to improve our understanding of the genotoxic potential of engineered NP is needed.

CONCLUSION

Advances in nanotechnology are behind the introduction of new products. The existing nanotoxicology studies have generally lagged behind these advances, but there is plenty of room in the area of nanotoxicology to create new nanotechnological products truly safe.

NP can enter the human body *via* different routes (skin, respiratory gastrointestinal tract, or parenteral administration). They have greater potential to travel through the organism than other materials of larger size. The translocation to the blood circulation allows the interaction between NP and biological elements. The various interactions and toxic effects of NP with fluids, cells, and tissues depend mainly on NP components and need to be considered (Table 2).

The toxic effects following exposure predominantly consist of inflammatory reactions. Tissue damage appears to be mediated by ROS generation and subsequent oxidative stress. The cytotoxic, genotoxic, and carcinogenic potential of NP is less apparent.

Table 2. Effects of NP on biological components

Nanomaterial	Cytotoxicity Mechanism
TiO2	Oxidative stress: ROS generation and glutathione depletion as a result of photoactivity and redox properties Cell membrane disruption leading to cell death Possible DNA damage Pulmonary inflammation and lung tissue injuries Decrease of cell viability
Zn/ZnO	ROS generation Dissolution and release of toxic metals Lysosomal damage Inflammation
Ag	Dissolution and release of metals that inhibit respiratory enzymes and ATP production ROS generation Cell membrane disruption and transport processes leading to cell death
Au	Disruption of protein conformation ROS generation
CdSe	Dissolution and release of toxic Cd ions
SiO2	ROS generation Protein unfolding Membrane disruption Toxic to fibroblast cells Amorphous are more toxic than crystalline SiO_2
Fe3O4	ROS generation Release of toxic metals Induction of cell autophagy
CeO2	ROS generation Protein aggregation Increased MDA levels Decrease in glutathione and α-tocopherol Bacterial cytotoxicity Adsorption onto bacterial outer membrane Decrease of cell viability

continued on following page

Table 2. Continued

Nanomaterial	Cytotoxicity Mechanism
SWCNT and MWCNT	Frustrated phagocytosis causes chronic tissue inflammation ROS generation due to the metal impurities (carbonyl and MDA generation, protein thiol reduction, etc) Oxidative damage of DNA Reduction of antioxidant system (glutathione depletion) Skin thickening Inflammation, fibrosis, and granuloma in lungs: -Granulomatous inflammation due to hydrophobic CNT aggregation -Interstitial pulmonary fibrosis due to fibroblast-mediated collagen production
Asbestos	Inflammation and fibrosis of lungs Mesothelioma
Fullerenes	ROS generation
Cationic nanospheres and dendrimers	Membrane damage, thinning, and leakage Damage to endosomal compartment
Co, Ni, and magnetic metallic NP	Liberation of toxic cations
Al2O3	ROS generation Proinflammatory reaction
Cu/CuO	ROS generation DNA damage
MoO3	Membrane disruption
Quantum dot	Cytotoxic with UV irradiation (ROS and radical generation) Membrane, mitochondrion, and Nucleus damage leading to cell death

The importance of studying the interactions of NP with biological systems and their resultant consequences on human health has attracted increasing attention over the last years. Understanding of the molecular and cellular mechanisms involved in NP toxicity, as well as the availability of sensitive methods for studying such interactions, will be of great help in preventing potential toxic effects of NP.

Three principal research directions can be foreseen. First, comparative testing should be done on a wide spectrum of cell cultures and organisms in order to have a broader base for comparison. Second, studies on the molecular mechanisms of nanotoxicity should be done because only such studies can provide detailed information on how these materials act and how to prevent their toxic effects. Third, the studies of NP-biological system interaction requires the development of new methods and techniques for monitoring NP in all steps of interaction, involving internalization mechanisms, migration within the cell and interaction with different organelles, penetration in the nucleus, translocation among tissues and organs, and the elimination of the nanoparticle.

Alternate assay methodologies need to be investigated to determine further the cytotoxic potential of NP. The application of genomics technologies can shed useful information on the mechanism of toxicity. The use of analytical chemistry along with the application of high-throughput biological assay procedures can help develop non-toxic NP. This, in conjunction with *in silico* studies, may be helpful to identify NP with desired physicochemical characteristics and biocompatibility.

Additionally, monitoring of the biodistribution, biotransformation, and excretion of NP is required, and long-term effects of exposure to NP must be more assessed. Therefore, strategies to extrapolate acute toxicity models for the prediction of chronic effects are thus needed using exposure levels relevant to human and environmental exposure.

Finally, scientist from different fields can play an important role in recognizing the hazards of NP so that health risks are reduced as the opportunities of nanotechnology are realized.

REFERENCES

Ahamed, M., Karns, M., Goodson, M., Rowe, J., Hussain, S. M., Schlager, J. J., & Hong, Y. (2008). DNA damage response to different surface chemistry of silver nanoparticles in mammalian cells. *Toxicology and Applied Pharmacology*, *233*(3), 404–410. doi:10.1016/j.taap.2008.09.015 PMID:18930072

Ambrosi, A., & Pumera, M. (2010). Regulatory Peptides Are Susceptible to Oxidation by Metallic Impurities within Carbon Nanotubes. *Chemistry (Weinheim an der Bergstrasse, Germany)*, *16*(6), 1786–1792. doi:10.1002/chem.200902534 PMID:20066697

An, H., Liu, Q., Ji, Q., & Jin, B. (2010). DNA binding and aggregation by carbon nanoparticles. *Biochemical and Biophysical Research Communications*, *393*(4), 571–576. doi:10.1016/j.bbrc.2010.02.006 PMID:20156419

Baan, R., Straif, K., Grosse, Y., Secretan, B., El Ghissassi, F., & Cogliano, V. (2006). Carcinogenicity of carbon black, titanium dioxide, and talc. *The Lancet Oncology*, *7*(4), 295–296. doi:10.1016/S1470-2045(06)70651-9 PMID:16598890

Babynin, E. V., Nuretdinov, I. A., Gubskaya, V. P., & Barabanshchikov, B. I. (2002). Study of Mutagenic Activity of Fullerene and Some of Its Derivatives Using His+ Reversions of Salmonella typhimurium as an Example. *Russian Journal of Genetics*, *38*(4), 359–363. doi:10.1023/A:1015237916596 PMID:12018160

Baeza-Squiban, A., Boland, S., Hussain, S., & Marano, F. (2009). *Health Effects of Nanoparticles. In General, Applied and Systems Toxicology*. John Wiley & Sons, Ltd.

Barillet, S., Jugan, M. L., Laye, M., Leconte, Y., Herlin-Boime, N., Reynaud, C., & Carriere, M. (2010). In vitro evaluation of SiC nanoparticles impact on A549 pulmonary cells: Cyto-, genotoxicity and oxidative stress. *Toxicology Letters*, *198*(3), 324–330. doi:10.1016/j.toxlet.2010.07.009 PMID:20655996

Barnes, C. A., Elsaesser, A., Arkusz, J., Smok, A., Palus, J., Lesniak, A., ... Howard, C. V. (2008). Reproducible comet assay of amorphous silica nanoparticles detects no genotoxicity. *Nano Letters*, *8*(9), 3069–3074. doi:10.1021/nl801661w PMID:18698730

Beer, C. (2016). Nanotoxicology and Regulatory Affairs. In K. A. Howard, T. Vorup-Jensen, & D. Peer (Eds.), *Nanomedicine: Nanotechnology, Biology, and Medicine* (pp. 279–310). New York, NY: Springer New York.

Beer, C., Foldbjerg, R., Hayashi, Y., Sutherland, D. S., & Autrup, H. (2012). Toxicity of silver nanoparticles - nanoparticle or silver ion? *Toxicology Letters*, *208*(3), 286–292. doi:10.1016/j.toxlet.2011.11.002 PMID:22101214

Bhabra, G., Sood, A., Fisher, B., Cartwright, L., Saunders, M., Evans, W. H., ... Case, C. P. (2009). Nanoparticles can cause DNA damage across a cellular barrier. *Nat Nano*, *4*(12), 876–883. doi:10.1038/nnano.2009.313 PMID:19893513

Carbon black, titanium dioxide, and talc. (2010). *IARC Monographs on the Evaluation of Carcinogenic Risks to Humans, 93*, 1-413.

Chen, M., & von Mikecz, A. (2005). Formation of nucleoplasmic protein aggregates impairs nuclear function in response to SiO2 nanoparticles. *Experimental Cell Research*, *305*(1), 51–62. doi:10.1016/j.yexcr.2004.12.021 PMID:15777787

Chen, T., Zang, J., Wang, H., Nie, H., Wang, X., Shen, Z., ... Jia, G. (2012). Water-soluble taurine-functionalized multi-walled carbon nanotubes induce less damage to mitochondria of RAW 264.7 cells. *Journal of Nanoscience and Nanotechnology*, *12*(10), 8008–8016. doi:10.1166/jnn.2012.6593 PMID:23421171

Choi, H. S., Ashitate, Y., Lee, J. H., Kim, S. H., Matsui, A., Insin, N., . . . Tsuda, A. (2010). *Rapid translocation of nanoparticles from the lung airspaces to the body*. Retrieved from https://www.nature.com/articles/nbt.1696#supplementary-information

Chu, M., Wu, Q., Yang, H., Yuan, R., Hou, S., Yang, Y., ... Sheng, L. (2010). Transfer of quantum dots from pregnant mice to pups across the placental barrier. *Small*, *6*(5), 670–678. doi:10.1002mll.200902049 PMID:20143348

Cui, D., Tian, F., Ozkan, C. S., Wang, M., & Gao, H. (2005). Effect of single wall carbon nanotubes on human HEK293 cells. *Toxicology Letters*, *155*(1), 73–85. doi:10.1016/j.toxlet.2004.08.015 PMID:15585362

Curtis, J., Greenberg, M., Kester, J., Phillips, S., & Krieger, G. (2006). Nanotechnology and nanotoxicology: A primer for clinicians. *Toxicological Reviews*, *25*(4), 245–260. doi:10.2165/00139709-200625040-00005 PMID:17288496

Dankovic, D., Kuempel, E., & Wheeler, M. (2007). An approach to risk assessment for TiO2. *Inhalation Toxicology*, *19*(sup1Suppl 1), 205–212. doi:10.1080/08958370701497754 PMID:17886069

De Jong, W. H., Hagens, W. I., Krystek, P., Burger, M. C., Sips, A. J., & Geertsma, R. E. (2008). Particle size-dependent organ distribution of gold nanoparticles after intravenous administration. *Biomaterials*, *29*(12), 1912–1919. doi:10.1016/j.biomaterials.2007.12.037 PMID:18242692

Dhawan, A., Taurozzi, J. S., Pandey, A. K., Shan, W., Miller, S. M., Hashsham, S. A., & Tarabara, V. V. (2006). Stable colloidal dispersions of C60 fullerenes in water: Evidence for genotoxicity. *Environmental Science & Technology*, *40*(23), 7394–7401. doi:10.1021/es0609708 PMID:17180994

Folkmann, J. K., Risom, L., Jacobsen, N. R., Wallin, H., Loft, S., & Moller, P. (2009). Oxidatively damaged DNA in rats exposed by oral gavage to C60 fullerenes and single-walled carbon nanotubes. *Environmental Health Perspectives*, *117*(5), 703–708. doi:10.1289/ehp.11922 PMID:19479010

Godugu, C., Singh, R. P., & Poduri, R. (2015). Nanotoxicology: Contemporary Issues and Future Directions. In P. V. Devarajan & S. Jain (Eds.), *Targeted Drug Delivery: Concepts and Design* (pp. 733–781). Cham: Springer International Publishing.

Gonzalez, L., Thomassen, L. C., Plas, G., Rabolli, V., Napierska, D., Decordier, I., ... Kirsch-Volders, M. (2010). Exploring the aneugenic and clastogenic potential in the nanosize range: A549 human lung carcinoma cells and amorphous monodisperse silica nanoparticles as models. *Nanotoxicology*, *4*(4), 382–395. doi:10.3109/17435390.2010.501913 PMID:20925446

Green, M., & Howman, E. (2005). Semiconductor quantum dots and free radical induced DNA nicking. Chemical Communications, (1), 121-123.

Guo, L., Von Dem Bussche, A., Buechner, M., Yan, A., Kane, A. B., & Hurt, R. H. (2008). Adsorption of essential micronutrients by carbon nanotubes and the implications for nanotoxicity testing. *Small*, *4*(6), 721–727. doi:10.1002mll.200700754 PMID:18504717

Gurr, J. R., Wang, A. S., Chen, C. H., & Jan, K. Y. (2005). Ultrafine titanium dioxide particles in the absence of photoactivation can induce oxidative damage to human bronchial epithelial cells. *Toxicology*, *213*(1-2), 66–73. doi:10.1016/j.tox.2005.05.007 PMID:15970370

Hackenberg, S., Friehs, G., Kessler, M., Froelich, K., Ginzkey, C., Koehler, C., ... Kleinsasser, N. (2011). Nanosized titanium dioxide particles do not induce DNA damage in human peripheral blood lymphocytes. *Environmental and Molecular Mutagenesis*, *52*(4), 264–268. doi:10.1002/em.20615 PMID:20740634

Hamad, I., Christy Hunter, A., Rutt, K. J., Liu, Z., Dai, H., & Moein Moghimi, S. (2008). Complement activation by PEGylated single-walled carbon nanotubes is independent of C1q and alternative pathway turnover. *Molecular Immunology*, *45*(14), 3797–3803. doi:10.1016/j.molimm.2008.05.020 PMID:18602161

He, L., Yang, L., Zhang, Z. R., Gong, T., Deng, L., Gu, Z., & Sun, X. (2009). In vitro evaluation of the genotoxicity of a family of novel MeO-PEG-poly(D,L-lactic-co-glycolic acid)-PEG-OMe triblock copolymer and PLGA nanoparticles. *Nanotechnology*, *20*(45), 455102. doi:10.1088/0957-4484/20/45/455102 PMID:19822931

Hillyer, J. F., & Albrecht, R. M. (2001). Gastrointestinal persorption and tissue distribution of differently sized colloidal gold nanoparticles. *Journal of Pharmaceutical Sciences*, *90*(12), 1927–1936. doi:10.1002/jps.1143 PMID:11745751

Hirano, S., Fujitani, Y., Furuyama, A., & Kanno, S. (2012). Macrophage receptor with collagenous structure (MARCO) is a dynamic adhesive molecule that enhances uptake of carbon nanotubes by CHO-K1 cells. *Toxicology and Applied Pharmacology*, *259*(1), 96–103. doi:10.1016/j.taap.2011.12.012 PMID:22209804

Hirano, S., Kanno, S., & Furuyama, A. (2008). Multi-walled carbon nanotubes injure the plasma membrane of macrophages. *Toxicology and Applied Pharmacology*, *232*(2), 244–251. doi:10.1016/j.taap.2008.06.016 PMID:18655803

Huang, S., Chueh, P. J., Lin, Y. W., Shih, T. S., & Chuang, S. M. (2009). Disturbed mitotic progression and genome segregation are involved in cell transformation mediated by nano-TiO2 long-term exposure. *Toxicology and Applied Pharmacology*, *241*(2), 182–194. doi:10.1016/j.taap.2009.08.013 PMID:19695278

Jain, K. K. (2017). Nanotoxicology. In K. K. Jain (Ed.), *The Handbook of Nanomedicine* (pp. 575–603). New York, NY: Springer New York. doi:10.1007/978-1-4939-6966-1_17

Kang, J. L., Moon, C., Lee, H. S., Lee, H. W., Park, E. M., Kim, H. S., & Castranova, V. (2008). Comparison of the biological activity between ultrafine and fine titanium dioxide particles in RAW 264.7 cells associated with oxidative stress. *Journal of Toxicology and Environmental Health. Part A., 71*(8), 478–485. doi:10.1080/15287390801906675 PMID:18338282

Kisin, E. R., Murray, A. R., Keane, M. J., Shi, X. C., Schwegler-Berry, D., Gorelik, O., ... Shvedova, A. A. (2007). Single-walled carbon nanotubes: Geno- and cytotoxic effects in lung fibroblast V79 cells. *Journal of Toxicology and Environmental Health. Part A., 70*(24), 2071–2079. doi:10.1080/15287390701601251 PMID:18049996

Krug, H. F., & Wick, P. (2011). Nanotoxicology: An interdisciplinary challenge. *Angewandte Chemie International Edition, 50*(6), 1260–1278. doi:10.1002/anie.201001037 PMID:21290492

Lowe, A. C., Hunter-Ellul, L. A., & Wilkerson, M. G. (2013). Nanotoxicology. In A. Nasir, A. Friedman, & S. Wang (Eds.), *Nanotechnology in Dermatology* (pp. 231–251). New York, NY: Springer New York. doi:10.1007/978-1-4614-5034-4_22

Lu, Z.-X., Zhang, Z.-L., Zhang, M.-X., Xie, H.-Y., Tian, Z.-Q., Chen, P., ... Pang, D.-W. (2005). Core/Shell Quantum-Dot-Photosensitized Nano-TiO2 Films: Fabrication and Application to the Damage of Cells and DNA. *The Journal of Physical Chemistry B, 109*(47), 22663–22666. doi:10.1021/jp054472h PMID:16853950

Lunov, O., Syrovets, T., Loos, C., Nienhaus, G. U., Mailander, V., Landfester, K., ... Simmet, T. (2011). Amino-Functionalized Polystyrene Nanoparticles Activate the NLRP3 Inflammasome in Human Macrophages. *ACS Nano*, 5. PMID:22111911

Magdolenova, Z., Collins, A., Kumar, A., Dhawan, A., Stone, V., & Dusinska, M. (2014). Mechanisms of genotoxicity. A review of in vitro and in vivo studies with engineered nanoparticles. *Nanotoxicology, 8*(3), 233–278. doi:10.3109/17435390.2013.773464 PMID:23379603

Manna, S. K., Sarkar, S., Barr, J., Wise, K., Barrera, E. V., Jejelowo, O., ... Ramesh, G. T. (2005). Single-Walled Carbon Nanotube Induces Oxidative Stress and Activates Nuclear Transcription Factor-κB in Human Keratinocytes. *Nano Letters, 5*(9), 1676–1684. doi:10.1021/nl0507966 PMID:16159204

Mecke, A., Uppuluri, S., Sassanella, T. M., Lee, D. K., Ramamoorthy, A., Baker, J. R. Jr, ... Banaszak Holl, M. M. (2004). Direct observation of lipid bilayer disruption by poly(amidoamine) dendrimers. *Chemistry and Physics of Lipids, 132*(1), 3–14. doi:10.1016/j.chemphyslip.2004.09.001 PMID:15530443

Michalet, X., Pinaud, F. F., Bentolila, L. A., Tsay, J. M., Doose, S., Li, J. J., ... Weiss, S. (2005). Quantum dots for live cells, in vivo imaging, and diagnostics. *Science, 307*(5709), 538–544. doi:10.1126cience.1104274 PMID:15681376

Mortimer, G. M., & Minchin, R. F. (2017). Chapter Twenty - Nanotoxicology and Nanovaccines A2 - Skwarczynski, Mariusz. In I. Toth (Ed.), *Micro and Nanotechnology in Vaccine Development* (pp. 373–392). William Andrew Publishing. doi:10.1016/B978-0-323-39981-4.00020-8

Mortimer, M., Kasemets, K., & Kahru, A. (2010). Toxicity of ZnO and CuO nanoparticles to ciliated protozoa Tetrahymena thermophila. *Toxicology*, *269*(2-3), 182–189. doi:10.1016/j.tox.2009.07.007 PMID:19622384

Muller, J., Decordier, I., Hoet, P. H., Lombaert, N., Thomassen, L., Huaux, F., ... Kirsch-Volders, M. (2008). Clastogenic and aneugenic effects of multi-wall carbon nanotubes in epithelial cells. *Carcinogenesis*, *29*(2), 427–433. doi:10.1093/carcin/bgm243 PMID:18174261

Neumann, S., Burkert, K., Kemp, R., Rades, T., Rod Dunbar, P., & Hook, S. (2014). Activation of the NLRP3 inflammasome is not a feature of all particulate vaccine adjuvants. *Immunology and Cell Biology*, *92*(6), 535–542. doi:10.1038/icb.2014.21 PMID:24687021

Ng, C.-T., Li, J. J., Bay, B.-H., & Yung, L.-Y. L. (2010). Current Studies into the Genotoxic Effects of Nanomaterials. *Journal of Nucleic Acids*, *2010*, 12. doi:10.4061/2010/947859 PMID:20936181

Oberdorster, G., Sharp, Z., Atudorei, V., Elder, A., Gelein, R., Kreyling, W., & Cox, C. (2004). Translocation of inhaled ultrafine particles to the brain. *Inhalation Toxicology*, *16*(6-7), 437–445. doi:10.1080/08958370490439597 PMID:15204759

Pacurari, M., Yin, X. J., Ding, M., Leonard, S. S., Schwegler-berry, D., Ducatman, B. S., ... Vallyathan, V. (2008). Oxidative and molecular interactions of multi-wall carbon nanotubes (MWCNT) in normal and malignant human mesothelial cells. *Nanotoxicology*, *2*(3), 155–170. doi:10.1080/17435390802318356

Pacurari, M., Yin, X. J., Zhao, J., Ding, M., Leonard, S. S., Schwegler-Berry, D., ... Vallyathan, V. (2008). Raw single-wall carbon nanotubes induce oxidative stress and activate MAPKs, AP-1, NF-kappaB, and Akt in normal and malignant human mesothelial cells. *Environmental Health Perspectives*, *116*(9), 1211–1217. doi:10.1289/ehp.10924 PMID:18795165

Park, E. J., Yi, J., Chung, K. H., Ryu, D. Y., Choi, J., & Park, K. (2008). Oxidative stress and apoptosis induced by titanium dioxide nanoparticles in cultured BEAS-2B cells. *Toxicology Letters*, *180*(3), 222–229. doi:10.1016/j.toxlet.2008.06.869 PMID:18662754

Patlolla, A. K., Hackett, D., & Tchounwou, P. B. (2015). Genotoxicity study of silver nanoparticles in bone marrow cells of Sprague–Dawley rats. *Food and Chemical Toxicology*, *85*, 52–60. doi:10.1016/j.fct.2015.05.005 PMID:26032631

Peter, C., & Hummer, G. (2005). Ion Transport through Membrane-Spanning Nanopores Studied by Molecular Dynamics Simulations and Continuum Electrostatics Calculations. *Biophysical Journal*, *89*(4), 2222–2234. doi:10.1529/biophysj.105.065946 PMID:16006629

Pumera, M., & Iwai, H. (2009). Multicomponent Metallic Impurities and Their Influence upon the Electrochemistry of Carbon Nanotubes. *The Journal of Physical Chemistry C*, *113*(11), 4401–4405. doi:10.1021/jp900069e

Pumera, M., Iwai, H., & Miyahara, Y. (2009). Bimetallic nickel-iron impurities within single-walled carbon nanotubes exhibit redox activity towards the oxidation of amino acids. *ChemPhysChem*, *10*(11), 1770–1773. doi:10.1002/cphc.200900355 PMID:19603449

Rajagopalan, P., Wudl, F., Schinazi, R. F., & Boudinot, F. D. (1996). Pharmacokinetics of a water-soluble fullerene in rats. *Antimicrobial Agents and Chemotherapy*, *40*(10), 2262–2265. PMID:8891126

Rao, D. B., Wong, B. A., McManus, B. E., McElveen, A. M., James, A. R., & Dorman, D. C. (2003). Inhaled iron, unlike manganese, is not transported to the rat brain via the olfactory pathway. *Toxicology and Applied Pharmacology*, *193*(1), 116–126. doi:10.1016/S0041-008X(03)00340-5 PMID:14613722

Rodriguez-Yañez, Y., Muñoz, B., & Albores, A. (2013). Mechanisms of toxicity by carbon nanotubes. *Toxicology Mechanisms and Methods*, *23*(3), 178–195. doi:10.3109/15376516.2012.754534 PMID:23193995

Ryman-Rasmussen, J. P., Riviere, J. E., & Monteiro-Riviere, N. A. (2007). Surface coatings determine cytotoxicity and irritation potential of quantum dot nanoparticles in epidermal keratinocytes. *The Journal of Investigative Dermatology*, *127*(1), 143–153. doi:10.1038j.jid.5700508 PMID:16902417

Salvador-Morales, C., Basiuk, E. V., Basiuk, V. A., Green, M. L., & Sim, R. B. (2008). Effects of covalent functionalization on the biocompatibility characteristics of multi-walled carbon nanotubes. *Journal of Nanoscience and Nanotechnology*, *8*(5), 2347–2356. doi:10.1166/jnn.2008.090 PMID:18572648

Salvador-Morales, C., Townsend, P., Flahaut, E., Vénien-Bryan, C., Vlandas, A., Green, M. L. H., & Sim, R. B. (2007). Binding of pulmonary surfactant proteins to carbon nanotubes; potential for damage to lung immune defense mechanisms. *Carbon*, *45*(3), 607–617. doi:10.1016/j.carbon.2006.10.011

Sandberg, W. J., Låg, M., Holme, J. A., Friede, B., Gualtieri, M., Kruszewski, M., ... Refsnes, M. (2012). Comparison of non-crystalline silica nanoparticles in IL-1β release from macrophages. *Particle and Fibre Toxicology*, *9*(1), 32. doi:10.1186/1743-8977-9-32 PMID:22882971

Sarkar, A., Ghosh, M., & Sil, P. C. (2014). Nanotoxicity: Oxidative stress mediated toxicity of metal and metal oxide nanoparticles. *Journal of Nanoscience and Nanotechnology*, *14*(1), 730–743. doi:10.1166/jnn.2014.8752 PMID:24730293

Sato, Y., Shibata, K., Kataoka, H., Ogino, S., Bunshi, F., Yokoyama, A., ... Tohji, K. (2005). Strict preparation and evaluation of water-soluble hat-stacked carbon nanofibers for biomedical application and their high biocompatibility: Influence of nanofiber-surface functional groups on cytotoxicity. *Molecular BioSystems*, *1*(2), 142–145. doi:10.1039/b501222h PMID:16880976

Sayes, C. M., Gobin, A. M., Ausman, K. D., Mendez, J., West, J. L., & Colvin, V. L. (2005). Nano-C60 cytotoxicity is due to lipid peroxidation. *Biomaterials*, *26*(36), 7587–7595. doi:10.1016/j.biomaterials.2005.05.027 PMID:16005959

Scipioni, R., Pumera, M., Boero, M., Miyahara, Y., & Ohno, T. (2010). Investigation of the Mechanism of Adsorption of β-Nicotinamide Adenine Dinucleotide on Single-Walled Carbon Nanotubes. *The Journal of Physical Chemistry Letters*, *1*(1), 122–125. doi:10.1021/jz9000714

Sera, N., Tokiwa, H., & Miyata, N. (1996). Mutagenicity of the fullerene C60-generated singlet oxygen dependent formation of lipid peroxides. *Carcinogenesis*, *17*(10), 2163–2169. doi:10.1093/carcin/17.10.2163 PMID:8895484

Sharma, V., Shukla, R. K., Saxena, N., Parmar, D., Das, M., & Dhawan, A. (2009). DNA damaging potential of zinc oxide nanoparticles in human epidermal cells. *Toxicology Letters*, *185*(3), 211–218. doi:10.1016/j.toxlet.2009.01.008 PMID:19382294

Shimizu, K., Uchiyama, A., Yamashita, M., Hirose, A., Nishimura, T., & Oku, N. (2013). Biomembrane damage caused by exposure to multi-walled carbon nanotubes. *The Journal of Toxicological Sciences*, *38*(1), 7–12. doi:10.2131/jts.38.7 PMID:23358135

Shimizu, M., Tainaka, H., Oba, T., Mizuo, K., Umezawa, M., & Takeda, K. (2009). Maternal exposure to nanoparticulate titanium dioxide during the prenatal period alters gene expression related to brain development in the mouse. *Particle and Fibre Toxicology*, *6*(1), 20–20. doi:10.1186/1743-8977-6-20 PMID:19640265

Shinohara, N., Matsumoto, K., Endoh, S., Maru, J., & Nakanishi, J. (2009). In vitro and in vivo genotoxicity tests on fullerene C60 nanoparticles. *Toxicology Letters*, *191*(2-3), 289–296. doi:10.1016/j.toxlet.2009.09.012 PMID:19772904

Shvedova, A. A., Castranova, V., Kisin, E. R., Schwegler-Berry, D., Murray, A. R., Gandelsman, V. Z., ... Baron, P. (2003). Exposure to carbon nanotube material: Assessment of nanotube cytotoxicity using human keratinocyte cells. *Journal of Toxicology and Environmental Health. Part A.*, *66*(20), 1909–1926. doi:10.1080/713853956 PMID:14514433

Singh, N. (2009). Conference scene - nanotoxicology: Health and environmental impacts. *Nanomedicine (London)*, *4*(4), 385–390. doi:10.2217/nnm.09.20 PMID:19505241

Sun, B., Wang, X., Ji, Z., Wang, M., Liao, Y.-P., Chang, C. H., ... Xia, T. (2015). NADPH Oxidase-Dependent NLRP3 Inflammasome Activation and its Important Role in Lung Fibrosis by Multiwalled Carbon Nanotubes. *Small*, *11*(17), 2087–2097. doi:10.1002mll.201402859 PMID:25581126

Tino, A., Ambrosone, A., Marchesano, V., & Tortiglione, C. (2014). Molecular Bases of Nanotoxicology. In Bio- and Bioinspired Nanomaterials (pp. 229-254): Wiley-VCH Verlag GmbH & Co. KGaA.

Tsuchiya, T., Oguri, I., Yamakoshi, Y. N., & Miyata, N. (1996). Novel harmful effects of [60]fullerene on mouse embryos in vitro and in vivo. *FEBS Letters*, *393*(1), 139–145. doi:10.1016/0014-5793(96)00812-5 PMID:8804443

Verones, B., & Oortgiesen, M. (2001). Neurogenic inflammation and particulate matter (PM) air pollutants. *Neurotoxicology*, *22*(6), 795–810. doi:10.1016/S0161-813X(01)00062-6 PMID:11829413

Wang, X., Guo, J., Chen, T., Nie, H., Wang, H., Zang, J., ... Jia, G. (2012). Multi-walled carbon nanotubes induce apoptosis via mitochondrial pathway and scavenger receptor. *Toxicology In Vitro*, *26*(6), 799–806. doi:10.1016/j.tiv.2012.05.010 PMID:22664788

Xia, T., Kovochich, M., Liong, M., Mädler, L., Gilbert, B., Shi, H., ... Nel, A. E. (2008). Comparison of the Mechanism of Toxicity of Zinc Oxide and Cerium Oxide Nanoparticles Based on Dissolution and Oxidative Stress Properties. *ACS Nano*, *2*(10), 2121–2134. doi:10.1021/nn800511k PMID:19206459

Yamashita, K., Yoshioka, Y., Higashisaka, K., Mimura, K., Morishita, Y., Nozaki, M., . . . Tsutsumi, Y. (2011). *Silica and titanium dioxide nanoparticles cause pregnancy complications in mice*. Retrieved from https://www.nature.com/articles/nnano.2011.41#supplementary-information

Yang, E.-J., Kim, S., Kim, J. S., & Choi, I.-H. (2012). Inflammasome formation and IL-1β release by human blood monocytes in response to silver nanoparticles. *Biomaterials*, *33*(28), 6858–6867. doi:10.1016/j.biomaterials.2012.06.016 PMID:22770526

Yuryev, M., Ferreira, M. P., Balasubramanian, V., Correia, A. M., Makila, E. M., Jokinen, V., ... Rivera, C. (2016). Active diffusion of nanoparticles of maternal origin within the embryonic brain. *Nanomedicine (London)*, *11*(19), 2471–2481. doi:10.2217/nnm-2016-0207 PMID:27622305

Zakharenko, L. P., Zakharov, I. K., Vasiunina, E. A., Karamysheva, T. V., Danilenko, A. M., & Nikiforov, A. A. (1997). Determination of the genotoxicity of fullerene C60 and fullerol using the method of somatic mosaics on cells of Drosophila melanogaster wing and SOS-chromotest. *Genetika*, *33*(3), 405–409. PMID:9244774

Zhang, Q., Xu, H., Zheng, S., Su, M., & Wang, J. (2015). Genotoxicity of mesoporous silica nanoparticles in human embryonic kidney 293 cells. *Drug Testing and Analysis*, *7*(9), 787–796. doi:10.1002/dta.1773 PMID:25583601

Zhao, J., & Castranova, V. (2011). Toxicology of nanomaterials used in nanomedicine. *Journal of Toxicology and Environmental Health. Part B, Critical Reviews*, *14*(8), 593–632. doi:10.1080/10937404.2011.615113 PMID:22008094

Zhong, Z., Zhai, Y., Liang, S., Mori, Y., Han, R., Sutterwala, F. S., & Qiao, L. (2013). *TRPM2 links oxidative stress to NLRP3 inflammasome activation*. Retrieved from https://www.nature.com/articles/ncomms2608#supplementary-information

Zhu, L., Chang, D. W., Dai, L., & Hong, Y. (2007). DNA damage induced by multiwalled carbon nanotubes in mouse embryonic stem cells. *Nano Letters*, *7*(12), 3592–3597. doi:10.1021/nl071303v PMID:18044946

Compilation of References

Aarthi, T., & Madras, G. (2007). Photocatalytic degradation of rhodamine dyes with nano-TiO2. *Industrial & Engineering Chemistry Research*, *46*(1), 7–14. doi:10.1021/ie060948n

Abburi, K. (2003). Adsorption of phenol and p-chlorophenol from their single and bisolute aqueous solutions on Amberlite XAD-16 resin. *Journal of Hazardous Materials*, *105*(1), 143–156. doi:10.1016/j.jhazmat.2003.08.004 PMID:14623424

Abdel Ghafar, H. H., Ali, G. A. M., Fouad, O. A., & Makhlouf, S. A. (2015). Enhancement of adsorption efficiency of methylene blue on $Co_3O_{4/}SiO_2$ nanocomposite. *Desalination and Water Treatment*, *53*(11), 2980–2989. doi:10.1080/19443994.2013.871343

Abdelghany, T. M., Al-Rajhi, A. M. H., Al Abboud, M. A., Alawlaqi, M. M., Ganash Magdah, A., Helmy, E. A. M., & Mabrouk, A. S. (2017). Recent Advances in Green Synthesis of Silver Nanoparticles and Their Applications: About Future Directions. A Review. *BioNanoScience*. doi:10.100712668-017-0413-3

Abdelkader, N. B., Bentouami, A., Derriche, Z., Bettahar, N., & de Ménorval, L. C. (2011). Synthesis and characterization of Mg–Fe layer double hydroxides and its application on adsorption of orange G from aqueous solution. *Chemical Engineering Journal*, *169*(1-3), 231–238. doi:10.1016/j.cej.2011.03.019

Abdul Salam, H., Sivaraj, R., & Venckatesh, R. (2014). Green synthesis and characterization of zinc oxide nanoparticles from Ocimum basilicum L. var. purpurascens Benth.-Lamiaceae leaf extract. *Materials Letters*, *131*, 16–18. doi:10.1016/j.matlet.2014.05.033

Abell, J. M., Hamilton, D. P., & Paterson, J. (2011). Reducing the external environmental costs of pastoral farming in New Zealand: Experiences from the Te Arawa lakes, Rotorua. *Australasian Journal of Environmental Management*, *18*(3), 139–154. doi:10.1080/14486563.2011.591520

Abid, A. D., Kanematsu, M., Young, T. M., & Kennedy, I. M. (2013). Arsenic removal from water using flame-synthesized iron oxide nanoparticles with variable oxidation states. *Aerosol Science and Technology*, *47*(2), 169–176. doi:10.1080/02786826.2012.735380 PMID:23645964

Absalan, G., Asadi, M., Kamran, S., Sheikhian, L., & Goltz, D. M. (2011). Removal of reactive red-120 and 4-(2-pyridylazo) resorcinol from aqueous samples by Fe 3 O 4 magnetic nanoparticles using ionic liquid as modifier. *Journal of Hazardous Materials*, *192*(2), 476–484. doi:10.1016/j.jhazmat.2011.05.046 PMID:21645968

Abu-Daabes, M. A., & Pinto, N. G. (2005). Synthesis and characterization of a nano-structured sorbent for the direct removal of mercury vapor from flue gases by chelation. *Chemical Engineering Science*, *60*(7), 1901–1910. doi:10.1016/j.ces.2004.11.028

Adeleye, A. S., Conway, J. R., Garner, K., Huang, Y., Su, Y., & Keller, A. A. (2016). Engineered nanomaterials for water treatment and remediation: Costs, benefits, and applicability. *Chemical Engineering Journal, 286*, 640–662. doi:10.1016/j.cej.2015.10.105

Aditya, D., Rohan, P., & Suresh, G. (n.d.). Nano-adsorbents for wastewater treatment: a review. *Research Journal of Chemical & Environmental Sciences, 15,* 1033–40.

Agarry, S. E., Ogunleye, O. O., & Aworanti, O. A. (2013). Biosorption equilibrium, kinetic and thermodynamic modelling of naphthalene removal from aqueous solution onto modified spent tea leaves. *Environmental Technology, 34*(7), 825–839. doi:10.1080/09593330.2012.720616 PMID:23837334

Agarwal, S., Tyagi, I., Gupta, V. K., Ghaedi, M., Masoomzade, M., Ghaedi, A., & Mirtamizdoust, B. (2016). Kinetics and thermodynamics of methyl orange adsorption from aqueous solutions-artificial neural network-particle swarm optimization modeling. *Journal of Molecular Liquids, 218*, 354–362. doi:10.1016/j.molliq.2016.02.048

Aguado, J., Arsuaga, J. M., & Arencibia, A. (2005). Adsorption of aqueous mercury (II) on propylthiolfunctionalized mesoporous silica obtained by cocondensation. *Industrial & Engineering Chemistry Research, 44*(10), 3665–3671. doi:10.1021/ie0487585

Aguado, J., Arsuaga, J. M., & Arencibia, A. (2008). Influence of synthesis conditions on mercury adsorption capacity of propylthiol functionalized SBA-15 obtained by co-condensation. *Microporous and Mesoporous Materials, 109*(1-3), 513–524. doi:10.1016/j.micromeso.2007.05.061

Aguinaldo, J. T. (2007). Integrated Precipitative Softening and Immersed Spiral Wound. *Conference Proceedings.*

Ahamed, M., Karns, M., Goodson, M., Rowe, J., Hussain, S. M., Schlager, J. J., & Hong, Y. (2008). DNA damage response to different surface chemistry of silver nanoparticles in mammalian cells. *Toxicology and Applied Pharmacology, 233*(3), 404–410. doi:10.1016/j.taap.2008.09.015 PMID:18930072

Ahluwalia, V. (2006). *Green chemistry: Environmentally benign reaction.* Ane Books Pvt Ltd.

Ahmad, A., Senapati, S., Khan, M. I., Kumar, R., & Sastry, M. (2003). Extracellular biosynthesis of monodisperse gold nanoparticles by a novel extremophilic actinomycete, Thermomonospora sp. *Langmuir, 19*(8), 3550–3553. doi:10.1021/la026772l

Ahmad, H., Kamarudin, S. K., Harsan, U. A., & Daud, W. R. W. (2010). Overview of hybrid membranes for direct methanol fuel cell applications. *International Journal of Hydrogen Energy, 35*(5), 2160–2175. doi:10.1016/j.ijhydene.2009.12.054

Ahmad, N., Sharma, S., Alam, M. K., Singh, V. N., Shamsi, S. F., Mehta, B. R., & Fatma, A. (2010). Rapid synthesis of silver nanoparticles using dried medicinal plant of basil. *Colloids and Surfaces. B, Biointerfaces, 81*(1), 81–86. doi:10.1016/j.colsurfb.2010.06.029 PMID:20656463

Ahmad, R., Nelson, P. N., & Kookana, R. S. (2006). The molecular composition of soil organic matter as determined by 13C NMR and elemental analyses and correlation with pesticide sorption. *European Journal of Soil Science, 57*(6), 883–893. doi:10.1111/j.1365-2389.2005.00784.x

Ahmaruzzaman, M., & Gayatri, S. L. (2010). Activated tea waste as a potential low-cost adsorbent for the removal of p-nitrophenol from waste water. *Journal of Chemical & Engineering Data, 55*(11), 4614–4623. doi:10.1021/je100117s

Ahmed, M. A., & Abdel Messih, M. F. (2017). Synthesis of metallic silver nanoparticles decorated mesoporous SnO2 for removal of methylene blue dye by coupling adsorption and photocatalytic processes. *Journal of Photochemistry and Photobiology A Chemistry, 346*, 77–88. doi:10.1016/j.jphotochem.2017.05.048

Ahmed, M. J. K., Ahmaruzzaman, M., & Reza, R. A. (2014). Lignocellulosic-derived modified agricultural waste: Development, characterization and implementation in sequestering pyridine from aqueous solutions. *Journal of Colloid and Interface Science*, *428*, 222–234. doi:10.1016/j.jcis.2014.04.049 PMID:24910057

Ahn, J., Chung, W.-J., Pinnau, I., & Guiver, M. D. (2008). Polysulfone/silica nanoparticle mixed-matrix membranes for gas separation. *Journal of Membrane Science*, *314*(1-2), 123–133. doi:10.1016/j.memsci.2008.01.031

Ai, L., & Jiang, J. (2012). Removal of methylene blue from aqueous solution with self-assembled cylindrical graphene–carbon nanotube hybrid. *Chemical Engineering Journal*, *192*, 156–163. doi:10.1016/j.cej.2012.03.056

Ain Samat, N., & Md Nor, R. (2013). Sol–gel synthesis of zinc oxide nanoparticles using Citrus aurantifolia extracts. *Ceramics International*, *39*, S545–S548. doi:10.1016/j.ceramint.2012.10.132

Ajayan, P. M. (2003). Bulk metal and ceramics nanocomposites. *Nanocomposite Science and Technology*, 1-76.

Ake, C. L., Wiles, M. C., Huebner, H. J., McDonald, T. J., Cosgriff, D., Richardson, M. B., & Phillips, T. D. (2003). Porous organoclay composite for the sorption of polycyclic aromatic hydrocarbons and pentachlorophenol from groundwater. *Chemosphere*, *51*(9), 835–844. doi:10.1016/S0045-6535(03)00040-7 PMID:12697173

Aksu, Z., & Yener, J. (2001). A comparative adsorption/biosorption study of mono-chlorinated phenols onto various sorbents. *Waste Management (New York, N.Y.)*, *21*(8), 695–702. doi:10.1016/S0956-053X(01)00006-X PMID:11699627

Akyildiz, I. F., & Jornet, J. M. (2010). Electromagnetic wireless nanosensor networks. *Nano Communication Networks*, *1*(1), 3–19. doi:10.1016/j.nancom.2010.04.001

Alberti, G., Casciola, M., & Costantino, U. (1983). Inorganic ion-exchange membranes made of acid salts of tetravalent metals. *Journal of Membrane Science*, *16*, 137–149. doi:10.1016/S0376-7388(00)81305-8

Alberti, G., Casciola, M., & Palombari, R. (2000). Inorganic-Organic Proton Conducting Membranes for Fuel Cells and Sensors at Medium Temperatures. *Journal of Membrane Science*, *172*(1-2), 233–239. doi:10.1016/S0376-7388(00)00332-X

Albo, J., Luis, P., & Irabien, A. (2010). Carbon Dioxide Capture from Flue Gases Using a Cross-Flow Membrane Contactor and 1-Ethyl-3-methylimidazolium Ionic Liquid. *Industrial & Engineering Chemistry Research*, *49*(21), 11045–11051. doi:10.1021/ie1014266

Alcántara, M. T., Gómez, J., & Sanromán, M. A. (2008). Combined treatment of PAHs contaminated soils using the sequence extraction with surfactant- electrochemical degradation. *Chemosphere*, *70*(8), 1438–1444. doi:10.1016/j.chemosphere.2007.08.070 PMID:17936331

Al-Degs, Y. S., El-Barghouthi, M. I., Issa, A. A., Khraisheh, M. A., & Walker, G. M. (2006). Sorption of Zn (II), Pb (II), and Co (II) using natural sorbents: Equilibrium and kinetic studies. *Water Research*, *40*(14), 2645–2658. doi:10.1016/j.watres.2006.05.018 PMID:16839582

Alén, B., Martınez-Pastor, J., Garcıa-Cristobal, A., González, L., & Garcıa, J. M. (2001). Optical transitions and excitonic recombination in InAs/InP self-assembled quantum wires. *Applied Physics Letters*, *78*(25), 4025–4027. doi:10.1063/1.1379991

Alexander, J. V., Neely, J. W., & Grulke, E. A. (2014). Effect of chemical functionalization on the mechanical properties of polypropylene hollow fiber membranes. *Journal of Polymer Science. Part B, Polymer Physics*, *52*(20), 1366–1373. doi:10.1002/polb.23572

Alghuthaymi, M. A., Almoammar, H., Rai, M., Said-Galiev, E., & Abd-Elsalam, K. A. (2015). Myconanoparticles: Synthesis and their role in phytopathogens management. *Biotechnology, Biotechnological Equipment*, *29*(2), 221–236. doi:10.1080/13102818.2015.1008194 PMID:26019636

Alhoshan, M., Alam, J., Dass, L. A., & Al-Homaidi, N. (2013). Fabrication of Polysulfone/ZnO Membrane: Influence of ZnO Nanoparticles on Membrane Characteristics. *Advances in Polymer Technology*, *32*(4), 1–7. doi:10.1002/adv.21369

Ali, G. A. M., Lih Teo, E. Y., Aboelazm, E. A. A., Sadegh, H., Memar, A. O. H., Shahryari-Ghoshekandi, R., & Chong, K. F. (2017). Capacitive performance of cysteamine functionalized carbon nanotubes. *Materials Chemistry and Physics*, *197*, 100–104. doi:10.1016/j.matchemphys.2017.05.019

Ali, G. A. M., Makhlouf, S. A., Yusoff, M. M., & Chong, K. F. (2015). Structural and electrochemical characteristics of graphene nanosheets as supercapacitor electrodes. *Reviews on Advanced Materials Science*, *40*(1), 35–43.

Ali, I. (2010). The Quest for Active Carbon Adsorbent Substitutes: Inexpensive Adsorbents for Toxic Metal Ions Removal from Wastewater. *Separation and Purification Reviews*, *39*(3-4), 95–171. doi:10.1080/15422119.2010.527802

Ali, I. (2012). New generation adsorbents for water treatment. *Chemical Reviews*, *112*(10), 5073–5091. doi:10.1021/cr300133d PMID:22731247

Ali, I., & Aboul-Enein, H. Y. (2004). *Chiral Pollutants: Distribution, Toxicity and Analysis by Chromatography and Capillary Electrophoresis*. Chichester, UK: John Wiley & Sons.

Ali, I., & Aboul-Enein, H. Y. (2006). *Instrumental Methods in Metal Ions Speciation: Chromatography, Capillary Electrophoresis and Electrochemistry*. New York: Taylor & Francis Ltd. doi:10.1201/9781420019407

Ali, I., Aboul-Enein, H. Y., & Gupta, V. K. (2009). *Nano Chromatography and Capillary Electrophoresis: Pharmaceutical and Environmental Analyses*. Hoboken, NJ: John Wiley & Sons.

Ali, I., Al-Othman, Z. A., & Alharbi, O. M. (2016). Uptake of pantoprazole drug residue from water using novel synthesized composite iron nano adsorbent. *Journal of Molecular Liquids*, *218*, 465–472. doi:10.1016/j.molliq.2016.02.088

Ali, I., AL-Othman, Z. A., & Alwarthan, A. (2016). Green synthesis of functionalized iron nano particles and molecular liquid phase adsorption of ametryn from water. *Journal of Molecular Liquids*, *221*, 1168–1174. doi:10.1016/j.molliq.2016.06.089

Ali, I., AL-Othman, Z. A., & Alwarthan, A. (2016). Molecular uptake of congo red dye from water on iron composite nano particles. *Journal of Molecular Liquids*, *224*, 171–176. doi:10.1016/j.molliq.2016.09.108

Ali, I., AL-Othman, Z. A., & Alwarthan, A. (2016). Synthesis of composite iron nano adsorbent and removal of ibuprofen drug residue from water. *Journal of Molecular Liquids*, *219*, 858–864. doi:10.1016/j.molliq.2016.04.031

Ali, I., Alothman, Z. A., & Alwarthan, A. (2017). Uptake of propranolol on ionic liquid iron nanocomposite adsorbent: Kinetic, thermodynamics and mechanism of adsorption. *Journal of Molecular Liquids*, *236*, 205–213. doi:10.1016/j.molliq.2017.04.028

Ali, I., Alothman, Z. A., & Sanagi, M. M. (2015). Green synthesis of iron nano-impregnated adsorbent for fast removal of fluoride from water. *Journal of Molecular Liquids*, *211*, 457–465. doi:10.1016/j.molliq.2015.07.034

Ali, I., Asim, M., & Khan, T. A. (2012). Low cost adsorbents for the removal of organic pollutants from waste water. *Journal of Environmental Management*, *113*, 170–183. doi:10.1016/j.jenvman.2012.08.028 PMID:23023039

Ali, K., Dwivedi, S., Azam, A., Saquib, Q., Al-Said, M. S., Alkhedhairy, A. A., & Musarrat, J. (2016). Aloe vera extract functionalized zinc oxide nanoparticles as nanoantibiotics against multi-drug resistant clinical bacterial isolates. *Journal of Colloid and Interface Science*, *472*, 145–156. doi:10.1016/j.jcis.2016.03.021 PMID:27031596

Almao, P. P. (2012). In situ upgrading of bitumen and heavy oils via nanocatalysis. *Canadian Journal of Chemical Engineering*, *90*(2), 320–329. doi:10.1002/cjce.21646

AlOmar, M. K., Alsaadi, M. A., Hayyan, M., Akib, S., Ibrahim, M., & Hashim, M. A. (2017a). Allyl triphenyl phosphonium bromide based DES-functionalized carbon nanotubes for the removal of mercury from water. *Chemosphere*, *167*(Supplement C), 44–52. doi:10.1016/j.chemosphere.2016.09.133 PMID:27710842

AlOmar, M. K., Alsaadi, M. A., Jassam, T. M., Akib, S., & Ali Hashim, M. (2017b). Novel deep eutectic solvent-functionalized carbon nanotubes adsorbent for mercury removal from water. *Journal of Colloid and Interface Science*, *497*(Supplement C), 413–421. doi:10.1016/j.jcis.2017.03.014 PMID:28314146

Alsohaimi, I. H., Wabaidur, S. M., Kumar, M., Khan, M. A., Alothman, Z. A., & Abdalla, M. A. (2015). Synthesis, characterization of PMDA/TMSPEDA hybrid nano-composite and its applications as an adsorbent for the removal of bivalent heavy metals ions. *Chemical Engineering Journal*, *270*, 9–21. doi:10.1016/j.cej.2015.02.008

Altavilla, C., & Ciliberto, E. (2016). *Inorganic nanoparticles: synthesis, applications, and perspectives*. CRC Press.

Altınışık, A., Gür, E., & Seki, Y. (2010). A natural sorbent, Luffa cylindrica for the removal of a model basic dye. *Journal of Hazardous Materials*, *179*(1-3), 658–664. doi:10.1016/j.jhazmat.2010.03.053 PMID:20378245

Alwin, S., Bhat, S. D., Sahu, A. K., Jalajakshi, A., Sridgar, P., Sridgar, S., & Shukla, A. K. (2011). Modified-pore-filled-PVDF-membrane electrolytes for direct methanol fuel cells. *Journal of the Electrochemical Society*, *158*(2), B91–B98. doi:10.1149/1.3518774

Ambashta, R. D., & Sillanpää, M. (2010). Water purification using magnetic assistance: A review. *Journal of Hazardous Materials*, *180*(1), 38–49. doi:10.1016/j.jhazmat.2010.04.105 PMID:20488616

Ambrosi, A., & Pumera, M. (2010). Regulatory Peptides Are Susceptible to Oxidation by Metallic Impurities within Carbon Nanotubes. *Chemistry (Weinheim an der Bergstrasse, Germany)*, *16*(6), 1786–1792. doi:10.1002/chem.200902534 PMID:20066697

Amin, M., Alazba, A., & Manzoor, U. (2014). A review of removal of pollutants from water/wastewater using different types of nanomaterials. *Advances in Materials Science and Engineering*.

Amin, N. K. (2009). Removal of direct blue-106 dye from aqueous solution using new activated carbons developed from pomegranate peel: Adsorption equilibrium and kinetics. *Journal of Hazardous Materials*, *165*(1), 52–62. doi:10.1016/j.jhazmat.2008.09.067 PMID:18986765

Amirinejad, M., Madaenia, S. S., & Navarrab, M. A. (2011). Preparation and characterization of phosphotungstic acid-derived salt/Nafion nanocomposite membranes for proton exchange membrane fuel cells. *Journal of Power Sources*, *196*(3), 988–998. doi:10.1016/j.jpowsour.2010.08.062

Amutha, C., Thanikaikarasan, S., Ramadas, V., Bahadur, S. A., Natarajan, B., & Kalyani, R. (2016). Synthesis, characterization and antibacterial efficiency of ZnO nanoparticles using rice as soft bio-template. *Optik - International Journal for Light and Electron Optics, 127*(10), 4281-4286. doi:10.1016/j.ijleo.2016.01.124

Anastas, P. T., & Zimmerman, J. B. (2012). Innovations in Green Chemistry and Green Engineering: Selected Entries from the Encyclopedia of Sustainability Science and Technology. Springer.

Anastas, P. T., Boethling, R., & Voutchkova-Kostal, A. (2014). *Handbook of Green Chemistry, Green Processes, Designing Safer Chemicals* (Vol. 9). John Wiley & Sons.

Anastas, P. T., & Warner, J. C. (2000). *Green chemistry: theory and practice*. Oxford university press.

Anbia, M., & Amirmahmoodi, S. (2011). Adsorption of phenolic compoundsfrom aqueous solutions using functionalized SBA-15 asa nano-sorbent. *Scientia Iranica*, *18*(3), 446–452. doi:10.1016/j.scient.2011.05.007

Anenberg, S. C., Horowitz, L. W., Tong, D. Q., & West, J. J. (2010). An estimate of the global burden of anthropogenic ozone and fine particulate matter on premature human mortality using atmospheric modeling. *Environmental Health Perspectives, 118*(9), 1189–1195. doi:10.1289/ehp.0901220 PMID:20382579

An, H., Liu, Q., Ji, Q., & Jin, B. (2010). DNA binding and aggregation by carbon nanoparticles. *Biochemical and Biophysical Research Communications, 393*(4), 571–576. doi:10.1016/j.bbrc.2010.02.006 PMID:20156419

Anilkumar, M. R., Nagaswarupa, H. P., Nagabhushana, H., Sharma, S. C., Vidya, Y. S., Anantharaju, K. S., ... Gurushantha, K. (2015). Bio-inspired route for the synthesis of spherical shaped MgO:Fe(3+) nanoparticles: Structural, photoluminescence and photocatalytic investigation. *Spectrochimica Acta. Part A: Molecular and Biomolecular Spectroscopy, 149*, 703–713. doi:10.1016/j.saa.2015.05.003 PMID:25988816

Anjum, M., Miandad, R., Waqas, M., Gehany, F., & Barakat, M. A. (2016). Remediation of wastewater using various nano-materials. *Arabian Journal of Chemistry*. doi:10.1016/j.arabjc.2016.10.004

Antochshuk, V., Olkhovyk, O., Jaroniec, M., Park, I.-S., & Ryoo, R. (2003). Benzoylthiourea modified mesoporous silica for mercury(II) removal. *Langmuir, 19*(7), 3031–3034. doi:10.1021/la026739z

Antonopoulou, M., Karagianni, P., & Konstantinou, I. K. (2016). Kinetic and mechanistic study of photocatalytic degradation of flame retardant Tris (1-chloro-2-propyl) phosphate (TCPP). *Applied Catalysis B: Environmental, 192*, 152–160. doi:10.1016/j.apcatb.2016.03.039

Apul, O. G., Wang, Q., Zhou, Y., & Karanfil, T. (2013). Adsorption of aromatic organic contaminants by graphene nanosheets: Comparison with carbon nanotubes and activated carbon. *Water Research, 47*(4), 1648–1654. doi:10.1016/j.watres.2012.12.031 PMID:23313232

Arakaki, L. N. H., Nunes, L. M., Simoni, J. A., & Airoldi, C. (2000). Ethyleneimine anchored on thiol-modified silica gel surface - Adsorption of divalent cations and calorimetric data. *Journal of Colloid and Interface Science, 228*(1), 46–51. doi:10.1006/jcis.2000.6842 PMID:10882492

Araki, S., Sato, H., Yokoyama, K., & Murata, K. (2000). Subclinical neurophysiological effects of lead: A review on peripheral, central, and autonomic nervous system effects in lead workers. *American Journal of Industrial Medicine, 37*(2), 193–204. doi:10.1002/(SICI)1097-0274(200002)37:2<193::AID-AJIM5>3.0.CO;2-J PMID:10615100

Arbain, R., Othman, M., & Palaniandy, S. (2011). Preparation of iron oxide nanoparticles by mechanical milling. *Minerals Engineering, 24*(1), 1–9. doi:10.1016/j.mineng.2010.08.025

Arbizzani, C., Donnadia, A., Pica, M., Saganappa, M., Varzi, A., Casciola, M., & Mastragostino, M. (2010). Methanol permeability and performance of Nafion-zirconium phosphate composite membranes in active and passive direct methanol fuel cells. *Journal of Power Sources, 195*(23), 7751–7756. doi:10.1016/j.jpowsour.2009.07.034

Arias, M., Barral, M., & Mejuto, J. (2002). Enhancement of copper and cadmium adsorption on kaolin by the presence of humic acids. *Chemosphere, 48*(10), 1081–1088. doi:10.1016/S0045-6535(02)00169-8 PMID:12227513

Aroguz, A. Z. (2006). Kinetics and thermodynamics of adsorption of azinphosmethyl from aqueous solution onto pyrolyzed (at 600 °C) ocean peat moss (Sphagnum sp.). *Journal of Hazardous Materials, 135*(1-3), 100–105. doi:10.1016/j.jhazmat.2005.11.027 PMID:16442228

Arora, M., Eddy, N. K., Mumford, K. A., Baba, Y., Perera, J. M., & Stevens, G. W. (2010). Surface modification of natural zeolite by chitosan and its use for nitrate removal in cold regions. *Cold Regions Science and Technology, 62*(2–3), 92–97. doi:10.1016/j.coldregions.2010.03.002

Arshadi, M., Foroughifard, S., Etemad Gholtash, J., & Abbaspourrad, A. (2015). Preparation of iron nanoparticles-loaded Spondias purpurea seed waste as an excellent adsorbent for removal of phosphate from synthetic and natural waters. *Journal of Colloid and Interface Science*, *452*, 69–77. doi:10.1016/j.jcis.2015.04.019 PMID:25919431

Ashbolt, N. J. (2004). Microbial contamination of drinking water and disease outcomes in developing regions. *Toxicology*, *198*(1–3), 229–238. doi:10.1016/j.tox.2004.01.030 PMID:15138046

Asmatullah, Q., Qureshi, S. N., & Shakoori, A. R. (1998). Hexavalent Chromium-induced Congenital Abnormalities in Chick Embryos. *Journal of Applied Toxicology*, *18*(3), 167–171. doi:10.1002/(SICI)1099-1263(199805/06)18:3<167::AID-JAT492>3.0.CO;2-8 PMID:9685044

Auimviriyavat, J., Changkhamchom, S., & Sirivat, A. (2011). Development of poly (ether ether ketone) (Peek) with inorganic filler for direct methanol fuel cells (DMFCS). *Industrial & Engineering Chemistry Research*, *50*(22), 12527–12533. doi:10.1021/ie2006005

Awwad, A. M., Albiss, B., & Ahmad, A. L. (2014). Green synthesis, characterization and optical properties of zinc oxide nanosheets using Olea europea leaf extract. *Advanced Materials Letters*, *5*(9), 520–524. doi:10.5185/amlett.2014.5575

Ayyasamy, P. M., Shanthi, K., Lakshmanaperumalsamy, P., Lee, S.-J., Choi, N.-C., & Kim, D.-J. (2007). Two-stage removal of nitrate from groundwater using biological and chemical treatments. *Journal of Bioscience and Bioengineering*, *104*(2), 129–134. doi:10.1263/jbb.104.129 PMID:17884658

Azam, A., Ahmed, A. S., Oves, M., Khan, M. S., Habib, S. S., & Memic, A. (2012). Antimicrobial activity of metal oxide nanoparticles against Gram-positive and Gram-negative bacteria: A comparative study. *International Journal of Nanomedicine*, *7*, 6003. doi:10.2147/IJN.S35347 PMID:23233805

Azam, A., Ahmed, A. S., Oves, M., Khan, M., & Memic, A. (2012). Size-dependent antimicrobial properties of CuO nanoparticles against Gram-positive and-negative bacterial strains. *International Journal of Nanomedicine*, *7*, 3527. doi:10.2147/IJN.S29020 PMID:22848176

Azamat, J., Sattary, B. S., Khataee, A., & Joo, S. W. (2015). Removal of a hazardous heavy metal from aqueous solution using functionalized graphene and boron nitride nanosheets: Insights from simulations. *Journal of Molecular Graphics & Modelling*, *61*, 13–20. doi:10.1016/j.jmgm.2015.06.012 PMID:26186492

Azam, M. A., Alias, F. M., Tack, L. W., Seman, R. N. A. R., & Taib, M. F. M. (2017). Electronic properties and gas adsorption behaviour of pristine, silicon-, and boron-doped (8, 0) single-walled carbon nanotube: A first principles study. *Journal of Molecular Graphics & Modelling*, *75*, 85–93. doi:10.1016/j.jmgm.2017.05.003 PMID:28531817

Azizi, S., Ahmad, M. B., Namvar, F., & Mohamad, R. (2014). Green biosynthesis and characterization of zinc oxide nanoparticles using brown marine macroalga Sargassum muticum aqueous extract. *Materials Letters*, *116*, 275–277. doi:10.1016/j.matlet.2013.11.038

Azizullah, A., Khattak, M. N. K., Richter, P., & Häder, D.-P. (2011). Water pollution in Pakistan and its impact on public health—a review. *Environment International*, *37*(2), 479–497. doi:10.1016/j.envint.2010.10.007 PMID:21087795

Baan, R., Straif, K., Grosse, Y., Secretan, B., El Ghissassi, F., & Cogliano, V. (2006). Carcinogenicity of carbon black, titanium dioxide, and talc. *The Lancet Oncology*, *7*(4), 295–296. doi:10.1016/S1470-2045(06)70651-9 PMID:16598890

Babel, S., & Kurniawan, T. A. (2003). Low-cost adsorbents for heavy metals uptake from contaminated water: A review. *Journal of Hazardous Materials*, *97*(1), 219–243. doi:10.1016/S0304-3894(02)00263-7 PMID:12573840

Babynin, E. V., Nuretdinov, I. A., Gubskaya, V. P., & Barabanshchikov, B. I. (2002). Study of Mutagenic Activity of Fullerene and Some of Its Derivatives Using His+ Reversions of Salmonella typhimurium as an Example. *Russian Journal of Genetics*, *38*(4), 359–363. doi:10.1023/A:1015237916596 PMID:12018160

Badruddoza, A. Z. M., Shawon, Z. B. Z., Tay, D. W. J., Hidajat, K., & Uddin, M. S. (2013). Endocrine disrupters and toxic metal ions removal by carboxymethyl-β-cyclodextrin polymer grafted onto magnetic nanoadsorbents. *Chemical Engineering Journal*, *27*(1), 69–73.

Baeza-Squiban, A., Boland, S., Hussain, S., & Marano, F. (2009). *Health Effects of Nanoparticles. In General, Applied and Systems Toxicology*. John Wiley & Sons, Ltd.

Bagherzade, G., Tavakoli, M. M., & Namaei, M. H. (2017). Green synthesis of silver nanoparticles using aqueous extract of saffron (Crocus sativus L.) wastages and its antibacterial activity against six bacteria. *Asian Pacific Journal of Tropical Biomedicine*, *7*(3), 227–233. doi:10.1016/j.apjtb.2016.12.014

Bahgat, M., Farghali, A., El Rouby, W., & Khedr, M. (2011). Synthesis and modification of multi-walled carbon nano-tubes (MWCNTs) for water treatment applications. *Journal of Analytical and Applied Pyrolysis*, *92*(2), 307–313. doi:10.1016/j.jaap.2011.07.002

Baijal, M., Singh, S., Shukla, R. N., & Sanwal, G. G. (1972). Enzymes Of The Banana Plant: Optimum Conditions For Extraction. *Phytochemistry*, *11*(3), 929–936. doi:10.1016/S0031-9422(00)88435-6

Bai, S., Shen, X., Zhong, X., Liu, Y., Zhu, G., Xu, X., & Chen, K. (2012). One-pot solvothermal preparation of magnetic reduced graphene oxide-ferrite hybrids for organic dye removal. *Carbon*, *50*(6), 2337–2346. doi:10.1016/j.carbon.2012.01.057

Bajpai, S. K., & Kumari, M. (2015). A green approach to prepare silver nanoparticles loaded gum acacia/poly(acrylate) hydrogels. *International Journal of Biological Macromolecules*, *80*, 177–188. doi:10.1016/j.ijbiomac.2015.06.048 PMID:26123815

Baker, C., Pradhan, A., Pakstis, L., Pochan, D. J., & Shah, S. I. (2005). Synthesis and antibacterial properties of silver nanoparticles. *Journal of Nanoscience and Nanotechnology*, *5*(2), 244–249. doi:10.1166/jnn.2005.034 PMID:15853142

Baker, R. W. (2004). *Overview of membrane science and technology* (2nd ed.). Membrane Technology and Applications.

Bala, N., Saha, S., Chakraborty, M., Maiti, M., Das, S., Basu, R., & Nandy, P. (2015). Green synthesis of zinc oxide nanoparticles using Hibiscus subdariffa leaf extract: Effect of temperature on synthesis, anti-bacterial activity and anti-diabetic activity. *RSC Advances*, *5*(7), 4993–5003. doi:10.1039/C4RA12784F

Balasubramanian, N., & Madhavan, K. (2001). Arsenic removal from Industrial Effluents through electrocoagulation. *Journal of Chemical Engineering & Tecnology*, *24*(5), 519–521. doi:10.1002/1521-4125(200105)24:5<519::AID-CEAT519>3.0.CO;2-P

Balayssac, S., Trefi, S., Gilard, V., Malet-Martino, M., Martino, R., & Delsuc, M.-A. (2009). 2D and 3D DOSY 1 H NMR, a useful tool for analysis of complex mixtures: Application to herbal drugs or dietary supplements for erectile dysfunction. *Journal of Pharmaceutical and Biomedical Analysis*, *50*(4), 602–612. doi:10.1016/j.jpba.2008.10.034 PMID:19108978

Balogh, L., Swanson, D. R., Tomalia, D. A., Hagnauer, G. L., & McManus, A. T. (2001). Dendrimer−Silver Complexes and Nanocomposites as Antimicrobial Agents. *Nano Letters*, *1*(1), 18–21. doi:10.1021/nl005502p

Balogh, L., Valluzzi, R., Laverdure, K. S., Gido, S. P., Hagnauer Bergmann, C. P., & Machado, F. M. (Eds.). (2015). *Carbon nanomaterials as adsorbents for environmental and biological applications*. New York City: Springer International Publishing.

Balogh, L., Valluzzi, R., Laverdure, K. S., Gido, S. P., Hagnauer, G. L., & Tomalia, D. A. (1999).. . *Journal of Nanoparticle Research*, *1*(3), 353–368. doi:10.1023/A:1010060404024

Bandara, N., Zeng, H., & Wu, J. (2013). Marine mussel adhesion: Biochemistry, mechanisms, and biomimetics. *Journal of Adhesion Science and Technology*, *27*(18-19), 2139–2162. doi:10.1080/01694243.2012.697703

Banerjee, R., Katsenovich, Y., Lagos, L., Senn, M., Naja, M., Balsamo, V., ... Li, C. (2010). Functional magnetic nanoshells integrated nanosensor for trace analysis of environmental uranium contamination. *Electrochimica Acta*, *55*(27), 7897–7902. doi:10.1016/j.electacta.2010.05.060

Banerjee, S. S., & Chen, D.-H. (2007). Fast removal of copper ions by gum arabic modified magnetic nano-adsorbent. *Journal of Hazardous Materials*, *147*(3), 792–799. doi:10.1016/j.jhazmat.2007.01.079 PMID:17321674

Bankar, A., Joshi, B., Kumar, A. R., & Zinjarde, S. (2010). Banana peel extract mediated synthesis of gold nanoparticles. *Colloids and Surfaces. B, Biointerfaces*, *80*(1), 45–50. doi:10.1016/j.colsurfb.2010.05.029 PMID:20620890

Barabadi, H., Honary, S., Mohammadi, M. A., Ahmadpour, E., Rahimi, M. T., Alizadeh, A., ... Saravanan, M. (2017). Green chemical synthesis of gold nanoparticles by using Penicillium aculeatum and their scolicidal activity against hydatid cyst protoscolices of Echinococcus granulosus. *Environmental Science and Pollution Research International*, *24*(6), 5800–5810. doi:10.100711356-016-8291-8 PMID:28054267

Barakat, M. A. (2011). New trends in removing heavy metals from industrial wastewater. *Arabian Journal of Chemistry*, *4*(4), 361–377. doi:10.1016/j.arabjc.2010.07.019

Barakat, M., Al-Hutailah, R., Hashim, M., Qayyum, E., & Kuhn, J. (2013). Titania-supported silver-based bimetallic nanoparticles as photocatalysts. *Environmental Science and Pollution Research International*, *20*(6), 3751–3759. doi:10.100711356-012-1306-1 PMID:23161501

Baran, E., & Hasirci, V. (2002). In vivo half life of nanoencapsulated L-asparaginase. *Journal of Materials Science. Materials in Medicine*, *13*(12), 1113–1121. doi:10.1023/A:1021125617828 PMID:15348652

Barillet, S., Jugan, M. L., Laye, M., Leconte, Y., Herlin-Boime, N., Reynaud, C., & Carriere, M. (2010). In vitro evaluation of SiC nanoparticles impact on A549 pulmonary cells: Cyto-, genotoxicity and oxidative stress. *Toxicology Letters*, *198*(3), 324–330. doi:10.1016/j.toxlet.2010.07.009 PMID:20655996

Barnes, C. A., Elsaesser, A., Arkusz, J., Smok, A., Palus, J., Lesniak, A., ... Howard, C. V. (2008). Reproducible comet assay of amorphous silica nanoparticles detects no genotoxicity. *Nano Letters*, *8*(9), 3069–3074. doi:10.1021/nl801661w PMID:18698730

Baruah, S., & Dutta, J. (2009). Nanotechnology applications in pollution sensing and degradation in agriculture: A review. *Environmental Chemistry Letters*, *7*(3), 191–204. doi:10.100710311-009-0228-8

Batley, G. E., Kirby, J. K., & McLaughlin, M. J. (2012). Fate and risks of nanomaterials in aquatic and terrestrial environments. *Accounts of Chemical Research*, *46*(3), 854–862. doi:10.1021/ar2003368 PMID:22759090

Bauer, B., Jones, D. J., & Roziere, J. (2000). Electrochemical characterization of sulfonated polyetherketone membranes. *Journal of New Materials for Electrochemical Systems*, *3*, 93–98.

Bayer, T., Bishop, S. R., Nishihara, M., Sasaki, K., & Lyth, S. M. (2014). Characterization of a graphene oxide membrane fuel cell. *Journal of Power Sources*, *272*, 239–247. doi:10.1016/j.jpowsour.2014.08.071

Bazrafshan, E., Mostafapour, F. K., Hosseini, A. R., Raksh Khorshid, A., & Mahvi, A. H. (2012). Decolorisation of reactive red 120 dye by using single-walled carbon nanotubes in aqueous solutions. *Journal of Chemistry*, 2013, 1-8. doi:10.1155/2013/938374

Bechhold, H. (1907). Kolloidstudien mit der Filtrationsmethode. *Zeitschrift für Physikalische Chemie*, *60*(1), 257–318. doi:10.1515/zpch-1907-0113

Beer, C. (2016). Nanotoxicology and Regulatory Affairs. In K. A. Howard, T. Vorup-Jensen, & D. Peer (Eds.), *Nanomedicine: Nanotechnology, Biology, and Medicine* (pp. 279–310). New York, NY: Springer New York.

Beer, C., Foldbjerg, R., Hayashi, Y., Sutherland, D. S., & Autrup, H. (2012). Toxicity of silver nanoparticles - nanoparticle or silver ion? *Toxicology Letters*, *208*(3), 286–292. doi:10.1016/j.toxlet.2011.11.002 PMID:22101214

Bello, B. A., Khan, S. A., Khan, J. A., Syed, F. Q., Mirza, M. B., Shah, L., & Khan, S. B. (2017). Anticancer, antibacterial and pollutant degradation potential of silver nanoparticles from *Hyphaene thebaica. Biochemical and Biophysical Research Communications*, *490*(3), 889–894. doi:10.1016/j.bbrc.2017.06.136 PMID:28648600

Benelli, G., & Lukehart, C. M. (2017). Special Issue: Applications of Green-Synthesized Nanoparticles in Pharmacology, Parasitology and Entomology. *Journal of Cluster Science*, *28*(1), 1–2. doi:10.100710876-017-1165-5

Benrabaa, R., Löfberg, A., Rubbens, A., Bordes-Richard, E., Vannier, R. N., & Barama, A. (2013). Structure, reactivity and catalytic properties of nanoparticles of nickel ferrite in the dry reforming of methane. *Catalysis Today*, *203*, 188–195. doi:10.1016/j.cattod.2012.06.002

Bergeson, L. L. (2013). *Sustainable nanomaterials: emerging governance systems*. ACS Publications.

Bernd, N. (2010). *Pollution prevention and treatment using nanotechnology*. Wiley Online Library. doi:10.1002/9783527628155.nanotech010

Betha, R., Behera, S. N., & Balasubramanian, R. (2014). 2013 Southeast Asian smoke haze: Fractionation of particulate-bound elements and associated health risk. *Environmental Science & Technology*, *48*(8), 4327–4335. doi:10.1021/es405533d PMID:24646334

Bhabra, G., Sood, A., Fisher, B., Cartwright, L., Saunders, M., Evans, W. H., ... Case, C. P. (2009). Nanoparticles can cause DNA damage across a cellular barrier. *Nat Nano*, *4*(12), 876–883. doi:10.1038/nnano.2009.313 PMID:19893513

Bhatnagar, A., & Sillanpää, M. (2014). Application of nanoadsorbents in water treatment. *Nanomaterials for Environmental Protection*, 237-247.

Bhatnagar, A. (2013). *Application of Adsorbents for Water Pollution Control*. Sharjah, UAE: Bentham Science Publishers.

Bhatnagar, A., Ji, M., Choi, Y. H., Jung, W., Lee, S. H., Kim, S. J., ... Kang, J. W. (2008). Removal of Nitrate from Water by Adsorption onto Zinc Chloride Treated Activated Carbon. *Separation Science and Technology*, *43*(4), 886–907. doi:10.1080/01496390701787461

Bhatnagar, A., & Sillanpää, M. (2011). A review of emerging adsorbents for nitrate removal from water. *Chemical Engineering Journal*, *168*(2), 493–504. doi:10.1016/j.cej.2011.01.103

Bhattacharya, A. K., Mandal, S. N., & Das, S. K. (2008). Heavy metals accumulation in water, sediment and tissues of different edible fishes in upper stretch of Gangetic West Bengal. *Trends in Applied Sciences Research*, *3*(1), 61–68. doi:10.3923/tasr.2008.61.68

Bhaumik, M., Agarwal, S., Gupta, V. K., & Maity, A. (2016). Enhanced removal of Cr (VI) from aqueous solutions using polypyrrole wrapped oxidized MWCNTs nanocomposites adsorbent. *Journal of Colloid and Interface Science*, *470*, 257–267. doi:10.1016/j.jcis.2016.02.054 PMID:26962976

Bhaumik, M., Maity, A., Srinivasu, V. V., & Onyango, M. S. (2011). Enhanced removal of Cr(VI) from aqueous solution using polypyrrole/Fe3O4 magnetic nanocomposite. *Journal of Hazardous Materials*, *190*(1), 381–390. doi:10.1016/j.jhazmat.2011.03.062 PMID:21497438

Bhave, R. (2012). *Inorganic Membranes Synthesis, Characteristics and Applications: Synthesis, characteristics, and applications*. Springer Science & Business Media.

Bhawana, P., & Fulekar, M. H. (2012). Nanotechnology: Remediation technologies to clean up the environmental pollutants. *Research Journal of Chemical Sciences*, *2*(2), 90–96.

Bhutta, Z. A., Sommerfeld, J., Lassi, Z. S., Salam, R. A., & Das, J. K. (2014). Global burden, distribution, and interventions for infectious diseases of poverty. *Infectious Diseases of Poverty*, *3*(1), 21. doi:10.1186/2049-9957-3-21 PMID:25110585

Bibi, M., Hashmi, M. Z., & Malik, R. N. (2015). Human exposure to arsenic in groundwater from Lahore district, Pakistan. *Environmental Toxicology and Pharmacology*, *39*(1), 42–52. doi:10.1016/j.etap.2014.10.020 PMID:25434761

Biesheuvel, P., Zhao, R., Porada, S., & Van der Wal, A. (2011). Theory of membrane capacitive deionization including the effect of the electrode pore space. *Journal of Colloid and Interface Science*, *360*(1), 239–248. doi:10.1016/j.jcis.2011.04.049 PMID:21592485

Bilgili, M. S. (2006). Adsorption of 4-chlorophenol from aqueous solutions by xad-4 resin: Isotherm, kinetic, and thermodynamic analysis. *Journal of Hazardous Materials*, *137*(1), 157–164. doi:10.1016/j.jhazmat.2006.01.005 PMID:16487655

Bilstad, T. (1997). Membrane operations. *Water Science and Technology*, *36*(2-3), 17–24.

Bina, B., Amin, M., Rashidi, A., & Pourzamani, H. (2012). Benzene and toluene removal by carbon nanotubes from aqueous solution. *Archives of Environmental Protection*, *38*(1), 3–25. doi:10.2478/v10265-012-0001-0

Bindhu, M. R., & Umadevi, M. (2013). Synthesis of monodispersed silver nanoparticles using Hibiscus cannabinus leaf extract and its antimicrobial activity. *Spectrochimica Acta. Part A: Molecular and Biomolecular Spectroscopy*, *101*, 184–190. doi:10.1016/j.saa.2012.09.031 PMID:23103459

Binning, G., & Quate, C. F. (1986). Article. *Physical Review Letters*, *56*, 930–933. doi:10.1103/PhysRevLett.56.930 PMID:10033323

Biskup, B., & Subotic, B. (2004). Kinetic analysis of the exchange processes between sodium ions from zeolite A and cadmium, copper and nickel ions from solutions. *Separation and Purification Technology*, *37*(1), 17–31. doi:10.1016/S1383-5866(03)00220-X

Biswas, P., & Wu, C.-Y. (2005). Nanoparticles and the Environment. *Journal of the Air & Waste Management Association*, *55*(6), 708–746. doi:10.1080/10473289.2005.10464656 PMID:16022411

Bogdanowicz, K. A., Tylkowski, B., & Giamberini, M. (2013). Preparation and characterization of light-sensitive microcapsules based on a liquid crystalline polyester. *Langmuir*, *29*(5), 1601–1608. doi:10.1021/la3038878 PMID:23245267

Boissiere, C., Larbot, A., & Prouzet, E. (2000). Synthesis of mesoporous MSU-X materials using inexpensive silica sources. *Chemistry of Materials*, *12*(7), 1937–1940. doi:10.1021/cm001012m

Boland, D. J., Brooker, M. I. H., Chippendale, G. M., Hall, N., Hyland, B. P. M., Johnston, R. D., . . . Turner, J. D. (2006). Forest Trees Of Australia (5th ed.). CSIRO Publishing.

Bonetto, L. R., Ferrarini, F., deMarco, C., Crespo, J. S., Guégan, R., & Giovanela, M. (2015). Removal of methyl violet 2B dye from aqueous solution using a magnetic composite as an adsorbent. *Journal of Water Process Engineering*, *6*, 11–20. doi:10.1016/j.jwpe.2015.02.006

Borase, H. P., Salunke, B. K., Salunkhe, R. B., Patil, C. D., Hallsworth, J. E., Kim, B. S., & Patil, S. V. (2014). Plant extract: A promising biomatrix for ecofriendly, controlled synthesis of silver nanoparticles. *Applied Biochemistry and Biotechnology*, *173*(1), 1–29. doi:10.100712010-014-0831-4 PMID:24622849

Bordes, P., Pollet, E., & Avérous, L. (2009). Nano-biocomposites: Biodegradable polyester/nanoclay systems. *Progress in Polymer Science*, *34*(2), 125–155. doi:10.1016/j.progpolymsci.2008.10.002

Bosman, A. W., Janssen, H. M., & Meijer, E. W. (1999). About dendrimers: Structure, physical properties, and applications. *Chemical Reviews*, *99*(7), 1665–1688. doi:10.1021/cr970069y PMID:11849007

Botes, M., & Cloete, T. E. (2010). The potential of nanofibers and nanobiocides in water purification. *Critical Reviews in Microbiology*, *36*(1), 68–81. doi:10.3109/10408410903397332 PMID:20088684

Bouwmeester, H., Dekkers, S., Noordam, M. Y., Hagens, W. I., Bulder, A. S., De Heer, C., ... Sips, A. J. (2009). Review of health safety aspects of nanotechnologies in food production. *Regulatory Toxicology and Pharmacology*, *53*(1), 52–62. doi:10.1016/j.yrtph.2008.10.008 PMID:19027049

Boxall, A. B., Tiede, K., & Chaudhry, Q. (2007). *Engineered nanomaterials in soils and water: how do they behave and could they pose a risk to human health?* Academic Press.

Brabec, E., Schulte, S., & Richards, P. L. (2002). Impervious surfaces and water quality: A review of current literature and its implications for watershed planning. *CPL Bibliography*, *16*(4), 499–514.

Bradl, H. B. (2005). *Heavy Metals in the Environment: Origin, Interaction and Remediation*. Amsterdam, The Netherlands: Elsevier.

Brar, S. K., Verma, M., Tyagi, R., & Surampalli, R. (2010). Engineered nanoparticles in wastewater and wastewater sludge–Evidence and impacts. *Waste Management (New York, N.Y.)*, *30*(3), 504–520. doi:10.1016/j.wasman.2009.10.012 PMID:19926463

Bréchignac, C., Houdy, P., & Lahmani, M. (2008). *Nanomaterials and Nanochemistry*. Springer Berlin Heidelberg.

Brett, R., Brent, C., Nanthi, S.,B., Santiago, M., Marc, G., Christopher, M., . . . Georgina, M. (2004). *Arsenic in the New Zealand environment*. Paper presented at the SuperSoil 2004: 3rd Australian New Zealand Soils Conference, Sydney, Australia.

Brillas, E., & Martínez-Huitle, C. A. (2015). Decontamination of wastewaters containing synthetic organic dyes by electrochemical methods. An updated review. *Applied Catalysis B: Environmental*, *166*, 603–643. doi:10.1016/j.apcatb.2014.11.016

Brook, R. D., Rajagopalan, S., Pope, C. A., Brook, J. R., Bhatnagar, A., Diez-Roux, A. V., & Peters, A. (2010). Particulate matter air pollution and cardiovascular disease. *Circulation*, *121*(21), 2331–2378. doi:10.1161/CIR.0b013e3181dbece1 PMID:20458016

Brumfiel, G. (2003). Nanotechnology: A little knowledge. *Nature, 424*(6946), 246-248.

Brumfiel, G. (2003). Nanotechnology: A little knowledge. *Nature*, *424*(6946), 246–248. doi:10.1038/424246a PMID:12867950

Brunner, T. J., Wick, P., Manser, P., Spohn, P., Grass, R. N., Limbach, L. K., ... Stark, W. J. (2006). In vitro cytotoxicity of oxide nanoparticles: Comparison to asbestos, silica, and the effect of particle solubility. *Environmental Science & Technology*, *40*(14), 4374–4381. doi:10.1021/es052069i PMID:16903273

Bundschuh, J., & Maity, J. P. (2015). Geothermal arsenic: Occurrence, mobility and environmental implications. *Renewable & Sustainable Energy Reviews*, *42*, 1214–1222. doi:10.1016/j.rser.2014.10.092

Burkhard, R., Deletic, A., & Craig, A. (2000). Techniques for water and wastewater management: A review of techniques and their integration in planning. *Urban Water*, *2*(3), 197–221. doi:10.1016/S1462-0758(00)00056-X

Burleigh, M. C., & Dai, S. (2004). Functionalized Nanoporous Adsorbents for Environmental Remediation. *Series on Chemical Engineering*, *4*, 756–771. doi:10.1142/9781860946561_0024

Calvo, J. I., Hernandez, A., Pradanos, P., Martınez, L., & Bowen, W. R. (1995). Pore size distributions in microporous membranes II. Bulk characterization of track-etched filters by air porometry and mercury porosimetry. *Journal of Colloid and Interface Science*, *176*(2), 467–478. doi:10.1006/jcis.1995.9944

Camargo, J. A., Alonso, A., & Salamanca, A. (2005). Nitrate toxicity to aquatic animals: A review with new data for freshwater invertebrates. *Chemosphere*, *58*(9), 1255–1267. doi:10.1016/j.chemosphere.2004.10.044 PMID:15667845

Cao, G. (2004). *Nanostructures & Nanomaterials: Synthesis, Properties & Applications*. Imperial College Press. doi:10.1142/p305

Cao, R., Brakenhielm, E., Pawliuk, R., Wariaro, D., Post, M. J., Wahlberg, E., ... Cao, Y. (2003). Angiogenic synergism, vascular stability and improvement of hind-limb ischemia by a combination of PDGF-BB and FGF-2. *Nature Medicine*, *9*(5), 604–613. doi:10.1038/nm848 PMID:12669032

Cappelletti, G., Ardizzone, S., Bianchi, C. L., Gialanella, S., Naldoni, A., Pirola, C., & Ragaini, V. (2008). Photodegradation of Pollutants in Air: Enhanced Properties of Nano-TiO2Prepared by Ultrasound. *Nanoscale Research Letters*, *4*(2), 97–105. doi:10.100711671-008-9208-3 PMID:20596293

Carbon black, titanium dioxide, and talc. (2010). *IARC Monographs on the Evaluation of Carcinogenic Risks to Humans*, *93*, 1-413.

Cardenas, S. A., Velazquez, T. G., Revilla, G. O., Cortez, M., & Perera, B. G. (2005). Adsorption of phenol and dichlorophenol from aqueous solutions by porous clay heterostructure. *Journal of the Mexican Chemical Society*, *49*(3), 287–291.

Carević, M. V., Abazović, N. D., Novaković, T. B., Pavlović, V. B., & Čomor, M. I. (2016). Zirconium dioxide nanopowders with incorporated Si4+ ions as efficient photocatalyst for degradation of trichlorophenol using simulated solar light. *Applied Catalysis B: Environmental*, *195*, 112–120. doi:10.1016/j.apcatb.2016.05.005

Carmen, Z., & Daniela, S. (2012). *Textile organic dyes–characteristics, polluting effects and separation/elimination procedures from industrial effluents–a critical overview.* Paper presented at the Organic Pollutants Ten Years After the Stockholm Convention-Environmental and Analytical Update. 10.5772/32373

Carretero, P., Molina, S., Lozano, A., de Abajo, J., Calvo, J. I., Prádanos, P., & Hernández, A. (2013). Liquid–liquid displacement porosimetry applied to several MF and UF membranes. *Desalination*, *327*, 14–23. doi:10.1016/j.desal.2013.08.001

Carrott, P., Carrott, M. R., & Roberts, R. (1991). Physical adsorption of gases by microporous carbons. *Colloids and Surfaces*, *58*(4), 385–400. doi:10.1016/0166-6622(91)80217-C

Carstea, E. M., Bridgeman, J., Baker, A., & Reynolds, D. M. (2016). Fluorescence spectroscopy for wastewater monitoring: A review. *Water Research*, *95*, 205–219. doi:10.1016/j.watres.2016.03.021 PMID:26999254

Carter, D., & Wing, J. (2013). The fuel cell industry review 2013. *Fuel Cell Today*, *1*, 1–46.

Cascone, M. G., Lazzeri, L., Carmignani, C., & Zhu, Z. (2002). Gelatin nanoparticles produced by a simple W/O emulsion as delivery system for methotrexate. *Journal of Materials Science. Materials in Medicine*, *13*(5), 523–526. doi:10.1023/A:1014791327253 PMID:15348607

Cassano, A., Donato, L., Conidi, C., & Drioli, E. (2008). Recovery of bioactive compounds in kiwifruit juice by ultrafiltration. *Innovative Food Science & Emerging Technologies*, *9*(4), 556–562. doi:10.1016/j.ifset.2008.03.004

Cassir, M., Ceballos, A. M., Ringuede, A., & Lair, V. (2016). Molten carbonate fuel cells. *Compendium of Hydrogen Energy*, *3*, 71–87. doi:10.1016/B978-1-78242-363-8.00003-7

Cedillo-González, E. I., Riccò, R., Montorsi, M., Montorsi, M., Falcaro, P., & Siligardi, C. (2014). Self-cleaning glass prepared from a commercial TiO 2 nano-dispersion and its photocatalytic performance under common anthropogenic and atmospheric factors. *Building and Environment*, *71*, 7–14. doi:10.1016/j.buildenv.2013.09.007

Cele, N. P. (2010). Carbon Nanotubes Based NafionComposite Membranes for Fuel Cell Applications. *Fuel Cells (Weinheim)*, *10*, 64–71.

Chae, K. J., Kim, K. Y., Choi, M. J., Yang, E., Kim, S., Ren, X., & Lee, M. (2014). Sulfonated polyether ether ketone (SPEEK)-based composite proton exchange membrane reinforced with nanofibers for microbial electrolysis cells. *Chemical Engineering Journal*, *254*, 393–398. doi:10.1016/j.cej.2014.05.145

Chandra, V., Park, J., Chun, Y., Lee, J. W., Hwang, I.-C., & Kim, K. S. (2010). Water-dispersible magnetite-reduced graphene oxide composites for arsenic removal. *ACS Nano*, *4*(7), 3979–3986. doi:10.1021/nn1008897 PMID:20552997

Chang, C. F., Chang, C. Y., Chen, K. H., Tsai, W. T., Shie, J. L., & Chen, Y. H. (2004a). Adsorption of Naphthalene on Zeolite from Aqueous Solution. *Journal of Colloid and Interface Science*, *277*(1), 29–34. doi:10.1016/j.jcis.2004.04.022 PMID:15276034

Chang, C., Lian, F., & Zhu, L. (2011). Simultaneous adsorption and degradation of γ-HCH by nZVI/Cu bimetallic nanoparticles with activated carbon support. *Environmental Pollution*, *159*(10), 2507–2514. doi:10.1016/j.envpol.2011.06.021 PMID:21763039

Chang, M.-C., Shu, H.-Y., Hsieh, W.-P., & Wang, M.-C. (2007). Remediation of soil contaminated with pyrene using ground nanoscale zero-valent iron. *Journal of the Air & Waste Management Association*, *57*(2), 221–227. doi:10.1080/10473289.2007.10465312 PMID:17355083

Chang, Y. C., & Chen, D. H. (2005). Adsorption Kinetics and Thermodynamics of Acid Dyes on a Carboxymethylated Chitosan-Conjugated Magnetic Nano-Adsorbent. *Macromolecular Bioscience*, *5*(3), 254–261. doi:10.1002/mabi.200400153 PMID:15768445

Chang, Y.-C., & Chen, D.-H. (2005). Preparation and adsorption properties of monodisperse chitosan-bound Fe 3 O 4 magnetic nanoparticles for removal of Cu (II) ions. *Journal of Colloid and Interface Science*, *283*(2), 446–451. doi:10.1016/j.jcis.2004.09.010 PMID:15721917

Chatterjee, S., Chatterjee, T., Lim, S.-R., & Woo, S. H. (2011). Effect of the addition mode of carbon nanotubes for the production of chitosan hydrogel core–shell beads on adsorption of Congo red from aqueous solution. *Bioresource Technology*, *102*(6), 4402–4409. doi:10.1016/j.biortech.2010.12.117 PMID:21277770

Chatterjee, S., Lee, M. W., & Woo, S. H. (2010). Adsorption of congo red by chitosan hydrogel beads impregnated with carbon nanotubes. *Bioresource Technology*, *101*(6), 1800–1806. doi:10.1016/j.biortech.2009.10.051 PMID:19962883

Chaturvedi, S., Dave, P. N., & Shah, N. (2012). Applications of nano-catalyst in new era. *Journal of Saudi Chemical Society*, *16*(3), 307–325. doi:10.1016/j.jscs.2011.01.015

Chatwal, G. R., & Anand, S. K. (2002). Instrumental method of Chemical Analysis: Thermal Methods (5th ed.). Himalaya Publishing House.

Chau, C. C., & Im, J. H. (1989). *U.S. Patent No. 4,874,568*. Washington, DC: U.S. Patent and Trademark Office.

Chaudhary, M., Mishra, S., & Kumar, A. (2017). Estimation of water pollution and probability of health risk due to imbalanced nutrients in River Ganga, India. *International Journal of River Basin Management*, *15*(1), 53–60. doi:10.1080/15715124.2016.1205078

Chauhan, A., & Chauhan, P. (2014). Article. *Journal of Analytical & Bioanalytical Techniques*, *5*(5), 1–5.

Chellaram, C., Murugaboopathi, G., John, A. A., Sivakumar, R., Ganesan, S., Krithika, S., & Priya, G. (2014). Significance of Nanotechnology in Food Industry. *APCBEE Procedia*, *8*(Supplement C), 109–113. doi:10.1016/j.apcbee.2014.03.010

Chella, S., Kollu, P., Komarala, E. V. P., Doshi, S., Saranya, M., Felix, S., ... Venugopal, V. (2015). Solvothermal synthesis of MnFe 2 O 4-graphene composite—Investigation of its adsorption and antimicrobial properties. *Applied Surface Science*, *327*, 27–36. doi:10.1016/j.apsusc.2014.11.096

Chen, C., Liang, B., Ogino, A., Wang, X., & Nagatsu, M. (2009). Oxygen functionalization of multiwall carbon nanotubes by microwave-excited surface-wave plasma treatment. *The Journal of Physical Chemistry. C, Nanomaterials and Interfaces*, *113*(18), 7659.

Chen, A. H., Liu, S. C., Chen, C. Y., & Chen, C.-Y. (2008). Comparative adsorption of Cu(II), Zn(II), and Pb(II) ions in aqueous solution on the crosslinked chitosan with epichlorohydrin. *Journal of Hazardous Materials*, *154*(1-3), 184–191. doi:10.1016/j.jhazmat.2007.10.009 PMID:18031930

Chen, C. Y., & Chiang, C. L. (2008). Preparation of cotton fibers with antibacterial silver nanoparticles. *Materials Letters*, *62*(21-22), 3607–3609. doi:10.1016/j.matlet.2008.04.008

Chen, C., & Wang, X. (2006). Adsorption of Ni (II) from aqueous solution using oxidized multiwall carbon nanotubes. *Industrial & Engineering Chemistry Research*, *45*(26), 9144–9149. doi:10.1021/ie060791z

Chen, D., Huang, F., Ren, G., Li, D., Zheng, M., Wang, Y., & Lin, Z. (2010). ZnS nano-architectures: Photocatalysis, deactivation and regeneration. *Nanoscale*, *2*(10), 2062–2064. doi:10.1039/c0nr00171f PMID:20689875

Chen, D., Li, Y., Zhang, J., Zhou, J. Z., Guo, Y., & Liu, H. (2012). Magnetic Fe3O4/ZnCr-layered double hydroxide composite with enhanced adsorption and photocatalytic activity. *Chemical Engineering Journal*, *185–186*, 120–126. doi:10.1016/j.cej.2012.01.059

Chen, D., Sharma, S. K., & Mudhoo, A. (2011). *Handbook on applications of ultrasound: sonochemistry for sustainability*. CRC Press.

Chen, F., Liu, S., & Yu, J. (2016). Efficient removal of gaseous formaldehyde in air using hierarchical titanate nanospheres with in situ amine functionalization. *Physical Chemistry Chemical Physics*, *18*(27), 18161–18168. doi:10.1039/C6CP03037H PMID:27331932

Chen, G. C., Shan, X. Q., Pei, Z. G., Wang, H. H., Zheng, L. R., Zhang, J., & Xie, Y. (2011). Adsorption of diuron and dichlobenil on multiwalled carbon nanotubes as affected by lead. *Journal of Hazardous Materials*, *188*(1-3), 156–163. doi:10.1016/j.jhazmat.2011.01.095 PMID:21324587

Cheng, L., Liu, S., & Dixon, K. (1998). Analysis of repair and mutagenesis of chromium-induced DNA damage in yeast, mammalian cells, and transgenic mice. *Environmental Health Perspectives*, *106*(Suppl 4), 1027–1032. doi:10.1289/ehp.98106s41027 PMID:9703488

Cheng, X., Huang, X., Wang, X., Zhao, B., Chen, A., & Sun, D. (2009). Phosphate adsorption from sewage sludge filtrate using zinc–aluminum layered double hydroxides. *Journal of Hazardous Materials, 169*(1–3), 958–964. doi:10.1016/j.jhazmat.2009.04.052 PMID:19443104

Cheng, Z., Tan, A. L. K., Tao, Y., Shan, D., Ting, K. E., & Yin, X. J. (2012). Synthesis and characterization of iron oxide nanoparticles and applications in the removal of heavy metals from industrial wastewater. *International Journal of Photoenergy*.

Chen, H., Li, J., Shao, D., Ren, X., & Wang, X. (2012). Poly (acrylic acid) grafted multiwall carbon nanotubes by plasma techniques for Co (II) removal from aqueous solution. *Chemical Engineering Journal, 210*, 475–481. doi:10.1016/j.cej.2012.08.082

Chen, H., Seiber, J. N., & Hotze, M. (2014). ACS Select on Nanotechnology in Food and Agriculture: A Perspective on Implications and Applications. *Journal of Agricultural and Food Chemistry, 62*(6), 1209–1212. doi:10.1021/jf5002588 PMID:24479582

Chen, J., Hamon, M. A., Hu, H., Chen, Y., Rao, A. M., Eklund, P. C., & Haddon, R. C. (1998). Solution properties of single-walled carbon nanotubes. *Science, 282*(5386), 95–98. doi:10.1126cience.282.5386.95 PMID:9756485

Chen, J., Jiang, J., Zhang, F., Yu, H., & Zhang, J. (2004). Cytotoxic environmentally relevant chlorophenols on L929 cells and their mechanisms. *Cell Biology and Toxicology, 20*(3), 183–196. doi:10.1023/B:CBTO.0000029468.89746.64 PMID:15250542

Chen, J., & Poon, C. (2009). Photocatalytic construction and building materials: From fundamentals to applications. *Building and Environment, 44*(9), 1899–1906. doi:10.1016/j.buildenv.2009.01.002

Chen, K., He, J., Li, Y., Cai, X., Zhang, K., Liu, T., ... Liu, J. (2017). Removal of cadmium and lead ions from water by sulfonated magnetic nanoparticle adsorbents. *Journal of Colloid and Interface Science, 494*(Supplement C), 307–316. doi:10.1016/j.jcis.2017.01.082 PMID:28161502

Chen, L., Yu, T. L., Lin, H. L., & Yeh, S. H. (2008). Nafion/PTFE and zirconium phosphate modified Nafion/PTFE composite membranes for direct methanol fuel cells. *Journal of Membrane Science, 307*(1), 10–20. doi:10.1016/j.memsci.2007.03.008

Chen, L., Zhao, X., Pan, B., Zhang, W., Hua, M., Lv, L., & Zhang, W. (2015). Preferable removal of phosphate from water using hydrous zirconium oxide-based nanocomposite of high stability. *Journal of Hazardous Materials, 284*, 35–42. doi:10.1016/j.jhazmat.2014.10.048 PMID:25463215

Chen, M., Bao, C., Cun, T., & Huang, Q. (2017). One-pot synthesis of ZnO/oligoaniline nanocomposites with improved removal of organic dyes in water: Effect of adsorption on photocatalytic degradation. *Materials Research Bulletin, 95*, 459–467. doi:10.1016/j.materresbull.2017.08.017

Chen, M., & von Mikecz, A. (2005). Formation of nucleoplasmic protein aggregates impairs nuclear function in response to SiO2 nanoparticles. *Experimental Cell Research, 305*(1), 51–62. doi:10.1016/j.yexcr.2004.12.021 PMID:15777787

Chen, T., Zang, J., Wang, H., Nie, H., Wang, X., Shen, Z., ... Jia, G. (2012). Water-soluble taurine-functionalized multi-walled carbon nanotubes induce less damage to mitochondria of RAW 264.7 cells. *Journal of Nanoscience and Nanotechnology, 12*(10), 8008–8016. doi:10.1166/jnn.2012.6593 PMID:23421171

Chen, X., & Mao, S. S. (2007). Titanium dioxide nanomaterials: Synthesis, properties, modifications, and applications. *Chemical Reviews, 107*(7), 2891–2959. doi:10.1021/cr0500535 PMID:17590053

Chen, Y., Guo, J., & Kim, H. (2010). Preparation of poly(vinylidene fluoride)/phosphotungstic acid composite nanofiber membranes by electrospinning for proton conductivity. *Reactive & Functional Polymers*, *70*(1), 69–74. doi:10.1016/j.reactfunctpolym.2009.10.006

Chen, Y., Wang, L., Jiang, S., & Yu, H. (2003). Study on Novel Antibacterial Polymer Materials (I) Preparation of Zeolite Antibacterial Agents and Antibacterial Polymer Composite and Their Antibacterial Properties. *Journal of Polymer Materials*, *20*(3), 279–284.

Chen, Z., Meng, H., Xing, G., Chen, C., Zhao, Y., Jia, G., ... Zhao, F. (2006). Acute toxicological effects of copper nanoparticles in vivo. *Toxicology Letters*, *163*(2), 109–120. doi:10.1016/j.toxlet.2005.10.003 PMID:16289865

Chequer, F. M. D., Oliveira, G. A. R. d., Ferraz, E. R. A. c., Cardoso, J. C., Zanoni, M. V. B., & Oliveira, D. P. d. (2013). Textile Dyes: Dyeing Process and Environmental Impact. In M. Günay (Ed.), Eco-Friendly Textile Dyeing and Finishing. Rijeka: InTech.

Che, S., Garcia-Bennett, A. E., Yokoi, T., Sakamoto, K., Kunieda, H., Terasaki, O., & Tatsumi, T. (2003). Mesoporous Silica of Novel Structures with Periodic Modulations Synthesized by Anionic Surfactant Templating Route. *Nature Materials*, *2*, 801–805. doi:10.1038/nmat1022 PMID:14634644

Chiang, L. F., & Doong, R. A. (2014). Cu–TiO_2 nanorods with enhanced ultraviolet- and visible-light photoactivity for bisphenol degradation. *Journal of Hazardous Materials*, *277*, 84–92. doi:10.1016/j.jhazmat.2014.01.047 PMID:24556011

Chien, H. C., Tsai, L. D., Huang, C. P., Kang, C. Y., Lin, J. N., & Chang, F. C. (2013). Sulfonated graphene oxide/Nafion composite membranes for high-performance direct methanol fuel cells. *International Journal of Hydrogen Energy*, *38*(31), 13792–13801. doi:10.1016/j.ijhydene.2013.08.036

Chirag, M. P. N. (2015). Nanotechnology: future of environmental pollution control. *International Journal on Recent Innovation Trends in Computing and Communication, 3*(2), 79-81.

Chithrani, B. D., Ghazani, A. A., & Chan, W. C. W. (2006). Determining the Size and Shape Dependence of Gold Nanoparticle Uptake into Mammalian Cells. *Nano Letters*, *6*(4), 662–668. doi:10.1021/nl052396o PMID:16608261

Choi, H. S., Ashitate, Y., Lee, J. H., Kim, S. H., Matsui, A., Insin, N., . . . Tsuda, A. (2010). *Rapid translocation of nanoparticles from the lung airspaces to the body*. Retrieved from https://www.nature.com/articles/nbt.1696#supplementary-information

Cho, K. Y., Jung, H. Y., Choi, N. S., Sung, S. J., Pak, J. K., Choi, J. H., & Sung, Y. E. (2005). A coated Nafion membrane with a PVDF copolymer/Nafion blend for direct methanol fuel cells (DMFCs). *Solid State Ionics*, *176*(39-40), 3027–3030. doi:10.1016/j.ssi.2005.09.048

Chong, M. N., Vimonses, V., Lei, S., Jin, B., Chow, C., & Saint, C. (2009). Synthesis and characterisation of novel titania impregnated kaolinite nano-photocatalyst. *Microporous and Mesoporous Materials*, *117*(1), 233–242. doi:10.1016/j.micromeso.2008.06.039

Choong, T. S. Y., Chuah, T. G., Robiah, Y., Gregory Koay, F. L., & Azni, I. (2007). Arsenic toxicity, health hazards and removal techniques from water: An overview. *Desalination*, *217*(1-3), 139–166. doi:10.1016/j.desal.2007.01.015

Choudhary, B. C., Paul, D., Gupta, T., Tetgure, S. R., Garole, V. J., Borse, A. U., & Garole, D. J. (2017). Photocatalytic reduction of organic pollutant under visible light by green route synthesized gold nanoparticles. *Journal of Environmental Sciences (China)*, *55*, 236–246. doi:10.1016/j.jes.2016.05.044 PMID:28477818

Choudhary, V. R., Patil, V. P., Jana, P., & Uphade, B. S. (2008). Nano-gold supported on Fe2O3: A highly active catalyst for low temperature oxidative destruction of methane green house gas from exhaust/waste gases. *Applied Catalysis A, General*, *350*(2), 186–190. doi:10.1016/j.apcata.2008.08.008

Chowdhury, S. R., & Yanful, E. K. (2010). Arsenic and chromium removal by mixed magnetite–maghemite nanoparticles and the effect of phosphate on removal. *Journal of Environmental Management*, *91*(11), 2238–2247. doi:10.1016/j.jenvman.2010.06.003 PMID:20598797

Chu, M., Wu, Q., Yang, H., Yuan, R., Hou, S., Yang, Y., ... Sheng, L. (2010). Transfer of quantum dots from pregnant mice to pups across the placental barrier. *Small*, *6*(5), 670–678. doi:10.1002mll.200902049 PMID:20143348

Cifuentes, Z., Custardoy, L., de la Fuente, J. M., Marquina, C., Ibarra, M. R., Rubiales, D., & Pérez-de-Luque, A. (2010). Absorption and translocation to the aerial part of magnetic carbon-coated nanoparticles through the root of different crop plants. *Journal of Nanobiotechnology*, *8*(1), 26. doi:10.1186/1477-3155-8-26 PMID:21059206

Cincotti, A., Mameli, A., Locci, A. M., Orru, R., & Cao, G. (2006). Heavy metals uptake by Sardinian natural zeolites: Experiment and modeling. *Industrial & Engineering Chemistry Research*, *45*(3), 1074–1084. doi:10.1021/ie050375z

Clean Air Act, NAAQS (National Ambient Air Quality Standards (NAAQS), Air and Radiation US EPA.

Clearfield, A. (1988). Role of Ion Exchange in Solid-state Chemistry. *Chemical Reviews*, *88*(1), 125–148. doi:10.1021/cr00083a007

Cloete, T. E. (2010). *Nanotechnology in water treatment applications*. Horizon Scientific Press.

Coats, A. W., & Redfren, J. P. (1963). Thermogravimetric Analysis. *RE:view*, *88*, 906–924.

Corriu, R. J. P., Mehdi, A., & Reye, C. (2005). Molecular chemistry and nanosciences: On the way to interactive materials. *Journal of Materials Chemistry*, *15*(40), 4285–4294. doi:10.1039/b504536n

Cortie, M., & Van der Lingen, E. (2002). *Catalytic gold nano-particles*. Paper presented at the Materials forum.

Costa, A. S., Romão, L. P. C., Araújo, B. R., Lucas, S. C. O., Maciel, S. T. A., Wisniewski, A. Jr, & Alexandre, M. R. (2012). Environmental strategies to remove volatile aromatic fractions (BTEX) from petroleum industry wastewater using biomass. *Bioresource Technology*, *105*, 31–39. doi:10.1016/j.biortech.2011.11.096 PMID:22178487

Coutanceau, C., Koffi, R. K., Leger, J. M., Marestin, K., Mercier, R., Nayoze, C., & Capron, P. (2006). Development of materials for mini DMFC working at room temperature for portable applications. *Journal of Power Sources*, *160*(1), 334–339. doi:10.1016/j.jpowsour.2006.01.073

Cuenya, B. R. (2010). Synthesis and catalytic properties of metal nanoparticles: Size, shape, support, composition, and oxidation state effects. *Thin Solid Films*, *518*(12), 3127–3150. doi:10.1016/j.tsf.2010.01.018

Cuenya, B. R., Baeck, S.-H., Jaramillo, T. F., & McFarland, E. W. (2003). Size-and support-dependent electronic and catalytic properties of Au0/Au3+ nanoparticles synthesized from block copolymer micelles. *Journal of the American Chemical Society*, *125*(42), 12928–12934. doi:10.1021/ja036468u PMID:14558841

Cui, D., Tian, F., Ozkan, C. S., Wang, M., & Gao, H. (2005). Effect of single wall carbon nanotubes on human HEK293 cells. *Toxicology Letters*, *155*(1), 73–85. doi:10.1016/j.toxlet.2004.08.015 PMID:15585362

Cui, H., Li, Q., Gao, S., & Shang, J. K. (2012). Strong adsorption of arsenic species by amorphous zirconium oxide nanoparticles. *Journal of Industrial and Engineering Chemistry*, *18*(4), 1418–1427. doi:10.1016/j.jiec.2012.01.045

Cui, L.-F., Yang, Y., Hsu, C.-M., & Cui, Y. (2009). Carbon– silicon core– shell nanowires as high capacity electrode for lithium ion batteries. *Nano Letters*, *9*(9), 3370–3374. doi:10.1021/nl901670t PMID:19655765

Cui, L., Geng, Q., Gong, C., Liu, H., Zheng, G., Wang, G., ... Wen, S. (2015). Novel sulfonated poly (ether ether ketone)/ silica coated carbon nanotubes high-performance composite membranes for direct methanol fuel cell. *Polymers for Advanced Technologies*, *26*(5), 457–464. doi:10.1002/pat.3473

Cullity, B. D., & Graham, C. D. (2009). *Introduction to Magnetic Materials*. John Wiley & Sons.

Cumming, G. S., & Allen, C. R. (2017). Protected areas as social-ecological systems: Perspectives from resilience and social-ecological systems theory. *Ecological Applications*, *27*(6), 1709–1717. doi:10.1002/eap.1584 PMID:28618079

Cuperus, F. P., & Smolders, C. A. (1991). Characterization of UF membranes: Membrane characteristics and characterization techniques. *Advances in Colloid and Interface Science*, *34*, 135–173. doi:10.1016/0001-8686(91)80049-P

Curtis, J., Greenberg, M., Kester, J., Phillips, S., & Krieger, G. (2006). Nanotechnology and nanotoxicology: A primer for clinicians. *Toxicological Reviews*, *25*(4), 245–260. doi:10.2165/00139709-200625040-00005 PMID:17288496

Da'na, E., & Sayari, A. (2010). Optimization of copper removal efficiency by adsorption on amine modified SBA-15: Experimental design methodology. *Chemical Engineering Journal*, *167*(1), 91–98. doi:10.1016/j.cej.2010.12.005

Da'na, E., & Sayari, A. (2011). Adsorption of copper on amine-functionalized SBA-15 prepared by cocondensation: Equilibrium properties. *Chemical Engineering Journal*, *166*(1), 445–453. doi:10.1016/j.cej.2010.11.016

Da'na, E., Silva, N. D., & Sayari, A. (2011). Adsorption of copper on amine-functionalized SBA-15 prepared by co-condensation: Kinetics properties. *Chemical Engineering Journal*, *166*(1), 454–459. doi:10.1016/j.cej.2010.11.017

Dadfarnia, S., Haji Shabani, A. M., Moradi, S. E., & Emami, S. (2015). Methyl red removal from water by iron based metal-organic frameworks loaded onto iron oxide nanoparticle adsorbent. *Applied Surface Science*, *330*, 85–93. doi:10.1016/j.apsusc.2014.12.196

Dai, Z., Qin, F., Zhao, H. P., Ding, J., Liu, Y. L., & Chen, R. (2016). Crystal Defect Engineering of Aurivillius Bi_2MoO_6 by Ce Doping for Increased Reactive Species Production in Photocatalysis. *ACS Catalysis*, *6*(5), 3180–3192. doi:10.1021/acscatal.6b00490

Dalali, N., Khoramnezhad, M., Habibizadeh, M., & Faraji, M. (2011). Magnetic removal of acidic dyes from waste waters using surfactant-coated magnetite nanoparticles: optimization of process by Taguchi method. *Proceedings of International Conference on Environmental and Agriculture Engineering IPCBEE.*

Dang, V. B. H., Doan, H. D., Dang-Vuc, T., & Lohi, A. (2009). Equilibrium and kinetics of biosorption of cadmium(II) and copper(II) ions by wheat straw. *Bioresource Technology*, *100*(1), 211–219. doi:10.1016/j.biortech.2008.05.031 PMID:18599289

Danilczuk, M., Lund, A., Sadlo, J., Yamada, H., & Michalik, J. (2006). Conduction electron spin resonance of small silver particles, *Spectrochimica Acta-art A. Molecular and Biomolecular Spectroscopy*, *63*(1), 189–191. doi:10.1016/j.saa.2005.05.002 PMID:15978868

Dankovic, D., Kuempel, E., & Wheeler, M. (2007). An approach to risk assessment for TiO2. *Inhalation Toxicology*, *19*(sup1Suppl 1), 205–212. doi:10.1080/08958370701497754 PMID:17886069

Daraei, P., Madaeni, S. S., Ghaemi, N., Salehi, E., Khadivi, M. A., Moradian, R., & Astinchap, B. (2012). Novel polyethersulfone nanocomposite membrane prepared by PANI/Fe3O4 nanoparticles with enhanced performance for Cu(II) removal from water. *Journal of Membrane Science*, *415-416*, 250–259. doi:10.1016/j.memsci.2012.05.007

Darvish Manesh, S., Tasselli, F., Jansen, J. C., Tocci, E., Bazzarelli, F., Bernardo, P., & Van der Bruggen, B. (2011). Preparation of solvent stable polyphenyl sulfone hollow fiber nanofiltration membranes. *Journal of Membrane Science*, *384*(1), 89-96.

Das, R. K., Gogoi, N., & Bora, U. (2011). Green synthesis of gold nanoparticles using Nyctanthes arbortristis flower extract. *Bioprocess and Biosystems Engineering*, *34*(5), 615–619. doi:10.100700449-010-0510-y PMID:21229266

Das, S., Sen, B., & Debnath, N. (2015). Recent trends in nanomaterials applications in environmental monitoring and remediation. *Environmental Science and Pollution Research International*, *22*(23), 18333–18344. doi:10.100711356-015-5491-6 PMID:26490920

Dave, P. N., & Chopda, L. V. (2014). Application of iron oxide nanomaterials for the removal of heavy metals. *Journal of Nanotechnology*.

Davydova, S. (2005). Heavy metals as toxicants in big cities. *Microchemical Journal*, *79*(1-2), 133–136. doi:10.1016/j.microc.2004.06.010

De Angelis, F., Pujia, A., Falcone, C., Iaccino, E., Palmieri, C., Liberale, C., ... De Laurentiis, A. (2010). Water soluble nanoporous nanoparticle for in vivo targeted drug delivery and controlled release in B cells tumor context. *Nanoscale*, *2*(10), 2230–2236. doi:10.1039/c0nr00161a PMID:20835434

De Beyer, J., Kollars, N., Edwards, N., & Cheung, H. (2004). *Research on Tobacco in China: An Annotated Bibilography of Research on Tobacco Use*. Health Effects, Policies, Farming and Industry.

De Jong, W. H., & Borm, P. J. A. (2008). Drug delivery and nanoparticles: Applications and hazards. *International Journal of Nanomedicine*, *3*(2), 133–149. doi:10.2147/IJN.S596 PMID:18686775

De Jong, W. H., Hagens, W. I., Krystek, P., Burger, M. C., Sips, A. J., & Geertsma, R. E. (2008). Particle size-dependent organ distribution of gold nanoparticles after intravenous administration. *Biomaterials*, *29*(12), 1912–1919. doi:10.1016/j.biomaterials.2007.12.037 PMID:18242692

De Martino, A., Iorio, M., Xing, B. S., & Capasso, R. (2012). Removal of 4-chloro-2-methylphenoxyacetic acid from water by sorption on carbon nanotubes and metal oxide nanoparticles. *RSC Advances*, *2*(13), 5693–5700. doi:10.1039/c2ra00038e

De Sales, P. F., Magriotis, Z. M., Rossi, M. A. L. S., Resende, R. F., & Nunes, C. A. (2015). Comparative analysis of tropaeolin adsorption onto raw and acid-treated kaolinite: Optimization by response surface methodology. *Journal of Environmental Management*, *151*, 144–152. doi:10.1016/j.jenvman.2014.12.008 PMID:25559496

De, D., Panda, P. K., Roy, M., Bhunia, S., & Jaman, A. I. (2013). Reinforcing effect of nanosilica on the properties of natural rubber/reclaimed ground rubber tire vulcanizates. *Polymer Engineering and Science*, *53*(2), 227–237. doi:10.1002/pen.23255

Deliyanni, E. A., Bakoyannakis, D. N., Zouboulis, A. I., & Matis, K. A. (2003). Sorption of As(V) ions by akaganeite-type nanocrystals. *Chemosphere*, *50*(1), 155–163. doi:10.1016/S0045-6535(02)00351-X PMID:12656241

Deluca, N. W., & Elabd, Y. A. (2006). Direct methanol fuel cell performance of Nafion®/poly(vinyl alcohol) blend membranes. *Journal of Power Sources*, *163*(1), 386–391. doi:10.1016/j.jpowsour.2006.09.009

Demirbas, A. (2009). Agricultural based activated carbons for the removal of dyes from aqueous solutions: A review. *Journal of Hazardous Materials*, *167*(1-3), 1–9. doi:10.1016/j.jhazmat.2008.12.114 PMID:19181447

Deng, J., Shao, Y. S., Gao, N. Y., Deng, Y., Tan, C. Q., Zhou, S. Q., & Hu, X. (2012a). Multiwalled carbon nanotubes as adsorbents for removal of herbicide diuron from aqueous solution. *Chemical Engineering Journal*, *193*, 339–347. doi:10.1016/j.cej.2012.04.051

Denis, P. A., & Iribarne, F. (2012). A first-principles study on the interaction between alkyl radicals and graphene. *Chemistry (Weinheim an der Bergstrasse, Germany)*, *18*(24), 7568–7574. doi:10.1002/chem.201103711 PMID:22532481

Deutsche, F. G. (2001). *Commission for the investigation of health hazards of chemical compounds in the work area MAK and BAT values 2001*. Report no. 37. Wiley-VCH, Weinheim.

Devi, R. S., & Gayathri, R. (2014). Green Synthesis of Zinc Oxide Nanoparticles by using Hibiscus rosa-sinensis. *International Journal of Current Engineering and Technology*, *4*(4).

Dhakras, P. A. (2011). Nanotechnology Applications in Water Purification and Waste Water Treatment: A Review. The Institute of Electrical and Electronics Engineers (professional association).

Dhawan, A., Taurozzi, J. S., Pandey, A. K., Shan, W., Miller, S. M., Hashsham, S. A., & Tarabara, V. V. (2006). Stable colloidal dispersions of C60 fullerenes in water: Evidence for genotoxicity. *Environmental Science & Technology*, *40*(23), 7394–7401. doi:10.1021/es0609708 PMID:17180994

Dhir, B., Sharmila, P., & Saradhi, P. P. (2009). Potential of aquatic macrophytes for removing contaminants from the environment. *Critical Reviews in Environmental Science and Technology*, *39*(9), 754–781. doi:10.1080/10643380801977776

Diallo, M. S., & Savage, N. (2005). Nanoparticles and water quality. *Journal of Nanoparticle Research*, *7*(4), 325–330. doi:10.100711051-005-8543-x

Diaz, C., Barrera, G., Segovia, M., Valenzuela, M. L., & Osiak, M. (2015). #x, & Dwyer, C. (2015). Crystallizing Vanadium Pentoxide Nanostructures in the Solid-State Using Modified Block Copolymer and Chitosan Complexes. *Journal of Nanomaterials*, *13*. doi:10.1155/2015/105157

Dil, E. A., Ghaedi, M., & Asfaram, A. (2017). The performance of nanorods material as adsorbent for removal of azo dyes and heavy metal ions: Application of ultrasound wave, optimization and modeling. *Ultrasonics Sonochemistry*, *34*, 792–802. doi:10.1016/j.ultsonch.2016.07.015 PMID:27773307

Dimitrova, P., Friedrich, K. A., Stimming, U., & Vogt, B. (2002). Modified Nafion-based membranes for use in direct methanol fuel cells. *Solid State Ionics*, *150*(1-2), 115–122. doi:10.1016/S0167-2738(02)00267-9

Ding, X., Ho, W., Shang, J., & Zhang, L. (2016). Self-doping promoted photocatalytic removal of no under visible light with bi2moo6: Indispensable role of superoxide ions. *Applied Catalysis B: Environmental*, *182*, 316–325. doi:10.1016/j.apcatb.2015.09.046

Ding, Y., & Bikson, B. (2010). Preparation and characterization of semi-crystalline poly (ether ether ketone) hollow fiber membranes. *Journal of Membrane Science*, *357*(1), 192–198. doi:10.1016/j.memsci.2010.04.021

Ditsch, A., Lindenmann, S., Laibinis, P. E., Wang, D. I., & Hatton, T. A. (2005). High-gradient magnetic separation of magnetic nanoclusters. *Industrial & Engineering Chemistry Research*, *44*(17), 6824–6836. doi:10.1021/ie048841s

Di, Z.-C., Ding, J., Peng, X.-J., Li, Y.-H., Luan, Z.-K., & Liang, J. (2006). Chromium adsorption by aligned carbon nanotubes supported ceria nanoparticles. *Chemosphere*, *62*(5), 861–865. doi:10.1016/j.chemosphere.2004.06.044 PMID:16403560

Donega, M. C., Liljeroth, P., & Vanmaekelbergh, D. (2005). Physicochemical evaluation of the hot-injection method, a synthesis route for monodisperse nanocrystals. *Small*, *1*(12), 1152–1162. doi:10.1002mll.200500239 PMID:17193409

Dong, C., Zhang, X., Cai, H., & Cao, C. (2014). Facile and one-step synthesis of monodisperse silver nanoparticles using gum acacia in aqueous solution. *Journal of Molecular Liquids*, *196*, 135–141. doi:10.1016/j.molliq.2014.03.009

Dong, H., Lin, B., Gilmore, K., Hou, T., Lee, S. T., & Li, Y. (2015). B 40 fullerene: An efficient material for CO_2 capture, storage and separation. *Current Applied Physics*, *15*(9), 1084–1089. doi:10.1016/j.cap.2015.06.008

Doole, G. J., Marsh, D., & Ramilan, T. (2013). Evaluation of agri-environmental policies for reducing nitrate pollution from New Zealand dairy farms accounting for firm heterogeneity. *Land Use Policy*, *30*(1), 57–66. doi:10.1016/j.landusepol.2012.02.007

Dubey, M., Bhadauria, S., & Kushwah, B. (2009). Green synthesis of nanosilver particles from extract of Eucalyptus hybrida (safeda) leaf. *Digest Journal of Nanomaterials and Biostructures*, *4*(3), 537–543.

Dubey, S. P., Dwivedi, A. D., Lahtinen, M., Lee, C., Kwon, Y. N., & Sillanpaa, M. (2013). Protocol for development of various plants leaves extract in single-pot synthesis of metal nanoparticles. *Spectrochimica Acta. Part A: Molecular and Biomolecular Spectroscopy*, *103*, 134–142. doi:10.1016/j.saa.2012.11.021 PMID:23257341

Dubey, S., Banerjee, S., Upadhyay, S. N., & Sharma, Y. C. (2017). Application of common nano-materials for removal of selected metallic species from water and wastewaters: A critical review. *Journal of Molecular Liquids*, *240*, 656–677. doi:10.1016/j.molliq.2017.05.107

Du, Q., Sun, J., Li, Y., Yang, X., Wang, X., Wang, Z., & Xia, L. (2014). Highly enhanced adsorption of congo red onto graphene oxide/chitosan fibers by wet-chemical etching off silica nanoparticles. *Chemical Engineering Journal*, *245*, 99–106. doi:10.1016/j.cej.2014.02.006

Duran, C., Ozdes, D., Gundogdu, A., & Senturk, H. B. (2011). Kinetics and isotherm analysis of basic dyes adsorption onto almond shell (Prunus dulcis) as a low cost adsorbent. *Journal of Chemical & Engineering Data*, *56*(5), 2136–2147. doi:10.1021/je101204j

Dutta, A. K., Maji, S. K., & Adhikary, B. (2014). γ-Fe 2 O 3 nanoparticles: An easily recoverable effective photo-catalyst for the degradation of rose bengal and methylene blue dyes in the waste-water treatment plant. *Materials Research Bulletin*, *49*, 28–34. doi:10.1016/j.materresbull.2013.08.024

Eapen, D. E., Suseendiran, S. R., & Rengaswamy, R. (2016). Phosphoric acid fuel cells. *Compendium of Hydrogen Energy*, *3*, 57–70. doi:10.1016/B978-1-78242-363-8.00002-5

Ebert, K., Fritsch, D., Koll, J., & Tjahjawiguna, C. (2004). Influence of inorganic fillers on the compaction behaviour of porous polymer based membranes. *Journal of Membrane Science*, *233*(1–2), 71–78. doi:10.1016/j.memsci.2003.12.012

Ebrahimian Pirbazari, A., Saberikhah, E., & Habibzadeh Kozani, S. S. (2014). Fe3O4–wheat straw: Preparation, characterization and its application for methylene blue adsorption. *Water Resources and Industry*, *7-8*(Supplement C), 23–37. doi:10.1016/j.wri.2014.09.001

Egger, S., Lehmann, R. P., Height, M. J., Loessner, M. J., & Schuppler, M. (2009). Antimicrobial properties of a novel silver-silica nanocomposite material. *Applied and Environmental Microbiology*, *75*(9), 2973–2976. doi:10.1128/AEM.01658-08 PMID:19270121

Ehiowemwenguan, G., Emoghene, A. O., & Inetianbor, J. E. (2014). Antibacterial and phytochemical analysis of Banana fruit peel. *IOSR Journal Of Pharmacy*, *4*(8), 18–25. doi:10.9790/3013-0408018025

Ehrampoush, M. H., Miria, M., Salmani, M. H., & Mahvi, A. H. (2015). Cadmium removal from aqueous solution by green synthesis iron oxide nanoparticles with tangerine peel extract. *Journal of Environmental Health Science & Engineering*, *13*(1), 84. doi:10.118640201-015-0237-4 PMID:26682059

Ekmekyapar, F., Aslan, A., Bayhan, Y. K., & Cakici, A. (2006). Biosorption of copper (II) by nonliving lichen biomass of Cladonia rangiformis hoffm. *Journal of Hazardous Materials*, *137*(1), 293–298. doi:10.1016/j.jhazmat.2006.02.003 PMID:16530938

El-Bendary, H., & El-Helaly, A. (2013). First record nanotechnology in agricultural: Silica nano-particles a potential new insecticide for pest control. *App. Sci. Report*, *4*(3), 241–246.

El-Latif, M. M. A., Ibrahim, A. M., Showman, M. S., & Hamide, R. R. A. (2013). Alumina/Iron Oxide nano composite for cadmium ions removal from aqueous solutions. *International Journal of Nonferrous Metallurgy*, *2*(02), 47–62. doi:10.4236/ijnm.2013.22007

Elmoubarki, R., Mahjoubi, F. Z., Tounsadi, H., Moustadraf, J., Abdennouri, M., Zouhri, A., ... Barka, N. (2015). Adsorption of textile dyes on raw and decanted Moroccan clays: Kinetics, equilibrium and thermodynamics. *Water Resources and Industry*, *9*, 16–29. doi:10.1016/j.wri.2014.11.001

Elnashaie, S. S., Danafar, F., & Rafsanjani, H. H. (2015). *From Nanotechnology to Nanoengineering Nanotechnology for Chemical Engineers*. Springer. doi:10.1007/978-981-287-496-2_2

Elnsar, T. A. S., Soliman, M. H., & Ayash, M. A.-E.-A. A. (2017). Modified Hydroxyapatite Adsorbent for Removal of Iron Dissolved in Water Wells in Sohag, Egypt. *Chemistry of Advanced Materials, 2*(1).

El-Temsah, Y. S., & Joner, E. J. (2013). Effects of nano-sized zero-valent iron (nZVI) on DDT degradation in soil and its toxicity to collembola and ostracods. *Chemosphere*, *92*(1), 131–137. doi:10.1016/j.chemosphere.2013.02.039 PMID:23522781

El-Temsah, Y. S., Sevcu, A., Bobcikova, K., Cernik, M., & Joner, E. J. (2016). DDT degradation efficiency and ecotoxicological effects of two types of nano-sized zero-valent iron (nZVI) in water and soil. *Chemosphere*, *144*, 2221–2228. doi:10.1016/j.chemosphere.2015.10.122 PMID:26598990

Elumalai, K., Velmurugan, S., Ravi, S., Kathiravan, V., & Ashokkumar, S. (2015b). Facile, eco-friendly and template free photosynthesis of cauliflower like ZnO nanoparticles using leaf extract of Tamarindus indica (L.) and its biological evolution of antibacterial and antifungal activities. *Spectrochim Acta A Mol Biomol Spectrosc, 136 Pt B*, 1052-1057. doi:10.1016/j.saa.2014.09.129

Elumalai, K., & Velmurugan, S. (2015). Green synthesis, characterization and antimicrobial activities of zinc oxide nanoparticles from the leaf extract of Azadirachta indica (L.). *Applied Surface Science*, *345*, 329–336. doi:10.1016/j.apsusc.2015.03.176

Elumalai, K., Velmurugan, S., Ravi, S., Kathiravan, V., & Adaikala Raj, G. (2015). Bio-approach: Plant mediated synthesis of ZnO nanoparticles and their catalytic reduction of methylene blue and antimicrobial activity. *Advanced Powder Technology*, *26*(6), 1639–1651. doi:10.1016/j.apt.2015.09.008

Elumalai, K., Velmurugan, S., Ravi, S., Kathiravan, V., & Ashokkumar, S. (2015a). Bio-fabrication of zinc oxide nanoparticles using leaf extract of curry leaf (Murraya koenigii) and its antimicrobial activities. *Materials Science in Semiconductor Processing*, *34*, 365–372. doi:10.1016/j.mssp.2015.01.048

Elumalai, K., Velmurugan, S., Ravi, S., Kathiravan, V., & Ashokkumar, S. (2015c). Green synthesis of zinc oxide nanoparticles using Moringa oleifera leaf extract and evaluation of its antimicrobial activity. *Spectrochimica Acta. Part A: Molecular and Biomolecular Spectroscopy*, *143*, 158–164. doi:10.1016/j.saa.2015.02.011 PMID:25725211

El-Zahhar, A. A., Awwad, N. S., & El-Katori, E. E. (2014). Removal of bromophenol blue dyes from industrial waste water by synthesizing polymer-clay composite. *Journal of Molecular Liquids*, *199*, 454–461. doi:10.1016/j.molliq.2014.07.034

Energy, U. S. D. (2000). *Fuel cell handbook* (5th ed.). Science Applications International Corporation E&G Services Parson Inc.

Ensink, J. H., Mahmood, T., van der Hoek, W., Raschid-Sally, L., & Amerasinghe, F. P. (2004). A nationwide assessment of wastewater use in Pakistan: An obscure activity or a vitally important one? *Water Policy*, *6*(3), 197–206.

Eppler, A. S., Rupprechter, G., Anderson, E. A., & Somorjai, G. A. (2000). Thermal and Chemical Stability and Adhesion Strength of Pt Nanoparticle Arrays Supported on Silica Studied by Transmission Electron Microscopy and Atomic Force Microscopy. *The Journal of Physical Chemistry B*, *104*(31), 7286–7292. doi:10.1021/jp0006429

Erkartal, M., Usta, H., Citir, M., & Sen, U. (2016). Proton conducting poly(vinyl alcohol) (PVA)/ poly (2-acrylamido-2-methylpropane sulfonic acid) (PAMPS)/ zeolitic imidazolate framework (ZIF) ternary composite membrane. *Journal of Membrane Science*, *499*, 156–16. doi:10.1016/j.memsci.2015.10.032

Ernhart, C. B. (1992). A critical review of low-level prenatal lead exposure in the human: 1. Effects on the fetus and newborn. *Reproductive Toxicology (Elmsford, N.Y.)*, *6*(1), 9–19. doi:10.1016/0890-6238(92)90017-N PMID:1562804

Esrafili, M. D. (2017). N_2O reduction over a fullerene-like boron nitride nanocage: A DFT study. *Physics Letters. [Part A]*, *381*(25), 2085–2091. doi:10.1016/j.physleta.2017.04.009

Ewecharoen, A., Thiravetyan, P., Wendel, E., & Bertagnolli, H. (2009). Nickel adsorption by sodium polyacrylate-grafted activated carbon. *Journal of Hazardous Materials*, *171*(1), 335–339. doi:10.1016/j.jhazmat.2009.06.008 PMID:19576692

Falah, M., MacKenzie, K. J. D., Knibbe, R., Page, S. J., & Hanna, J. V. (2016). New composites of nanoparticle Cu (I) oxide and titania in a novelinorganic polymer (geopolymer) matrix for destruction of dyes andhazardous organic pollutants. *Journal of Hazardous Materials*, *318*, 772–782. doi:10.1016/j.jhazmat.2016.06.016 PMID:27329791

Fan, L., Zhang, S., Zhang, X., Zhou, H., Lu, Z., & Wang, S. (2015). Removal of arsenic from simulation wastewater using nano-iron/oyster shell composites. *Journal of Environmental Management*, *156*, 109–114. doi:10.1016/j.jenvman.2015.03.044 PMID:25839745

Fan, Q., Sirkar, K. K., & Michniak, B. (2008). Iontophoretic transdermal drug delivery system using a conducting polymeric membrane. *Journal of Membrane Science*, *321*(2), 240–249. doi:10.1016/j.memsci.2008.04.061

FAO. (2010). *Information System on Water and Agriculture*. Retrieved from http://www.fao.org/nr/water/aquastat/water_use/index6.stm

Fatimah, I. (2016). Green synthesis of silver nanoparticles using extract of Parkia speciosa Hassk pods assisted by microwave irradiation. *Journal of Advanced Research*, *7*(6), 961–969. doi:10.1016/j.jare.2016.10.002 PMID:27857843

Fei, B., Qian, B., Yang, Z., Wang, R., Liu, W. C., Mak, C. L., & Xin, J. H. (2008). Coating carbon nanotubes by spontaneous oxidative polymerization of dopamine. *Carbon*, *46*(13), 1795–1797. doi:10.1016/j.carbon.2008.06.049

Feldheim, D., & Foss, C. (2002). *Metal nanoparticles: synthesis, characterization, and applications*. Boca Raton, FL: CRC Press.

Feldmann, C. (2003). Polyol-Mediated Synthesis of Nanoscale Functional Materials. *Advanced Functional Materials*, *13*(2), 101–107. doi:10.1002/adfm.200390014

Fendler, J. H. (2007). *Nanoparticles and Nanostructured Films: Preparation, Characterization, and Applications*. Wiley-VCH Verlag GmbH.

Feng, C. Y., Khulbe, K. C., Matsuura, T., & Ismail, A. F. (2013). Recent progresses in polymeric hollow fiber membrane preparation, characterization and applications. *Separation and Purification Technology*, *111*, 43–71. doi:10.1016/j.seppur.2013.03.017

Feng, J. T., Zhu, J. W., Lv, W., Li, J. J., & Yan, W. (2015). Effect of hydroxyl group of carboxylic acids on the adsorption of acid red G and methylene blue on TiO2. *Chemical Engineering Journal*, *269*, 316–322. doi:10.1016/j.cej.2015.01.109

Feng, J., Zhang, D.-D., Liu, Y.-F., Bai, Y., Chen, Q.-D., Liu, S.-Y., & Sun, H.-B. (2010). Magnetic nanofilm of Fe_3O_4 for highly efficient organic light-emitting devices. *The Journal of Physical Chemistry C*, *114*(14), 6718–6721. doi:10.1021/jp9122503

Feng, L., Cao, M., Ma, X., Zhu, Y., & Hu, C. (2012). Superparamagnetic high-surface-area Fe_3O_4 nanoparticles as adsorbents for arsenic removal. *Journal of Hazardous Materials*, *217*, 439–446. doi:10.1016/j.jhazmat.2012.03.073 PMID:22494901

Feng, Q. L., Wu, J., Chen, G. Q., Cui, F. Z., Kim, T. N., & Kim, J. O. (2000). A mechanistic study of the antibacterial effect of silver ions on *Escherichia coli* and Staphylococcus aureus. *Journal of Biomedical Materials Research*, *52*(4), 662–668. doi:10.1002/1097-4636(20001215)52:4<662::AID-JBM10>3.0.CO;2-3 PMID:11033548

Feng, Q., Xu, J., Dong, H., Li, S., & Wei, Y. (2000). Synthesis of Polystyrene-Silica Hybrid Mesoporous Materials via the Nonsurfactant-template Sol-gel Process. *Journal of Materials Chemistry*, *10*(11), 2490–2494. doi:10.1039/b003170o

Feng, Y. F., Yang, F., Wang, Y. Q., Ma, L., Wu, Y. H., Kerr, P. G., & Yang, L. Z. (2011). Basic dye adsorption onto an agro-based waste material-sesame hull (Sesamum indicum L.). *Bioresource Technology*, *102*(22), 10280–10285. doi:10.1016/j.biortech.2011.08.090 PMID:21962534

Feng, Y., Gong, J.-L., Zeng, G.-M., Niu, Q.-Y., Zhang, H.-Y., Niu, C.-G., ... Yan, M. (2010). Adsorption of Cd (II) and Zn (II) from aqueous solutions using magnetic hydroxyapatite nanoparticles as adsorbents. *Chemical Engineering Journal*, *162*(2), 487–494. doi:10.1016/j.cej.2010.05.049

Feng, Y., Han, G., Chung, T. S., Weber, M., Widjojo, N., & Maletzko, C. (2017). Effects of polyethylene glycol on membrane formation and properties of hydrophilic sulfonated polyphenylenesulfone (sPPSU) membranes. *Journal of Membrane Science*, *531*, 27–35. doi:10.1016/j.memsci.2017.02.040

Ferreira, A., Cemlyn-Jones, J., & Cordeiro, C. R. (2013). Nanoparticles, nanotechnology and pulmonary nanotoxicology. [English Edition]. *Revista Portuguesa de Pneumologia*, *19*(1), 28–37. doi:10.1016/j.rppneu.2012.09.003 PMID:23265236

Fihri, A., Bouhrara, M., Nekoueishahraki, B., Basset, J.-M., & Polshettiwar, V. (2011). Nanocatalysts for Suzuki cross-coupling reactions. *Chemical Society Reviews*, *40*(10), 5181–5203. doi:10.1039/c1cs15079k PMID:21804997

Fine, G. F., Cavanagh, L. M., Afonja, A., & Binions, R. (2010). Metal oxide semi-conductor gas sensors in environmental monitoring. *Sensors (Basel)*, *10*(6), 5469–5502. doi:10.3390100605469 PMID:22219672

Folke, C., Hahn, T., Olsson, P., & Norberg, J. (2005). Adaptive governance of social-ecological systems. *Annual Review of Environment and Resources*, *30*(1), 441–473. doi:10.1146/annurev.energy.30.050504.144511

Folkmann, J. K., Risom, L., Jacobsen, N. R., Wallin, H., Loft, S., & Moller, P. (2009). Oxidatively damaged DNA in rats exposed by oral gavage to C60 fullerenes and single-walled carbon nanotubes. *Environmental Health Perspectives*, *117*(5), 703–708. doi:10.1289/ehp.11922 PMID:19479010

Fotiou, T., Triantis, T. M., Kaloudis, T., O'Shea, K. E., Dionysiou, D. D., & Hiskia, A. (2016). Assessment of the roles of reactive oxygen species in the UV and visible light photocatalytic degradation of cyanotoxins and water taste and odor compounds using $CeTiO_2$. *Water Research*, *90*, 52–61. doi:10.1016/j.watres.2015.12.006 PMID:26724439

Fox, M. (1988). Photocatalytic Oxidation of Organic Substances. In Photocatalysis and Environment: Trends and Applications. New York: Academic Publishers.

Freeman, B. D., & Pinnau, I. (Eds.). (1999). *Polymer membranes for gas and vapor separation: chemistry and materials science*. American Chemical Society. doi:10.1021/bk-1999-0733

Freundlich, H. M. F. (1906). Über die adsorption in Lösungen. Z. *Journal of Physical Chemistry*, *57*, 385–470.

Fryxell, G. E. (2006). The synthesis of functional mesoporous materials. *Inorganic Chemistry Communications*, *9*(11), 1141–1150. doi:10.1016/j.inoche.2006.06.012

Fryxell, G. E., Lin, Y., Fiskum, S., Birnbaum, J. C., Wu, H., Kemner, K., & Kelly, S. (2005). Actinide Sequestration Using Self-Assembled Monolayers on Mesoporous Supports (SAMMS). *Environmental Science & Technology*, *39*(5), 1324–1331. doi:10.1021/es049201j PMID:15787373

Fryxell, G. E., Mattigod, S. V., Lin, Y., Wu, H., Fiskum, S., Parker, K., ... Feng, X. (2007). Design and synthesis of self-assembled monolayers on mesoporous supports (SAMMS): The importance of ligand posture in functional nanomaterials. *Journal of Materials Chemistry*, *17*(28), 2863–2874. doi:10.1039/b702422c

Fu, F., & Wang, Q. (2011). Removal of heavy metal ions from wastewaters: A review. *Journal of Environmental Management*, *92*(3), 407–418. doi:10.1016/j.jenvman.2010.11.011 PMID:21138785

Fukumoto, T., Yoshioka, T., Nagasawa, H., Kanezashi, M., & Tsuru, T. (2014). Development and gas permeation properties of microporous amorphous TiO 2–ZrO 2–organic composite membranes using chelating ligands. *Journal of Membrane Science*, *461*, 96–105. doi:10.1016/j.memsci.2014.02.031

Gaballah, I., & Kilbertus, G. (1998). Recovery of heavy metal ions through decontamination of synthetic solutions and industrial effluents using modified barks. *Journal of Geochemical Exploration*, *62*(1–3), 241–286. doi:10.1016/S0375-6742(97)00068-X

Galdames, A., Mendoza, A., Orueta, M., de Soto García, I., Sánchez, M., Virto, I., & Vilas, J. (2017). *Development of new remediation technologies for contaminated soils based on the application of zero-valent iron nanoparticles and bioremediation with compost*. Resource-Efficient Technologies.

Gan, P. P., & Li, S. F. Y. (2012). Potential of plant as a biological factory to synthesize gold and silver nanoparticles and their applications. *Reviews in Environmental Science and Biotechnology*, *11*(2), 169–206. doi:10.100711157-012-9278-7

Gan, P. P., Ng, S. H., Huang, Y., & Li, S. F. Y. (2012). Green synthesis of gold nanoparticles using palm oil mill effluent (POME): A low-cost and eco-friendly viable approach. *Bioresource Technology*, *113*, 132–135. doi:10.1016/j.biortech.2012.01.015 PMID:22297042

Gao, B., Gao, Y., & Li, Y. (2010). Preparation and chelation adsorption property of composite chelating material poly(amidoxime)/SiO_2 towards heavy metal ions. *Chemical Engineering Journal*, *158*(3), 542–549. doi:10.1016/j.cej.2010.01.046

Gao, C., Zhang, W., Li, H., Lang, L., & Xu, Z. (2008). Controllable fabrication of mesoporous MgO with various morphologies and their absorption performance for toxic pollutants in water. *Crystal Growth & Design*, *8*(10), 3785–3790. doi:10.1021/cg8004147

Gao, H., Sun, Y., Zhou, J., Xu, R., & Duan, H. (2013). Mussel-inspired synthesis of polydopamine-functionalized graphene hydrogel as reusable adsorbents for water purification. *ACS Applied Materials & Interfaces*, *5*(2), 425–432. doi:10.1021/am302500v PMID:23265565

Gao, L. X. (2004). Current Situation and Improved Measures of Iindoor Air Cleaner. *Period of Haerbin Industry University*, *36*(2), 199–201.

Gao, P., Feng, Y. J., Zhang, Z. H., Liu, J. F., & Ren, N. Q. (2011). Comparison of competitive and synergetic adsorption of three phenolic compounds on river sediment. *Environmental Pollution*, *159*(10), 2876–2881. doi:10.1016/j.envpol.2011.04.047 PMID:21636192

Gao, W., Majumder, M., Alemany, L. B., Narayanan, T. N., Ibarra, M. A., Pradhan, B. K., & Ajayan, P. M. (2011). Engineered graphite oxide materials for application in water purification. *ACS Applied Materials & Interfaces*, *3*(6), 1821–1826. doi:10.1021/am200300u PMID:21568266

Gao, Z., Pereira, T., Choi, O., Wang, Y., & Hahn, H. T. (2006). Surface functionalized alumina nanoparticle filled polymeric nanocomposites with enhanced mechanical properties. *Journal of Materials Chemistry*, *16*(27), 2800–2808. doi:10.1039/b603020c

Garcia, A., Ferreira, L., Leitão, A., & Rodrigues, A. (1999). Binary Adsorption of Phenol and m-Cresol Mixtures onto a Polymeric Adsorbent. *Adsorption*, *5*(4), 359–368. doi:10.1023/A:1008904631147

Geetha, D., Kavitha, S., & Ramesh, P. S. (2015). A novel bio-degradable polymer stabilized Ag/TiO2 nanocomposites and their catalytic activity on reduction of methylene blue under natural sun light. *Ecotoxicology and Environmental Safety*, *121*, 126–134. doi:10.1016/j.ecoenv.2015.04.042 PMID:25976106

Gehrke, I., Geiser, A., & Somborn-Schulz, A. (2015). Innovations in nanotechnology for water treatment. *Nanotechnology, Science and Applications*, *8*, 1–17. doi:10.2147/NSA.S43773 PMID:25609931

Geng, Q. J., Guo, Q. J., Cao, C. Q., & Wang, H. Q. (2008). Investigation into photocatalytic degradation of gaseous benzene in a circulated photocatalytic reactor (CPCR). *Chemical Engineering & Technology*, *31*(7), 1023–1030. doi:10.1002/ceat.200700510

Ge, P., Li, F., & Zhang, B. (2010). Synthesis of modified mesoporous materials and comparative studies of removal of heavy metal from aqueous solutions. *Polish Journal of Environmental Studies*, *19*, 301–308.

Geyikçi, F. (2013). Adsorption of acid blue 161 (AB 161) dye from water by multi-walled carbon nanotubes. *Fullerenes, Nanotubes, and Carbon Nanostructures*, *21*(7), 579–593. doi:10.1080/1536383X.2011.643428

Ghaedi, M., Ghobadzadeh, P., Nasiri Kokhdan, S., & Soylak, M. (2013). Oxidized multiwalled carbon nanotubes as adsorbents for kinetic and equilibrium study of removal of 5-(4-dimethyl amino benzylidene) rhodanine. *Arabian Journal for Science & Engineering, 38*(7).

Ghaedi, M., Haghdoust, S., Kokhdan, S. N., Mihandoost, A., Sahraie, R., & Daneshfar, A. (2012a). Comparison of activated carbon, multiwalled carbon nanotubes, and cadmium hydroxide nanowire loaded on activated carbon as adsorbents for kinetic and equilibrium study of removal of Safranine O. *Spectroscopy Letters*, *45*(7), 500–510. doi:10.1080/00387010.2011.641058

Ghaedi, M., Hassanzadeh, A., & Kokhdan, S. N. (2011a). Multiwalled carbon nanotubes as adsorbents for the kinetic and equilibrium study of the removal of alizarin red S and morin. *Journal of Chemical & Engineering Data*, *56*(5), 2511–2520. doi:10.1021/je2000414

Ghaedi, M., Khajehsharifi, H., Yadkuri, A. H., Roosta, M., & Asghari, A. (2012b). Oxidized multiwalled carbon nanotubes as efficient adsorbent for bromothymol blue. *Toxicological and Environmental Chemistry*, *94*(5), 873–883. doi:10.1080/02772248.2012.678999

Ghaedi, M., & Kokhdan, S. N. (2012). Oxidized multiwalled carbon nanotubes for the removal of methyl red (MR): Kinetics and equilibrium study. *Desalination and Water Treatment*, *49*(1-3), 317–325. doi:10.1080/19443994.2012.719355

Ghaedi, M., Shokrollahi, A., Hossainian, H., & Kokhdan, S. N. (2011b). Comparison of activated carbon and multiwalled carbon nanotubes for efficient removal of eriochrome cyanine R (ECR): Kinetic, isotherm, and thermodynamic study of the removal process. *Journal of Chemical & Engineering Data*, *56*(7), 3227–3235. doi:10.1021/je200331u

Ghaedi, M., Shokrollahi, A., Tavallali, H., Shojaiepoor, F., Keshavarz, B., Hossainian, H., ... Purkait, M. K. (2011c). Activated carbon and multiwalled carbon nanotubes as efficient adsorbents for removal of arsenazo(III) and methyl red from waste water. *Toxicological and Environmental Chemistry*, *93*(3), 438–449. doi:10.1080/02772248.2010.540244

Ghodake, G. S., Deshpande, N. G., Lee, Y. P., & Jin, E. S. (2010). Pear fruit extract-assisted room-temperature biosynthesis of gold nanoplates. *Colloids and Surfaces. B, Biointerfaces*, *75*(2), 584–589. doi:10.1016/j.colsurfb.2009.09.040 PMID:19879738

Gholami, M., Nassehinia, H. R., Jonidi-Jafari, A., Nasseri, S., & Esrafili, A. (2014). Comparison of Benzene & Toluene removal from synthetic polluted air with use of Nano photocatalytic TiO_2 /ZNO process. *Journal of Environmental Health Science & Engineering*, *12*(1), 45. doi:10.1186/2052-336X-12-45 PMID:24499601

Ghorai, T. K., Biswas, S. K., & Pramanik, P. (2008). Surface modification with both phosphorylcholine and stearyl groups to adjust hydrophilicity and hydrophobicity. *Applied Surface Science*, *254*, 498–7504.

Ghosh, S., Patil, S., Ahire, M., Kitture, R., Gurav, D. D., Jabgunde, A. M., . . . Chopade, B. A. (2012). Gnidia glauca flower extract mediated synthesis of gold nanoparticles and evaluation of its chemocatalytic potential. *Journal of Nanobiotechnology, 10*(17), 1-9.

Ghosh, S., Patil, S., Ahire, M., Kitture, R., Kale, S., Pardesi, K., ... Chopade, B. A. (2012). Synthesis of silver nanoparticles using Dioscorea bulbifera tuber extract and evaluation of its synergistic potential in combination with antimicrobial agents. *International Journal of Nanomedicine*, *7*, 483–496. doi:10.2147/IJN.S24793 PMID:22334779

Gier, S., & Johns, W. D. (2000). Heavy metal-adsorption on micas and clay minerals studied by X–ray photoelectron spectroscopy. *Applied Clay Science*, *16*(5-6), 289–299. doi:10.1016/S0169-1317(00)00004-1

Gilroy, K. D., Ruditskiy, A., Peng, H.-C., Qin, D., & Xia, Y. (2016). Bimetallic nanocrystals: Syntheses, properties, and applications. *Chemical Reviews*, *116*(18), 10414–10472. doi:10.1021/acs.chemrev.6b00211 PMID:27367000

Glusker, J. P., Katz, A. K., & Bock, C. W. (1999). Metal Ions In Biological Systems. *The Rigaku Journal*, *16*(2), 8–17.

Godugu, C., Singh, R. P., & Poduri, R. (2015). Nanotoxicology: Contemporary Issues and Future Directions. In P. V. Devarajan & S. Jain (Eds.), *Targeted Drug Delivery: Concepts and Design* (pp. 733–781). Cham: Springer International Publishing.

Goh, K., Setiawan, L., Wei, L., Si, R., Fane, A. G., Wang, R., & Chen, Y. (2015). Preparation of graphene oxide-based hydrogels as efficient dye adsorbents for wastewater treatment. *Nanoscale Research Letters*, *10*(1), 272–281. doi:10.118611671-015-0931-2 PMID:26123269

Goldoni, A., Larciprete, R., Petaccia, L., & Lizzit, S. (2003). Single-wall carbon nanotube interaction with gases: Sample contaminants and environmental monitoring. *Journal of the American Chemical Society*, *125*(37), 11329–11333. doi:10.1021/ja034898e PMID:16220955

Gomathi, M., Rajkumar, P., Prakasam, A., & Ravichandran, K. (2017). *Green synthesis of silver nanoparticles using Datura stramonium leaf extract and assessment of their antibacterial activity*. Resource-Efficient Technologies.

Gonçalves, J. R. (2016). The Soil and Groundwater Remediation with Zero Valent Iron Nanoparticles. *Procedia Engineering, 143*, 1268–1275. doi:10.1016/j.proeng.2016.06.122

Gong, J., Chen, L., Zeng, G., Long, F., Deng, J., Niu, Q., & He, X. (2012). Shellac-coated iron oxide nanoparticles for removal of cadmium(II) ions from aqueous solution. *Journal of Environmental Sciences (China)*, *24*(7), 1165–1173. doi:10.1016/S1001-0742(11)60934-0 PMID:23513435

Gong, J.-L., Wang, B., Zeng, G.-M., Yang, C.-P., Niu, C.-G., Niu, Q.-Y., ... Liang, Y. (2009). Removal of cationic dyes from aqueous solution using magnetic multi-wall carbon nanotube nanocomposite as adsorbent. *Journal of Hazardous Materials*, *164*(2), 1517–1522. doi:10.1016/j.jhazmat.2008.09.072 PMID:18977077

Gonzalez, L., Thomassen, L. C., Plas, G., Rabolli, V., Napierska, D., Decordier, I., ... Kirsch-Volders, M. (2010). Exploring the aneugenic and clastogenic potential in the nanosize range: A549 human lung carcinoma cells and amorphous monodisperse silica nanoparticles as models. *Nanotoxicology*, *4*(4), 382–395. doi:10.3109/17435390.2010.501913 PMID:20925446

Gonze, E., Commenges, N., Gonthier, Y., & Bernis, A. (2003). High frequency ultrasound as a pre- or a post-oxidation for paper mill wastewaters and landfill leachate treatment. *Chemical Engineering Journal*, *92*(1–3), 215–225. doi:10.1016/S1385-8947(02)00258-9

Gopalakrishnan, A., Krishnan, R., Thangavel, S., Venugopal, G., & Kim, S.-J. (2015). Removal of heavy metal ions from pharma-effluents using graphene-oxide nanosorbents and study of their adsorption kinetics. *Journal of Industrial and Engineering Chemistry*, *30*, 14–19. doi:10.1016/j.jiec.2015.06.005

Gottschalk, F., & Nowack, B. (2011). The release of engineered nanomaterials to the environment. *Journal of Environmental Monitoring*, *13*(5), 1145–1155. doi:10.1039/c0em00547a PMID:21387066

Graydon, J. W., Zhang, X., Kirk, D. W., & Jia, C. Q. (2009). Sorption and stability of mercury on activated carbon for emission control. *Journal of Hazardous Materials*, *168*(2), 978–982. doi:10.1016/j.jhazmat.2009.02.118 PMID:19327890

Green, M., & Howman, E. (2005). Semiconductor quantum dots and free radical induced DNA nicking. Chemical Communications, (1), 121-123.

Gregorio-Jauregui, K. M., Pineda, M. G., Rivera-Salinas, J. E., Hurtado, G., Saade, H., Martinez, J. L., ... López, R. G. (2012). One-step method for preparation of magnetic nanoparticles coated with chitosan. *Journal of Nanomaterials*, *2012*, 4. doi:10.1155/2012/813958

Grey, D., Garrick, D., Blackmore, D., Kelman, J., Muller, M., & Sadoff, C. (2013). Water security in one blue planet: twenty-first century policy challenges for science. *Philosophical Transactions of the Royal Society A*, *371*(2002), 20120406.

Grieger, K. D., Hjorth, R., Rice, J., Kumar, N., & Bang, J. (2015). *Nano-remediation: tiny particles cleaning up big environmental problems.* Blog entry for IUCN.

Gubin, S. P. (2009). *Magnetic Nanoparticles*. Wiley. doi:10.1002/9783527627561

Gubin, S. P., Koksharov, Y. A., Khomutov, G., & Yurkov, G. Y. e. (2005). Magnetic nanoparticles: Preparation, structure and properties. *Russian Chemical Reviews*, *74*(6), 489–520. doi:10.1070/RC2005v074n06ABEH000897

Guillen, G. R., Pan, Y., Li, M., & Hoek, E. M. (2011). Preparation and characterization of membranes formed by nonsolvent induced phase separation: A review. *Industrial & Engineering Chemistry Research*, *50*(7), 3798–3817. doi:10.1021/ie101928r

Gunalan, S., Sivaraj, R., & Rajendran, V. (2012). Green synthesized ZnO nanoparticles against bacterial and fungal pathogens. *Progress in Natural Science: Materials International*, *22*(6), 693–700. doi:10.1016/j.pnsc.2012.11.015

Gündoğan, R., Acemioğlu, B., & Alma, M. H. (2004). Copper (II) adsorption from aqueous solution by herbaceous peat. *Journal of Colloid and Interface Science*, *269*(2), 303–309. doi:10.1016/S0021-9797(03)00762-8 PMID:14654388

Guo, D., Xie, G., & Luo, J. (2013). Mechanical properties of nanoparticles: Basics and applications. *Journal of Physics. D, Applied Physics*, *47*(1), 013001. doi:10.1088/0022-3727/47/1/013001

Guo, J., Zhang, Q., Cai, Z., & Zhao, K. (2016). Preparation and dye filtration property of electrospun polyhydroxybutyrate–calcium alginate/carbon nanotubes composite nanofibrous filtration membrane. *Separation and Purification Technology*, *161*, 69–79. doi:10.1016/j.seppur.2016.01.036

Guo, L., Liu, Q., Li, G., Shi, J., Liu, J., Wang, T., & Jiang, G. (2012). A mussel-inspired polydopamine coating as a versatile platform for the in situ synthesis of graphene-based nanocomposites. *Nanoscale*, *4*(19), 5864–5867. doi:10.1039/c2nr31547e PMID:22945453

Guo, L., Von Dem Bussche, A., Buechner, M., Yan, A., Kane, A. B., & Hurt, R. H. (2008). Adsorption of essential micronutrients by carbon nanotubes and the implications for nanotoxicity testing. *Small*, *4*(6), 721–727. doi:10.1002mll.200700754 PMID:18504717

Guo, Z. B., Liu, J., & Liu, F. L. (2015). Selective adsorption of p-CNB and o-CNB in silica-coating HZSM-5 zeolite. *Microporous and Mesoporous Materials*, *213*, 8–13. doi:10.1016/j.micromeso.2015.04.006

Gupta, A., Rai, D. K., Pandey, R. S., & Sharma, B. (2009). Analysis of some heavy metals in the riverine water, sediments and fish from river Ganges at Allahabad. *Environmental Monitoring and Assessment*, *157*(1), 449–458. doi:10.100710661-008-0547-4 PMID:18850290

Gupta, A., Vidyarthi, S. R., & Sankararamakrishnan, N. (2014). Enhanced sorption of mercury from compact fluorescent bulbs and contaminated water streams using functionalized multiwalled carbon nanotubes. *Journal of Hazardous Materials*, *274*(Supplement C), 132–144. doi:10.1016/j.jhazmat.2014.03.020 PMID:24780855

Gupta, S. K., Behari, J., & Kesari, K. K. (2006). Low Frequencies Ultrasonic Treatment of Sludge. *Asian Journal of Water, Environment and Pollution*, *3*(2), 101–105.

Gupta, V. K., Agarwal, S., Sadegh, H., Ali, G. A. M., Bharti, A. K., & Hamdy, A. S. (2017). Facile route synthesis of novel graphene oxide-β-cyclodextrin nanocomposite and its application as adsorbent for removal of toxic bisphenol A from the aqueous phase. *Journal of Molecular Liquids*, *237*, 466–472. doi:10.1016/j.molliq.2017.04.113

Gupta, V. K., Agarwal, S., & Saleh, T. A. (2011). Synthesis and characterization of alumina-coated carbon nanotubes and their application for lead removal. *Journal of Hazardous Materials*, *185*(1), 17–23. doi:10.1016/j.jhazmat.2010.08.053 PMID:20888691

Gupta, V. K., Ali, I., Saleh, T. A., Nayak, A., & Agarwal, S. (2012). Chemical treatment technologies for waste-water recycling—an overview. *RSC Advances*, *2*(16), 6380–6388. doi:10.1039/c2ra20340e

Gupta, V. K., Gupta, M., & Sharma, S. (2001). Process development for the removal of lead and chromium from aqueous solutions using red mud-an aluminium industry waste. *Water Research*, *35*(5), 1125–1134. doi:10.1016/S0043-1354(00)00389-4 PMID:11268832

Gupta, V. K., Kumar, R., Nayak, A., Saleh, T. A., & Barakat, M. (2013). Adsorptive removal of dyes from aqueous solution onto carbon nanotubes: A review. *Advances in Colloid and Interface Science*, *193*, 24–34. doi:10.1016/j.cis.2013.03.003 PMID:23579224

Gupta, V. K., & Saleh, T. A. (2013). Sorption of pollutants by porous carbon,carbon nanotubes and fullerene- An overview. *Environmental Science and Pollution Research International*, *20*(5), 2828–2843. doi:10.100711356-013-1524-1 PMID:23430732

Gupta, V. K., Tyagi, I., Sadegh, H., Shahryari-Ghoshekandi, R., Makhlouf, A. S. H., & Maazinejad, B. (2015). Nanoparticles as adsorbent; a positive approach for removal of noxious metal ions: A review. *Science. Technology and Development, 34*(3), 195–214. doi:10.3923td.2015.195.214

Gupta, V., & Nayak, A. (2012). Cadmium removal and recovery from aqueous solutions by novel adsorbents prepared from orange peel and Fe 2 O 3 nanoparticles. *Chemical Engineering Journal, 180*, 81–90. doi:10.1016/j.cej.2011.11.006

Gurr, J. R., Wang, A. S., Chen, C. H., & Jan, K. Y. (2005). Ultrafine titanium dioxide particles in the absence of photoactivation can induce oxidative damage to human bronchial epithelial cells. *Toxicology, 213*(1-2), 66–73. doi:10.1016/j.tox.2005.05.007 PMID:15970370

Güzel, F., Sayğılı, H., Sayğılı, G. A., & Koyuncu, F. (2014). Elimination of anionic dye by using nanoporous carbon prepared from an industrial biowaste. *Journal of Molecular Liquids, 194*, 130–140. doi:10.1016/j.molliq.2014.01.018

Habeeb, O. A., Kanthasamy, R., Ali, G. A. M., Sethupathi, S., & Yunus, R. B. M. (2017a). Hydrogen sulfide emission sources, regulations, and removal techniques: A review. *Reviews in Chemical Engineering, 0*(0). doi:10.1515/revce-2017-0004

Habeeb, O. A., Ramesh, K., Ali, G. A. M., & Yunus, R. M. (2017b). Experimental design technique on removal of hydrogen sulfide using CaO-eggshells dispersed onto palm kernel shell activated carbon: Experiment, optimization, equilibrium and kinetic studies. *Journal of Wuhan University of Technology-Mater. Sci. Ed., 32*(2), 305–320. doi:10.100711595-017-1597-7

Habeeb, O. A., Ramesh, K., Ali, G. A. M., & Yunus, R. M. (2017c). Isothermal modelling based experimental study of dissolved hydrogen sulfide adsorption from waste water using eggshell based activated carbon. *The Malaysian Journal of Analytical Sciences, 21*(2), 334–345. doi:10.17576/mjas-2017-2102-08

Habeeb, O. A., Ramesh, K., Ali, G. A. M., & Yunus, R. M. (2017d). Low-cost and eco-friendly activated carbon from modified palm kernel shell for hydrogen sulfide removal from wastewater: Adsorption and kinetic studies. *Desalination and Water Treatment, 84*, 205–214. doi:10.5004/dwt.2017.21196

Hackenberg, S., Friehs, G., Kessler, M., Froelich, K., Ginzkey, C., Koehler, C., ... Kleinsasser, N. (2011). Nanosized titanium dioxide particles do not induce DNA damage in human peripheral blood lymphocytes. *Environmental and Molecular Mutagenesis, 52*(4), 264–268. doi:10.1002/em.20615 PMID:20740634

Hadavifar, M., Bahramifar, N., Younesi, H., & Li, Q. (2014). Adsorption of mercury ions from synthetic and real wastewater aqueous solution by functionalized multi-walled carbon nanotube with both amino and thiolated groups. *Chemical Engineering Journal, 237*(Supplement C), 217–228. doi:10.1016/j.cej.2013.10.014

Hagen, A., & Hertel, T. (2003). Quantitative analysis of optical spectra from individual single-wall carbon nanotubes. *AIP Conference Proceedings*. 10.1063/1.1628016

Haham, H., Grinblat, J., Sougrati, M.-T., Stievano, L., & Margel, S. (2015). Engineering of iron-based magnetic activated carbon fabrics for environmental remediation. *Materials (Basel), 8*(7), 4593–4607. doi:10.3390/ma8074593 PMID:28793459

Hajipour, M. J., Fromm, K. M., Ashkarran, A. A., de Aberasturi, D. J., de Larramendi, I. R., Rojo, T., ... Mahmoudi, M. (2012). Antibacterial properties of nanoparticles. *Trends in Biotechnology, 30*(10), 499–511. doi:10.1016/j.tibtech.2012.06.004 PMID:22884769

Hakami, O., Zhang, Y., & Banks, C. J. (2012). Thiol-functionalised mesoporous silica-coated magnetite nanoparticles for high efficiency removal and recovery of Hg from water. *Water Research, 46*(12), 3913–3922. doi:10.1016/j.watres.2012.04.032 PMID:22608609

Halek, F. S., Farahani, S. K., & Hosseini, S. M. (2015). Fabrication of poly(ether sulfone) based mixed matrix membranes modified by TiO2 nanoparticles for purification of biodiesel produced from waste cooking oils. *Korean Journal of Chemical Engineering*, *33*(2), 629–637. doi:10.100711814-015-0161-4

Hamad, I., Christy Hunter, A., Rutt, K. J., Liu, Z., Dai, H., & Moein Moghimi, S. (2008). Complement activation by PEGylated single-walled carbon nanotubes is independent of C1q and alternative pathway turnover. *Molecular Immunology*, *45*(14), 3797–3803. doi:10.1016/j.molimm.2008.05.020 PMID:18602161

Handy, R. D., Owen, R., & Valsami-Jones, E. (2008). The ecotoxicology of nanoparticles and nanomaterials: Current status, knowledge gaps, challenges, and future needs. *Ecotoxicology (London, England)*, *17*(5), 315–325. doi:10.100710646-008-0206-0 PMID:18408994

Hanley, N., Shogren, J. F., & White, B. (2007). *Environmental Economics* (2nd ed.). Basingstoke, UK: Palgrave Macmillan.

Han, W., Fu, F., Cheng, Z., Tang, B., & Wu, S. (2016). Studies on the optimum conditions using acid-washed zero-valent iron/aluminum mixtures in permeable reactive barriers for the removal of different heavy metal ions from wastewater. *Journal of Hazardous Materials*, *302*, 437–446. doi:10.1016/j.jhazmat.2015.09.041 PMID:26521089

Han, Y.-F., Chen, F., Zhong, Z., Ramesh, K., Chen, L., & Widjaja, E. (2006). Controlled Synthesis, Characterization, and Catalytic Properties of Mn_2O_3 and Mn_3O_4 Nanoparticles Supported on Mesoporous Silica SBA-15. *The Journal of Physical Chemistry B*, *110*(48), 24450–24456. doi:10.1021/jp064941v PMID:17134200

Hao, Y.-M., Man, C., & Hu, Z.-B. (2010). Effective removal of Cu (II) ions from aqueous solution by amino-functionalized magnetic nanoparticles. *Journal of Hazardous Materials*, *184*(1), 392–399. doi:10.1016/j.jhazmat.2010.08.048 PMID:20837378

Hao, Y., Meng, G., Ye, C., Zhang, X., & Zhang, L. (2005). Kinetics-Driven Growth of Orthogonally Branched Single-Crystalline Magnesium Oxide Nanostructures. *The Journal of Physical Chemistry B*, *109*(22), 11204–11208. doi:10.1021/jp0505451 PMID:16852367

Hariharan, C. (2006). Photocatalytic degradation of organic contaminants in water by ZnO nanoparticles: Revisited. *Applied Catalysis A, General*, *304*, 55–61. doi:10.1016/j.apcata.2006.02.020

Hassan, M. L., Kassem, N. F., & Abd El-Kader, A. H. (2010). Novel Zr(IV)/sugar beet pulp composite for removal of sulfate and nitrate anions. *Journal of Applied Polymer Science*, *117*(4), 2205–2212. doi:10.1002/app.32063

Hatakeyama, T., & Quinn, F. X. (1999). Thermal Analysis Fundamental and Applications to Polymer Science: Thermogravimetry (2nd ed.). John Wiley and Sons Publications.

Hatakeyama, K., Razaul, K. M., Ogata, C., Tateishi, H., Taniguchi, T., Koinuma, M., ... Matsumoto, Y. (2014). Optimization of proton conductivity in graphene oxide by filling sulfate ions. *Chemical Communications*, *50*(93), 14527–14530. doi:10.1039/C4CC07273A PMID:25302574

Hatton, B., Landskron, K., Whitnall, W., Perovic, D., & Ozin, G. A. (2005). Past, Present, and Future of Periodic Mesoporous Organosilicas The PMOs. *Accounts of Chemical Research*, *38*(4), 305–312. doi:10.1021/ar040164a PMID:15835877

Haverkamp, R. G., & Marshall, A. T. (2008). The mechanism of metal nanoparticle formation in plants: Limits on accumulation. *Journal of Nanoparticle Research*, *11*(6), 1453–1463. doi:10.100711051-008-9533-6

He, F., Zhao, D., & Paul, C. (2010). Field assessment of carboxymethyl cellulose stabilized iron nanoparticles for in situ destruction of chlorinated solvents in source zones. *Water Research*, *44*(7), 2360–2370. doi:10.1016/j.watres.2009.12.041 PMID:20106501

He, L., Yang, L., Zhang, Z. R., Gong, T., Deng, L., Gu, Z., & Sun, X. (2009). In vitro evaluation of the genotoxicity of a family of novel MeO-PEG-poly(D,L-lactic-co-glycolic acid)-PEG-OMe triblock copolymer and PLGA nanoparticles. *Nanotechnology*, *20*(45), 455102. doi:10.1088/0957-4484/20/45/455102 PMID:19822931

Hench, L. L., & West, J. K. (1990). The sol-gel process. *Chemical Reviews*, *90*(1), 33–72. doi:10.1021/cr00099a003

Herlekar, M., Barve, S., & Kumar, R. (2014). Plant-Mediated Green Synthesis of Iron Nanoparticles. *Journal of Nanoparticles*, *2014*, 1–9. doi:10.1155/2014/140614

Herrero, M. A., & Prato, M. (2008). Recent advances in the covalent functionalization of carbon nanotubes. *Molecular Crystals and Liquid Crystals (Philadelphia, Pa.)*, *483*(1), 21–32. doi:10.1080/15421400801898033

Hillyer, J. F., & Albrecht, R. M. (2001). Gastrointestinal persorption and tissue distribution of differently sized colloidal gold nanoparticles. *Journal of Pharmaceutical Sciences*, *90*(12), 1927–1936. doi:10.1002/jps.1143 PMID:11745751

Hinds, W. C. (1998). *Aerosol Technology* (2nd ed.). New York: John. Wiley & Sons.

Hirano, S., Fujitani, Y., Furuyama, A., & Kanno, S. (2012). Macrophage receptor with collagenous structure (MARCO) is a dynamic adhesive molecule that enhances uptake of carbon nanotubes by CHO-K1 cells. *Toxicology and Applied Pharmacology*, *259*(1), 96–103. doi:10.1016/j.taap.2011.12.012 PMID:22209804

Hirano, S., Kanno, S., & Furuyama, A. (2008). Multi-walled carbon nanotubes injure the plasma membrane of macrophages. *Toxicology and Applied Pharmacology*, *232*(2), 244–251. doi:10.1016/j.taap.2008.06.016 PMID:18655803

Hlihor, R. M., & Gavrilescu, M. (2009). Removal of some environmentally relevant heavy Metals using low-cost natural sorbents. *Environmental Engineering and Management Journal*, *8*(2), 353–372.

Hoag, G. E., Collins, J. B., Holcomb, J. L., Hoag, J. R., Nadagouda, M. N., & Varma, R. S. (2009). Degradation of bromothymol blue by 'greener'nano-scale zero-valent iron synthesized using tea polyphenols. *Journal of Materials Chemistry*, *19*(45), 8671–8677. doi:10.1039/b909148c

Hoffmann, D., Hoffmann, I., & El-Bayoumy, K. (2001). The less harmful cigarette: A controversial issue. A tribute to Ernst L. Wynder. *Chemical Research in Toxicology*, *14*(7), 767–790. doi:10.1021/tx000260u PMID:11453723

Hoffmann, F., Cornelius, M., Morell, J., & Froeba, M. (2006). Silica-based Mesoporous organic-inorganic hybrid materials. *Angewandte Chemie International Edition*, *45*(20), 3216–3251. doi:10.1002/anie.200503075 PMID:16676373

Ho, H. L., Chan, W. K., Blondy, A., Yeung, K. L., & Schrotter, J.-C. (2012). Experiment and modeling of advanced ozone membrane reactor for treatment of organic endocrine disrupting pollutants in water. *Catalysis Today*, *193*(1), 120–127. doi:10.1016/j.cattod.2012.03.059

Holdren, J. P., Smith, K. R., Kjellstrom, T., Streets, D., Wang, X., & Fischer, S. (2000). *Energy, the environment and health*. New York: United Nations Development Programme.

Holmberg, Z. B., Li, W., Wang, X., Munoz, W., & Yan, R. Y. (2006). Nafion/Zeolite Nanocomposite Membrane by in Situ Crystallization for a Direct Methanol Fuel Cell. *Chemistry of Materials*, *18*(24), 5669–5675. doi:10.1021/cm060841q

Hong, J., Peralta-Videa, J. R., Rico, C., Sahi, S., Viveros, M. N., Bartonjo, J., ... Gardea-Torresdey, J. L. (2014). Evidence of translocation and physiological impacts of foliar applied CeO2 nanoparticles on cucumber (Cucumis sativus) plants. *Environmental Science & Technology*, *48*(8), 4376–4385. doi:10.1021/es404931g PMID:24625209

Hornyak, G. L., Moore, J. J., Tibbals, H. F., & Dutta, J. (2008). *Fundamentals of Nanotechnology*. CRC Press Taylor & Francis Group.

Hosseini, S. J., Kokhdan, S. N., Ghaedi, A. M., & Moosavian, S. S. (2011). Comparison of multiwalled carbon nanotubes and activated carbon for efficient removal of methyl orange: Kinetic and thermodynamic investigation. *Fresenius Environmental Bulletin*, *20*, 219–234.

Howe, J. M., Fultz, B., & Miao, S. (2012). Characterization of Materials. John Wiley & Sons, Inc.

Howes, P. D., Chandrawati, R., & Stevens, M. M. (2014). Colloidal nanoparticles as advanced biological sensors. *Science*, *346*(6205), 1247390. doi:10.1126cience.1247390 PMID:25278614

Hristovski, K., Baumgardner, A., & Westerhoff, P. (2007). Selecting metal oxide nanomaterials for arsenic removal in fixed bed columns: From nanopowders to aggregated nanoparticle media. *Journal of Hazardous Materials*, *147*(1), 265–274. doi:10.1016/j.jhazmat.2007.01.017 PMID:17254707

Hsu, S., & Lu, C. (2007). Modification of Single-walled Carbon Nanotubes for Enhancing Isopropyl Alcohol Vapor Adsorption from Air Streams. *Separation Science and Technology*, *42*(12), 2751–2766. doi:10.1080/01496390701515060

Hua, M., Zhang, S., Pan, B., Zhang, W., Lv, L., & Zhang, Q. (2012). Heavy metal removal from water/wastewater by nanosized metal oxides: A review. *Journal of Hazardous Materials*, *211*, 317–331. doi:10.1016/j.jhazmat.2011.10.016 PMID:22018872

Huang, F. Y. (2014). *H_2-Cl_2 proton exchange membrane fuel cells, fuel cell assemblies including the same and systems for cogeneration of electricity and HCL.* US Patent no. US9437895 B2.

Huang, H., Zhang, J., Jiang, L., & Zang, Z. (2017). Preparation of cubic Cu2O nanoparticles wrapped by reduced graphene oxide for the efficient removal of rhodamine B. *Journal of Alloys and Compounds*, *718*, 112–115. doi:10.1016/j.jallcom.2017.05.132

Huang, J., & Wan, Q. (2009). Gas sensors based on semiconducting metal oxide one-dimensional nanostructures. *Sensors (Basel)*, *9*(12), 9903–9924. doi:10.339091209903 PMID:22303154

Huang, S., Chueh, P. J., Lin, Y. W., Shih, T. S., & Chuang, S. M. (2009). Disturbed mitotic progression and genome segregation are involved in cell transformation mediated by nano-TiO2 long-term exposure. *Toxicology and Applied Pharmacology*, *241*(2), 182–194. doi:10.1016/j.taap.2009.08.013 PMID:19695278

Huang, Y. C., Fan, R., Grusak, M. A., Sherrier, J. D., & Huang, C. (2014). Effects of nano-ZnO on the agronomically relevant Rhizobium–legume symbiosis. *The Science of the Total Environment*, *497*, 78–90. doi:10.1016/j.scitotenv.2014.07.100 PMID:25124056

Huang, Y., & Keller, A. A. (2013). Magnetic Nanoparticle Adsorbents for Emerging Organic Contaminants. *ACS Sustainable Chemistry & Engineering*. doi:10.1021c400047q

Huang, Z., Wang, X., & Yang, D. (2015). Adsorption of Cr (VI) in wastewater using magnetic multi-wall carbon nanotubes. *Water Science and Engineering*, *8*(3), 226–232. doi:10.1016/j.wse.2015.01.009

Huber, D. L. (2005). Synthesis, properties, and applications of iron nanoparticles. *Small*, *1*(5), 482–501. doi:10.1002mll.200500006 PMID:17193474

Hudson, S. P., Padera, R. F., Langer, R., & Kohane, D. S. (2008). The biocompatibility of mesoporous silicates. *Biomaterials*, *29*(30), 4045–4055. doi:10.1016/j.biomaterials.2008.07.007 PMID:18675454

Hu, H., Wang, Z., & Pan, L. (2010). Synthesis of monodisperse Fe_3O_4@ silica core–shell microspheres and their application for removal of heavy metal ions from water. *Journal of Alloys and Compounds*, *492*(1), 656–661. doi:10.1016/j.jallcom.2009.11.204

Hui, L., Xiao, D., Hua, H., Rui, L., & Zuo, P. (2013). Adsorption behavior and adsorption mechanism of Cu (II) ions on amino-functionalized magnetic nanoparticles. *Transactions of Nonferrous Metals Society of China*, *23*(9), 2657–2665. doi:10.1016/S1003-6326(13)62782-X

Hu, J., Chen, G., & Lo, I. M. C. (2005). Removal and recovery of Cr(VI) from wastewater by maghemite nanoparticles. *Water Research*, *39*(18), 4528–4536. doi:10.1016/j.watres.2005.05.051 PMID:16146639

Hu, J., Chen, G., & Lo, I. M. C. (2006). Selective Removal of Heavy Metals from Industrial Wastewater Using Maghemite Nanoparticle: Performance and Mechanisms. *Journal of Environmental Engineering*, *132*(7), 709–715. doi:10.1061/(ASCE)0733-9372(2006)132:7(709)

Hu, J., Lo, I. M. C., & Chen, G. (2007). Comparative study of various magnetic nanoparticles for Cr(VI) removal. *Separation and Purification Technology*, *56*(3), 249–256. doi:10.1016/j.seppur.2007.02.009

Hu, J., Lo, I., & Chen, G. (2004). Removal of Cr (VI) by magnetite. *Water Science and Technology*, *50*(12), 139–146. PMID:15686014

Hu, Q., Wujcik, E. K., Kelarakis, A., Cyriac, J., & Gong, X. (2017). Carbon-Based Nanomaterials as Novel Nanosensors. *Journal of Nanomaterials*.

Hussain, F., Hojjati, M., Okamoto, M., & Gorga, R. E. (2006). Review article: Polymer-matrix Nanocomposites, Processing, Manufacturing, and Application: An Overview. *Journal of Composite Materials*, *40*(17), 1511–1575. doi:10.1177/0021998306067321

Hussein, A. K. (2016). Applications of nanotechnology to improve the performance of solar collectors–Recent advances and overview. *Renewable & Sustainable Energy Reviews*, *62*, 767–792. doi:10.1016/j.rser.2016.04.050

Hutton, G., Haller, L., & Bartram, J. (2007). *Economic and Health Effects of Increasing Coverage of Low Cost Household Drinking Water Supply and Sanitation Interventions*. World Health Organization.

Hu, Y., Feng, G., Li, G., Cao, G., & Chen, X. (2005). Application of Active Carbon and Nano-titanium Dioxide photocatalysis web in air-conditioning system. *Indoor Air*, (Supplement, 15), 161–162.

Ian, S. (2012). Nanotechnology in water and air pollution treatment. *Journal Environmental Technology Reviews*, *1*(1), 136–148. doi:10.1080/21622515.2012.733966

Ibrahim, H. M. M. (2015). Green synthesis and characterization of silver nanoparticles using banana peel extract and their antimicrobial activity against representative microorganisms. *Journal of Radiation Research and Applied Sciences*, *8*(3), 265–275. doi:10.1016/j.jrras.2015.01.007

Ichikawa, T., Takahara, K., Shimoda, K., Seita, Y., & Emi, M. (1987). *U.S. Patent No. 4,708,800*. Washington, DC: U.S. Patent and Trademark Office.

Ihsanullah, H. A., Asmaly, T. A., Saleh, T., Laoui, V. K., Gupta, V. K., & Atieh, M. A. (2015). Enhanced adsorption of phenols from liquids by aluminum oxide/carbon nanotubes: Comprehensive study from synthesis to surface properties. *Journal of Molecular Liquids*, *206*, 176–182. doi:10.1016/j.molliq.2015.02.028

Iijima, S. (1991). Helical microtubules of graphitic carbon. *Nature*, *354*(6348), 56–58. doi:10.1038/354056a0

Im, H.-J., Yost, T. L., Yang, Y., Bramlett, J. M., Yu, X., Fagan, B. C., ... Xue, Z.-L. (2006). ACS Symp. Ser.: Vol. 943. *Organofunctional Sol-Gel Materials for Toxic Metal Separation*. Nuclear Waste Management, Accomplishments of the Environmental Management Science Program. doi:10.1021/bk-2006-0943.ch012

Inagaki, S., Guan, S., Fukushima, Y., Ohsuna, T., & Terasaki, O. (2000). Nanoporous Materials II, Proceedings of the Conference on Access in Nanoporous Materials, 2000. *Studies in Surface Science and Catalysis*, *129*, 155. doi:10.1016/S0167-2991(00)80209-8

Inoue, Y., Hoshino, M., Takahashi, H., Noguchi, T., Murata, T., Kanzaki, Y., ... Sasatsu, M. (2002). Bactericidal activity of Ag-zeolite mediated by reactive oxygen species under aerated conditions. *Journal of Inorganic Biochemistry*, *92*(1), 37–42. doi:10.1016/S0162-0134(02)00489-0 PMID:12230986

International Agency for Research on Cancer (IARC). (2009). IARC monographs on the evaluation of carcinogenic risks to humans, Vols 1-97 (until 2009). IARC Publications.

Iravani, S. (2011). Green synthesis of metal nanoparticles using plants. *Green Chemistry*, *13*(10), 2638–2650. doi:10.1039/c1gc15386b

Iravani, S., Korbekandi, H., Mirmohammadi, S., & Zolfaghari, B. (2014). Synthesis of silver nanoparticles: Chemical, physical and biological methods. *Research in Pharmaceutical Sciences*, *9*(6), 385. PMID:26339255

Ishibashi, K., Fujishima, A., Watanabe, T., & Hashimoto, K. (2000). Quantum yields of active oxidative species formed on TiO2 photocatalyst. *Journal of Photochemistry and Photobiology A Chemistry*, *134*(1-2), 139–142. doi:10.1016/S1010-6030(00)00264-1

Ismail, R. A., Sulaiman, G. M., Abdulrahman, S. A., & Marzoog, T. R. (2015). Antibacterial activity of magnetic iron oxide nanoparticles synthesized by laser ablation in liquid. *Materials Science and Engineering C*, *53*, 286–297. doi:10.1016/j.msec.2015.04.047 PMID:26042717

Issa, B., Obaidat, I. M., Albiss, B. A., & Haik, Y. (2013). Magnetic nanoparticles: Surface effects and properties related to biomedicine applications. *International Journal of Molecular Sciences*, *14*(11), 21266–21305. doi:10.3390/ijms141121266 PMID:24232575

Itoh, H., & Sugimoto, T. (2003). Systematic control of size, shape, structure, and magnetic properties of uniform magnetite and maghemite particles. *Journal of Colloid and Interface Science*, *265*(2), 283–295. doi:10.1016/S0021-9797(03)00511-3 PMID:12962662

IUPAC Recommendations, . (1985). *Pure and Applied Chemistry*, *57*, 603. doi:10.1351/pac198557040603

Iwamoto, T., & Ishigaki, T. (2013). *Fabrication of iron oxide nanoparticles using laser ablation in liquids.* Paper presented at the Journal of Physics: Conference Series. 10.1088/1742-6596/441/1/012034

Izumi, Y., Hisano, K., Hida, T. (1999). Acid catalysis of silica-included heteropolyacid in polar reaction media. *Applied Catalysis, A, 181*, 277-282.

Jabeen, H., Chandra, V., Jung, S., Lee, J. W., Kim, K. S., & Kim, S. B. (2011). Enhanced Cr(vi) removal using iron nanoparticle decorated graphene. *Nanoscale*, *3*(9), 3583–3585. doi:10.1039/c1nr10549c PMID:21814702

Jacob, S. J. P., Finub, J., & Narayanan, A. (2012). Synthesis of silver nanoparticles using Piper longum leaf extracts and its cytotoxic activity against Hep-2 cell line. *Colloids and Surfaces. B, Biointerfaces*, *91*, 212–214. doi:10.1016/j.colsurfb.2011.11.001 PMID:22119564

Jain, K. K. (2017). Nanotoxicology. In K. K. Jain (Ed.), *The Handbook of Nanomedicine* (pp. 575–603). New York, NY: Springer New York. doi:10.1007/978-1-4939-6966-1_17

Jalil, A. A., Triwahyono, S., Adam, S. H., Rahim, N. D., Aziz, M. A. A., Hairom, N. H. H., ... Mohamadiah, M. K. A. (2010). Adsorption of methyl orange from aqueous solution onto calcined Lapindo volcanic mud. *Journal of Hazardous Materials*, *181*(1-3), 755–762. doi:10.1016/j.jhazmat.2010.05.078 PMID:20538408

Jal, P. K., Patel, S., & Mishra, B. K. (2004). Chemical modification of silica surface by immobilization of functional groups for extractive concentration of metal ions. *Talanta*, *62*(5), 1005–1028. doi:10.1016/j.talanta.2003.10.028 PMID:18969392

Jang, J. H., Lee, J., Jung, S.-Y., Choi, D.-C., Won, Y.-J., Ahn, K. H., ... Lee, C.-H. (2015). Correlation between particle deposition and the size ratio of particles to patterns in nano-and micro-patterned membrane filtration systems. *Separation and Purification Technology*, *156*, 608–616. doi:10.1016/j.seppur.2015.10.056

Jang, Y. J., Simer, C., & Ohm, T. (2006). Comparison of zinc oxide nanoparticles and its nano-crystalline particles on the photocatalytic degradation of methylene blue. *Materials Research Bulletin*, *41*(1), 67–77. doi:10.1016/j.materresbull.2005.07.038

Jaroniec, M., & Olkhovyk, O. (2006). Synthesis and Characterization of Ordered Mesoporous Materials for Removal of Heavy Metal Ions. *NATO Science Series II.*, *228*, 325–336.

Jeon, E. K., Seo, E., Lee, E., Lee, W., Um, M. K., & Kim, B. S. (2013). Mussel-inspired green synthesis of silver nanoparticles on graphene oxide nanosheets for enhanced catalytic applications. *Chemical Communications (Cambridge)*, *49*(33), 3392–3394. doi:10.1039/c3cc00115f PMID:23435315

Jeon, H. J., Kim, J. S., Kim, T. G., Kim, J. H., Yu, W. R., & Youk, J. H. (2008). Preparation of poly(ε-caprolactone)-based polyurethane nanofibers containing silver nanoparticles. *Applied Surface Science*, *254*(18), 5886–5890. doi:10.1016/j.apsusc.2008.03.141

Jiang, J., Oberdörster, G., & Biswas, P. (2009). Characterization of size, surface charge, and agglomeration state of nanoparticle dispersions for toxicological studies. *Journal of Nanoparticle Research*, *11*(1), 77–89. doi:10.100711051-008-9446-4

Jian, M., Liu, B., Zhang, G., Liu, R., & Zhang, X. (2015). Adsorptive removal of arsenic from aqueous solution by zeolitic imidazolate framework-8 (ZIF-8) nanoparticles. *Colloids and Surfaces. A, Physicochemical and Engineering Aspects*, *465*, 67–76. doi:10.1016/j.colsurfa.2014.10.023

Jiles, D. (1998). *Introduction to Magnetism and Magnetic Materials*. Chapman & Hall.

Jimenez-Solomon, M. F., Gorgojo, P., Munoz-Ibanez, M., & Livingston, A. G. (2013). Beneath the surface: Influence of supports on thin film composite membranes by interfacial polymerization for organic solvent nanofiltration. *Journal of Membrane Science*, *448*, 102–113. doi:10.1016/j.memsci.2013.06.030

Jin, Y., Liu, F., Tong, M., & Hou, Y. (2012). Removal of arsenate by cetyltrimethylammonium bromide modified magnetic nanoparticles. *Journal of Hazardous Materials*, *227-228*(Supplement C), 461–468. doi:10.1016/j.jhazmat.2012.05.004 PMID:22703733

Johnson, R. F., Manjreker, T. G., & Halligan, J. E. (1973). Removal of oil from water surfaces by sorption on unstructured fibers. *Environmental Science & Technology*, *7*(5), 439–443. doi:10.1021/es60077a003 PMID:22283537

Jonker, M. T. O., & Koelmans, A. A. (2002). Sorption of polycyclic aromatic hydrocarbons. *Journal of Molecular Catalysis. B, Enzymatic*, *66*, 302–310.

Joo, S. H., & Zhao, D. (2016). Environmental dynamics of metal oxide nanoparticles in heterogeneous systems: A review. *Journal of Hazardous Materials*. doi:10.1016/j.jhazmat.2016.02.068 PMID:26961405

Jordan, N., Ritter, A., Foerstendorf, H., Scheinost, A., Weiß, S., Heim, K., ... Reuther, H. (2013). Adsorption mechanism of selenium (VI) onto maghemite. *Geochimica et Cosmochimica Acta*, *103*, 63–75. doi:10.1016/j.gca.2012.09.048

Joseph, L., Heo, J., Park, Y. G., Flora, J. R. V., & Yoon, Y. (2011a). Adsorption of bisphenol A and 17 alphaethinyl estradiol on single walled carbon nanotubes from seawater and brackish water. *Desalination*, *281*, 68–74. doi:10.1016/j.desal.2011.07.044

Jothivenkatachalam, K., Prabhu, S., Nithya, A., & Jeganathan, K. (2014). Facile synthesis of WO_3 with reduced particle size on zeolite and enhanced photocatalytic activity. *RSC Advances*, *4*(41), 21221–21229. doi:10.1039/C4RA01376J

Jung, D. H., Chob, S. Y., Pecka, D. H., Shina, D. R., & Kim, J. S. (2002). Performance evaluation of a Nafion/silicon oxide hybrid membrane for direct methanol fuel cell. *Journal of Power Sources*, *106*(1-2), 173–177. doi:10.1016/S0378-7753(01)01053-9

Kadirvelu, K., Thamaraiselvi, K., & Namasivayam, C. (2001). Adsorption of nickel (II) from aqueous solution onto activated carbon prepared from coirpith. *Separation and Purification Technology*, *24*(3), 497–505. doi:10.1016/S1383-5866(01)00149-6

Kalfa, O. M., Yalçınkaya, Ö., & Türker, A. R. (2009). Synthesis of nano B_2O_3/TiO_2 composite material as a new solid phase extractor and its application to preconcentration and separation of cadmium. *Journal of Hazardous Materials*, *166*(1), 455–461. doi:10.1016/j.jhazmat.2008.11.112 PMID:19135793

Kamaraj, R., Davidson, D. J., Sozhan, G., & Vasudevan, S. (2014). An in situ electrosynthesis of metal hydroxides and their application for adsorption of 4-chloro-2-methylphenoxyacetic acid (MCPA) from aqueous solution. *Journal of Environmental Chemical Engineering*, *2*(4), 2068–2077. doi:10.1016/j.jece.2014.08.027

Kamarudin, S. K., Ahmad, F., & Daud, W. R. W. (2009). Overview on application of direct methanol fuel cell (DMFC) for portable electronic devices. *International Journal of Hydrogen Energy*, *34*(16), 6902–6916. doi:10.1016/j.ijhydene.2009.06.013

Kang, J. L., Moon, C., Lee, H. S., Lee, H. W., Park, E. M., Kim, H. S., & Castranova, V. (2008). Comparison of the biological activity between ultrafine and fine titanium dioxide particles in RAW 264.7 cells associated with oxidative stress. *Journal of Toxicology and Environmental Health. Part A.*, *71*(8), 478–485. doi:10.1080/15287390801906675 PMID:18338282

Kar, D., Sur, P., Mandai, S. K., Saha, T., & Kole, R. K. (2008). Assessment of heavy metal pollution in surface water. *International Journal of Environmental Science and Technology*, *5*(1), 119–124. doi:10.1007/BF03326004

Karkare, M. (2008). *Nanotechnology: Fundamentals and Applications: I.K.* International Publishing House Pvt. Limited.

Kashmiri, Z., & Mankar, S. (2014). Free radicals and oxidative stress in bacteria. *International Journal of Current Microbiology and Applied Sciences*, *3*, 34–40.

Kasithevar, M., Periakaruppan, P., Muthupandian, S., & Mohan, M. (2017). Antibacterial efficacy of silver nanoparticles against multi-drug resistant clinical isolates from post-surgical wound infections. *Microbial Pathogenesis*, *107*, 327–334. doi:10.1016/j.micpath.2017.04.013 PMID:28411059

Kašpar, J., Fornasiero, P., & Hickey, N. (2003). Automotive catalytic converters: Current status and some perspectives. *Catalysis Today*, *77*(4), 419–449. doi:10.1016/S0920-5861(02)00384-X

Kassaee, M. Z., Motamedi, E., Mikhak, A., & Rahnemaie, R. (2011). Nitrate removal from water using iron nanoparticles produced by arc discharge vs. reduction. *Chemical Engineering Journal*, *166*(2), 490–495. doi:10.1016/j.cej.2010.10.077

Katiyar, A., Yadav, S., Smirniotis, P. G., & Pinto, N. G. (2006). Synthesis of ordered large pore SBA-15 spherical particles for adsorption of biomolecules. *Journal of Chromatography. A*, *1122*(1-2), 13–20. doi:10.1016/j.chroma.2006.04.055 PMID:16716334

Katz, M. G., & Baruch, G. (1986). New insights into the structure of microporous membranes obtained using a new pore size evaluation method. *Desalination, 58*(3), 199–211. doi:10.1016/0011-9164(86)87004-7

Kaur, A., & Kansal, S. K. (2016). Bi_2WO_6 nanocuboids: An efficient visible light active photocatalyst for the degradation of levofloxacin drug in aqueous phase. *Chemical Engineering Journal, 302*, 194–203. doi:10.1016/j.cej.2016.05.010

Kawasaki, N., Matsushige, K., Komatsu, K., Kohzu, A., Nara, F. W., Ogishi, F., ... Imai, A. (2011). Fast and precise method for HPLC-size exclusion chromatography with UV and TOC (NDIR) detection: Importance of multiple detectors to evaluate the characteristics of dissolved organic matter. *Water Research, 45*(18), 6240–6248. doi:10.1016/j.watres.2011.09.021 PMID:21959091

Kazi, T. G., Arain, M. B., Baig, J. A., Jamali, M. K., Afridi, H. I., Jalbani, N., ... Niaz, A. (2009). The correlation of arsenic levels in drinking water with the biological samples of skin disorders. *The Science of the Total Environment, 407*(3), 1019–1026. doi:10.1016/j.scitotenv.2008.10.013 PMID:19027142

Kebede, A., Singh, A. K., Rai, P. K., Giri, N. K., Rai, A. K., Watal, G., & Gholap, A. V. (2013). Controlled synthesis, characterization, and application of iron oxide nanoparticles for oral delivery of insulin. *Lasers in Medical Science, 28*(2), 579–587. doi:10.100710103-012-1106-3 PMID:22581389

Kesharwani, J., Yoon, K. Y., Hwang, J., & Rai, M. (2009). Phytofabrication of Silver Nanoparticles by Leaf Extract of Datura metel: Hypothetical Mechanism Involvedin Synthesis. *Journal of Bionanoscience, 3*(1), 1–6. doi:10.1166/jbns.2009.1008

Keyhanian, F., Shariati, S., Faraji, M., & Hesabi, M. (2016). Magnetite nanoparticles with surface modification for removal of methyl violet from aqueous solutions. *Arabian Journal of Chemistry, 9*(Supplement 1), S348–S354. doi:10.1016/j.arabjc.2011.04.012

Khajeh, M., Laurent, S., & Dastafkan, K. (2013). Nanoadsorbents: Classification, preparation, and applications (with emphasis on aqueous media). *Chemical Reviews, 113*(10), 7728–7768. doi:10.1021/cr400086v PMID:23869773

Khajeh, M., & Sanchooli, E. (2011). Synthesis and evaluation of silver nanoparticles material for solid phase extraction of cobalt from water samples. *Applied Nanoscience, 1*(4), 205–209. doi:10.100713204-011-0028-x

Khalid, M., Joly, G., Renaud, A., & Magnoux, P. (2004). Removal of Phenol from Water by Adsorption Using Zeolites. *Journal of Industrial and Engineering Chemistry Research, 43*, 5275-5280.

Khallaf, H., Oladeji, I. O., & Chow, L. (2008). Optimization of chemical bath deposited CdS thin films using nitrilotriacetic acid as a complexing agent. *Thin Solid Films, 516*(18), 5967–5973. doi:10.1016/j.tsf.2007.10.079

Khandanlou, R., Ahmad, M. B., Masoumi, H. R. F., Shameli, K., Basri, M., & Kalantari, K. (2015). Rapid adsorption of copper (II) and lead (II) by rice straw/Fe3O4 nanocomposite: Optimization, equilibrium isotherms, and adsorption kinetics study. *PLoS One, 10*(3), e0120264. doi:10.1371/journal.pone.0120264 PMID:25815470

Khani, A., & Mirzaei, M. (2008). *Comparative study of nitrate removal from aqueous solution using powder activated carbon and carbon nanotubes.* Paper presented at the 2nd International IUPAC Conference on Green Chemistry, Russia.

Khan, I., Saeed, K., & Khan, I. (2017). Nanoparticles: Properties, applications and toxicities. *Arabian Journal of Chemistry*. doi:10.1016/j.arabjc.2017.05.011

Khani, R., Sobhani, S., & Beyki, M. H. (2016). Highly selective and efficient removal of lead with magnetic nanoadsorbent: Multivariate optimization, isotherm and thermodynamic studies. *Journal of Colloid and Interface Science, 466*, 198–205. doi:10.1016/j.jcis.2015.12.027 PMID:26724702

Khanna, V. K. (2016). *Nanosensors: Physical, Chemical, and Biological.* CRC Press.

Khan, S., & Malik, A. (2014). Environmental and Health Effects of Textile Industry Wastewater. In A. Malik, E. Grohmann, & R. Akhtar (Eds.), *Environmental Deterioration and Human Health: Natural and anthropogenic determinants* (pp. 55–71). Dordrecht: Springer Netherlands. doi:10.1007/978-94-007-7890-0_4

Khan, T. A., Nazir, M., Ali, I., & Kumar, A. (2013). Removal of Chromium(VI) from aqueous solution using guar gum–nano zinc oxide biocomposite adsorbent. *Arabian Journal of Chemistry*. doi:10.1016/j.arabjc.2013.08.019

Kharissova, O. V., Dias, H. V., Kharisov, B. I., Perez, B. O., & Perez, V. M. (2013). The greener synthesis of nanoparticles. *Trends in Biotechnology*, *31*(4), 240–248. doi:10.1016/j.tibtech.2013.01.003 PMID:23434153

Khataee, A. R., Vafaei, F., & Jannatkhah, M. (2013). Biosorption of three textile dyes from contaminated water by filamentous green algal Spirogyra sp.: Kinetic, isotherm and thermodynamic studies. *International Biodeterioration & Biodegradation*, *83*, 33–40. doi:10.1016/j.ibiod.2013.04.004

Khavarian, M., Chai, S.-P., & Mohamed, A. R. (2013). Carbon dioxide conversion over carbon-based nanocatalysts. *Journal of Nanoscience and Nanotechnology*, *13*(7), 4825–4837. doi:10.1166/jnn.2013.7569 PMID:23901504

Khayet, M., & Matsuura, T. (2003). Determination of surface and bulk pore sizes of flat-sheet and hollow-fiber membranes by atomic force microscopy, gas permeation and solute transport methods. *Desalination*, *158*(1), 57–64. doi:10.1016/S0011-9164(03)00433-8

Khin, M. M., Nair, A. S., Babu, V. J., Murugan, R., & Ramakrishna, S. (2012). A review on nanomaterials for environmental remediation. *Energy & Environmental Science*, *5*(8), 8075–8109. doi:10.1039/c2ee21818f

Kickelbick, G. (2004). Hybrid inorganic-organic mesoporous materials. *Angewandte Chemie International Edition*, *43*(24), 3102–3104. doi:10.1002/anie.200301751 PMID:15199556

Kilpimaa, S., Runtti, H., Kangas, T., Lassi, U., & Kuokkanen, T. (2014). Removal of phosphate and nitrate over a modified carbon residue from biomass gasification. *Chemical Engineering Research & Design*, *92*(10), 1923–1933. doi:10.1016/j.cherd.2014.03.019

Kilpimaa, S., Runtti, H., Kangas, T., Lassi, U., & Kuokkanen, T. (2015). Physical activation of carbon residue from biomass gasification: Novel sorbent for the removal of phosphates and nitrates from aqueous solution. *Journal of Industrial and Engineering Chemistry*, *21*, 1354–1364. doi:10.1016/j.jiec.2014.06.006

Kim, D. J., Jo, M. J., & Nam, S. Y. (2015). A review of polymer–nanocomposite electrolyte membranes for fuel cell application. *Journal of Industrial and Engineering Chemistry*, *21*, 36–52. doi:10.1016/j.jiec.2014.04.030

Kim, J. S., Kuk, E., & Yu, K. N. (2007). Antimicrobial effects of silver nanoparticles Nanomedicine. *Nanotechnology. Biology, and Medicine*, *3*(1), 95–101.

Kim, J., Choi, S. W., Lee, J. H., Chung, Y., & Byun, Y. T. (2016). Gas sensing properties of defect-induced single-walled carbon nanotubes. *Sensors and Actuators. B, Chemical*, *228*, 688–692. doi:10.1016/j.snb.2016.01.094

Kim, J., Rheem, Y., Yoo, B., Chong, Y., Bozhilov, K. N., Kim, D., ... Myung, N. V. (2010). Peptide-mediated shape- and size-tunable synthesis of gold nanostructures. *Acta Biomaterialia*, *6*(7), 2681–2689. doi:10.1016/j.actbio.2010.01.019 PMID:20083240

Kim, K.-J., Sung, W. S., Suh, B. K., Moon, S.-K., Choi, J.-S., Kim, J. G., & Lee, D. G. (2009). Antifungal activity and mode of action of silver nano-particles on Candida albicans. *Biometals*, *22*(2), 235–242. doi:10.100710534-008-9159-2 PMID:18769871

Kim, S. S., Pauly, T. R., & Pinnavaia, T. J. (2000). Non-ionic surfactant assembly of wormhole silica molecular sieves from water soluble silicates. *Chemical Communications*, (10), 835–836. doi:10.1039/b001012j

Kim, S. W., Son, S. U., Lee, S. I., Hyeon, T., & Chung, Y. K. (2000). Cobalt on mesoporous silica: The first heterogeneous Pauson-Khand catalyst. *Journal of the American Chemical Society*, *122*(7), 1550–1551. doi:10.1021/ja9939237

Kim, Y.-J., & Choi, J.-H. (2012). Selective removal of nitrate ion using a novel composite carbon electrode in capacitive deionization. *Water Research*, *46*(18), 6033–6039. doi:10.1016/j.watres.2012.08.031 PMID:22980574

Kim, Y., & Yi, J. (2004). Advances in environmental technologies via the application of mesoporous materials. *Journal of Industrial and Engineering Chemistry*, *10*(1), 41–51.

Kipp, J. (2004). The role of solid nanoparticle technology in the parenteral delivery of poorly water-soluble drugs. *International Journal of Pharmaceutics*, *284*(1), 109–122. doi:10.1016/j.ijpharm.2004.07.019 PMID:15454302

Kisin, E. R., Murray, A. R., Keane, M. J., Shi, X. C., Schwegler-Berry, D., Gorelik, O., ... Shvedova, A. A. (2007). Single-walled carbon nanotubes: Geno- and cytotoxic effects in lung fibroblast V79 cells. *Journal of Toxicology and Environmental Health. Part A.*, *70*(24), 2071–2079. doi:10.1080/15287390701601251 PMID:18049996

Kobya, M., Demirbas, E., Senturk, E., & Ince, M. (2005). Adsorption of heavy metal ions from aqueous solutions by activated carbon prepared from apricot stone. *Bioresource Technology*, *96*(13), 1518–1521. doi:10.1016/j.biortech.2004.12.005 PMID:15939281

Kochameshki, M. G., Marjani, A., Mahmoudian, M., & Farhadi, K. (2017). Grafting of diallyldimethylammonium chloride on graphene oxide by RAFT polymerization for modification of nanocomposite polysulfone membranes using in water treatment. *Chemical Engineering Journal*, *309*, 206–221. doi:10.1016/j.cej.2016.10.008

Koksharov, Y. (2009). *Magnetism of Nanoparticles: Effects of Size, Shape, and Interactions*. Wiley-VCH Verlag GmbH & Co. KGaA

Konicki, W., Pełech, I., Mijowska, E., & Jasińska, I. (2012). Adsorption of anionic dye direct red 23 onto magnetic multi-walled carbon nanotubes-Fe3C nanocomposite: Kinetics, equilibrium and thermodynamics. *Chemical Engineering Journal*, *210*, 87–95. doi:10.1016/j.cej.2012.08.025

Korngold, E., Belayer, N., & Aronvol, L. (2001). Removal of arsenic from drinking water by anion exchangers. *Desalination*, *141*(1), 81–84. doi:10.1016/S0011-9164(01)00391-5

Kotas, J., & Stasicka, Z. (2000). Chromium occurrence in the environment and methods of its speciation. *Environmental Pollution*, *107*(3), 263–283. doi:10.1016/S0269-7491(99)00168-2 PMID:15092973

Kotov, N. A. (2006). Materials science: Carbon sheet solutions. *Nature*, *442*(7100), 254–255. doi:10.1038/442254a PMID:16855576

Kousha, M., Tavakoli, S., Daneshvar, E., Vazirzadeh, A., & Bhatnagar, A. (2015). Central composite design optimization of acid blue 25 dye biosorption using shrimp shell biomass. *Journal of Molecular Liquids*, *207*, 266–273. doi:10.1016/j.molliq.2015.03.046

Kowalczyk, P., & Holyst, R. (2008). Efficient adsorption of super greenhouse gas (tetrafluoromethane) in carbon nanotubes. *Environmental Science & Technology*, *42*(8), 2931–2936. doi:10.1021/es071306+ PMID:18497146

Krasteva, N., Besnard, I., Guse, B., Bauer, R. E., Müllen, K., Yasuda, A., & Vossmeyer, T. (2002). Self-assembled gold nanoparticle/dendrimer composite films for vapor sensing applications. *Nano Letters*, *2*(5), 551–555. doi:10.1021/nl020242s

Kraytsberg, A., & Eli, Y. E. (2014). Review of Advanced Materials for Proton Exchange Membrane Fuel Cells. *Energy & Fuels*, *28*(12), 7303–7330. doi:10.1021/ef501977k

Kreyling, W., Semmler, M., Erbe, F., Mayer, P., Takenaka, S., Schulz, H., ... Ziesenis, A. (2002). Translocation of ultrafine insoluble iridium particles from lung epithelium to extrapulmonary organs is size dependent but very low. *Journal of Toxicology and Environmental Health. Part A., 65*(20), 1513–1530. doi:10.1080/00984100290071649 PMID:12396866

Krishnan, K. A., & Haridas, A. (2008). Removal of phosphate from aqueous solutions and sewage using natural and surface modified coir pith. *Journal of Hazardous Materials, 152*(2), 527–535. doi:10.1016/j.jhazmat.2007.07.015 PMID:17706344

Krug, H. F., & Wick, P. (2011). Nanotoxicology: An interdisciplinary challenge. *Angewandte Chemie International Edition, 50*(6), 1260–1278. doi:10.1002/anie.201001037 PMID:21290492

Kruk, M., Jaroniec, M., & Sayari, A. (2000). New insights into pore-size expansion of mesoporous silicates using long-chain amines. *Microporous and Mesoporous Materials, 35*, 545–553. doi:10.1016/S1387-1811(99)00249-8

Krystofova, O., Shestivska, V., Galiova, M. V., Novotny, K., Kaiser, J., Zehnalek, J., ... Kizek, R. (2009). Sunflower Plants as Bioindicators of Environmental Pollution with Lead (II) Ions. *Sensors (Basel), 9*(7), 5040–5505. doi:10.339090705040 PMID:22346686

Kukovecz, A., Balogi, Z., Konya, Z., Toba, M., Lentz, P., Niwa, S.-I., ... Kiricsi, I. (2002). Synthesis, characterisation and catalytic applications of sol–gel derived silica–phosphotungstic acid composites. *Applied Catalysis A, General, 228*(1), 83–94. doi:10.1016/S0926-860X(01)00969-3

Kumar, B., Smita, K., Cumbal, L., & Debut, A. (2017). Green synthesis of silver nanoparticles using Andean blackberry fruit extract. *Saudi Journal of Biological Sciences, 24*(1), 45–50. doi:10.1016/j.sjbs.2015.09.006 PMID:28053570

Kumari, L., Li, W. Z., Vannoy, C. H., Leblanc, R. M., & Wang, D. Z. (2009). Synthesis, characterization and optical properties of Mg(OH)2 micro-/nanostructure and its conversion to MgO. *Ceramics International, 35*(8), 3355–3364. doi:10.1016/j.ceramint.2009.05.035

Kumar, M., & Puri, A. (2012). A review of permissible limits of drinking water. *Indian Journal of Occupational and Environmental Medicine, 16*(1), 40. doi:10.4103/0019-5278.99696 PMID:23112507

Kumar, P. V., Pammi, S., Kollu, P., Satyanarayana, K., & Shameem, U. (2014). Green synthesis and characterization of silver nanoparticles using Boerhaavia diffusa plant extract and their anti bacterial activity. *Industrial Crops and Products, 52*, 562–566. doi:10.1016/j.indcrop.2013.10.050

Kumar, R., Kumar, G., Akhtar, M., & Umar, A. (2015). Sonophotocatalytic degradation of methyl orange using ZnO nano-aggregates. *Journal of Alloys and Compounds, 629*, 167–172. doi:10.1016/j.jallcom.2014.12.232

Kumar, R., Mamlouk, M., & Scott, K. (2014). Sulfonated polyether ether ketone-sulfonated graphene oxide composite membranes for polymer electrolyte fuel cells. *RSC Advances, 4*(2), 617–623. doi:10.1039/C3RA42390E

Kumar, V. S., Nagaraja, B. M., Shashikala, V., Padmasri, A. H., Madhavendra, S. S., Raju, B. D., & Rao, K. S. R. (2004). Highly efficient Ag/C catalyst prepared by electro-chemical deposition method in controlling microorganisms in water. *Journal of Molecular Catalysis A Chemical, 223*(1-2), 313–319. doi:10.1016/j.molcata.2003.09.047

Kumar, V., & Yadav, S. K. (2009). Plant-mediated synthesis of silver and gold nanoparticles and their applications. *Journal of Chemical Technology and Biotechnology (Oxford, Oxfordshire), 84*(2), 151–157. doi:10.1002/jctb.2023

Kuppusamy, P., Yusoff, M. M., Maniam, G. P., & Govindan, N. (2016). Biosynthesis of metallic nanoparticles using plant derivatives and their new avenues in pharmacological applications–An updated report. *Saudi Pharmaceutical Journal, 24*(4), 473–484. doi:10.1016/j.jsps.2014.11.013 PMID:27330378

Kuppusamy, S., Thavamani, P., Megharaj, M., Venkateswarlu, K., Lee, Y. B., & Naidu, R. (2016). Potential of Melaleuca diosmifolia leaf as a low-cost adsorbent for hexavalent chromium removal from contaminated water bodies. *Process Safety and Environmental Protection*, *100*, 173–182. doi:10.1016/j.psep.2016.01.009

Kurian, M., & Nair, D. S. (2015). Heterogeneous Fenton behavior of nano nickel zinc ferrite catalysts in the degradation of 4-chlorophenol from water under neutral conditions. *Journal of Water Process Engineering*, *8*, 37–49. doi:10.1016/j.jwpe.2014.10.011

Kurniawan, T. A., & Lo, W. (2009). Removal of refractory compounds from stabilized landfill leachate using an integrated H_2O_2 oxidation and granular activated carbon (GAC) adsorption treatment. *Water Research*, *43*(16), 4079–4091. doi:10.1016/j.watres.2009.06.060 PMID:19695663

Kuroki, V., Bosco, G. E., Fadini, P. S., Mozeto, A. A., Cestari, A. R., & Carvalho, W. A. (2014). Use of a La(III)-modified bentonite for effective phosphate removal from aqueous media. *Journal of Hazardous Materials*, *274*, 124–131. doi:10.1016/j.jhazmat.2014.03.023 PMID:24769849

Kvitek, L., Panáček, A., Soukupova, J., Kolář, M., Večeřová, R., Prucek, R., ... Zbořil, R. (2008). Effect of surfactants and polymers on stability and antibacterial activity of silver nanoparticles (NPs). *The Journal of Physical Chemistry C*, *112*(15), 5825–5834. doi:10.1021/jp711616v

Kwon, S. G., & Hyeon, T. (2011). Formation Mechanisms of Uniform Nanocrystals via Hot-Injection and Heat-Up Methods. *Small*, *7*(19), 2685–2702. doi:10.1002mll.201002022 PMID:21805623

Lai, C. Y., Groth, A., Gray, S., & Duke, M. (2014). Preparation and characterization of poly (vinylidene fluoride)/nanoclay nanocomposite flat sheet membranes for abrasion resistance. *Water Research*, *57*, 56–66. doi:10.1016/j.watres.2014.03.005 PMID:24698723

Lai, L., Xie, Q., Chi, L., Gu, W., & Wu, D. (2016). Adsorption of phosphate from water by easily separable Fe3O4@SiO2 core/shell magnetic nanoparticles functionalized with hydrous lanthanum oxide. *Journal of Colloid and Interface Science*, *465*, 76–82. doi:10.1016/j.jcis.2015.11.043 PMID:26641568

Lala, N. L., Ramaseshan, R., Bojun, L., Sundarrajan, S., Barhate, R. S., Ying-jun, L., & Ramakrishna, S. (2007). Fabrication of nanofibers with antimicrobial functionality used as filters protection against bacterial contaminants. *Biotechnology and Bioengineering*, *97*(6), 1357–1365. doi:10.1002/bit.21351 PMID:17274060

Lal, R. (2016). Environmental Sustainability. In R. Lal, D. Kraybill, D. O. Hansen, B. R. Singh, T. Mosogoya, & L. O. Eik (Eds.), *Climate Change and Multi-Dimensional Sustainability in African Agriculture: Climate Change and Sustainability in Agriculture* (pp. 3–11). Cham: Springer International Publishing. doi:10.1007/978-3-319-41238-2_1

Langmuir, I. (1916). The constitution and fundamental properties of solids and liquids. *Journal of the American Chemical Society*, *38*(11), 2221–2295. doi:10.1021/ja02268a002

Lata, S., & Samadder, S. R. (2016). Removal of arsenic from water using nano adsorbents and challenges: A review. *Journal of Environmental Management*, *166*, 387–406. doi:10.1016/j.jenvman.2015.10.039 PMID:26546885

Lateef, A., Nazir, R., Jamil, N., Alam, S., Shah, R., Khan, M. N., & Saleem, M. (2016). Synthesis and characterization of zeolite based nano–composite: An environment friendly slow release fertilizer. *Microporous and Mesoporous Materials*, *232*, 174–183. doi:10.1016/j.micromeso.2016.06.020

Latorre, C. H., Méndez, J. A., García, J. B., Martín, S. G., & Crecente, R. M. P. (2012). Carbon nanotubes as solid-phase extraction sorbents prior to atomic spectrometric determination of metal species: A review. *Analytica Chimica Acta*, *749*, 16–35. doi:10.1016/j.aca.2012.09.001 PMID:23036463

Law, M., Kind, H., Messer, B., Kim, F., & Yang, P. (2002). Photochemical sensing of NO2 with SnO2 nanoribbon nanosensors at room temperature. *Angewandte Chemie*, *114*(13), 2511–2514. doi:10.1002/1521-3757(20020703)114:13<2511::AID-ANGE2511>3.0.CO;2-N PMID:12203608

Lazaridis, N. K., Matis, K. A., & Webb, M. (2001). Flotation of metal-loaded clay anion exchangers. Part I: The case of chromates. *Chemosphere*, *42*(4), 373–378. doi:10.1016/S0045-6535(00)00143-0 PMID:11100788

Lebeau, B., Gaslain, F., Fernandez-Martin, C., & Babonneau, F. (2009). In V. Valtchev, S. Mintova, & M. Tsapatsis (Eds.), *Ordered Porous Solids* (pp. 283–308). Oxford, UK: Elsevier Ltd. doi:10.1016/B978-0-444-53189-6.00011-1

Lebeau, B., Patarin, J., & Sanchez, C. (2004). Design and Properties of Hierarchically Organized Hybrid Organic-Inorganic Nanocomposites. *Advances in Technology of Materials and Materials Processing*, *6*, 298–307.

Lee, C. H., Shie, J. L., Tsai, C. Y., Yang, Y. T., & Chang, C. Y. (2013). Photocatalytic decomposition of indoor air pollution using dye-sensitized TiO_2 induced by anthocyanin and ru complexes. *Journal of Clean Energy Technologies*, *1*(2), 115–119. doi:10.7763/JOCET.2013.V1.27

Lee, H., Dellatore, S. M., Miller, W. M., & Messersmith, P. B. (2007). Mussel-Inspired Surface Chemistry for Multifunctional Coatings. *Science*, *318*(5849), 426–430. doi:10.1126cience.1147241 PMID:17947576

Lee, J. K., Li, W., & Manthiram, A. (2009). Poly(arylene ether sulfone)s containing pendant sulfonic acid groups as membrane materials for direct methanol fuel. *Journal of Membrane Science*, *330*(1-2), 73–79. doi:10.1016/j.memsci.2008.12.043

Lee, J. W., Shim, W. G., & Moon, H. (2004). Adsorption equilibrium and kinetics for capillary condensation of trichloroethylene on MCM-41 and MCM-48. *Microporous and Mesoporous Materials*, *73*(3), 109–119. doi:10.1016/j.micromeso.2004.04.020

Lee, J., & Pandey, B. D. (2012). Bio-processing of solid wastes and secondary resources for metal extraction–a review. *Waste Management (New York, N.Y.)*, *32*(1), 3–18. doi:10.1016/j.wasman.2011.08.010 PMID:21925857

Lee, K. S. (2011). Three ways of linking laboratory endeavours to the realm of policies. *European Journal of the History of Economic Thought*, *18*(5), 755–776. doi:10.1080/09672567.2011.616593

Lee, S. Y., Woo, H. S., Song, J. M., Sohn, J. Y., & Shin, J. (2016). Preparation of cross-linked SPEEK-zirconium phosphate hybrid membranes by electron beam irradiation. *Polymer Composites*, *2016*. doi:10.1002/pc.24147

Lee, Y. G., Park, J. H., Oh, C., Oh, S. G., & Kim, Y. C. (2007). Preparation of highly monodispersed hybrid silica spheres using a one-step sol-gel reaction in aqueous solution. *Langmuir*, *23*(22), 10875–10878. doi:10.1021/la702462b PMID:17902722

Lee, Y.-C., & Yang, J.-W. (2012). Self-assembled flower-like TiO_2 on exfoliated graphite oxide for heavy metal removal. *Journal of Industrial and Engineering Chemistry*, *18*(3), 1178–1185. doi:10.1016/j.jiec.2012.01.005

Lenntech, B. V. (2011). *Lead and water.* Water Treatment Solutions. Available at: http://www.lenntech.com/periodic/water/lead/lead-and-water.htm

Leonard, P., Hearty, S., Brennan, J., Dunne, L., Quinn, J., Chakraborty, T., & O'Kennedy, R. (2003). Advances in biosensors for detection of pathogens in food and water. *Enzyme and Microbial Technology*, *32*(1), 3–13. doi:10.1016/S0141-0229(02)00232-6

Levine, T. R., Kim, R. K., Park, H. S., & Hughes, M. (2006). Deception detection accuracy is a predictable linear function of message veracity base-rate: A formal test of Park and Levine's probability model. *Communication Monographs*, *73*(3), 243–260. doi:10.1080/03637750600873736

Liang, H. W., Wang, L., Chen, P. Y., Lin, H. T., Chen, L. F., He, D., & Yu, S. H. (2010). Carbonaceous nanofiber membranes for selective filtration and separation of nanoparticles. *Advanced Materials*, *22*(42), 4691–4695. doi:10.1002/adma.201001863 PMID:20859940

Liang, X., Zhang, F., Feng, W., Zou, X., & Zhu, G. (2013). From metal–organic framework (MOF) to MOF–polymer composite membrane: Enhancement of low-humidity proton conductivity. *Chemical Science (Cambridge)*, *4*(3), 983–992. doi:10.1039/C2SC21927A

Liao, M. H., Wu, K. Y., & Chen, D. H. (2005). Fast Adsorption of Crystal Violet on Polyacrylic Acid-Bound Magnetic Nanoparticles. *Separation Science and Technology*, *39*(7), 1563–1575. doi:10.1081/SS-120030802

Liau, S. Y., Read, D. C., Pugh, W. J., Furr, J. R., & Russell, A. D. (1997). Interaction of silver nitrate with readily identifiable groups: Relationship to the antibacterial action of silver ions. *Letters in Applied Microbiology*, *25*(4), 279–283. doi:10.1046/j.1472-765X.1997.00219.x PMID:9351278

Li, B., Cao, H., & Yin, G. (2011a). $Mg(OH)_2$ @ reduced graphene oxide composite for removal of dyes from water. *Journal of Materials Chemistry*, *21*(36), 13765–13768. doi:10.1039/c1jm13368c

Li, C., Dong, Y., Yang, J., Li, Y., & Huang, C. (2014). Modified nano-graphite/Fe_3O_4 composite as efficient adsorbent for the removal of methyl violet from aqueous solution. *Journal of Molecular Liquids*, *196*, 348–356. doi:10.1016/j.molliq.2014.04.010

Lide, D. R. (2005). *Handbook of Chemistry and Physics* (86th ed.). Cleveland, OH: Chemical Rubber Company.

Li, G., Gao, S., Zhang, G., & Zhang, X. (2014). Enhanced adsorption of phosphate from aqueous solution by nanostructured iron(III)–copper(II) binary oxides. *Chemical Engineering Journal*, *235*, 124–131. doi:10.1016/j.cej.2013.09.021

Li, H., & Nogami, M. (2002). Pore-controlled proton conducting silica films. *Advanced Materials*, *14*(12), 912–914. doi:10.1002/1521-4095(20020618)14:12<912::AID-ADMA912>3.0.CO;2-L

Li, J., Lu, Y., Ye, Q., Cinke, M., Han, J., & Meyyappan, M. (2003). Carbon nanotube sensors for gas and organic vapor detection. *Nano Letters*, *3*(7), 929–933. doi:10.1021/nl034220x

Li, J., Wang, L., Qi, T., Zhou, Y., Liu, C., Chu, J., & Zhang, Y. (2008). Different N-containing functional groups modified mesoporous adsorbents for Cr(VI) sequestration: Synthesis, characterization and comparison. *Microporous and Mesoporous Materials*, *110*(2-3), 442–450. doi:10.1016/j.micromeso.2007.06.033

Limmongkon, Y., Johns, J., & Chaerntanyarak, L. (2013). Preparation of a TiO2-coated photocatalytic air filter for use with an electrostatic air filter pack for xylene removal. *ScienceAsia*, *39*(3), 284–293. doi:10.2306cienceasia1513-1874.2013.39.284

Lim, T.-C., & Ramakrishna, S. (2006). A conceptual review of nanosensors. *Zeitschrift fur Naturforschung. Section A. Physical Sciences*, *61*(7-8), 402–412. doi:10.1515/zna-2006-7-815

Li, N., Zheng, M., Chang, X., Ji, G., Lu, H., Xue, L., ... Cao, J. (2011b). Preparation of magnetic $CoFe_2O_4$-functionalized graphene sheets via a facile hydrothermal method and their adsorption properties. *Journal of Solid State Chemistry*, *184*(4), 953–958. doi:10.1016/j.jssc.2011.01.014

Lin, D., & Xing, B. (2007). Phytotoxicity of nanoparticles: Inhibition of seed germination and root growth. *Environmental Pollution*, *150*(2), 243–250. doi:10.1016/j.envpol.2007.01.016 PMID:17374428

Linda, Z., Yong, G. L., Martin, H., & Eric, H. (2006). Removal of VOCs by photo catalysis process using adsorption enhanced TiO_2–SiO_2 catalyst. *Chemical Engineering and Processing*, *45*(11), 959–964. doi:10.1016/j.cep.2006.01.014

Lingamdinne, L. P., Koduru, J. R., Choi, Y.-L., Chang, Y.-Y., & Yang, J.-K. (2016). Studies on removal of Pb (II) and Cr (III) using graphene oxide based inverse spinel nickel ferrite nano-composite as sorbent. *Hydrometallurgy*, *165*, 64–72. doi:10.1016/j.hydromet.2015.11.005

Lin, J., Zhan, Y., Wang, H., Chu, M., Wang, C., He, Y., & Wang, X. (2017). Effect of calcium ion on phosphate adsorption onto hydrous zirconium oxide. *Chemical Engineering Journal*, *309*, 118–129. doi:10.1016/j.cej.2016.10.001

Lin, S., Hsiao, R., & Juang, R. (2006). Removal of soluble organics from water by a hybrid process of clay adsorption and membrane filtration. *Journal of Hazardous Materials*, *135*(1-3), 134–140. doi:10.1016/j.jhazmat.2005.11.030 PMID:16359788

Lin, S., Lu, D., & Liu, Z. (2012). Removal of arsenic contaminants with magnetic γ-Fe2O3 nanoparticles. *Chemical Engineering Journal*, *211-212*(Supplement C), 46–52. doi:10.1016/j.cej.2012.09.018

Lin, Y.-F., Chen, H.-W., Chen, Y.-C., & Chiou, C.-S. (2013). Application of magnetite modified with polyacrylamide to adsorb phosphate in aqueous solution. *Journal of the Taiwan Institute of Chemical Engineers*, *44*(1), 45–51. doi:10.1016/j.jtice.2012.09.005

Li, Q., Mahendra, S., Lyon, D. Y., Brunet, L., Liga, M. V., Li, D., & Alvarez, P. J. J. (2008). Antimicrobial nanomaterials for water disinfection and microbial control: Potential applications and implications. *Water Research*, *42*(18), 4591–4602. doi:10.1016/j.watres.2008.08.015 PMID:18804836

Li, Q., Xie, R., Li, Y. W., Mintz, E. A., & Shang, J. K. (2007). Enhanced visible-light-induced photocatalytic disinfection of E. coli by carbon-sensitized nitrogen-doped titanium oxide. *Environmental Science & Technology*, *41*(14), 5050–5056. doi:10.1021/es062753c PMID:17711222

Li, S. Z., Li, X. Y., Cui, Z. F., & Wang, D. Z. (2004). Application of ultrafiltration to improve the extraction of antibiotics. *Separation and Purification Technology*, *34*(1-3), 115–123. doi:10.1016/S1383-5866(03)00185-0

Liu, A. M., Hidajat, K., Kawi, S., & Zhao, D. Y. (2000a). A New Class of Hybrid Mesoporous Materials with functionalized Organic Monolayers for selective Adsorption of Heavy Metal ions. *Chemical Communications*, *13*(13), 1145–1146. doi:10.1039/b0026611

Liu, C.-H., Chuang, Y.-H., Chen, T.-Y., Tian, Y., Li, H., Wang, M.-K., & Zhang, W. (2015). Mechanism of arsenic adsorption on magnetite nanoparticles from water: Thermodynamic and spectroscopic studies. *Environmental Science & Technology*, *49*(13), 7726–7734. doi:10.1021/acs.est.5b00381 PMID:26055623

Liu, C., Hsu, P. C., Lee, H. W., Ye, M., Zheng, G., Liu, N., & Cui, Y. (2015). Transparent air filter for high-efficiency PM 2.5 capture. *Nature Communications*, *6*(1), 6205. doi:10.1038/ncomms7205 PMID:25683688

Liu, F., Hashim, N. A., Liu, Y., Moghareh Abed, M. R., & Li, K. (2011). Review: Progress in the production and modification of PVDF membranes. *Journal of Membrane Science*, *375*(1-2), 1–27. doi:10.1016/j.memsci.2011.03.014

Liu, H., Sun, X., Yin, C., & Hu, C. (2008). Removal of phosphate by mesoporous ZrO2. *Journal of Hazardous Materials*, *151*(2-3), 616–622. doi:10.1016/j.jhazmat.2007.06.033 PMID:17658689

Liu, J., Shin, Y., Nie, Z., Chang, J. H., Wang, L.-Q., Fryxell, G. E., ... Exarhos, G. J. (2000b). Molecular Assembly in Ordered Mesoporosity: A New Class of Highly Functional Nanoscale Materials. *The Journal of Physical Chemistry A*, *104*(36), 8328–8339. doi:10.1021/jp0009812

Liu, J., Shin, Y., Wang, L.-Q., Exarhos, G. J., Chang, J. H., Fryxell, G. E., ... Samuels, W. D. (2003). In S. L. Scott, C. M. Crudden, & C. W. Jones (Eds.), *Nanostructured Catalysts* (pp. 297–310). New York: Kluwer Academic/Plenum Publishers. doi:10.1007/978-0-387-30641-4_13

Liu, J., Zhao, Z., & Jiang, G. (2008). Coating Fe3O4 magnetic nanoparticles with humic acid for high efficient removal of heavy metals in water. *Environmental Science & Technology*, *42*(18), 6949–6954. doi:10.1021/es800924c PMID:18853814

Liu, R., Guo, Y., Odusote, G., Qu, F., & Priestley, R. D. (2013). Core-shell Fe3O4 polydopamine nanoparticles serve multipurpose as drug carrier, catalyst support and carbon adsorbent. *ACS Applied Materials & Interfaces*, *5*(18), 9167–9171. doi:10.1021/am402585y PMID:24010676

Liu, T., Li, B., Hao, Y., & Yao, Z. (2014). MoO 3-nanowire membrane and Bi 2 Mo 3 O 12/MoO 3 nano-heterostructural photocatalyst for wastewater treatment. *Chemical Engineering Journal*, *244*, 382–390. doi:10.1016/j.cej.2014.01.070

Liu, X., Guan, Y., Ma, Z., & Liu, H. (2004). Surface modification and characterization of magnetic polymer nanospheres prepared by miniemulsion polymerization. *Langmuir*, *20*(23), 10278–10282. doi:10.1021/la0491908 PMID:15518525

Liu, X., & Zhang, L. (2015). Removal of phosphate anions using the modified chitosan beads: Adsorption kinetic, isotherm and mechanism studies. *Powder Technology*, *277*, 112–119. doi:10.1016/j.powtec.2015.02.055

Liu, X., Zhao, X. F., Zhu, Y., & Zhang, F. Z. (2013). Experimental and theoretical investigation into the elimination of organic pollutants from solution by layered double hydroxides. *Applied Catalysis B: Environmental*, *140–141*, 241–248. doi:10.1016/j.apcatb.2013.04.008

Li, W.-R., Xie, X.-B., Shi, Q.-S., Zeng, H.-Y., You-Sheng, O.-Y., & Chen, Y.-B. (2010). Antibacterial activity and mechanism of silver nanoparticles on Escherichia coli. *Applied Microbiology and Biotechnology*, *85*(4), 1115–1122. doi:10.100700253-009-2159-5 PMID:19669753

Li, X., Xu, H., Chen, Z.-S., & Chen, G. (2011). Biosynthesis of Nanoparticles by Microorganisms and Their Applications. *Journal of Nanomaterials*, *16*. doi:10.1155/2011/270974

Li, Y., Chen, B., & Zhu, L. (2010). Enhanced sorption of polycyclic aromatic hydrocarbons from aqueous solution by modified pine bark. *Bioresource Technology*, *101*(19), 7307–7313. doi:10.1016/j.biortech.2010.04.088

Li, Y., Du, Q., Liu, T., Peng, X., Wang, J., Sun, J., ... Xia, Y. (2013). Comparative study of methylene blue dye adsorption onto activated carbon, graphene oxide, and carbon nanotubes. *Chemical Engineering Research & Design*, *91*(2), 361–368. doi:10.1016/j.cherd.2012.07.007

Li, Y.-f., Wang, B.-h., & Huang, W.-x. (2003). Study on bactericidal performance of nano-scale ZnO modified interior wall paint. *Paint and Coatings Industry-China*, *33*(8), 3–5.

Li, Y.-H., Ding, J., Luan, Z., Di, Z., Zhu, Y., Xu, C., ... Wei, B. (2003a). Competitive adsorption of Pb^{2+}, Cu^{2+} and Cd^{2+} ions from aqueous solutions by multiwalled carbon nanotubes. *Carbon*, *41*(14), 2787–2792. doi:10.1016/S0008-6223(03)00392-0

Li, Y.-H., Wang, S., Luan, Z., Ding, J., Xu, C., & Wu, D. (2003b). Adsorption of cadmium (II) from aqueous solution by surface oxidized carbon nanotubes. *Carbon*, *41*(5), 1057–1062. doi:10.1016/S0008-6223(02)00440-2

Loeb, S., & Sourirajan, S. (1962). *Sea water demineralization by means of an osmotic membrane*. Academic Press.

Lofrano, G., Carotenuto, M., Libralato, G., Domingos, R. F., Markus, A., Dini, L., ... Meric, S. (2016). Polymer functionalized nanocomposites for metals removal from water and wastewater: An overview. *Water Research*, *92*, 22–37. doi:10.1016/j.watres.2016.01.033 PMID:26827255

Lofrano, G., Libralato, G., & Brown, J. (2017). *Nanotechnologies for Environmental Remediation: Applications and Implications*. Springer International Publishing. doi:10.1007/978-3-319-53162-5

Long, F., Zhu, A., & Shi, H. (2013). Recent advances in optical biosensors for environmental monitoring and early warning. *Sensors (Basel)*, *13*(10), 13928–13948. doi:10.3390131013928 PMID:24132229

Long, R. Q., & Yang, R. T. (2001). Carbon nanotubes as a superior sorbent for nitrogen oxides. *Industrial & Engineering Chemistry Research*, *40*(20), 4288–4291. doi:10.1021/ie000976k

Lopatin, G., Yen, L. Y., & Rogers, R. R. (1989). *U.S. Patent No. 4,874,567*. Washington, DC: U.S. Patent and Trademark Office.

Lowe, A. C., Hunter-Ellul, L. A., & Wilkerson, M. G. (2013). Nanotoxicology. In A. Nasir, A. Friedman, & S. Wang (Eds.), *Nanotechnology in Dermatology* (pp. 231–251). New York, NY: Springer New York. doi:10.1007/978-1-4614-5034-4_22

Lubick, N., & Betts, K. (2008). *Silver socks have cloudy lining| Court bans widely used flame retardant*. ACS Publications.

Lu, C., & Liu, C. (2006). Removal of nickel (II) from aqueous solution by carbon nanotubes. *Journal of Chemical Technology and Biotechnology (Oxford, Oxfordshire)*, *81*(12), 1932–1940. doi:10.1002/jctb.1626

Lu, C.-Y., Wey, M.-Y., & Chen, L.-I. (2007). Application of polyol process to prepare AC-supported nanocatalyst for VOC oxidation. *Applied Catalysis A, General*, *325*(1), 163–174. doi:10.1016/j.apcata.2007.03.030

Lue, S. J., Pai, Y. L., Shih, C. M., Wu, M. C., & Lai, S. M. (2015). Novel bilayer well-aligned Nafion/graphene oxide composite membranes prepared using spin coating method for direct liquid fuel cells. *Journal of Membrane Science*, *493*, 212–223. doi:10.1016/j.memsci.2015.07.007

Lufrano, F., Baglio, V., Stati, P., Antonucci, V., & Arico, A. S. (2013). Performance analysis of polymer electrolyte membranes for direct methanol fuel cells. *Journal of Power Sources*, *243*, 519–534. doi:10.1016/j.jpowsour.2013.05.180

Lufrano, F., Squardrito, G., Patti, A., & Passalacqua, E. (2000). Sulfonated polysulfone as promising membranes for polymer electrolyte fuel cells. *Journal of Applied Polymer Science*, *77*(6), 1250–1257. doi:10.1002/1097-4628(20000808)77:6<1250::AID-APP9>3.0.CO;2-R

Lukman, A. I., Gong, B., Marjo, C. E., Roessner, U., & Harris, A. T. (2011). Facile synthesis, stabilization, and anti-bacterial performance of discrete Ag nanoparticles using Medicago sativa seed exudates. *Journal of Colloid and Interface Science*, *353*(2), 433–444. doi:10.1016/j.jcis.2010.09.088 PMID:20974473

Luna, F. M. T., Pontes-Filho, A. A., Trindade, E. D., Silva, I. J., Azevedo, D. C., & Cavalcante, C. L. (2008). Removal of aromatic compounds from mineral naphthenic oil by adsorption. *Industrial & Engineering Chemistry Research*, *47*(9), 3207–3212. doi:10.1021/ie071476v

Lunov, O., Syrovets, T., Loos, C., Nienhaus, G. U., Mailander, V., Landfester, K., ... Simmet, T. (2011). Amino-Functionalized Polystyrene Nanoparticles Activate the NLRP3 Inflammasome in Human Macrophages. *ACS Nano*, 5. PMID:22111911

Luo, Z., Zhu, S., Liu, Z., Liu, J., Huo, M., & Yang, W. (2015). Study of phosphate removal from aqueous solution by zinc oxide. *Journal of Water and Health*, *13*(3), 704–713. doi:10.2166/wh.2015.210 PMID:26322756

Luque, R., Varma, R. S., & Clark, J. H. (2012). *Sustainable Preparation of Metal Nanoparticles: Methods and Applications*. Royal Society of Chemistry. doi:10.1039/9781849735469

Lu, S. G., Bai, S. Q., Zhu, L., & Shan, H. D. (2009). Removal mechanism of phosphate from aqueous solution by fly ash. *Journal of Hazardous Materials*, *161*(1), 95–101. doi:10.1016/j.jhazmat.2008.02.123 PMID:18434007

Luximon-Ramma, A., Bahorun, T., Soobrattee, M. A., & Aruoma, O. I. (2002). Antioxidant Activities of Phenolic, Proanthocyanidin, and Flavonoid Components in Extracts of Cassia fistula. *Journal of Agricultural and Food Chemistry*, *50*(18), 5042–5047. doi:10.1021/jf0201172 PMID:12188605

Lu, Y., Li, J., Han, J., Ng, H.-T., Binder, C., Partridge, C., & Meyyappan, M. (2004). Room temperature methane detection using palladium loaded single-walled carbon nanotube sensors. *Chemical Physics Letters*, *391*(4), 344–348. doi:10.1016/j.cplett.2004.05.029

Lu, Z.-X., Zhang, Z.-L., Zhang, M.-X., Xie, H.-Y., Tian, Z.-Q., Chen, P., ... Pang, D.-W. (2005). Core/Shell Quantum-Dot-Photosensitized Nano-TiO2 Films: Fabrication and Application to the Damage of Cells and DNA. *The Journal of Physical Chemistry B*, *109*(47), 22663–22666. doi:10.1021/jp054472h PMID:16853950

Lynge, M. E., Westen, R., Postmab, A., & Stadler, B. (2011). Polydopamine—a nature-inspired polymer coating for biomedical science. *Nanoscale*, *3*(12), 4916–4928. doi:10.1039/c1nr10969c PMID:22024699

Maaz, K., Mumtaz, A., Hasanain, S. K., & Ceylan, A. (2007). Synthesis and magnetic properties of cobalt ferrite (CoFe2O4) nanoparticles prepared by wet chemical route. *Journal of Magnetism and Magnetic Materials*, *308*(2), 289–295. doi:10.1016/j.jmmm.2006.06.003

Machado, F. M., Bergmann, C. P., Fernandes, T. H., Lima, E. C., Royer, B., Calvete, T., & Fagan, S. B. (2011). Adsorption of Reactive Red M-2BE dye from water solutions by multi-walled carbon nanotubes and activated carbon. *Journal of Hazardous Materials*, *192*(3), 1122–1131. doi:10.1016/j.jhazmat.2011.06.020 PMID:21724329

Machado, S., Stawiński, W., Slonina, P., Pinto, A., Grosso, J., Nouws, H., ... Delerue-Matos, C. (2013). Application of green zero-valent iron nanoparticles to the remediation of soils contaminated with ibuprofen. *The Science of the Total Environment*, *461*, 323–329. doi:10.1016/j.scitotenv.2013.05.016 PMID:23738986

Machida, M., Mochimaru, T., & Tatsumoto, H. (2006). Lead (II) adsorption onto the graphene layer of carbonaceous materials in aqueous solution. *Carbon*, *44*(13), 2681–2688. doi:10.1016/j.carbon.2006.04.003

Madadrang, C. J., Kim, H. Y., Gao, G., Wang, N., Zhu, J., Feng, H., ... Hou, S. (2012). Adsorption behavior of EDTA-graphene oxide for Pb (II) removal. *ACS Applied Materials & Interfaces*, *4*(3), 1186–1193. doi:10.1021/am201645g PMID:22304446

Maensiri, S., Laokul, P., Klinkaewnarong, J., Phokha, S., Promarak, V., & Seraphin, S. (2008). Indium oxide (In2O3) nanoparticles using Aloe vera plant extract: Synthesis and optical properties. *Optoelectronics And Advanced Materials*, *2*(3), 161–165.

Magdolenova, Z., Collins, A., Kumar, A., Dhawan, A., Stone, V., & Dusinska, M. (2014). Mechanisms of genotoxicity. A review of in vitro and in vivo studies with engineered nanoparticles. *Nanotoxicology*, *8*(3), 233–278. doi:10.3109/17435390.2013.773464 PMID:23379603

Mageshwari, K., Mali, S. S., Sathyamoorthy, R., & Patil, P. S. (2013). Template-free synthesis of MgO nanoparticles for effective photocatalytic applications. *Powder Technology*, *249*, 456–462. doi:10.1016/j.powtec.2013.09.016

Ma, H., Wang, H., & Na, C. (2015). Microwave-assisted optimization of platinum-nickel nanoalloys for catalytic water treatment. *Applied Catalysis B: Environmental*, *163*, 198–204. doi:10.1016/j.apcatb.2014.07.062

Mahmoodi, N. M., Abdi, J., & Bastani, D. (2014). Direct dyes removal using modified magnetic ferrite nanoparticle. *Journal of Environmental Health Science & Engineering*, *12*(1), 96–96. doi:10.1186/2052-336X-12-96 PMID:24991427

Maiti, J., Kakati, N., Lee, S. H., Jee, S. H., & Yoon, Y. S. (2011). PVA nano composite membrane for DMFC application. *Solid State Ionics*, *201*(1), 21–26. doi:10.1016/j.ssi.2011.07.013

Ma, J., Zhang, L. Z., Wang, Y. H., Lei, S. L., Luo, X. B., Chen, S. H., ... Au, C. T. (2014). Mechanism of 2,4-dinitrophenol photocatalytic degradation by ζ-Bi2O3/Bi2MoO6 composites under solar and visible light irradiation. *Chemical Engineering Journal*, *251*, 371–380. doi:10.1016/j.cej.2014.04.085

Majedi, S. M., Kelly, B. C., & Lee, H. K. (2014). Combined effects of water temperature and chemistry on the environmental fate and behavior of nanosized zinc oxide. *The Science of the Total Environment*, *496*, 585–593. doi:10.1016/j.scitotenv.2014.07.082 PMID:25108799

Makarov, V. V., Love, A. J., Sinitsyna, O. V., Makarova, S. S., Yaminsky, I. V., Taliansky, M. E., & Kalinina, N. O. (2014). "Green" Nanotechnologies: Synthesis of Metal Nanoparticles Using Plant. *Acta naturae, 6*(20), 35-44.

Makhluf, S., Dror, R., Nitzan, Y., Abramovich, Y., Jelinek, R., & Gedanken, A. (2005). Microwave-assisted synthesis of nanocrystalline MgO and its use as a bacteriocide. *Advanced Functional Materials*, *15*(10), 1708–1715. doi:10.1002/adfm.200500029

Mak, S.-Y., & Chen, D.-H. (2004). Fast adsorption of methylene blue on polyacrylic acid-bound iron oxide magnetic nanoparticles. *Dyes and Pigments*, *61*(1), 93–98. doi:10.1016/j.dyepig.2003.10.008

Malik, P., Shankar, R., Malik, V., Sharma, N., & Mukherjee, T. K. (2014). Green Chemistry Based Benign Routes for Nanoparticle Synthesis. *Journal of Nanoparticles*, *14*. doi:10.1155/2014/302429

Maliyekkal, S. M., Lisha, K. P., & Pradeep, T. (2010). A novel cellulose–manganese oxide hybrid material by in situ soft chemical synthesis and its application for the removal of Pb (II) from water. *Journal of Hazardous Materials*, *181*(1), 986–995. doi:10.1016/j.jhazmat.2010.05.112 PMID:20638968

Maliyekkal, S. M., Sreeprasad, T., Krishnan, D., Kouser, S., Mishra, A. K., Waghmare, U. V., & Pradeep, T. (2013). Graphene: A reusable substrate for unprecedented adsorption of pesticides. *Small*, *9*(2), 273–283. doi:10.1002mll.201201125 PMID:23001848

Manna, S. K., Sarkar, S., Barr, J., Wise, K., Barrera, E. V., Jejelowo, O., ... Ramesh, G. T. (2005). Single-Walled Carbon Nanotube Induces Oxidative Stress and Activates Nuclear Transcription Factor-κB in Human Keratinocytes. *Nano Letters*, *5*(9), 1676–1684. doi:10.1021/nl0507966 PMID:16159204

Manoli, E., Samara, C., Konstantinou, I., & Albanis, T. (2000). Polycyclic aromatic hydrocarbons in the bulk precipitation and surface waters of Northern Greece. *Chemosphere*, *41*(12), 1845–1855. doi:10.1016/S0045-6535(00)00134-X PMID:11061306

Maphutha, S., Moothi, K., Meyyappan, M., & Iyuke, S. E. (2013). A carbon nanotube-infused polysulfone membrane with polyvinyl alcohol layer for treating oil-containing waste water. *Scientific Reports*, *3*(1), 1509. doi:10.1038rep01509 PMID:23518875

Marina, P. E., Ali, G. A. M., See, L. M., Teo, E. Y. L., Ng, E.-P., & Chong, K. F. (2016). In situ growth of redox-active iron-centered nanoparticles on graphene sheets for specific capacitance enhancement. *Arabian Journal of Chemistry*. doi:10.1016/j.arabjc.2016.02.006

Martins, Eggers, & Evertz. (2010). *Applications of nanomaterials in environmental engineering*. Environmental Research of the Federal Ministry of the Environment, Nature Conservation and Safety, UBA-FB001337.

Martinson, C. A., & Reddy, K. J. (2009). Adsorption of arsenic(III) and arsenic(V) by cupric oxide nanoparticles. *Journal of Colloid and Interface Science*, *336*(2), 406–411. doi:10.1016/j.jcis.2009.04.075 PMID:19477461

Masciangioli, T., & Zhang, W.-X. (2003). *Peer reviewed: environmental technologies at the nanoscale*. ACS Publications.

Maslin, B. R., & McDonald, M. W. (2004). AcaciaSearch- Evaluation of Acacia as a woody crop option for southern Australia. Rural Industries Research and Development Corporation; Dept. Conservation and Land Management.

Masui, T., Ozaki, T., Machida, K.-i., & Adachi, G.-y. (2000). Preparation of ceria–zirconia sub-catalysts for automotive exhaust cleaning. *Journal of Alloys and Compounds*, *303-304*(Supplement C), 49–55. doi:10.1016/S0925-8388(00)00603-4

Ma, X. M., Tsige, M., Uddin, S., & Talapatra, S. (2011b). Application of carbon nanotubes for removing organic contaminants from water. *Materials Express*, *1*(3), 183–200. doi:10.1166/mex.2011.1023

Mayo, J. T., Yavuz, C., Yean, S., Cong, L., Shipley, H., Yu, W., ... Colvin, V. L. (2007). The effect of nanocrystalline magnetite size on arsenic removal. *Science and Technology of Advanced Materials*, *8*(1-2), 71–75. doi:10.1016/j.stam.2006.10.005

McBride, M. B., & MartÍnez, C. E. (2000). Copper Phytotoxicity in a Contaminated Soil: Remediation Tests with Adsorptive Materials. *Environmental Science & Technology*, *34*(20), 4386–4391. doi:10.1021/es0009931

Mckeen, J. C., Yan, Y. S., & Davis, M. E. (2008). Proton conductivity of acid-functionalized zeolite beta, MCM-41 and MCM-48: Effect of acid strength. *Chemistry of Materials*, *20*(16), 5122–5124. doi:10.1021/cm801418r

McMurray, T. A., Byrne, J. A., Dunlop, P. S. M., & McAdams, E. T. (2005). Photocatalytic and electrochemically assisted photocatalytic oxidation of formic acid on TiO_2 films under UVA and UVB irradiation. *Journal of Applied Electrochemistry*, *35*(7-8), 723–731. doi:10.100710800-005-1397-1

McNicholas, T. P., Zhao, K., Yang, C., Hernandez, S. C., Mulchandani, A., Myung, N. V., & Deshusses, M. A. (2011). Sensitive detection of elemental mercury vapor by gold-nanoparticle-decorated carbon nanotube sensors. *The Journal of Physical Chemistry C*, *115*(28), 13927–13931. doi:10.1021/jp203662w PMID:21922036

Mecke, A., Uppuluri, S., Sassanella, T. M., Lee, D. K., Ramamoorthy, A., Baker, J. R. Jr, ... Banaszak Holl, M. M. (2004). Direct observation of lipid bilayer disruption by poly(amidoamine) dendrimers. *Chemistry and Physics of Lipids*, *132*(1), 3–14. doi:10.1016/j.chemphyslip.2004.09.001 PMID:15530443

Mercier, L., & Bell, V. (2000). Chapter. In H. Mostaghaci (Ed.), Environment Conscious Materials: Ecomaterials (pp. 599–606). Montreal, Canada: Canadian Institute of Mining, Metallurgy and Petroleum.

Mey-Marom, A., & Katz, M. G. (1986). Measurement of active pore size distribution of microporous membranes-a new approach. *Journal of Membrane Science*, *27*(2), 119–130. doi:10.1016/S0376-7388(00)82049-9

Michalet, X., Pinaud, F. F., Bentolila, L. A., Tsay, J. M., Doose, S., Li, J. J., ... Weiss, S. (2005). Quantum dots for live cells, in vivo imaging, and diagnostics. *Science*, *307*(5709), 538–544. doi:10.1126cience.1104274 PMID:15681376

Mickelson, E., Chiang, I., Zimmerman, J., Boul, P., Lozano, J., Liu, J., ... Margrave, J. (1999). Solvation of fluorinated single-wall carbon nanotubes in alcohol solvents. *The Journal of Physical Chemistry B*, *103*(21), 4318–4322. doi:10.1021/jp9845524

Milani, N., McLaughlin, M. J., Stacey, S. P., Kirby, J. K., Hettiarachchi, G. M., Beak, D. G., & Cornelis, G. (2012). Dissolution kinetics of macronutrient fertilizers coated with manufactured zinc oxide nanoparticles. *Journal of Agricultural and Food Chemistry*, *60*(16), 3991–3998. doi:10.1021/jf205191y PMID:22480134

Miller, G., van Schaik, D., & Adelaide, R. (2008). *Nanotechnology in food and agriculture*. Radio Adelaide.

Mioc, U. B., Milonjic, S. K., Malovic, D., Stamenkovic, V., Colomband, P., Mitrovic, M. M., & Dimitrijevic, R. (1997). Structure and proton conductivity of 12-tungstophosphoric acid doped silica. *Solid State Ionics*, *97*(1-4), 239–246. doi:10.1016/S0167-2738(97)00089-1

Mirkin, C. A., Letsinger, R. L., Mucic, R. C., & Storhoff, J. J. (1996). A DNA-based method for rationally assembling nanoparticles into macroscopic materials. *Nature, 382*(6592), 607-609.

Mirkin, C. A., Letsinger, R. L., Mucic, R. C., & Storhoff, J. J. (1996). A DNA-based method for rationally assembling nanoparticles into macroscopic materials. *Nature*, *382*(6592), 607–609. doi:10.1038/382607a0 PMID:8757129

Mirzaei, H., & Darroudi, M. (2017). Zinc oxide nanoparticles: Biological synthesis and biomedical applications. *Ceramics International*, *43*(11, Part B), 907–914. doi:10.1016/j.ceramint.2016.10.051

Mishra, P. C., & Patel, R. K. (2009). Use of agricultural waste for the removal of nitrate-nitrogen from aqueous medium. *Journal of Environmental Management*, *90*(1), 519–522. doi:10.1016/j.jenvman.2007.12.003 PMID:18294755

Mitrano, D. M., Motellier, S., Clavaguera, S., & Nowack, B. (2015). Review of nanomaterial aging and transformations through the life cycle of nano-enhanced products. *Environment International*, *77*, 132–147. doi:10.1016/j.envint.2015.01.013 PMID:25705000

Mittal, A. K., Chisti, Y., & Banerjee, U. C. (2013). Synthesis of metallic nanoparticles using plant extracts. *Biotechnology Advances*, *31*(2), 346–356. doi:10.1016/j.biotechadv.2013.01.003 PMID:23318667

Mizutani, T., Arai, K., Miyamoto, M., & Kimura, Y. (2006). Application of silica-containing nano-composite emulsion to wall paint: A new environmentally safe paint of high performance. *Progress in Organic Coatings*, *55*(3), 276–283. doi:10.1016/j.porgcoat.2005.12.001

Mohamed, E. F. (2017). Nanotechnology: Future of Environmental Air Pollution Control. *Environmental Management and Sustainable Development*, *6*(2), 429–454. doi:10.5296/emsd.v6i2.12047

Mohammadian, H., Heidari, M. B., & Esfandiarijahromi, E. (2013). *Analysis the Different Types of Nano Sensors on Based of Structure and it's Applications on Nano Electronics.* Paper presented at the 2nd International Conference on Computer and Communication Technologies (ICCCT'2013), Pattaya, Thailand.

Mohanpuria, P., Rana, N. K., & Yadav, S. K. (2008). Biosynthesis of nanoparticles: Technological concepts and future applications. *Journal of Nanoparticle Research*, *10*(3), 507–517. doi:10.100711051-007-9275-x

Mohapatra, D., Mishra, S., & Sutar, N. (2010). Banana And Its By Product Utilisation: An Overview. *Journal of Scientific and Industrial Research*, *69*, 323–329.

Mohmood, I., Lopes, C. B., Lopes, I., Ahmad, I., Duarte, A. C., & Pereira, E. (2013). Nanoscale materials and their use in water contaminants removal—a review. *Environmental Science and Pollution Research International*, *20*(3), 1239–1260. doi:10.100711356-012-1415-x PMID:23292223

Molinari, R., Poerio, T., Cassano, R., Picci, N., & Argurio, P. (2004). Copper (II) removal from wastewaters by a new synthesized selective extractant and SLM viability. *Industrial & Engineering Chemistry Research*, *43*(2), 623–628. doi:10.1021/ie030392t

Moniz, E. J. (2010). *Nanotechnology for the energy challenge.* John Wiley & Sons.

Moody, N., & Barron, A. R. (2015). *UV-Visible Spectroscopy of Noble Metal Nanoparticles.* OpenStax-CNX module: m55548.

Mook, W. T., Chakrabarti, M. H., Aroua, M. K., Khan, G. M. A., Ali, B. S., Islam, M. S., & Abu Hassan, M. A. (2012). Removal of total ammonia nitrogen (TAN), nitrate and total organic carbon (TOC) from aquaculture wastewater using electrochemical technology: A review. *Desalination*, *285*, 1–13. doi:10.1016/j.desal.2011.09.029

Moorthy, S. K., Ashok, C. H., Rao, K. V., & Viswanathan, C. (2015). Synthesis and Characterization of Mgo Nanoparticles by Neem Leaves through Green Method. *Materials Today: Proceedings*, *2*(9), 4360–4368. doi:10.1016/j.matpr.2015.10.027

Moradi, O., Sadegh, H., Shahryari-Ghoshekandi, R., & Norouzi, M. (2015). Application of carbon nanotubes in nanomedicine: new medical approach for tomorrow. In Handbook of Research on Diverse Applications of Nanotechnology in Biomedicine, Chemistry, and Engineering (pp. 90–128). Hershey, PA: IGI Global.

Moradi, O. (2013). Adsorption behavior of basic red 46 by single-walled carbon nanotubes surfaces. *Fullerenes, Nanotubes, and Carbon Nanostructures*, *21*(4), 286–301. doi:10.1080/1536383X.2011.572317

Moradi, O., & Zare, K. (2011). Adsorption of Pb (II), Cd (II) and Cu (II) ions in aqueous solution on SWCNTs and SWCNT–COOH surfaces: Kinetics studies. *Fullerenes, Nanotubes, and Carbon Nanostructures*, *19*(7), 628–652. doi: 10.1080/1536383X.2010.504955

Moradi, O., Zare, K., Monajjemi, M., Yari, M., & Aghaie, H. (2010). The studies of equilibrium and thermodynamic adsorption of Pb (II), Cd (II) and Cu (II) Ions from aqueous solution onto SWCNTs and SWCNT–COOH Surfaces. *Fullerenes, Nanotubes, and Carbon Nanostructures*, *18*(3), 285–302. doi:10.1080/15363831003783005

Morelli, J. (2011). Environmental Sustainability: A Definition for Environmental Professionals. *Journal of Environmental Sustainability*, *1*(1), 9. doi: 10.14448/jes.01.0002

Morelli, J. (2013). Environmental sustainability: A definition for environmental professionals. *Journal of Environmental Sustainability*, *1*(1), 2.

Morillo, D., Perez, G., & Valiente, M. (2015). Efficient arsenic(V) and arsenic(III) removal from acidic solutions with Novel Forager Sponge-loaded superparamagnetic iron oxide nanoparticles. *Journal of Colloid and Interface Science*, *453*, 132–141. doi:10.1016/j.jcis.2015.04.048 PMID:25982936

Morones, J. R., Elechiguerra, J. L., Camacho, A., Holt, K., Kouri, J. B., Ramírez, J. T., & Yacaman, M. J. (2005). T-bactericidal effect of silver nanoparticles. *Nanotechnology*, *16*(10), 2346–2353. doi:10.1088/0957-4484/16/10/059 PMID:20818017

Mortimer, G. M., & Minchin, R. F. (2017). Chapter Twenty - Nanotoxicology and Nanovaccines A2 - Skwarczynski, Mariusz. In I. Toth (Ed.), *Micro and Nanotechnology in Vaccine Development* (pp. 373–392). William Andrew Publishing. doi:10.1016/B978-0-323-39981-4.00020-8

Mortimer, M., Kasemets, K., & Kahru, A. (2010). Toxicity of ZnO and CuO nanoparticles to ciliated protozoa Tetrahymena thermophila. *Toxicology*, *269*(2-3), 182–189. doi:10.1016/j.tox.2009.07.007 PMID:19622384

Mostafa, S., Behafarid, F., Croy, J. R., Ono, L. K., Li, L., Yang, J. C., ... Cuenya, B. R. (2010). Shape-Dependent Catalytic Properties of Pt Nanoparticles. *Journal of the American Chemical Society*, *132*(44), 15714–15719. doi:10.1021/ja106679z PMID:20949968

Mrówczyński, R., Turcu, R., Leostean, C., Scheidt, H. A., & Liebscher, J. (2013). New versatile polydopamine coated functionalized magnetic nanoparticles. *Materials Chemistry and Physics*, *138*(1), 295–302. doi:10.1016/j.matchemphys.2012.11.059

Mthombeni, N. H., Mbakop, S., & Onyango, M. S. (2015). Magnetic zeolite-polymer composite as an adsorbent for the remediation of wastewaters containing vanadium. *International Journal of Environmental Sciences and Development*, *6*(8), 602–605. doi:10.7763/IJESD.2015.V6.665

MubarakAli, D., Thajuddin, N., Jeganathan, K., & Gunasekaran, M. (2011). Plant extract mediated synthesis of silver and gold nanoparticles and its antibacterial activity against clinically isolated pathogens. *Colloids and Surfaces. B, Biointerfaces*, *85*(2), 360–365. doi:10.1016/j.colsurfb.2011.03.009 PMID:21466948

Mubeen, S., Zhang, T., Chartuprayoon, N., Rheem, Y., Mulchandani, A., Myung, N. V., & Deshusses, M. A. (2009). Sensitive detection of H2S using gold nanoparticle decorated single-walled carbon nanotubes. *Analytical Chemistry*, *82*(1), 250–257. doi:10.1021/ac901871d PMID:19968250

Mueller, N. C., Braun, J., Bruns, J., Černík, M., Rissing, P., Rickerby, D., & Nowack, B. (2012). Application of nanoscale zero valent iron (NZVI) for groundwater remediation in Europe. *Environmental Science and Pollution Research International*, *19*(2), 550–558. doi:10.100711356-011-0576-3 PMID:21850484

Mueller, N. C., & Nowack, B. (2010). Nanoparticles for remediation: Solving big problems with little particles. *Elements*, *6*(6), 395–400. doi:10.2113/gselements.6.6.395

Muff, J., & Søgaard, E. G. (2010). Electrochemical degradation of PAH compounds in process water: A kinetic study on model solutions and a proof of concept study on runoff water from harbour sediment purification. *Water Science and Technology*, *61*(8), 2043–2051. doi:10.2166/wst.2010.129 PMID:20389002

Mukherjee, P., Ahmad, A., Mandal, D., Senapati, S., Sainkar, S. R., Khan, M. I., ... Kumar, R. (2001). Fungus-mediated synthesis of silver nanoparticles and their immobilization in the mycelial matrix: A novel biological approach to nanoparticle synthesis. *Nano Letters*, *1*(10), 515–519. doi:10.1021/nl0155274

Mukhopadhyay, S. S. (2014). Nanotechnology in agriculture: Prospects and constraints. *Nanotechnology, Science and Applications*, *7*, 63–71. doi:10.2147/NSA.S39409 PMID:25187699

Muller, J., Decordier, I., Hoet, P. H., Lombaert, N., Thomassen, L., Huaux, F., ... Kirsch-Volders, M. (2008). Clastogenic and aneugenic effects of multi-wall carbon nanotubes in epithelial cells. *Carcinogenesis*, *29*(2), 427–433. doi:10.1093/carcin/bgm243 PMID:18174261

Mulvaney, D. (2011). *Green Technology: An A-to-Z Guide*. SAGE Publications. doi:10.4135/9781412973793.n84

Murali, K. R., Swarna, L. M., & Nakkeeran, E. (2014). Nanoparticle-Membrane Filtration of Vehicular Exhaust to Reduce Air Pollution-A review. *International Research Journal of Environment Sciences*, *3*(4), 82–86.

Muralikisrishnan, R., Swarnalakshmi, M., & Nakkeeran, E. (2014). Nanoparticle-Membrane Filtration of Vehicular Exhaust to Reduce Air Polution- A Review. *International Research Journal of Environment Sciences, 3*(4), 82-86. Retrieved from http://www.isca.in/IJENS/Archive/v3/i4/12.ISCA-IRJEvS-2014-45.pdf

Murcia, M. D., Gómez, M., Gómez, E., Gómez, J. L., & Christofi, N. (2011). Photodegradation of congo red using XeBr, KrCl and Cl_2 barrier discharge excilamps: A kinetics study. *Desalination*, *281*, 364–371. doi:10.1016/j.desal.2011.08.011

Murphy, C. J. (2008). Sustainability as an emerging design criterion in nanoparticle synthesis and applications. *Journal of Materials Chemistry*, *18*(19), 2173–2176. doi:10.1039/b717456j

Murphy, C. J., Sau, T. K., Gole, A. M., Orendorff, C. J., Gao, J., Gou, L., ... Li, T. (2005, July). Anisotropic Metal Nanoparticles: Synthesis, Assembly, and Optical Applications. *The Journal of Physical Chemistry B*, *109*(29), 13857–13870. doi:10.1021/jp0516846 PMID:16852739

Murugan, R., & Parimelazhagan, T. (2014). Comparative evaluation of different extraction methods for antioxidant and anti-inflammatory properties from Osbeckia parvifolia Arn. – An in vitro approach. *Journal of King Saud University - Science, 26*(4), 267-275. doi:10.1016/j.jksus.2013.09.006

Muthukumaran, C., Sivakumar, V. M., & Thirumarimurugan, M. (2016). Adsorption isotherms and kinetic studies of crystal violet dye removal from aqueous solution using surfactant modified magnetic nanoadsorbent. *Journal of the Taiwan Institute of Chemical Engineers*, *63*, 354–362. doi:10.1016/j.jtice.2016.03.034

Mutin, P. H., & Vioux, A. (2009). Nonhydrolytic Processing of Oxide-Based Materials: Simple Routes to Control Homogeneity, Morphology, and Nanostructure. *Chemistry of Materials*, *21*(4), 582–596. doi:10.1021/cm802348c

Nabeela, F., Azizullah, A., Bibi, R., Uzma, S., Murad, W., Shakir, S. K., ... Häder, D.-P. (2014). Microbial contamination of drinking water in Pakistan—a review. *Environmental Science and Pollution Research International*, *21*(24), 13929–13942. doi:10.100711356-014-3348-z PMID:25056753

Nadafi, K., Mesdaghinia, A., Nabizadeh, R., Younesian, M., & Rad, M. J. (2011). The combination and optimization study on RB29 dye removal from water by peroxy acid and single-wall carbon nanotubes. *Desalination and Water Treatment*, *27*(1-3), 237–242. doi:10.5004/dwt.2011.1980

Nadagouda, M. N., Desai, I., Cruz, C., & Yang, D. J. (2012). Novel Pd based catalyst for the removal of organic and emerging contaminants. *RSC Advances*, *2*(19), 7540. doi:10.1039/c2ra20562a

Nadejda, K., Isabelle, B., Berit, G., Roland, E., Bauer, K. M., Akio, Y., & Tobias, V. (2002). Self-Assembled Gold Nanoparticle/Dendrimer Composite Film for Vapor Sensing Application. *Nanotechology Letter*, *2*(5), 551–555.

Nadim, F., Hoag, G. E., Liu, S., Carley, R. J., & Zack, P. (2000). Detection and remediation of soil aquifer systems contaminated with petroleum products: An overview. *Journal of Petroleum Science Engineering*, *26*(1-4), 169–178. doi:10.1016/S0920-4105(00)00031-0

Nagaraja, P., Yathirajan, H., Raju, C., Vasantha, R., Nagendra, P., & Kumar, M. (2003). 3-aminophenol as a novel coupling agent for the spectrophotometric determination of sulfonamide derivatives. *Il Farmaco*, *58*(12), 1295–1300. doi:10.1016/S0014-827X(03)00093-4 PMID:14630242

Nagarale, R. K., Gohil, G. S., & Shahi, V. K. (2006). Recent developments on ion-exchange membranes and electro-membrane processes. *Advances in Colloid and Interface Science*, *119*(2-3), 97–130. doi:10.1016/j.cis.2005.09.005 PMID:16325751

Nagarale, R. K., Gohil, G. S., Shahi, V. K., & Rangarajan, R. (2004). Organic-inorganic hybrid membrane: Thermally stable cation-exchange membrane prepared by sol-gel method. *Macromolecules*, *37*(26), 10023–10030. doi:10.1021/ma048404p

Nagarale, R. K., Shin, W., & Singh, P. K. (2010). Progress in ionic organic-inorganic composite membranes for fuel cell applications. *Polymer Chemistry*, *1*(4), 388–408. doi:10.1039/B9PY00235A

Nagaveni, K., Sivalingam, G., Hegde, M., & Madras, G. (2004a). Photocatalytic degradation of organic compounds over combustion-synthesized nano-TiO2. *Environmental Science & Technology*, *38*(5), 1600–1604. doi:10.1021/es034696i PMID:15046366

Nagaveni, K., Sivalingam, G., Hegde, M., & Madras, G. (2004b). Solar photocatalytic degradation of dyes: High activity of combustion synthesized nano TiO 2. *Applied Catalysis B: Environmental*, *48*(2), 83–93. doi:10.1016/j.apcatb.2003.09.013

Naghash, A., & Nezamzadeh-Ejhieh, A. (2015). Comparison of the efficiency of modified clinoptilolite with HDTMA and HDP surfactants for the removal of phosphate in aqueous solutions. *Journal of Industrial and Engineering Chemistry*, *31*, 185–191. doi:10.1016/j.jiec.2015.06.022

Nair, C.I., & Jayachandran, K., & Shashidhar. (2008). Biodegradation of Phenol. *African Journal of Biotechnology*, *7*(25), 4951–4958.

Namasivayam, C., & Sangeetha, D. (2004). Equilibrium and kinetic studies of adsorption of phosphate onto ZnCl2 activated coir pith carbon. *Journal of Colloid and Interface Science*, *280*(2), 359–365. doi:10.1016/j.jcis.2004.08.015 PMID:15533408

Narayanan, K. B., & Sakthivel, N. (2010). Biological synthesis of metal nanoparticles by microbes. *Advances in Colloid and Interface Science*, *156*(1), 1–13. doi:10.1016/j.cis.2010.02.001 PMID:20181326

Narayanan, K. B., & Sakthivel, N. (2011). Green synthesis of biogenic metal nanoparticles by terrestrial and aquatic phototrophic and heterotrophic eukaryotes and biocompatible agents. *Advances in Colloid and Interface Science*, *169*(2), 59–79. doi:10.1016/j.cis.2011.08.004 PMID:21981929

Narayanan, K. B., & Sakthivel, N. (2011). Synthesis and characterization of nano-gold composite using Cylindrocladium floridanum and its heterogeneous catalysis in the degradation of 4-nitrophenol. *Journal of Hazardous Materials*, *189*(1), 519–525. doi:10.1016/j.jhazmat.2011.02.069 PMID:21420237

Nassar, N. N. (2010). Rapid removal and recovery of Pb (II) from wastewater by magnetic nanoadsorbents. *Journal of Hazardous Materials*, *184*(1), 538–546. doi:10.1016/j.jhazmat.2010.08.069 PMID:20837379

Nassehinia, H. R., Gholami, M., Jafari, A. J., & Esrafily, A. (2013). Nano Photocatalytic Process Application of ZnO Nanoparticle and UV on Benzene Removal from Synthetic Air. *Asian Journal of Chemistry*, *25*(6), 3427–3430.

Natale, F., Erto, A., Lancia, A., & Musmarra, D. (2008). Experimental and modelling analysis of As (V) ions adsorption on granular activated carbon. *Water Research*, *42*(8), 2007–2016. doi:10.1016/j.watres.2007.12.008 PMID:18222519

National Toxicology Program (NTP). (2000). *Toxicology and carcinogenesis studies of naphthalene (CAS no. 91-20-3) in F344/N rats (inhalation studies).* Technical report series no. 91-20-3. NIH Publication no. 01-4434. US Department of Health and Human Services, Public Health Service, National Institutes of Health.

Navarro Abel, E., Portales Rosario, F., Sun-Kou María, R., & Llanos Bertha, P. (2008). Effect of pH on phenol biosorption by marine seaweeds. *Journal of Hazardous Materials*, *156*(1–3), 405–411. doi:10.1016/j.jhazmat.2007.12.039 PMID:18249490

Navarro, E., Piccapietra, F., Wagner, B., Marconi, F., Kaegi, R., Odzak, N., ... Behra, R. (2008). Toxicity of silver nanoparticles to Chlamydomonas reinhardtii. *Environmental Science & Technology*, *42*(23), 8959–8964. doi:10.1021/es801785m PMID:19192825

Neburchilov, V., Martin, J., Wang, H., & Zhang, J. (2007). A review of polymer electrolyte membranes for direct methanol fuel cells. *Journal of Power Sources*, *160*(2), 221–238. doi:10.1016/j.jpowsour.2007.03.044

Nethaji, S., Sivasamy, A., & Mandal, A. B. (2013). Preparation and characterization of corn cob activated carbon coated with nano-sized magnetite particles for the removal of Cr(VI). *Bioresource Technology*, *134*(Supplement C), 94–100. doi:10.1016/j.biortech.2013.02.012 PMID:23500565

Neumann, S., Burkert, K., Kemp, R., Rades, T., Rod Dunbar, P., & Hook, S. (2014). Activation of the NLRP3 inflammasome is not a feature of all particulate vaccine adjuvants. *Immunology and Cell Biology*, *92*(6), 535–542. doi:10.1038/icb.2014.21 PMID:24687021

Newalkar, B. L., Komarneni, S., & Katsuki, H. (2000). Rapid synthesis of mesoporous SBA-15 molecular sieve by a microwave-hydrothermal process. *Chemical Communications*, (23): 2389–2390. doi:10.1039/b0074411

Newalkar, B. L., Olanrewaju, J., & Komarneni, S. (2001). Direct synthesis of titanium-substituted mesoporous SBA-15 molecular sieveunder microwave-hydrothermal conditions. *Chemistry of Materials*, *13*(2), 552–557. doi:10.1021/cm000748g

Newkome, G. R., Moorefield, C. N., & Vögtle, F. (1996). Nomenclature. *Dendritic Molecules: Concepts, Syntheses. Perspectives - Gerontological Nursing Association*, 37–47.

Neyaz, N., & Siddiqui, W. (2015). Removal of Cu (II) by modified magnetite nanocomposite as a nanosorbent. *International Journal of Science and Research*, *4*, 1868–1873.

Ngah, W. S., & Fatinathan, S. (2008). Adsorption of Cu(II) ions in aqueous solution using chitosan beads, chitosan-GLA beads and chitosan-alginate beads. *Chemical Engineering Journal*, *143*(1-3), 62–72. doi:10.1016/j.cej.2007.12.006

Ngah, W. W., Teong, L., & Hanafiah, M. (2011). Adsorption of dyes and heavy metal ions by chitosan composites: A review. *Carbohydrate Polymers*, *83*(4), 1446–1456. doi:10.1016/j.carbpol.2010.11.004

Ng, C.-T., Li, J. J., Bay, B.-H., & Yung, L.-Y. L. (2010). Current Studies into the Genotoxic Effects of Nanomaterials. *Journal of Nucleic Acids*, *2010*, 12. doi:10.4061/2010/947859 PMID:20936181

Nguyen, M. T., Kryachko, E. S., & Vanquickenborne, L. G. (2003-d), The Chemistry of Phenols. In Z. Rappoport (Ed.), *General and theoretical aspects of phenols*. John Wiley and Sons, Ltd.

Nie, L. R., Lu, J., Zhang, W., He, A., & Yao, S. (2015). Ionic liquid-modified silica gel as adsorbents for adsorption and separation of water-soluble phenolic acids from Salvia militiorrhiza Bunge. *Seperation and Purification Technology*. .2015.01.03710.1016/j.seppur

Niederberger, M. (2007). Nonaqueous Sol–Gel Routes to Metal Oxide Nanoparticles. *Accounts of Chemical Research*, *40*(9), 793–800. doi:10.1021/ar600035e PMID:17461544

Niederberger, M., & Pinna, N. (2009). *Metal Oxide Nanoparticles in Organic Solvents: Synthesis, Formation, Assembly and Application*. London: Springer-Verlag. doi:10.1007/978-1-84882-671-7

Ni, F., He, J., Wang, Y., & Luan, Z. (2015). Preparation and characterization of a cost-effective red mud/polyaluminum chloride composite coagulant for enhanced phosphate removal from aqueous solutions. *Journal of Water Process Engineering*, *6*, 158–165. doi:10.1016/j.jwpe.2015.04.003

Njoku, V. O., Islam, M. A., Asif, M., & Hameed, B. H. (2014). Utilization of sky fruit husk agricultural waste to produce high quality activated carbon for the herbicide bentazon adsorption. *Chemical Engineering Journal*, *251*, 183–191. doi:10.1016/j.cej.2014.04.015

Nogueira, A. E., Giroto, A. S., Neto, A. B. S., & Ribeiro, C. (2016). CuO synthesized by solvothermal method as a high capacity adsorbent for hexavalent chromium. *Colloids and Surfaces. A, Physicochemical and Engineering Aspects*, *498*, 161–167. doi:10.1016/j.colsurfa.2016.03.022

Nora'aini, A., Wahab Mohammad, A., Jusoh, A., Hasan, M. R., Ghazali, N., & Kamaruzaman, K. (2005). Treatment of aquaculture wastewater using ultra-low pressure asymmetric polyethersulfone (PES) membrane. *Desalination*, *185*(1-3), 317–326. doi:10.1016/j.desal.2005.03.084

Noreen, S., Bhatti, H. N., Nausheen, S., Sadaf, S., & Ashfaq, M. (2013). Batch and fixed bed adsorption study for the removal of drimarine black CL-B dye from aqueous solution using a lignocellulosic waste: A cost affective adsorbent. *Industrial Crops and Products*, *50*, 568–579. doi:10.1016/j.indcrop.2013.07.065

O'Connell, D. W., Birkinshaw, C., & O'Dwyer, T. F. (2008). Heavy metal adsorbents prepared from the modification of cellulose: A review. *Bioresource Technology*, *99*(15), 6709–6724. doi:10.1016/j.biortech.2008.01.036 PMID:18334292

O'Hayre, R., Cha, S. W., Colella, W., & Prinz, F. B. (2016). *Fuel Cell fundamentals*. John Wiley and Sons. doi:10.1002/9781119191766

Oberdorster, G., Sharp, Z., Atudorei, V., Elder, A., Gelein, R., Kreyling, W., & Cox, C. (2004). Translocation of inhaled ultrafine particles to the brain. *Inhalation Toxicology*, *16*(6-7), 437–445. doi:10.1080/08958370490439597 PMID:15204759

Ogata, F., Ueda, A., Tanei, S., Imai, D., & Kawasaki, N. (2016). Simultaneous removal of phosphate and nitrite ions from aqueous solutions using modified soybean waste. *Journal of Industrial and Engineering Chemistry*, *35*, 287–294. doi:10.1016/j.jiec.2016.01.007

Ohe, K., Nagae, Y., Nakamura, S., & Baba, Y. (2003). Removal of nitrate anion by carbonaceous materials prepared from bamboo and coconut shell. *Journal of Chemical Engineering of Japan*, *36*(4), 511–515. doi:10.1252/jcej.36.511

Ohura, T., Amagai, T., Fusaya, M., & Matsushita, H. (2003). Polycyclic aromatic hydrocarbons in indoor and outdoor environments and factors affecting their concentrations. *Environmental Science & Technology*, *38*(1), 77–83. doi:10.1021/es030512o PMID:14740720

Oliveira, M., & Machado, A. (2013). Preparation of polymer-based nanocomposites by different routes. *Nanocomposites: synthesis, characterization and applications*, 1-22.

Olkhovyk, O., Antochshuk, V., & Jaroniec, M. (2004). Benzoylthiourea-modified MCM-48 mesoporous silica for mercury(II) adsorption from aqueous solutions. *Colloids and Surfaces. A, Physicochemical and Engineering Aspects*, *236*(1-3), 69–72. doi:10.1016/j.colsurfa.2003.12.032

Ondreiovic, G., & Sirota, A. (1995). Selective binding and removal of guests in a microporous metal-organic framework. *Nature*, *378*, 14.

Organization, W. H. (2012). *The Global Report for Research on Infectious Diseases of Poverty*. World Health Organization.

Orphanides, A., Goulas, V., & Gekas, V. (2013). Effect of Drying Method on the Phenolic Content and Antioxidant Capacity of Spearmint. *Czech Journal of Food Sciences*, *31*(5), 509–513. doi:10.17221/526/2012-CJFS

Ouraipryvan, P., Sreethawong, T., & Chavadej, S. (2009). Synthesis of crystalline MgO nanoparticle with mesoporous-assembled structure via a surfactant-modified sol–gel process. *Materials Letters*, *63*(21), 1862–1865. doi:10.1016/j.matlet.2009.05.068

Owlad, Aroua, M. K., Daud, W. A. W., & Baroutian, S. (2009). Removal of Hexavalent Chromium-Contaminated Water and Wastewater: A Review. *Water, Air, and Soil Pollution*, *200*(1), 59–77. doi:10.100711270-008-9893-7

Ozay, O., Ekici, S., Baran, Y., Aktas, N., & Sahiner, N. (2009). Removal of toxic metal ions with magnetic hydrogels. *Water Research*, *43*(17), 4403–4411. doi:10.1016/j.watres.2009.06.058 PMID:19625066

Pacheco, S., Medina, M., Valencia, F., & Tapia, J. (2006). Removal of inorganic mercury from polluted water using structured nanoparticles. *Journal of Environmental Engineering*, *132*(3), 342–349. doi:10.1061/(ASCE)0733-9372(2006)132:3(342)

Pacurari, M., Yin, X. J., Ding, M., Leonard, S. S., Schwegler-berry, D., Ducatman, B. S., ... Vallyathan, V. (2008). Oxidative and molecular interactions of multi-wall carbon nanotubes (MWCNT) in normal and malignant human mesothelial cells. *Nanotoxicology*, *2*(3), 155–170. doi:10.1080/17435390802318356

Pacurari, M., Yin, X. J., Zhao, J., Ding, M., Leonard, S. S., Schwegler-Berry, D., ... Vallyathan, V. (2008). Raw single-wall carbon nanotubes induce oxidative stress and activate MAPKs, AP-1, NF-kappaB, and Akt in normal and malignant human mesothelial cells. *Environmental Health Perspectives*, *116*(9), 1211–1217. doi:10.1289/ehp.10924 PMID:18795165

Padmini, E., & Miranda, L. R. (2013). Nanocatalyst from sol–sol doping of TiO2 with Vanadium and Cerium and its application for 3,4 Dichloroaniline degradation using visible light. *Chemical Engineering Journal*, *232*(Supplement C), 249–258. doi:10.1016/j.cej.2013.07.083

Pal, S., Tak, Y. K., & Song, J. M. (2007). Does the antibacterial activity of silver nanoparticles depend on the shape of the nanoparticle? A study of the gram-negative bacterium Escherichia coli. *Applied and Environmental Microbiology*, *73*(6), 1712–1720. doi:10.1128/AEM.02218-06 PMID:17261510

Panáček, A., Kvítek, L., Prucek, R., Kolář, M., Večeřová, R., Pizúrová, N., ... Zbořil, R. (2006). Silver colloid nanoparticles: Synthesis, characterization, and their antibacterial activity. *The Journal of Physical Chemistry B*, *110*(33), 16248–16253. doi:10.1021/jp063826h PMID:16913750

Pan, B. C., Xiong, Y., Li, A. M., Chen, J. L., Zhang, Q. X., & Jin, X. Y. (2002). Adsorption of aromatic acids on an aminated hypercrosslinked macroporous polymer. *Reactive & Functional Polymers*, *53*(2), 63–72. doi:10.1016/S1381-5148(02)00123-2

Pan, B., Han, F., Nie, G., Wu, B., He, K., & Lu, L. (2014). New Strategy To Enhance Phosphate Removal from Water by Hydrous Manganese Oxide. *Environmental Science & Technology*, *48*(9), 5101–5107. doi:10.1021/es5004044 PMID:24730751

Pan, B., Pan, B., Zhang, W., Lv, L., Zhang, Q., & Zheng, S. (2009). Development of polymeric and polymer-based hybrid adsorbents for pollutants removal from waters. *Chemical Engineering Journal*, *151*(1), 19–29. doi:10.1016/j.cej.2009.02.036

Pan, B., & Xing, B. (2008). Adsorption mechanisms of organic chemicals on carbon nano-tubes. *Environmental Science & Technology*, *42*(24), 9005–9013. doi:10.1021/es801777n PMID:19174865

Pandey, B., & Fulekar, M. H. (2012). Nanotechnology: Remediation Technologies to clean up the Environmental pollutants, *Research. Journal of Chemical. The Sciences*, *2*(2), 90–96.

Pandey, J., Mir, F. Q., & Shukla, A. (2014). Synthesis of silica immobilized phosphotungstic acid (Si-PWA)-poly(vinyl alcohol) (PVA) composite ion-exchange membrane for direct methanol fuel cell. *International Journal of Hydrogen Energy*, *39*(17), 9473–9481. doi:10.1016/j.ijhydene.2014.03.237

Pandey, J., & Tankal, B. R. (2016). Performance of the vanadium redox-flow battery (VRB) for Si PWA/PVA nanocomposite membrane. *Journal of Solid State Electrochemistry*, *8*(8), 2259–2265. doi:10.100710008-016-3244-1

Pang, Y., Zeng, G., Tang, L., Zhang, Y., Liu, Y., Lei, X., ... Xie, G. (2011). PEI-grafted magnetic porous powder for highly effective adsorption of heavy metal ions. *Desalination*, *281*, 278–284. doi:10.1016/j.desal.2011.08.001

Panigrahi, S., Kundu, S., Ghosh, S. K., Nath, S., & Pal, T. (2004). General method of synthesis for metal nanoparticles. *Journal of Nanoparticle Research*, *6*(4), 411–414. doi:10.100711051-004-6575-2

Pan, J. H., Dou, H., Xiong, Z., Xu, C., Ma, J., & Zhao, X. S. (2010). Porous photocatalysts for advanced water purifications. *Journal of Materials Chemistry*, *20*(22), 4512–4528. doi:10.1039/b925523k

Pan, Z. F., Chen, R., An, L., & Li, Y. S. (2017). Alkaline anion exchange membrane fuel cells for cogeneration of electricity and valuable chemicals. *Journal of Power Sources*, *365*, 430–455. doi:10.1016/j.jpowsour.2017.09.013

Parida, P., Lolage, M., Angal, A., & Rautaray, D. (2017). Iron Oxide Nanoparticles to Remove Arsenic from Water. In S. Ranjan, N. Dasgupta, & E. Lichtfouse (Eds.), *Nanoscience in Food and Agriculture 4* (pp. 279–299). Cham: Springer International Publishing. doi:10.1007/978-3-319-53112-0_10

Park, J., Joo, J. C., Kwon, S.G., Jang, Y., & Hyeon, T. (2007). Synthesis of Monodisperse Spherical Nanocrystals. Anegwandte Chemie International Edition, 46, 4630- 4660.

Park, E. J., Yi, J., Chung, K. H., Ryu, D. Y., Choi, J., & Park, K. (2008). Oxidative stress and apoptosis induced by titanium dioxide nanoparticles in cultured BEAS-2B cells. *Toxicology Letters*, *180*(3), 222–229. doi:10.1016/j.toxlet.2008.06.869 PMID:18662754

Park, H. B., Jung, C. H., Lee, Y. M., Hill, A. J., Pas, S. J., Mudie, S. T., & Cookson, D. J. (2007). Polymers with cavities tuned for fast selective transport of small molecules and ions. *Science*, *318*(5848), 254–258. doi:10.1126cience.1146744 PMID:17932294

Park, J.-H., Gu, L., Von Maltzahn, G., Ruoslahti, E., Bhatia, S. N., & Sailor, M. J. (2009). Biodegradable luminescent porous silicon nanoparticles for in vivo applications. *Nature Materials*, *8*(4), 331–336. doi:10.1038/nmat2398 PMID:19234444

Park, J., Joo, J., Kwon, S. G., Jang, Y., & Hyeon, T. (2007). Synthesis of monodisperse spherical nanocrystals. *Angewandte Chemie International Edition*, *46*(25), 4630–4660. doi:10.1002/anie.200603148 PMID:17525914

Park, S. G., Chae, K. J., & Lee, M. (2017). A sulfonated poly(arylene ether sulfone)/polyimide nanofiber composite proton exchange membrane for microbial electrolysis cell application under the coexistence of diverse competitive cations and protons. *Journal of Membrane Science*, *540*, 165–173. doi:10.1016/j.memsci.2017.06.048

Park, Y. (2014). New paradigm shift for the green synthesis of antibacterial silver nanoparticles utilizing plant extracts. *Toxicological Research*, *30*(3), 169–178. doi:10.5487/TR.2014.30.3.169 PMID:25343010

Park, Y., Hong, Y., Weyers, A., Kim, Y., & Linhardt, R. (2011). Polysaccharides and phytochemicals: A natural reservoir for the green synthesis of gold and silver nanoparticles. *IET Nanobiotechnology / IET*, *5*(3), 69–78. doi:10.1049/iet-nbt.2010.0033 PMID:21913788

Particle Sizing using image. (2013). Retrieved from http://mesa.ac.nz/2013/03/particle-sizing-using-imagej/

Patel, J., Patel, O., & Raval, B. (2012). *Green Chemistry: New Avenues in Chemical Research: Focus in Healthcare*. LAP LAMBERT Academic Publishing.

Patlolla, A. K., Hackett, D., & Tchounwou, P. B. (2015). Genotoxicity study of silver nanoparticles in bone marrow cells of Sprague–Dawley rats. *Food and Chemical Toxicology*, *85*, 52–60. doi:10.1016/j.fct.2015.05.005 PMID:26032631

Patnukao, P., Kongsuwan, A., & Pavasant, P. (2008). Batch studies of adsorption of copper and lead on activated carbon from Eucalyptus camaldulensis Dehn. bark. *Journal of Environmental Sciences (China)*, *20*(9), 1028–1034. doi:10.1016/S1001-0742(08)62145-2 PMID:19143307

Pavličević, J., Ikonić, B., Bera, O., Vukić, N., & Budinski-Simendić, J. (2016). Environmental nanotechnology: benefits and risk assessment. *Conference: 10. International Conference Risk and safety engineering*, *3*, 129-134.

Peighambardoust, S. J., Rowshanzamir, S., & Amjadi, M. (2010). Review of the proton exchange membranes for fuel cell applications. *International Journal of Hydrogen Energy*, *35*(17), 9349–9384. doi:10.1016/j.ijhydene.2010.05.017

Pena, M. E., Korfiatis, G. P., Patel, M., Lippincott, L., & Meng, X. (2005). Adsorption of As (V) and As (III) by nanocrystalline titanium dioxide. *Water Research*, *39*(11), 2327–2337. doi:10.1016/j.watres.2005.04.006 PMID:15896821

Pena, M., Meng, X., Korfiatis, G. P., & Jing, C. (2006). Adsorption mechanism of arsenic on nanocrystalline titanium dioxide. *Environmental Science & Technology*, *40*(4), 1257–1262. doi:10.1021/es052040e PMID:16572784

Pendergast, M. M., & Hoek, E. M. (2011). A review of water treatment membrane nanotechnologies. *Energy & Environmental Science*, *4*(6), 1946–1971. doi:10.1039/c0ee00541j

Pendergast, M. T. M., Nygaard, J. M., Ghosh, A. K., & Hoek, E. M. V. (2010). Using nanocomposite materials technology to understand and control reverse osmosis membrane compaction. *Desalination*, *261*(3), 255–263. doi:10.1016/j.desal.2010.06.008

Peng, H. B., Pan, B., Wu, M., Liu, Y., Zhang, D., & Xing, B. S. (2012). Adsorption of ofloxacin and norfloxacin on carbon nanotubes: Hydrophobicity- and structure-controlled process. *Journal of Hazardous Materials*, *233*, 89–96. doi:10.1016/j.jhazmat.2012.06.058 PMID:22819959

Penza, M., Cassano, G., Aversa, P., Cusano, A., Consales, M., Giordano, M., & Nicolais, L. (2006). Acoustic and optical VOCs sensors incorporating carbon nanotubes. *IEEE Sensors Journal*, *6*(4), 867–875. doi:10.1109/JSEN.2006.877974

Penza, M., Rossi, R., Alvisi, M., Cassano, G., & Serra, E. (2009). Functional characterization of carbon nanotube networked films functionalized with tuned loading of Au nanoclusters for gas sensing applications. *Sensors and Actuators. B, Chemical*, *140*(1), 176–184. doi:10.1016/j.snb.2009.04.008

Pérez, E., Ayele, L., Getachew, G., Fetter, G., Bosch, P., Mayoral, A., & Díaz, I. (2015). Removal of chromium(VI) using nano-hydrotalcite/SiO2 composite. *Journal of Environmental Chemical Engineering*, *3*(3), 1555–1561. doi:10.1016/j.jece.2015.05.009

Pérez-Quintanilla, D., Del Hierro, I., Fajardo, M., & Sierra, I. (2006). Preparation of 2-mercaptobenzothiazole-derivatized mesoporous silica and removal of Hg(II) from aqueous solution. *Journal of Environmental Monitoring C*, *8*(1), 214–222. doi:10.1039/B507983G PMID:16395482

Perreault, F., Fonseca de Faria, A., & Elimelech, M. (2015). Environmental applications of graphene-based nanomaterials. *Chemical Society Reviews*, *44*(16), 5861–5896. doi:10.1039/C5CS00021A PMID:25812036

Petala, E., Dimos, K., Douvalis, A., Bakas, T., Tucek, J., Zbořil, R., & Karakassides, M. A. (2013). Nanoscale zero-valent iron supported on mesoporous silica: Characterization and reactivity for Cr (VI) removal from aqueous solution. *Journal of Hazardous Materials*, *261*, 295–306. doi:10.1016/j.jhazmat.2013.07.046 PMID:23959249

Petcharoen, K., & Sirivat, A. (2012). Synthesis and characterization of magnetite nanoparticles via the chemical co-precipitation method. *Materials Science and Engineering B*, *177*(5), 421–427. doi:10.1016/j.mseb.2012.01.003

Peter, C., & Hummer, G. (2005). Ion Transport through Membrane-Spanning Nanopores Studied by Molecular Dynamics Simulations and Continuum Electrostatics Calculations. *Biophysical Journal*, *89*(4), 2222–2234. doi:10.1529/biophysj.105.065946 PMID:16006629

Philippot, K., & Serp, P. (2012). *Concepts in nanocatalysis*. Nanomaterials in Catalysis.

Phiwdang, K., Suphankij, S., Mekprasart, W., & Pecharapa, W. (2013). Synthesis of CuO nanoparticles by precipitation method using different precursors. *Energy Procedia*, *34*, 740–745. doi:10.1016/j.egypro.2013.06.808

Phuoc, T. X., Howard, B. H., Martello, D. V., Soong, Y., & Chyu, M. K. (2008). Synthesis of Mg(OH)2, MgO, and Mg nanoparticles using laser ablation of magnesium in water and solvents. *Optics and Lasers in Engineering*, *46*(11), 829–834. doi:10.1016/j.optlaseng.2008.05.018

Pineda, M. G., Torres, S., López, L. V., Enríquez-Medrano, F. J., De León, R. D., Fernández, S., ... López, R. G. (2014). Chitosan-coated magnetic nanoparticles prepared in one-step by precipitation in a high-aqueous phase content reverse microemulsion. *Molecules (Basel, Switzerland)*, *19*(7), 9273–9287. doi:10.3390/molecules19079273 PMID:24991759

Plaper, A., Jenko-Brinovec, S., Premzl, A., Kos, J., & Raspor, P. (2002). Genotoxicity of trivalent chromium in bacterial cells. Possible effects on DNA topology. *Chemical Research in Toxicology*, *15*(7), 943–949. doi:10.1021/tx010096q PMID:12119005

Polshettiwar, V., Asefa, T., & Hutchings, G. (2013). *Nanocatalysis: Synthesis and Applications*. Wiley. doi:10.1002/9781118609811

Polshettiwar, V., Luque, R., Fihri, A., Zhu, H., Bouhrara, M., & Basset, J.-M. (2011). Magnetically Recoverable Nanocatalysts. *Chemical Reviews*, *111*(5), 3036–3075. doi:10.1021/cr100230z PMID:21401074

Pomiès, M., Choubert, J. M., Wisniewski, C., & Coquery, M. (2013). Modelling of micropollutant removal in biological wastewater treatments: A review. *The Science of the Total Environment*, *443*, 733–748. doi:10.1016/j.scitotenv.2012.11.037 PMID:23228719

Pompeo, F., & Resasco, D. E. (2002). Water solubilization of single-walled carbon nanotubes by functionalization with glucosamine. *Nano Letters*, *2*(4), 369–373. doi:10.1021/nl015680y

Prasad, T. N., Kambala, V. S. R., & Naidu, R. (2013). Phyconanotechnology: Synthesis of silver nanoparticles using brown marine algae Cystophora moniliformis and their characterisation. *Journal of Applied Phycology*, *25*(1), 177–182. doi:10.100710811-012-9851-z

Prasse, C., Stalter, D., Schulte-Oehlmann, U., Oehlmann, J., & Ternes, T. A. (2015). Spoilt for choice: A critical review on the chemical and biological assessment of current wastewater treatment technologies. *Water Research*, *87*, 237–270. doi:10.1016/j.watres.2015.09.023 PMID:26431616

Pretty, J. N., Mason, C. F., Nedwell, D. B., Hine, R. E., Leaf, S., & Dils, R. (2003). Environmental costs of freshwater eutrophication in England and Wales. *Environmental Science & Technology*, *37*(2), 201–208. doi:10.1021/es020793k PMID:12564888

Prieto, G., Zečević, J., Friedrich, H., De Jong, K. P., & De Jongh, P. E. (2013). Towards stable catalysts by controlling collective properties of supported metal nanoparticles. *Nature Materials*, *12*(1), 34–39. doi:10.1038/nmat3471 PMID:23142841

Pulido, J. M. O., Martinez, J. R. C., Stoller, M., & Ferez, A. M. (2017). *Organic/TiO2 Nanocomposite Membranes: Recent Developments, Springer International Publishing AG 2017*. Organic–Inorganic Composite Polymer Electrolyte Membranes; doi:10.1007/978-3-319-52739-02

Pumera, M., & Iwai, H. (2009). Multicomponent Metallic Impurities and Their Influence upon the Electrochemistry of Carbon Nanotubes. *The Journal of Physical Chemistry C*, *113*(11), 4401–4405. doi:10.1021/jp900069e

Pumera, M., Iwai, H., & Miyahara, Y. (2009). Bimetallic nickel-iron impurities within single-walled carbon nanotubes exhibit redox activity towards the oxidation of amino acids. *ChemPhysChem*, *10*(11), 1770–1773. doi:10.1002/cphc.200900355 PMID:19603449

Pummakarnchana, O., Tripathi, N., & Dutta, J. (2005). Air pollution monitoring and GIS modeling: A new use of nanotechnology based solid state gas sensors. *Science and Technology of Advanced Materials*, *6*(3), 251–255. doi:10.1016/j.stam.2005.02.003

Pungor, E. (1995). A Practical Guide to Instrumental Analysis. Academic Press.

Pyrzyńska, K., & Bystrzejewski, M. (2010). Comparative study of heavy metal ions sorption onto activated carbon, carbon nanotubes, and carbon-encapsulated magnetic nanoparticles. *Colloids and Surfaces. A, Physicochemical and Engineering Aspects*, *362*(1), 102–109. doi:10.1016/j.colsurfa.2010.03.047

Qian, H., Peng, X., Han, X., Ren, J., Sun, L., & Fu, Z. (2013). Comparison of the toxicity of silver nanoparticles and silver ions on the growth of terrestrial plant model Arabidopsis thaliana. *Journal of Environmental Sciences (China)*, *25*(9), 1947–1956. doi:10.1016/S1001-0742(12)60301-5 PMID:24520739

Qian, W., Shang, Y., Fang, M., Wang, S., Xie, X., Wang, J., ... Mao, Z. (2012). Sulfonated polybenzimidazole/zirconium phosphate composite membrane for high temperature applications. *International Journal of Hydrogen Energy*, *37*(17), 12919–12924. doi:10.1016/j.ijhydene.2012.05.076

Qijin, G., Qingjie, G., & Changing, C. (2008). Investigation into photo-catalytic degradation of gaseous benzene in a circulated photocatalytic reactor (CPCR). *Chemical Engineering & Technology*, *31*(7), 1–9.

Qi, M., Zhang, K., Li, S., Wu, J., Pham-Huy, C., Diao, X., ... He, H. (2016). Superparamagnetic Fe 3 O 4 nanoparticles: Synthesis by a solvothermal process and functionalization for a magnetic targeted curcumin delivery system. *New Journal of Chemistry*, *40*(5), 4480–4491. doi:10.1039/C5NJ02441B

Qin, S., Liu, X., & Dong, L. (2017). *Advances in Application of Nanomaterials in Life Science*. Academic Press.

Qiu, L., Peng, Y., Liu, B., Lin, B., Peng, Y., Malik, M. J., & Yan, F. (2012). Polypyrrole nanotube-supported gold nanoparticles: An efficient electrocatalyst for oxygen reduction and catalytic reduction of 4-nitrophenol. *Applied Catalysis A, General*, *413*, 230–237. doi:10.1016/j.apcata.2011.11.013

Qu, J., Luo, C., & Hou, J. (2011). Synthesis of ZnO nanoparticles from Zn-hyperaccumulator (Sedum alfredii Hance) plants. *Micro & Nano Letters*, *6*(3), 174. doi:10.1049/mnl.2011.0004

Qu, J., Yuan, X., Wang, X., & Shao, P. (2011). Zinc accumulation and synthesis of ZnO nanoparticles using Physalis alkekengi L. *Environmental Pollution*, *159*(7), 1783–1788. doi:10.1016/j.envpol.2011.04.016 PMID:21549461

Qu, S., Huang, F., Yu, S., Chen, G., & Kong, J. (2008). Magnetic removal of dyes from aqueous solution using multi-walled carbon nanotubes filled with Fe 2 O 3 particles. *Journal of Hazardous Materials*, *160*(2), 643–647. doi:10.1016/j.jhazmat.2008.03.037 PMID:18430510

Qu, X., Alvarez, P. J., & Li, Q. (2013). Applications of nanotechnology in water and wastewater treatment. *Water Research*, *47*(12), 3931–3946. doi:10.1016/j.watres.2012.09.058 PMID:23571110

Qu, X., Brame, J., Li, Q., & Alvarez, P. J. (2013). Nanotechnology for a safe and sustainable water supply: Enabling integrated water treatment and reuse. *Accounts of Chemical Research*, *46*(3), 834–843. doi:10.1021/ar300029v PMID:22738389

Qu, Z., Yan, L., Li, L., Xu, J., Liu, M., Li, Z., & Yan, N. (2014). Ultraeffective ZnS nanocrystals sorbent for mercury (II) removal based on size-dependent cation exchange. *ACS Applied Materials & Interfaces*, *6*(20), 18026–18032. doi:10.1021/am504896w PMID:25299972

Rafiq, Z., Nazir, R., Durr-e-Shahwar, Shah, M. R., & Ali, S. (2014). Utilization of magnesium and zinc oxide nano-adsorbents as potential materials for treatment of copper electroplating industry wastewater. *Journal of Environmental Chemical Engineering*, *2*(1), 642–651. doi:10.1016/j.jece.2013.11.004

Rahman, A. K. M. R., Hossain, S. M., & Akramuzzaman, M. M. (2010). Distribution of heavy metals in rice plant cultivated in industrial effluent receiving soil. *Environment Asia*, *3*(2), 15–19.

Rahmani, A., Ghaffari, H., & Samadi, M. (2011). A comparative study on arsenic (III) removal from aqueous solution using nano and micro sized zero-valent iron. *Journal of Environmental Health Science & Engineering*, *8*(2), 157–166.

Rahmani, H. R., & Rezaei, M. (2007). Effluent of industrial effluents of Zob- Ahan on soil, water and vegetable plants. *Journal of Applied Sciences (Faisalabad)*, 72519–72522.

Rai, M., & Duran, N. (2011). *Metal Nanoparticles in Microbiology*. Springer. doi:10.1007/978-3-642-18312-6

Rai, M., & Ingle, A. (2012). Role of nanotechnology in agriculture with special reference to management of insect pests. *Applied Microbiology and Biotechnology*, *94*(2), 287–293. doi:10.100700253-012-3969-4 PMID:22388570

Rai, M., & Yadav, A. (2013). Plants as potential synthesiser of precious metal nanoparticles: Progress and prospects. *IET Nanobiotechnology / IET*, *7*(3), 117–124. doi:10.1049/iet-nbt.2012.0031 PMID:24028810

Rai, U. N., Prasad, D., Verma, S., Upadhyay, A. K., & Singh, N. K. (2012). Biomonitoring of metals in Ganga water at different ghats of Haridwar: Implications of constructed wetland for sewage detoxification. *Bulletin of Environmental Contamination and Toxicology*, *89*(4), 805–810. doi:10.100700128-012-0766-9 PMID:22869393

Rajabi, H., Ghaemi, N., Madaeni, S. S., Daraei, P., Astinchap, B., Zinadini, S., & Razavizadeh, S. H. (2015). Nano-ZnO embedded mixed matrix polyethersulfone (PES) membrane: Influence of nanofiller shape on characterization and fouling resistance. *Applied Surface Science*, *349*, 66–77. doi:10.1016/j.apsusc.2015.04.214

Rajagopalan, P., Wudl, F., Schinazi, R. F., & Boudinot, F. D. (1996). Pharmacokinetics of a water-soluble fullerene in rats. *Antimicrobial Agents and Chemotherapy*, *40*(10), 2262–2265. PMID:8891126

Rajeswari, A., Amalraj, A., & Pius, A. (2015). Removal of phosphate using chitosan-polymer composites. *Journal of Environmental Chemical Engineering*, *3*(4), 2331–2341. doi:10.1016/j.jece.2015.08.022

Rajiv, P., Rajeshwari, S., & Venckatesh, R. (2013). Bio-fabrication of zinc oxide nanoparticles using leaf extract of Parthenium hysterophorus L. and its size-dependent antifungal activity against plant fungal pathogens. *Spectrochimica Acta. Part A: Molecular and Biomolecular Spectroscopy*, *112*, 384–387. doi:10.1016/j.saa.2013.04.072 PMID:23686093

Raj, L. F. A., & e, J. (2015). Biosynthesis and Characterization of Zinc Oxide Nanoparticles Using Root Extract of Zingiber Officinale. *Oriental Journal of Chemistry*, *31*(1), 51–56. doi:10.13005/ojc/310105

Ramana, D. V., Yu, J. S., & Seshaiah, K. (2013). Silver nanoparticles deposited multiwalled carbon nanotubes for removal of Cu (II) and Cd (II) from water: Surface, kinetic, equilibrium, and thermal adsorption properties. *Chemical Engineering Journal*, *223*, 806–815. doi:10.1016/j.cej.2013.03.001

Ramani, V., Kunz, H. R., & Fenton, J. M. (2005). Stabilized heteropolyacid/Nafion composite membranes for elevated temperature/low relative humidity PEFC operation. *Electrochimica Acta*, *50*(5), 1181–1187. doi:10.1016/j.electacta.2004.08.015

Ramanujam, K., & Sundrarajan, M. (2014). Antibacterial effects of biosynthesized MgO nanoparticles using ethanolic fruit extract of Emblica officinalis. *Journal of Photochemistry and Photobiology. B, Biology*, *141*, 296–300. doi:10.1016/j.jphotobiol.2014.09.011 PMID:25463681

Ramaswamy, P., Wong, N. E., Benjamin, S. G., & Shimizu, G. K. H. (2015). A water stable magnesium MOF that conducts proton over 10^{-2} Scm^{-1}. *Journey of the American Chemical Society.*, *137*(24), 7640–7643. doi:10.1021/jacs.5b04399 PMID:26053659

Ramaswamy, P., Wong, N. E., & Shimizu, G. K. H. (2014). MOFs as proton conductors- challenges and opportunities. *Chemical Society Reviews*, *43*(16), 5913–5932. doi:10.1039/C4CS00093E PMID:24733639

Ramesh, M., Anbuvannan, M., & Viruthagiri, G. (2015). Green synthesis of ZnO nanoparticles using Solanum nigrum leaf extract and their antibacterial activity. *Spectrochim Acta A Mol Biomol Spectrosc, 136 Pt B*, 864-870. doi:10.1016/j.saa.2014.09.105

Ramesha, G. K., Kumara, A. V., Muralidhara, H. B., & Sampath, S. S. (2011). Graphene and graphene oxide as effective adsorbents toward anionic and cationic dyes. *Journal of Colloid and Interface Science*, *361*(1), 270–277. doi:10.1016/j.jcis.2011.05.050 PMID:21679961

Ramesh, P., Rajendran, A., & Meenakshisundaram, M. (2014). Green Syntheis of Zinc Oxide Nanoparticles Using Flower Extract Cassia Auriculata. *Journal of Nanoscience and Nanotechnology*, *2*(1), 41–45.

Ramkumar, V. S., Pugazhendhi, A., Gopalakrishnan, K., Sivagurunathan, P., Saratale, G. D., Dung, T. N. B., & Kannapiran, E. (2017). Biofabrication and characterization of silver nanoparticles using aqueous extract of seaweed Enteromorpha compressa and its biomedical properties. *Biotechnology Reports (Amsterdam, Netherlands)*, *14*, 1–7. doi:10.1016/j.btre.2017.02.001 PMID:28459002

Ramsden, J. (2011). *Nanotechnology: An Introduction.* William Andrew.

Rao, C. N. R., Müller, A., & Cheetham, A. K. (2006). *The chemistry of nanomaterials: synthesis, properties and applications*. John Wiley & Sons.

Rao, D. B., Wong, B. A., McManus, B. E., McElveen, A. M., James, A. R., & Dorman, D. C. (2003). Inhaled iron, unlike manganese, is not transported to the rat brain via the olfactory pathway. *Toxicology and Applied Pharmacology*, *193*(1), 116–126. doi:10.1016/S0041-008X(03)00340-5 PMID:14613722

Rao, G. P., Lu, C., & Su, F. (2007). Sorption of divalent metal ions from aqueous solution by carbon nanotubes: A review. *Separation and Purification Technology*, *58*(1), 224–231. doi:10.1016/j.seppur.2006.12.006

Rao, K. G., Ashok, C., Rao, K. V., & Chakra, C. S., & A, A. (2015). Eco-Friendly Synthesis of MgO Nanoparticles from Orange Fruit Waste. *International Journal of Advanced Research in Physical Science*, *2*(3), 1–6.

Rao, M. M., Ramesh, A., Rao, G. P. C., & Seshaiah, K. (2006). Removal of copper and cadmium from the aqueous solutions by activated carbon derived from Ceiba pentandra hulls. *Journal of Hazardous Materials*, *129*(1), 123–129. PMID:16191464

Rao, M., Parwate, A. V., & Bhole, A. G. (2002). Removal of Cr 6^+ and Ni 2^+ from aqueous solution using bagasse and fly ash. *Waste Management (New York, N.Y.)*, *22*(7), 821–830. doi:10.1016/S0956-053X(02)00011-9 PMID:12365785

Raoof, J., Hosseini, S. R., & Rezaee, S. (2014). Preparation of Pt/poly(2-methoxyaniline)/multiwalled carbon nanotube nanocomposite and its application for electrocatalytic oxidation of methanol. *Journal of Molecular Liquids*, *200*, 196–204. doi:10.1016/j.molliq.2014.09.043

Rathinam, A., Rao, J. R., & Nair, B. U. (2011). Adsorption of phenol onto activated carbon from seaweed: Determination of the optimal experimental parameters using factorial design. *Journal of Taiwan Institute of Chemical Engineers*, *42*(6), 952–956. doi:10.1016/j.jtice.2011.04.003

Ratner, M. A., & Ratner, D. (2003). *Nanotechnology: A Gentle Introduction to the Next Big Idea*. Prentice Hall.

Raut, S., Thorat, P. V., & Thakre, R. (2015). Green Synthesis of Zinc Oxide (ZnO) Nanoparticles Using Ocimum Tenuiflorum Leaves. *International Journal of Science and Research*, *4*(5), 2319–7064.

Rauwel, P., Küünal, S., Ferdov, S., & Rauwel, E. (2015). A Review on the Green Synthesis of Silver Nanoparticles and Their Morphologies Studied via TEM. *Advances in Materials Science and Engineering*, *2015*, 1–9. doi:10.1155/2015/682749

Raval, N. P., Shah, P. U., & Shah, N. K. (2016). Adsorptive removal of nickel (II) ions from aqueous environment: A review. *Journal of Environmental Management*, *179*, 1–20. doi:10.1016/j.jenvman.2016.04.045 PMID:27149285

Ravanchi, M. T., Kaghazchi, T., & Kargari, A. (2009). Application of membrane separation processes in petrochemical industry: A review. *Desalination*, *235*(1), 199–244. doi:10.1016/j.desal.2007.10.042

Ravichandran, R. (2010). Nanotechnology Applications in Food and Food Processing: Innovative Green Approaches, Opportunities and Uncertainties for Global Market. *International Journal of Green Nanotechnology: Physics and Chemistry*, *1*(2), 72–P96. doi:10.1080/19430871003684440

Ray, P. C. (2010). Size and shape dependent second order nonlinear optical properties of nanomaterials and its application in biological and chemical sensing. *Chemical Reviews*, *110*(9), 5332–5365. doi:10.1021/cr900335q PMID:20469927

Recillas, S., García, A., González, E., Casals, E., Puntes, V., Sanchez, A., & Font, X. (2011). Use of CeO_2, TiO_2 and Fe_3O_4 nanoparticles for the removal of lead from water: Toxicity of nanoparticles and derived compounds. *Desalination*, *277*(1), 213–220. doi:10.1016/j.desal.2011.04.036

Reddy, A., Madhavi, V., Reddy, K. G., & Madhavi, G. (2012). Remediation of chlorpyrifos-contaminated soils by laboratory-synthesized zero-valent nano iron particles: Effect of pH and aluminium salts. *Journal of Chemistry*.

Reetz, M. T., Maase, M., Schilling, T., & Tesche, B. (2000). Computer Image Processing of Transmission Electron Micrograph Pictures as a Fast and Reliable Tool To Analyze the Size of Nanoparticles. *The Journal of Physical Chemistry B*, *104*(37), 8779–8781. doi:10.1021/jp000328e

Renuka, N., Sood, A., Ratha, S. K., Prasanna, R., & Ahluwalia, A. S. (2013). Evaluation of microalgal consortia for treatment of primary treated sewage effluent and biomass production. *Journal of Applied Phycology*, *25*(5), 1529–1537. doi:10.100710811-013-9982-x

Ren, X., Chen, C., Nagatsu, M., & Wang, X. (2011). Carbon nanotubes as adsorbents in environmental pollution management: A review. *Chemical Engineering Journal*, *170*(2), 395–410. doi:10.1016/j.cej.2010.08.045

Repo, E., Warchoł, J. K., Bhatnagar, A., & Sillanpää, M. (2011). Heavy metals adsorption by novel EDTA-modified chitosan-silica hybrid materials. *Journal of Colloid and Interface Science*, *358*(1), 261–267. doi:10.1016/j.jcis.2011.02.059 PMID:21440904

Reynolds, T. D. (1977). *Unit operations and processes in environmental engineering*. Brooks.

Rezaee, A., Godini, H., Dehestani, S., & Khavanin, A. (2008). Application of impregnated almond shell activated carbon by zinc and zinc sulfate for nitrate removal from water. *Iranian Journal of Environmental Health Sciences & Engineering*, *5*, 125–130.

Rezaee, R., Nasseri, S., Mahvi, A. H., Nabizadeh, R., Mousavi, S. A., Rashidi, A., ... Nazmara, S. (2015). Fabrication and characterization of a polysulfone-graphene oxide nanocomposite membrane for arsenate rejection from water. *Journal of Environmental Health Science & Engineering*, *13*(1), 61–71. doi:10.118640201-015-0217-8 PMID:26301096

Richard, C., Balavoine, F., Schultz, P., Ebbesen, T. W., & Mioskowski, C. (2003). Supramolecular self-assembly of lipid derivatives on carbon nanotubes. *Science*, *300*(5620), 775–778. doi:10.1126cience.1080848 PMID:12730595

Rickerby, D. G. (2007). Nanotechnology Nanotechnologies and nanoparticles in remediation and pollution prevention. *Chemistry Today*, *25*(6), 24.

Rickerby, D., & Morrison, M. (2007). Nanotechnology and the environment: A European perspective. *Science and Technology of Advanced Materials*, *8*(1), 19–24. doi:10.1016/j.stam.2006.10.002

Ridoutt, B. G., & Pfister, S. (2010). A revised approach to water footprinting to make transparent the impacts of consumption and production on global freshwater scarcity. *Global Environmental Change*, *20*(1), 113–120. doi:10.1016/j.gloenvcha.2009.08.003

Ridoutt, B. G., & Pfister, S. (2013). A new water footprint calculation method integrating consumptive and degradative water use into a single stand-alone weighted indicator. *The International Journal of Life Cycle Assessment*, *18*(1), 204–207. doi:10.100711367-012-0458-z

Rikukawa, M., & Sanui, K. (2000). Proton-conducting polymer electrolyte membranes based on hydrocarbon polymers. *Progress in Polymer Science*, *25*(10), 1463–1502. doi:10.1016/S0079-6700(00)00032-0

Riu, J., Maroto, A., & Rius, F. X. (2006). Nanosensors in environmental analysis. *Talanta*, *69*(2), 288–301. doi:10.1016/j.talanta.2005.09.045 PMID:18970568

Robert, C., Valle, K., & Sanchez, C. (2011). Design and properties of functional hybrid organic–inorganic membranes for fuel cells. *Chemical Society Reviews*, *40*(2), 961–1005. doi:10.1039/c0cs00144a PMID:21218233

Robinson, B., Clothier, B., Bolan, N. S., Mahimairaja, S., Greven, M., Moni, C., . . . Milne, G. (2004). *Arsenic in the New Zealand environment.* Paper presented at the SuperSoil 2004: 3rd Australian New Zealand Soils Conference, Sydney, Australia.

Rockstr¨om, J. (2003). Water for food and nature in drought-prone tropics: Vapour shift in rain-fed agriculture. *Philosophical Transactions of the Royal Society of London. Series B, Biological Sciences*, *358*(1440), 1997–2009. doi:10.1098/rstb.2003.1400 PMID:14728794

Rodriguez-Yañez, Y., Muñoz, B., & Albores, A. (2013). Mechanisms of toxicity by carbon nanotubes. *Toxicology Mechanisms and Methods*, *23*(3), 178–195. doi:10.3109/15376516.2012.754534 PMID:23193995

Roduner, E. (2006). Size matters: Why nanomaterials are different. *Chemical Society Reviews*, *35*(7), 583–592. doi:10.1039/b502142c PMID:16791330

Rogers, K. (2006). Recent advances in biosensor techniques for environmental monitoring. *Analytica Chimica Acta*, *568*(1), 222–231. doi:10.1016/j.aca.2005.12.067 PMID:17761264

Roostaei, N., & Tezel, F. H. (2004). Removal of phenol from aqueous solutions by adsorption. *Journal of Environmental Management*, *70*(2), 157–164. doi:10.1016/j.jenvman.2003.11.004 PMID:15160741

Ruelle, B., Peeterbroeck, S., Godfroid, T., Bittencourt, C., Hecq, M., Snyders, R., & Dubois, P. (2012). Selective grafting of primary amines onto carbon nanotubes via free-radical treatment in microwave plasma post-discharge. *Polymers*, *4*(1), 296–315. doi:10.3390/polym4010296

Ruffini-Castiglione, M., & Cremonini, R. (2009). Article. *Caryologia*, *62*, 161–165. doi:10.1080/00087114.2004.10589681

Ruffmann, B., Silva, H., & Schulte, B. (2003). Organic/inorganic composite membranes for application in DMFC. *Solid State Ionics*, *162*, 269–275. doi:10.1016/S0167-2738(03)00240-6

Ruthven, D. M. (1984). *Principles of Adsorption and Adsorption Processes* (1st ed.). ohn Wiley and Sons, Inc.

Ryman-Rasmussen, J. P., Riviere, J. E., & Monteiro-Riviere, N. A. (2007). Surface coatings determine cytotoxicity and irritation potential of quantum dot nanoparticles in epidermal keratinocytes. *The Journal of Investigative Dermatology*, *127*(1), 143–153. doi:10.1038j.jid.5700508 PMID:16902417

Saad, S. A., Isa, K. M., & Bahari, R. (2010). Chemically modified sugarcane bagasse as a potentially low-cost biosorbent for dye removal. *Desalination*, *264*(1-2), 123–128. doi:10.1016/j.desal.2010.07.015

Sadegh, H., Ali, G. A. M., Abbasi, Z., & Nadagoud, M. N. (2017a). Adsorption of Ammonium Ions onto Multi-Walled Carbon Nanotubes. *Studia Universitatis Babes-Bolyai. Chemia*, *62*(2), 233–245. doi:10.24193ubbchem.2017.2.18

Sadegh, H., Ali, G. A. M., Gupta, V. K., Makhlouf, A. S. H., Shahryari-Ghoshekandi, R., Nadagouda, M. N., ... Megiel, E. (2017b). The role of nanomaterials as effective adsorbents and their applications in wastewater treatment. *Journal of Nanostructure in Chemistry*, *7*(1), 1–14. doi:10.100740097-017-0219-4

Sadegh, H., Ali, G. A. M., Makhlouf, A. S. H., Chong, K. F., Alharbi, N. S., Agarwal, S., & Gupta, V. K. (2018). MWCNTs-Fe3O4 nanocomposite for Hg(II) high adsorption efficiency. *Journal of Molecular Liquids*, *258*, 345–353. doi:10.1016/j.molliq.2018.03.012

Sadegh, H., & Shahryari-ghoshekandi, R. (2015). Functionalization of carbon nanotubes and its application in nanomedicine: A review. *Nanomedicine Journal*, *2*(4), 231–248.

Sadegh, H., Shahryari-ghoshekandi, R., Agarwal, S., Tyagi, I., Asif, M., & Gupta, V. K. (2015). Microwave-assisted removal of malachite green by carboxylate functionalized multi-walled carbon nanotubes: Kinetics and equilibrium study. *Journal of Molecular Liquids*, *206*, 151–158. doi:10.1016/j.molliq.2015.02.007

Sadegh, H., Zare, K., Maazinejad, B., Shahryari-Ghoshekandi, R., Tyagi, I., Agarwal, S., & Gupta, V. K. (2016). Synthesis of MWCNT-COOH-Cysteamine composite and its application for dye removal. *Journal of Molecular Liquids*, *215*, 221–228. doi:10.1016/j.molliq.2015.12.042

Sahaa, R. K., Acharyaa, S., Shovon, S. S. H., & Royb, P. (2013). Medicinal activities of the leaves of Musa sapientum var. sylvesteris in vitro. *Asian Pacific Journal of Tropical Biomedicine*, *3*(6), 476–482. doi:10.1016/S2221-1691(13)60099-4 PMID:23730561

Saha, J., Begum, A., Mukherjee, A., & Kumar, S. (2017). A novel green synthesis of silver nanoparticles and their catalytic action in reduction of Methylene Blue dye. *Sustainable Environment Research*, *27*(5), 245–250. doi:10.1016/j.serj.2017.04.003

Sahiner, N., Demirci, S., Sahiner, M., Yilmaz, S., & Al-Lohedan, H. (2015). The use of superporous p(3-acrylamidopropyl) trimethyl ammonium chloride cryogels for removal of toxic arsenate anions. *Journal of Environmental Management*, *152*, 66–74. doi:10.1016/j.jenvman.2015.01.023 PMID:25617870

Sahu, A. K., Selvarani, S., Pitchumani, G., Sridhar, P., Shukla, A. K., Narayanan, N., ... Chandrakumar, N. (2008). PVA-PSSA Membrane with Interpenetrating Networks and its Methanol Crossover Mitigating Effect in DMFCs. *Journal of the Electrochemical Society*, *155*(7), B686–B695. doi:10.1149/1.2912040

Saikia, N., Sarma, J., Borah, J. M., & Mahiuddin, S. (2013). Adsorption of 3,4-dihydroxybenzoic acid onto hematite surface in aqueous medium: Importance of position of phenolic –OH groups and understanding of the same using catechol as an auxiliary model. *Journal of Colloid and Interface Science*, *398*, 227–233. doi:10.1016/j.jcis.2013.02.028 PMID:23518302

Salam, M. A. (2013). Coating carbon nanotubes with crystalline manganese dioxide nanoparticles and their application for lead ions removal from model and real water. *Colloids and Surfaces. A, Physicochemical and Engineering Aspects*, *419*, 69–79. doi:10.1016/j.colsurfa.2012.11.064

Salam, M. A., El-Shishtawy, R. M., & Obaid, A. Y. (2014). Synthesis of magnetic multi-walled carbon nanotubes/magnetite/chitin magnetic nanocomposite for the removal of Rose Bengal from real and model solution. *Journal of Industrial and Engineering Chemistry*, *20*(5), 3559–3567. doi:10.1016/j.jiec.2013.12.049

Salam, M. A., Fageeh, O., Al-Thabaiti, S. A., & Obaid, A. Y. (2015). Removal of nitrate ions from aqueous solution using zero-valent iron nanoparticles supported on high surface area nanographenes. *Journal of Molecular Liquids*, *212*, 708–715. doi:10.1016/j.molliq.2015.09.029

Saleh, T. A., Al-Saadi, A. A., & Gupta, V. K. (2014). Carbonaceous adsorbent prepared from waste tires: Experimental and computational evaluations of organic dyemethyl orange. *Journal of Molecular Liquids*, *191*, 85–91. doi:10.1016/j.molliq.2013.11.028

Salem Attia, T. M., Hu, X. L., & Yin, D. Q. (2014). Synthesised magnetic nanoparticles coated zeolite (MNCZ) for the removal of arsenic (As) from aqueous solution. *Journal of Experimental Nanoscience*, *9*(6), 551–560. doi:10.1080/17458080.2012.677549

Salman, J. M., Njoku, V. O., & Hameed, B. H. (2011). Adsorption of pesticides from aqueous solution onto banana stalk activated carbon. *Chemical Engineering Journal*, *174*(1), 41–48. doi:10.1016/j.cej.2011.08.026

Salvador-Morales, C., Basiuk, E. V., Basiuk, V. A., Green, M. L., & Sim, R. B. (2008). Effects of covalent functionalization on the biocompatibility characteristics of multi-walled carbon nanotubes. *Journal of Nanoscience and Nanotechnology*, *8*(5), 2347–2356. doi:10.1166/jnn.2008.090 PMID:18572648

Salvador-Morales, C., Townsend, P., Flahaut, E., Vénien-Bryan, C., Vlandas, A., Green, M. L. H., & Sim, R. B. (2007). Binding of pulmonary surfactant proteins to carbon nanotubes; potential for damage to lung immune defense mechanisms. *Carbon*, *45*(3), 607–617. doi:10.1016/j.carbon.2006.10.011

Sanchez, C., Boissiere, C., Grosso, D., Laberty, C., & Nicole, L. (2008). Design, Synthesis, and Properties of Inorganic and Hybrid Thin Films Having Periodically Organized Nanoporosity. *Chemistry of Materials*, *20*(3), 682–737. doi:10.1021/cm702100t

Sanchez-Mendieta, V., & Vilchis-Nestor, A. R. (2012). *Green synthesis of noble metal (Au, Ag, Pt) nanoparticles, assisted by plant-extracts. In Noble Metals*. InTech. doi:10.5772/34335

Sandberg, W. J., Låg, M., Holme, J. A., Friede, B., Gualtieri, M., Kruszewski, M., ... Refsnes, M. (2012). Comparison of non-crystalline silica nanoparticles in IL-1β release from macrophages. *Particle and Fibre Toxicology*, *9*(1), 32. doi:10.1186/1743-8977-9-32 PMID:22882971

Sangeetha, G., Rajeshwari, S., & Venckatesh, R. (2011). Green synthesis of zinc oxide nanoparticles by aloe barbadensis miller leaf extract: Structure and optical properties. *Materials Research Bulletin*, *46*(12), 2560–2566. doi:10.1016/j.materresbull.2011.07.046

Sanghi, R., & Singh, V. (2012). *Green Chemistry for Environmental Remediation*. Wiley.

Santhosh, C., Kollu, P., Doshi, S., Sharma, M., Bahadur, D., Vanchinathan, M. T., ... Grace, A. N. (2014). Adsorption, photodegradation and antibacterial study of graphene–Fe 3 O 4 nanocomposite for multipurpose water purification application. *RSC Advances*, *4*(54), 28300–28308. doi:10.1039/C4RA02913E

Santhosh, C., Velmurugan, V., Jacob, G., Jeong, S. K., Grace, A. N., & Bhatnagar, A. (2016). Role of nanomaterials in water treatment applications: A review. *Chemical Engineering Journal*, *306*, 1116–1137. doi:10.1016/j.cej.2016.08.053

Santos, C., Albuquerque, A., Sampaio, F., & Keyson, D. (2013). Nanomaterials with antimicrobial properties: applications in health sciences. *Microbial Pathogens and Strategies for Combating Them: Science, Technology and Education, 4*(2).

Santra, S., Zhang, P., Wang, K., Tapec, R., & Tan, W. (2001). Conjugation of biomolecules with luminophore-doped silica nanoparticles for photostable biomarkers. *Analytical Chemistry*, *73*(20), 4988–4993. doi:10.1021/ac010406+ PMID:11681477

Sanz, R., Calleja, G., Arencibia, A., & Sanz-Pérez, E. S. (2012). Amino functionalized mesostructured SBA-15 silica for CO_2 capture: Exploring the relation between the adsorption capacity and the distribution of amino groups by TEM. *Microporous and Mesoporous Materials*, *158*, 309–317. doi:10.1016/j.micromeso.2012.03.053

Saratale, R. G., Saratale, G. D., Shin, H. S., Jacob, J. M., Pugazhendhi, A., Bhaisare, M., & Kumar, G. (2017). New insights on the green synthesis of metallic nanoparticles using plant and waste biomaterials: Current knowledge, their agricultural and environmental applications. *Environmental Science and Pollution Research International*. doi:10.100711356-017-9912-6 PMID:28815433

Sarkar, A., Ghosh, M., & Sil, P. C. (2014). Nanotoxicity: Oxidative stress mediated toxicity of metal and metal oxide nanoparticles. *Journal of Nanoscience and Nanotechnology*, *14*(1), 730–743. doi:10.1166/jnn.2014.8752 PMID:24730293

Sarkar, S. K., Saha, M., Takada, H., Bhattacharya, A., Mishra, P., & Bhattacharya, B. (2007). Water quality management in the lower stretch of the river Ganges, east coast of India: An approach through environmental education. *Journal of Cleaner Production*, *15*(16), 1559–1567. doi:10.1016/j.jclepro.2006.07.030

Sarkar, T., Gao, Y., & Mulchandani, A. (2013). Carbon nanotubes-based label-free affinity sensors for environmental monitoring. *Applied Biochemistry and Biotechnology*, *170*(5), 1011–1025. doi:10.100712010-013-0233-z PMID:23653139

Satapanajaru, T., Anurakpongsatorn, P., Pengthamkeerati, P., & Boparai, H. (2008). Remediation of atrazine-contaminated soil and water by nano zerovalent iron. *Water, Air, and Soil Pollution*, *192*(1-4), 349–359. doi:10.100711270-008-9661-8

Sathishkumar, M., Sneha, K., Won, S. W., Cho, C. W., Kim, S., & Yun, Y. S. (2009). Cinnamon zeylanicum bark extract and powder mediated green synthesis of nano-crystalline silver particles and its bactericidal activity. *Colloids and Surfaces. B, Biointerfaces*, *73*(2), 332–338. doi:10.1016/j.colsurfb.2009.06.005 PMID:19576733

Sathyamoorthy, R., Mageshwari, K., Mali, S. S., Priyadharshini, S., & Patil, P. S. (2013). Effect of organic capping agent on the photocatalytic activity of MgO nanoflakes obtained by thermal decomposition route. *Ceramics International*, *39*(1), 323–330. doi:10.1016/j.ceramint.2012.06.028

Sato, Y., Shibata, K., Kataoka, H., Ogino, S., Bunshi, F., Yokoyama, A., ... Tohji, K. (2005). Strict preparation and evaluation of water-soluble hat-stacked carbon nanofibers for biomedical application and their high biocompatibility: Influence of nanofiber-surface functional groups on cytotoxicity. *Molecular BioSystems*, *1*(2), 142–145. doi:10.1039/b501222h PMID:16880976

Savage, N., & Diallo, M. S. (2005). Nanomaterials and water purification: Opportunities and challenges. *Journal of Nanoparticle Research*, *7*(4), 331–342. doi:10.100711051-005-7523-5

Sayari, A., & Hamoudi, S. (2001). Periodic mesoporous silica-based organic-inorganic nano-composite materials. *Chemistry of Materials*, *13*(10), 3151–3168. doi:10.1021/cm011039l

Sayes, C. M., Gobin, A. M., Ausman, K. D., Mendez, J., West, J. L., & Colvin, V. L. (2005). Nano-C60 cytotoxicity is due to lipid peroxidation. *Biomaterials*, *26*(36), 7587–7595. doi:10.1016/j.biomaterials.2005.05.027 PMID:16005959

Scanlon, B. R., Jolly, I., Sophocleous, M., & Zhang, L. (2007). Global impacts of conversions from natural to agricultural ecosystems on water resources: Quantity versus quality. *Water Resources Research, 43*(3). doi:10.1029/2006wr005486

Schaffer, B., Hohenester, U., Trügler, A., & Hofer, F. (2009). High-resolution surface plasmon imaging of gold nanoparticles by energy-filtered transmission electron microscopy. *Physical Review B: Condensed Matter and Materials Physics*, *79*(4), 041401. doi:10.1103/PhysRevB.79.041401

Schauer, J. J., Kleeman, M. J., Cass, G. R., & Simoneit, B. R. T. (2001). Measurement of emissions from air pollution sources, 3. C1–C29 organic compounds from fireplace combustion of wood. *Environmental Science & Technology*, *35*(9), 1716–1728. doi:10.1021/es001331e PMID:11355184

Schlupp, M., Weil, T., Berresheim, A. J., Wiesler, U. M., Bargon, J., & Müllen, K. (2001). Polyphenylene dendrimers as sensitive and selective sensor layers. *Angewandte Chemie International Edition*, *40*(21), 4011–4015. doi:10.1002/1521-3773(20011105)40:21<4011::AID-ANIE4011>3.0.CO;2-C PMID:12404477

Schwarzenbach, R. P., Escher, B. I., Fenner, K., Hofstetter, T. B., Johnson, C. A., Von Gunten, U., & Wehrli, B. (2006). The challenge of micropollutants in aquatic systems. *Science*, *313*(5790), 1072–1077. doi:10.1126cience.1127291 PMID:16931750

Scipioni, R., Pumera, M., Boero, M., Miyahara, Y., & Ohno, T. (2010). Investigation of the Mechanism of Adsorption of β-Nicotinamide Adenine Dinucleotide on Single-Walled Carbon Nanotubes. *The Journal of Physical Chemistry Letters*, *1*(1), 122–125. doi:10.1021/jz9000714

Seaton, A., & Donaldson, K. (2005). Nanoscience, nanotoxicology, and the need to think small. *Lancet*, *365*(9463), 923–924. doi:10.1016/S0140-6736(05)71061-8 PMID:15766981

Seepana, M. M., Pandey, J., & Shukla, A. (2017). Design and synthesis of highly stable poly(tetrafluoroethylene-zirconium phosphate (PTFE-ZrP) ion-exchange membrane for vanadium redox flow battery (VRFB). *Ionics*, *23*(6), 1471–1480. doi:10.100711581-016-1967-8

See-Toh, Y. H., Silva, M., & Livingston, A. (2008). Controlling molecular weight cut-off curves for highly solvent stable organic solvent nanofiltration (OSN) membranes. *Journal of Membrane Science*, *324*(1), 220–232. doi:10.1016/j.memsci.2008.07.023

Seigler, D. S. (2003). Phytochemistry of Acacia—sensu lato. *Biochemical Systematics and Ecology*, *31*(8), 845–873. doi:10.1016/S0305-1978(03)00082-6

Selid, P. D., Xu, H., Collins, E. M., Striped Face-Collins, M., & Zhao, J. X. (2009). Sensing mercury for biomedical and environmental monitoring. *Sensors (Basel)*, *9*(7), 5446–5459. doi:10.339090705446 PMID:22346707

Sellers, K., Mackay, C., Bergeson, L. L., Clough, S. R., Hoyt, M., Chen, J., & Hamblen, J. (2008). *Nanotechnology and the Environment*. CRC Press.

Selvi, K., Pattabhi, S., & Kadirvelu, K. (2001). Removal of Cr (VI) from aqueous solution by adsorption onto activated carbon. *Bioresource Technology*, *80*(1), 87–89. doi:10.1016/S0960-8524(01)00068-2 PMID:11554606

Sen, D., Ghosh, A. K., Majumder, S., Bindal, R. C., & Tiwari, P. K. (2014). Novel polysulfone-spray-dried silica composite membrane for water purification: Preparation, characterization and performance evaluation. *Separation and Purification Technology*, *123*, 79–86. doi:10.1016/j.seppur.2013.12.019

Sen, T. K., Afroze, S., & Ang, H. (2011). Equilibrium, kinetics and mechanism of removal of methylene blue from aqueous solution by adsorption onto pine cone biomass of Pinus radiate. *Water, Air, and Soil Pollution*, *218*(1-4), 499–515. doi:10.100711270-010-0663-y

Sepeur, S. (2008). *Nanotechnology: technical basics and applications*. Hannover: Vincentz.

Sera, N., Tokiwa, H., & Miyata, N. (1996). Mutagenicity of the fullerene C60-generated singlet oxygen dependent formation of lipid peroxides. *Carcinogenesis*, *17*(10), 2163–2169. doi:10.1093/carcin/17.10.2163 PMID:8895484

Seredych, M., & Bandosz, T. J. (2012). Manganese oxide and graphite oxide/MnO_2 composites as reactive adsorbents of ammonia at ambient conditions. *Microporous and Mesoporous Materials*, *150*, 55–63. doi:10.1016/j.micromeso.2011.09.010

Serrano, E., Rus, G., & Garcia-Martinez, J. (2009). Nanotechnology for sustainable energy. *Renewable & Sustainable Energy Reviews*, *13*(9), 2373–2384. doi:10.1016/j.rser.2009.06.003

Sevcenco, A. M., Paravidino, M., Vrouwenvelder, J. S., Wolterbeek, H. T., van Loosdrecht, M. C., & Hagen, W. R. (2015). Phosphate and arsenate removal efficiency by thermostable ferritin enzyme from Pyrococcus furiosus using radioisotopes. *Water Research*, *76*, 181–186. doi:10.1016/j.watres.2015.02.054 PMID:25817554

Seyfi, B., Baghalha, M., & Kazemian, H. (2009). Modified LaCoO 3 nano-perovskite catalysts for the environmental application of automotive CO oxidation. *Chemical Engineering Journal*, *148*(2), 306–311. doi:10.1016/j.cej.2008.08.041

Shahbazi, A., Younesi, H., & Badiei, A. (2012). Batch and fixed-bed column adsorption of Cu(II), Pb(II) and Cd(II) from aqueous solution onto functionalised SBA-15 mesoporous silica. *Canadian Journal of Chemical Engineering*, *91*(4), 739–750. doi:10.1002/cjce.21691

Shah, J., Jan, M. R., Khan, M., & Amir, S. (2016). Removal and recovery of cadmium from aqueous solutions using magnetic nanoparticle-modified sawdust: Kinetics and adsorption isotherm studies. *Desalination and Water Treatment*, *57*(21), 9736–9744. doi:10.1080/19443994.2015.1030777

Shah, M. A., & Qurashi, A. (2009). Novel surfactant-free synthesis of MgO nanoflakes. *Journal of Alloys and Compounds*, *482*(1-2), 548–551. doi:10.1016/j.jallcom.2009.04.129

Shah, R. K., Boruah, F., & Parween, N. (2015). Synthesis and Characterization of ZnO Nanoparticles using Leaf Extract of Camellia sinesis and Evaluation of their Antimicrobial Efficacy. *International Journal of Current Microbiology and Applied Sciences*, *4*(8), 444–450.

Shahverdi, A., Shakibaie, M., & Nazari, P. (2011). Basic and practical procedures for microbial synthesis of nanoparticles. In M. Rai & N. Duran (Eds.), *Metal nanoparticles in microbiology*. Berlin: Springer. doi:10.1007/978-3-642-18312-6_8

Shamaila, S., Sajjad, A. K. L., Farooqi, S. A., Jabeen, N., Majeed, S., & Farooq, I. (2016). Advancements in nanoparticle fabrication by hazard free eco-friendly green routes. *Applied Materials Today*, *5*, 150–199. doi:10.1016/j.apmt.2016.09.009

Shankar, S. S., Ahmad, A., & Sastry, M. (2003). Geranium Leaf Assisted Biosynthesis of Silver Nanoparticles. *Biotechnology Progress*, *19*(6), 1627–1631. doi:10.1021/bp034070w PMID:14656132

Shankar, S. S., Rai, A., Ahmad, A., & Sastry, M. (2004). Rapid synthesis of Au, Ag, and bimetallic Au core–Ag shell nanoparticles using Neem (Azadirachta indica) leaf broth. *Journal of Colloid and Interface Science*, *275*(2), 496–502. doi:10.1016/j.jcis.2004.03.003 PMID:15178278

Shankar, S. S., Rai, A., Ahmad, A., & Sastry, M. (2005). Controlling the Optical Properties of Lemongrass Extract Synthesized Gold Nanotriangles and Potential Application in Infrared-Absorbing Optical Coatings. *Chemistry of Materials*, *17*(3), 566–572. doi:10.1021/cm048292g

Shankar, S., Jaiswal, L., Aparna, R., & Prasad, R. (2014). Synthesis, characterization, in vitro biocompatibility, and antimicrobial activity of gold, silver and gold silver alloy nanoparticles prepared from Lansium domesticum fruit peel extract. *Materials Letters*, *137*, 75–78. doi:10.1016/j.matlet.2014.08.122

Shan, R. R., Yan, L. G., Yang, K. Y., Yu, S. J., Hao, Y. F., Yu, H. Q., & Du, B. (2014). Magnetic Fe3O4/MgAl- LDH composite for effective removal of three red dyes from aqueous solution. *Chemical Engineering Journal*, *252*, 38–46. doi:10.1016/j.cej.2014.04.105

Shariati-Rad, M., Irandoust, M., Amri, S., Feyzi, M., & Ja'fari, F. (2014). Magnetic solid phase adsorption, preconcentration and determination of methyl orange in water samples using silica coated magnetic nanoparticles and central composite design. *International Nano Letters*, *4*(4), 91–101. doi:10.100740089-014-0124-5

Sharma, J. K., Akhtar, M. S., Ameen, S., Srivastava, P., & Singh, G. (2015). Green synthesis of CuO nanoparticles with leaf extract of Calotropis gigantea and its dye-sensitized solar cells applications. *Journal of Alloys and Compounds*, *632*, 321–325. doi:10.1016/j.jallcom.2015.01.172

Sharma, K., Ko, E. Y., Assefa, A. D., Ha, S., Nile, S. H., Lee, E. T., & Park, S. W. (2015). Temperature-dependent studies on the total phenolics, flavonoids, antioxidant activities, and sugar content in six onion varieties. *Journal of Food and Drug Analysis*, *23*(2), 243–252. doi:10.1016/j.jfda.2014.10.005 PMID:28911379

Sharma, R., & Gupta, N. (2015). Green HRM: An Innovative Approach to Environmental Sustainability. *Proceeding of the Twelfth AIMS International Conference on Management.*

Sharma, S. S., & Dietz, K. (2009). The relationship between metal toxicity and cellular redox imbalance. *Trends in Plant Science, 14*(1), 43–45. doi:10.1016/j.tplants.2008.10.007 PMID:19070530

Sharma, V. K., Yngard, R. A., & Lin, Y. (2009). Silver nanoparticles: Green synthesis and their antimicrobial activities. *Advances in Colloid and Interface Science, 145*(1–2), 83–96. doi:10.1016/j.cis.2008.09.002 PMID:18945421

Sharma, V., Shukla, R. K., Saxena, N., Parmar, D., Das, M., & Dhawan, A. (2009). DNA damaging potential of zinc oxide nanoparticles in human epidermal cells. *Toxicology Letters, 185*(3), 211–218. doi:10.1016/j.toxlet.2009.01.008 PMID:19382294

Sharma, Y. C. (2011). Adsorption characteristics of a low-cost activated carbon for the reclamation of colored effluents containing malachite green. *Journal of Chemical & Engineering Data, 56*(3), 478–484. doi:10.1021/je1008922

Sharma, Y. C., Prasad, G., & Rupainwar, D. C. (1992). Heavy metal pollution of river Ganga in Mirzapur, India. *The International Journal of Environmental Studies, 40*(1), 41–53. doi:10.1080/00207239208710712

Sharma, Y. C., Srivastava, V., Singh, V. K., Kaul, S. N., & Weng, C. H. (2009). Nano-adsorbents for the removal of metallic pollutants from water and wastewater. *Environmental Technology, 30*(6), 583–609. doi:10.1080/09593330902838080 PMID:19603705

Shawabkeh, R. A., & Abu-Nameh, E. S. M. (2007). Absorption of Phenol and Methylene Blue by Activated Carbon from Pecan Shells. *Colloid Journal, 69*(3), 355–359. doi:10.1134/S1061933X07030143

Shen, H., Tao, S., Wang, R., Wang, B., Shen, G., Li, W., ... Sun, K. (2011). Global time trends in PAH emissions from motor vehicles. *Atmospheric Environment, 45*(12), 2067–2073. doi:10.1016/j.atmosenv.2011.01.054 PMID:24198716

Sheta, A. S., Falatah, A. M., Al-Sewailem, M. S., Khaled, E. M., & Sallam, A. S. H. (2003). Sorption characteristics of zinc and iron by natural zeolite and bentonite. *Microporous and Mesoporous Materials, 61*(1), 127–136. doi:10.1016/S1387-1811(03)00360-3

Shimizu, K., Uchiyama, A., Yamashita, M., Hirose, A., Nishimura, T., & Oku, N. (2013). Biomembrane damage caused by exposure to multi-walled carbon nanotubes. *The Journal of Toxicological Sciences, 38*(1), 7–12. doi:10.2131/jts.38.7 PMID:23358135

Shimizu, M., Tainaka, H., Oba, T., Mizuo, K., Umezawa, M., & Takeda, K. (2009). Maternal exposure to nanoparticulate titanium dioxide during the prenatal period alters gene expression related to brain development in the mouse. *Particle and Fibre Toxicology, 6*(1), 20–20. doi:10.1186/1743-8977-6-20 PMID:19640265

Shinohara, N., Matsumoto, K., Endoh, S., Maru, J., & Nakanishi, J. (2009). In vitro and in vivo genotoxicity tests on fullerene C60 nanoparticles. *Toxicology Letters, 191*(2-3), 289–296. doi:10.1016/j.toxlet.2009.09.012 PMID:19772904

Shirmardi, M., Mesdaghinia, A., Mahvi, A. H., Nasseri, S., & Nabizadeh, R. (2012). Kinetics and equilibrium studies on adsorption of acid red 18 (Azo-Dye) using multiwall carbon nanotubes (MWCNTs) from aqueous solution. *Journal of Chemistry, 9*(4), 2371–2383.

Shirsath, D. S., & Shirivastava, V. S. (2015). Adsorptive removal of heavy metals by magnetic nanoadsorbent: An equilibrium and thermodynamic study. *Applied Nanoscience, 5*(8), 927–935. doi:10.100713204-014-0390-6

Shirsat, M. D., Sarkar, T., Kakoullis, J. Jr, Myung, N. V., Konnanath, B., Spanias, A., & Mulchandani, A. (2012). Porphyrin-functionalized single-walled carbon nanotube chemiresistive sensor arrays for VOCs. *The Journal of Physical Chemistry C, 116*(5), 3845–3850. doi:10.1021/jp210582t PMID:22393460

Shi, S., Fan, Y. W., & Huang, Y. M. (2013). Facile low temperature hydrothermal synthesis of magnetic mesoporous carbon nanocomposite for adsorption removal of ciprofloxacin antibiotics. *Industrial & Engineering Chemistry Research, 52*(7), 2604–2612. doi:10.1021/ie303036e

Shokri, A., Rabiee, F., & Mahanpoor, K. (2017). Employing a novel nanocatalyst (Mn/Iranian hematite) for oxidation of SO2 pollutant in aqueous environment. *International Journal of Environmental Science and Technology, 14*(11), 2485–2494. doi:10.100713762-017-1346-7

Short, J. L. (1993). Membrane Industry News. Westford, MA: Academic Press.

Shrivastava, S., Bera, T., Roy, A., Singh, G., Ramachand, P., & Dash, D. (2007). Characterization of enhanced antibacterial effect of novel silver nanoparticles. *Nanotechnology, 18*(22), 225103. doi:10.1088/0957-4484/18/22/225103

Shumilla, J. A., Broderick, R. J., Wang, Y., & Barchowsky, A. (1999). Chromium(VI) Inhibits the Transcriptional Activity of Nuclear Factor-B by Decreasing the Interaction of p65 with cAMP-responsive Element-binding Protein-binding Protein. *The Journal of Biological Chemistry, 274*(51), 36207–36212. doi:10.1074/jbc.274.51.36207 PMID:10593907

Shvedova, A. A., Castranova, V., Kisin, E. R., Schwegler-Berry, D., Murray, A. R., Gandelsman, V. Z., ... Baron, P. (2003). Exposure to carbon nanotube material: Assessment of nanotube cytotoxicity using human keratinocyte cells. *Journal of Toxicology and Environmental Health. Part A., 66*(20), 1909–1926. doi:10.1080/713853956 PMID:14514433

Siboni, M. S., Samadi, M. T., Yang, J. K., & Lee, S. M. (2011). Photocatalytic reduction of Cr(VI) and Ni(II) in aqueous solution by synthesized nanoparticle ZnO under ultraviolet light irradiation: A kinetic study. *Environmental Technology, 32*(13-14), 1573–1579. doi:10.1080/09593330.2010.543933 PMID:22329148

Siddique, H., Rundquist, E., Bhole, Y., Peeva, L. G., & Livingston, A. G. (2014). Mixed matrix membranes for organic solvent nanofiltration. *Journal of Membrane Science, 452*, 354–366. doi:10.1016/j.memsci.2013.10.012

Si, J., & Yang, H. (2011). Preparation and characterization of bio-compatible Fe3O4@Polydopamine spheres with core/shell nanostructure. *Materials Chemistry and Physics, 128*(3), 519–524. doi:10.1016/j.matchemphys.2011.03.039

Simon, I., Bârsan, N., Bauer, M., & Weimar, U. (2001). Micromachined metal oxide gas sensors: Opportunities to improve sensor performance. *Sensors and Actuators. B, Chemical, 73*(1), 1–26. doi:10.1016/S0925-4005(00)00639-0

Simsek, E. B., Duranoglu, D., & Beker, U. (2012). Heavy metal adsorption by magnetic hybrid-sorbent: An experimental and theoretical approach. *Separation Science and Technology, 47*(9), 1334–1340. doi:10.1080/01496395.2012.672845

Singh, S., Barick, K., & Bahadur, D. (2011). Novel and efficient three dimensional mesoporous ZnO nanoassemblies for envirnomental remediation. *International Journal of Nanoscience, 10*(4-5), 1001-1005.

Singh, A. K., Talat, M., Singh, D. P., & Srivastava, O. N. (2010). Biosynthesis of gold and silver nanoparticles by natural precursor clove and their functionalization with amine group. *Journal of Nanoparticle Research, 12*(5), 1667–1675. doi:10.100711051-009-9835-3

Singhal, R., Gangadhar, B., Basu, H., Manisha, V., Naidu, G., & Reddy, A. (2012). Remediation of malathion contaminated soil using zero valent iron nano-particles. *American Journal of Analytical Chemistry, 3*(01), 76–82. doi:10.4236/ajac.2012.31011

Singh, M., & Tlali, S. (2017). Effects of size and shape on the specific heat, melting entropy and enthalpy of nanomaterials. *Journal of Taibah University for Science, 11*(6), 922–929. doi:10.1016/j.jtusci.2016.09.011

Singh, N. (2009). Conference scene - nanotoxicology: Health and environmental impacts. *Nanomedicine (London), 4*(4), 385–390. doi:10.2217/nnm.09.20 PMID:19505241

Singh, N., Singh, S. P., Gupta, V., Yadav, H. K., Ahuja, T., Tripathy, S. S., & Rashmi. (2013). A process for the selective removal of arsenic from contaminated water using acetate functionalized zinc oxide nanomaterials. *Environmental Progress & Sustainable Energy*, *32*(4), 1023–1029. doi:10.1002/ep.11698

Singh, R., Manickam, N., Mudiam, M. K. R., Murthy, R. C., & Misra, V. (2013). An integrated (nano-bio) technisque for degradation of γ-HCH contaminated soil. *Journal of Hazardous Materials*, *258*, 35–41. doi:10.1016/j.jhazmat.2013.04.016 PMID:23692681

Singh, S., Barick, K., & Bahadur, D. (2013). Functional oxide nanomaterials and nanocomposites for the removal of heavy metals and dyes. *Nanomaterials and Nanotechnology*, *3*, 20. doi:10.5772/57237

Singh, T. S., & Pant, K. (2004). Equilibrium, kinetics and thermodynamic studies for adsorption of As (III) on activated alumina. *Separation and Purification Technology*, *36*(2), 139–147. doi:10.1016/S1383-5866(03)00209-0

Sinha, R., Peters, U., Cross, A. J., Kulldorff, M., Weissfeld, J. L., Pinsky, P. F., ... Colorectal, T. O. C. P. (2005). Meat, meat cooking methods and preservation, and risk for colorectal adenoma. *Cancer Research*, *65*(17), 8034–8041. doi:10.1158/0008-5472.CAN-04-3429 PMID:16140978

Si, S., & Mandal, T. K. (2007). Tryptophan-based peptides to synthesize gold and silver nanoparticles: A mechanistic and kinetic study. *Chemistry*, *13*(11), 3160–3168. doi:10.1002/chem.200601492 PMID:17245786

Sivalingam, G., Nagaveni, K., Hegde, M., & Madras, G. (2003). Photocatalytic degradation of various dyes by combustion synthesized nano anatase TiO 2. *Applied Catalysis B: Environmental*, *45*(1), 23–38. doi:10.1016/S0926-3373(03)00124-3

Skubal, L. R., McArthur, A. L., & Newville, M. (2008). TiO_2 films for self-detection and decontamination. *International Journal of Photoenergy*, *2008*, 1–6. doi:10.1155/2008/343714

Slade, S., Campbell, S. A., Ralph, T. R., & Walsh, F. C. (2002). Ionic Conductivity of an Extruded Nafion 1100 EW Series of Membranes. *Journal of the Electrochemical Society*, *149*(12), A1556–A1564. doi:10.1149/1.1517281

Slowing, I. I., Trewyn, B. G., & Lin, V. S. Y. (2007). Mesoporous silica nanoparticles for intracellular delivery of membrane-impermeable proteins. *Journal of the American Chemical Society*, *129*(28), 8845–8849. doi:10.1021/ja0719780 PMID:17589996

Smitha, B., Sridhar, S., & Khan, A. A. (2005). Solid polymer electrolyte membrane for fuel cell applications. *Journal of Membrane Science*, *259*(1-2), 10–26. doi:10.1016/j.memsci.2005.01.035

Soler-Illia, G. J. de A.A., Sanchez, C., Lebeau, B., & Patarin, J. (2002). Chemical strategies to design textured materials: From microporous and mesoporous oxides to nanonetworks and hierarchical structures. *Chemical Reviews*, *102*(11), 4093–4138. doi:10.1021/cr0200062 PMID:12428985

Sondi, I., & Salopek-Sondi, B. (2004). Silver nanoparticles as antimicrobial agent: A case study on E. coli as a model for Gramnegative bacteria. *Journal of Colloid and Interface Science*, *275*(1), 177–182. doi:10.1016/j.jcis.2004.02.012 PMID:15158396

Sopyan, I. (2007). Kinetic analysis on photocatalytic degradation of gaseous acetaldehyde, ammonia and hydrogen sulfide on nanosized porous TiO_2 films. *Science and Technology of Advanced Materials*, *8*(1), 33–39. doi:10.1016/j.stam.2006.10.004

Spadaro, J. A., Berger, T. J., Barranco, S. D., Chapin, S. E., & Becker, R. O. (1974). Antibacterial effects of silver electrodes with weak direct current. *Antimicrobial Agents and Chemotherapy*, *6*(5), 637–642. doi:10.1128/AAC.6.5.637 PMID:15825319

Srivastava, N., & Majumder, C. (2008). Novel biofiltration methods for the treatment of heavy metals from industrial wastewater. *Journal of Hazardous Materials*, *151*(1), 1–8. doi:10.1016/j.jhazmat.2007.09.101 PMID:17997034

Srivastava, R. R., Kim, M., Lee, J., & Ilyas, S. (2015). Liquid–liquid extraction of rhenium (VII) from an acidic chloride solution using Cyanex 923. *Hydrometallurgy*, *157*, 33–38. doi:10.1016/j.hydromet.2015.07.011

Stafiej, A., & Pyrzynska, K. (2007). Adsorption of heavy metal ions with carbon nanotubes. *Separation and Purification Technology*, *58*(1), 49–52. doi:10.1016/j.seppur.2007.07.008

Steduto, P., Faurès, J.-M., Hoogeveen, J., Winpenny, J. T., & Burke, J. J. (2012). Coping with Water Scarcity: An Action Framework for Agricultural and Food Security. Food and Agricultural Organization of the United Nations.

Steele, B. C., & Heinzel, H. A. (2001). Materials for fuel cell technologies. *Nature*, *414*(6861), 345–352. doi:10.1038/35104620 PMID:11713541

Steffen, W., Crutzen, P. J., & McNeill, J. R. (2007). The Anthropocene: Are humans now overwhelming the great forces of nature. *Ambio*, *36*(8), 614–621. doi:10.1579/0044-7447(2007)36[614:TAAHNO]2.0.CO;2 PMID:18240674

Steffen, W., Persson, Å., Deutsch, L., Zalasiewicz, J., Williams, M., Richardson, K., ... Svedin, U. (2011). The Anthropocene: From Global Change to Planetary Stewardship. *Ambio*, *40*(7), 739–761. doi:10.100713280-011-0185-x PMID:22338713

Stein, A., Melde, B. J., & Schroden, R. C. (2000). Hybrid Inorganic-Organic Mesoporous Silicates-Nanoscopic Reactors Coming of Age. *Advanced Materials*, *12*(19), 1403–1419. doi:10.1002/1521-4095(200010)12:19<1403::AID-ADMA1403>3.0.CO;2-X

Steinfeld, H., Gerber, P., Wassenaar, T., Castel, V., Rosales, M., & Haan, C. d. (2006). *Livestock's Long Shadow environmental issues and options.* Retrieved from Room, Italy: Stott, K. J., & Gourley, C. J. P. (2016). Intensification, nitrogen use and recovery in grazing-based dairy systems. *Agricultural Systems*, *144*, 101–112. doi:10.1016/j.agsy.2016.01.003

Stenzel, M. H. (1993). Remove organics by activated carbon adsorption. *Chemical Engineering Progress*, *89*(4).

Stone, V., Nowack, B., Baun, A., van den Brink, N., von der Kammer, F., Dusinska, M., ... Fernandes, T. F. (2010). Nanomaterials for environmental studies: Classification, reference material issues, and strategies for physico-chemical characterisation. *The Science of the Total Environment*, *408*(7), 1745–1754. doi:10.1016/j.scitotenv.2009.10.035 PMID:19903569

Strasser, P., Koh, S., Anniyev, T., Greeley, J., More, K., Yu, C., . . . Nilsson, A. (2010). Lattice-strain control of the activity in dealloyed core–shell fuel cell catalysts. *Nat Chem, 2*(6), 454-460. Retrieved from http://www.nature.com/nchem/journal/v2/n6/suppinfo/nchem.623_S1.html

Strathmann, H. (1985). *Production of microporous media by phase inversion processes.* Academic Press.

Strathmann, H. (2004). Ion-Exchange Membrane Separation Processes. *Elsevier Science Technology*, *9*, 10–30.

Suchithra, P. S., Vazhayal, L., Mohamed, A. P., & Ananthakumar, S. (2012). Mesoporous organic-inorganic hybrid aerogels through ultrasonic assisted sol-gel intercalation of silica-PEG in bentonite for effective removal of dyes, volatile organic pollutants and petroleum products from aqueous solution. *Chemical Engineering Journal*, *200–202*, 589–600. doi:10.1016/j.cej.2012.06.083

Su, F., Lu, C., Cnen, W., Bai, H., & Hwang, J. F. (2009). Capture of CO_2 from flue gas via multi-walled carbon nanotubes. *The Science of the Total Environment*, *407*(8), 3017–3023. doi:10.1016/j.scitotenv.2009.01.007 PMID:19201012

Sugg, T., Tholstrup, C., Taylor, R., & Staggers, J. (2012). *The development and application of nanoporous materials.* Texas A&M University.

Su, H., Li, Y., Li, X.-Y., & Wong, K. S. (2009). Optical and electrical properties of Au nanoparticles in two-dimensional networks: An effective cluster model. *Optics Express*, *17*(24), 22223–22234. doi:10.1364/OE.17.022223 PMID:19997469

Sui, G., Zhong, W., Yang, X., Yu, Y., & Zhao, S. (2008). Preparation and properties of natural rubber composites reinforced with pretreated carbon nanotubes. *Polymers for Advanced Technologies*, *19*(11), 1543–1549.

Sui, Z. Y., Meng, Q. H., Zhang, X. T., Ma, R., & Cao, B. (2012b). Green synthesis of carbon nanotube-graphene hybrid aerogels and their use as versatile agents for water purification. *Journal of Materials Chemistry*, *22*(18), 8767–8771. doi:10.1039/c2jm00055e

Sullivan, C., Tyrer, M., Cheeseman, C. R., & Graham, N. J. D. (2010). Disposal of water treatment wastes containing arsenic — A review. *The Science of the Total Environment*, *408*(8), 1770–1778. doi:10.1016/j.scitotenv.2010.01.010 PMID:20153878

Sun, B., Wang, X., Ji, Z., Wang, M., Liao, Y.-P., Chang, C. H., ... Xia, T. (2015). NADPH Oxidase-Dependent NLRP3 Inflammasome Activation and its Important Role in Lung Fibrosis by Multiwalled Carbon Nanotubes. *Small*, *11*(17), 2087–2097. doi:10.1002mll.201402859 PMID:25581126

Sun, J., Wang, X., Sun, J., Sun, R., Sun, S., & Qiao, L. (2006). Photocatalytic degradation and kinetics of Orange G using nano-sized Sn (IV)/TiO 2/AC photocatalyst. *Journal of Molecular Catalysis A Chemical*, *260*(1), 241–246. doi:10.1016/j.molcata.2006.07.033

Sun, S., Murray, C. B., Weller, D., Folks, L., & Moser, A. (2000). Monodisperse FePt Nanoparticles and Ferromagnetic FePt Nanocrystal Superlattices. *Science*, *287*(5460), 1989–1992. doi:10.1126cience.287.5460.1989 PMID:10720318

Suresh, D., Nethravathi, P. C., Udayabhanu, Rajanaika, H., Nagabhushana, H., & Sharma, S. C. (2015). Green synthesis of multifunctional zinc oxide (ZnO) nanoparticles using Cassia fistula plant extract and their photodegradative, antioxidant and antibacterial activities. *Materials Science in Semiconductor Processing*, *31*, 446–454. doi:10.1016/j.mssp.2014.12.023

Sureshkumar, M., Lee, P.-N., & Lee, C.-K. (2011). Stepwise assembly of multimetallic nanoparticles via self-polymerized polydopamine. *Journal of Materials Chemistry*, *21*(33), 12316. doi:10.1039/c1jm11914a

Su, S., Wu, W., Gao, J., Lu, J., & Fan, C. (2012). Nanomaterials-based sensors for applications in environmental monitoring. *Journal of Materials Chemistry*, *22*(35), 18101–18110. doi:10.1039/c2jm33284a

Su, Y., Cui, H., Li, Q., Gao, S., & Shang, J. K. (2013). Strong adsorption of phosphate by amorphous zirconium oxide nanoparticles. *Water Research*, *47*(14), 5018–5026. doi:10.1016/j.watres.2013.05.044 PMID:23850213

Su, Y., Yang, W., Sun, W., Li, Q., & Shang, J. K. (2015). Synthesis of mesoporous cerium–zirconium binary oxide nanoadsorbents by a solvothermal process and their effective adsorption of phosphate from water. *Chemical Engineering Journal*, *268*, 270–279. doi:10.1016/j.cej.2015.01.070

Sweetman, L. J., Nghiem, L., Chironi, I., Triani, G., Panhuis, M. I. H., & Ralph, S. F. (2012). Synthesis, properties and water permeability of SWNT buckypapers. *Journal of Materials Chemistry*, *22*(27), 13800–13810. doi:10.1039/c2jm31382k

Szubzda, B., Szmaja, A., Ozimek, M., & Mazurkiewicz, S. (2017). Polymer membranes as separators for supercapacitors. Applied Physics A. *Materials and Processing*, *117*, 1801–1809.

Szymanska, K., Zouboulis, A. I., & Zamboulis, D. (2014). Hybrid ozonation–microfiltration system for the treatment of surface water using ceramic membrane. *Journal of Membrane Science*, *468*, 163–171. doi:10.1016/j.memsci.2014.05.056

Tahir, L., Nazir, R., & Khan, N. (2013). *Nanoherbal antiseptic medicine/bandage for effective and safe treatment of external wounds*. Paper presented at the Nanotechnology 2013: Bio Sensors, Instruments, Medical, Environment and Energy, Washington, DC.

Taka, A. L., Pillay, K., & Mbianda, X. Y. (2017). Nanosponge cyclodextrin polyurethanes and their modification with nanomaterials for the removal of pollutants from waste water: A review. *Carbohydrate Polymers*, *159*, 94–107. doi:10.1016/j.carbpol.2016.12.027 PMID:28038758

Takeshita, K., Watanabe, K., Nakano, Y., & Watanabe, M. (2003). Solvent extraction of Cd(II) and Zn(II) with the organophosphorus extractant D2EHPA and the aqueous nitrogen-donor ligand TPEN. *Hydrometallurgy*, *70*(1-3), 63–71. doi:10.1016/S0304-386X(03)00046-X

Tang, H., Pan, M., & Jiang, S. P. (2010). One- step synthesized HPW/meso-silica inorganic proton exchange membranes for fuel cells. *Chemical Communications*, *46*(24), 4351–4353. doi:10.1039/c003129a PMID:20464019

Tang, W.-W., Zeng, G.-M., Gong, J.-L., Liu, Y., Wang, X.-Y., Liu, Y.-Y., ... Tu, D.-Z. (2012). Simultaneous adsorption of atrazine and Cu (II) from wastewater by magnetic multi-walled carbon nanotube. *Chemical Engineering Journal*, *211*, 470–478. doi:10.1016/j.cej.2012.09.102

Tan, K. A., Morad, N., Teng, T. T., Norli, I., & Panneerselvam, P. (2012). Removal of Cationic Dye by Magnetic Nanoparticle (Fe3O4) Impregnated onto Activated Maize Cob Powder and Kinetic Study of Dye Waste Adsorption. *APCBEE Procedia*, *1*(Supplement C), 83–89. doi:10.1016/j.apcbee.2012.03.015

Tan, Y. N., Lee, J. Y., & Wang, D. I. C. (2010). Uncovering the Design Rules for Peptide Synthesis of Metal Nanoparticles. *Journal of the American Chemical Society*, *132*(16), 5677–5686. doi:10.1021/ja907454f PMID:20355728

Tarabara, V. V. (2009). Multifunctional nanomaterial-enabled membranes for water treatment. *Nanotechnology Applications for Clean Water*, *5*, 59-75.

Tarigh, G. D., & Shemirani, F. (2013). Magnetic multi-wall carbon nanotube nanocomposite as an adsorbent for preconcentration and determination of lead (II) and manganese (II) in various matrices. *Talanta*, *115*, 744–750. doi:10.1016/j.talanta.2013.06.018 PMID:24054657

Tasaltin, C., & Basarir, F. (2014). Preparation of flexible VOC sensor based on carbon nanotubes and gold nanoparticles. *Sensors and Actuators. B, Chemical*, *194*, 173–179. doi:10.1016/j.snb.2013.12.063

Tavlarides, L. L., & Lee, J. S. (2001). Synthesis and application of functionalized organo-ceramic selective adsorbents. *Solvent Extraction and Ion Exchange*, *14*, 169–227.

Tawabini, B. S., Al-Khaldi, S. F., Khaled, M. M., & Atieh, M. A. (2011). Removal of arsenic from water by iron oxide nanoparticles impregnated on carbon nanotubes. *Journal of Environmental Science and Health. Part A*, *46*(3), 215–223. PMID:21279891

Taylor, J. M., Dawson, K. W., & Shimizu, G. K. H. (2013). A Water-Stable Metal–Organic Framework with Highly Acidic Pores for Proton-Conducting Applications. *Journal of the American Chemical Society*, *135*(4), 1193–1196. doi:10.1021/ja310435e PMID:23305324

Thayumanavan, B., & Sadasivam, S. (1984). Physicochemical basis for the preferential uses of certain rice varieties. *Qual Plant Plant Foods Hum Nutr*, *34*(4), 253–259. doi:10.1007/BF01126554

Theron, J., Cloete, T. E., & de Kwaadsteniet, M. (2010). Current Molecular and Emerging Nanobiotechnology Approaches for. *Nanotechnology in Water Treatment Applications*, 39.

Theron, J., Walker, J., & Cloete, T. (2008). Nanotechnology and water treatment: Applications and emerging opportunities. *Critical Reviews in Microbiology*, *34*(1), 43–69. doi:10.1080/10408410701710442 PMID:18259980

Theron, S., Zussman, E., & Yarin, A. (2004). Experimental investigation of the governing parameters in the electrospinning of polymer solutions. *Polymer*, *45*(6), 2017–2030. doi:10.1016/j.polymer.2004.01.024

Tian, H., Peng, J., Lv, T., Sun, C., & He, H. (2018). Preparation and performance study of MgFe2O4/metal–organic framework composite for rapid removal of organic dyes from water. *Journal of Solid State Chemistry*, *257*, 40–48. doi:10.1016/j.jssc.2017.09.017

Tian, Y., Gao, B., Morales, V. L., Chen, H., Wang, Y., & Li, H. (2013). Removal of sulfamethoxazole and sulfapyridine by carbon nanotubes in fixed-bed columns. *Chemosphere*, *90*(10), 2597–2605. doi:10.1016/j.chemosphere.2012.11.010 PMID:23232047

Timonen, K. L., Vanninen, E., De Hartog, J., Ibald-Mulli, A., Brunekreef, B., Gold, D. R., & Tarkiainen, T. (2006). Effects of ultrafine and fine particulate and gaseous air pollution on cardiac autonomic control in subjects with coronary artery disease: The ULTRA study. *Journal of Exposure Science & Environmental Epidemiology*, *16*(4), 332–341. doi:10.1038j.jea.7500460 PMID:16205787

Tingey-Holyoak, J. L. (2014). Water sharing risk in agriculture: Perceptions of farm dam management accountability in Australia. *Agricultural Water Management*, *145*, 123–133. doi:10.1016/j.agwat.2014.02.011

Tino, A., Ambrosone, A., Marchesano, V., & Tortiglione, C. (2014). Molecular Bases of Nanotoxicology. In Bio- and Bioinspired Nanomaterials (pp. 229-254): Wiley-VCH Verlag GmbH & Co. KGaA.

Tiwari, D. K., Behari, J., & Sen, P. (2008). Application of Nanoparticles in Waste Water Treatment. *World Applied Sciences Journal*, *3*(3), 417–433.

Toyomoto, K., Higuchi, A., Osada, Y., & Nakagawa, T. (1992). *Membrane Science and Technology*. Academic Press.

Tran, H. V., Tran, L. D., & Nguyen, T. N. (2010). Preparation of chitosan/magnetite composite beads and their application for removal of Pb(II) and Ni(II) from aqueous solution. *Materials Science and Engineering C*, *30*(2), 304–310. doi:10.1016/j.msec.2009.11.008

Tratnyek, P. G., & Johnson, R. L. (2006). Nanotechnologies for environmental cleanup. *Nano Today*, *1*(2), 44–48. doi:10.1016/S1748-0132(06)70048-2

Tricoli, V., & Nannetti, F. (2003). Zeolite-Nafion composites as ion conducting membrane materials. *Electrochimica Acta*, *48*(18), 2625–2633. doi:10.1016/S0013-4686(03)00306-2

Tripathy, A., Raichur, A. M., Chandrasekaran, N., Prathna, T., & Mukherjee, A. (2010). Process variables in biomimetic synthesis of silver nanoparticles by aqueous extract of Azadirachta indica (Neem) leaves. *Journal of Nanoparticle Research*, *12*(1), 237–246. doi:10.100711051-009-9602-5

Tsang, S. C., Caps, V., Paraskevas, I., Chadwick, D., & Thompsett, D. (2004). Magnetically Separable, Carbon-Supported Nanocatalysts for the Manufacture of Fine Chemicals. *Angewandte Chemie*, *116*(42), 5763–5767. doi:10.1002/ange.200460552 PMID:15495206

Tseng, J.-Y., Chang, C.-Y., Chang, C.-F., Chen, Y.-H., Chang, C.-C., Ji, D.-R., ... Chiang, P.-C. (2009). Kinetics and equilibrium of desorption removal of copper from magnetic polymer adsorbent. *Journal of Hazardous Materials*, *171*(1), 370–377. doi:10.1016/j.jhazmat.2009.06.030 PMID:19595507

Tsibranska, I. H., & Tylkowski, B. (2013). Concentration of ethanolic extracts from Sideritis ssp. L. by nanofiltration: Comparison of dead-end and cross-flow modes. *Food and Bioproducts Processing*, *91*(2), 169–174. doi:10.1016/j.fbp.2012.09.004

Tsibranska, I., Tylkowski, B., Kochanov, R., & Alipieva, K. (2011). Extraction of biologically active compounds from Sideritis ssp. L. *Food and Bioproducts Processing*, *89*(4), 273–280. doi:10.1016/j.fbp.2010.10.004

Tsuchiya, T., Oguri, I., Yamakoshi, Y. N., & Miyata, N. (1996). Novel harmful effects of [60]fullerene on mouse embryos in vitro and in vivo. *FEBS Letters*, *393*(1), 139–145. doi:10.1016/0014-5793(96)00812-5 PMID:8804443

Tsujimoto, Y., Matsushita, Y., Yu, S., Yamaura, K., & Uchikoshi, T. (2015). Size dependence of structural, magnetic, and electrical properties in corundum-type Ti2O3 nanoparticles showing insulator–metal transition. *Journal of Asian Ceramic Societies*, *3*(3), 325–333. doi:10.1016/j.jascer.2015.06.007

Tsuru, T., Nakasuji, T., Oka, M., Kanezashi, M., & Yoshioka, T. (2011). Preparation of hydrophobic nanoporous methylated SiO 2 membranes and application to nanofiltration of hexane solutions. *Journal of Membrane Science*, *384*(1), 149–156. doi:10.1016/j.memsci.2011.09.018

Tsuru, T., Narita, M., Shinagawa, R., & Yoshioka, T. (2008). Nanoporous titania membranes for permeation and filtration of organic solutions. *Desalination*, *233*(1), 1–9. doi:10.1016/j.desal.2007.09.021

Tsuzuki, T., & McCormick, P. G. (2004). Mechanochemical synthesis of nanoparticles. *Journal of Materials Science*, *39*(16), 5143–5146. doi:10.1023/B:JMSC.0000039199.56155.f9

Tungittiplakorn, W., Lion, L. W., Cohen, C., & Kim, J.-Y. (2004). Engineered polymeric nanoparticles for soil remediation. *Environmental Science & Technology*, *38*(5), 1605–1610. doi:10.1021/es0348997 PMID:15046367

Tung, K. L., Chang, K. S., Wu, T. T., Lin, N. J., Lee, K. R., & Lai, J. Y. (2014). Recent advances in the characterization of membrane morphology. *Current Opinion in Chemical Engineering*, *4*, 121–127. doi:10.1016/j.coche.2014.03.002

Turkevich, J., Stevenson, P. C., & Hillier, J. (1951). A study of the nucleation and growth processes in the synthesis of colloidal gold. *Discussions of the Faraday Society*, *11*, 55–75. doi:10.1039/df9511100055

Tuzen, M., & Soylak, M. (2007). Multiwalled carbon nanotubes for speciation of chromium in environmental samples. *Journal of Hazardous Materials*, *147*(1), 219–225. doi:10.1016/j.jhazmat.2006.12.069 PMID:17258392

Tylkowski, B., Carosio, F., Castañeda, J., Alongi, J., García-Valls, R., Malucelli, G., & Giamberini, M. (2013). Permeation behavior of polysulfone membranes modified by fully organic layer-by-layer assemblies. *Industrial & Engineering Chemistry Research*, *52*(46), 16406–16413. doi:10.1021/ie402942g

Tylkowski, B., Peris, S., Giamberini, M., Garcia-Valls, R., Reina, J. A., & Ronda, J. C. (2010). Light-induced switching of the wettability of novel asymmetrical poly (vinyl alcohol)-co-ethylene membranes blended with azobenzene polymers. *Langmuir*, *26*(18), 14821–14829. doi:10.1021/la101809r PMID:20799708

Tylkowski, B., Trusheva, B., Bankova, V., Giamberini, M., Peev, G., & Nikolova, A. (2010). Extraction of biologically active compounds from propolis and concentration of extract by nanofiltration. *Journal of Membrane Science*, *348*(1), 124–130. doi:10.1016/j.memsci.2009.10.049

Tylkowski, B., Tsibranska, I., Kochanov, R., Peev, G., & Giamberini, M. (2011). Concentration of biologically active compounds extracted from Sideritis ssp. L. by nanofiltration. *Food and Bioproducts Processing*, *89*(4), 307–314. doi:10.1016/j.fbp.2010.11.003

U.S. Department of the Interior. (n.d.). *U.S. Geological Survey*. Retrieved from http://water.usgs.gov/edu/earthwherewater.html

Ünal, M. Ü. (2007). Properties of polyphenol oxidase from Anamur banana (Musa cavendishii). *Food Chemistry*, *100*(3), 909–913. doi:10.1016/j.foodchem.2005.10.048

UNESCO. (2003). *Water for People, Water for Life*. United Nations World Water Development Report. Retrieved from www.unesdoc.unesco.org

Ungureanu, G., Santos, S., Boaventura, R., & Botelho, C. (2015). Arsenic and antimony in water and wastewater: Overview of removal techniques with special reference to latest advances in adsorption. *Journal of Environmental Management*, *151*, 326–342. doi:10.1016/j.jenvman.2014.12.051 PMID:25585146

UNICEF/WHO. (2010). *Progress on Sanitation and Drinking Water: 2010 Update*. WHO/UNICEF Joint Monitoring Programme for Water Supply and Sanitation.

United Nations Development Programme. (2006). *Coping with water scarcity; Challenge of the twenty-first century* [Press release]. UNDP.

United Nations Secretary-General's High-Level Panel on Global Sustainability. (2012). Resilient People, Resilient Planet. A Future Worth Choosing. New York: UN GSP.

Uricanu, V., Donescu, D., Banu, A. G., Serban, S., Olteanu, M., & Dudau, M. (2004). Organic-inorganic hybrids made from polymerizable precursors. *Materials Chemistry and Physics*, *85*(1), 120–130. doi:10.1016/j.matchemphys.2003.12.024

Usman, M. S., El Zowalaty, M. E., Shameli, K., Zainuddin, N., Salama, M., & Ibrahim, N. A. (2013). Synthesis, characterization, and antimicrobial properties of copper nanoparticles. *International Journal of Nanomedicine*, *8*, 4467. PMID:24293998

Uzun, I., & Güzel, F. (2000). Adsorption of some heavy metal ions from aqueous solution by activated carbon and comparison of percent adsorption results of activated carbon with those of some other adsorbents. *Turkish Journal of Chemistry*, *24*, 291–297.

Uzunova-Bujnova, M., Todorovska, R., Milanova, M., Kralchevska, R., & Todorovsky, D. (2009). On the spray-drying deposition of TiO_2 photocatalytic films. *Applied Surface Science*, *256*(3), 830–837. doi:10.1016/j.apsusc.2009.08.069

Vaiano, V., Sacco, O., Sannino, D., & Ciambelli, P. (2015). Photocatalytic removal of spiramycin from wastewater under visible light with N-doped TiO2 photocatalysts. *Chemical Engineering Journal*, *261*, 3–8. doi:10.1016/j.cej.2014.02.071

Vajnhandl, S., & Valh, J. V. (2014). The status of water reuse in European textile sector. *Journal of Environmental Management*, *141*, 29–35. doi:10.1016/j.jenvman.2014.03.014 PMID:24768832

Valadez-Blanco, R., & Livingston, A. G. (2009). Solute molecular transport through polyimide asymmetric organic solvent nanofiltration (OSN) membranes and the effect of membrane-formation parameters on mass transfer. *Journal of Membrane Science*, *326*(2), 332–342. doi:10.1016/j.memsci.2008.10.016

Valderrama, C., Cortina, J. L., Farran, A., Gamisans, X., & Lao, C. (2007). Kinetics of Sorption of polyaromatic Hydrocarbons onto Granular Activated Carbon and Macronet Hyper-Cross-Linked Polymers (MN200). *Journal of Colloid and Interface Science*, *310*(1), 35–46. doi:10.1016/j.jcis.2007.01.039 PMID:17367802

Van der Bruggen, B., & Vandecasteele, C. (2003). Removal of pollutants from surface water and groundwater by nanofiltration: Overview of possible applications in the drinking water industry. *Environmental Pollution*, *122*(3), 435–445. doi:10.1016/S0269-7491(02)00308-1 PMID:12547533

Vanathi, P., Rajiv, P., Narendhran, S., Rajeshwari, S., Rahman, P. K. S. M., & Venckatesh, R. (2014). Biosynthesis and characterization of phyto mediated zinc oxide nanoparticles: A green chemistry approach. *Materials Letters*, *134*, 13–15. doi:10.1016/j.matlet.2014.07.029

Vani Padmaja, G. (2016). Nanomaterials: A Measure for Air Pollutant Remedial Technologies. *International Journal of Science and Research*, *5*(10), 1307–1310.

Varanasi, P., Fullana, A., & Sidhu, S. (2007). Remediation of PCB contaminated soils using iron nano-particles. *Chemosphere*, *66*(6), 1031–1038. doi:10.1016/j.chemosphere.2006.07.036 PMID:16962632

Varma, R. S. (2012). Greener approach to nanomaterials and their sustainable applications. *Current Opinion in Chemical Engineering*, *1*(2), 123–128. doi:10.1016/j.coche.2011.12.002

Varma, S., Sarode, D., Wakale, S., Bhanvase, B., & Deosarkar, M. (2013). Removal of nickel from waste water using graphene nanocomposite. *International Journal of Chemical and Physical Sciences*, *2*, 132–139.

Venkatachalam, N., Palanichamy, M., & Murugesan, V. (2007). Sol–gel preparation and characterization of alkaline earth metal doped nano TiO 2: Efficient photocatalytic degradation of 4-chlorophenol. *Journal of Molecular Catalysis A Chemical*, *273*(1), 177–185. doi:10.1016/j.molcata.2007.03.077

Venkatesham, V., Madhu, G., Satyanarayana, S., & Preetham, H. (2013). Adsorption of lead on gel combustion derived nano ZnO. *Procedia Engineering*, *51*, 308–313. doi:10.1016/j.proeng.2013.01.041

Verma, A. K., Dash, R. R., & Bhunia, P. (2012). A review on chemical coagulation/flocculation technologies for removal of colour from textile wastewaters. *Journal of Environmental Management*, *93*(1), 154–168. doi:10.1016/j.jenvman.2011.09.012 PMID:22054582

Verones, B., & Oortgiesen, M. (2001). Neurogenic inflammation and particulate matter (PM) air pollutants. *Neurotoxicology*, *22*(6), 795–810. doi:10.1016/S0161-813X(01)00062-6 PMID:11829413

Vidal, M., Domínguez, J., & Luís, A. (2011). Spatial and temporal patterns of polycyclic aromatic hydrocarbons (PAHs) in eggs of a coastal bird from northwestern Iberia after a major oil spill. *The Science of the Total Environment*, *409*(13), 2668–2673. doi:10.1016/j.scitotenv.2011.03.025 PMID:21482431

Vijayaraghavan, K., Jegan, J., Palanivelu, K., & Velan, M. (2005). Biosorption of cobalt (II) and nickel (II) by seaweeds: Batch and column studies. *Separation and Purification Technology*, *44*(1), 53–59. doi:10.1016/j.seppur.2004.12.003

Vimala, K., Samba Sivudu, K., Murali Mohan, Y., Sreedhar, B., & Mohana Raju, K. (2009). Controlled silver nanoparticles synthesis in semi-hydrogel networks of poly(acrylamide) and carbohydrates: A rational methodology for antibacterial application. *Carbohydrate Polymers*, *75*(3), 463–471. doi:10.1016/j.carbpol.2008.08.009

Vinhal, J. O., Lage, M. R., Carneiro, J. W. M., Lima, C. F., & Cassella, R. J. (2015). Modeling, kinetic, and equilibrium characterization of paraquat adsorption onto polyurethane foamusing the ion-pairing technique. *Journal of Environmental Management*, *156*, 200–208. doi:10.1016/j.jenvman.2015.03.022 PMID:25846000

Vinu, A., Hossain, K. Z., & Ariga, K. (2005). Recent advances in functionalization of mesoporous silica. *Journal of Nanoscience and Nanotechnology*, *5*(3), 347–371. doi:10.1166/jnn.2005.089 PMID:15913241

Vossmeyer, T. B., Guse, I., Besnard, R. E., Bauer, K., Müllen, A., & Yasuda, A. (2002). Gold nanoparticle/polyphenylene dendrimer composite films: Preparation and vapor sensing properties. *Advanced Materials*, *14*(3), 238–242. doi:10.1002/1521-4095(20020205)14:3<238::AID-ADMA238>3.0.CO;2-#

Wainright, J. S., Wang, J. T., Weng, D., Savinell, R. F., & Litt, M. (1995). Acid doped polybenzimidazoles a new polymer electrolyte. *Journal of the Electrochemical Society*, *142*(7), L121–L123. doi:10.1149/1.2044337

Walcarius, A., Etienne, M., Sayen, S., & Lebeau, B. (2003). Grafted silicas in electroanalysis: Amorphous versus ordered mesoporous materials. *Electroanalysis*, *15*(5-6), 414–421. doi:10.1002/elan.200390048

Wang, Q., Xiong, Y., & Lou, L. (2015). Biosensor for detection of pesticide residue. *American Journal of Nano Research and Application*, *3*(1), 18-22.

Wang, A. A., Li, S. X., Teng, Y., Liu, W. X., Wu, L. H., Zhang, H. B., ... Christie, P. (2013). Adsorption and desorption characteristics of diphenylarsenicals in two contrasting soils. *Journal of Environmental Sciences (China)*, *25*(6), 1172–1179. doi:10.1016/S1001-0742(12)60148-X PMID:24191607

Wang, C., Luo, H., Zhang, Z., Wu, Y., Zhang, J., & Chen, S. (2014). Removal of As(III) and As(V) from aqueous solutions using nanoscale zero valent iron-reduced graphite oxide modified composites. *Journal of Hazardous Materials*, *268*, 124–131. doi:10.1016/j.jhazmat.2014.01.009 PMID:24480523

Wang, H., Dong, Z., & Na, C. (2013a). Hierarchical carbon nanotube membrane-supported gold nanoparticles for rapid catalytic reduction of p-nitrophenol. *ACS Sustainable Chemistry & Engineering*, *1*(7), 746–752. doi:10.1021c400048m

Wang, H., Wang, Y., Chen, J., Wang, S., Cheng, J., & Zhou, D. (2009). Application of modified nano-particle black carbon for the remediation of soil heavy metal pollution. *Zhongguo Huanjing Kexue*, *29*(4), 431–436.

Wang, H., Yuan, X., Wu, Y., Zeng, G., Chen, X., Leng, L., ... Li, H. (2015). Facile synthesis of amino-functionalized titanium metal-organic frameworks and their superior visible-light photocatalytic activity for Cr (VI) reduction. *Journal of Hazardous Materials*, *286*, 187–194. doi:10.1016/j.jhazmat.2014.11.039 PMID:25585267

Wang, L., Li, J., Jiang, Q., & Zhao, L. (2012). Water-soluble Fe 3 O 4 nanoparticles with high solubility for removal of heavy-metal ions from waste water. *Dalton Transactions (Cambridge, England)*, *41*(15), 4544–4551. doi:10.1039/c2dt11827k PMID:22358186

Wang, L., Wu, X. L., Xu, W. H., Huang, X. J., Liu, J. H., & Xu, A. W. (2012). Stable organic-inorganic hybrid of polyaniline/α-zirconium phosphate for efficient removal of organic pollutants in water environment. *ACS Applied Materials & Interfaces*, *4*(5), 2686–2692. doi:10.1021/am300335e PMID:22545781

Wang, L., Zhang, J., & Wang, A. (2011). Fast removal of methylene blue from aqueous solution by adsorption onto chitosan-g-poly (acrylic acid)/attapulgite composite. *Desalination*, *266*(1-3), 33–39. doi:10.1016/j.desal.2010.07.065

Wang, P., Menzies, N. W., Lombi, E., McKenna, B. A., Johannessen, B., Glover, C. J., ... Kopittke, P. M. (2013). Fate of ZnO nanoparticles in soils and cowpea (Vigna unguiculata). *Environmental Science & Technology*, *47*(23), 13822–13830. doi:10.1021/es403466p PMID:24195448

Wang, S. H., Chou, T. C., & Liu, C. C. (2003). Nano-crystalline tungsten oxide NO_2 sensor. *Sensors and Actuators. B, Chemical*, *94*(3), 343–351. doi:10.1016/S0925-4005(03)00383-6

Wang, S., Gao, B., Li, Y., Creamer, A. E., & He, F. (2016). Adsorptive removal of arsenate from aqueous solutions by biochar supported zero-valent iron nanocomposite: Batch and continuous flow tests. *Journal of Hazardous Materials*. doi:10.1016/j.jhazmat.2016.01.052 PMID:26852252

Wang, S., Ng, C. W., Wang, W., Li, Q., & Hao, Z. (2012a). Synergistic and competitive adsorption of organic dyes on multiwalled carbon nanotubes. *Chemical Engineering Journal*, *197*, 34–40. doi:10.1016/j.cej.2012.05.008

Wang, S., Ng, C. W., Wang, W., Li, Q., & Li, L. (2012b). A comparative study on the adsorption of acid and reactive dyes on multiwall carbon nanotubes in single and binary dye systems. *Journal of Chemical & Engineering Data*, *57*(5), 1563–1569. doi:10.1021/je3001552

Wang, S., Sun, H., Ang, H. M., & Tadé, M. O. (2013). Adsorptive remediation of environmental pollutants using novel graphene-based nanomaterials. *Chemical Engineering Journal*, *226*, 336–347. doi:10.1016/j.cej.2013.04.070

Wang, T., Jin, X., Chen, Z., Megharaj, M., & Naidu, R. (2014). Simultaneous removal of Pb(II) and Cr(III) by magnetite nanoparticles using various synthesis conditions. *Journal of Industrial and Engineering Chemistry*, *20*(5), 3543–3549. doi:10.1016/j.jiec.2013.12.047

Wang, T., Lin, J., Chen, Z., Megharaj, M., & Naidu, R. (2014). Green synthesized iron nanoparticles by green tea and eucalyptus leaves extracts used for removal of nitrate in aqueous solution. *Journal of Cleaner Production*, *83*, 413–419. doi:10.1016/j.jclepro.2014.07.006

Wang, W., & Gu, B. (2005). Self-assembly of Two- and Three-Dimensional Particle Arrays by Manipulating Hydrophobicity of Silica Nanoparticles. *The Journal of Physical Chemistry B*, *109*(47), 22175–22180. doi:10.1021/jp053692s PMID:16853885

Wang, W., Gu, B., & Liang, L. (2007). Effect of anionic surfactants on synthesis and self-assembling of silica colloidal nanoparticles. *Journal of Colloid and Interface Science*, *313*(1), 169–173. doi:10.1016/j.jcis.2007.04.042 PMID:17512000

Wang, W., Gu, B., Liang, L. Y., & Hamilton, W. (2003, April). (2003-b), Fabrication of Two and Three-Dimensional Silica Nanocolloidal Particle Arrays. *The Journal of Physical Chemistry B*, *107*(15), 3400–3404. doi:10.1021/jp0221800

Wang, W., Gu, B., Liang, L. Y., & Hamilton, W. (2003, November). (2003-a), Fabrication of Near Infrared Photonic Crystals using Highly Monodispersed Submicrometer SiO_2 Spheres. *The Journal of Physical Chemistry B*, *107*(44), 12113–12117. doi:10.1021/jp034161b

Wang, W., Qiao, X., Chen, J., Tan, F., & Li, H. (2009). Influence of titanium doping on the structure and morphology of MgO prepared by coprecipitation method. *Materials Characterization*, *60*(8), 858–862. doi:10.1016/j.matchar.2009.02.002

Wang, X., Cai, W., Liu, S., Wang, G., Wu, Z., & Zhao, H. (2013b). ZnO hollow microspheres with exposed porous nanosheets surface: Structurally enhanced adsorption towards heavy metal ions. *Colloids and Surfaces. A, Physicochemical and Engineering Aspects*, *422*, 199–205. doi:10.1016/j.colsurfa.2013.01.031

Wang, X., Chan, J. C. C., Tseng, Y. H., & Cheng, S. (2006). Synthesis, characterization and catalytic activity of ordered SBA-15 materials containing high loading of diamine functional groups. *Microporous and Mesoporous Materials*, *95*(1-3), 57–65. doi:10.1016/j.micromeso.2006.05.003

Wang, X., Guo, J., Chen, T., Nie, H., Wang, H., Zang, J., ... Jia, G. (2012). Multi-walled carbon nanotubes induce apoptosis via mitochondrial pathway and scavenger receptor. *Toxicology In Vitro*, *26*(6), 799–806. doi:10.1016/j.tiv.2012.05.010 PMID:22664788

Wang, X., Guo, Y., Yang, L., Han, M., Zhao, J., & Cheng, X. (2012). Nanomaterials as sorbents to remove heavy metal ions in wastewater treatment. *Journal of Environmental & Analytical Toxicology*, *2*(7), 1000154. doi:10.4172/2161-0525.1000154

Wang, X., Han, H., Liu, X., Gu, X., Chen, K., & Lu, D. (2012). Multi-walled carbon nanotubes can enhance root elongation of wheat (Triticum aestivum) plants. *Journal of Nanoparticle Research*, *14*(6), 841. doi:10.100711051-012-0841-5

Wang, X., & Xing, B. (2007). Sorption of organic contaminants by biopolymer-derived chars. *Environmental Science & Technology*, *41*(24), 8342–8348. doi:10.1021/es071290n PMID:18200861

Wang, Y., & Xia, Y. (2004). Bottom-up and top-down approaches to the synthesis of monodispersed spherical colloids of low melting-point metals. *Nano Letters*, *4*(10), 2047–2050. doi:10.1021/nl048689j

Wan, Y., & Zhao, D. Y. (2007). On the Controllable Soft-Templating Approach to Mesoporous Silicates. *Chemical Reviews*, *107*(7), 2821–2860. doi:10.1021/cr068020s PMID:17580976

Washmon-Kriel, L., Jimenez, V. L., & Balkus, K. J. Jr. (2000). Cytochrome c immobilization into meoporous molecular sieves. *Journal of Molecular Catalysis. B, Enzymatic*, *10*(5), 453–469. doi:10.1016/S1381-1177(99)00123-X

Watarai, H. (2006). *Trend of Research and Development for Magnesium Alloys-Reducing the Weight of Structural Materialsin Motor Vehicles*. NISTEP Science & Technology Foresight Center.

Wee, J. H. (2007). Feasibility study on direct methanol fuel cells for laptop, computers based on a cost comparison with lithium ion batteries. *Journal of Power Sources*, *173*(1), 424–436. doi:10.1016/j.jpowsour.2007.04.084

Wei, H., Deng, S., Huang, Q., Nie, Y., Wang, B., Huang, J., & Yu, G. (2013). Regenerable granular carbon nanotubes/alumina hybrid adsorbents for diclofenac sodium and carbamazepine removal from aqueous solution. *Water Research*, *47*(12), 4139–4147. doi:10.1016/j.watres.2012.11.062 PMID:23579087

Weiner, R. G., & Skrabalak, S. E. (2015). Metal Dendrimers: Synthesis of Hierarchically Stellated Nanocrystals by Sequential Seed-Directed Overgrowth. *Angewandte Chemie International Edition*, *54*(4), 1181–1184. doi:10.1002/anie.201409966 PMID:25424302

Wei, Q., Yang, D., Fan, M., & Harris, H. G. (2013). Applications of nanomaterial-based membranes in pollution control. *Critical Reviews in Environmental Science and Technology*, *43*(22), 2389–2438. doi:10.1080/10643389.2012.672066

Wei, Y., Chen, S., Kowalczyk, B., Huda, S., Gray, T. P., & Grzybowski, B. A. (2010). Synthesis of Stable, Low-Dispersity Copper Nanoparticles and Nanorods and Their Antifungal and Catalytic Properties. *The Journal of Physical Chemistry C*, *114*(37), 15612–15616. doi:10.1021/jp1055683

Wei, Y., Feng, Q., Xu, J., Dong, H., Qiu, K.-Y., Jansen, S. A., ... Ong, K. (2000). Polymethacrylate-Silica Hybrid Nanoporous Materials: A Bridge between Inorganic and Polymeric Molecular Sieves. *Advanced Materials*, *12*(19), 1448–1450. doi:10.1002/1521-4095(200010)12:19<1448::AID-ADMA1448>3.0.CO;2-W

Wen, Z., Zhang, Y., & Dai, C. (2014). Removal of phosphate from aqueous solution using nanoscale zerovalent iron (nZVI). *Colloids and Surfaces. A, Physicochemical and Engineering Aspects*, *457*, 433–440. doi:10.1016/j.colsurfa.2014.06.017

Weon, S., & Choi, W. (2016). TiO_2 Nanotubes with Open Channels as Deactivation-Resistant Photocatalyst for the Degradation of Volatile Organic Compounds. *Environmental Science & Technology*, *50*(5), 2556–2563. doi:10.1021/acs.est.5b05418 PMID:26854616

Wheeler, S. A., Zuo, A., & Loch, A. (2015). Watering the farm: Comparing organic and conventional irrigation water use in the Murray–Darling Basin, Australia. *Ecological Economics*, *112*, 78–85. doi:10.1016/j.ecolecon.2015.02.019

WHO/UNICEF. (2012). Joint Monitoring Programme for Water Supply and Sanitation. Progress on Sanitation and Drinking-water. (NLM classification: WA 670)

Wickramasinghe, H. K., Martin, Y., & Williams, C. C. (1987). Article . *Journal of Applied Physics*, *61*, 4723. doi:10.1063/1.338807

Wight, A. P., & Davis, M. E. (2002). Design and Preparation of Organic−Inorganic Hybrid Catalysts. *Chemical Reviews*, *102*(10), 3589–3614. doi:10.1021/cr010334m PMID:12371895

Wikipedia. (2018a). *Differential Scanning Calorimetry*. Retrieved from https://en.wikipedia.org/wiki/Differential_scanning_calorimetry

Wikipedia. (2018b). *Vibrating Sample Magnetometer*. Retrieved from https://en.wikipedia.org/wiki/Vibrating_sample_magnetometer

Wikipedia. (2018c). *Ultraviolet Visible Spectroscopy*. Retrieved from https://en.wikipedia.org/wiki/Ultraviolet%E2%80%93visible_spectroscopy

Williams, T. D., Gensberg, K., Minchin, S. D., & Chipman, J. K. (2003). A DNA expression array to detect toxic stress response in European flounder (*Platichthys flesus*). *Aquatic Toxicology (Amsterdam, Netherlands)*, *65*(2), 141–157. doi:10.1016/S0166-445X(03)00119-X PMID:12946615

Woehrle, G. H., Hutchison, J. E., Ozkar, S., & Finke, R. G. (2006). Article . *Turkish Journal of Chemistry*, *30*, 1–13.

Worch, E. (2012). *Adsorption Technology in Water Treatment, Fundamentals, Processes, and Modeling.* Gottingen: GmbH & Co. KG. doi:10.1515/9783110240238

World Economic Forum. (2015). *Global Risks. 10.* Retrieved from http://www3.weforum.org/docs/WEF_Global_Risks_2015_Report15.pdf

World Health Organization (WHO) and UNICEF. (2013). *Progress on Sanitation and Drinking-Water.* Geneva, Switzerland: World Health Organization.

World Health Organization. (2015). *Drinking-Water-Fact Sheet No. 391.* WHO.

Wu, Yu, Liu, Dionysiou, Yang, Wang, & Liu. (2017). *Novel Fluorinated Bi_2MoO_6 Nanocrystals for Efficient Photocatalytic Removal of Water Organic Pollutants under Different Light Source Illumination.* Academic Press.

Wu, B., Lin, X., Ge, L., Wu, L., & Xu, T. (2013). A novel route for preparing highly proton conductive membrane materials with metal-organic frameworks. *Chemical Communications*, *49*(2), 143–145. doi:10.1039/C2CC37045J PMID:23108195

Wu, P., Li, S., Ju, L., Zhu, N., Wu, J., Li, P., & Dang, Z. (2012). Mechanism of the reduction of hexavalent chromium by organo-montmorillonite supported iron nanoparticles. *Journal of Hazardous Materials*, *219*, 283–288. doi:10.1016/j.jhazmat.2012.04.008 PMID:22521796

Wu, S., Deng, F., Wei, H., Huang, J., Wang, X., Hao, Y., & Guo, X. (2014). Association of cardiopulmonary health effects with source-appointed ambient fine particulate in Beijing, China: A combined analysis from the Healthy Volunteer Natural Relocation (HVNR) study. *Environmental Science & Technology*, *48*(6), 3438–3448. doi:10.1021/es404778w PMID:24521469

Wu, T., Cai, X., Tan, S., Li, H., Liu, J., & Yang, W. (2011). Adsorption characteristics of acrylonitrile, p-toluenesulfonic acid, 1-naphthalenesulfonic acid and methyl blue on graphene in aqueous solutions. *Chemical Engineering Journal*, *173*(1), 144–149. doi:10.1016/j.cej.2011.07.050

Wu, Y., Yi, N., Huang, L., Zhang, T., Fang, S., Chang, H., ... Chen, Y. (2015). Three-dimensionally bonded spongy graphene material with super compressive elasticity and near-zero Poisson's ratio. *Nature Communications*, *6*(1), 6141–6149. doi:10.1038/ncomms7141 PMID:25601131

Wuyts, N., De Waele, D., & Swennen, R. (2006). Extraction and partial characterization of polyphenol oxidase from banana (Musa acuminata Grande naine) roots. *Plant Physiology and Biochemistry*, *44*(5-6), 308–314. doi:10.1016/j.plaphy.2006.06.005 PMID:16814556

Wu, Z. J., Han, H., Han, W., Kim, B., Ahn, K. H., & Lee, K. (2007). Controlling the hydrophobicity of submicrometer silica spheres via surface modification for nanocomposite applications. *Langmuir*, *23*(14), 7799–7803. doi:10.1021/la700386z PMID:17559243

Wycisk, R., Chisholm, J., Lee, J., Lin, J., & Pintauro, P. N. (2006). Direct methanol fuel cell membranes from Nafion–polybenzimidazole blends. *Journal of Power Sources*, *163*(1), 9–17. doi:10.1016/j.jpowsour.2005.11.056

Xiao, Q., Sun, Y., Zhang, J., & Li, Q. (2015). Size-dependent of chromium (VI) adsorption on nano α-Fe2O3 surface. *Applied Surface Science*, *356*, 18–23. doi:10.1016/j.apsusc.2015.08.005

Xia, S., Zhang, L., Pan, G., Qian, P., & Ni, Z. (2015). Photocatalytic degradation of methylene blue with a nanocomposite system: Synthesis, photocatalysis and degradation pathways. *Physical Chemistry Chemical Physics*, *17*(7), 5345–5351. doi:10.1039/C4CP03877K PMID:25611297

Xia, T., Kovochich, M., Liong, M., Mädler, L., Gilbert, B., Shi, H., ... Nel, A. E. (2008). Comparison of the Mechanism of Toxicity of Zinc Oxide and Cerium Oxide Nanoparticles Based on Dissolution and Oxidative Stress Properties. *ACS Nano*, *2*(10), 2121–2134. doi:10.1021/nn800511k PMID:19206459

Xie, J., Lin, Y., Li, C., Wu, D., & Kong, H. (2015). Removal and recovery of phosphate from water by activated aluminum oxide and lanthanum oxide. *Powder Technology*, *269*, 351–357. doi:10.1016/j.powtec.2014.09.024

Xie, J., Wang, Z., Fang, D., Li, C., & Wu, D. (2014). Green synthesis of a novel hybrid sorbent of zeolite/lanthanum hydroxide and its application in the removal and recovery of phosphate from water. *Journal of Colloid and Interface Science*, *423*, 13–19. doi:10.1016/j.jcis.2014.02.020 PMID:24703662

Xie, J., Wang, Z., Wu, D., & Kong, H. (2014). Synthesis and properties of zeolite/hydrated iron oxide composite from coal fly ash as efficient adsorbent to simultaneously retain cationic and anionic pollutants from water. *Fuel*, *116*, 71–76. doi:10.1016/j.fuel.2013.07.126

Xie, Y., Yan, B., Xu, H., Chen, J., Liu, Q., Deng, Y., & Zeng, H. (2014). Highly regenerable mussel-inspired Fe(3)O(4)@ polydopamine-Ag core-shell microspheres as catalyst and adsorbent for methylene blue removal. *ACS Applied Materials & Interfaces*, *6*(11), 8845–8852. doi:10.1021/am501632f PMID:24787615

Xiong, J., He, Z., Mahmood, Q., Liu, D., Yang, X., & Islam, E. (2008). Phosphate removal from solution using steel slag through magnetic separation. *Journal of Hazardous Materials*, *152*(1), 211–215. doi:10.1016/j.jhazmat.2007.06.103 PMID:17703877

Xiu, Z. M., Ma, J., & Alvarez, P. J. J. (2011). Differential effect of common ligands and molecular oxygen on antimicrobial activity of silver nanoparticles versus silver ions. *Environmental Science & Technology*, *45*(20), 9003–9008. doi:10.1021/es201918f PMID:21950450

Xiu, Z. M., Zhang, Q. B., Puppala, H. L., Colvin, V. L., & Alvarez, P. J. J. (2012). Negligible particle-specific antibacterial activity of silver nanoparticles. *Nano Letters*, *12*(8), 4271–4275. doi:10.1021/nl301934w PMID:22765771

Xi, Y., Sun, Z., Hreid, T., Ayoko, G. A., & Frost, R. L. (2014). Bisphenol A degradation enhanced by air bubbles via advanced oxidation using in situ generated ferrous ions from nano zero-valent iron/palygorskite composite materials. *Chemical Engineering Journal*, *247*, 66–74. doi:10.1016/j.cej.2014.02.077

Xu, K., Deng, T., Liu, J., & Peng, W. (2010). Study on the phosphate removal from aqueous solution using modified fly ash. *Fuel*, *89*(12), 3668–3674. doi:10.1016/j.fuel.2010.07.034

Xu, K., Deng, T., Liu, J., & Peng, W. (2012). Phosphate Removal from Digested Sludge Supernatant Using Modified Fly Ash. *Water Environment Research*, *84*(5), 411–416. doi:10.2175/106143012X13347678384323 PMID:22852426

Xu, M. F., Lin, S., Chen, X. M., & Peng, Y. Z. (2006). Studies on characteristics of nanostructure of N-TiO_2 thin films and photo-bactericidal action. *Journal of Zhejiang University. Science. B.*, *7*(7), 586–590. doi:10.1631/jzus.2006.B0586 PMID:16773734

Xu, M., Wang, H., Lei, D., Qu, D., Zhai, Y., & Wang, Y. (2013). Removal of Pb (II) from aqueous solution by hydrous manganese dioxide: Adsorption behavior and mechanism. *Journal of Environmental Sciences (China)*, *25*(3), 479–486. doi:10.1016/S1001-0742(12)60100-4 PMID:23923420

Xu, P., Zeng, G. M., Huang, D. L., Feng, C. L., Hu, S., Zhao, M. H., ... Xie, G. X. (2012). Use of iron oxide nanomaterials in wastewater treatment: A review. *The Science of the Total Environment*, *424*, 1–10. doi:10.1016/j.scitotenv.2012.02.023 PMID:22391097

Xu, S., Lv, Y., Zeng, X., & Cao, D. (2017). ZIF-Derived Nitrogen-doped porous carbons as highly efficient adsorbents for removal of organic compounds from wastewater. *Chemical Engineering Journal*, *323*, 502–511. doi:10.1016/j.cej.2017.04.093

Xu, W., Liu, C., Xue, X., Su, Y., Lv, Y., Xing, W., & Lu, T. (2004). New proton exchange membranes based on poly(vinyl alcohol) for DMFCs. *Solid State Ionics*, *171*(1-2), 121–127. doi:10.1016/j.ssi.2004.04.009

Xu, X., Gao, B., Jin, B., & Yue, Q. (2016). Removal of anionic pollutants from liquids by biomass materials: A review. *Journal of Molecular Liquids*, *215*, 565–595. doi:10.1016/j.molliq.2015.12.101

Xu, X., Lin, L., Papelis, C., Myint, M., Cath, T. Y., & Xu, P. (2015). Use of drinking water treatment solids for arsenate removal from desalination concentrate. *Journal of Colloid and Interface Science*, *445*, 252–261. doi:10.1016/j.jcis.2014.12.090 PMID:25622050

Xu, Y., Li, C., Zhu, X., Huang, W. E., & Zhang, D. (2014). Application of magnetic nanoparticles in drinking water purification. *Environmental Engineering and Management Journal*, *13*(8).

Yamanaka, M., Hara, K., & Kudo, J. (2005). Bactericidal actions of a silver ion solution on *Escherichia coli*, studied by energy filtering transmission electron microscopy and proteomic analysis. *Applied and Environmental Microbiology*, *71*(11), 7589–7593. doi:10.1128/AEM.71.11.7589-7593.2005 PMID:16269810

Yamashita, K., Yoshioka, Y., Higashisaka, K., Mimura, K., Morishita, Y., Nozaki, M., . . . Tsutsumi, Y. (2011). *Silica and titanium dioxide nanoparticles cause pregnancy complications in mice*. Retrieved from https://www.nature.com/articles/nnano.2011.41#supplementary-information

Yang, B., Geng, P., & Chen, G. (2015). One-dimensional structured IrO 2 nanorods modified membrane for electrochemical anti-fouling in filtration of oily wastewater. *Separation and Purification Technology*, *156*, 931–941. doi:10.1016/j.seppur.2015.10.040

Yang, E.-J., Kim, S., Kim, J. S., & Choi, I.-H. (2012). Inflammasome formation and IL-1β release by human blood monocytes in response to silver nanoparticles. *Biomaterials*, *33*(28), 6858–6867. doi:10.1016/j.biomaterials.2012.06.016 PMID:22770526

Yang, L., Tang, B., & Wu, P. (2015). Metal–organic framework–graphene oxide composites: A facile method to highly improve the proton conductivity of PEMs operated under low humidity. *Journal of Materials Chemistry. A, Materials for Energy and Sustainability*, *3*(31), 15838–15842. doi:10.1039/C5TA03507D

Yang, Q., Sha, J., Wang, L., Wang, J., & Yang, D. (2006). MgO nanostructures synthesized by thermal evaporation. *Materials Science and Engineering C*, *26*(5-7), 1097–1101. doi:10.1016/j.msec.2005.09.082

Yang, S.-T., Chang, Y., Wang, H., Liu, G., Chen, S., Wang, Y., ... Cao, A. (2010). Folding/aggregation of graphene oxide and its application in Cu^{2+} removal. *Journal of Colloid and Interface Science*, *351*(1), 122–127. doi:10.1016/j.jcis.2010.07.042 PMID:20705296

Yang, W., Ding, P., Zhou, L., Yu, J., Chen, X., & Jiao, F. (2013). Preparation of diamine modified mesoporous silica on multi-walled carbon nanotubes for the adsorption of heavy metals in aqueous solution. *Applied Surface Science*, *282*, 38–45. doi:10.1016/j.apsusc.2013.05.028

Yang, X., Cao, C., Erickson, L., Hohn, K., Maghirang, R., & Klabunde, K. (2008). Synthesis of visible-light-active TiO_2-based photocatalysts by carbon and nitrogen doping. *Journal of Catalysis*, *260*(1), 128–133. doi:10.1016/j.jcat.2008.09.016

Yang, Y., Ma, J., Qin, Q., & Zhai, X. (2007). Degradation of nitrobenzene by nano-TiO 2 catalyzed ozonation. *Journal of Molecular Catalysis A Chemical*, *267*(1), 41–48. doi:10.1016/j.molcata.2006.09.010

Yang, Z., Ye, Z., Xu, Z., & zhao, B. (2009). Effect of the morphology on the optical properties of ZnO nanostructures. *Physica E, Low-Dimensional Systems and Nanostructures*, *42*(2), 116–119. doi:10.1016/j.physe.2009.09.010

Yan, H., Yang, L., Yang, Z., Yang, H., Li, A., & Cheng, R. (2012). Preparation of chitosan/poly (acrylic acid) magnetic composite microspheres and applications in the removal of copper (II) ions from aqueous solutions. *Journal of Hazardous Materials*, *229*, 371–380. doi:10.1016/j.jhazmat.2012.06.014 PMID:22749139

Yao, N., & Yeung, K. L. (2011). Investigation of the performance of TiO_2 photocatalytic coatings. *Chemical Engineering Journal*, *167*(1), 13–21. doi:10.1016/j.cej.2010.11.061

Yeo, Z. Y., Chai, S. P., Zhu, P. W., & Mohamed, A. R. (2014). An overview: Synthesis of thin films/membranes of metal organic frameworks and its gas separation performances. *RSC Advances*, *4*(97), 54322–54334. doi:10.1039/C4RA08884K

Ye, Q., Zhou, F., & Liu, W. (2011). Bioinspired catecholic chemistry for surface modification. *Chemical Society Reviews*, *40*(7), 4244–4258. doi:10.1039/c1cs15026j PMID:21603689

Yin, J., & Deng, B. (2015). Polymer-matrix nanocomposite membranes for water treatment. *Journal of Membrane Science*, *479*, 256–275. doi:10.1016/j.memsci.2014.11.019

Yin, J., Zhu, G., & Deng, B. (2013). Multi-walled carbon nanotubes (MWNTs)/polysulfone (PSU) mixed matrix hollow fiber membranes for enhanced water treatment. *Journal of Membrane Science*, *437*, 237–248. doi:10.1016/j.memsci.2013.03.021

Yi, X., Wenzhong, W., Yitai, Q., Li, Y., & Zhiwen, C. (1996). Deposition and microstructural characterization of MgO thin films by a spray pyrolysis method. *Surface and Coatings Technology*, *82*(3), 291–293. doi:10.1016/0257-8972(95)02762-9

Yoshitake, H. (2005). Highly-controlled synthesis of organic layers on mesoporous silica: Their structure and application to toxic ion adsorptions. *New Journal of Chemistry*, *29*(9), 1107–1117. doi:10.1039/b504957a

You, X., Farran, A., Guaya, D., Valderrama, C., Soldatov, V., & Cortina, J. L. (2016). Phosphate removal from aqueous solutions using a hybrid fibrous exchanger containing hydrated ferric oxide nanoparticles. *Journal of Environmental Chemical Engineering*, *4*(1), 388–397. doi:10.1016/j.jece.2015.11.032

Yu, C., Yu, Y., & Zhao, D. (2000). Highly ordered large caged cubic mesoporous silica structures template by triblock PEO–PBO–PEO copolymer. *Journal of Chemical Communications*, (7), 575-576.

Yuan, P., Fan, M., Yang, D., He, H., Liu, D., Yuan, A., ... Chen, T. (2009). Montmorillonite-supported magnetite nanoparticles for the removal of hexavalent chromium [Cr(VI)] from aqueous solutions. *Journal of Hazardous Materials*, *166*(2), 821–829. doi:10.1016/j.jhazmat.2008.11.083 PMID:19135796

Yuan, P., Liu, D., Fan, M., Yang, D., Zhu, R., Ge, F., ... He, H. (2010). Removal of hexavalent chromium [Cr(VI)] from aqueous solutions by the diatomite-supported/unsupported magnetite nanoparticles. *Journal of Hazardous Materials*, *173*(1), 614–621. doi:10.1016/j.jhazmat.2009.08.129 PMID:19748178

Yu, B., Zeng, J., Gong, L., Zhang, M., Zhang, L., & Chen, X. (2007). Investigation of the photocatalytic degradation of organochlorine pesticides on a nano-TiO 2 coated film. *Talanta*, *72*(5), 1667–1674. doi:10.1016/j.talanta.2007.03.013 PMID:19071814

Yu, C. L., Fan, Q. Z., Xie, Y., Chen, J. C., Shu, Q., & Yu, J. C. (2012). Sonochemical fabrication of novel square-shaped F doped TiO2 nanocrystals with enhanced performance in photocatalytic degradation of phenol. *Journal of Hazardous Materials*, *237–238*, 38–45. doi:10.1016/j.jhazmat.2012.07.072 PMID:22947181

Yu, C. L., Wu, Z., Liu, R. Y., He, H. B., Fan, W. H., & Xue, S. S. (2016). The effects of Gd3þ doping on the physical structure and photocatalytic performance of Bi2MoO6 nanoplate crystals. *Journal of Physics and Chemistry of Solids, 93*, 7–13. doi:10.1016/j.jpcs.2016.02.008

Yu, C., Wu, Z., Liu, R., Dionysiou, D. D., Yang, K., Wang, C., & Liu, H. (2017). Novel Fluorinated Bi_2MoO_6 nanocrystals for efficient photocatalytic removal of water organic pollutants under different light source illumination. *Applied Catalysis B: Environmental, 209*, 1–11. doi:10.1016/j.apcatb.2017.02.057

Yu, F., Chen, J., Chen, L., Huai, J., Gong, W., Yuan, Z., ... Ma, J. (2012). Magnetic carbon nanotubes synthesis by Fenton's reagent method and their potential application for removal of azo dye from aqueous solution. *Journal of Colloid and Interface Science, 378*(1), 175–183. doi:10.1016/j.jcis.2012.04.024 PMID:22564767

Yu, H., Zhang, K., & Rossi, C. (2007). Experimental study of the photocatalytic degradation of formaldehyde in indoor air using a nano-particulate titanium dioxide photocatalyst. *Indoor and Built Environment, 16*(6), 529–537. doi:10.1177/1420326X07083513

Yu, J. C., Lin, J., & Kwok, R. W. (1998). Ti1-x Zr x O2 Solid Solutions for the Photocatalytic Degradation of Acetone in Air. *The Journal of Physical Chemistry B, 102*(26), 5094–5098. doi:10.1021/jp980332e

Yu, J.-G., Zhao, X.-H., Yang, H., Chen, X.-H., Yang, Q., Yu, L.-Y., ... Chen, X.-Q. (2014). Aqueous adsorption and removal of organic contaminants by carbon nanotubes. *The Science of the Total Environment, 482*, 241–251. doi:10.1016/j.scitotenv.2014.02.129 PMID:24657369

Yuming, D., Kun, H., Lin, Y., & Aimin, Z. (2007). A facile route to controlled synthesis of Co_3O_4 nanoparticles and their environmental catalytic properties. *Nanotechnology, 18*(43), 435602. doi:10.1088/0957-4484/18/43/435602

Yuryev, M., Ferreira, M. P., Balasubramanian, V., Correia, A. M., Makila, E. M., Jokinen, V., ... Rivera, C. (2016). Active diffusion of nanoparticles of maternal origin within the embryonic brain. *Nanomedicine (London), 11*(19), 2471–2481. doi:10.2217/nnm-2016-0207 PMID:27622305

Yuvakkumar, R., Suresh, J., Nathanael, A. J., Sundrarajan, M., & Hong, S. I. (2014). Novel green synthetic strategy to prepare ZnO nanocrystals using rambutan (Nephelium lappaceum L.) peel extract and its antibacterial applications. *Materials Science and Engineering C, 41*, 17–27. doi:10.1016/j.msec.2014.04.025 PMID:24907732

Yu, W., Zhan, S., Shen, Z., Zhou, Q., & Yang, D. (2017). Efficient removal mechanism for antibiotic resistance genes from aquatic environments by graphene oxide nanosheet. *Chemical Engineering Journal, 313*, 836–846. doi:10.1016/j.cej.2016.10.107

Zäch, M., Hägglund, C., Chakarov, D., & Kasemo, B. (2006). Nanoscience and nanotechnology for advanced energy systems. *Current Opinion in Solid State and Materials Science, 10*(3), 132–143. doi:10.1016/j.cossms.2007.04.004

Zaitseva, N., Zaitsev, V., & Walcarius, A. (2013). Chromium(VI) removal via reduction–sorption on bi-functional silica adsorbents. *Journal of Hazardous Materials, 250–251*, 454–461. doi:10.1016/j.jhazmat.2013.02.019 PMID:23500426

Zakharenko, L. P., Zakharov, I. K., Vasiunina, E. A., Karamysheva, T. V., Danilenko, A. M., & Nikiforov, A. A. (1997). Determination of the genotoxicity of fullerene C60 and fullerol using the method of somatic mosaics on cells of Drosophila melanogaster wing and SOS-chromotest. *Genetika, 33*(3), 405–409. PMID:9244774

Zakrzewska-Trznadel, G. (2006). Membrane processes for environmental protection: Applications in nuclear technology. *Nukleonika, 51*, 101–111.

Zambianchi, M., Durso, M., Liscio, A., Treossi, E., Bettini, C., Capobianco, M. L., … Melucci, M. (2017). *Graphene oxide doped polysulfone membrane adsorbers for the removal of organic contaminants from water*. DOI: 10.1016/j.cej.2017.05.143

Zang, L. (2011). *Energy efficiency and renewable energy through nanotechnology*. Springer.

Zan, L., Wang, S., Fa, W., Hu, Y., Tian, L., & Deng, K. (2006). Solid-phase photocatalytic degradation of polystyrene with modified nano-TiO 2 catalyst. *Polymer*, *47*(24), 8155–8162. doi:10.1016/j.polymer.2006.09.023

Zaporotskova, I. V., Boroznina, N. P., Parkhomenko, Y. N., & Kozhitov, L. V. (2016). Carbon nanotubes: Sensor properties. A review. *Modern Electronic Materials*, *2*(4), 95–105. doi:10.1016/j.moem.2017.02.002

Zare-Dorabei, R., Ferdowsi, S. M., Barzin, A., & Tadjarodi, A. (2016). Highly efficient simultaneous ultrasonic-assisted adsorption of Pb (II), Cd (II), Ni (II) and Cu (II) ions from aqueous solutions by graphene oxide modified with 2, 2′-dipyridylamine: Central composite design optimization. *Ultrasonics Sonochemistry*, *32*, 265–276. doi:10.1016/j.ultsonch.2016.03.020 PMID:27150770

Zare, K., Najafi, F., Sadegh, H., & ghoshekandi, R. (2013). Studies of ab initio and Monte Carlo simulation on interaction of fluorouracil anticancer drug with carbon nanotube. *Journal of Nanostructure in Chemistry*, *3*(1), 71. doi:10.1186/2193-8865-3-71

Zargar, B., Parham, H., & Hatamie, A. (2009). Fast removal and recovery of amaranth by modified iron oxide magnetic nanoparticles. *Chemosphere*, *76*(4), 554–557. doi:10.1016/j.chemosphere.2009.02.065 PMID:19345980

Zayed, M. F., Eisa, W. H., & Shabaka, A. A. (2012). Malva parviflora extract assisted green synthesis of silver nanoparticles. *Spectrochimica Acta. Part A: Molecular and Biomolecular Spectroscopy*, *98*, 423–428. doi:10.1016/j.saa.2012.08.072 PMID:23010627

Zeledón-Toruño, Z. C., Lao-Luque, C., de las Heras, F. X. C., & Sole-Sardans, M. (2007). Removal of PAHs from water using an immature coal (leonardite). *Chemosphere*, *67*(3), 505–512. doi:10.1016/j.chemosphere.2006.09.047 PMID:17109923

Zeleňák, V., Badaničová, M., Halamová, D., Čejka, J., Zukal, A., Murafa, N., & Goerigk, G. (2008). Amine-modified ordered mesoporous silica: Effect of pore size on carbon dioxide capture. *Chemical Engineering Journal*, *144*(2), 336–342. doi:10.1016/j.cej.2008.07.025

Zelenay, P., & Piela, P. (2004). Researchers redefine the DMFC roadmap. *The Fuel Cell Review*, *1*, 17–23.

Zelmanov, G., & Semiat, R. (2015). The influence of competitive inorganic ions on phosphate removal from water by adsorption on iron (Fe+3) oxide/hydroxide nanoparticles-based agglomerates. *Journal of Water Process Engineering*, *5*, 143–152. doi:10.1016/j.jwpe.2014.06.008

Zeng, H., Hwang, D. S., Israelachvili, J. N., & Waite, J. H. (2010). Strong reversible Fe3+-mediated bridging between dopa-containing protein films in water. *Proceedings of the National Academy of Sciences of the United States of America*, *107*(29), 12850–12853. doi:10.1073/pnas.1007416107 PMID:20615994

Zenon Environmental, B. V. (1995). Ultrafiltration combined with a bioreactor for the treatment of tannery effluents. *World Leather*, 43-5.

Zewail, T. M., & El-Garf, S. A. M. (2010). Preparation of agriculture residue based adsorbents for heavy metal removal. *Desalination and Water Treatment*, *22*(1-3), 363–370. doi:10.5004/dwt.2010.1245

Zhang, C. L., Qiao, G. L., Zhao, F., & Wang, Y. (2011). Thermodynamic and kinetic parameters of ciprofloxacin adsorption onto modified coal fly ash from aqueous solution. *Journal of Molecular Liquids*, *163*(1), 53–56. doi:10.1016/j.molliq.2011.07.005

Zhang, D., Berry, J. P., Zhu, D., Wang, Y., Chen, Y., Jiang, B., ... Davison, P. A. (2015). Magnetic nanoparticle-mediated isolation of functional bacteria in a complex microbial community. *The ISME Journal*, *9*(3), 603–614. doi:10.1038/ismej.2014.161 PMID:25191996

Zhang, J., Zhai, S., Li, S., Xiao, Z., Song, Y., An, Q., & Tian, G. (2013). Pb (II) removal of Fe_3O_4@ SiO_2–NH_2 core–shell nanomaterials prepared via a controllable sol–gel process. *Chemical Engineering Journal*, *215*, 461–471. doi:10.1016/j.cej.2012.11.043

Zhang, L. Y., Zhang, H. Y., Guo, W., & Tian, Y. L. (2014). Removal of malachite green and crystal violet cationic dyes from aqueous solution using activated sintering process red mud. *Applied Clay Science*, *93–94*, 85–93. doi:10.1016/j.clay.2014.03.004

Zhang, L., Cheng, Z., Guo, X., Jiang, X., & Liu, R. (2014). Process optimization, kinetics and equilibrium of orange G and acid orange 7 adsorptions onto chitosan/surfactant. *Journal of Molecular Liquids*, *197*, 353–367. doi:10.1016/j.molliq.2014.06.007

Zhang, L., & Fang, M. (2010). Nanomaterials in pollution trace detection and environmental improvement. *Nano Today*, *5*(2), 128–142. doi:10.1016/j.nantod.2010.03.002

Zhang, M., He, F., Zhao, D., & Hao, X. (2011). Degradation of soil-sorbed trichloroethylene by stabilized zero valent iron nanoparticles: Effects of sorption, surfactants, and natural organic matter. *Water Research*, *45*(7), 2401–2414. doi:10.1016/j.watres.2011.01.028 PubMed

Zhang, M., He, X., Chen, L., & Zhang, Y. (2010). Preparation of IDA-Cu functionalized core–satellite Fe3O4/polydopamine/Au magnetic nanocomposites and their application for depletion of abundant protein in bovine blood. *Journal of Materials Chemistry*, *20*(47), 10696. doi:10.1039/c0jm01336f

Zhang, M., Li, J., Li, H., Li, Y., & Shen, W. (2009). Morphology-dependent redox and catalytic properties of CeO2 nanostructures: Nanowires, nanorods and nanoparticles. *Catalysis Today*, *148*(1-2), 179–183. doi:10.1016/j.cattod.2009.02.016

Zhang, Q. Y., Deng, S. D., Yu, G., & Huang, J. (2011). Removal of perfluorooctane sulfonate from aqueous solution by crosslinked chitosan beads: Sorption kinetics and uptake mechanism. *Bioresource Technology*, *102*(3), 2265–2271. doi:10.1016/j.biortech.2010.10.040 PMID:21044835

Zhang, Q., Xu, H., Zheng, S., Su, M., & Wang, J. (2015). Genotoxicity of mesoporous silica nanoparticles in human embryonic kidney 293 cells. *Drug Testing and Analysis*, *7*(9), 787–796. doi:10.1002/dta.1773 PMID:25583601

Zhang, Q., Xu, R., Xu, P., Chen, R., He, Q., Zhong, J., & Gu, X. (2014). Performance study of ZrO 2 ceramic microfiltration membranes used in pretreatment of DMF wastewater. *Desalination*, *346*, 1–8. doi:10.1016/j.desal.2014.05.006

Zhang, R., Jing, J., Tao, J., Hsu, S. C., Wang, G., Cao, J., & Shen, Z. (2013). Chemical characterization and source apportionment of PM 2.5 in Beijing: Seasonal perspective. *Atmospheric Chemistry and Physics*, *13*(14), 7053–7074. doi:10.5194/acp-13-7053-2013

Zhang, S., Niu, H., Cai, Y., Zhao, X., & Shi, Y. (2010). Arsenite and arsenate adsorption on coprecipitated bimetal oxide magnetic nanomaterials: MnFe2O4 and CoFe2O4. *Chemical Engineering Journal*, *158*(3), 599–607. doi:10.1016/j.cej.2010.02.013

Zhang, S., Zhang, Y., Bi, G., Liu, J., Wang, Z., Xu, Q., ... Li, X. (2014). Mussel-inspired polydopamine biopolymer decorated with magneticnanoparticles for multiple pollutants removal. *Journal of Hazardous Materials*, *270*, 27–34. doi:10.1016/j.jhazmat.2014.01.039 PMID:24525161

Zhang, T., Mubeen, S., Myung, N. V., & Deshusses, M. A. (2008). Recent progress in carbon nanotube-based gas sensors. *Nanotechnology*, *19*(33), 332001. doi:10.1088/0957-4484/19/33/332001 PMID:21730614

Zhang, X. B., Zhang, L., Hu, J. S., & Huang, X. H. (2016). Facile hydrothermal synthesis and improved photocatalytic activities of Zn^{2+} doped Bi_2MoO_6 nanosheets. *RSC Advances*, *6*(38), 32349–32357. doi:10.1039/C6RA06972J

Zhang, X., Yang, B., Dai, Z., & Luo, C. (2012). The gas response of hydroxyl modified SWCNTs and carboxyl modified SWCNTs to H_2S and SO_2. *Przeglad Elektrotechniczny*, *88*(7B), 311–314.

Zhang, Y., Yuan, S., Feng, X., Li, H., Zhou, J., & Wang, B. (2016). Preparation of Nanofibrous Metal–Organic Framework Filters for Efficient Air Pollution Control. *Journal of the American Chemical Society*, *138*(18), 5785–5788. doi:10.1021/jacs.6b02553 PMID:27090776

Zhang, Z., Li, M., Chen, W., Zhu, S., Liu, N., & Zhu, L. (2010). Immobilization of lead and cadmium from aqueous solution and contaminated sediment using nano-hydroxyapatite. *Environmental Pollution*, *158*(2), 514–519. doi:10.1016/j.envpol.2009.08.024 PMID:19783084

Zhao, C., Pelaez, M., Dionysiou, D. D., Pillai, S. C., Byrne, J. A., & O'Shea, K. E. (2014). UV and visible light activated TiO2 photocatalysis of 6-hydroxymethyl uracil, a model compound for the potent cyanotoxin cylindrospermopsin. *Catalysis Today*, *224*, 70–76. doi:10.1016/j.cattod.2013.09.042

Zhao, G., Li, J., Ren, X., Chen, C., & Wang, X. (2011a). Few-layered graphene oxide nanosheets as superior sorbents for heavy metal ion pollution management. *Environmental Science & Technology*, *45*(24), 10454–10462. doi:10.1021/es203439v PMID:22070750

Zhao, G., Ren, X., Gao, X., Tan, X., Li, J., Chen, C., ... Wang, X. (2011b). Removal of Pb (II) ions from aqueous solutions on few-layered graphene oxide nanosheets. *Dalton Transactions (Cambridge, England)*, *40*(41), 10945–10952. doi:10.1039/c1dt11005e PMID:21918761

Zhao, G., & Stevens, S. E. Jr. (1998). Multiple parameters for the comprehensive evaluation of the susceptibility of *Escherichia coli* to the silver ion. *Biometals*, *11*(1), 27–32. doi:10.1023/A:1009253223055 PMID:9450315

Zhao, G., Wen, T., Yang, X., Yang, S., Liao, J., Hu, J., ... Wang, X. (2012). Preconcentration of U (VI) ions on few-layered graphene oxide nanosheets from aqueous solutions. *Dalton Transactions (Cambridge, England)*, *41*(20), 6182–6188. doi:10.1039/c2dt00054g PMID:22473651

Zhao, J., & Castranova, V. (2011). Toxicology of nanomaterials used in nanomedicine. *Journal of Toxicology and Environmental Health. Part B, Critical Reviews*, *14*(8), 593–632. doi:10.1080/10937404.2011.615113 PMID:22008094

Zhao, W., Mou, Q., Zhang, X., Shi, J., Sun, S., & Zhao, C. (2013). Preparation and characterization of sulfonated polyethersulfone membranes by a facile approach. *European Polymer Journal*, *49*(3), 738–751. doi:10.1016/j.eurpolymj.2012.11.018

Zhao, X., Chen, X., & Hu, J. (2017). Composition-dependent dual halide anion-doped bismuth terephthalate hybrids for enhanced pollutants removal. *Microporous and Mesoporous Materials*, *244*, 284–290. doi:10.1016/j.micromeso.2016.11.003

Zhao, X., Qu, J. H., Liu, H. J., Qiang, Z. M., Liu, R. P., & Hu, C. Z. (2009). Photoelectrochemical degradation of anti-inflammatory pharmaceuticals at Bi2MoO6–boron-doped diamond hybrid electrode under visible light irradiation. *Applied Catalysis B: Environmental*, *91*(1-2), 539–545. doi:10.1016/j.apcatb.2009.06.025

Zhao, Y. L., & Stoddart, J. F. (2009). Noncovalent functionalization of single-walled carbon nanotubes. *Accounts of Chemical Research*, *42*(8), 1161–1171. doi:10.1021/ar900056z PMID:19462997

Zheng, S., Jiang, W. J., Cai, Y., Dionysiou, D. D., & O'Shea, K. E. (2014). Adsorption and photocatalytic degradation of aromatic organoarsenic compounds in TiO2 suspension. *Catalysis Today*, *224*, 83–88. doi:10.1016/j.cattod.2013.09.040

Zheng, Y., & Wang, A. (2012). Ag nanoparticle-entrapped hydrogel as promising material for catalytic reduction of organic dyes. *Journal of Materials Chemistry*, *22*(32), 16552. doi:10.1039/c2jm32774k

Zhong, Z., Zhai, Y., Liang, S., Mori, Y., Han, R., Sutterwala, F. S., & Qiao, L. (2013). *TRPM2 links oxidative stress to NLRP3 inflammasome activation.* Retrieved from https://www.nature.com/articles/ncomms2608#supplementary-information

Zhou, K., Wang, R., Xu, B., & Li, Y. (2006). Synthesis, characterization and catalytic properties of CuO nanocrystals with various shapes. *Nanotechnology*, *17*(15), 3939–3943. doi:10.1088/0957-4484/17/15/055

Zhou, N., Hao, Z., Zhao, X., Maharjan, S., Zhu, S., Song, Y., & Lu, L. (2015). A novel fluorescent retrograde neural tracer: Cholera toxin B conjugated carbon dots. *Nanoscale*, *7*(38), 15635–15642. doi:10.1039/C5NR04361A PMID:26285001

Zhou, S., Shao, Y., Gao, N., Deng, J., & Tan, C. (2013). Equilibrium, kinetic, and thermodynamic studies on the adsorption of triclosan onto multi-walled carbon nanotubes. *CLEAN–Soil, Air. Water (Basel)*, *41*(6), 539–547.

Zhou, X., Lee, S., Xu, Z., & Yoon, J. (2015). Recent progress on the development of chemosensors for gases. *Chemical Reviews*, *115*(15), 7944–8000. doi:10.1021/cr500567r PMID:25651137

Zhou, Y.-T., Nie, H.-L., Branford-White, C., He, Z.-Y., & Zhu, L.-M. (2009). Removal of Cu 2+ from aqueous solution by chitosan-coated magnetic nanoparticles modified with α-ketoglutaric acid. *Journal of Colloid and Interface Science*, *330*(1), 29–37. doi:10.1016/j.jcis.2008.10.026 PMID:18990406

Zhu, H., Jiang, R., Xiao, L., Chang, Y., Guan, Y., Li, X., & Zeng, G. (2009). Photocatalytic decolorization and degradation of Congo Red on innovative crosslinked chitosan/nano-CdS composite catalyst under visible light irradiation. *Journal of Hazardous Materials*, *169*(1), 933–940. doi:10.1016/j.jhazmat.2009.04.037 PMID:19477069

Zhu, H., Jia, Y., Wu, X., & Wang, H. (2009). Removal of arsenite from drinking water by activated carbon supported nano zero-valent iron. *Huan jing ke xue=. Huan Jing Ke Xue*, *30*(6), 1644–1648. PMID:19662844

Zhu, H., Jia, Y., Yao, S., Wu, X., & Wang, S. (2009). Removal of arsenate from drinking water by activated carbon supported nano zero-valent iron. *Huan jing ke xue=. Huan Jing Ke Xue*, *30*(12), 3562–3567. PMID:20187387

Zhu, L., Chang, D. W., Dai, L., & Hong, Y. (2007). DNA damage induced by multiwalled carbon nanotubes in mouse embryonic stem cells. *Nano Letters*, *7*(12), 3592–3597. doi:10.1021/nl071303v PMID:18044946

Zhu, L.-P., Yu, J.-Z., Xu, Y.-Y., Xi, Z.-Y., & Zhu, B.-K. (2009). Surface modification of PVDF porous membranes via poly(DOPA) coating and heparin immobilization. *Colloids and Surfaces. B, Biointerfaces*, *69*(1), 152–155. doi:10.1016/j.colsurfb.2008.11.011 PMID:19117736

Zhu, Q. L., & Xu, Q. (2014). Metal–organic framework composites. *Chemical Society Reviews*, *43*(16), 5468–5512. doi:10.1039/C3CS60472A PMID:24638055

Zhu, W. Z., & Deevi, S. C. (2003). A review on the status of anode materials for solid oxide fuel cells. *Materials Science and Engineering*, *362*(1-2), 228–239. doi:10.1016/S0921-5093(03)00620-8

Zylstra, G. J., Bang, S. W., Newman, L. M., & Perry, L. L. (2000). *Biodegradation of Nitroaromatic Compounds and Explosives. In Microbial Degradation of Mononitrophenols and Mononitrobenzoates* (pp. 145–160). Boca Raton, FL: Lewis Publishers.

About the Contributors

Rabia Nazir is currently working as Senior Scientific Officer in a reputable R&D organization named as Pakistan Council of Scientific and Industrial Research (PCSIR) Laboratories Complex, Lahore, Pakistan. Nanotechnology is her specific field of interest and she is handling several R&D projects in that area pertaining to development of nanomaterials and nanocomposites with application in water and wastewater treatment, pharmaceuticals, health care etc. She is also the member of R&D committee and Commercialization committee of PCSIR where her work is to participate in planning, arranging and supporting programs for need based & market oriented research & development projects. She has completed 10 different projects that helped local industry and academia; utilizing the indigenous resources. She is an author of two books on Nanotechnology and has contributed two chapters in two other books published on Nanotechnology. She has 44 research publications in local and international reputed journals. She has supervised 42 research students in her career spanning over 16 years.

* * *

Matloob Ahmad is working as 'Assistant Professor' in Department of Chemistry, Government College University, Faisalabad, Pakistan since Dec. 2011. His research interests are in heterocyclic compounds (organic chemistry). He has published around forty research articles dealing the synthesis of various families of organic compounds, their structure elucidation and medicinal importance. Moreover, he has keen research interests in water treatment techniques.

Gomaa A. M. Ali works as an Assistant Professor at Chemistry Department, Faculty of Science, Al-Azhar University Egypt. He obtained his B.Sc. (2006) and M.Sc. (2012) in Physical Chemistry from Al-Azhar University, Faculty of Science. He was awarded his Ph.D. (2015) in Advanced Nanomaterials for Energy Storage from UMP, Malaysia. His research interests are: Materials Science, Nanocomposites, Humidity Sensing, Graphene, Supercapacitors, and Water Treatment. So far, he has 41 published papers, 2 patents and 13 medals (gold, platinum, and diamond).

Sana Aslam is working as 'Assistant Professor' in Department of Chemistry, Government College Women University, Faisalabad since July 2013. Her research interests are in organic synthesis. She is working on antiviral compounds and on the synthesis of codrugs.

Abderrezak Khelfi is a toxicologist and the head of the department of research and development in the National Center of Toxicology of Algiers. He studied pharmacy and became specialized in toxicology. He received his PhD degree from the University of medical sciences of Algiers. He is a member of the Algerian Society of Toxicology. He published different papers on the effect of xenobiotics, pollutants and contaminants on human body. He has contributed extensively to the world medical literature with publications appearing in the Journal of clinical and analytical toxicology, annales pharmaceutiques françaises and journal of pharmacy and nutrition sciences.

Mintu Maan Dutta was born in Nazira, Sibsagar, Assam, India in May 10, 1992. He obtained his B.Sc. degree from Cotton College in 2012 and M.Sc (Physical Chemistry) degree in 2014, from Gauhati University, Assam, India. He is carrying out his Ph.D. work in the Department of Chemistry, Gauhati University. His current research interest includes the "Use of magnetic nanoparticles for various organic transformations".

Zahra Mahmoodi is a PhD student in Department of Chemistry, Science and Research Branch, Islamic Azad University, Tehran, Iran.

Saima Nasreen is currently working as Assistant Professor in Department of Environmental Sciences at The Women University Multan. She has done one year Interim Placement as Assistant Professor in Department of Environmental Sciences at International Islamic University Islamabad. She has done her PhD in Environmental Sciences from Fatima Jinnah Women University Rawalpindi, Pakistan. She has conducted her PhD research work at University of Maryland, College Park USA and a part at University of Malaya, Malaysia as an exchange student. She has presented papers Nationally and Internationally and 14 International research publications are on her credit.

Hamid Jafari Nia is a PhD student in Environmental Engineering in University of Politechnic Catalonia.

Durr-e-Shahwar Noman was born and raised in Lahore, the second largest city in Pakistan. Her parents ensured that Durr-e- Shahwar, her sister, and her two brothers were given the best opportunities for education, and she stayed in her home town to attend primary school through to university until she left in 2016 to begin her PhD studies. Durr-e- Shahwar attended the University of the Punjab (PU), Lahore - the oldest public university in Pakistan - where she completed a BSc, MSc and in 2009 an MPhil, all in Environmental Science. Throughout her studies she was acknowledged as the best student in her class and received several awards and scholarships. Since 2007, when she completed her MSc, Durr-e- Shahwar has been a Lecturer at several academic and research organizations in Lahore, including PU, Kinnaird College for Women and the City District Government, where she also helped develop environmental monitoring systems. She has authored numerous research papers in international journals and written books in her field of expertize – nanotechnology and the application of advanced oxidation processes for wastewater treatment. The opportunity to study overseas towards a PhD was enabled by an International Post Graduate Research Scholarship from the Australian Government's Department of Education and Training, and Durr-e- Shahwar's research continues her investigations at UniSA into methods of treating wastewater using engineered nanoparticles. Water is a scarce resource that needs to be effectively managed in Pakistan, and she expects her research will enhance her skills in the physio-chemical characterization of nanomaterials and evaluation of their properties to resolve water pollution challenges. After completing her PhD, Durr-e- Shahwar plans to return to PU, where she hopes to play a leading role in water treatment research. She also expects to contribute to policies for tackling environmental issues faced by many people in her country.

Sumaira Nosheen is a PhD Scholar and Mphil in Polymer Technology from University of the Punjab, Pakistan. She has been serving P.C.S.I.R as Scientist for the last nine years. She has more than eight publications in different fields of science and engineering.

Jay Pandey is PhD in Chemical Engineering from Indian Institute of Technology (IIT) Delhi, New Delhi, India and Postdoc from Van ‘t Hoff Institute for Molecular Sciences (HIMS), University of Amsterdam, The Netherlands. His research interests include solid state electro-chemistry, advanced functional energy materials (membranes, perovskites and electro-catalysts), energy storage and conversion devices (fuel cells and redox flow batteries) and biomass derived bio-fuel. Currently he is Principal Scientist at Sardar Patel Renewable Energy Research Institute (SPRERI) Anand, Gujrat, India. So far Dr. Pandey has more than 25 research publications in peer reviewed journals and international conference proceedings.

Uzaira Rafique did her Ph.D. in 2000 in the field of Polymer Chemistry and postdoctoral on Commonwealth fellowship at Imperial College, London in 2009. She joined the public sector first Women University of Pakistan in 1999 and developed Department of Environmental Sciences. She is currently working as Professor and Dean for Faculty of Science and Technology. She is a member of different national and international professional bodies. Dr. Rafique presented her research work not only nationally but also internationally on different forums. Dr. Rafique has fifty plus publications on her credit.

Phool Shahzadi is PhD scholar and Silver Medalist from Minhaj university in M.Sc.and M.Phil in Chemistry. The young scientist is serving as scientific officer at P.C.S.I.R Laboratories Complex Lahore since 2009.She has more than eleven publications, three international books, one full length chapter and filed three patents.

Index

A

adsorption 5, 9, 29, 77-78, 112, 114, 126-131, 133, 136, 143-156, 158, 173, 177-184, 187, 201-202, 204, 206-210, 212, 226-228, 230-232, 242-245, 247-248, 252, 268, 272-275, 279, 282, 285, 319
adverse 38, 172, 242, 344-345, 350
AFM 52, 63-64, 79, 307
air-remediation 136
as 1-3, 5-7, 9, 12, 14, 17, 29-39, 44-45, 47-50, 52-57, 60, 62, 66, 68-70, 73, 75-79, 81-83, 87-94, 103-107, 109, 111-116, 121-122, 124-126, 128-129, 131-135, 143-148, 150-158, 171-184, 186, 201-204, 206-211, 220-223, 225-232, 241-250, 252-253, 268-270, 272-275, 277-280, 307, 309-314, 317-321, 323, 325-327, 329-335, 344-350, 352-358, 360-361

B

blood-brain barrier 344-345

C

carbon nano tubes 126, 137, 273-274
COD 35, 173, 269, 307
Cr 14, 144, 150-151, 154-156, 177-178, 207-208, 227, 229-231, 249, 307, 352

D

dermal 344-346
DLS 307
DMAC 307
DNA 18, 92, 177, 246, 344-345, 350, 353, 356-358
DSC 69-73, 226
DTA 73-75, 226

E

EDS 60, 62-63, 307
EDX 57, 272, 307
effects 28-32, 35, 37-38, 53, 81, 87, 123, 127, 144, 153, 172, 175, 181, 207, 211, 220, 223-224, 229, 242, 269-270, 344-345, 348-351, 353-361
EG 307
environment 1-2, 7, 12, 15, 17-18, 28, 31, 34, 36-39, 46, 52, 81-83, 85-90, 92-94, 113-114, 116, 121, 123, 133, 144, 157, 172-174, 176-177, 179, 182, 202, 209, 220-221, 223, 241, 268-269, 274, 279-280, 309, 335, 345, 350
environmental monitoring 3, 6
environmental pollution 114, 121, 181, 202
environmental remediation 7, 12, 31, 82, 121, 146, 174, 177, 243
exposure 5-6, 37, 94, 123, 133-134, 144, 151-152, 181, 269, 278, 344-345, 350, 354-356, 358-359, 361

F

FT-IR 272, 307
fuel cell 221, 308-314, 318, 321, 323, 325, 327, 329-335
functionalized hybrids 171, 179-182

G

GA 280, 307
green route 81-82, 90-91
green synthesis 81-82, 87-88, 250, 253, 269, 275, 280

H

health 38-39, 81, 92, 94, 115, 121-123, 129, 143-144, 151, 172, 177, 201, 204, 208, 221, 223-224, 229, 241-242, 268, 344, 351, 356, 358, 360-361

heavy metals 5-6, 9, 12, 14, 39, 113, 116, 126, 146-148, 151-154, 156-158, 171, 174-175, 201-202, 207, 220-229, 231-232, 247, 274, 285
HMO 307

I

inflammatory 344-345, 352-355, 359
ingestion 179, 344-345
inhalation 18, 344-345, 355

M

magnesium oxide 74, 226, 272, 279
magnetic nanomaterials 201, 203-204, 209
membranes 15, 103-114, 116-117, 203, 228, 243, 245-246, 269, 273, 307-309, 312-314, 316-319, 321-332, 334-335, 355-356
metal oxides 5, 28, 124, 126, 133, 146-147, 154, 157, 210, 228-229, 231, 252, 273-274, 309, 318-319
mg 128, 134, 147, 150-156, 179, 204, 206-211, 223, 230-231, 279, 307, 321
microorganisms 36, 81, 87, 89, 114, 210, 275, 333
mitochondria 344-345, 353-354, 357
MMMs 279, 307
MNP 209, 230, 278, 307
MONPs 274-275, 280, 307

N

Nafion 309, 312-316, 323, 325-327, 330-335
nano particles 86, 115, 221, 225, 227, 247, 274
Nanoadsorbents 9, 146-147, 156-157, 208, 274
nano-catalysis 137
nanocomposite membrane 279, 285, 318, 323, 325, 327, 330, 333, 335
nano-dendrimers 124, 135-136
nanomaterials 1-4, 6-7, 9, 15, 17-18, 28-38, 44, 53, 57, 78-79, 81-83, 87-88, 92, 94, 114, 116, 126, 143-144, 146-148, 153-155, 158, 173, 175, 201-204, 207, 209-210, 220, 241-243, 245-246, 252, 270-272, 344
nanoparticles 12, 17-18, 28, 30-39, 44, 46, 55-56, 60, 69, 81-85, 87-92, 94, 115-116, 126, 144, 146-147, 157, 173, 204, 206-211, 226-227, 229, 231, 241-243, 245-247, 249-250, 252-253, 268-269, 271-272, 274-275, 277-278, 307, 311, 325, 344-346, 349, 351
nanoscience 87, 144, 173
Nano-Sensors 114, 124
nanotechnology 1-3, 7, 15, 17-18, 28, 38, 44, 79, 82-85, 87, 91-94, 103, 113-116, 121, 124, 126, 136-137, 144, 171, 173, 202, 220-222, 228, 232, 268-271, 273, 275, 344, 359, 361
Nitrogen Sorption 182
NPs 4-6, 10, 14, 17, 34, 36-37, 39, 91-92, 204, 245-246, 269-280, 307
nZVI 12, 14, 307

O

organic pollutants 9, 12, 14-15, 92, 147-148, 155-156, 171, 181-182, 207, 221-222, 241-242, 244-245, 247, 249, 273
oxidative stress 224-225, 344-345, 350-353, 355, 357, 359

P

PDA 247, 307
PES 307, 317, 321
phenols 175, 179-182, 184, 247, 250, 252
Photocatalysis 132, 246-247
plants 37, 57, 82, 87-89, 92-93, 103, 114, 121, 129, 179, 242, 269, 275, 277-278, 280
pollutants 3, 7, 9, 12, 14-15, 17, 36, 83, 85, 92, 113, 115, 121-124, 126-129, 131, 133-137, 145-148, 155-156, 171-177, 181-182, 184, 201-203, 207, 212, 221-223, 226, 229, 241-245, 247, 249, 252-253, 269-270, 273-274, 279-280, 282
pollution prevention 15, 17, 116, 136
proton conductivity 308-309, 313-314, 316-321, 323, 325-332, 335
PVP 307, 330

R

removal 7, 9, 14, 17, 34, 92, 116, 126-127, 129, 131, 133-137, 144, 146-148, 150-158, 171-175, 177-182, 201, 203-204, 206-212, 220-222, 225-226, 228-232, 241-245, 247, 249-250, 252, 268-270, 273, 279, 282, 285

S

SEM 52, 57-60, 79, 109, 112, 129-131, 211, 226,

246, 272, 307
stability 15, 18, 29, 66, 106, 115, 153, 174, 181, 208, 231, 276, 279, 308-309, 313, 316, 318-321, 323, 325, 328, 332, 335
structure property relationship 174-175
surface modification 125, 129, 137, 144-146, 148, 206, 209, 231, 350

T

TEM 52-55, 57, 60, 79, 91, 112, 226, 272, 307
TGA 66-68, 75, 226
toxicity 18, 36-39, 94, 146, 148, 151, 157, 179, 211, 221, 223, 226, 246, 270, 344-345, 348-355, 360-361

V

VSM, 75

W

Waste water treatment 226, 242
wastewater 12, 77, 92, 104, 114-116, 143-144, 146-148, 151, 153-158, 172-174, 176, 179, 201-204, 207-208, 212, 220-223, 225, 227-229, 231-232, 242-246, 249, 268-270, 273, 279-280, 282, 333
wastewater treatment 104, 114, 144, 147, 157-158, 173, 202-203, 212, 220-222, 225, 227-228, 268-270, 273, 279-280, 282
water treatment 91-92, 113, 144, 151, 155, 158, 201-203, 222, 225-226, 228, 232, 242, 249, 252-253, 280, 285

X

XRD 47-52, 79, 226-227, 272, 307, 319

Z

zinc oxide 91, 124, 128, 133, 173, 226, 231, 272, 278-279
Zn 14, 35, 144, 151-152, 154, 156, 176, 207, 227, 230, 246, 277-278, 307

www.ingramcontent.com/pod-product-compliance
Lightning Source LLC
LaVergne TN
LVHW080532300926
849162LV00007B/1591
* 9 7 8 1 5 2 2 5 5 7 4 5 6 *